CAMBRIDGE LIBRARY COLLECTION

Books of enduring scholarly value

Mathematical Sciences

From its pre-historic roots in simple counting to the algorithms powering modern desktop computers, from the genius of Archimedes to the genius of Einstein, advances in mathematical understanding and numerical techniques have been directly responsible for creating the modern world as we know it. This series will provide a library of the most influential publications and writers on mathematics in its broadest sense. As such, it will show not only the deep roots from which modern science and technology have grown, but also the astonishing breadth of application of mathematical techniques in the humanities and social sciences, and in everyday life.

The Collected Mathematical Papers

Arthur Cayley (1821-1895) was a key figure in the creation of modern algebra. He studied mathematics at Cambridge and published three papers while still an undergraduate. He then qualified as a lawyer and published about 250 mathematical papers during his fourteen years at the Bar. In 1863 he took a significant salary cut to become the first Sadleirian Professor of Pure Mathematics at Cambridge, where he continued to publish at a phenomenal rate on nearly every aspect of the subject, his most important work being in matrices, geometry and abstract groups. In 1882 he spent five months at Johns Hopkins University, and in 1883 became president of the British Association for the Advancement of Science. Publication of his Collected Papers - 967 papers in 13 volumes plus an index volume - began in 1889 and was completed after his death under the editorship of his successor in the Sadleirian Chair. This volume contains 93 papers mostly published between 1878 and 1883, including a series of articles written for the ninth edition of the Encyclopaedia Britannica.

The Collected Mathematical Papers

VOLUME 11

ARTHUR CAYLEY

CAMBRIDGE
UNIVERSITY PRESS

CAMBRIDGE UNIVERSITY PRESS

Cambridge New York Melbourne Madrid Cape Town Singapore São Paolo Delhi

Published in the United States of America by Cambridge University Press, New York

www.cambridge.org
Information on this title: www.cambridge.org/9781108005036

This edition first published 1896
This digitally printed version 2009

ISBN 978-1-108-00503-6

MATHEMATICAL PAPERS.

London: C. J. CLAY and SONS,
CAMBRIDGE UNIVERSITY PRESS WAREHOUSE,
AVE MARIA LANE.
Glasgow: 263, ARGYLE STREET.

Leipzig: F. A. BROCKHAUS.
New York: THE MACMILLAN CO.

CAMBRIDGE:
PRINTED BY J. AND C. F. CLAY,
AT THE UNIVERSITY PRESS.

ADVERTISEMENT.

THE present volume contains 93 papers numbered 706 to 798, published, with the exception of one series, for the most part in the years 1878 to 1883. This series is constituted by the articles which Professor Cayley wrote for the *Encyclopædia Britannica* between the years 1878 and 1888; it seemed desirable to place these together in the same volume, in spite of the departure from the chronological arrangement which governs the sequence of the papers in the volumes generally. The Syndics of the University Press desire to acknowledge their obligation to Messrs Adam and Charles Black, Publishers of the ninth Edition of the *Encyclopædia Britannica*, for their courteous consent in allowing these articles to be included in the *Collected Mathematical Papers*. Exact references to the volumes, from which the articles are extracted, will be found in the Table of Contents.

The frontispiece to the present volume is a reproduction by Mr A. G. Dew-Smith, of Trinity College, of a photograph of Professor Cayley which he made in the year 1885. The Syndics of the Press desire to acknowledge their obligation to Mr Dew-Smith.

The Table for the eleven volumes is

Vol.	I.	Numbers	1	to	100,
,,	II.	,,	101	,,	158,
,,	III.	,,	159	,,	222,
,,	IV.	,,	223	,,	299,
,,	V.	,,	300	,,	383,
,,	VI.	,,	384	,,	416,
,,	VII.	,,	417	,,	485,
,,	VIII.	,,	486	,,	555,
,,	IX.	,,	556	,,	629,
,,	X.	,,	630	,,	705,
,,	XI.	,,	706	,,	798.

A. R. FORSYTH.

21 *November*, 1896.

CONTENTS.

[An Asterisk means that the paper is not printed in full.]

CLASSIFICATION.

ANALYSIS.

706.

ON THE DISTRIBUTION OF ELECTRICITY ON TWO SPHERICAL SURFACES.

[From the *Philosophical Magazine*, vol. v. (1878), pp. 54—60.]

IN the two memoirs "Sur la distribution de l'électricité à la surface des corps conducteurs," *Mém. de l'Inst.* 1811, Poisson considers the question of the distribution of electricity upon two spheres: viz. if the radii be a, b, and the distance of the centres be c (where $c > a+b$, the spheres being exterior to each other), and the potentials within the two spheres respectively have the constant values h and g, then—for Poisson's $f\left(\frac{x}{a}\right)$ writing $\phi(x)$, and for his $F\left(\frac{x}{b}\right)$ writing $\Phi(x)$—the question depends on the solution of the functional equations

$$a\phi(x) + \frac{b^2}{c-x}\Phi\left(\frac{b^2}{c-x}\right) = h,$$

$$\frac{a^2}{c-x}\phi\left(\frac{a^2}{c-x}\right) + b\Phi(x) = g,$$

where of course the x of either equation may be replaced by a different variable.

It is proper to consider the meaning of these equations: for a point on the axis, at the distance x from the centre of the first sphere, or say from the point A, the potential of the electricity on this spherical surface is $a\phi x$ or $\frac{a^2}{x}\phi\left(\frac{a^2}{x}\right)$, according as the point is interior or exterior; and, similarly, if x now denote the distance from the centre of the second sphere (or, say, from the point B), then the potential of the electricity on this spherical surface is $b\Phi x$ or $\frac{b^2}{x}\Phi\left(\frac{b^2}{x}\right)$, according as the point is interior or exterior; $\phi(x)$ is thus the same function of (x, a, b) that $\Phi(x)$ is of

(x, b, a). Hence, first, for a point interior to the sphere A, if x denote the distance from A, and therefore $c-x$ the distance of the same point from B, the potential of the point in question is

$$= a\phi x + \frac{b^2}{c-x}\Phi\left(\frac{b^2}{c-x}\right);$$

and, secondly, for a point interior to the sphere B, if x denote the distance from B and therefore $c-x$ the distance of the same point from A, the potential of the point is

$$= \frac{a^2}{c-x}\phi\left(\frac{a^2}{c-x}\right) + b\Phi(x).$$

The two equations thus express that the potentials of a point interior to A and of a point interior to B are $=h$ and g respectively.

It is to be added that the potential of an exterior point, distances from the points A and $B = x$ and $c-x$ respectively, is

$$= \frac{a^2}{x}\phi\left(\frac{a^2}{x}\right) + \frac{b^2}{c-x}\Phi\left(\frac{b^2}{c-x}\right);$$

and that, by the known properties of Legendre's coefficients, when the potential upon an axial point is given, it is possible to pass at once to the expression for the potential of a point not on the axis, and also to the expression for the electrical density at a point on the two spherical surfaces respectively. The determination of the functions $\phi(x)$ and $\Phi(x)$ gives thus the complete solution of the question.

I obtain Poisson's solution by a different process as follows:—Consider the two functions

$$\frac{a^2(c-x)}{c^2-b^2-cx}, \quad = \frac{\mathrm{a}x+\mathrm{b}}{\mathrm{c}x+\mathrm{d}}, \text{ suppose,}$$

and

$$\frac{b^2(c-x)}{c^2-a^2-cx}, \quad = \frac{\alpha x+\beta}{\gamma x+\delta}, \text{ suppose;}$$

and let the nth functions be

$$\frac{\mathrm{a}_n x+\mathrm{b}_n}{\mathrm{c}_n x+\mathrm{d}_n} \text{ and } \frac{\alpha_n x+\beta_n}{\gamma_n x+\delta_n}$$

respectively.

Observing that the values of the coefficients are

$$\left(\begin{array}{cc}\mathrm{a}, & \mathrm{b}\\ \mathrm{c}, & \mathrm{d}\end{array}\right) = \left(\begin{array}{cc}-a^2, & a^2c\\ -c, & c^2-b^2\end{array}\right), \text{ and } \left(\begin{array}{cc}\alpha, & \beta\\ \gamma, & \delta\end{array}\right) = \left(\begin{array}{cc}-b^2, & b^2c\\ -c, & c^2-a^2\end{array}\right),$$

so that we have

$$\mathrm{a}+\mathrm{d} = \alpha+\delta, \quad = c^2-a^2-b^2, \qquad \mathrm{ad}-\mathrm{bc} = \alpha\delta-\beta\gamma, \quad = a^2b^2,$$

and consequently that the two equations

$$\frac{(\lambda+1)^2}{\lambda} = \frac{(\mathrm{a}+\mathrm{d})^2}{\mathrm{ad}-\mathrm{bc}}, \quad \frac{(\lambda+1)^2}{\lambda} = \frac{(\alpha+\delta)^2}{\alpha\delta-\beta\gamma},$$

are in fact one and the same equation

$$\frac{(\lambda+1)^2}{\lambda} = \frac{(c^2-a^2-b^2)^2}{a^2b^2}$$

for the determination of λ, then (by a theorem which [686, 687] I have recently obtained) we have the following equations for the coefficients

$$\begin{pmatrix} \mathrm{a}_n, & \mathrm{b}_n \\ \mathrm{c}_n, & \mathrm{d}_n \end{pmatrix}, \quad \begin{pmatrix} \alpha_n, & \beta_n \\ \gamma_n, & \delta_n \end{pmatrix}$$

of the nth functions; viz. these are:—

$$\mathrm{a}_n x + \mathrm{b}_n = \frac{1}{\lambda^2-1}\left(\frac{\mathrm{a}+\mathrm{d}}{\lambda+1}\right)^{n-1} \{(\lambda^{n+1}-1)(\mathrm{a}x+\mathrm{b}) + (\lambda^n-\lambda)(-\mathrm{d}x+\mathrm{b})\},$$

$$\mathrm{c}_n x + \mathrm{d}_n = \quad ,, \quad ,, \quad \{(\lambda^{n+1}-1)(\mathrm{c}x+\mathrm{d}) + (\lambda^n-\lambda)(\ \mathrm{c}x-\mathrm{a})\};$$

and similarly

$$\alpha_n x + \beta_n = \frac{1}{\lambda^2-1}\left(\frac{\alpha+\delta}{\lambda+1}\right)^{n-1} \{(\lambda^{n+1}-1)(\alpha x+\beta) + (\lambda^n-\lambda)(-\delta x+\beta)\},$$

$$\gamma_n x + \delta_n = \quad ,, \quad ,, \quad \{(\lambda^{n+1}-1)(\gamma x+\ \delta) + (\lambda^n-\lambda)(\ \gamma x-\alpha)\}.$$

Observe that these equations give, as they ought to do,

$$\mathrm{a}_0 x + \mathrm{b}_0 = x, \quad \mathrm{c}_0 x + \mathrm{d}_0 = 1, \quad \mathrm{a}_1 x + \mathrm{b}_1 = \mathrm{a}x + \mathrm{b}, \quad \mathrm{c}_1 x + \mathrm{d}_1 = \mathrm{c}x + \mathrm{d};$$

and similarly

$$\alpha_0 x + \beta_0 = x, \quad \gamma_0 x + \delta_0 = 1, \quad \alpha_1 x + \beta_1 = \alpha x + \beta, \quad \gamma_1 x + \delta_1 = \gamma x + \delta.$$

Substituting in the first two equations $\frac{a^2}{c-x}$ in place of x, and in the second two equations $\frac{b^2}{c-x}$ in place of x, we obtain the following results which will be useful:—

$$\mathrm{a}_n a^2 + \mathrm{b}_n (c-x) = a^2(\gamma_n x + \delta_n),$$

$$\mathrm{c}_n a^2 + \mathrm{d}_n (c-x) = \frac{1}{b^2}(\alpha_{n+1} x + \beta_{n+1}),$$

$$\alpha_n b^2 + \beta_n (c-x) = b^2(\mathrm{c}_n x + \mathrm{d}_n),$$

$$\gamma_n b^2 + \delta_n (c-x) = \frac{1}{a^2}(\mathrm{a}_{n+1} x + \mathrm{b}_{n+1}),$$

the last two of which are obtained from the first two by a mere interchange of letters; it will therefore be sufficient to prove the first and second equations.

For the first equation we have

$$\mathrm{a}_n a^2 + \mathrm{b}_n (c-x) = \frac{1}{\lambda^2-1}\left(\frac{\mathrm{a}+\mathrm{d}}{\lambda+1}\right)^{n-1} \{(\lambda^{n+1}-1)[\mathrm{a}a^2 + \mathrm{b}(c-x)] + (\lambda^n-\lambda)[-\mathrm{d}a^2 + \mathrm{b}(c-x)]\},$$

where the term in $\{.\}$ is

$$=(\lambda^{n+1}-1)\,[-a^4+a^2c\,(c-x)]+(\lambda^n-\lambda)\,[a^2\,(b^2-c^2)+a^2c\,(c-x)]\,;$$

viz. this is

$$=a^2\,\{(\lambda^{n+1}-1)\,(c^2-a^2-cx)+(\lambda^n-\lambda)\,(b^2-cx)\}\,;$$

or it is

$$=a^2\,\{(\lambda^{n+1}-1)\,(\gamma x+\delta)\qquad+(\lambda^n-\lambda)\,(\gamma x-\alpha)\},$$

whence the relation in question.

The proof of the second equation is a little more complicated. We have

$$\mathrm{c}_n a^2+\mathrm{d}_n\,(c-x)=\frac{1}{\lambda^2-1}\left(\frac{\mathrm{a}+\mathrm{d}}{\lambda+1}\right)^{n-1}\{(\lambda^{n+1}-1)\,[ca^2+\mathrm{d}\,(c-x)]+(\lambda^n-\lambda)\,[ca^2-\mathrm{a}\,(c-x)]\},$$

where the term in $\{\ \}$ is

$$=(\lambda^{n+1}-1)\,[-ca^2+(c^2-b^2)\,(c-x)]+(\lambda^n-\lambda)\,[-ca^2+a^2\,(c-x)].$$

Comparing this with

$$\alpha_{n+1}x+\beta_{n+1}=\frac{1}{\lambda^2+1}\left(\frac{\alpha+\delta}{\lambda+1}\right)^n\{(\lambda^{n+2}-1)\,(\alpha x+\beta)+(\lambda^{n+1}-\lambda)\,(-\delta x+\beta)\},$$

where the term in $\{\ \}$ is

$$=(\lambda^{n+2}-1)\,[b^2\,(c-x)]+(\lambda^{n+1}-\lambda)\,[-c\,(c^2-a^2-b^2)+(c^2-a^2)\,(c-x)],$$

it is to be observed that the quotient of the two terms in $\{\ \}$ is in fact a constant; this is most easily verified as follows. Dividing the first of them by the second, we have a quotient which when $x=c$ is

$$\frac{(\lambda^{n+1}-1)\,(-ca^2)+(\lambda^n-\lambda)\,(-ca^2)}{(\lambda^{n+1}-\lambda)\,\{-c\,(c^2-a^2-b^2)\}},\quad=\frac{a^2\,(\lambda^{n+1}-1+\lambda^n-\lambda)}{(\lambda^{n+1}-\lambda)\,(c^2-a^2-b^2)},\quad=\frac{a^2\,(\lambda+1)}{(c^2-a^2-b^2)\,\lambda},$$

and when $x=0$ is

$$\frac{(\lambda^{n+1}-1)\,c\,(c^2-a^2-b^2)}{(\lambda^{n+2}-1)\,b^2c+(\lambda^{n+1}-\lambda)\,b^2c},\quad=\frac{(\lambda^{n+1}-1)\,(c^2-a^2-b^2)}{(\lambda^{n+2}-1+\lambda^{n+1}-\lambda)\,b^2},\quad=\frac{c^2-a^2-b^2}{b^2\,(\lambda+1)}:$$

these two values are equal by virtue of the equation which defines λ; and hence the quotient of the two linear functions having equal values for $x=c$ and $x=0$, has always the same value; say it is $=\dfrac{c^2-a^2-b^2}{b^2(\lambda+1)}$. Hence, observing that $\mathrm{a}+\mathrm{d}=\alpha+\delta$, $=c^2-a^2-b^2$, the quotient, $\mathrm{c}_n a^2+\mathrm{d}_n\,(c-x)$ divided by $\alpha_{n+1}x+\beta_{n+1}$, is

$$=\frac{\lambda+1}{c^2-a^2-b^2}\cdot\frac{c^2-a^2-b^2}{b^2\,(\lambda+1)},\quad=\frac{1}{b^2}\,;$$

or we have the required equation

$$\mathrm{c}_n a^2+\mathrm{d}_n\,(c-x)=\frac{1}{b^2}\,(\alpha_{n+1}x+\beta_{n+1}).$$

Considering now the functional equations, suppose for the moment that g is $=0$; the two equations may be satisfied by assuming

$$\phi(x) = \quad h\left\{\frac{1}{\mathrm{c}_0 x + \mathrm{d}_0} + \frac{\omega}{\mathrm{c}_1 x + \mathrm{d}_1} + \ldots\right\} L,$$

$$\Phi(x) = -h\left\{\frac{\omega}{\alpha_1 x + \beta_1} + \frac{\omega^2}{\alpha_2 x + \beta_2} + \ldots\right\} M.$$

We in fact, from the foregoing relations, at once obtain

$$\frac{a^2}{c-x}\phi\frac{a^2}{c-x} = \quad h\left\{\frac{\omega}{\alpha_1 x + \beta_1} + \frac{\omega^2}{\alpha_2 x + \beta_2}\ldots\right\}\frac{a^2b^2L}{\omega},$$

$$\frac{b^2}{c-x}\Phi\frac{b^2}{c-x} = -h\left\{\frac{\omega}{\mathrm{c}_1 x + \mathrm{d}_1} + \frac{\omega^2}{\mathrm{c}_2 x + \mathrm{d}_2}\ldots\right\} M.$$

To satisfy the first equation we must have $M = aL$; viz. this being so, the equation becomes

$$a\phi x + \frac{b^2}{c-x}\Phi\left(\frac{b^2}{c-x}\right) = \frac{aLh}{\mathrm{c}_0 x + \mathrm{d}_0};$$

or, since $\mathrm{c}_0 x + \mathrm{d}_0 = 1$, the equation will be satisfied if only $aL = 1$, whence also $M = 1$. And the second equation will be satisfied if only $\dfrac{a^2b^2L}{\omega} = bM$; viz. substituting for L, M their value, we find $\omega = ab$.

Supposing, in like manner, that $h = 0$, g retaining its proper value, we find a like solution for the two equations; and by simply adding the solutions thus obtained, we have a solution of the original two equations

$$a\phi(x) + \frac{b^2}{c-x}\Phi\left(\frac{b^2}{c-x}\right) = h,$$

$$\frac{a^2}{c-x}\phi\left(\frac{a^2}{c-x}\right) + b\Phi(x) = g;$$

viz. the solution is

$$\phi(x) = \quad \frac{h}{a}\left\{\frac{1}{\mathrm{c}_0 x + \mathrm{d}_0} + \frac{ab}{\mathrm{c}_1 x + \mathrm{d}_1} + \ldots\right\} - g\left\{\frac{ab}{\mathrm{a}_1 x + \mathrm{b}_1} + \frac{(ab)^2}{\mathrm{a}_2 x + \mathrm{b}_2} + \ldots\right\}$$

$$\Phi(x) = -h\left\{\frac{ab}{\alpha_1 x + \beta_1} + \frac{(ab)^2}{\alpha_2 x + \beta_2} + \ldots\right\} + \frac{g}{b}\left\{\frac{1}{\gamma_0 x + \delta_0} + \frac{ab}{\gamma_1 x + \delta_1} + \ldots\right\}.$$

We have a general solution containing an arbitrary constant P by adding to the foregoing values for ϕx a term

$$= \frac{Pb(a-b)}{\sqrt{a^2(c-x) - x(c^2 - b^2 - cx)}},$$

and for Φx a term

$$= \frac{Pa(b-a)}{\sqrt{b^2(c-x) - x(c^2 - a^2 - cx)}},$$

as may be easily verified if we observe that the function

$$a^2(c-x) - x(c^2 - b^2 - cx),$$

writing therein $\dfrac{a^2}{c-x}$ for x, becomes

$$= \frac{a^2}{(c-x)^2}\{b^2(c-x) - x(c^2 - a^2 - cx)\}:$$

and similarly that

$$b^2(c-x) - x(c^2 - a^2 - cx),$$

writing therein $\dfrac{b^2}{c-x}$ for x, becomes

$$= \frac{b^2}{(c-x)^2}\{a^2(c-x) - x(c^2 - b^2 - cx)\}.$$

More generally, the terms to be added are for ϕx a term as above, where P denotes a function of x which remains unaltered when x is changed into $\dfrac{a^2(c-x)}{c^2 - b^2 - cx}$, and for Φx a term as above with P' instead of P, where P' denotes what P becomes when x is changed into $\dfrac{a^2}{c-x}$. But these additional terms vanish for the electrical problem, and the correct values of ϕx, Φx are the particular values given above.

It is to be remarked that the function

$$\frac{a^2(c-x)}{c^2 - b^2 - cx} \text{ is } = \frac{a^2}{c - \dfrac{b^2}{c-x}};$$

viz. considering x as the distance of a point X from A, then taking the image of X in regard to the sphere B, and again the image of this image in regard to the sphere A, the function in question is the distance of this second image from A. And similarly the function

$$\frac{b^2(c-x)}{c^2 - a^2 - cx} \text{ is } = \frac{b^2}{c - \dfrac{a^2}{c-x}};$$

viz. considering here x as the distance of the point X from B, then taking the image of X in regard to the sphere A, and again the image of this image in regard to the sphere B, the function in question is the distance of this second image from B. It thus appears that Poisson's solution depends upon the successive images of X in regard to the spheres B and A alternately, and also on the successive images of X in regard to the spheres A and B alternately. This method of images is in fact employed in Sir W. Thomson's paper "On the Mutual Attraction or Repulsion between two Electrified Spherical Conductors," *Phil. Mag.*, April and August, 1853.

707.

ON THE COLOURING OF MAPS.

[From the *Proceedings of the Royal Geographical Society*, vol. I., no. 4 (1879), pp. 259—261.]

THE theorem that four colours are sufficient for any map, is mentioned somewhere by the late Professor De Morgan, who refers to it as a theorem known to map-makers. To state the theorem in a precise form, let the term "area" be understood to mean a simply or multiply connected* area: and let two areas, if they touch along a line, be said to be "attached" to each other; but if they touch only at a point or points, let them be said to be "appointed" to each other. For instance, if a circular area be divided by radii into sectors, then each sector is attached to the two contiguous sectors, but it is appointed to the several other sectors. The theorem then is, that if an area be partitioned in any manner into areas, these can be, with four colours only, coloured in such wise that in every case two attached areas have distinct colours; appointed areas may have the same colour. Detached areas may in a map represent parts of the same country, but this relation is not in anywise attended to: the colours of such detached areas will be the same, or different, as the theorem may require.

It is easy to see that four colours are wanted; for instance, we have a circle divided into three sectors, the whole circle forming an *enclave* in another area; then we require three colours for the three sectors, and a fourth colour for the surrounding area: if the circle were divided into four sectors, then for these two colours would

* An area is "connected" when every two points of the area can be joined by a continuous line lying wholly within the area; the area within a non-intersecting closed curve, or say an area having a single boundary, is "simply connected"; but if besides the exterior boundary there is one or more than one interior boundary (that is, if there is within the exterior boundary one or more than one *enclave* not belonging to the area), then the area is "multiply connected." The theorem extends to multiply connected areas, but there is no real loss of generality in taking, and we may for convenience take the areas of the theorem to be each of them a simply connected area.

be sufficient, and taking a third colour for the surrounding area, three colours only would be wanted; and so in general according as the number of sectors is even or odd, three colours or four colours are wanted. And in any tolerably simple case it can be seen that four colours are sufficient. But I have not succeeded in obtaining a general proof: and it is worth while to explain wherein the difficulty consists. Supposing a system of n areas coloured according to the theorem with four colours only, if we add an $(n+1)$th area, it by no means follows that we can *without altering the original colouring* colour this with one of the four colours. For instance, if the original colouring be such that the four colours all present themselves in the exterior boundary of the n areas, and if the new area be an area enclosing the n areas, then there is not any one of the four colours available for the new area.

The theorem, if it is true at all, is true under more stringent conditions. For instance, if in any case the figure includes four or more areas meeting in a point (such as the sectors of a circle), then if (introducing a new area) we place at the point a small circular area, cut out from and attaching itself to each of the original sectorial areas, it must according to the theorem be possible with four colours only to colour the new figure; and this implies that it must be possible to colour the original figure so that only three colours (or it may be two) are used for the sectorial areas. And in precisely the same way (the theorem is in fact really the same) it must be possible to colour the original figure in such wise that only three colours (or it may be two) present themselves in the exterior boundary of the figure.

But now suppose that the theorem *under these more stringent conditions* is true for n areas: say that it is possible with four colours only, to colour the n areas in such wise that not more than three colours present themselves in the external boundary: then it might be easy to prove that the $n+1$ areas could be coloured with four colours only: but this would be insufficient for the purpose of a general proof; it would be necessary to show further that the $n+1$ areas could be with the four colours only coloured *in accordance with the foregoing boundary condition;* for without this we cannot from the case of the $n+1$ areas pass to the next case of $n+2$ areas. And so in general, whatever more stringent conditions we import into the theorem as regards the n areas, it is necessary to show not only that the $n+1$ areas can be coloured with four colours only, but that they can be coloured in accordance with the more stringent conditions. As already mentioned, I have failed to obtain a proof.

708.

NOTE SUR LA THÉORIE DES COURBES DE L'ESPACE.

[From the *Compte Rendu de l'Association Française pour l'Avancement des Sciences* (1880), pp. 135—139.]

En considérant dans l'espace une courbe d'espèce donnée, déterminée au moyen d'un nombre suffisant de points, la courbe n'est pas déterminée uniquement; mais on a par les points un certain nombre de telles courbes. Par exemple, la courbe unicursale d'ordre $2p$ dépend, comme on voit sans peine, de $8p$ constantes et sera ainsi déterminée par $4p$ points (le cas $p=1$ est une exception): on ne connaît pas, je pense, le nombre des courbes par les $4p$ points; mais pour le cas particulier $p=2$ (c'est-à-dire pour une courbe quartique de seconde espèce, ou autrement dit, une courbe excubo-quartique) ce nombre est $=4$: théorème démontré par moi depuis longtemps par des considérations géométriques. (Voir Salmon, *Geometry of three dimensions*, 3e éd. 1874, p. 319.) Ce n'est que dernièrement que j'ai considéré la question analytique, de trouver les équations d'une courbe excubo-quartique qui passe par 8 points donnés; et même j'ai pris pour les 8 points une disposition qui n'est pas tout à fait générale: l'investigation elle-même, et la forme du résultat, m'ont paru assez intéressantes pour que je les soumette à l'Association.

En considérant sur une courbe excubo-quartique 4 points donnés, le plan passant par 3 quelconques de ces points rencontre la courbe dans un seul point; et l'on obtient ainsi encore 4 points sur la courbe: voilà mon système de 8 points donnés, savoir en partant de 4 points quelconques, je prends un point quelconque dans chacun des plans qui passent par 3 de ces points, et j'obtiens ainsi les autres 4 points. Et par un tel système de 8 points, je cherche à faire passer une courbe de l'espèce dont il s'agit.

En prenant $x=0$, $y=0$, $z=0$, $w=0$, pour les équations des plans du tétraèdre formé par les 4 premiers points, les coordonnées de ces points seront $(1, 0, 0, 0)$, $(0, 1, 0, 0)$, $(0, 0, 1, 0)$, $(0, 0, 0, 1)$: et pour les coordonnées des 4 autres points, je prends $(0, y_1, z_1, w_1)$, $(x_2, 0, z_2, w_2)$, $(x_3, y_3, 0, w_3)$, $(x_4, y_4, z_4, 0)$.

Les équations de la courbe sont $x : y : z : w = P : Q : R : S$, où P, Q, R, S sont des fonctions (*)$(\theta, 1)^4$ d'un paramètre variable θ; il s'agit de faire passer une telle courbe par les 8 points.

Je prends α, β, γ, δ, a, b, c, d pour les valeurs du paramètre θ qui correspondent aux 8 points respectivement.

Pour que la courbe passe par les premiers 4 points, il faut et il suffit que les équations soient de la forme

$$x : y : z : w = A\,\frac{\theta - a}{\theta - \alpha} : B\,\frac{\theta - b}{\theta - \beta} : C\,\frac{\theta - c}{\theta - \gamma} : D\,\frac{\theta - d}{\theta - \delta};$$

les conditions pour les autres 4 points seront alors

$$\begin{array}{l} \;.\;\; y_1 : z_1 : w_1 = \quad . \quad B\,\dfrac{a-b}{a-\beta} : C\,\dfrac{a-c}{a-\gamma} : D\,\dfrac{a-d}{a-\delta}, \\[2ex] x_2 \;.\; : z_2 : w_2 = A\,\dfrac{b-a}{b-\alpha} \quad . \quad : C\,\dfrac{b-c}{b-\gamma} : D\,\dfrac{b-d}{b-\delta}, \\[2ex] x_3 : y_3 \;.\; : w_3 = A\,\dfrac{c-a}{c-\alpha} : B\,\dfrac{c-b}{c-\beta} \quad . \quad : D\,\dfrac{c-d}{c-\delta}, \\[2ex] x_4 : y_4 : z_4 \;.\; = A\,\dfrac{d-a}{d-\alpha} : B\,\dfrac{d-b}{d-\beta} : C\,\dfrac{d-c}{d-\gamma} \quad . \;\; . \end{array}$$

Évidemment il y a deux équations qui donnent la valeur de $B : C$, et qui servent ainsi pour éliminer cette quantité. De cette manière on obtient six équations que j'écris comme voici:

$$\lambda = \frac{y_1 z_4}{y_4 z_1} = \frac{a-b\,.\,d-c}{a-c\,.\,d-b} \cdot \frac{a-\gamma\,.\,d-\beta}{a-\beta\,.\,d-\gamma},$$

$$\mu = \frac{w_1 y_3}{y_1 w_3} = \frac{a-d\,.\,c-b}{a-b\,.\,c-d} \cdot \frac{a-\beta\,.\,c-\delta}{a-\delta\,.\,c-\beta},$$

$$\nu = \frac{z_1 w_2}{z_2 w_1} = \frac{a-c\,.\,b-d}{a-d\,.\,b-c} \cdot \frac{a-\delta\,.\,b-\gamma}{a-\gamma\,.\,b-\delta},$$

$$\varpi = \frac{z_2 x_4}{z\,x_2} = \frac{b-c\,.\,d-a}{b-a\,.\,d-c} \cdot \frac{b-\alpha\,.\,d-\gamma}{b-\gamma\,.\,d-\alpha},$$

$$\kappa = \frac{x_2 w_3}{x_3 w_2} = \frac{b-a\,.\,c-d}{b-d\,.\,c-a} \cdot \frac{b-\delta\,.\,c-\alpha}{b-\alpha\,.\,c-\delta},$$

$$\rho = \frac{x_3 y_4}{x_4 y_3} = \frac{c-a\,.\,d-b}{c-b\,.\,d-a} \cdot \frac{c-\beta\,.\,d-\alpha}{c-\alpha\,.\,d-\beta};$$

savoir λ, μ, ν, ϖ, κ, ρ dénotent ici les quantités données $\lambda = \dfrac{y_1 z_4}{y_4 z_1}$, etc. Le nombre des équations indépendantes est 5, car l'on a identiquement $\lambda\mu\nu\varpi\kappa\rho = 1$. Je remarque que l'on peut faire sur le paramètre θ une transformation linéaire quelconque $(h\theta + i) : (j\theta + k)$, et introduire ainsi 3 constantes arbitraires; on peut donc prendre à volonté 3 valeurs du paramètre θ, c'est-à-dire les valeurs de 3 quelconques des quantités $\alpha, \beta, \gamma, \delta, a, b, c, d$; et cela étant les 5 équations donneront les valeurs des autres 5 quantités. Si au moyen des équations on élimine $\alpha, \beta, \gamma, \delta$, on obtient entre a, b, c, d une équation qui sera, comme on va voir, de l'ordre 4 par rapport à chacune de ces quantités: en prenant comme données a, b, c il y aura donc 4 valeurs de d; et pour l'une quelconque de ces valeurs, celles de $\alpha, \beta, \gamma, \delta$ seront déterminées uniquement: il y aura ainsi 4 courbes qui passent chacune par les 8 points; ce qui est le théorème dont il s'agit.

J'introduis, pour abréger, la notation

$$\begin{array}{cccccc} a-d, & b-d, & c-d, & b-c, & c-a, & a-b, \\ =\mathrm{f}, & \mathrm{g}, & \mathrm{h}, & \mathrm{a}, & \mathrm{b}, & \mathrm{c}: \end{array}$$

on a donc identiquement

$$\mathrm{a}, \mathrm{b}, \mathrm{c} = \mathrm{g}-\mathrm{h},\ \mathrm{h}-\mathrm{f},\ \mathrm{f}-\mathrm{g},$$

$$\mathrm{a}+\mathrm{b}+\mathrm{c}=0,$$

$$\mathrm{fa}+\mathrm{gb}+\mathrm{hc}=0.$$

Les équations prennent ainsi la forme

$$\lambda = -\frac{\mathrm{hc}}{\mathrm{gb}}\,\frac{a-\gamma\,.\,d-\beta}{a-\beta\,.\,d-\gamma}, \text{ etc.};$$

ou, en introduisant pour plus de commodité, les symboles

$$L, \quad M, \quad N, \quad P, \quad Q, \quad R,$$

pour désigner respectivement

$$-\frac{\mathrm{gb}}{\mathrm{hc}}\lambda, \quad -\frac{\mathrm{hc}}{\mathrm{fa}}\mu, \quad -\frac{\mathrm{fa}}{\mathrm{gb}}\nu, \quad -\frac{\mathrm{hc}}{\mathrm{fa}}\varpi, \quad -\frac{\mathrm{gb}}{\mathrm{hc}}\kappa, \quad -\frac{\mathrm{fa}}{\mathrm{gb}}\rho,$$

les équations seront

$$L=\frac{a-\gamma\,.\,d-\beta}{a-\beta\,.\,d-\gamma},$$

$$M=\frac{a-\beta\,.\,c-\delta}{a-\delta\,.\,c-\beta},$$

$$N=\frac{a-\delta\,.\,b-\gamma}{a-\gamma\,.\,b-\delta},$$

$$P=\frac{b-\alpha\,.\,d-\gamma}{b-\gamma\,.\,d-\alpha},$$

$$Q=\frac{b-\delta\,.\,c-\alpha}{b-\alpha\,.\,c-\delta},$$

$$R=\frac{c-\beta\,.\,d-\alpha}{c-\alpha\,.\,d-\beta},$$

avec la relation identique $LMNPQR=1$; il s'agit entre ces 5 équations d'éliminer $\alpha, \beta, \gamma, \delta$.

J'écris $\alpha = a - \phi$, les facteurs $b - \alpha$, $c - \alpha$, $d - \alpha$ de P, Q, R deviennent ainsi respectivement $-\mathrm{c} + \phi$, $\mathrm{g} + \phi$, $-\mathrm{f} + \phi$; cela étant, les valeurs de P, Q, R servent à exprimer β, γ, δ en fonction de ϕ: substituant ces valeurs de β, γ, δ dans celles de L, M, N, on obtient sans peine

$$L = -\frac{\mathrm{h}}{\mathrm{g}P}\,\frac{\mathrm{f}(\mathrm{c}-\phi) + \mathrm{c}P(-\mathrm{f}+\phi)}{\mathrm{b}(-\mathrm{f}+\phi) + \mathrm{f}R(\mathrm{b}+\phi)},$$

$$M = -\frac{\mathrm{a}}{\mathrm{h}R}\,\frac{\mathrm{b}(-\mathrm{f}+\phi) + \mathrm{f}R(\mathrm{b}+\phi)}{\mathrm{c}(\mathrm{b}+\phi) + \mathrm{b}Q(-\mathrm{c}+\phi)},$$

$$N = -\frac{\mathrm{g}}{\mathrm{a}Q}\,\frac{\mathrm{c}(\mathrm{b}+\phi) + \mathrm{b}Q(-\mathrm{c}+\phi)}{\mathrm{f}(\mathrm{c}-\phi) + \mathrm{c}P(-\mathrm{f}+\phi)},$$

valeurs qui donnent, comme cela doit être, $LMNPQR = 1$: il faut entre ces équations éliminer ϕ.

En rétablissant λ, μ, ν, ϖ, κ, ρ au lieu de L, M, N, P, Q, R, ces équations deviennent

$$\xi = \frac{\mathrm{g}}{\mathrm{h}}\lambda\varpi = \frac{X + Y\phi}{X_1 + Y_1\phi},$$

$$\eta = \frac{\mathrm{h}}{\mathrm{a}}\,\mu\rho = \frac{X_1 + Y_1\phi}{X_2 + Y_2\phi},$$

$$\zeta = \frac{\mathrm{a}}{\mathrm{g}}\,\kappa\nu = \frac{X_2 + Y_2\phi}{X + Y\phi},$$

(évidemment $\xi\eta\zeta = 1$), où j'écris ξ, η, ζ pour dénoter les expressions $\frac{\mathrm{b}}{\mathrm{c}}\lambda\varpi$, etc., et où les valeurs des coefficients X, Y, etc., sont

$$\begin{aligned} X &= \mathrm{fc}\,(\mathrm{fa} + \varpi\mathrm{hc}), & Y &= -\mathrm{f}^2\mathrm{a} - \varpi\mathrm{hc}^2, \\ X_1 &= \mathrm{fb}\,(\mathrm{gb} + \rho\mathrm{fa}), & Y_1 &= \mathrm{gb}^2 + \rho\mathrm{f}^2\mathrm{a}, \\ X_2 &= \mathrm{bc}\,(\mathrm{hc} + \kappa\mathrm{gb}), & Y_2 &= \mathrm{hc}^2 - \kappa\mathrm{gb}^2. \end{aligned}$$

Les deux premières équations donnent

$$\xi\eta\,(X_1Y_2 - X_2Y_1) + \eta\,(X_2Y - XY_2) + XY_1 - XY_1 = 0,$$

ou, ce qui est la même chose,

$$\frac{1}{\zeta}(X_1Y_2 - X_2Y_1) + \eta\,(X_2Y - XY_2) + XY_1 - XY_1 = 0,$$

et l'on n'a qu'à substituer la valeur de ces coefficients.

On a

$$\begin{aligned} X_1Y_2 - X_2Y_1 &= \mathrm{fb}\,(\mathrm{gb} + \rho\mathrm{fa})\,(\mathrm{hc}^2 - \kappa\mathrm{gb}^2) - \mathrm{bc}\,(\mathrm{hc} + \kappa\mathrm{gb})\,(-\,\mathrm{gb}^2 + \rho\mathrm{f}^2\mathrm{a}) \\ &= \mathrm{fghb}^2\mathrm{c}^2 - \mathrm{fg}^2\mathrm{b}^4\kappa + \mathrm{f}^2\mathrm{habc}^2\rho - \mathrm{f}^2\mathrm{gab}^3\kappa\rho + \mathrm{ghb}^3\mathrm{c}^2 + \mathrm{g}^2\mathrm{b}^4\mathrm{c}\kappa - \mathrm{f}^2\mathrm{habc}^2\rho - \mathrm{f}^2\mathrm{gab}^2\mathrm{c}\kappa\rho \\ &= \mathrm{ghb}^2\mathrm{c}^2\,(\mathrm{f} + \mathrm{b}) + \mathrm{g}^2\mathrm{b}^4\,(-\,\mathrm{f} + \mathrm{c})\,\rho - \mathrm{f}^2\mathrm{gab}^2\,(\mathrm{b} + \mathrm{c})\,\kappa\rho \\ &= \mathrm{ghb}^2\mathrm{c}^2\mathrm{h} + \mathrm{g}^2\mathrm{b}^4\,(-\,\mathrm{g})\,\rho + \mathrm{f}^2\mathrm{gab}^2\mathrm{a}\kappa\rho \\ &= \mathrm{gb}^2\,(\mathrm{h}^2\mathrm{c}^2 - \mathrm{g}^2\mathrm{b}^2\kappa + \mathrm{a}^2\mathrm{f}^2\kappa\rho)\,; \end{aligned}$$

et de même

$$XY_2 - X_2Y = \mathrm{hc}^2(\mathrm{f}^2\mathrm{a}^2 - \mathrm{h}^2\mathrm{c}^2\varpi + \mathrm{g}^2\mathrm{b}^2\varpi\kappa),$$

$$XY_1 - X_1Y = \mathrm{f}^2\mathrm{a}(\mathrm{g}^2\mathrm{b}^2 - \mathrm{f}^2\mathrm{a}^2\rho + \mathrm{h}^2\mathrm{c}^2\varpi\rho).$$

Donc

$$\frac{\mathrm{g}}{\mathrm{a}}\frac{1}{\kappa\nu}\mathrm{gb}^2(\mathrm{h}^2\mathrm{c}^2 - \mathrm{g}^2\mathrm{b}^2\kappa + \mathrm{f}^2\mathrm{a}^2\kappa\rho)$$

$$+\frac{\mathrm{h}}{\mathrm{a}}\mu\rho\,\mathrm{hc}^2(\mathrm{f}^2\mathrm{a}^2 - \mathrm{h}^2\mathrm{c}^2\varpi + \mathrm{g}^2\mathrm{b}^2\varpi\kappa)$$

$$+\quad \mathrm{f}^2\mathrm{a}(\mathrm{g}^2\mathrm{b}^2 - \mathrm{f}^2\mathrm{a}^2\rho + \mathrm{h}^2\mathrm{c}^2\varpi\rho) = 0,$$

ou enfin en multipliant par $-\mathrm{a}\nu$, et dans un terme $-\mathrm{g}^2\mathrm{b}^2\mathrm{h}^2\mathrm{c}^2\mu\nu\rho\varpi\kappa$, au lieu de $\mu\nu\rho\varpi\kappa$ écrivant $\frac{1}{\lambda}$, l'équation devient

$$(\mathrm{fa})^4\nu\rho + (\mathrm{gb})^4 + (\mathrm{hc})^4\frac{1}{\lambda\kappa} - (\mathrm{gb})^2(\mathrm{hc})^2\left(\frac{1}{\kappa}+\frac{1}{\lambda}\right)$$

$$-(\mathrm{hc})^2(\mathrm{fa})^2\nu\rho(\varpi+\mu) - (\mathrm{fa})^2(\mathrm{gb})^2(\nu+\rho) = 0,$$

ou, comme on peut l'écrire,

$$\left(\nu\rho,\ 1,\ \frac{1}{\lambda\kappa},\ -\left(\frac{1}{\kappa}+\frac{1}{\lambda}\right),\ -\nu\rho(\varpi+\mu),\ -(\nu+\rho)\right)((\mathrm{fa})^2,\ (\mathrm{gb})^2,\ (\mathrm{hc})^2)^2 = 0.$$

C'est la deuxième d'un système de trois équations équivalentes; savoir, en multipliant par $\frac{1}{\nu\rho}$ et en réduisant par $\lambda\mu\nu\varpi\kappa\rho = 1$, on obtient la première forme: et, en multipliant par $\lambda\kappa$ et réduisant de même, on obtient la troisième forme: le système est

$$\left(1,\ \frac{1}{\nu\rho},\ \mu\varpi,\ -\mu\varpi(\lambda+\kappa),\ -(\mu+\varpi),\ -\left(\frac{1}{\nu}+\frac{1}{\rho}\right)\right)((\mathrm{fa})^2,\ (\mathrm{gb})^2,\ (\mathrm{hc})^2)^2 = 0,$$

$$\left(\nu\rho,\ 1,\ \frac{1}{\lambda\kappa},\ -\left(\frac{1}{\kappa}+\frac{1}{\lambda}\right),\ -\nu\rho(\mu+\varpi),\ -(\nu+\rho)\right)(\ ,,\quad ,,\quad ,,\ )^2 = 0,$$

$$\left(\frac{1}{\mu\varpi},\ \lambda\kappa,\ 1,\ -(\lambda+\kappa),\ -\left(\frac{1}{\mu}+\frac{1}{\varpi}\right),\ -\lambda\kappa(\nu+\rho)\right)(\ ,,\quad ,,\quad ,,\ )^2 = 0.$$

En écrivant $\mathrm{hc} = -\mathrm{fa} - \mathrm{gb}$, on obtient une équation de la forme $(*)(\mathrm{fa},\ \mathrm{gb})^4 = 0$, savoir une équation quartique pour avoir fa : gb, c'est-à-dire, le rapport anharmonique $(a-d)(b-c) : (b-d)(c-a)$: en considérant a, b, c comme données, il y a donc 4 valeurs de d: et l'on a déjà vu que les valeurs α, β, γ, δ sont données rationnellement en fonctions de a, b, c, d: le théorème est donc démontré.

Cambridge, juillet, 1880.

709.

ON THE NUMBER OF CONSTANTS IN THE EQUATION OF A SURFACE $PS-QR=0$.

[From the *Tidsskrift for Mathematik*, Ser. 4, t. IV. (1880), pp. 145—148.]

THE very important results contained in Mr H. Valentiner's paper "Nogle Sætninger om fuldstændige Skjæringskurver mellem to Flader" may be considered from a somewhat different point of view, and established in a more simple manner, as follows*.

Assuming throughout $n \geqq p+q$, $p \geqq q$, and moreover that P, Q, R, S denote functions of the coordinates (x, y, z, w) of the orders p, q, $n-q$, $n-p$ respectively: then the equation of a surface of the order n containing the curve of intersection of two surfaces of the orders p and q respectively, is

$$\begin{vmatrix} P, & Q \\ R, & S \end{vmatrix} = 0,$$

so that the number of constants in the equation of a surface of the order n satisfying the condition in question is in fact the number of constants contained in an equation of the last-mentioned form. Writing for shortness

$$a_p = \tfrac{1}{6}(p+1)(p+2)(p+3)-1, \; = \tfrac{1}{6}p(p^2+6p+11),$$

the number of constants contained in a function of the order p is $=a_p+1$; or if we take one of the coefficients (for instance that of x^p) to be unity, then the number

* Idet vi med stor Glæde optage Prof. Cayley's simple Forklaring af den Reduktion af Konstanttallet i Ligningen $PS-QR=0$, som Hr. Valentiner havde paavist (*Tidsskr. f. Math.* 1879, S. 22), skulle vi dog bemærke, at Grunden til, at dennes Bevis er bleven saa vanskeligt, er den, at han tillige har villet bevise, at der ikke finder nogen *yderligere* Reduktion Sted.

of the remaining constants is $=a_p$; viz. a_p is the number of constants in the equation of a surface of the order p. As regards the surface in question

$$\begin{vmatrix} P, & Q \\ R, & S \end{vmatrix} = 0,$$

we may it is clear take P, Q, R each with a coefficient unity as above, but in the remaining function S, the coefficient must remain arbitrary: the apparent number of constants is thus $= a_p + a_q + a_{n-p} + a_{n-q} + 1$; but there is a deduction from this number.

The equation may in fact be written in the form

$$\begin{vmatrix} P+\alpha Q, & Q \\ R+\alpha S+\beta P+\alpha\beta Q, & S+\beta Q \end{vmatrix} = 0,$$

where α represents an arbitrary function of the order $p-q$, and β an arbitrary function of the degree $n-p-q$: we thus introduce $(a_{p-q}+1)+(a_{n-p-q}+1)$, $= a_{p-q}+a_{n-p-q}+2$, constants, and by means of these we can impose the like number of arbitrary relations upon the constants originally contained in the functions P, Q, R, S respectively (say we can reduce to zero this number $a_{p-q}+a_{n-p-q}+2$ of the original constants): hence the real number of constants is

$$\begin{aligned} & a_p + a_q + a_{n-p} + a_{n-q} + 1 - (a_{p-q} + a_{n-p-q} + 2), \\ & = a_p + a_q + a_{n-p} + a_{n-q} - a_{p-q} - a_{n-p-q} - 1 \\ & = \omega \text{ suppose}; \end{aligned}$$

viz. this is the required number in the case $n > p+q$, $p > q$.

If however $n = p+q$, or $p = q$, or if these relations are both satisfied, then there is a further deduction of 1, 1, or 2: in fact, calling the last-mentioned determinant $\begin{vmatrix} P', & Q' \\ R', & S' \end{vmatrix}$, then the four cases are

$$n > p+q,\ p > q,\quad \begin{vmatrix} P', & Q' \\ R', & S' \end{vmatrix} = \begin{vmatrix} P', & Q' \\ R', & S' \end{vmatrix}$$

$$n = p+q,\ p > q,\quad \begin{vmatrix} P', & Q' \\ R', & S' \end{vmatrix} = \begin{vmatrix} P'+kR', & Q'+kS' \\ R', & S' \end{vmatrix}$$

$$n > p+q,\ p = q,\quad \begin{vmatrix} P', & Q' \\ R', & S' \end{vmatrix} = \begin{vmatrix} P', & Q'+kP' \\ R', & S'+kR' \end{vmatrix}$$

$$n = p+q,\ p = q,\quad \begin{vmatrix} P', & Q' \\ R', & S' \end{vmatrix} = \begin{vmatrix} P'+kR', & Q'+lP'+kS'+klR' \\ R', & S'+lR' \end{vmatrix}$$

where k, l denote arbitrary constants: these, like the constants of α and β, may be used to impose arbitrary relations upon the original constants of P, Q, R, S; and hence the number of constants is $=\omega$, $\omega-1$, $\omega-1$, $\omega-2$ in the four cases respectively; where as above

$$\omega = a_p + a_q + a_{n-p} + a_{n-q} - a_{p-q} - a_{n-p-q} - 1.$$

If $n=4$, there is in each of the four cases one system of values of p, q; viz. the cases are

$p, q =$		
2 1	No. $= a_2 + a_1 + a_2 + a_3 - a_1 - a_1 - 1 =$	$9+3+\ 9+19-3-3-1, =33,$
3 1	" $a_3 + a_1 + a_1 + a_3 - a_2 - a_0 - 2 =$	$19+3+\ 3+19-9-0-2, =33,$
1 1	" $a_1 + a_1 + a_3 + a_3 - a_0 - a_2 - 2 =$	$3+3+19+19-0-9-2, =33,$
2 2	" $a_2 + a_2 + a_2 + a_2 - a_0 - a_0 - 3 =$	$9+9+\ 9+\ 9-0-0-3, =33,$

and the number of constants is in each case $=33$. This is easily verified: in the first case we have a quartic surface containing a conic, the plane of the conic is therefore a quadruple tangent plane; and the existence of such a plane is 1 condition. In the second case the surface contains a plane cubic; the plane of this cubic is a triple tangent plane, having the points of contact in a line; and this is 1 condition. In the third case the surface contains a line, which is 1 condition: hence in each of these cases the number of constants is $34-1, =33$. In the fourth case, where the surface contains a quadriquadric curve, we repeat in some measure the general reasoning: the quadriquadric curve contains 16 constants, and we have thus 16 as the number of constants really contained in the equations $P=0$, $Q=0$ of the quadriquadric curve: the equation $PS-QR=0$, contains in addition $9+10, =19$ constants, but writing it in the form $P(S+kQ)-Q(R+kP)=0$, we have a diminution $=1$, or the number apparently is $16+19-1, =34$. But the quadriquadric curve is one of a singly infinite series $P+lR=0$, $Q+lS=0$ of such curves, and we have on this account a diminution $=1$; the number of constants is thus $34-1, =33$ as above: the reasoning is, in fact, the same as for the case of a plane passing through a line; the line contains 4 constants, hence the plane, quà arbitrary plane through the line, would contain $1+4, =5$ constants; but the line being one of a doubly infinite system of lines on the plane the number is really $5-2, =3$, as it should be.

Cambridge, 2nd Sept., 1880.

710.

ON A DIFFERENTIAL EQUATION.

[From *Collectanea Mathematica: in memoriam Dominici Chelini*, (Milan, Hoepli, 1881), pp. 17—26.]

IN the Memoir on hypergeometric series, *Crelle*, t. XV. (1836), Kummer in effect considers a differential equation

$$\frac{(a'z^2+2b'z+c')\,dz^2}{z^2(z-1)^2}=\frac{(ax^2+2bx+c)\,dx^2}{x^2(x-1)^2},$$

viz. he seeks for solutions of an equation of this form which also satisfy a certain differential equation of the third order. The coefficients a, b, c are either all arbitrary, or they are two or one of them, arbitrary; but this last case (or say the case where the function of x is the completely determinate function $x^2+2bx+c$) is scarcely considered: a', b', c' are regarded as determinable in terms of a, b, c; and z is to be found as a function of x independent of a, b, c: so that when these coefficients are arbitrary, the equation breaks up into three equations, and when two of the coefficients are arbitrary, it breaks up into two equations, satisfied in each case by the same value of z; and the value of z is thus determined without any integration: these cases will be considered in the sequel, but they are of course included in the general case where the coefficients a, b, c are regarded as having any given values whatever.

Writing for shortness $X=ax^2+2bx+c$, in general the integral

$$\int\frac{N\,dx}{D\sqrt{X}},$$

where D is the product of any number n of distinct linear factors $x-p$, and N is a rational and integral function of x of the order n at most, and therefore also the integral

$$\int\frac{N\sqrt{X}\,dx}{D}=\int\frac{NX\,dx}{D\sqrt{X}},$$

where N is now of the order $n-2$ at most, is expressible as the logarithm of a quasi-algebraical function, that is, a function containing powers the exponents of which are incommensurable (for instance, $x^{\sqrt{2}}$ is a quasi-algebraical function): in fact, the integral is of the form

$$\int\left(M+\frac{A}{x-p}+\frac{B}{x-q}+\ldots\right)\frac{dx}{\sqrt{X}},$$

where each term is separately integrable,

$$\int\frac{dx}{\sqrt{X}}=\frac{1}{\sqrt{a}}\log\{ax+b+\sqrt{a}\,.\,\sqrt{X}\},$$

$$\int\frac{dx}{(x-p)\sqrt{X}}=-\frac{1}{\sqrt{P}}\log\left\{\frac{(ap+b)\,x+(bp+c)+\sqrt{P}\,.\,\sqrt{X}}{x-p}\right\},$$

where P is written to denote $ap^2+2bp+c$: the integral is thus $=\log\Omega$, where Ω is a product of factors

$$ax+b+\sqrt{a}\,.\,\sqrt{X},\quad \frac{(ap+b)\,x+(bp+c)+\sqrt{P}\,.\,\sqrt{X}}{x-p},\ \text{etc.},$$

raised to powers $\frac{M}{\sqrt{a}}$, $\frac{-A}{\sqrt{P}}$, etc.: hence, if we have a differential equation

$$\frac{N'dz}{D'\sqrt{Z}}=\frac{Ndx}{D\sqrt{X}},\quad\text{or}\quad\frac{N'\sqrt{Z}dz}{D'}=\frac{N\sqrt{X}dx}{D},$$

where $Z\,(=a'z^2+2b'z+c')$, and N', D' are functions of z such as X, N, D are of x; then, taking $\log C$ for the constant of integration, the general integral is

$$\log\Omega'=\log C+\log\Omega:$$

viz. we have the quasi-algebraical integral $\Omega'-C\Omega=0$.

The constants a, b, c, p, q, ... etc. may be such that the exponents are rational, and the integral is then algebraical: in particular, for the differential equation

$$\frac{\sqrt{z^2+14z+1}\,dz}{z\,(z-1)}=\frac{\sqrt{x^2+14x+1}\,dx}{x\,(x-1)},$$

the general integral is in the first instance obtained in the form

$$\frac{(z+1+\sqrt{Z})\,(z-1)^2}{\sqrt{z}\,(2z+2+\sqrt{Z})^2}=C\frac{(x+1+\sqrt{X})\,(x-1)^2}{\sqrt{x}\,(2x+2+\sqrt{X})^2}$$

which, observing that $(2x+2)^2-X=3\,(x-1)^2$, may also be written

$$\frac{(z+1)\,(z^2-34z+1)+Z\sqrt{Z}}{\sqrt{z}\,(z-1)^2}=C\frac{(x+1)\,(x^2-34x+1)+X\sqrt{X}}{\sqrt{x}\,(x-1)^2}.$$

I had previously obtained the solution

$$z = \left(\frac{1-\sqrt[4]{x}}{1+\sqrt[4]{x}}\right)^4,$$

and I wish to show that this is, in fact, the particular integral belonging to the value $C=1$ of the constant of integration: for this purpose I proceed to rationalise the general integral as regards z.

Writing for a moment

$$P = (z+1)(z^2-34z+1),$$

$$Q = (z^2+14z+1)\sqrt{z^2+14z+1},$$

$$R = M\sqrt{z}\,(z-1)^2,$$

where

$$M = C\,\frac{(x+1)(x^2-34x+1)+(x^2+14x+1)\sqrt{x^2+14x+1}}{\sqrt{x}\,(x-1)^2},$$

the integral is $P+Q+R=0$; or rationalising, it is

$$(P^2-Q^2)^2 - 2R^2(P^2+Q^2) + R^4 = 0;$$

we have

$$P^2 = (1,\ -66,\ 1023,\ 2180,\ 1023,\ -66,\ 1 \between z,\ 1)^6,$$

$$Q^2 = (1,\ 42,\ 591,\ 2828,\ 591,\ 42,\ 1 \between z,\ 1)^6,$$

and thence

$$P^2-Q^2 = (0,\ -108,\ 432,\ -648,\ 432,\ -108,\ 0 \between z,\ 1)^6,$$

$$= -108z(z-1)^4;$$

$$P^2+Q^2 = 2(1,\ -12,\ 807,\ 2504,\ 807,\ -12,\ 1 \between z,\ 1)^6.$$

Writing the equation in the form

$$\tfrac{1}{2}(P^2+Q^2) - \tfrac{1}{4}\left\{R^2 + \frac{(P^2-Q^2)^2}{R^2}\right\} = 0,$$

it thus becomes

$$(1,\ -12,\ 807,\ 2504,\ 807,\ -12,\ 1 \between z,\ 1)^6 - z(z-1)^4\left\{M^2 + \frac{(108)^2}{M^2}\right\} = 0,$$

where M has its above-mentioned value; and if we now assume $C=1$, then

$$M = \frac{(x+1)(x^2-34x+1)+(x^2+14x+1)\sqrt{x^2+14x+1}}{\sqrt{x}\,(x-1)^2},$$

$$\frac{108}{M} = \frac{(x+1)(x^2-34x+1)-(x^2+14x+1)\sqrt{x^2+14x+1}}{\sqrt{x}\,(x-1)^2},$$

and thence

$$M^2 + \frac{(108)^2}{M^2},\ = \left(M - \frac{108}{M}\right)^2 + 216,\ = 4\,\frac{(x+1)^2(x^2-34x+1)^2}{x(x-1)^4} + 216,$$

$$= \frac{4}{x(x-1)^4}\cdot(1,\ -12,\ 807,\ 2504,\ 807,\ -12,\ 1 \between x,\ 1)^6:$$

and the rationalised equation is

$$(1,\ -12,\ 807,\ 2504,\ 807,\ -12,\ 1\between z,\ 1)^6$$
$$-\frac{z(z-1)^4}{x(x-1)^4}(1,\ -12,\ 807,\ 2504,\ 807,\ -12,\ 1\between x,\ 1)^6=0.$$

This is a sextic equation in z, of the form

$$z^3+\frac{1}{z^3}+\lambda\left(z^2+\frac{1}{z^2}\right)+\mu\left(z+\frac{1}{z}\right)+\nu=0,$$

where

$$\lambda,\ \mu,\ \nu=-12-\Omega,\quad 807+4\Omega,\quad 2504-6\Omega,$$

if Ω denote the function of x which enters into the equation; and writing $z+\frac{1}{z}=\theta$, this becomes

$$\theta^3-3\theta+\lambda(\theta^2-2)+\mu\theta+\nu=0.$$

But the equation in z is satisfied by the value $z=x$, and therefore the equation in θ by the value $\theta=x+\frac{1}{x}=\alpha$ suppose, we have therefore

$$\alpha^3-3\alpha+\lambda(\alpha^2-2)+\mu\alpha+\nu=0,$$

and thence subtracting, and throwing out the factor $\theta-\alpha$,

$$\theta^2+\theta\alpha+\alpha^2-3+\lambda(\theta+\alpha)+\mu=0;$$

viz. writing for λ, μ, α their values, this is

$$\theta^2+\theta\left(x+\frac{1}{x}-12-\Omega\right)+x^2-1+\frac{1}{x^2}-\left(x+\frac{1}{x}\right)(12+\Omega)+807+4\Omega=0,$$

or, what is the same thing,

$$\theta^2+\theta\left(x-12+\frac{1}{x}-\Omega\right)+x^2-12x+806-\frac{12}{x}+\frac{1}{x^2}-\left(x-4-\frac{1}{x}\right)\Omega=0,$$

where

$$\Omega=\frac{1}{x(x-1)^4}(1,\ -12,\ 807,\ 2504,\ 807,\ -12,\ 1\between x,\ 1)^6.$$

Hence in the quadric equation, the coefficients, each multiplied by $(x-1)^4$, are

$$(x-1)^4\left(x-12+\frac{1}{x}\right)-\frac{1}{x}(1,\ -12,\ 807,\ 2504,\ 807,\ -12,\ 1\between x,\ 1)^6,$$

and

$$(x-1)^4\left(x^2-12x+806-\frac{12}{x}+\frac{1}{x^2}\right)$$
$$-\frac{1}{x}\left(x-4+\frac{1}{x}\right)(1,\ -12,\ 807,\ 2504,\ 807,\ -12,\ 1\between x,\ 1)^6,$$

which are respectively rational and integral quartic functions of x; and, writing for θ its value, the equation finally is

$$\left(z+\frac{1}{z}\right)^2-4\left(z+\frac{1}{z}\right)\frac{(1,\ 188,\ 646,\ 188,\ 1\between x,\ 1)^4}{(x-1)^4}+4\,\frac{(1,\ -644,\ 3334,\ -644,\ 1\between x,\ 1)^4}{(x-1)^4}=0$$

Writing

$$\xi = \sqrt[4]{x},\quad A = \frac{1-\xi}{1+\xi},\quad B = \frac{1+\xi}{1-\xi},\quad C = \frac{1-i\xi}{1+i\xi},\quad D = \frac{1+i\xi}{1-i\xi},\qquad (i = \sqrt{-1} \text{ as usual}),$$

this is

$$(z - A^4)(z - B^4)(z - C^4)(z - D^4) = 0,$$

or, what is the same thing,

$$\left\{z + \frac{1}{z} - (A^4 + B^4)\right\}\left\{z + \frac{1}{z} - (C^4 + D^4)\right\} = 0,$$

that is,

$$\left(z + \frac{1}{z}\right)^2 - \left(z + \frac{1}{z}\right)(A^4 + B^4 + C^4 + D^4) + (A^4 + B^4)(C^4 + D^4) = 0;$$

for we have

$$\tfrac{1}{2}(A^4 + B^4) = \frac{(1,\ \ 28,\ 70,\ \ 28,\ 1\,)\!(\,\xi^2,\ 1)^4}{(\xi^2 - 1)^4},$$

$$\tfrac{1}{2}(C^4 + D^4) = \frac{(1,\ -28,\ 70,\ -28,\ 1\,)\!(\,\xi^2,\ 1)^4}{(\xi^2 + 1)^4}.$$

And substituting these values, the coefficients will be rational functions of ξ^4, that is, of x, and it is easy to verify that they have in fact their foregoing values.

It thus appears that for $C = 1$, besides the values x and $\frac{1}{x}$, we have for z only the values

$$\left(\frac{1-\xi}{1+\xi}\right)^4,\quad \left(\frac{1+\xi}{1-\xi}\right)^4,\quad \left(\frac{1-i\xi}{1+i\xi}\right)^4,\quad \left(\frac{1+i\xi}{1-i\xi}\right)^4;$$

viz. that the only solution is

$$z = \left(\frac{1-\sqrt[4]{x}}{1+\sqrt[4]{x}}\right)^4.$$

The example shows that although the differential equation

$$\frac{\sqrt{a'z^2 + 2b'z + c'}\,dz}{z(z-1)} = \frac{\sqrt{ax^2 + 2bx + c}\,dx}{x(x-1)}$$

can be integrated generally in a quasi-algebraical or algebraical form as above, yet we cannot from the general solution deduce, at once or easily, the various particular integrals comprised therein: nor can we find for what values of the constants a, b, c and a', b', c' the differential equation admits of a simple solution, or say of a solution where z is expressed as an explicit (irrational) function of x.

In the cases considered by Kummer there is a second (or it may be also a third) differential equation of the like form, the equations being each of them satisfied by the same value of z: hence eliminating the differentials dx, dz, the relation between x and z is of the form

$$\frac{P'}{Q'} = \frac{P}{Q},$$

where P, Q are quadric functions of x; P', Q' quadric functions of z. But P and Q may contain a common factor, and the integral is then expressible in the form $x=\frac{P'}{Q'}$, the quotient of two quadric functions of z; or P' and Q' may have a common factor, and the integral is then expressible in the form $z=\frac{P}{Q}$, the quotient of two quadric functions of x; or there may be a common factor of P, Q, and also a common factor of P' and Q', and the integral is then of the form $z=\frac{L}{M}$, the quotient of two linear functions of x.

In the general case the differential equation is

$$\frac{\lambda(aP'+bQ')\,dz^2}{z^2(z-1)^2}=\frac{(aP+bQ)\,dx^2}{x^2(x-1)^2},$$

where a, b are arbitrary constants, λ is a constant the value of which can in each particular case be at once determined; so when the integral is $z=\frac{P}{Q}$, the differential equation is

$$\frac{\lambda\,(az+b)\,dz^2}{z^2(z-1)^2}=\frac{(aP+bQ)\,dx^2}{x^2(x-1)^2},$$

where a, b are arbitrary constants, but λ is now a linear function of z the value of which can in each particular case be at once determined. When the integral is $z=\frac{L}{M}$, the differential equation is

$$\frac{\lambda\,(az^2+2bz+c)\,dz^2}{z^2(z-1)^2}=\frac{(aL^2+2bLM+cM^2)\,dx^2}{x^2(x-1)^2}$$

containing the three arbitrary constants a, b, c; λ is a constant the value of which can be at once determined.

There are in all 6 integrals of the form $z=\frac{L}{M}$, for which the differential equation contains three arbitrary constants: 18 integrals of the form $z=\frac{P}{Q}$ $\left(\text{and of course the same number of integrals of the form } x=\frac{P'}{Q'}\right)$, and 9 integrals of the form $\frac{P}{Q}=\frac{P'}{Q'}$, for all of which the differential equation contains two arbitrary constants. It is to be remarked that Kummer, considering the values of z as a function of x, obtains the 72 rational and irrational values mentioned in his equations (31), (35), (36), (37), (38), and (39): but the 72 values are made up as follows, viz. the 18 values of z as a rational function of x, the 36 irrational values obtained from the 18 expressions of x as a rational function of z, and the 18 irrational values of z obtained from the 9 integrals in which neither of the variables is a rational function of the other: $18+36+18=72$.

The several integrals together with the expressions of the functions

$$a'z^2 + 2b'z + c' \quad \text{and} \quad ax^2 + 2bx + c$$

which enter into the differential equation are as follows:

	$z =$	$a'z^2 + 2b'z + c' =$	$ax^2 + 2bx + c =$
1.	x	$az^2 + 2bz + c$	$ax^2 + 2bx + c$
	$1 - x$	,,	$a(x-1)^2 - 2b(x-1) + c$
	$\dfrac{1}{x}$	,,	$a + 2bx + cx^2$
	$\dfrac{1}{1-x}$	,,	$a - 2b(x-1) + c(x-1)^2$
	$\dfrac{x}{x-1}$	,,	$ax^2 + 2bx(x-1) + c(x-1)^2$
	$\dfrac{x-1}{x}$	,,	$a(x-1)^2 + 2bx(x-1) + cx^2$
2.	$\left(\dfrac{x+1}{x-1}\right)^2$	$az^2 + bz$	$a(x+1)^2 + b(x-1)^2$
	$(2x-1)^2$	,,	$a(2x-1)^2 + b$
	$\left(\dfrac{x-2}{x}\right)^2$	,,	$a(x-2)^2 + bx^2$
	$\dfrac{(x+1)^2}{4x}$	,,	$a(x+1)^2 + 4bx$
	$\dfrac{(2x-1)^2}{4x(x-1)}$	,,	$a(2x-1)^2 + 4bx(x-1)$
	$-\dfrac{(x-2)^2}{4(x-1)}$	,,	$a(x-2)^2 - 4b(x-1)$
3.	$\left(\dfrac{x-1}{x+1}\right)^2$	$bz + c$	$b(x-1)^2 + c(x+1)^2$
	$\left(\dfrac{1}{2x-1}\right)^2$	,,	$b + c(2x-1)^2$
	$\left(\dfrac{x}{x-2}\right)^2$	,,	$bx^2 + c(x-2)^2$
	$\dfrac{4x}{(x+1)^2}$	,,	$4bx + c(x+1)^2$
	$\dfrac{4x(x-1)}{(2x-1)^2}$	,,	$4bx(x-1) + c(2x-1)^2$
	$-\dfrac{4(x-1)}{(x-2)^2}$	,,	$-4b(x-1) + c(x-2)^2$

	$z=$	$a'z^2+2b'z+c'=$	$ax^2+2bx+c=$
4.	$-\frac{(x-1)^2}{4x}$	$az^2-(a+c)z+c$	$a(x-1)^2+4cx$
	$-4x(x-1)$	,,	$4ax(x-1)+c$
	$\frac{4(x-1)}{x^2}$	,,	$-4a(x-1)+cx^2$
	$-\frac{4x}{(x-1)^2}$	,,	$4ax+c(x-1)^2$
	$\frac{-1}{4x(x-1)}$	,,	$a+4cx(x-1)$
	$\frac{x^2}{4(x-1)}$	,,	$ax^2-4c(x-1)$

5.
6. same as 2, 3, 4 interchanging x and z.
7.

	$z=$	$a'z^2+2b'z+c'=$	$ax^2+2bx+c=$
8.	$\frac{(z-1)^2}{4z}=\frac{4x}{(x-1)^2}$	$a(z-1)^2+4bz$	$4ax+b(x-1)^2$
	$\frac{z^2}{4(z-1)}=\frac{4(x-1)}{x^2}$	$az^2+4b(z-1)$	$-4a(x-1)-bx^2$
	$4z(z-1)=\frac{1}{4x(x-1)}$	$4az(z-1)+b$	$a+4bx(x-1)$
9.	$\frac{(z-1)^2}{4z}=4x(x-1)$	$a(z-1)^2+4bz$	$4ax(x-1)+b$
	$\frac{z^2}{4(z-1)}=-4x(x-1)$	$az^2-4b(z-1)$	$4ax(x-1)+b$
	$\frac{(z-1)^2}{4z}=-\frac{4(x-1)}{x^2}$	$a(z-1)^2+4bz$	$-4a(x-1)+bx^2$
10.	$4z(z-1)=\frac{(x-1)^2}{4x}$	$4az(z-1)+b$	$a(x-1)^2+4bx$
	$4z(z-1)=-\frac{x^2}{4(x-1)}$	$4az(z-1)+b$	$ax^2-4b(x-1)$
	$\frac{4(z-1)}{z^2}=-\frac{(x-1)^2}{4x}$	$-4a(z-1)+bz^2$	$a(x-1)^2+4bx$

The six functions of the set (1), that is,

$$x,\quad 1-x,\quad \frac{1}{x},\quad \frac{1}{1-x},\quad \frac{x}{x-1},\quad \frac{x-1}{x},$$

form a group; and by operating with the substitutions of this group, and of the like group

$$z,\quad 1-z,\quad \frac{1}{z},\quad \frac{1}{1-z},\quad \frac{z}{z-1},\quad \frac{z-1}{z},$$

upon any value of z in the sets (2), (3), (4), for instance upon $z=\left(\frac{x+1}{x-1}\right)^2$, we form all the 18 functions of these sets.

In any one of these sets (2), (3), and (4), comparing two forms (the same or different), for instance in the set (2), writing y for z and then in one form z for x,

$$y=\left(\frac{x+1}{x-1}\right)^2 \text{ and } y=\left(\frac{z+1}{z-1}\right)^2, \text{ whence } \left(\frac{x+1}{x-1}\right)^2=\left(\frac{z+1}{z-1}\right)^2,$$

or

$$y=\left(\frac{x+1}{x-1}\right)^2 \text{ and } y=\frac{(z+1)^2}{4z}, \text{ whence } \left(\frac{x+1}{x-1}\right)^2=\frac{(z+1)^2}{4z},$$

we obtain either the equations of the set (1) or those of the sets (8), (9) and (10); and whether we use the set (2), (3) or (4), the only new equations obtained are thus the 9 equations of the sets (8), (9) and (10). These several equations present themselves however in different forms: for instance, instead of the equation

$$\frac{(z-1)^2}{4z}=\frac{4x}{(x-1)^2},$$

we may obtain

$$\frac{(z+1)^2}{4z}=\left(\frac{x+1}{x-1}\right)^2.$$

If, to get rid of this variety of form, we multiply out the denominators, the 9 equations are

$$\begin{aligned}
0 &= x^2z^2-2x^2z-2xz^2+x^2-12xz+z^2-2x-2z+1,\\
0 &= x^2z^2-16xz+16x+16z-16,\\
0 &= 16x^2z^2-16x^2z-16xz^2+16xz-1,\\
0 &= x^2z^2-2x^2z+x^2+16xz-16z,\\
0 &= 16x^2z-16xz-z^2+2z-1,\\
0 &= 16x^2z-16x^2-16xz+z^2+16x,\\
0 &= x^2z^2-2xz^2+16xz+z^2-16x,\\
0 &= 16xz^2-x^2-16xz+2x+1,\\
0 &= 16xz^2+x^2-16xz-16z^2+16z.
\end{aligned}$$

These 9 equations are derivable all from any one of them by the changes of the set (1) upon x and z.

Cambridge, 3rd June, 1879.

711.

ON A DIAGRAM CONNECTED WITH THE TRANSFORMATION OF ELLIPTIC FUNCTIONS.

[From the *Report of the British Association for the Advancement of Science*, (1881), p. 534.]

THE diagram relates to a known theorem, and is constructed as follows. Consider the infinite half-plane $y = +$; draw in it, centre the origin and radius unity, a semicircle; and draw the infinite half-lines $x = -\frac{1}{2}$, and $x = \frac{1}{2}$; then we have a region included between the lines, but exterior to the semicircle. The region in question may be regarded as a curvilinear triangle, with the angles 60°, 60°, and 0°. The region may be moved parallel to itself in the direction of the axis of x, through the distance 1; say this is a "displacement"; or we may take the "image" of the region in regard to the semicircle. Performing any number of times, and in any order, these two operations of making the displacement and of taking the image, we obtain a new region, which is always a curvilinear triangle (bounded by circular arcs) and having the angles 60°, 60°, 0°; and the theorem is that the whole series of the new regions thus obtained completely covers, without interstices or overlapping, the infinite half-plane. The number of regions is infinite, and the size of the successive regions diminishes very rapidly. The diagram was a coloured one, exhibiting the regions obtained by a few of the successive operations.

The analytical theorem is that the whole series of transformations, ω into $\dfrac{\alpha\omega+\beta}{\gamma\omega+\delta}$, where α, β, γ, δ are integers such that $\alpha\delta - \beta\gamma = 1$, can be obtained by combination of the transformations ω into $\omega + 1$ and ω into $-\dfrac{1}{\omega}$.

712.

A PARTIAL DIFFERENTIAL EQUATION CONNECTED WITH THE SIMPLEST CASE OF ABEL'S THEOREM.

[From the *Report of the British Association for the Advancement of Science*, (1881), pp. 534, 535.]

CONSIDER a given cubic curve cut by a line in the points (x_1, y_1), (x_2, y_2), (x_3, y_3); taking the first and second points at pleasure, these determine uniquely the third point. Analytically, the equation of the curve determines y_1 as a function of x_1, and y_2 as a function of x_2: writing in the equation

$$x_3 = \lambda x_1 + (1-\lambda) x_2, \quad y_3 = \lambda y_1 + (1-\lambda) y_2,$$

we have λ by a simple equation, and thence x_3; viz. x_3 is found as a function of x_1, x_2, and of the nine constants of the equation. Hence forming the derived equations (in regard to x_1, x_2) of the first, second, and third orders, we have $(1+2+3+4=)$ 10 equations from which to eliminate the 9 constants; x_3, considered as a function of x_1 and x_2, thus satisfies a partial differential equation of the third order, independent of the particular cubic curve.

To obtain this equation it is only necessary to observe that we have, by Abel's theorem,

$$\frac{dx_1}{X_1} + \frac{dx_2}{X_2} + \frac{dx_3}{X_3} = 0,$$

where X_1 is a given function of x_1 and y_1, that is, of x_1; X_2 and X_3 are the like functions of x_2 and x_3 respectively. Hence, considering x_3 as a function of x_1 and x_2, we have

$$\frac{dx_3}{dx_1} = -\frac{X_3}{X_1}, \quad \frac{dx_3}{dx_2} = -\frac{X_3}{X_2},$$

and consequently

$$\frac{dx_3}{dx_1} \div \frac{dx_3}{dx_2} = \frac{X_2}{X_1};$$

where X_2, X_1 are functions of x_2, x_1 respectively: hence taking the logarithm and differentiating successively with regard to x_1 and x_2, we have

$$\frac{d}{dx_1}\frac{d}{dx_2}\log\left(\frac{dx_3}{dx_1} \div \frac{dx_3}{dx_2}\right) = 0,$$

which is the required partial differential equation of the third order.

This differential equation has a simple geometrical signification. Consider three consecutive positions of the line meeting the cubic curve in the points 1, 2, 3; 1′, 2′, 3′; 1″, 2″, 3″ respectively: *quà* equation of the third order, the equation should in effect determine 3″ by means of the other points. And, in fact, the three positions of the line constitute a cubic curve; the nine points are thus the intersections of two cubic curves, or, say, they are an "ennead" of points; any eight of the points thus determine uniquely the ninth point.

713.

ADDITION TO MR ROWE'S MEMOIR ON ABEL'S THEOREM.

[From the *Philosophical Transactions of the Royal Society of London*, vol. 172, Part III. (1881), pp. 751—758. Received May 27,—Read June 10, 1880.]

IN Abel's general theorem y is an irrational function of x determined by an equation $\chi(y) = 0$, or say $\chi(x, y) = 0$, of the order n as regards y: and it was shown by him that the sum of any number of the integrals considered may be reduced to a sum of γ integrals; where γ is a determinate number depending only on the form of the equation $\chi(x, y) = 0$, and given in his equation (62), [*Œuvres Complètes*, (1881), t. I. p. 168]: viz. if, solving the equation so as to obtain from it developments of y in descending series of powers of x, we have*

$$\begin{array}{llll} n_1\mu_1 & \text{series each of the form} & & y = x^{\frac{m_1}{\mu_1}} + \ldots, \\ n_2\mu_2 & \quad ,, & ,, & y = x^{\frac{m_2}{\mu_2}} + \ldots, \\ \vdots & & & \vdots \\ n_k\mu_k & \quad ,, & ,, & y = x^{\frac{m_k}{\mu_k}} + \ldots, \end{array}$$

* The several powers of x have coefficients: the form really is $y = A_1 x^{\frac{m_1}{\mu_1}} + \ldots$, which is regarded as representing the μ_1 different values of y obtained by giving to the radical $x^{\frac{1}{\mu_1}}$ each of its μ_1 values, and the corresponding values to the radicals which enter into the coefficients of the series: and (so understanding it) the meaning is that there are n_1 such series each representing μ_1 values of y. It is assumed that the series contains *only* the radical $x^{\frac{1}{\mu_1}}$, that is, the indices after the leading index $\frac{m_1}{\mu_1}$ are $\frac{m_1 - 1}{\mu_1}$, $\frac{m_1 - 2}{\mu_1}$, ...; a series such as $y = A_1 x^{\frac{4}{3}} + B_1 x^{\frac{3}{5}} + \ldots$, depending on the two radicals $x^{\frac{1}{3}}$, $x^{\frac{1}{5}}$ represents 15 different values, and would be written $y = A_1 x^{\frac{20}{15}} + \ldots$, or the values of m_1 and μ_1 would be 20 and 15 respectively: in a case like this where $\frac{m_1}{\mu_1}$ is not in its least terms, the number of values of the leading coefficient A_1 is equal, not to μ_1, but to a submultiple of μ_1. But the case is excluded by Abel's assumption that $\frac{m_1}{\mu_1}$, $\frac{m_2}{\mu_2}$, ..., are fractions each of them in its least terms.

(so that $n = n_1\mu_1 + n_2\mu_2 + \ldots + n_k\mu_k$), then γ is a determinate function of n_1, m_1, μ_1; n_2, m_2, μ_2; ...; n_k, m_k, μ_k.

Mr Rowe has expressed Abel's γ in the following form, viz. assuming

$$\frac{m_1}{\mu_1} > \frac{m_2}{\mu_2} > \ldots > \frac{m_k}{\mu_k},$$

then this expression is

$$\gamma = \sum_{s>r} n_r m_r n_s \mu_s + \tfrac{1}{2}\Sigma n^2 m\mu - \tfrac{1}{2}\Sigma nm - \tfrac{1}{2}\Sigma n - \tfrac{1}{2}n + 1,$$

or, what is the same thing, for n writing its value $\Sigma n\mu$,

$$\gamma = \sum_{s>r} n_r m_r n_s \mu_s + \tfrac{1}{2}\Sigma n^2 m\mu - \tfrac{1}{2}\Sigma nm - \tfrac{1}{2}\Sigma n\mu - \tfrac{1}{2}\Sigma n + 1,$$

where in the first sum r, s have each of them the values 1, 2, ..., k, subject to the condition $s > r$; in each of the other sums n, m, and μ are considered as having the suffix r, which has the values 1, 2, ..., k.

It is a leading result in Riemann's theory of the Abelian integrals that γ is the deficiency (Geschlecht) of the curve represented by the equation $\chi(x, y) = 0$: and it must consequently be demonstrable *à posteriori* that the foregoing expression for γ is in fact = deficiency of curve $\chi(x, y) = 0$. I propose to verify this by means of the formulæ given in my paper "On the Higher Singularities of a Plane Curve," *Quart. Math. Jour.*, vol. VII., (1866), pp. 212—223, [374].

It is necessary to distinguish between the values of $\frac{m}{\mu}$ which are >, =, and < 1; and to fix the ideas I assume $k = 7$, and

$$\frac{m_1}{\mu_1},\ \frac{m_2}{\mu_2},\ \frac{m_3}{\mu_3}, \text{ each } > 1,$$

$$\frac{m_4}{\mu_4} = 1; \text{ say } m_4 = \mu_4 = \lambda, \text{ and } n_4 = \theta;$$

$$\frac{m_5}{\mu_5},\ \frac{m_6}{\mu_6},\ \frac{m_7}{\mu_7}, \text{ each } < 1,$$

but it will be easily seen that the reasoning is quite general. I use Σ' to denote a sum in regard to the first set of suffixes 1, 2, 3, and Σ'' to denote a sum in regard to the second set of suffixes 5, 6, 7. The foregoing value of n is thus

$$n = \Sigma' n\mu + \lambda\theta + \Sigma'' n\mu.$$

Introducing a third coordinate z for homogeneity, the equation $\chi(x, y) = 0$ of the curve will be

$$0 = \left(yz^{\frac{m_1}{\mu_1}-1} - x^{\frac{m_1}{\mu_1}}\right)^{n_1\mu_1} \ldots \left(y - x^{\frac{\lambda}{\lambda}}\right)^{\lambda\theta} \left(y - x^{\frac{m_5}{\mu_5}} z^{1-\frac{m_5}{\mu_5}}\right) \ldots,$$

where it is to be observed that $(\quad)^{n_1\mu_1}$ is written to denote the product of $n_1\mu_1$ different series each of the form $yz^{\frac{m_1}{\mu_1}-1} - A_1 x^{\frac{m_1}{\mu_1}} - \ldots$; these divide themselves into n_1

groups, each a product of μ_1 series; and in each such product the μ_1 coefficients A_1 are in general the μ_1 values of a function containing a radical $a^{\frac{1}{\mu_1}}$ and are thus different from each other: it is in what follows in effect assumed not only that this is so, but that all the $n_1\mu_1$ coefficients A_1 are different from each other*: the like remarks apply to the other factors. It applies in particular to the term $\left(y - x^{\frac{\lambda}{\lambda}}\right)^{\lambda\theta}$, viz. it is assumed that the coefficients A in the $\lambda\theta$ series $y = Ax^{\frac{\lambda}{\lambda}} + \ldots$ are all of them different from each other. These assumptions as to the leading coefficients really imply Abel's assumption that $\frac{m_1}{\mu_1}, \ldots, \frac{m_k}{\mu_k}$ are all of them fractions in their least terms, and in particular that $\frac{\lambda}{\lambda}$ is a fraction in its least terms, viz. that $\lambda = 1$: I retain however for convenience the general value λ, putting it ultimately $= 1$.

In the product of the several infinite series, the terms containing negative powers all disappear of themselves; and the product is a rational and integral function $F(x, y, z)$ of the coordinates, which on putting therein $z = 1$ becomes $= \chi(x, y)$. The equation of the curve thus is $F(x, y, z) = 0$; and the order is

$$= \frac{m_1}{\mu_1} n_1\mu_1 + \ldots + \lambda\theta + n_5\mu_5 + \ldots, \; = m_1n_1 + \ldots + \lambda\theta + n_5\mu_5 + \ldots;$$

viz. if K is the order of the curve $\chi(x, y) = 0$, then $K = \Sigma' nm + \lambda\theta + \Sigma'' n\mu$.

The curve has singularities (singular points) at infinity, that is, on the line $z = 0$: viz.—

First, a singularity at ($z = 0$, $x = 0$), where the tangent is $x = 0$, and which, writing for convenience $y = 1$, is denoted by the function

$$\left(z - x^{\frac{m_1}{m_1 - \mu_1}}\right)^{n_1(m_1 - \mu_1)} \ldots;$$

where observe that the expressed factor indicates n_1 branches $\left(z - x^{\frac{m_1}{m_1-\mu_1}}\right)^{m_1 - \mu_1}$, or say $n_1(m_1 - \mu_1)$ partial branches $z - x^{\frac{m_1}{m_1 - \mu_1}}$, that is, $n_1(m_1 - \mu_1)$ partial branches $z = A_1 x^{\frac{m_1}{m_1 - \mu_1}} + \ldots$, with in all $n_1(m_1 - \mu_1)$ distinct values of A_1: and the like as regards the unexpressed factors with the suffixes 2 and 3.

Secondly, a singularity at ($z = 0$, $y = 0$), where the tangent is $y = 0$, and which, writing for convenience $x = 1$, is denoted by the function

$$\left(z - y^{\frac{\mu_5}{\mu_5 - m_5}}\right)^{n_5(\mu_5 - m_5)} \ldots;$$

* This assumption is virtually made by Abel, (*l. c.*) p. 162, in the expression "alors on aura en général, excepté quelques cas particuliers que je me dispense de considérer: $h(y' - y'') = hy'$, &c.": viz. the meaning is that the degree of y' being greater than or equal to that of y'', then the degree of $y' - y''$ is equal to that of y'': of course when the degrees are equal, this implies that the coefficients of the two leading terms must be unequal.

where observe that the expressed factor indicates n_5 branches $\left(z-y^{\frac{\mu_5}{\mu_5-m_5}}\right)^{\mu_5-m_5}$, or say $n_5(\mu_5-m_5)$ partial branches $z-y^{\frac{\mu_5}{\mu_5-m_5}}$, that is, $n_5(\mu_5-m_5)$ partial branches $z=A_5y^{\frac{\mu_5}{\mu_5-m_5}}+\ldots$, with in all $n_5(\mu_5-m_5)$ distinct values of A_5: and the like as regards the unexpressed factors with the suffixes 6 and 7.

Thirdly, singularities at the θ points $(z=0,\ y-Ax=0)$, A having here θ distinct values, at any one of which the tangent is $y-Ax=0$, and which are denoted by the function

$$\left(y-x^{\frac{\lambda}{\lambda}}\right)^{\lambda\theta}:$$

but in the case ultimately considered λ is $=1$; and these are then the θ ordinary points at infinity, $(z=0,\ y-Ax=0)$.

According to the theory explained in my paper above referred to, these several singularities are together equivalent to a certain number $\delta'+\kappa'$ of nodes and cusps; viz. we have

$$\delta'=\tfrac{1}{2}M-\tfrac{3}{2}\Sigma(\alpha-1),$$

$$\kappa'=\qquad\Sigma(\alpha-1),$$

hence

$$\delta'+\kappa'=\tfrac{1}{2}M-\tfrac{1}{2}\Sigma(\alpha-1).$$

Assuming that there are no other singularities, the deficiency

$$\tfrac{1}{2}(K-1)(K-2)-\delta'-\kappa'$$

is

$$=\tfrac{1}{2}(K-1)(K-2)-\tfrac{1}{2}M+\tfrac{1}{2}\Sigma(\alpha-1).$$

This should be equal to the before-mentioned value of γ; viz. we ought to have

$$(K-1)(K-2)-M+\Sigma(\alpha-1)=2\sum_{s>r}n_rm_rn_s\mu_s+\Sigma n^2m\mu-\Sigma nm-\Sigma n\mu-\Sigma n+2,$$

or, as it will be convenient to write it,

$$M=K^2-3K+\Sigma(\alpha-1)-2\sum_{s>r}n_rm_rn_s\mu_s-\Sigma n^2m\mu+\Sigma nm+\Sigma n\mu+\Sigma n,$$

which is the equation which ought to be satisfied by the values of M and $\Sigma(\alpha-1)$ calculated, according to the method of my paper, for the foregoing singularities of the curve.

We have as before

$$K=\Sigma' nm+\Sigma'' n\mu+\theta\lambda.$$

The term $\sum_{s>r}n_rm_rn_s\mu_s$, written at length, is

$$\begin{aligned}
= \quad & n_1m_1\,(n_2\mu_2+n_3\mu_3+\theta\lambda+n_5\mu_5+n_6\mu_6+n_7\mu_7)\\
+\, & n_2m_2\,(\qquad\quad n_3\mu_3+\theta\lambda+n_5\mu_5+n_6\mu_6+n_7\mu_7)\\
+\, & n_3m_3\,(\qquad\qquad\qquad \theta\lambda+n_5\mu_5+n_6\mu_6+n_7\mu_7)\\
+\, & \theta\lambda\,(\qquad\qquad\qquad\qquad\quad n_5\mu_5+n_6\mu_6+n_7\mu_7)\\
+\, & n_5m_5\,(\qquad\qquad\qquad\qquad\qquad\quad n_6\mu_6+n_7\mu_7)\\
+\, & n_6m_6\,(\qquad\qquad\qquad\qquad\qquad\qquad\quad n_7\mu_7),
\end{aligned}$$

which is

$$=\sum_{s>r}'n_r m_r n_s \mu_s + \theta\lambda\,(\Sigma' nm + \Sigma'' n\mu) + \Sigma' nm\,.\,\Sigma'' n\mu + \sum_{s>r}'' n_r m_r n_s \mu_s.$$

We have moreover

$$\begin{aligned} \Sigma n^2 m\mu &= \Sigma' n^2 m\mu + \theta^2\lambda^2 + \Sigma'' n^2 m\mu,\\ \Sigma nm &= \Sigma' nm \;\; + \theta\lambda \;\; + \Sigma'' nm,\\ \Sigma n\mu &= \Sigma' n\mu \;\; + \theta\lambda \;\; + \Sigma'' n\mu,\\ \Sigma n &= \Sigma' n \;\; + \theta \;\; + \Sigma'' n. \end{aligned}$$

We next calculate $\Sigma(\alpha - 1)$.

For the singularity

$$\left(z - x^{\frac{m_1}{m_1 - \mu_1}}\right)^{n_1(m_1 - \mu_1)} \ldots,$$

each branch $\left(z - x^{\frac{m_1}{m_1-\mu_1}}\right)^{m_1-\mu_1}$ gives $\alpha = m_1 - \mu_1$, and the value of $\Sigma(\alpha - 1)$ for this singularity is

$$n_1(m_1 - \mu_1 - 1) + n_2(m_2 - \mu_2 - 1) + n_3(m_3 - \mu_3 - 1),$$

which is

$$= \Sigma' nm - \Sigma' n\mu - \Sigma' n.$$

For the singularity

$$\left(z - y^{\frac{\mu_5}{\mu_5 - m_5}}\right)^{n_5(\mu_5 - m_5)} \ldots,$$

each branch $\left(z - y^{\frac{\mu_5}{\mu_5-m_5}}\right)^{\mu_5-m_5}$ gives $\alpha = \mu_5 - m_5$, and the value of $\Sigma(\alpha - 1)$ for this singularity is

$$n_5(\mu_5 - m_5 - 1) + n_6(\mu_6 - m_6 - 1) + n_7(\mu_7 - m_7 - 1),$$

which is

$$= \Sigma'' n\mu - \Sigma'' nm - \Sigma'' n.$$

For each of the θ singularities

$$\left(y - x^{\frac{\lambda}{\lambda}}\right)^{\lambda\theta},$$

we have $\alpha = \lambda$ and the value of $\Sigma(\alpha - 1)$ is $= \theta(\lambda - 1)$: this is $= 0$ for the value $\lambda = 1$, which is ultimately attributed to λ.

The complete value of $\Sigma(\alpha - 1)$ is thus

$$= \Sigma' nm - \Sigma'' nm - \Sigma' n\mu + \Sigma'' n\mu - \Sigma' n - \Sigma'' n + \theta\lambda - \theta.$$

Substituting all these values, we have

$$\begin{aligned} M = {} & (\Sigma' nm + \Sigma'' n\mu)^2 + 2\theta\lambda\,(\Sigma' nm + \Sigma'' n\mu) + (\theta\lambda)^2\\ & - 3\,(\Sigma' nm + \Sigma'' n\mu) - 3\theta\lambda\\ & + \Sigma' nm - \Sigma'' nm - \Sigma' n\mu + \Sigma'' n\mu - \Sigma' n - \Sigma'' n + \theta\lambda - \theta\\ & - 2\sum_{s>r}' n_r m_r n_s \mu_s - 2\theta\lambda\,(\Sigma' nm + \Sigma'' n\mu) - 2\Sigma' nm\,.\,\Sigma'' n\mu - 2\sum_{s>r}'' n_r m_r n_s \mu_s\\ & - \Sigma' n^2 m\mu - (\theta\lambda)^2 - \Sigma'' n^2 m\mu\\ & + \Sigma' nm + \theta\lambda + \Sigma'' nm\\ & + \Sigma' n\mu + \theta\lambda + \Sigma'' n\mu\\ & + \Sigma' n + \theta + \Sigma'' n, \end{aligned}$$

or, reducing,

$$M = (\Sigma' nm)^2 - \Sigma' nm - \Sigma' n^2 m\mu - 2\Sigma'_{s>r} n_r m_r n_s \mu_s$$
$$+ (\Sigma'' n\mu)^2 - \Sigma'' n\mu - \Sigma'' n^2 m\mu - 2\Sigma''_{s>r} n_r m_r n_s \mu_s;$$

and it is to be shown that the two lines of this expression are in fact the values of M belonging to the singularities

$$\left(z - x^{\frac{m_1}{m_1-\mu_1}}\right)^{n_1(m_1-\mu_1)} \ldots, \text{ and } \left(z - y^{\frac{\mu_5}{\mu_5-m_5}}\right)^{n_5(\mu_5-m_5)} \ldots,$$

respectively. We assume $\lambda = 1$, and there is thus no singularity $\left(y - x^{\frac{\lambda}{\lambda}}\right)^{\lambda\theta}$.

I recall that, considering the several partial branches which meet at a singular point, M denotes the sum of the number of the intersections of each partial branch by every other partial branch: so that for each pair of partial branches the intersections are to be counted *twice*. Supposing that the tangent is $x = 0$, and that for any two branches we have $z_1 = A_1 x^{p_1}$, $z_2 = A_2 x^{p_2}$ (where p_1, p_2 are each equal to or greater than 1), then if $p_2 = p_1$, and $z_1 - z_2 = (A_1 - A_2) x^{p_1}$ where $A_1 - A_2$ not $= 0$ (an assumption which has been already made as regards the cases about to be considered), then the number of intersections is taken to be $= p_1$; and if p_1 and p_2 are unequal, then *taking* p_2 *to be the greater of them*, the leading term of $z_1 - z_2$ is $= A_1 x^{p_1}$, and the number of intersections is taken to be $= p_1$; viz. in the case of unequal exponents, it is equal to the smaller exponent.

Consider now the singularity $\left(z - x^{\frac{m_1}{m_1-\mu_1}}\right)^{n_1(m_1-\mu_1)} \ldots$; and first the intersections of a partial branch $z - x^{\frac{m_1}{m_1-\mu_1}}$ by each of the remaining $n_1(m_1 - \mu_1) - 1$ partial branches of the same set: the number of intersections with any one of these is $= \dfrac{m_1}{m_1 - \mu_1}$; and consequently the number with all of them is $= \dfrac{m_1}{m_1 - \mu_1} [n_1(m_1 - \mu_1) - 1]$. But we obtain this same number from each of the $n_1(m_1 - \mu_1)$ partial branches, and thus the whole number is

$$n_1(m_1 - \mu_1) \frac{m_1}{m_1 - \mu_1} [n_1(m_1 - \mu_1) - 1], \; = n_1 m_1 [n_1(m_1 - \mu_1) - 1].$$

Taking account of the other sets, each with itself, the whole number of such intersections is

$$n_1 m_1 [n_1(m_1 - \mu_1) - 1] + n_2 m_2 [n_2(m_2 - \mu_2) - 1] + n_3 m_3 [n_3(m_3 - \mu_3) - 1],$$

which is

$$= \Sigma' n^2 m^2 - \Sigma' n^2 m\mu - \Sigma' nm.$$

Observe now that $\frac{m_1}{\mu_1} > \frac{m_2}{\mu_2}$, that is, $\frac{\mu_1}{m_1} < \frac{\mu_2}{m_2}$, and that, these being each < 1, we thence have $1 - \frac{\mu_1}{m_1} > 1 - \frac{\mu_2}{m_2}$, that is, $\frac{m_1 - \mu_1}{m_1} > \frac{m_2 - \mu_2}{m_2}$: and we thus have

$$\frac{m_1}{m_1 - \mu_1} < \frac{m_2}{m_2 - \mu_2} < \frac{m_3}{m_3 - \mu_3}.$$

Considering now the intersections of partial branches of the two sets

$$\left(z - x^{\frac{m_1}{m_1 - \mu_1}}\right)^{n_1 (m_1 - \mu_1)} \text{ and } \left(z - x^{\frac{m_2}{m_2 - \mu_2}}\right)^{n_2 (m_2 - \mu_2)}$$

respectively, a partial branch $z - x^{\frac{m_1}{m_1 - \mu_1}}$ gives with each partial branch of the other set a number $= \frac{m_1}{m_1 - \mu_1}$; and in this way taking each partial branch of each set, the number is

$$n_1 (m_1 - \mu_1) \,.\, n_2 (m_2 - \mu_2) \,.\, \frac{m_1}{m_1 - \mu_1}, \quad = n_1 m_1 n_2 (m_2 - \mu_2);$$

and thus for all the sets the number is

$$= n_1 m_1 n_2 (m_2 - \mu_2) + n_1 m_1 n_3 (m_3 - \mu_3) + n_2 m_2 n_3 (m_3 - \mu_3),$$

which is

$$= \Sigma' n_r m_r n_s m_s - \sum_{s>r}' n_r m_r n_s \mu_s,$$

where in the first sum the Σ' refers to each pair of values of the suffixes. But the intersections are to be taken twice; the number thus is

$$= 2\Sigma' n_r m_r n_s m_s - 2\sum_{s>r}' n_r m_r n_s \mu_s.$$

Adding the foregoing number

$$\Sigma' n^2 m^2 - \Sigma' n^2 m\mu - \Sigma' nm,$$

the whole number for the singularity in question is

$$= (\Sigma' nm)^2 - \Sigma' nm - \Sigma' n^2 m\mu - 2\sum_{s>r}' n_r m_r n_s \mu_s.$$

Similarly for the singularity $\left(z - y^{\frac{\mu_5}{\mu_5 - m_5}}\right)^{n_5 (\mu_5 - m_5)} \ldots$; taking each set with itself, the number of intersections is

$$n_5 \mu_5 [n_5 (\mu_5 - m_5) - 1] + n_6 \mu_6 [n_6 (\mu_6 - m_6) - 1] + n_7 \mu_7 [n_7 (\mu_7 - m_7) - 1],$$

which is

$$= \Sigma'' n^2 \mu^2 - \Sigma'' n^2 m\mu - \Sigma'' n\mu.$$

We have here $\frac{m_5}{\mu_5} > \frac{m_6}{\mu_6}$; each of these being less than 1, we have $1 - \frac{m_5}{\mu_5} < 1 - \frac{m_6}{\mu_6}$, that is, $\frac{\mu_5 - m_5}{\mu_5} < \frac{\mu_6 - m_6}{\mu_6}$, or $\frac{\mu_5}{\mu_5 - m_5} > \frac{\mu_6}{\mu_6 - m_6}$; and so

$$\frac{\mu_7}{\mu_7 - m_7} < \frac{\mu_6}{\mu_6 - m_6} < \frac{\mu_5}{\mu_5 - m_5}.$$

Hence considering the two sets

$$\left(z - y^{\frac{\mu_5}{\mu_5 - m_5}}\right)^{n_5(\mu_5 - m_5)} \text{ and } \left(z - y^{\frac{\mu_6}{\mu_6 - m_6}}\right)^{n_6(\mu_6 - m_6)},$$

a partial branch of the first set gives with a partial branch of the second set $\frac{\mu_6}{\mu_6 - m_6}$ intersections: and the number thus obtained is

$$n_5(\mu_5 - m_5) \,.\, n_6(\mu_6 - m_6) \,.\, \frac{\mu_6}{\mu_6 - m_6}, \; = n_5 n_6 \mu_6 (\mu_5 - m_5).$$

For all the sets the number is

$$n_5 n_6 \mu_6 (\mu_5 - m_5) + n_5 n_7 \mu_7 (\mu_5 - m_5) + n_6 n_7 \mu_7 (\mu_6 - m_6)$$

or taking this twice, the number is

$$= 2\Sigma'' n_r \mu_r n_s \mu_s - 2\underset{s>r}{\Sigma''} n_r m_r n_s \mu_s$$

where in the first sum the Σ'' refers to each pair of suffixes. Adding the foregoing value

$$\Sigma'' n^2 \mu^2 - \Sigma'' n^2 m \mu - \Sigma'' n \mu,$$

the whole number for the singularity in question is

$$= (\Sigma'' n\mu)^2 - \Sigma'' n\mu - \Sigma'' n^2 m\mu - 2\underset{s>r}{\Sigma''} n_r m_r n_s \mu_s;$$

and the proof is thus completed.

Referring to the foot-note (ante, p. 31), I remark that the theorem $\gamma =$ deficiency, is absolute, and applies to a curve with any singularities whatever: in a curve which has singularities not taken account of in Abel's theory, the "quelques cas particuliers que je me dispense de considérer," the singularities not taken account of give rise to a diminution in the deficiency of the curve, and also to an equal diminution of the value of γ as determined by Abel's formula; and the actual deficiency will be $=$ Abel's $\gamma -$ such diminution, that is, it will be $=$ true value of γ.

714.

VARIOUS NOTES.

[From the *Messenger of Mathematics*, vol. VII. (1878), pp. 69, 115, 124, 125.]

An Identity.

THE following remarkable identity is given under a slightly different form by Gauss, *Werke*, t. III., p. 424,

$$1+\left(\frac{\frac{1}{2}}{1}\right)^3 x+\left(\frac{\frac{1}{2}\cdot\frac{3}{2}}{1\,.\,2}\right)^3 x^2+\left(\frac{\frac{1}{2}\cdot\frac{3}{2}\cdot\frac{5}{2}}{1\,.\,2\,.\,3}\right)^3 x^3+\&c.$$

$$=\left\{1+\left(\frac{\frac{1}{4}}{1}\right)^2 x+\left(\frac{\frac{1}{4}\cdot\frac{5}{4}}{1\,.\,2}\right)^2 x^2+\left(\frac{\frac{1}{4}\cdot\frac{5}{4}\cdot\frac{9}{4}}{1\,.\,2\,.\,3}\right)^2 x^3+\&c.\right\}^2.$$

On two related quadric functions.

Assume

$$\phi x = a^2(c-x) - x(c^2-b^2-cx),$$
$$\psi x = b^2(c-x) - x(c^2-a^2-cx):$$

then

$$\phi\left(\frac{a^2}{c-x}\right) = \frac{a^2}{(c-x)^2}\psi x,$$

$$\psi\left(\frac{b^2}{c-x}\right) = \frac{b^2}{(c-x)^2}\phi x.$$

In the first of these for x write $\dfrac{b^2}{c-x}$; then

$$\phi\left\{\frac{a^2(c-x)}{c^2-b^2-cx}\right\} = \frac{a^2(c-x)^2}{(c^2-b^2-cx)^2}\,\frac{b^2}{(c-x)^2}\,\phi x = \frac{a^2 b}{(c^2-b^2-cx)^2}\,\phi(x).$$

A Trigonometrical Identity.

$$\begin{aligned} &\cos(b-c)\cos(b+c+d)+\cos a\cos(a+d) \\ &=\cos(c-a)\cos(c+a+d)+\cos b\cos(b+d) \\ &=\cos(a-b)\cos(a+b+d)+\cos c\cos(c+d) \\ &=\cos a\cos(a+d)+\cos b\cos(b+d)+\cos c\cos(c+d)-\cos d. \end{aligned}$$

Extract from a Letter.

"I wish to construct a correspondence such as

$$(x+iy)^3+(x+iy)=X+iY,$$

or, say, for greater convenience

$$4(x+iy)^3-3(x+iy)=X+iY;$$

viz. if

$$x+iy=\cos u,$$

then

$$X+iY=\cos 3u.$$

Suppose $3u_0$ is a value of $3u$ corresponding to a given value of $X+iY$, then the three values of $x+iy$ are of course $\cos u_0$, $\cos\left(u_0 \pm \frac{2\pi}{3}\right)$; but I am afraid that the calculation of u_0, even with cosh and sinh tables, would be very laborious. Writing

$$X+iY=R(\cos\Theta+i\sin\Theta),$$

the intervals for Θ might be 5°, 10° or even 15°, those of R, say 0·1 from 0 to 2, and then 0·5 up to 4 or 5; and 2 places of decimals would be quite sufficient; but even this would probably involve a great mass of calculation.

It has occurred to me that perhaps a geometrical solution might be found for the equation $X+iY=\cos 3u$."

October 31, 1877.

715.

NOTE ON A SYSTEM OF ALGEBRAICAL EQUATIONS.

[From the *Messenger of Mathematics*, vol. VII. (1878), pp. 17, 18.]

ASSUME

$$x+y+z=P,$$
$$yz+zx+xy=Q,$$
$$xyz=R,$$
$$A=x(nyz+Q)-w^2(mx+P),$$
$$B=y(nzx+Q)-w^2(my+P),$$
$$C=z(nxy+Q)-w^2(mz+P),$$
$$\Theta=-mnR+PQ.$$

Then

$$\begin{aligned}&(mz+P)B-(my+P)C\\&\quad=(myz+Py)(nzx+Q)-(myz+Pz)(nxy+Q)\\&\quad=myz(nzx+Q-nxy-Q)+Pnxyz+PQy-Pnxyz-PQz\\&\quad=mnxyz(z-y)-PQ(z-y)\\&\quad=(z-y)\{mnxyz-PQ\}=(y-z)\Theta;\end{aligned}$$

whence, identically,

$$(mz+P)B-(my+P)C=(y-z)\Theta,$$
$$(mx+P)C-(mz+P)A=(z-x)\Theta,$$
$$(my+P)A-(mx+P)B=(x-y)\Theta.$$

Hence any two of the equations $A=0$, $B=0$, $C=0$ imply the third equation.

We have

$$A = x\{(n+1)yz + zx + xy\} - w^2\{(m+1)x + (y+z)\}$$
$$= (x^2 - w^2)(y+z) - x[(m+1)w^2 - (n+1)yz],$$

and similarly for B and C. The three equations therefore are

$$\frac{x}{x^2 - w^2} = \frac{y+z}{(m+1)w^2 - (n+1)yz},$$

$$\frac{y}{y^2 - w^2} = \frac{z+x}{(m+1)w^2 - (n+1)zx},$$

$$\frac{z}{z^2 - w^2} = \frac{x+y}{(m+1)w^2 - (n+1)xy};$$

and any two of these equations imply the third equation.

716.

AN ILLUSTRATION OF THE THEORY OF THE ϑ-FUNCTIONS.

[From the *Messenger of Mathematics*, vol. VII. (1878), pp. 27—32.]

IF X be a given quartic function of x, and if u, or for convenience a constant multiple αu, be the value of the integral $\int \frac{dx}{\sqrt{(X)}}$ taken from a given inferior limit to the superior limit x; then, conversely, x is expressible as a function of u, viz. it is expressible in terms of ϑ-functions of u, where ϑu, or say $\vartheta(u, \mathfrak{F})$ ($\mathfrak{F}$ a parameter upon which the function depends), is given by definition as the sum of a series of exponentials of u; and it is possible from the assumed equation $\alpha u = \int \frac{dx}{\sqrt{(X)}}$, and the definition of ϑu, to obtain by general theory the actual formulæ for the determination of x as such a function of u.

I propose here to obtain these formulæ, in the case where X is a product of real factors, in a less scientific manner, by connecting the function ϑu (as given by such definition) with Jacobi's function Θ, and by reducing the integral $\int \frac{dx}{\sqrt{(X)}}$ by a linear substitution to the form of an elliptic integral; the object being merely to obtain for the case in question the actual formulæ for the expression of x in terms of ϑ-functions of u.

The definition of ϑu or, when the parameter is expressed, $\vartheta(u, \mathfrak{F})$ is

$$\vartheta u = \Sigma (-)^s e^{-\mathfrak{F}s^2 + 2isu},$$

where s has all positive or negative integer values, zero included, from $-\infty$ to $+\infty$ (that is, from $-S$ to $+S$, $S=\infty$); the parameter $\mathfrak{F}$, or (if imaginary) its real part, must be positive.

Evidently ϑu is an even function: $\vartheta(-u) = \vartheta u$. Moreover, it is at once seen that we have

$$\vartheta(u+\pi) = \vartheta u,\ \vartheta(u+i\mathfrak{F}) = -e^{\mathfrak{F}-2iu}\,\vartheta u,$$

whence also

$$\vartheta(u+m\pi+ni\mathfrak{F}),$$

where m and n are any positive or negative integers, is the product of ϑu by an exponential factor, or say simply that it is a multiple of ϑu.

Writing $u = -\tfrac{1}{2}i\mathfrak{F}$, we have $\vartheta(-\tfrac{1}{2}i\mathfrak{F}) = \vartheta(\tfrac{1}{2}i\mathfrak{F})$, that is,

$$\vartheta(\tfrac{1}{2}i\mathfrak{F}) = 0,$$

and therefore also

$$\vartheta\{m\pi + (n+\tfrac{1}{2})i\mathfrak{F}\} = 0.$$

The above properties are general, but if $\mathfrak{F}$ be real, then k, K, K', q being as in Jacobi (consequently k being real, positive, and less than 1, and K and K' real and positive), and assuming $\mathfrak{F} = \dfrac{\pi K'}{K}$, or, what is the same thing,

$$q\left(= e^{-\frac{\pi K'}{K}}\right) = e^{-\mathfrak{F}},$$

the function ϑ is given in terms of Jacobi's Θ by the equation $\vartheta u = \Theta\left(\dfrac{2Ku}{\pi}\right)$; or, what is the same thing, $\Theta u = \vartheta\left(\dfrac{\pi u}{2K}\right)$.

We hence at once obtain expressions of the elliptic functions $\operatorname{sn} u$, $\operatorname{cn} u$, $\operatorname{dn} u$ in terms of ϑ, viz. these are

$$\operatorname{sn} u = \frac{-i}{\sqrt{k}}\, e^{-\frac{\pi}{4K}(K'-2iu)}\,\vartheta\left(\frac{\pi u}{2K}+\tfrac{1}{2}i\mathfrak{F}\right) \div \vartheta\left(\frac{\pi u}{2K}\right),$$

$$\operatorname{cn} u = \sqrt{\left(\frac{k'}{k}\right)}\, e^{-\frac{\pi}{4K}(K'-2iu)}\,\vartheta\left(\frac{\pi u}{2K}+\tfrac{1}{2}\pi+\tfrac{1}{2}i\mathfrak{F}\right) \div \vartheta\left(\frac{\pi u}{2K}\right),$$

$$\operatorname{dn} u = \sqrt{k'}\,\vartheta\left(\frac{\pi u}{2K}+\tfrac{1}{2}\pi\right) \div \vartheta\left(\frac{\pi u}{2K}\right).$$

Consider now the integral

$$\int_a \frac{dx}{\sqrt{\{(-)\,x-a\,.\,x-b\,.\,x-c\,.\,x-d\}}},\ = \int_a \frac{dx}{\sqrt{(X)}} \text{ suppose,}$$

where a, b, c, d are taken to be real, and in the order of increasing magnitude, viz. it is assumed that $b-a$, $c-a$, $d-a$, $c-b$, $d-b$, $d-c$ are all positive; x considered as the variable under the integral sign is always real; when it is between a and b or between c and d, X is positive, and we assume that $\sqrt{(X)}$ denotes the positive value of the radical; but if x is between b and c, X is negative, and we assume that the sign of $\sqrt{(X)}$ is taken so that $\dfrac{1}{\sqrt{(X)}}$ is equal to a positive multiple of i, and this being so the integral is taken from the inferior limit a to the superior limit x, which is real.

Take x a linear function of y, such that for

$$x = a,\ b,\ c,\ d,$$

$$y = 0,\ 1,\ \frac{1}{k^2},\ \infty\text{, respectively,}$$

so that, x increasing continuously from a to d, y will increase continuously from 0 to ∞. We have

$$k^2 = \frac{b-a\,.\,d-c}{d-b\,.\,c-a},$$

$$y = \frac{b-d}{b-a}\,\frac{x-a}{x-d},$$

$$1-y = \frac{d-a}{b-a}\,\frac{x-b}{x-d},$$

$$1-k^2y = \frac{d-a}{c-a}\,\frac{x-c}{x-d};$$

and, thence,

$$\sqrt{(y\,.\,1-y\,.\,1-k^2y)} = \frac{d-a}{c-a}\sqrt{\left(\frac{d-b}{c-a}\right)}\cdot\frac{\sqrt{(X)}}{(x-d)^2},$$

where $\sqrt{\left(\dfrac{d-b}{c-a}\right)}$ is taken to be positive, and the sign of $\sqrt{(X)}$ is fixed as above. Then for y between 0 and 1 or $> \dfrac{1}{k^2}$, $y\,.\,1-y\,.\,1-k^2y$ will be positive, and $\sqrt{(y\,.\,1-y\,.\,1-k^2y)}$ will also be positive; but y being between 1 and $\dfrac{1}{k^2}$, $y\,.\,1-y\,.\,1-k^2y$ will be negative, and the sign of the radical is such that $\dfrac{1}{\sqrt{(y\,.\,1-y\,.\,1-k^2y)}}$ is a positive multiple of i.

We have moreover

$$dy = \frac{d-a}{b-a}(d-b)\frac{dx}{(x-d)^2};$$

and therefore

$$\frac{dy}{\sqrt{(y\,.\,1-y\,.\,1-k^2y)}} = \sqrt{(d-b\,.\,c-a)}\,\frac{dx}{\sqrt{(X)}},$$

where $\sqrt{(d-b\,.\,c-a)}$ is positive; or, say,

$$\int_0 \frac{dy}{\sqrt{(y\,.\,1-y\,.\,1-k^2y)}} = \sqrt{(d-b\,.\,c-a)}\int_a \frac{dx}{\sqrt{(X)}}.$$

Hence, writing $y = z^2 = \text{sn}^2 u$, we have

$$2u = \sqrt{(d-b\,.\,c-a)} \int_a \frac{dx}{\sqrt{(X)}},$$

and it is to be further noticed that to

$$x = a,\ b,\ c,\ d,$$

correspond

$$\text{sn}\, u = 0,\ 1,\ \frac{1}{k},\ \infty,$$

or we may say

$$u = 0,\quad K,\quad K + iK',\quad 2K + iK'.$$

Writing for shortness

$$\frac{2}{\sqrt{(d-b\,.\,c-a)}} = \alpha,$$

we have

$$\alpha u = \int_a \frac{dx}{\sqrt{(X)}};$$

and moreover

$$\alpha K = \int_a^b \frac{dx}{\sqrt{(X)}},$$

$$\alpha (K + iK') = \int_a^c \frac{dx}{\sqrt{(X)}},$$

$$\alpha (2K + iK') = \int_a^d \frac{dx}{\sqrt{(X)}},$$

or if for a moment we write

$$\int_0^a \frac{dx}{\sqrt{(X)}} = A, \text{ \&c.},$$

then these equations are

$$\alpha K = B - A,$$

$$\alpha (K + iK') = C - A,$$

$$\alpha (2K + iK') = D - A.$$

Hence $B + C - 2A = D - A$, that is, $A - B - C + D = 0$, or $B - A = D - C$, that is,

$$\int_a^b \frac{dx}{\sqrt{(X)}} = \int_c^d \frac{dx}{\sqrt{(X)}},$$

where observe as before that $x = a$ to $x = b$, or $x = c$ to $x = d$, X is positive, and the radical $\sqrt{(X)}$ is taken to be positive.

We have also

$$\alpha K = B - A = \int_a^b \frac{dx}{\sqrt{(X)}},$$

$$\alpha i K' = C - B = \int_b^c \frac{dx}{\sqrt{(X)}},$$

where, as before, from b to c, X is negative, and the sign of the radical is such that $\dfrac{1}{\sqrt{(X)}}$ is a positive multiple of i; the last formula may be more conveniently written

$$\alpha K' = \int_b^c \frac{dx}{\sqrt{(-X)}},$$

where, from b to c, $-X$ is positive, and $\sqrt{(-X)}$ is also taken to be positive.

Collecting the results, we have

$$\int_a \frac{dx}{\sqrt{(X)}} = \alpha u, \quad \alpha = \frac{2}{\sqrt{(d-b\,.\,c-a)}}, \quad k^2 = \frac{b-a\,.\,d-c}{d-b\,.\,c-a},$$

and also

$$k'^2 = \frac{d-a\,.\,c-b}{d-b\,.\,c-a},$$

and then conversely

$$x = \frac{a(d-b) + d(b-a)\,\text{sn}^2 u}{(d-b) + (b-a)\,\text{sn}^2 u};$$

or, what is the same thing,

$$\text{sn}^2 u = \frac{b-d\,.\,x-a}{b-a\,.\,x-d},$$

$$\text{cn}^2 u = \frac{d-a\,.\,x-b}{b-a\,.\,x-d},$$

$$\text{dn}^2 u = \frac{d-a\,.\,x-c}{c-a\,.\,x-d};$$

where, in place of the elliptic functions we are to substitute their ϑ-values; it will be recollected that $\mathfrak{F}$, the parameter of the ϑ-functions, has the value

$$\mathfrak{F}\left(=\frac{\pi K'}{K}\right) = \pi \int_b^c \frac{dx}{\sqrt{(-X)}} \div \int_a^b \frac{dx}{\sqrt{(X)}},$$

and, as before,

$$K = \frac{1}{\alpha}\int_a^b \frac{dx}{\sqrt{(X)}}.$$

Hence, finally, α, k, k', K, $\mathfrak{F}$ denoting given functions of a, b, c, d, if as above

$$\int_a \frac{dx}{\sqrt{(X)}} = \alpha u,$$

we have conversely

$$\frac{b-d\,.\,x-a}{b-a\,.\,x-d} = -\frac{1}{k}\, e^{-\frac{1}{4}\mathfrak{F} + \frac{i\pi u}{2K}}\, \vartheta^2\left(\frac{\pi u}{2K} + \tfrac{1}{2}i\mathfrak{F}\right) \div \vartheta^2 \frac{\pi u}{2K},$$

$$\frac{d-a\,.\,x-b}{b-a\,.\,x-d} = \frac{k'}{k}\, e^{-\frac{1}{4}\mathfrak{F} + \frac{i\pi u}{2K}}\, \vartheta^2\left(\frac{\pi u}{2K} + \tfrac{1}{2}\pi + \tfrac{1}{2}i\mathfrak{F}\right) \div \vartheta^2 \frac{\pi u}{2K},$$

$$\frac{d-a\,.\,x-c}{c-a\,.\,x-d} = k'\, \vartheta^2\left(\frac{\pi u}{2K} + \tfrac{1}{2}\pi\right) \div \vartheta^2 \frac{\pi u}{2K},$$

which are the formulæ in question.

The problem is to obtain them (and that in the more general case where a, b, c, d have any given imaginary values) directly from the assumed equation

$$\int_a \frac{dx}{\sqrt{(X)}} = \alpha u,$$

and from the foregoing definition of the function ϑ.

It may be recalled that the function ϑu is a doubly infinite product

$$\vartheta u = \Pi\Pi \left\{1 - \frac{u}{m\pi + (n + \tfrac{1}{2})\, i\mathfrak{F}}\right\};$$

m and n positive or negative integers from $-\infty$ to $+\infty$; I purposely omit all further explanations as to limits; or, what is the same thing,

$$\vartheta \frac{\pi u}{2K} = \Pi\Pi \left\{1 - \frac{u}{2mK + (2n+1)\, iK'}\right\};$$

and consequently that, disregarding constant and exponential factors, the foregoing expressions of

$$\frac{b-d \,.\, x-a}{b-a \,.\, x-d},\quad \frac{d-a \,.\, x-b}{b-a \,.\, x-d},\quad \frac{d-a \,.\, x-c}{c-a \,.\, x-d},$$

are the squares of the expressions $\frac{X}{W}, \frac{Y}{W}, \frac{Z}{W}$, where X, Y, Z, W are respectively of the form

$$u\Pi\Pi\left\{1 + \frac{u}{(m,\ n)}\right\},\quad \Pi\Pi\left\{1 + \frac{u}{(\overline{m},\ n)}\right\},$$

$$\Pi\Pi\left\{1 + \frac{u}{(\overline{m},\ \overline{n})}\right\},\quad \Pi\Pi\left\{1 + \frac{u}{(m,\ \overline{n})}\right\},$$

where $(m,\ n) = 2mK + 2niK'$, and the stroke over the m or the n denotes that the $2m$ or the $2n$ (as the case may be) is to be changed into $2m+1$ or $2n+1$. But this is a transformation which has apparently no application to the ϑ-functions of more than one variable.

717.

ON THE TRIPLE THETA-FUNCTIONS.

[From the *Messenger of Mathematics*, vol. VII. (1878), pp. 48—50.]

As a specimen of mathematical notation, viz. of the notation which appears to me the easiest to *read* and also to *print*, I give the definition and demonstration of the fundamental properties of the triple theta-functions.

Definition.

$$\vartheta(U,\ V,\ W) = \Sigma \exp. \Theta,$$

where

$$\Theta = (A,\ B,\ C,\ F,\ G,\ H)(l,\ m,\ n)^2 + 2(U,\ V,\ W)(l,\ m,\ n),$$

Σ denoting the sum in regard to all positive and negative integer values from $-\infty$ to $+\infty$ (zero included) of l, m, n respectively.

$\vartheta(U,\ V,\ W)$ is considered as a function of the arguments $(U,\ V,\ W)$, and it depends also on the parameters $(A,\ B,\ C,\ F,\ G,\ H)$.

First Property. $\vartheta(U,\ V,\ W) = 0$, for

$$U = \tfrac{1}{2}\{x\pi i + (A,\ H,\ G)(\alpha,\ \beta,\ \gamma)\},$$
$$V = \tfrac{1}{2}\{y\pi i + (H,\ B,\ F)(\alpha,\ \beta,\ \gamma)\},$$
$$W = \tfrac{1}{2}\{z\pi i + (G,\ F,\ C)(\alpha,\ \beta,\ \gamma)\},$$

x, y, z, α, β, γ being any positive or negative integer numbers, such that $\alpha x + \beta y + \gamma z$ = odd number.

Demonstration. It is only necessary to show that to each term of ϑ there corresponds a second term, such that the indices of the two exponentials differ by an odd multiple of πi.

Taking l, m, n as the integers which belong to the one term, those belonging to the other term are

$$-(l+\alpha),\quad -(m+\beta),\quad -(n+\gamma),$$

(where observe that one at least of the numbers α, β, γ being odd, this system of values is not in any case identical with l, m, n). The two exponents then are

$$\Theta, =(A, B, C, F, G, H)(l, m, n)^2+2(U, V, W)(l, m, n),$$

and

$$\Theta', =(A, B, C, F, G, H)(l+\alpha, m+\beta, n+\gamma)^2-2(U, V, W)(l+\alpha, m+\beta, n+\gamma);$$

viz. the value of Θ' is

$$\begin{aligned} = \quad & (A, B, C, F, G, H)(l, m, n)^2+(A, B, C, F, G, H)(\alpha, \beta, \gamma)^2 \\ & +2(A, B, C, F, G, H)(l, m, n)(\alpha, \beta, \gamma) \\ & -2(U, V, W)(l+\alpha, m+\beta, n+\gamma), \end{aligned}$$

and we then have

$$\begin{aligned} \Theta'-\Theta = & \, 2(A, B, C, F, G, H)(l, m, n)(\alpha, \beta, \gamma) \\ & +(A, B, C, F, G, H)(\alpha, \beta, \gamma)^2 \\ & -2(U, V, W)(2l+\alpha, 2m+\beta, 2n+\gamma). \end{aligned}$$

Substituting herein for U, V, W their values, the last term is

$$\begin{aligned} = & -\{(2l+\alpha)x+(2m+\beta)y+(2n+\gamma)z\} \\ & -2(A, B, C, F, G, H)(l, m, n)(\alpha, \beta, \gamma) \\ & - \quad (A, B, C, F, G, H)(\alpha, \beta, \gamma)^2, \end{aligned}$$

and thence

$$\Theta'-\Theta=-\{(2l+\alpha)x+(2m+\beta)y+(2n+\gamma)z\}\,\pi i,$$

which proves the theorem.

As to the notation, remark that, after (A, B, C, F, G, H) has been once written out in full, we may instead of

$$(A, B, C, F, G, H)(l, m, n)^2, \text{ \&c., write } (A, \ldots)(l, m, n)^2, \text{ \&c.,}$$

and that we may use the like abbreviations

$(A, \ldots)(l, m, n)$,	to denote	$(A, H, G)(l, m, n)$	respectively,
$(H, \ldots)(l, m, n)$,	,,	$(H, B, F)(l, m, n)$	,, ,
$(G, \ldots)(l, m, n)$,	,,	$(G, F, C)(l, m, n)$	,, .

These are not only abbreviations, but they make the formulæ actually clearer, as bringing them into a smaller compass; and I accordingly use them in the demonstration which follows.

Second Property. If U_1, V_1, W_1 denote

$$U + x\pi i + (A, H, G)(\alpha, \beta, \gamma),$$
$$V + y\pi i + (H, B, F)(\alpha, \beta, \gamma),$$
$$W + z\pi i + (G, F, C)(\alpha, \beta, \gamma),$$

respectively, where x, y, z, α, β, γ are any positive or negative integers (zero values admissible), then

$$\vartheta(U_1, V_1, W_1) = \exp.\{-(A, B, C, F, G, H)(\alpha, \beta, \gamma)^2\}.\exp.\{-2(\alpha U + \beta V + \gamma W)\}.\vartheta(U, V, W),$$

or say

$$= \exp.\{-(A, \ldots)(\alpha, \beta, \gamma)^2\}.\exp.\{-2(\alpha U + \beta V + \gamma W)\}.\vartheta(U, V, W).$$

Demonstration. Writing $\vartheta(U_1, V_1, W_1) = \Sigma.\exp.\Theta_1$, then in the expression of Θ_1 we may in place of l, m, n write $l - \alpha$, $m - \beta$, $n - \gamma$; we thus obtain

$$\begin{aligned}\Theta_1 = (A, \ldots)(l-\alpha, m-\beta, n-\gamma)^2 + \{&(l-\alpha)[U + x\pi i + (A, \ldots)(\alpha, \beta, \gamma)] \\ &+ (m-\beta)[V + y\pi i + (H, \ldots)(\alpha, \beta, \gamma)] \\ &+ (n-\gamma)[W + z\pi i + (G, \ldots)(\alpha, \beta, \gamma)]\},\end{aligned}$$

which is

$$\begin{aligned} = \ & (A, \ldots)(l, m, n)^2 \\ & + 2(lU + mV + nW) + 2(lx + my + nz)\pi i + 2(A, \ldots)(l, m, n)(\alpha, \beta, \gamma) \\ & \qquad - 2(A, \ldots)(l, m, n)(\alpha, \beta, \gamma) \\ & - 2(\alpha U + \beta V + \gamma W) - 2(\alpha x + \beta y + \gamma z)\pi i - 2(A, \ldots)(\alpha, \beta, \gamma)^2 \\ & \qquad + (A, \ldots)(\alpha, \beta, \gamma)^2,\end{aligned}$$

which is

$$\begin{aligned} = \ & (A, \ldots)(l, m, n)^2 + 2(lU + mV + nW) \\ & - (A, \ldots)(\alpha, \beta, \gamma)^2 - 2(\alpha U + \beta V + \gamma W) \\ & + 2[(l-\alpha)x + (m-\beta)y + (n-\gamma)z]\pi i.\end{aligned}$$

Hence, rejecting the last line, which (as an even multiple of πi) leaves the exponential unaltered, we see that $\vartheta(U_1, V_1, W_1)$ is $= \vartheta(U, V, W)$ multiplied by the factor

$$\exp.\{-(A, \ldots)(\alpha, \beta, \gamma)^2\}.\exp.\{-2(\alpha U + \beta V + \gamma W)\},$$

which is the theorem in question.

In many cases a formula, which belongs to an indefinite number s of letters, is most easily intelligible when written out for three letters, but it is sometimes convenient to speak of the s letters $l, m, \ldots, n$, or even the s letters $l, \ldots, n$, and to write out the formulæ accordingly.

718.

ADDITION TO MR GENESE'S NOTE ON THE THEORY OF ENVELOPES.

[From the *Messenger of Mathematics*, vol. VII. (1878), pp. 62, 63.]

THE example, although simple, is an instructive one. Introducing z, μ for homogeneity, the equation is

$$\lambda^2 y(y - bz) + 2\lambda\mu xy + \mu^2 x(x - az) = 0,$$

giving the envelope

$$xy[(x - az)(y - bz) - xy] = 0;$$

that is,

$$xy(bx + ay - abz)z = 0;$$

viz. we have thus the four lines

$$x = 0, \quad y = 0, \quad \frac{x}{a} + \frac{y}{b} - z = 0, \quad z = 0.$$

Writing these values successively in the equation of the curve, we find respectively

$$\lambda^2 y(y - bz) = 0,$$

$$\mu^2 x(x - az) = 0,$$

$$(b\lambda - a\mu)^2 \frac{xy}{ab} = 0,$$

$$(\lambda y + \mu x)^2 = 0;$$

viz. in each case the equation in λ, μ has (as it should have) two equal roots; but in the first three cases the values are *constant;* viz. we find $\lambda = 0$, $\mu = 0$, $b\lambda - a\mu = 0$, respectively; and the curves $x = 0$, $y = 0$, $\frac{x}{a} + \frac{y}{b} - z = 0$, are for this reason not proper envelopes.

It is to be remarked that writing in the equation of the parabola these values $\lambda=0$, $\mu=0$, $b\lambda-a\mu=0$ successively, we find respectively

$$x(x-az)=0,$$

$$y(y-bz)=0,$$

$$(bx+ay)(bx+ay-abz)=0;$$

viz. in each case the parabola reduces itself to a pair of lines, one of the given lines and a line parallel thereto through the intersection of the other two lines; the parabola thus becomes a curve having a dp on the line at infinity.

In the fourth case $z=0$, the equation in λ, μ is $(\lambda y+\mu x)^2=0$, giving a variable value $\lambda\div\mu=-x\div y$; hence $z=0$, the line at infinity is a proper envelope.

The true geometrical result is that the envelope consists of the three points A, B, C, and the line at infinity; a point *quâ* curve of the order 0 and class 1 is not representable by a single equation in point-coordinates, and hence the peculiarity in the form of the analytical result.

719.

SUGGESTION OF A MECHANICAL INTEGRATOR FOR THE CALCULATION OF $\int(Xdx + Ydy)$ ALONG AN ARBITRARY PATH*.

[From the *Messenger of Mathematics*, vol. VII. (1878), pp. 92—95; *British Association Report*, 1877, pp. 18—20.]

I CONSIDER an integral $\int(Xdx + Ydy)$, where X, Y are each of them a given function of the variables (x, y); $Xdx + Ydy$ is thus not in general an exact differential; but assuming a relation between (x, y), that is, a path of the integral, there is in effect one variable only, and the integral becomes calculable. I wish to show how for any given values of the functions X, Y, but for an arbitrary path, it is possible to construct a mechanism for the calculation of the integral: viz. a mechanism such that, a point D thereof being moved in a plane along a path chosen at pleasure, the corresponding value of the integral shall be exhibited on a dial.

The mechanism (for convenience I speak of it as actually existing) consists of a square block or inverted box, the upper horizontal face whereof is taken as the plane of xy, the equations of its edges being $y=0$, $y=1$, $x=0$, $x=1$ respectively. In the wall faces represented by these equations, we have the endless bands A, A', B, B' respectively; and in the plane of xy, a driving point D, the coordinates of which are (x, y), and a regulating point R, mechanically connected with D, in suchwise that the coordinates of R are always the given functions X, Y of the coordinates of D†; the nature of the mechanical connexion will of course depend upon the particular functions X, Y.

This being so, D drives the bands A and B in such manner that, to the given motions dx, dy of D, correspond a motion dx of the band A and a motion dy of

* Read at the British Association Meeting at Plymouth, August 20, 1877.

† It might be convenient to have as the coordinates of R, not X, Y but ξ, η, determinate functions of X, Y respectively.

the band B; A drives A' with a velocity-ratio depending on the position of the regulator R in suchwise that, the coordinates of R being X, Y, then to the motion dx of A corresponds a motion Xdx of A'; and, similarly, B drives B' with a velocity-ratio depending on the position of R, in suchwise that to the motion dy of B corresponds a motion Ydy of B'. Hence, to the motions dx, dy of the driver D, there correspond the motions Xdx and Ydy of the bands A' and B' respectively; the band A' drives a hand or index, and the band B' drives in the contrary sense a graduated dial, the hand and dial rotating independently of each other about a common centre; the increased reading of the hand on the dial is thus $= Xdx + Ydy$; and supposing the original reading to be zero, and the driver D to be moved from its original position along an arbitrary path to any other position whatever, the reading on the dial will be the corresponding value of the integral $\int(Xdx + Ydy)$.

It is obvious that we might, by means of a combination of two such mechanisms, calculate the value of an integral $\int f(u)\,du$ along an arbitrary path of the complex variable u, $= x + iy$; in fact, writing $f(x+iy) = P + iQ$, the differential is

$$(P+iQ)(dx+idy), \quad = Pdx - Qdy + i(Qdx + Pdy);$$

and we thus require the calculation of the two integrals

$$\int(Pdx - Qdy) \text{ and } \int(Qdx + Pdy),$$

each of which is an integral of the above form. Taking for the path a closed curve, it would be very curious to see the machine giving a value zero or a value different from zero, according as the path did not include or included within it a critical point; it seems to me that this discontinuity would really exhibit itself without the necessity of any change in the setting of the machine.

The ordinary modes of establishing a continuously-variable velocity-ratio between two parts of a machine depend upon friction; and, in particular, this is the case in Prof. James Thomson's mechanical integrator—there is thus of course a limitation of the driving power. It seems to me that a variable velocity-ratio, the variation of which is practically although not strictly continuous, might be established by means of toothed wheels (and so with unlimited driving power) in the following manner.

Consider a revolving wheel A, which by means of a link BC, pivoted to a point B of the wheel A and a point C of a toothed wheel or arc D, communicates a reciprocating motion to D; the extent of this reciprocating motion depending on the distance of B from the centre of A, which distance, or say the half-throw, is assumed to be variable. Here during a half-revolution of A, D moves in one direction, say upwards; and during the other half-revolution of A, D moves in the other direction, say downwards; the extent of these equal and opposite motions varying with the throw. Suppose then that D works a pinion E, the centre of which is not absolutely fixed but is so connected with A that during the first half-revolution of A (or while D is moving upwards), E is in gear with D, and during the second half-revolution of A, or while

D is moving downwards, E is out of gear with D; the continuous rotation of A will communicate an intermittent rotation to E, in such manner nevertheless that, to each entire revolution of A or rotation through the angle 2π, there will (the throw remaining constant) correspond a rotation of E through the angle $n\,.\,2\pi$, where the coefficient n depends upon the throw*. And evidently if A be driven by a wheel A', the angular velocity of which is $\frac{1}{\lambda}$ times that of A, then to a rotation of A' through each angle $\frac{2\pi}{\lambda}$, there will correspond an entire revolution of A, and therefore, as before, a rotation of E through the determinate angle $n\,.\,2\pi$; hence, λ being sufficiently large to each increment of rotation of A', there corresponds in E an increment of rotation which is $n\lambda$ times the first-mentioned increment; viz. E moves (intermittently and possibly also with some "loss of time" on E coming successively in gear and out of gear with D, or in beats as explained) with an angular velocity which is $= n\lambda$ times the angular velocity of A'. And thus the throw (and therefore n) being variable, the velocity-ratio $n\lambda$ is also variable.

We may imagine the wheel A as carrying upon it a piece L sliding between guides, which piece L carries the pivot B of the link BC, and works by a rack on a toothed wheel α concentric with A, but capable of rotating independently thereof. Then if α rotates along with A, as if forming one piece therewith, it will act as a clamp upon L, keeping the distance of B from the centre of A, that is, the half-throw, constant; whereas, if α has given to it an angular velocity different from that of A, the effect will be to vary the distance in question; that is, to vary the half-throw, and consequently the velocity-ratio of A and E. And, in some such manner, substituting for A and E the bands A and A' of the foregoing description, it might be possible to establish between these bands the required variable velocity-ratio.

* If instead of the wheel or arc D with a reciprocating circular motion, we have a double rack D with a reciprocating rectilinear motion, such that the wheel E is placed between the two racks, and is in gear on the one side with one of them when the rack is moving upwards, and on the other side with the other of them when the rack is moving downwards; then the continuous circular motion of A will communicate to E a continuous circular motion, not of course uniform, but such that to each entire revolution of A or rotation through the angle 2π, there will correspond a rotation of E through an angle $n\,.\,2\pi$ as before. This is in fact a mechanical arrangement made use of in a mangle, the double rack being there the follower instead of the driver.

720.

NOTE ON ARBOGAST'S METHOD OF DERIVATIONS.

[From the *Messenger of Mathematics*, vol. VII. (1878), p. 158.]

It is an injustice to Arbogast to speak of his *first* method, as Arbogast's method*. There is really nothing in this, it is the straightforward process of expanding

$$\phi\left(a + bx + \frac{1}{1\,.\,2}cx^2 + \ldots\right)$$

by the differentiation of ϕu, writing $a, b, c, d, \ldots$ in place of $u, \frac{du}{dx}, \frac{d^2u}{dx^2}, \frac{d^3u}{dx^3}$, &c. or say in place of u, u', u'', u''', &c. respectively; thus

$$\phi a,\ \phi' a\,.\,b,\ \tfrac{1}{2}\{\phi' a\,.\,c + \phi'' a\,.\,b^2\},\quad \tfrac{1}{6}\left\{\begin{array}{l}\phi' a\,.\,d + \phi'' a\,.\ bc \\ \qquad\qquad + \phi'' a\,.\,2bc + \phi''' a\,.\,b^3\end{array}\right\}$$

$$= \tfrac{1}{6}\ \{\phi' a\,.\,d + \phi'' a\,.\,3bc + \phi''' a\,.\,b^3\},\ \&\text{c.},$$

and in subsequent terms the number of additions necessary for obtaining the numerical coefficients increases with great rapidity.

That which is specifically Arbogast's method, is his *second* method, viz. here the coefficients of the successive powers of x in the expansion of $\phi(a + bx + cx^2 + dx^3 + \ldots)$, are obtained by the rule of the last and the last but one; thus we have

$$\phi a,\ \phi' a\,.\,b,\ \phi' a\,.\,c + \phi'' a\,.\,\tfrac{1}{2}b^2,\quad \phi' a\,.\,d + \phi'' a\,.\,bc + \phi''' a\,.\,\tfrac{1}{6}b^3,\ \&\text{c.},$$

where each numerical coefficient is found directly, without an addition in any case.

* See *Messenger of Mathematics*, vol. VII. (1878), pp. 142, 143.

721.

FORMULÆ INVOLVING THE SEVENTH ROOTS OF UNITY.

[From the *Messenger of Mathematics*, vol. VII. (1878), pp. 177—182.]

LET ω be an imaginary cube root of unity, $\omega^2+\omega+1=0$, or say $\omega=\frac{1}{2}\{-1+i\sqrt{(3)}\}$; $\alpha^3=-7(1+3\omega)$, $\beta^3=-7(1+3\omega^2)$, values giving $\alpha^3\beta^3=343$, and the cube roots α, β being such that $\alpha\beta=7$; then $\alpha+\beta$, $=\alpha+\frac{7}{\alpha}$, is a three-valued function $\left(\text{since changing the root } \omega \text{ we merely interchange } \alpha \text{ and } \frac{7}{\alpha}\right)$; and if r be an imaginary seventh root of unity, then

$$\begin{aligned} 3(r+r^6) &= \alpha+\beta-1,\\ 3(r^2+r^5) &= \omega\alpha+\omega^2\beta-1,\\ 3(r^4+r^3) &= \omega^2\alpha+\omega\beta-1. \end{aligned}$$

Any one of these formulæ gives the other two; for observe that we have $\alpha^3=-\alpha\beta(1+3\omega)$, $\beta^3=-\alpha\beta(1+3\omega^2)$, that is, $\alpha^2=-\beta(1+3\omega)$, $\beta^2=-\alpha(1+3\omega^2)$; hence, starting for instance with the first formula, we deduce

$$\begin{aligned} 9(r^2+r^5+2) &= \alpha^2+2\alpha\beta+\beta^2-2\alpha-2\beta+1,\\ &= -\beta(1+3\omega)+14-\alpha(1+3\omega^2)-2\alpha-2\beta+1,\\ &= -\alpha(3+3\omega^2)-\beta(3+3\omega)+15,\\ &= 3\omega\alpha+3\omega^2\beta+15, \end{aligned}$$

that is,

$$3(r^2+r^5)=\omega\alpha+\omega^2\beta-1;$$

and in like manner by squaring each side of this we have the third formula

$$3(r^4+r^3)=\omega^2\alpha+\omega\beta-1.$$

The foregoing formulæ apply to the combinations $r+r^6$, r^2+r^5, r^4+r^3 of the seventh roots of unity, but we may investigate the theory for the roots themselves r, r^2, r^3, r^4, r^5, r^6. These depend on the new radical $\sqrt{(-7)}$ or $i\sqrt{(7)}$; introducing instead hereof X, Y, where

$$X = \tfrac{1}{2}\{-1+i\sqrt{(7)}\},$$
$$Y = \tfrac{1}{2}\{-1-i\sqrt{(7)}\},$$

then if

$$A^3 = 6 + 3\omega X + (1+3\omega^2)\,Y,$$
$$B^3 = 6 + 3\omega^2 X + (1+3\omega\,)\,Y,$$

where

$$AB = i\sqrt{(7)},$$

we have (Lagrange, *Équations Numériques*, p. 294),

$$3r = X + A + B.$$

I found that, in order to bring this into connexion with the foregoing formula, $3(r+r^6) = \alpha+\beta-1$, where as before $\alpha^3 = -7(1+3\omega)$, $\beta^3 = -7(1+3\omega^2)$, $\alpha\beta = 7$, it is necessary that B, A should be linear multiples of α, β respectively, the coefficients being rational functions of ω, X; and that the actual relations are

$$B = \frac{\alpha}{7}\{4-\omega+X(1-2\omega)\},$$

$$A = \frac{\beta}{7}\{5+\omega+X(3+2\omega)\};$$

in verification of which, it may be remarked that these equations give

$$AB = \frac{\alpha\beta}{49}\{(20-\omega-\omega^2)+X(17-4\omega-4\omega^2)+X^2(3-4\omega-4\omega^2)\},$$

viz. in virtue of the equation $\omega^2+\omega+1=0$, the term in $\{\ \}$ is $=21+21X+7X^2$, $=7(X^2+3X+3)$, or since $X^2+X+2=0$, this is $=7(2X+1)$, $=7i\sqrt{(7)}$; the equation thus is $7AB = \alpha\beta\,.\,i\sqrt{(7)}$, which is true in virtue of $AB = i\sqrt{(7)}$ and $\alpha\beta = 7$. The same relations may also be written

$$-\alpha = B(\omega^2+X),$$
$$-\beta = A(\omega\ +X).$$

I found in the first instance

$$3r\ = X + A + B,$$
$$3r^6 = -1 - X + A(\omega^2 - X) + B(\omega - X),$$
$$3r^2 = X + \omega^2 A + \omega B,$$
$$3r^5 = -1 - X + A(\omega - \omega^2 X) + B(\omega^2 - \omega X),$$
$$3r^4 = X + \omega A + \omega^2 B,$$
$$3r^3 = -1 - X + A(1 - \omega X) + B(1 - \omega^2 X),$$

which in fact gave the foregoing formulæ

$$3(r+r^6) = -1 + \alpha + \beta,$$
$$3(r^2+r^5) = -1 + \omega\alpha + \omega^2\beta,$$
$$3(r^4+r^3) = -1 + \omega^2\alpha + \omega\beta.$$

But there is a want of symmetry in these expressions for r, r^2, &c., inasmuch as the values of r, r^2, r^4 are of a different form from those of r^6, r^5, r^3; to obtain the proper forms, we must for A, B substitute their values in terms of α, β, and we thus obtain

$$3r = X + \frac{\alpha}{7}\{4 - \omega + X(1 - 2\omega)\} + \frac{\beta}{7}\{5 + \omega + X(3 + 2\omega)\},$$

$$3r^6 = -1 - X + \frac{\alpha}{7}\{3 + \omega + X(-1 + 2\omega)\} + \frac{\beta}{7}\{2 - \omega + X(-3 - 2\omega)\},$$

$$3r^2 = X + \frac{\alpha}{7}\{1 + 5\omega + X(2 + 3\omega)\} + \frac{\beta}{7}\{-4 - 5\omega + X(-1 - 3\omega)\},$$

$$3r^5 = -1 - X + \frac{\alpha}{7}\{-1 + 2\omega + X(-2 - 3\omega)\} + \frac{\beta}{7}\{-3 - 2\omega + X(1 + 3\omega)\},$$

$$3r^4 = X + \frac{\alpha}{7}\{-5 - 4\omega + X(-3 - \omega)\} + \frac{\beta}{7}\{-1 + 4\omega + X(-2 + \omega)\},$$

$$3r^3 = -1 - X + \frac{\alpha}{7}\{-2 - 3\omega + X(3 + \omega)\} + \frac{\beta}{7}\{1 + 3\omega + X(2 - \omega)\};$$

viz. each of the imaginary seventh roots is thus expressed as a linear function of the cubic radicals α, β (involving ω under the radical signs) with coefficients which are functions of ω, X.

Recollecting the equations $\alpha^2 = -\beta(1+3\omega)$, $\beta^2 = -\alpha(1+3\omega^2)$, $\alpha\beta = 7$; $\omega^2+\omega+1=0$, $X^2+X+2=0$; it is clear that, starting for instance from the equation for $3r$, and squaring each side of the equation, we should, after proper reductions, obtain for $9r^2$ an expression of the like form; viz. we thus in fact obtain the expression for $3r^2$; then from the expressions of $3r$ and $3r^2$, multiplying together and reducing, we should obtain the expression for $3r^3$; and so on; viz. from any one of the six equations we can in this manner obtain the remaining five equations.

At the time of writing what precedes I did not recollect Jacobi's paper "Ueber die Kreistheilung und ihre Anwendung auf die Zahlentheorie," *Berliner Monatsber.*, (1837) and *Crelle*, t. XXX. (1846), pp. 166—182; [*Ges. Werke*, t. VI. pp. 254—274]. The starting-point is the following theorem: if x be a root of the equation $\dfrac{x^p-1}{x-1}=0$, p a prime number, and if g is a prime root of p, and

$$F(\alpha) = x + \alpha x^g + \alpha^2 x^{g^2} + \ldots + \alpha^{p-1} x^{g^{p-2}},$$

where α is any root of $\dfrac{\alpha^{p-1}-1}{\alpha-1}=0$, we have

$$F(\alpha^m)\,F(\alpha^n) = \psi(\alpha)\,F(\alpha^{m+n}),$$

where $\psi(\alpha)$ is a rational and integral function of α with integral coefficients; or, what is the same thing, if α and β be any two roots of the above-mentioned equation, then

$$F(\alpha) F(\beta) = \psi(\alpha, \beta) F(\alpha\beta),$$

where $\psi(\alpha, \beta)$ is a rational and integral function of α, β with integral coefficients. As regards the proof of this, it may be remarked that, writing x^g for x, $F(\alpha)$, $F(\beta)$, and $F(\alpha\beta)$ become respectively $\alpha^{-1}F(\alpha)$, $\beta^{-1}F(\beta)$, $(\alpha\beta)^{-1}F(\alpha\beta)$; hence, $F(\alpha)F(\beta) \div F(\alpha\beta)$ remains unaltered, and it thus appears that the function in question is expressible rationally in terms of the *adjoint* quantities α and β. With this explanation the following extract will be easily intelligible:

"The true form (never yet given) of the roots of the equation $x^p - 1 = 0$ is as follows: The roots, as is known, can easily be expressed by mere addition of the functions $F(\alpha)$. If λ is a factor of $p-1$ and $\alpha^\lambda = 1$, then it is further known that $\{F(\alpha)\}^\lambda$ is a mere function of α. But it is only necessary to know those values of $F(\alpha)$ for which λ is *the power of a prime number*. For suppose $\lambda\lambda'\lambda''\ldots$ is a factor of $p-1$; further let λ, λ', λ'', ... be powers of different prime numbers, and α, α', α'', ... prime λth, λ'th, λ''th, ... roots of unity, then

$$F(\alpha\alpha'\alpha''\ldots) = \frac{F(\alpha)F(\alpha')F(\alpha'')\ldots}{\psi(\alpha, \alpha', \alpha'', \ldots)}$$

where $\psi(\alpha, \alpha', \alpha'', \ldots)$ denotes a rational and integral function of α, α', α'', ... with integral coefficients. Hence, considering always the $(p-1)$th roots of unity as given, there are contained in the expression for x only radicals, the exponents of which are powers of prime numbers, and products of such radicals. But if λ is a power of a prime number, $= \mu^n$, suppose, the corresponding function $F(\alpha)$ can be found as follows: Assume

$$F(\alpha) F(\alpha^i) = \psi_i(\alpha) F(\alpha^{i+1}):$$

then

$$F(\alpha) \ = \sqrt[\mu]{\{\psi_1(\alpha)\ \psi_2(\alpha)\ \ldots \psi_{\mu-1}(\alpha)\ \ F(\alpha^\mu)\}},$$

$$F(\alpha^\mu) = \sqrt[\mu]{\{\psi_1(\alpha^\mu)\psi_2(\alpha^\mu)\ldots\psi_{\mu-1}(\alpha^\mu)F(\alpha^{\mu^2})\}},$$

and so on, up to

$$F(\alpha^{\mu^{n-1}}) = \sqrt[\mu]{\{\psi_1(\alpha^{\mu^{n-1}})\psi_2(\alpha^{\mu^{n-1}})\ldots\psi_{\mu-1}(\alpha^{\mu^{n-1}})(-)^{\frac{p-1}{\mu}} p\}},"$$

so that the formulæ contain ultimately μth roots only. It is remarked in a footnote that, when $n=1$, the $\mu-1$ functions can always be reduced to one-sixth part in number, and that by an induction continued as far as $\mu = 31$, Jacobi had found that all the functions ψ could be expressed by means of the values of a single one of these functions.

"The $\mu-1$ functions determine, not only the values of all the magnitudes under the radical signs, but also the mutual dependence of the radicals themselves. For replacing α by the different powers of α, one can by means of the values so obtained for these functions rationally express all the $\mu^n - 1$ functions $F(\alpha^i)$ by means of the powers of $F(\alpha)$; since all the $\mu^n - 1$ magnitudes $\{F(\alpha)\}^i \div F(\alpha^i)$ are each of them equal to a product of several of the functions $\psi(\alpha)$. Herein consists one of the great advantages of the method over that of Gauss, since in this the discovery of the mutual dependency of the different radicals requires a special investigation, which, on account of its laboriousness, is scarcely practicable for even small primes; whereas the introduction of the functions ψ gives simultaneously the quantities under the radical signs, and the mutual dependency of the radicals. The formation of the functions ψ is obtained by a very simple algorithm, which requires only that one should, from the table for the residues of g^m, form another table giving $g^{m'} = 1 + g^m \pmod{p}$, [see Table IV. of the Memoir]. According to these rules one of my auditors [Rosenhain] in a Prize-Essay of the [Berlin] *Academy* has completely solved the equations $x^p - 1 = 0$ for all the prime numbers p up to 103."

I am endeavouring to procure the Prize-Essay just referred to. As an example—which however is too simple a one to fully bring out Jacobi's method, and its difference from that of Gauss—consider the equation for the fifth roots of unity, $x^4 + x^3 + x^2 + x + 1 = 0$. According to Gauss, we have $x + x^4$ and $x^2 + x^3$, the roots of the equation $u^2 + u - 1 = 0$; say $x + x^4 = \frac{1}{2}\{-1 + \sqrt{(5)}\}$, $x^2 + x^3 = \frac{1}{2}\{-1 - \sqrt{(5)}\}$. The first of these, combined with $x \cdot x^4 = 1$, gives $x - x^4 = \sqrt{[-\frac{1}{2}\{5 + \sqrt{(5)}\}]}$; and thence $4x = -1 + \sqrt{(5)} + \sqrt{[-2\{5 + \sqrt{(5)}\}]}$; if from the second of them, combined with $x^2 \cdot x^3 = 1$, we were in like manner to obtain the values of x^2 and x^3, it would be necessary to investigate the signs to be given to the radicals, in order that the values so obtained for x^2 and x^3 might be consistent with the value just found for x. For the Jacobian process, observing that a prime fourth root of unity is $\alpha = i$, and writing for shortness F_1, F_2, F_3, F_4 to denote $F(\alpha)$, $F(\alpha^2)$, $F(\alpha^3)$, $F(\alpha^4)$ respectively, these functions are

$$\begin{aligned} F_1 &= x - x^4 + i(x^2 - x^3), \\ F_2 &= x + x^4 - (x^2 + x^3), \\ F_3 &= x - x^4 - i(x^2 - x^3), \\ F_4 &= x + x^4 + x^2 + x^3, \end{aligned}$$

viz. we have $F_4 = -1$, $F_2^2 = 5$, or say $F_2 = \sqrt{(5)}$, $F_1^2 = -(1+2i)F_2$, $= -(1+2i)\sqrt{(5)}$; and similarly $F_3^2 = -(1-2i)F_2$, $= -(1-2i)\sqrt{(5)}$; but also $F_1F_3 = -5$, so that the values $F_1 = \sqrt{\{-(1+2i)\sqrt{(5)}\}}$, $F_3 = \sqrt{\{-(1-2i)\sqrt{(5)}\}}$, must be taken consistently with this last equation $F_1F_3 = \sqrt{(5)}$. The values of F_1, F_2, F_3, F_4 being thus known, the four equations then give simultaneously x, x^4, x^2, x^3, these values being of course consistent with each other. It may be remarked that the form in which x presents itself is

$$4x = -1 + \sqrt{(5)} + \sqrt{\{-(1+2i)\sqrt{(5)}\}} + \sqrt{\{-(1-2i)\sqrt{(5)}\}},$$

with the before-mentioned condition as to the last two radicals; with this condition we, in fact, have

$$\sqrt{\{-(1+2i)\sqrt{(5)}\}} + \sqrt{\{-(1-2i)\sqrt{(5)}\}} = \sqrt{[-2\{5 + \sqrt{(5)}\}]},$$

as is at once verified by squaring the two sides.

722.

A PROBLEM IN PARTITIONS.

[From the *Messenger of Mathematics*, vol. VII. (1878), pp. 187, 188.]

TAKE for instance 6 letters; a partition into 3's, such as *abc . def* contains the 6 duads *ab, ac, bc, de, df, ef.* A partition into 2's such as *ab . cd . ef* contains the 3 duads *ab, cd, ef.* Hence if there are α partitions into 3's, and β partitions into 2's, and these contain all the duads each once and only once, $6\alpha + 3\beta = 15$, or $2\alpha + \beta = 5$. The solutions of this last equation are ($\alpha = 0$, $\beta = 5$), ($\alpha = 1$, $\beta = 3$), ($\alpha = 2$, $\beta = 1$), and it is at once seen that the first two sets give solutions of the partition problem, but that the third set gives no solution; thus we have

$\alpha = 0,\quad \beta = 5$	$\alpha = 1,\quad \beta = 3$
ab . cd . ef	*abc . def*
ac . be . df	*ad . be . cf*
ad . bf . ce	*ae . bf . cd*
ae . bd . cf	*af . bd . ce.*
af . bc . de	

Similarly for any other number of letters, for instance 15; if we have α partitions into 5's and β partitions into 3's, then, if these contain all the duads, $4\alpha + 2\beta = 14$, or what is the same $2\alpha + \beta = 7$; if $\alpha = 0$, $\beta = 7$, the partition problem can be solved (this is in fact the problem of the 15 school-girls): but can it be solved for any other values (and if so which values) of α, β? Or again for 30 letters; if we have α partitions into 5's, β partitions into 3's and γ partitions into 2's; then, if these contain all the duads, $4\alpha + 2\beta + \gamma = 29$; and the question is for what values of α, β, γ, does the partition-problem admit of solution.

The question is important from its connexion with the theory of groups, but it seems to be a very difficult one.

I take the opportunity of mentioning the following theorem: two non-commutative symbols α, β, which are such that $\beta\alpha = \alpha^2\beta^2$ cannot give rise to a group made up of symbols of the form $\alpha^p\beta^q$. In fact, the assumed relation gives $\beta\alpha^2 = \alpha^2\beta\alpha^2\beta^2$; and hence, if $\beta\alpha^2$ be of the form in question, $= \alpha^x\beta^y$ suppose, we have

$$\alpha^x\beta^y = \alpha^2 \,.\, \alpha^x\beta^y \,.\, \beta^2, \quad = \alpha^{x+2}\beta^{y+2};$$

that is, $1 = \alpha^2\beta^2$, and thence $\beta\alpha = 1$, that is, $\beta = \alpha^{-1}$, viz. the symbols are commutative, and the only group is that made up of the powers of α.

723.

VARIOUS NOTES.

[From the *Messenger of Mathematics*, vol. VIII. (1879), pp. 45—46, 126, 127.]

An Algebraical Identity: p. 45.

Let a, b, c, f, g, h be the differences of four quantities α, β, γ, δ, say

$$a,\ b,\ c,\ f,\ g,\ h = \beta-\gamma,\ \gamma-\alpha,\ \alpha-\beta,\ \alpha-\delta,\ \beta-\delta,\ \gamma-\delta;$$

then

$$\begin{array}{rrrr} . & h & -g & +a=0, \\ -h & . & +f & +b=0, \\ g & -f & . & +c=0, \\ -a & -b & -c & .=0. \end{array}$$

Now Cauchy's identity

$$(a+b)^7-a^7-b^7=7ab(a+b)(a^2+ab+b^2)^2,$$

putting therein $a+b=-c$, becomes

$$a^7+b^7+c^7 = 7abc\,(\ bc+ca+ab)^2;$$

hence we have

$$\begin{array}{rrrrl} . & h^7 & -g^7 & +a^7 & =-7agh\,(-ga+ah-hg)^2, \\ -h^7 & . & +f^7 & +b^7 & =-7bhf\,(-hb+bf-fh)^2, \\ g^7 & -f^7 & . & +c^7 & =-7cfg\,(-fc+cg-gf)^2, \\ -a^7 & -b^7 & -c^7 & . & =-7abc\,(\ bc+ca+ab)^2; \end{array}$$

whence, adding,

$$agh\,(-ga+ah-hg)^2+bhf\,(-hb+bf-fh)^2+cfg\,(-fc+cg-gf)^2+abc\,(bc+ca+ab)^2=0,$$

or, as this may also be written,

$$agh\,(g^2+h^2+a^2)^2+bhf\,(h^2+f^2+b^2)^2+cfg\,(f^2+g^2+c^2)^2+abc\,(a^2+b^2+c^2)^2=0,$$

an identity if a, b, c, f, g, h denote their values in terms of $\alpha, \beta, \gamma, \delta$.

Note on a Definite Integral: p. 126.

The integral

$$J=\int_0^1 \frac{k^2x^2dx}{\sqrt{(1-x^2\,.\,1-k^2x^2)}},$$

used by Weierstrass, is at once seen to be $=K-E$; but the proof that the other integral

$$J'=\int_1^{\frac{1}{k}} \frac{k^2x^2dx}{\sqrt{(x^2-1\,.\,1-k^2x^2)}}$$

is $=E'$ is not so immediate.

We have

$$\frac{d}{dy}\frac{y\sqrt{(1-y^2)}}{\sqrt{(1-k^2y^2)}}=\frac{1-2y^2+k^2y^4}{(1-y^2)^{\frac{1}{2}}(1-k^2y^2)^{\frac{3}{2}}},$$

and thence

$$0=\int_0^1 \frac{(1-2y^2+k^2y^4)\,dy}{(1-y^2)^{\frac{1}{2}}(1-k^2y^2)^{\frac{3}{2}}};$$

viz. replacing the numerator by

$$-\frac{k'^2}{k^2}+\frac{1}{k^2}(1-k^2y^2)^2,$$

this becomes

$$0=-\frac{k'^2}{k^2}\int_0^1 \frac{dy}{(1-y^2)^{\frac{1}{2}}(1-k^2y^2)^{\frac{3}{2}}}+\frac{1}{k^2}\int_0^1 \frac{(1-k^2y^2)^{\frac{1}{2}}\,dy}{(1-y^2)^{\frac{1}{2}}},$$

that is,

$$\int_0^1 \frac{dy}{(1-y^2)^{\frac{1}{2}}(1-k^2y^2)^{\frac{3}{2}}}=\frac{1}{k'^2}E;$$

or, writing k' for k,

$$\int_0^1 \frac{dy}{(1-y^2)^{\frac{1}{2}}(1-k'^2y^2)^{\frac{3}{2}}}=\frac{1}{k^2}E'.$$

The integral J' writing therein $x=\dfrac{1}{\sqrt{(1-k'^2y^2)}}$ becomes

$$J'=k^2\int_0^1 \frac{dy}{(1-y^2)^{\frac{1}{2}}(1-k'^2y^2)^{\frac{3}{2}}},$$

viz. its value is thus $=E'$.

On a Formula in Elliptic Functions: p. 127.

Writing $\text{en}\,u = \dfrac{\text{cn}\,u}{\text{dn}\,u}$, then the formulæ p. 63 of my *Elliptic Functions* give

$$\text{sn}\,(u+v) = \frac{T-T'}{C-C'}, \quad \text{en}\,(u+v) = \frac{B+B'}{C-C'};$$

and, substituting for T, T', B, B', and C, C' their values, we obtain

$$\text{sn}\,(u+v) = \frac{\text{sn}\,u\,\text{en}\,v + \text{sn}\,v\,\text{en}\,u}{1 + k^2\,\text{sn}\,u\,\text{en}\,u\,\text{sn}\,v\,\text{en}\,v},$$

$$\text{en}\,(u+v) = \frac{\text{en}\,u\,\text{en}\,v - \text{sn}\,u\,\text{sn}\,v}{1 - k^2\,\text{sn}\,u\,\text{en}\,u\,\text{sn}\,v\,\text{en}\,v},$$

formulæ which, as regards their numerators, correspond precisely with the formulæ,

$$\sin\,(u+v) = \sin u \cos v + \sin v \cos u$$

and

$$\cos\,(u+v) = \cos u \cos v - \sin u \sin v,$$

of the circular functions, and which in fact reduce themselves to these on putting $k=0$.

The foregoing formulæ, putting therein $k^2 = -1$, are the formulæ given by Gauss, *Werke*, t. III., p. 404, for the lemniscate functions $\sin\text{lemn}\,(a \pm b)$ and $\cos\text{lemn}\,(a \pm b)$; where it is to be observed that these notations do not represent a sine and a cosine, but they are related as the sn and en, viz. that

$$\cos\text{lemn}\,a = \sqrt{(1 - \sin\text{lemn}^2 a)} \div \sqrt{(1 + \sin\text{lemn}^2 a)}.$$

724.

ON THE DEFORMATION OF A MODEL OF A HYPERBOLOID.

[From the *Messenger of Mathematics*, vol. VIII. (1879), pp. 51, 52.]

THE following is a solution of Mr Greenhill's problem set in the Senate-House Examination, January 14, 1878.

"Prove that, if a model of a hyperboloid of one sheet be constructed of rods representing the generating lines, jointed at the points of crossing; then if the model be deformed it will assume the form of a confocal hyperboloid, and prove that the trajectory of a point on the model will be orthogonal to the system of confocal hyperboloids."

Let (x_1, y_1, z_1), (x_2, y_2, z_2) be points on the generating line of

$$\frac{x^2}{a^2}+\frac{y^2}{b^2}-\frac{z^2}{c^2}=1,$$

then

$$\frac{x_1^2}{a^2}+\frac{y_1^2}{b^2}-\frac{z_1^2}{c^2}=1,$$

$$\frac{x_2^2}{a^2}+\frac{y_2^2}{b^2}-\frac{z_2^2}{c^2}=1,$$

$$\frac{x_1x_2}{a^2}+\frac{y_1y_2}{b^2}-\frac{z_1z_2}{c^2}=1;$$

or, what is the same thing, if

$$\frac{x_1}{a},\ \frac{y_1}{b},\ \frac{z_1}{c}=p_1,\ q_1,\ r_1;\quad \frac{x_2}{a},\ \frac{y_2}{b},\ \frac{z_2}{c}=p_2,\ q_2,\ r_2;$$

then

$$p_1^2+q_1^2-r_1^2=1,$$
$$p_2^2+q_2^2-r_2^2=1,$$
$$p_1p_2+q_1q_2-r_1r_2=1.$$

Similarly, if (ξ_1, η_1, ζ_1), (ξ_2, η_2, ζ_2) be points on generating line of

$$\frac{\xi^2}{\alpha^2}+\frac{\eta^2}{\beta^2}-\frac{\zeta^2}{\gamma^2}=1,$$

and if

$$\frac{\xi_1}{\alpha},\ \frac{\eta_1}{\beta},\ \frac{\zeta_1}{\gamma}=\mathrm{p}_1,\ \mathrm{q}_1,\ \mathrm{r}_1;\quad \frac{\xi_2}{\alpha},\ \frac{\eta_2}{\beta},\ \frac{\zeta_2}{\gamma}=\mathrm{p}_2,\ \mathrm{q}_2,\ \mathrm{r}_2;$$

then

$$\begin{aligned}\mathrm{p}_1^2+\mathrm{q}_1^2-\mathrm{r}_1^2&=1,\\ \mathrm{p}_2^2+\mathrm{q}_2^2-\mathrm{r}_2^2&=1,\\ \mathrm{p}_1\mathrm{p}_2+\mathrm{q}_1\mathrm{q}_2-\mathrm{r}_1\mathrm{r}_2&=1.\end{aligned}$$

Hence if (x_1, y_1, z_1), (ξ_1, η_1, ζ_1) be corresponding points on the two surfaces, that is, if

$$\frac{x_1}{a},\ \frac{y_1}{b},\ \frac{z_1}{c}=\frac{\xi_1}{\alpha},\ \frac{\eta_1}{\beta},\ \frac{\zeta_1}{\gamma},\ =p_1,\ q_1,\ r_1,$$

and similarly, if (x_2, y_2, z_2), (ξ_2, η_2, ζ_2) are corresponding points, that is, if

$$\frac{x_2}{a},\ \frac{y_2}{b},\ \frac{z_2}{c}=\frac{\xi_2}{\alpha},\ \frac{\eta_2}{\beta},\ \frac{\zeta_2}{\gamma}=p_2,\ q_2,\ r_2;$$

then we have, as before, the system of three equations

$$\begin{aligned}p_1^2+q_1^2-r_1^2&=1,\\ p_2^2+q_2^2-r_2^2&=1,\\ p_1p_2+q_1q_2-r_1r_2&=1.\end{aligned}$$

Then if the two surfaces are confocal, that is, if

$$\alpha^2,\ \beta^2,\ -\gamma^2=a^2+h,\ b^2+h,\ -c^2+h,$$

we shall have

$$(x_1-x_2)^2+(y_1-y_2)^2+(z_1-z_2)^2=(\xi_1-\xi_2)^2+(\eta_1-\eta_2)^2+(\zeta_1-\zeta_2)^2.$$

For this equation is

$$a^2(p_1-p_2)^2+b^2(q_1-q_2)^2+c^2(r_1-r_2)^2=\alpha^2(p_1-p_2)^2+\beta^2(q_1-q_2)^2+\gamma^2(r_1-r_2)^2,$$

that is,

$$(p_1-p_2)^2+(q_1-q_2)^2-(r_1-r_2)^2=0,$$

an equation which is obviously true in virtue of the above system of three equations.

Hence, if on confocal surfaces

$$\frac{x^2}{a^2}+\frac{y^2}{b^2}-\frac{z^2}{c^2}=1,\quad \frac{\xi^2}{a^2+h}+\frac{\eta^2}{b^2+h}-\frac{\zeta^2}{c^2-h}=1,$$

we take two points P_1, P_2 on the first, and Q_1, Q_2 the corresponding points on the second; then P_1, P_2 being on a generating line of the first surface, Q_1, Q_2 will be on a generating line of the second surface, and P_1P_2 will be $=Q_1Q_2$. The same is evidently true for the quadrilaterals $P_1P_2P_3P_4$ and $Q_1Q_2Q_3Q_4$, where P_1P_2, P_2P_3, P_3P_4, P_4P_1 are generating lines on the first surface: and therefore Q_1Q_2, Q_2Q_3, Q_3Q_4, Q_4Q_1 are generating lines on the second surface, which proves the theorem.

725.

NEW FORMULÆ FOR THE INTEGRATION OF $\frac{dx}{\surd X}+\frac{dy}{\surd Y}=0$.

[From the *Messenger of Mathematics*, vol. VIII. (1879), pp. 60—62.]

I HAVE found in regard to the differential equation

$$\frac{dx}{\surd(a-x\,.\,b-x\,.\,c-x\,.\,d-x)}+\frac{dy}{\surd(a-y\,.\,b-y\,.\,c-y\,.\,d-y)}=0,$$

a system of formulæ analogous to those given, p. 63, of my *Treatise on Elliptic Functions*, for the values of $\text{sn}\,(u+v)$, $\text{cn}\,(u+v)$, $\text{dn}\,(u+v)$. Writing for shortness

$$\text{a}\,,\ \text{b}\,,\ \text{c}\,,\ \text{d}\ =a-x,\ b-x,\ c-x,\ d-x,$$
$$\text{a}_1,\ \text{b}_1,\ \text{c}_1,\ \text{d}_1=a-y,\ b-y,\ c-y,\ d-y,$$

and $(bc,\ ad)$ to denote the determinant

$$\begin{vmatrix} 1, & x+y, & xy \\ 1, & b+c, & bc \\ 1, & a+d, & ad \end{vmatrix},$$

and $(cd,\ ab)$, $(bd,\ ac)$ to denote the like determinants; then the formulæ are

$$\sqrt{\left(\frac{a-z}{d-z}\right)}=\frac{\surd(a-b\,.\,a-c)\,\{\surd(\text{a}\text{d}\text{b}_1\text{c}_1)+\surd(\text{a}_1\text{d}_1\text{b}\text{c})\}}{(bc,\ ad)},$$

$$=\frac{\surd(a-b\,.\,a-c)\,(x-y)}{\surd(\text{a}\text{d}\text{b}_1\text{c}_1)-\surd(\text{a}_1\text{d}_1\text{b}\text{c})},$$

$$=\frac{\surd(a-b\,.\,a-c)\,\{\surd(\text{a}\text{b}\text{c}_1\text{d}_1)+\surd(\text{a}_1\text{b}_1\text{c}\text{d})\}}{(a-c)\,\surd(\text{b}\text{d}\text{b}_1\text{d}_1)-(b-d)\,\surd(\text{a}\text{c}\text{a}_1\text{c}_1)},$$

$$=\frac{\surd(a-b\,.\,a-c)\,\{\surd(\text{a}\text{c}\text{b}_1\text{d}_1)+\surd(\text{a}_1\text{c}_1\text{b}\text{d})\}}{(a-b)\,\surd(\text{c}\text{d}\text{c}_1\text{d}_1)-(c-d)\,\surd(\text{a}\text{b}\text{a}_1\text{b}_1)},$$

$$\sqrt{\left(\frac{b-z}{d-z}\right)}=\frac{\sqrt{\left(\frac{a-b}{a-d}\right)}\{(a-c)\surd(\mathrm{bdb_1d_1})+(b-d)\surd(\mathrm{aca_1c_1})\}}{(bc,\ ad)},$$

$$=\frac{\sqrt{\left(\frac{a-b}{a-d}\right)}\{\surd(\mathrm{abc_1d_1})-\surd(\mathrm{a_1b_1cd})\}}{\surd(\mathrm{adb_1c_1})-\surd(\mathrm{a_1d_1bc})},$$

$$=\frac{\sqrt{\left(\frac{a-b}{a-d}\right)}(cd,\ ab)}{(a-c)\surd(\mathrm{bdb_1d_1})-(b-d)\surd(\mathrm{aca_1c_1})},$$

$$=\frac{\sqrt{\left(\frac{a-b}{a-d}\right)}\{(a-d)\surd(\mathrm{bcb_1c_1})+(b-c)\surd(\mathrm{ada_1d_1})\}}{(a-b)\surd(\mathrm{cdc_1d_1})-(c-d)\surd(\mathrm{aba_1b_1})},$$

$$\sqrt{\left(\frac{c-z}{d-z}\right)}=\frac{\sqrt{\left(\frac{a-c}{a-d}\right)}\{(a-b)\surd(\mathrm{cdc_1d_1})+(c-d)\surd(\mathrm{aba_1b_1})\}}{(bc,\ ad)}.$$

$$=\frac{\sqrt{\left(\frac{a-c}{a-d}\right)}\{\surd(\mathrm{acb_1d_1})-\surd(\mathrm{a_1c_1bd})\}}{\surd(\mathrm{adb_1c_1})-\surd(\mathrm{a_1d_1bc})},$$

$$=\frac{\sqrt{\left(\frac{a-c}{a-d}\right)}\{(a-d)\surd(\mathrm{bcb_1c_1})-(b-c)\surd(\mathrm{ada_1d_1})\}}{(a-c)\surd(\mathrm{bdb_1d_1})-(b-d)\surd(\mathrm{aca_1c_1})},$$

$$=\frac{\sqrt{\left(\frac{a-c}{a-d}\right)}(bd,\ ac)}{(a-b)\surd(\mathrm{cdc_1d_1})-(c-d)\surd(\mathrm{aba_1b_1})}.$$

The twelve equations are equivalent to each other, each giving z as one and the same function of x, y; and regarding z as a constant of integration, any one of the equations is a form of the integral of the proposed differential equation.

Writing in the formulæ $x=a,\ b,\ c,\ d$ successively, the formulæ become

$x=a$,	$x=b$,	$x=c$,	$x=d$,
$\frac{a-z}{d-z}=\frac{\mathrm{a_1}}{\mathrm{d_1}}$,	$-\frac{c-a}{d-b}\frac{\mathrm{b_1}}{\mathrm{c_1}}$,	$-\frac{b-a}{d-c}\frac{\mathrm{c_1}}{\mathrm{b_1}}$,	$\frac{a-b\,.\,a-c}{d-b\,.\,d-c}\frac{\mathrm{d_1}}{\mathrm{a_1}}$,
$\frac{b-z}{d-z}=\frac{\mathrm{b_1}}{\mathrm{d_1}}$,	$-\frac{c-b}{d-a}\frac{\mathrm{a_1}}{\mathrm{c_1}}$,	$\frac{b-a\,.\,b-c}{d-a\,.\,d-c}\frac{\mathrm{d_1}}{\mathrm{b_1}}$,	$-\frac{a-b}{d-c}\frac{\mathrm{c_1}}{\mathrm{a_1}}$,
$\frac{c-z}{d-z}=\frac{\mathrm{c_1}}{\mathrm{d_1}}$,	$\frac{c-a\,.\,c-b}{d-a\,.\,d-b}\frac{\mathrm{d_1}}{\mathrm{c_1}}$,	$-\frac{b-c}{d-a}\frac{\mathrm{a_1}}{\mathrm{b_1}}$,	$-\frac{a-c}{d-b}\frac{\mathrm{b_1}}{\mathrm{a_1}}$,

viz. in the first case we have $z=y$, and in each of the other cases z equal to a linear function $\frac{\alpha y+\beta}{\gamma y+\delta}$ of y.

Cambridge, July 3, 1878.

726.

A FORMULA BY GAUSS FOR THE CALCULATION OF LOG 2 AND CERTAIN OTHER LOGARITHMS.

[From the *Messenger of Mathematics*, vol. VIII. (1879), pp. 125, 126.]

GAUSS has given, *Werke*, t. II., p. 501, a formula which is in effect as follows:

$$2^{196} = 10^{59}\left(\frac{1025}{1024}\right)^5\left(\frac{1048576}{1048575}\right)^8\left(\frac{6560}{6561}\right)^3\left(\frac{15624}{15625}\right)^8\left(\frac{9801}{9800}\right)^4,$$

viz. this is

$$= 2^{59}\,.\,5^{59}\left(\frac{5^2\,.\,41}{2^{10}}\right)^5\left(\frac{2^{20}}{5^2\,.\,3\,.\,11\,.\,31\,.\,41}\right)^8\left(\frac{5\,.\,2^5\,.\,41}{3^8}\right)^3\left(\frac{2^3\,.\,3^2\,.\,7\,.\,31}{5^6}\right)^8\left(\frac{3^4\,.\,11^2}{2^3\,.\,5^2\,.\,7^2}\right)^4,$$

where on the right-hand side the several prime factors have the indices following, viz.

2,	index is	$(59+160+15+24-50-12)=196$,
3	„	$(16+16-8-24)=0$,
5	„	$(59+10+3-16-48-8)=0$,
7	„	$(8-8)=0$,
11	„	$(8-8)=0$,
31	„	$(8-8)=0$,
41	„	$(5+3-8)=0$,

or the right-hand side is $=2^{196}$ as it should be. The value of log 2 calculated from $2^{196}=10^{59}$ is $\log 2 = \frac{59}{196} = \cdot 301020$, viz. there is an error of a unit in fifth place of decimals. The actual value of 2^{196} has been given me by Mr Glaisher:

$$2^{196} = 10043\ 36277\ 66186\ 89222\ 13726\ 30771$$
$$32266\ 26576\ 37687\ 11142\ 45522\ 06336.^*$$

Supposing log 2 calculated by the form, we then have

$$41 = \left(\tfrac{1025}{1024}\right)2^{12} \div 10^2, \text{ giving } \log 41,$$

and

$$3^8 = 10\,.\,\tfrac{6561}{6560}\,.\,2^4\,.\,41, \text{ giving } \log 3;$$

and formulæ may be obtained proper for the calculation of the logarithms of $\frac{11}{7}$, 11 . 31, and 7 . 31.

* The value was deduced by Mr Glaisher from Mr Shanks's value of 2^{193} in his *Rectification of the Circle*, (1853), p. 90.

727.

EQUATION OF THE WAVE-SURFACE IN ELLIPTIC COORDINATES.

[From the *Messenger of Mathematics*, vol. VIII. (1879), pp. 190, 191.]

THE equation of the wave-surface

$$\frac{ax^2}{x^2+y^2+z^2-a}+\frac{by^2}{x^2+y^2+z^2-b}+\frac{cz^2}{x^2+y^2+z^2}=0,$$

when transformed to coordinates p, q, r, such that

$$\frac{x^2}{-a+p}+\frac{y^2}{-b+p}+\frac{z^2}{-c+p}=1,$$

$$\frac{x^2}{-a+q}+\frac{y^2}{-b+q}+\frac{z^2}{-c+q}=1,$$

$$\frac{x^2}{-a+r}+\frac{y^2}{-b+r}+\frac{z^2}{-c+r}=1;$$

(that is, to the elliptic coordinates belonging to the quadric surface $\frac{x^2}{-a}+\frac{y^2}{-b}+\frac{z^2}{-c}=1$), assumes the form

$$(q+r-a-b-c)(r+p-a-b-c)(p+q-a-b-c)=0,$$

(Senate-House Problem, January 14, 1879).

In fact, p, q, r are the roots of the equation

$$\frac{x^2}{-a+u}+\frac{y^2}{-b+u}+\frac{z^2}{-c+u}=1;$$

we have therefore

$$\begin{aligned}(u-p)(u-q)(u-r)&=(u-a)(u-b)(u-c)\\ &\quad -x^2(u-b)(u-c)-y^2(u-c)(u-a)-z^2(u-a)(u-b);\end{aligned}$$

whence, writing for shortness

$$\begin{aligned} A &= a+b+c \quad , & P &= p+q+r, \\ B &= bc+ca+ab, & Q &= qr+rp+pq, \\ C &= abc \quad , & R &= pqr, \end{aligned}$$

we have

$$\begin{aligned} x^2 + y^2 + z^2 &= P - A, \\ (b+c)\,x^2 + (c+a)\,y^2 + (a+b)\,z^2 &= Q - B, \\ bcx^2 + cay^2 + abz^2 &= R - C, \end{aligned}$$

and thence also

$$\begin{aligned} a\,(b+c)\,x^2 + b\,(c+a)\,y^2 + c\,(a+b)\,z^2 &= B\,(P-A) - (R-C), \\ ax^2 + by^2 + cz^2 &= A\,(P-A) - (Q-B). \end{aligned}$$

The equation of the wave-surface is

$$abc - \{a\,(b+c)\,x^2 + b\,(c+a)\,y^2 + c\,(a+b)\,z^2\} + (x^2+y^2+z^2)(ax^2+by^2+cz^2) = 0.$$

By the formulæ just obtained, this is

$$C - [B\,(P-A) - (R-C)] + (P-A)\,[A\,(P-A) - (Q-B)] = 0,$$

that is,

$$A^3 - 2A^2P + A\,(P^2+Q) - (PQ-R) = 0,$$

that is,

$$\{A - (q+r)\}\,\{A - (r+p)\}\,\{A - (p+q) = 0,$$

or, substituting for A its value $a+b+c$, and reversing the sign of each factor, we have the formula in question.

It is easy to see that, taking a, b, c to be each positive, $(a > b > c)$, and assuming also $p > q > r$, we obtain the different real points of space by giving to these coordinates respectively the different real values from ∞ to a, a to b, and b to c respectively. Hence

	greatest,	least value, is
$q+r$,	$a+b$,	$a+c$,
$r+p$,	∞ ,	$a+c$,
$p+q$,	∞ ,	$a+b$,

so that $r+p$, $p+q$, may be either of them $=a+b+c$, but $q+r$ cannot be $=a+b+c$, that is, $q+r=a+b+c$ does not belong to any real point on the wave-surface. We can only have $r+p$ and $p+q$ each $=a+b+c$, if $p=a+c$, $q=r=b$, and these values belong as is easily shown to the nodes on the wave-surface; hence, the equations $r+p=a+b+c$ and $p+q=a+b+c$ being satisfied simultaneously only at the nodes of the surface, must belong to the two sheets respectively. And it can be shown that $p+r=a+b+c$ belongs to the external sheet, and $p+q=a+b+c$ belongs to the internal sheet. In fact, for the point $(0, 0, \sqrt{a})$, which is on the external sheet, we have $p=a+c$, $q=a$, $r=b$, and therefore $p+r=a+b+c$: for the point $(0, 0, \sqrt{b})$, which is on the internal sheet, either

$$(p=b+c,\ q=a,\ r=b) \quad \text{or} \quad (p=a,\ q=b+c,\ r=c),$$

according as $b+c>a$ or $b+c<a$: but in each case

$$p+q=a+b+c.$$

728.

A THEOREM IN ELLIPTIC FUNCTIONS.

[From the *Proceedings of the London Mathematical Society*, vol. x. (1879), pp. 43—48. Read January 8, 1879.]

THE theorem is as follows:

If $u+v+r+s=0$, then

$$-k'^2 \operatorname{sn} u \operatorname{sn} v \operatorname{sn} r \operatorname{sn} s + \operatorname{cn} u \operatorname{cn} v \operatorname{cn} r \operatorname{cn} s - \frac{1}{k^2} \operatorname{dn} u \operatorname{dn} v \operatorname{dn} r \operatorname{dn} s = -\frac{k'^2}{k^2}.$$

It is easy to see that, if a linear relation exists between the three products, then it must be this relation: for the relation must be satisfied on writing therein $v=-u$, $s=-r$, and the only linear relation connecting $\operatorname{sn}^2 u \operatorname{sn}^2 r$, $\operatorname{cn}^2 u \operatorname{cn}^2 r$, $\operatorname{dn}^2 u \operatorname{dn}^2 r$ is the relation in question

$$-k'^2 \operatorname{sn}^2 u \operatorname{sn}^2 r + \operatorname{cn}^2 u \operatorname{cn}^2 r - \frac{1}{k^2} \operatorname{dn}^2 u \operatorname{dn}^2 r = -\frac{k'^2}{k^2}.$$

A demonstration of the theorem was recently communicated to me by Mr Glaisher; and this led me to the somewhat more general theorem

$$\begin{aligned} &-k'^2 \operatorname{sn}(\alpha+\beta) \operatorname{sn}(\alpha-\beta) \operatorname{sn}(\gamma+\delta) \operatorname{sn}(\gamma-\delta) \\ &+\quad \operatorname{cn}(\alpha+\beta) \operatorname{cn}(\alpha-\beta) \operatorname{cn}(\gamma+\delta) \operatorname{cn}(\gamma-\delta) \\ &-\frac{1}{k^2} \operatorname{dn}(\alpha+\beta) \operatorname{dn}(\alpha-\beta) \operatorname{dn}(\gamma+\delta) \operatorname{dn}(\gamma-\delta) \\ &= -\frac{k'^2}{k^2} - \frac{2k'^2 (\operatorname{sn}^2\alpha - \operatorname{sn}^2\gamma)(\operatorname{sn}^2\beta - \operatorname{sn}^2\delta)}{1-k^2\operatorname{sn}^2\alpha \operatorname{sn}^2\beta \,.\, 1-k^2\operatorname{sn}^2\gamma\operatorname{sn}^2\delta}. \end{aligned}$$

In fact, writing herein $\alpha+\gamma=0$, that is, $\gamma=-\alpha$, the right-hand side becomes $=0$; and the arcs on the left-hand side are $\alpha+\beta$, $\alpha-\beta$, $-\alpha+\delta$, $-\alpha-\delta$, which represent any four arcs the sum of which is $=0$.

Writing in the last-mentioned equation x, y, z, w for the sn's of α, β, γ, δ respectively, also

$$\begin{aligned}
P &= x^2-y^2, & P_1 &= z^2-w^2,\\
Q &= 1-x^2-y^2+k^2x^2y^2, & Q_1 &= 1-z^2-w^2+k^2z^2w^2,\\
R &= 1-k^2x^2-k^2y^2+k^2x^2y^2, & R_1 &= 1-k^2z^2-k^2w^2+k^2z^2w^2,\\
D &= 1-k^2x^2y^2, & D_1 &= 1-k^2z^2w^2,
\end{aligned}$$

the equation is

$$-k'^2\frac{PP_1}{DD_1}+\frac{QQ_1}{DD_1}-\frac{1}{k^2}\frac{RR_1}{DD_1}=-\frac{k'^2}{k^2}-\frac{2k'^2(x^2-z^2)(y^2-w^2)}{DD_1},$$

that is,

$$-k'^2PP_1+QQ_1-\frac{1}{k^2}RR_1+\frac{k'^2}{k^2}DD_1+2k'^2(x^2-z^2)(y^2-w^2)=0.$$

It is easy to verify that the terms of the orders 0, 1, 2, 3 and 4 in x^2, y^2, z^2, w^2 separately destroy each other; for instance, for the terms of the order 2, we have

$$\begin{aligned}
&-k'^2(x^2-y^2)(z^2-w^2)+\{(x^2+y^2)(z^2+w^2)+k^2(x^2y^2+z^2w^2)\}\\
&-\frac{1}{k^2}\{k^4(x^2+y^2)(z^2+w^2)+k^2(x^2y^2+z^2w^2)\}\\
&+\frac{k'^2}{k^2}\{-k^2(x^2y^2+z^2w^2)\}+2k'^2(x^2-z^2)(y^2-w^2)=0,
\end{aligned}$$

that is,

$$\begin{aligned}
&-k'^2(x^2-y^2)(z^2-w^2)+(1-k^2)(x^2+y^2)(z^2+w^2)\\
&\qquad+(k^2-1-k'^2)(x^2y^2+z^2w^2)+2k'^2(x^2-z^2)(y^2-w^2)=0\,;
\end{aligned}$$

or, omitting the factor k'^2, this is

$$-(x^2-y^2)(z^2-w^2)+(x^2+y^2)(z^2+w^2)-2(x^2y^2+z^2w^2)+2(x^2-z^2)(y^2-w^2)=0,$$

as it should be.

The theorem in its original form was obtained by me as follows: using the elliptic coordinates p, q, r, such that

$$\frac{x^2}{a+p}+\frac{y^2}{b+p}+\frac{z^2}{c+p}=1,$$

$$\frac{x^2}{a+q}+\frac{y^2}{b+q}+\frac{z^2}{c+q}=1,$$

$$\frac{x^2}{a+r}+\frac{y^2}{b+r}+\frac{z^2}{c+r}=1\,;$$

or, what is the same thing,

$$-\beta\gamma x^2 = a+p\,.\,a+q\,.\,a+r,$$

$$-\gamma\alpha y^2 = b+p\,.\,b+q\,.\,b+r,$$

$$-\alpha\beta z^2 = c+p\,.\,c+q\,.\,c+r,$$

where α, β, γ denote $b-c$, $c-a$, $a-b$ respectively; then, treating r as a constant, the coordinates x, y, z will belong to a point on the ellipsoid

$$\frac{x^2}{a+r}+\frac{y^2}{b+r}+\frac{z^2}{c+r}=1,$$

and the differential equation of the right lines upon this surface is

$$\frac{dp}{\sqrt{a+p\,.\,b+p\,.\,c+p}}=\frac{dq}{\sqrt{a+q\,.\,b+q\,.\,c+q}}.$$

Take x_0, y_0, z_0 the coordinates of a point on the surface, and p_0, q_0 the corresponding values of p, q, so that

$$-\beta\gamma x_0^2 = a+p_0\,.\,a+q_0\,.\,a+r,$$

$$-\gamma\alpha y_0^2 = b+p_0\,.\,b+q_0\,.\,b+r,$$

$$-\alpha\beta z_0^2 = c+p_0\,.\,c+q_0\,.\,c+r,$$

then the equation of the tangent plane at the point (x_0, y_0, z_0) is

$$\frac{xx_0}{a+r}+\frac{yy_0}{b+r}+\frac{zz_0}{c+r}=1,$$

or, substituting for x^2, x_0^2, &c., their values, we have

$$-\frac{\beta\gamma xx_0}{a+r}=\sqrt{a+p\,.\,a+q\,.\,a+p_0\,.\,a+q_0},\ \&\text{c.},$$

and consequently the equation of the tangent plane is

$$\alpha\sqrt{a+p\,.\,a+q\,.\,a+p_0\,.\,a+q_0}+\beta\sqrt{b+p\,.\,b+q\,.\,b+p_0\,.\,b+q_0}$$
$$+\gamma\sqrt{c+p\,.\,c+q\,.\,c+p_0\,.\,c+q_0}=-\alpha\beta\gamma,$$

the equation of a plane intersecting the ellipsoid in a pair of lines; hence this equation (containing in appearance the two arbitrary constants p_0 and q_0) is the integral of the proposed differential equation.

Writing

$$\text{sn}^2 u = A\,(a+p),\quad \text{cn}^2 u = B\,(b+p),\quad \text{dn}^2 u = C\,(c+p),$$

the values of A, B, C, k are determined; and, assuming for q, p_0, q_0 the like forms with the arguments v, u_0, v_0, the differential equation becomes $du=dv$, having the integral $u-u_0=v-v_0$; while the foregoing integral equation, on reducing the constant coefficients contained therein, takes the form

$$\begin{aligned}
&-k'^2 \operatorname{sn} u \operatorname{sn} v \operatorname{sn} u_0 \operatorname{sn} v_0 \\
&+\quad \operatorname{cn} u \operatorname{cn} v \operatorname{cn} u_0 \operatorname{cn} v_0 \\
&-\frac{1}{k^2} \operatorname{dn} u \operatorname{dn} v \operatorname{dn} u_0 \operatorname{dn} v_0 \\
=&-\frac{k'^2}{k^2};
\end{aligned}$$

viz. this equation holds good if $u-u_0=v-v_0$. And by a change of signs we have the theorem.

If, as above, $u+v+r+s=0$, the theorem gives a linear relation between the three products $\operatorname{sn} u \operatorname{sn} v \operatorname{sn} r \operatorname{sn} s$, $\operatorname{cn} u \operatorname{cn} v \operatorname{cn} r \operatorname{cn} s$, $\operatorname{dn} u \operatorname{dn} v \operatorname{dn} r \operatorname{dn} s$, and regarding at pleasure the sn's, the cn's, or the dn's as rational, one of these products will be rational while the other two will be each of them a quadric radical; and hence, rationalising, we obtain an equation which contains the product in question linearly, and contains besides only the squares of the sn's, cn's, or dn's; that is, we have three such equations containing the three products respectively. Bringing to one side the terms which contain the product, and again squaring, we obtain an equation involving only the squares of the sn's, cn's, or dn's; but the three equations thus obtained represent, it is clear, one and the same rational equation, which may be expressed as an equation between the squares of the sn's, or of the cn's, or of the dn's, at pleasure. This equation may be obtained, as I will show, from the ordinary addition-equations of the elliptic functions, but it is not obvious how to obtain from them the three equations involving the products respectively, and these last have the advantage of being of a degree which is the half of the equation which involves only the squared functions.

Write x, y, z, w for $\operatorname{sn} u$, $\operatorname{sn} v$, $\operatorname{sn} r$, $\operatorname{sn} s$ respectively; then, writing

$$\begin{aligned}
A &= x\sqrt{1-y^2 \,.\, 1-k^2y^2}, & \alpha &= z\sqrt{1-w^2 \,.\, 1-k^2w^2}, \\
A' &= y\sqrt{1-x^2 \,.\, 1-k^2x^2}, & \alpha' &= w\sqrt{1-z^2 \,.\, 1-k^2z^2}, \\
P &= x^2-y^2, & \varpi &= z^2-w^2, \\
D &= 1-k^2x^2y^2, & \delta &= 1-k^2z^2w^2,
\end{aligned}$$

we have

$$\operatorname{sn}(u+v) = -\operatorname{sn}(r+s),$$

that is,

$$\frac{A+A'}{D} = \frac{P}{A-A'} = -\frac{\alpha+\alpha'}{\delta} = -\frac{\varpi}{\alpha-\alpha'},$$

and consequently

$$\begin{aligned}
D\varpi &= -(\alpha-\alpha')(A+A'), \\
P\delta &= -(\alpha+\alpha')(A-A');
\end{aligned}$$

whence

$$D\varpi - P\delta = 2(A\alpha' - A'\alpha),$$

that is,
$$(z^2-w^2)(1-k^2x^2y^2)-(x^2-y^2)(1-k^2z^2w^2)$$
$$=2\{xw\sqrt{1-y^2\,.\,1-k^2y^2\,.\,1-z^2\,.\,1-k^2z^2}-yz\sqrt{1-x^2\,.\,1-k^2x^2\,.\,1-w^2\,.\,1-k^2w^2}\}.$$

Rationalising, we obtain, as mentioned above, an equation containing only the squares x^2, y^2, z^2, w^2; it therefore is of a degree twice that of the equation containing the product $xyzw$. I worked out in this way the equation in (x^2, y^2, z^2, w^2), but the calculation was lost, and the easier way of obtaining it is obviously by means of the equation involving $xyzw$.

We have, by the theorem,
$$\begin{aligned}&-k'^2\,xyzw\\&+\sqrt{1-x^2\,.\,1-y^2\,.\,1-z^2\,.\,1-w^2}\\&-\frac{1}{k^2}\sqrt{1-k^2x^2\,.\,1-k^2y^2\,.\,1-k^2z^2\,.\,1-k^2w^2}=-\frac{k'^2}{k^2},\end{aligned}$$
that is,
$$\begin{aligned}k'^2(1-k^2xyzw)=k^2\sqrt{1-x^2\,.\,1-y^2\,.\,1-z^2\,.\,1-w^2}\\-\sqrt{1-k^2x^2\,.\,1-k^2y^2\,.\,1-k^2z^2\,.\,1-k^2w^2};\end{aligned}$$
and then, writing
$$\begin{aligned}P&=x^2+y^2+z^2+w^2,\\Q&=x^2y^2+x^2z^2+x^2w^2+y^2z^2+y^2w^2+z^2w^2,\\R&=x^2y^2z^2+x^2y^2w^2+x^2z^2w^2+y^2z^2w^2,\\S&=x^2y^2z^2w^2,\end{aligned}$$
and using $\sqrt{S}$ to denote the rational function $xyzw$, we have
$$\begin{aligned}&k'^4(1-2k^2\sqrt{S}+k^4S)\\&\quad=k^4(1-P+Q-R+S)\\&\quad\quad+1-k^2P+k^4Q-k^6R+k^8S\\&\quad\quad-2k^2\sqrt{(1-P+Q-R+S)(1-k^2P+k^4Q-k^6R+k^8S)};\end{aligned}$$
or, if for a moment the radical is called $\sqrt{\Delta}$, then the factor k^2 divides out, and the equation becomes
$$2\sqrt{\Delta}=2-(1+k^2)P+2k^2Q-(k^2+k^4)R+2k^4S+2k'^4\sqrt{S},$$
whence
$$\begin{aligned}&4(1-P+Q-R+S)(1-k^2P+k^4Q-k^6R+k^8S)\\&-\{2-(1+k^2)P+2k^2Q-(k^2+k^4)R+2k^4S\}^2-4k'^8S\\&=-2k^4\sqrt{S}\{2-(1+k^2)P+2k^2Q-(k^2+k^4)R+2k^4S\}.\end{aligned}$$

The factor k'^4 divides out; omitting it, we have
$$\begin{aligned}&4Q-P^2-4(1+k^2)R+16k^2S+2k^2PR-4(k^2+k^4)PS-k^4R^2+4k^4QS\\&\quad=-2\sqrt{S}\{2-(1+k^2)P+2k^2Q-(k^2+k^4)R+2k^4S\},\end{aligned}$$
or, as this may also be written,
$$\begin{aligned}&\{(-P^2+4Q-4R)+k^2(-4R+2PR+16S-4PS)+k^4(-R^2+4QS-PS)\}\\&\quad=-2\sqrt{S}\{2-P+k^2(-P+2Q-R)+k^4(-R+2S)\},\end{aligned}$$
which is the required rational equation involving the product of the sn's.

729.

ON A THEOREM RELATING TO CONFORMABLE FIGURES.

[From the *Proceedings of the London Mathematical Society*, vol. X. (1879), pp. 143—146. Read May 8, 1879.]

CONSIDER two plane figures, say the figure of the points P referred to axes Ox, Oy, and that of the points P' referred to axes Ox', Oy'; and let x, y be the coordinates of P, and x', y' those of P'. If the figures correspond to each other in any manner whatever, P and P' being corresponding points, then we have x', y' each of them a function of x, y; and we may consider the second figure as derived from the first by altering the distance OP in the ratio $\sqrt{x'^2+y'^2} \div \sqrt{x^2+y^2}$, and by rotating it through the angle $\tan^{-1}\frac{y'}{x'} - \tan^{-1}\frac{y}{x}$; say by the Extension $\sqrt{x'^2+y'^2} \div \sqrt{x^2+y^2}$, and by the Rotation $\tan^{-1}\frac{y'}{x'} - \tan^{-1}\frac{y}{x}$; where the Extension and the Rotation are each of them a determinate function of x, y, the coordinates of P.

Passing from the point P to a consecutive point Q, the coordinates of which are $x+dx$, $y+dy$ (the ratio $dy \div dx$ being arbitrary), then the coordinates of the corresponding point Q' will be $x'+dx'$, $y'+dy'$, where

$$dx' = \frac{dx'}{dx}dx + \frac{dx'}{dy}dy, \quad dy' = \frac{dy'}{dx}dx + \frac{dy'}{dy}dy.$$

Writing $\frac{dy'}{dx'}$ and $\frac{dy}{dx}$ instead of $dy' \div dx'$ and $dy \div dx$, the expressions

$$\sqrt{dx'^2+dy'^2} \div \sqrt{dx^2+dy^2}, \text{ and } \tan^{-1}\frac{dy'}{dx'} - \tan^{-1}\frac{dy}{dx},$$

will in general have values depending upon that of the arbitrary ratio $dy : dx$. But they may be independent of this ratio; viz. this is the case when x', y' are functions of x, y such that

$$\frac{dx'}{dy} = -\frac{dy'}{dx}, \quad \frac{dy'}{dy} = \frac{dx'}{dx};$$

and the two figures are then conformable (or conjugate) figures; that is, figures similar as regards corresponding infinitesimal elements of area. We have, in this case,

$$\sqrt{dx'^2+dy'^2} \div \sqrt{dx^2+dy^2}, \text{ and } \tan^{-1}\frac{dy'}{dx'} - \tan^{-1}\frac{dy}{dx},$$

each a determinate function of x, y, the coordinates of P; and we pass from the element PQ to the corresponding element $P'Q'$ by altering the length in the ratio $\sqrt{dx'^2+dy'^2} \div \sqrt{dx^2+dy^2}$, and rotating the element through the angle $\tan^{-1}\frac{dy'}{dx'} - \tan^{-1}\frac{dy}{dx}$; say, this ratio and this angle are the Auxesis and the Streblosis respectively, these being, as already mentioned, functions of x, y only.

Considering now any two conformable figures, say the figure of the points P, and that of the points P'; we have the theorem that we can from the first figure obtain a third conformable figure by means of an Auxesis and a Streblosis which are respectively equal to the Extension and the Rotation by which the second figure is derived from the first.

In fact, if in the three figures respectively we take x, y, x', y', and x'', y'', for the coordinates of the corresponding points P, P', P'', the first and second figures are conformable: and we have therefore

$$\frac{dx'}{dy} = -\frac{dy'}{dx}, \quad \frac{dy'}{dy} = \frac{dx'}{dx};$$

the third figure is to have the Auxesis $\sqrt{x'^2+y'^2} \div \sqrt{x^2+y^2}$, and the Streblosis

$$\tan^{-1}\frac{y'}{x'} - \tan^{-1}\frac{y}{x};$$

viz. writing r for $\sqrt{x^2+y^2}$, we ought to have

$$dx'' = \frac{xx'+yy'}{r^2}\,dx - \frac{xy'-x'y}{r^2}\,dy,$$

$$dy'' = \frac{xy'-x'y}{r^2}\,dx + \frac{xx'+yy'}{r^2}\,dy;$$

and it is therefore to be shown that there exist x'', y'' functions of x, y satisfying these relations; for, this being so, we have

$$\frac{dx''}{dy} = -\frac{dy''}{dx}, \quad \frac{dy''}{dy} = \frac{dx''}{dx};$$

and the third figure is thus conformable with the first.

Writing, for shortness,

$$A = \frac{xx'+yy'}{r^2}, \quad B = -\frac{xy'-x'y}{r^2},$$

the equations are

$$dx'' = Adx - Bdy,$$
$$dy'' = Bdx + Ady;$$

or the conditions for the existence of the functions x'', y'' are

$$\frac{dA}{dy} + \frac{dB}{dx} = 0, \quad \frac{dA}{dx} - \frac{dB}{dy} = 0.$$

We, in fact, have

$$\frac{dA}{dy} + \frac{dB}{dx} = \frac{1}{r^2}\left\{x\left(\frac{dx'}{dy} + \frac{dy'}{dx}\right) + y\left(\frac{dy'}{dy} - \frac{dx'}{dx}\right) + 2y'\right\} - \frac{2}{r^4}\{(xx' + yy')\,y + (xy' - x'y)\,x\}$$
$$= \frac{2y'}{r^2} - \frac{2}{r^4}(x^2 + y^2)\,y' = 0;$$

and similarly

$$\frac{dA}{dx} - \frac{dB}{dy} = \frac{1}{r^2}\left\{x\left(\frac{dx'}{dx} - \frac{dy'}{dy}\right) + y\left(\frac{dy'}{dx} + \frac{dx'}{dy}\right) + 2x'\right\} - \frac{2}{r^4}\{(xx' + yy')\,x - (xy' - x'y)\,y\}$$
$$= \frac{2x'}{r^2} - \frac{2}{r^4}(x^2 + y^2)\,x' = 0;$$

which proves the theorem.

The theorem is closely connected with the theory of the function of an imaginary variable; for, writing the conditions for the conformable figures in the form

$$\frac{dx'}{dx} = \frac{dy'}{dy} = F, \quad \frac{dx'}{dy} = -\frac{dy'}{dx} = -G,$$

we have

$$dx' = Fdx - Gdy,$$
$$dy' = Gdx - Fdy;$$

that is,

$$dx' + idy' = (F + iG)(dx + idy):$$

whence $F + iG$ is a function of $x + iy$, and then by integration $x' + iy'$ is also a function of $x + iy$. In one point of view, any function such as $\phi(x, y) + i\psi(x, y)$ is a function of $x + iy$, for the quantity $x + iy$ is only known by means of its real components x, y; that is, knowing $x + iy$, we know x, y, and therefore also

$$\phi(x, y) + i\psi(x, y);$$

and Cauchy, adopting this definition, introduced the expression "fonction monogène" of $x + iy$, to denote that which is in the more restricted (and the ordinary) sense termed a function of $x + iy$. And MM. Briot and Bouquet, in their "Théorie des fonctions elliptiques" (Paris, 1875), although not using Cauchy's expression *fonction monogène*, but the simple term *fonction*, do this under the qualification stated p. 3:—"Dans tout ce qui suit, nous ne nous occuperons que des fonctions qui admettent une dérivée." Now, a function admitting of a derivative (that is, in the ordinary sense, a function) of the imaginary variable z, $=x+iy$, is a function such that, for a consecutive value z', $=x+iy+dx+idy$, we have

$$\frac{f(z')-f(z)}{z'-z}$$

= a quantity independent of the ratio of the real components dx, dy of the increment $dx+idy$ of the imaginary variable. Or, what is the same thing, writing $f(z)=x'+iy'$, the condition in order that $x'+iy'$ may be a function of $x+iy$ is

$$dx'+idy'=(F+iG)(dx+idy),$$

where F and G are functions of x and y. It is not part of the condition that $F+iG$ shall be a function of $x+iy$, and it is only a long way further on that the authors prove that this is the case (see the definition of a "*fonction holomorphe,*" p. 14; and the proof, p. 137). The last-mentioned equation

$$dx'+idy'=(F+iG)(dx+idy),$$

where F and G are only assumed to be functions of x and y, has, if we represent $x+iy$ by means of the point P with coordinates (x, y), and in like manner $x'+iy'$ by means of the point P' with coordinates (x', y'), the geometrical interpretation that the figures of the points P and P' are conformable figures, that is, figures similar as regards their infinitesimal elements. The foregoing theorem in regard to the Auxesis and the Streblosis is that we can, by means of F and G, construct a third conformable figure,—in fact, the Auxesis and the Streblosis are $=\sqrt{F^2+G^2}$ and $\tan^{-1}\frac{G}{F}$ respectively; and, using these as an Extension and a Rotation, we have the third conformable figure $x''+iy''=(F+iG)(x+iy)$; that is, $(F+iG)(x+iy)$, and therefore also $F+iG$, is a function of $x+iy$,—and we have thus the derivative of a function of $x+iy$ as itself a function of $x+iy$.

It is to be remarked that, although the theorem of the Auxesis and the Streblosis, considered as a property of conformable figures, is not by any means geometrically self-evident, yet the foregoing analytical proof is only a proof conducted by means of real quantities, of what (admitting the theory of imaginary quantities) is in fact self-evident; viz. the analytical conclusion really is that, F, G denoting functions of x, y, then, if $dx'+idy'=(F+iG)(dx+idy)$, that is, if $(F+iG)(dx+idy)$ be a complete differential, then $F+iG$ is a function of $x+iy$.

730.

[ADDITION TO MR SPOTTISWOODE'S PAPER "ON THE TWENTY-ONE COORDINATES OF A CONIC IN SPACE."]

[From the *Proceedings of the London Mathematical Society*, vol. x. (1879), pp. 194—196.]

WRITE

$$\begin{aligned} U &= (a, b, c, d, f, g, h, l, m, n \between x, y, z, t)^2, \\ U_0 &= (\quad\quad ,, \quad\quad \between \xi, \eta, \zeta, \omega)^2, \\ W &= (\quad\quad ,, \quad\quad \between x, y, z, t \between \xi, \eta, \zeta, \omega), \\ P &= (\alpha, \beta, \gamma, \delta \between x, y, z, t), \\ P_0 &= (\alpha, \beta, \gamma, \delta \between \xi, \eta, \zeta, \omega). \end{aligned}$$

Then the equation of the cone, having for its vertex the arbitrary point $(\xi, \eta, \zeta, \omega)$, and passing through the conic $U=0$, $P=0$, is

$$UP_0^2 - 2WPP_0 + U_0P^2 = 0.$$

Or if, to put the coefficients ξ, η, ζ, ω in evidence, we write for a moment

$$\begin{aligned} A &= (a, h, g, l \between x, y, z, t), \\ B &= (h, b, f, m \between \quad ,, \quad), \\ C &= (g, f, c, n \between \quad ,, \quad), \\ D &= (l, m, n, d \between \quad ,, \quad), \end{aligned}$$

and therefore

$$W = A\xi + B\eta + C\zeta + D\omega;$$

then the equation is

$$U(\alpha\xi + \beta\eta + \gamma\zeta + \delta\omega)^2 - 2P(\alpha\xi + \beta\eta + \gamma\zeta + \delta\omega)(A\xi + B\eta + C\zeta + D\omega)$$
$$+ P^2(a, b, c, d, f, g, h, l, m, n \between \xi, \eta, \zeta, \omega)^2 = 0.$$

And if we expand first in ξ, η, ζ, ω, and then in x, y, z, t, the final result is

	x^2	y^2	z^2	t^2	yz	zx	xy	xt	yt	zt	
ξ^2		C	B	F	$2A'$				$2L$	$2L'$	$= 0.$
$+\ \eta^2$	C		A	G		$2B'$		$2M'$		$2M$	
$+\ \zeta^2$	B	A		H			$2C'$	$2N$	$2N'$		
$+\ \omega^2$	F	G	H		$2F'$	$2G'$	$2H'$				
$+\ \eta\zeta$	$2A'$			$2F'$	$-2A$	$-2C'$	$-2B'$	$2(Q-R)$	$-2M$	$-2N'$	
$+\ \zeta\xi$		$2B'$		$2G'$	$-2C'$	$-2B$	$-2A'$	$-2L'$	$2(R-P)$	$-2N$	
$+\ \xi\eta$			$2C'$	$2H'$	$-2B'$	$-2A'$	$-2C$	$-2L$	$-2M'$	$2(P-Q)$	
$+\ \xi\omega$		$2M'$	$2N$		$2(Q-R)$	$-2L'$	$-2L$	$-2F$	$-2H'$	$-2G'$	
$+\ \eta\omega$	$2L$		$2N'$		$-2M$	$-2(R-P)$	$-2M'$	$-2H'$	$-2G$	$-2F'$	
$+\ \zeta\omega$	$2L'$	$2M$			$-2N'$	$-2N$	$2(P-Q)$	$-2G'$	$-2F'$	$-2H$	

In particular, if $\eta=0$, $\zeta=0$, $\omega=0$, then we have the foregoing equation $X=0$; and the like for the equations $Y=0$, $Z=0$, and $W=0$ respectively.

Take a, b, c, f, g, h for the six coordinates of the line through the points

$$\left|\begin{matrix} x, & y, & z, & t \\ \xi, & \eta, & \zeta, & \omega \end{matrix}\right| ;$$

that is, write

$$\begin{aligned} \mathrm{a} &= y\zeta - z\eta, & \mathrm{f} &= x\omega - t\xi, \\ \mathrm{b} &= z\xi - x\zeta, & \mathrm{g} &= y\omega - t\eta, \\ \mathrm{c} &= x\eta - y\xi, & \mathrm{h} &= z\omega - t\zeta,. \end{aligned}$$

where, of course,

$$\mathrm{af} + \mathrm{bg} + \mathrm{ch} = 0.$$

Then the foregoing equation of the cone is

$$\left.\begin{array}{l} A\mathrm{a}^2 \ + \ B\mathrm{b}^2 \ + \ C\mathrm{c}^2 \ + \ F\mathrm{f}^2 \ + \ G\mathrm{g}^2 \ + \ H\mathrm{h}^2 \\ -\,2A'\mathrm{bc} - 2B'\mathrm{ca} - 2C'\mathrm{ab} + 2F'\mathrm{gh} + 2G'\mathrm{hf} + 2H'\mathrm{fg} \\ +\,2P\mathrm{af} \ + 2M\mathrm{ag} - 2N'\mathrm{ah} \\ -\,2L'\mathrm{bf} \ + 2Q\mathrm{bg} \ + 2N\mathrm{bh} \\ +\,2L\mathrm{cf} \ - 2M'\mathrm{cg} + 2R\mathrm{ch} \end{array}\right\} = 0.$$

And this may be regarded as the equation of the *conic* in terms of the twenty-one coordinates of the conic, and of the six coordinates of an arbitrary line meeting the conic. It is, in fact, the general form of the equation given in the paper—Cayley, "On a new Analytical Representation of a Curve in Space," *Quart. Math. Jour.*, vol. III. (1860), [284; this Collection, vol. IV. p. 453].

731.

ON THE BINOMIAL EQUATION $x^p - 1 = 0$; TRISECTION AND QUARTISECTION.

[From the *Proceedings of the London Mathematical Society*, vol. XI. (1880), pp. 4—17. Read November 13, 1879.]

THE solution of the binomial equation $x^p - 1 = 0$, p a prime number, or, say rather, the equation

$$x^{p-1} + x^{p-2} + \ldots + x + 1 = 0,$$

depends upon the Jacobian function

$$F\alpha = x^1 + \alpha x^g + \ldots + \alpha^{p-2} x^{g^{p-2}},$$

where g is a prime root of p, α any root whatever of the equation $u^{p-1} - 1 = 0$. Taking e a factor of $p-1$, and f for the complementary factor (that is, $p - 1 = ef$), then, if for α we write α^f, or, what is the same thing, taking α^f, $= \beta$, a root of $u^e - 1 = 0$, we have

$$F\beta = X_0 + \beta X_1 + \ldots + \beta^{e-1} X_{e-1},$$

where $X_0, X_1, \ldots, X_{e-1}$ denote each of them a period or sum of f, $= \frac{p-1}{e}$, roots, viz.

$$\begin{aligned} X_0 \;&= (1, \quad g^e, \quad \ldots, g^{(f-1)e}), \\ X_1 \;&= (g^e, \quad g^{e+1}, \quad \ldots, g^{(f-1)e+1}), \\ &\vdots \\ X_{e-1} &= (g^{e-1}, \quad g^{2e-1}, \quad \ldots, g^{fe-1}) \end{aligned}$$

(read $X_0 = x^1 + x^{g^e} + \ldots + x^{g^{(f-1)e}}$, and so for the other functions).

We have, of course, $F(1)$, $= X_0 + X_1 + \ldots + X_{e-1}$, the sum of all the roots $= -1$; and, further, the general property that any rational and integral function of these periods is expressible as a sum

$$a_0 X_0 + a_1 X_1 + \ldots + a_{e-1} X_{e-1}$$

with known coefficients

$$a_0, a_1, \ldots, a_{e-1}.$$

The several cases $e = 2, 3, 4, \ldots$ may be termed those of the bisection, trisection, quartisection, &c., of the equation; viz.

$e = 2$, there are two periods, X, Y, and $F(-1) = X - Y$;

$e = 3$, three periods, X, Y, Z, and $F\gamma = X + \gamma Y + \gamma^2 Z$, if γ is a root of $u^3 - 1 = 0$;

$e = 4$, four periods, X, Y, Z, W, and $F\delta = X + \delta Y + \delta^2 Z + \delta^3 W$, if δ be a root of $u^4 - 1 = 0$.

It is sufficient to attend to the prime roots γ and δ of the equations

$$u^3 - 1 = 0, \quad u^4 - 1 = 0,$$

respectively; for, if γ or δ be $= 1$, we have simply $F(1)$, $= -1$; and if δ be $= -1$, then the function is $F(-1)$, $= X + Z - (Y + W)$, where $X + Z$ and $Y + W$ are the periods for the bisection. The prime roots δ are of course i and $-i$, and we have

$$F(i) \;\; = X + iY - Z - iW,$$
$$F(-i) = X - iY - Z + iW,$$

respectively.

As regards the bisection, it is known that $(X - Y)^2 = (-)^{\frac{p-1}{4}} p$, which is $+p$ or $-p$, according as p is $\equiv 1$ or 3, mod. 4; and the values of X, Y are thus determined. In what follows, I consider the cases $e = 3$ and $e = 4$ of the trisection and the quartisection respectively.

It is to be remembered that, not the division into periods, but the order of the periods, depends on the choice of g, a prime root at pleasure of p; and, in what follows, I select the prime root used in Reuschle's *Tafeln complexer Primzahlen welche aus Wurzeln der Einheit gebildet sind* (4to, Berlin, 1875): viz. these are

$p = 3, 5, 7, 11, 13, 17, 19, 23, 29, 31, 37, 41, 43, 47, 53,$
$59, 61, 67, 71, 73, 79, 83, 89, 97,$

$g = 2, 2, 5, 2, 2, 3, 2, -2, 2, 3, 2, 6, 3, 10, 2,$
$2, 2, 2, 62, 5, 3, 2, 30, 10,$

where I quote the whole series, although I am here only concerned with the values of p which are $\equiv 1$ (mod. 3), or $\equiv 1$ (mod. 4).

The periods are consequently those of Reuschle, viz. X, Y, Z are his η_0, η_1, η_2, and X, Y, Z, W his η_0, η_1, η_2, η_3: they can of course, without referring to his work, be easily recalculated, but it is, I think, convenient to have for his values of g the series of residues such as are given (for differently selected values of g) in Jacobi's *Canon Arithmeticus* (4to, Berlin, 1839); and I have accordingly taken out of Reuschle, and annex, such a table.

For instance, $p = 13$, the powers of g are 1, 2, 4, 8, 3, 6, 12, 11, 9, 5, 10, 7; and, by writing these down in order in columns of 3 or of 4,

1	8	12	5		1	3	9
2	3	11	10		2	6	5
4	6	9	7		4	12	10
					8	11	7

we have the periods X, Y, Z or X, Y, Z, W, belonging to the trisection and the quartisection of $p = 13$.

I further remark that the equations which I am concerned with are all given in Reuschle, but in a somewhat different form; thus, $p = 13$, quartisection (see p. 13), he has

$$\eta_0^2 = \eta_1 + 2\eta_2, \quad \eta_0\eta_1 = -1 - \eta_2, \quad \eta_0\eta_2 = 3 + \eta_1 + \eta_3, \quad \eta_0\eta_3 = -1 - \eta_1,$$

(where observe that here and in every case the value of $\eta_0\eta_3$ is at once obtained from that of $\eta_0\eta_1$ by a mere cyclical interchange of the suffixes, so that the last equation is in fact superfluous); the other equations, using $\eta_0 + \eta_1 + \eta_2 + \eta_3 = -1$ to eliminate any constant term which occurs, give my values

$$\begin{aligned} X^2 &= (\;\;0,\;\; 1,\;\; 2,\;\; 0)(X, Y, Z, W), \\ XY &= (\;\;1,\;\; 1,\;\; 0,\;\; 1)(\quad " \quad), \\ XZ &= (-3, -2, -3, -2)(\quad " \quad). \end{aligned}$$

Similarly, in the case of a trisection, the equation for $\eta_0\eta_2$ is superfluous, and the other equations give my values of X^2 and XY.

Reuschle gives also, and I take from him, the cubic and the quartic equations (such as $p = 13$, $\eta^3 + \eta^2 - 4\eta + 1 = 0$, $\eta^4 + \eta^3 + 2\eta^2 - 4\eta + 3 = 0$), which determine the periods in the trisections and the quartisections respectively.

Many of the results obtained accord with, and furnish exemplifications of general theorems contained in Jacobi's memoir, "Ueber die Kreistheilung und ihre Anwendung auf die Zahlentheorie," *Crelle*, t. XXX. (1846), pp. 166—189; [*Ges. Werke*, t. VI. pp. 254—274].

Trisection, $e = 3$; $p \equiv 1$ (*mod.* 3).

We have three periods X, Y, Z; and we thence obtain

$$\begin{aligned} X^2 &= (a, b, c)(X, Y, Z), \\ XY &= (f, g, h)(\quad " \quad), \end{aligned}$$

the coefficients a, b, c, f, g, h being determinate integers. And, by cyclical interchanges, we obtain equations which may be written

$$\begin{aligned} X^2 &= a, b, c, \\ Y^2 &= c, a, b, \\ Z^2 &= b, c, a, \\ XY &= f, g, h, \\ YZ &= h, f, g, \\ ZX &= g, h, f; \end{aligned}$$

viz. here and elsewhere the coefficients a, b, c are written to denote the sum

$$aX + bY + cZ.$$

It is easy to see that

$$f + g + h = \tfrac{1}{3}(p - 1);$$

in fact, a period contains $\frac{1}{3}(p-1)$ terms, and in two consecutive periods X, Y, there are no terms the product of which is unity; hence XY contains $\frac{1}{9}(p-1)^2$ terms, each a power of x, and the sum $XY+YZ+ZX$ contains $\frac{1}{3}(p-1)^2$ such terms, being in fact the sum $X+Y+Z$ taken $\frac{1}{3}(p-1)$ times; whence the relation in question.

Hence also

$$YZ+ZX+XY=-\tfrac{1}{3}(p-1).$$

From the equation $X+Y+Z=-1$, multiplying by X, and for X^2, XY, XZ substituting their values, we obtain an expression

$$(a+f+g+1)X+(b+g+h)Y+(c+h+f)Z,$$

which must identically vanish; viz. the three coefficients must be each of them $=0$; or we must have

$$\begin{aligned} a &= -f-g-1, \\ b &= -g-h, \\ c &= -h-f; \end{aligned}$$

so that, taking f, g, h as known, the other coefficients a, b, c are given in terms of them. The equations give

$$a+b+c=-2(f+g+h)-1.$$

We have $X.YZ=Y.ZX$; that is, $X(h, f, g)=Y(g, h, f)$; or, substituting for X^2, XY, &c. their values,

$$\begin{aligned} h(a, b, c) = {} & g(f, g, h) \\ +f(f, g, h) & +h(c, a, b) \\ +g(g, h, f) & +f(h, f, g); \end{aligned}$$

that is,

$$\begin{aligned} ah+f^2+g^2 &= gf+ch+fh, \\ bh+fg+gh &= g^2+ah+f^2, \\ ch+fh+fg &= gh+bh+fg, \end{aligned}$$

equations which reduce themselves to the single equation

$$gh+hf+fg+h=f^2+g^2+h^2;$$

and this is the only relation obtainable by consideration of the three equal values

$$X.YZ, \quad Y.ZX, \quad Z.XY.$$

Moreover, this equation being satisfied, the six functions in the three equations become each of them $=fg-h^2$; or we have

$$XYZ=(fg-h^2, \ fg-h^2, \ fg-h^2);$$

that is,

$$XYZ=h^2-fg.$$

We have

$$\begin{aligned} F\gamma . F\gamma^2 &= X^2+Y^2+Z^2-YZ-ZX-XY \\ &= (a+b+c-f-g-h)(X+Y+Z) \\ &= -(a+b+c)+(f+g+h) \\ &= 3(f+g+h)+1; \end{aligned}$$

that is,

$$F\gamma \,.\, F\gamma^2 = p.$$

We have, moreover,

$$\begin{aligned}(F\gamma)^2 = {} & X^2 + 2YZ + \gamma(Z^2 + 2XY) + \gamma^2(Y^2 + 2ZX) \\ = {} & [(a,\ b,\ c) + 2(h,\ f,\ g)] \\ & + \gamma\,[(b,\ c,\ a) + 2(f,\ g,\ h)] \\ & + \gamma^2[(c,\ a,\ b) + 2(g,\ h,\ f)],\end{aligned}$$

which is

$$= \{(a + 2h) + \gamma(b + 2f) + \gamma^2(c + 2g)\}\,(X + \gamma^2 Y + \gamma Z),$$

as is at once seen by comparing the coefficients of X, Y, Z respectively.

Hence, writing

$$\begin{aligned}& a + 2h + \gamma(b + 2f) + \gamma^2(c + 2g) \\ & = a + 2h + \gamma(b + 2f) - (1 + \gamma)(c + 2g) \\ & = A + B\gamma,\end{aligned}$$

we have

$$\begin{aligned}A &= a + 2h - c - 2g = 3h - 3g - 1, \\ B &= b + 2f - c - 2g = 3f - 3g.\end{aligned}$$

We have

$$(F\gamma)^2 = (A + B\gamma)\,F\gamma^2,$$

and thence, writing γ^2 for γ,

$$(F\gamma^2)^2 = (A + B\gamma^2)\,F\gamma,$$

equations which give

$$F\gamma \,.\, F\gamma^2, \ = p, \ = (A + B\gamma)(A + B\gamma^2);$$

or, say $p = A^2 - AB + B^2$; viz. p has the complex factor

$$A + B\gamma, \ = 3h - 3g - 1 + \gamma(3f - 3g).$$

Hence also

$$\begin{aligned}(F\gamma)^3 &= p(A + B\gamma), \\ (F\gamma^2)^3 &= p(A + B\gamma^2),\end{aligned}$$

and, as before,

$$F\gamma \,.\, F\gamma^2 = p;$$

which equations determine $F\gamma$, $F\gamma^2$, and from these and $F(1) = -1$ we obtain the periods X, Y, Z; we have thus, in fact, the solution of the cubic equation which gives these periods. We have already found the coefficients of this cubic equation, viz.

$$X + Y + Z = -1, \quad YZ + ZX + XY = -\tfrac{1}{3}(p - 1), \quad XYZ = h^2 - fg;$$

the equation thus is

$$\eta^3 + \eta^2 - \tfrac{1}{3}(p - 1)\,\eta + (fg - h^2) = 0.$$

As already remarked, the values of a, b, c; f, g, h, and the equations in η, are in effect given in Reuschle; the complex factors of p, as given p. 1 ($7 = 2\gamma - 3\gamma^2$, &c.), when reduced to the form $A + B\gamma$, are not identical with the $A + B\gamma$ of the foregoing theory; viz. this $A + B\gamma$ is not Reuschle's selected primary form. I give, in the annexed table for the primes 7, 13, ..., to 97, the values from Reuschle of a, b, c; f, g, h, and of the coefficients of the η-equation; also the values of A and B derived from f, g, h by the foregoing formulæ. It will be seen that all the values are consistent with the theory.

TABLE FOR THE TRISECTION.

p	$a,$ $f,$	$b,$ $g,$	c h	$\eta^3+\eta^2+$ η^1	η^0	A	B	Page in Reuschle
7	− 2 1	− 1 0	− 2 1	− 2	− 1	2	3	p. 6
13	− 4 1	− 3 2	− 2 1	− 4	− 1	− 4	− 3	p. 15
19	− 4 1	− 5 2	− 4 3	− 6	− 7	2	− 3	p. 26
31	− 7 4	− 6 2	− 8 4	− 10	− 8	5	6	p. 45
37	− 8 5	− 10 4	− 7 3	− 12	11	− 4	3	p. 54
43	− 11 6	− 8 4	− 10 4	− 14	8	− 1	6	p. 69
61	− 14 5	− 13 8	− 15 7	− 20	− 9	− 4	− 9	p. 97
67	− 16 9	− 13 6	− 16 7	− 22	5	2	9	p. 105
73	− 16 6	− 18 9	− 15 9	− 24	− 27	− 1	− 9	p. 128
79	− 20 9	− 17 10	− 16 7	− 26	41	− 10	− 3	p. 138
97	− 20 10	− 23 9	− 22 13	− 32	− 79	11	3	p. 168

Quartisection, $e = 4$; $p \equiv 1$ (*mod.* 4).

We have four periods X, Y, Z, W; and we obtain

$$\begin{aligned} X^2 &= (a, b, c, d)(X, Y, Z, W), \\ XY &= (f, g, h, k)(\quad " \quad), \\ XZ &= (l, m, l, m)(\quad " \quad), \end{aligned}$$

the coefficients being determinate integers. It can be shown that $l + m = \frac{1}{8}(p-1)$ or $-\frac{1}{8}(3p+1)$ according as $p \equiv 1$ or 5 (mod. 8). And then, by cyclical interchanges,

$$\begin{aligned} X^2 &= a, b, c, d, \\ Y^2 &= d, a, b, c, \\ Z^2 &= c, d, a, b, \\ W^2 &= b, c, d, a, \\ XY &= f, g, h, k, \\ YZ &= k, f, g, h, \\ ZX &= h, k, f, g, \\ XW &= g, h, k, f, \\ XZ &= l, m, l, m, \\ YW &= m, l, m, l. \end{aligned}$$

We have, in like manner as for the trisection,

$$f + g + h + k = \tfrac{1}{4}(p-1),$$

and so also the expression for

$$\Sigma XY, \; = XY + XZ + XW + YZ + YW + ZW$$

is

$$= -(f + g + h + k + l + m) = -\tfrac{1}{4}(p-1) - l - m\,;$$

and, in virtue of the foregoing value of $l + m$, this is $= -\frac{3}{8}(p-1)$ or $\frac{1}{8}(p+3)$ according as $p \equiv 1$ or 5 (mod. 8).

Again, from the equation $X + Y + Z + W = -1$, multiplying by X and reducing,

$$\begin{aligned} a &= -1 - f - g - l, \\ b &= \quad\;\; - g - h - m, \\ c &= \quad\;\; - h - k - l, \\ d &= \quad\;\; - k - f - m, \end{aligned}$$

and thence

$$a + b + c + d = -1 - 2(f + g + h + k) - 2(l + m),$$

and

$$a - b + c - d = -1 + 2(m - l).$$

We have

$$X.YZ=Y.ZX=Z.XY,$$

that is,

$$X(k,\ f,\ g,\ h)=Y(l,\ m,\ l,\ m)=Z(f,\ g,\ h,\ k),$$

and thence

$$\begin{array}{lll} k(a,\ b,\ c,\ d)= & l\ (f,\ g,\ h,\ k)= & f(l,\ m,\ l,\ m) \\ +f(f,\ g,\ h,\ k) & +m(d,\ a,\ b,\ c) & +g(k,\ f,\ g,\ h) \\ +g(l,\ m,\ l,\ m) & +l\ (k,\ f,\ g,\ h) & +h(c,\ d,\ a,\ b) \\ +h(g,\ h,\ k,\ f) & +m(m,\ l,\ m,\ l) & +k(h,\ k,\ f,\ g), \end{array}$$

that is,

$$\begin{aligned} ka+f^2+gl+gh &= lf+md+lk+m^2=lf+gk+ch+kh, \\ kb+fg+gm+h^2 &= lg+am+lf+ml=fm+fg+hd+k^2, \\ kc+fh+gl+hk &= lh+mb+lg+m^2=fl+g^2+ah+kf, \\ kd+fk+gm+fh &= kl+mc+lh+lm=fm+gh+bh+gk, \end{aligned}$$

in which equations $a,\ b,\ c,\ d$ may be regarded as having their foregoing values.

One of these equations is

$$kc+fh+gl+hk=lf+g^2+ah+kf,$$

that is,

$$-k(h+k+l)+fh+gl+hk=lf+g^2-h(f+g+l+1)+kf,$$

or, reducing,

$$l(g+h-f-k)=g^2+k^2-2hf-hg-h+kf,$$

which gives l.

Again, another equation is

$$kb+fg+gm+h^2=fm+fg+hd+k^2,$$

that is,

$$-k(g+h+m)+fg+gm+h^2=fm+fg-h(k+f+m)+k^2,$$

or, reducing,

$$m(g+h-f-k)=k^2-h^2+gk-hf,$$

which gives m.

And we have also

$$md+lk+m^2=gk+ch+kh,$$

that is,

$$-m(k+f+m)+lk+m^2=gk+kh-h(h+k+l),$$

or, reducing,

$$l(k+h)-m(f+k)=gk-h^2.$$

Substituting herein for $l,\ m$ their values, we have

$$(k+h)[g^2+k^2-2hf-hg+kf-h]-(f+k)[k^2-h^2+gk-hf]+(h^2-gk)[g+h-f-k]=0.$$

In this equation the only terms of the second order are $-h(h+k)$, which contain the factor h; the terms of the third order contain this same factor h, and throwing it out, and reducing, the equation is found to be

$$(g-k)^2+(h-f)^2=h+k,$$

or, as it may also be written,

$$g^2+k^2-2hf-h+(h^2+f^2-2gk-k)=0;$$

and the foregoing values of l, m are

$$l=\frac{(g^2+k^2-2hf-h)-(gh-kf)}{g+h-k-f},$$

$$m=\frac{k^2-h^2+gk-hf}{g+h-k-f};$$

and by means of these three equations all the foregoing equations are satisfied.

We have

$$\begin{aligned} FiFi^3 &= (X-Z)^2+(Y-W)^2 \\ &= X^2+Y^2+Z^2+W^2-2(XZ+YW) \\ &= -(a+b+c+d)+2(l+m); \end{aligned}$$

or, substituting for a, b, c, d, this is

$$=1+2(f+g+h+k)+4(l+m),$$

viz. it is

$$=\tfrac{1}{2}(p+1)+4(l+m);$$

or, substituting for $l+m$ its before-mentioned value, then, according as $p\equiv 1$ or 5 (mod. 8), the value is $=p$ or $-p$; that is, we have

$$FiFi^3=(-)^{\frac{p-1}{4}}p.$$

Again, we have

$$\begin{aligned} (Fi)^2 &= (X+iY-Z-iW)^2 \\ &= X^2-Y^2+Z^2-W^2-2XZ+2YW+2i(XY-YZ+ZW-WX) \\ &= \{a-b+c-d+2(m-l)+2(f-g+h-k)i\}(X-Y+Z-W) \\ &= (A+Bi)F(-1), \end{aligned}$$

where

$$\begin{aligned} A &= a-b+c-d+2(m-l), \;=-1+4(m-l), \\ B &= 2(f-g+h-k); \end{aligned}$$

or, since $X-Y+Z-W=F(-1)$, this equation is

$$(Fi)^2=(A+Bi)F(-1):$$

and similarly

$$(Fi^3)^2=(A-Bi)F(-1).$$

Moreover

$$[F(-1)]^2=(-)^{\frac{p-1}{2}}p, \ =p;$$

and we have therefore

$$(\pm p)^2=(A^2+B^2)p,$$

that is,

$$A^2+B^2=p;$$

or the expression $A+Bi$ determined as above is a complex factor of p.

We may investigate the quartic equation for the determination of the periods X, Y, Z, W. The values of $X+Y+Z+W$ and $XY+XZ+XW+YZ+YW+ZW$ are already known: for the next coefficient $XYZ+XYW+XZW+YZW$, we have $XYZ=(\alpha, \beta, \gamma, \delta)$, where each of the coefficients α, β, γ, δ is given under three different forms: the values of YZW, ZWX, WXY are $(\delta, \alpha, \beta, \gamma)$, $(\gamma, \delta, \alpha, \beta)$, $(\beta, \gamma, \delta, \alpha)$; and the required sum therefore is

$$(\alpha+\beta+\gamma+\delta)(X+Y+Z+W), \ =-(\alpha+\beta+\gamma+\delta).$$

Taking the first expressions of these coefficients respectively, we have

$$\begin{aligned}\alpha+\beta+\gamma+\delta= \quad & k(a+b+c+d)\\ & +f(f+g+h+k)\\ & +2g(l+m)\\ & +h(f+g+h+k),\\ =k\{-1-\tfrac{1}{2}(p-1)-2(l+m)\}&+(f+h)\{\tfrac{1}{4}(p-1)\}+2g(l+m),\\ =2(g-k)(l+m)+\tfrac{1}{4}&(f+h)(p-1)-\tfrac{1}{2}k(p+1).\end{aligned}$$

We find $XYZW$ most readily as the product of XZ and YW; we thus obtain

$$\begin{aligned}XYZW &= lm(X^2+Y^2+Z^2+W^2+2XZ+2YW)+(l^2+m^2)(XY+XW+YZ+ZW),\\ &= lm(-a-b-c-d-2l-2m)-(l^2+m^2)(f+g+h+k),\\ &= lm\{1+2(f+g+h+k)\}-(l^2+m^2)(f+g+h+k);\end{aligned}$$

or, substituting for $f+g+h+k$ its value $\tfrac{1}{4}(p-1)$, this is

$$lm-\tfrac{1}{4}(l-m)^2(p-1), \ =\tfrac{1}{4}\{(l+m)^2-(l-m)^2p\}.$$

Hence the required equation, having roots X, Y, Z, W, is

$$\begin{aligned}&\eta^4+\eta^3\\ &-\eta^2\{\tfrac{1}{4}(p-1)+l+m\}\\ &+\eta\ \{\tfrac{1}{4}(f+h)(p-1)-\tfrac{1}{2}k(p+1)+2(g-k)(l+m)\}\\ &+lm-\tfrac{1}{4}(l-m)^2(p-1)\\ &=0,\end{aligned}$$

where, for the sake of having a single formula, I have retained $l+m$ in place of its value $=-\tfrac{3}{8}(p-1)$ or $\tfrac{1}{8}(p+3)$ according as $p\equiv 1$ or 5 (mod. 8).

We thus have the following:—

TABLE FOR THE QUARTISECTION.

p	a f l	b g m	c h	d k	η^2	$\eta^4+\eta^3+$ η^1	η^0	A	B	Page in Reuschle
5	0	1	0	0	1	1	1	− 1	− 2	p. 2
	0	0	0	1						
	− 1	− 1								
13	0	1	2	0	2	− 4	3	3	− 2	p. 13
	1	1	0	1						
	− 3	− 2								
17	− 4	− 2	− 3	− 4	− 6	− 1	1	− 1	4	p. 19
	2	0	1	1						
	1	1								
29	2	3	0	2	4	20	23	− 5	− 2	p. 36
	1	1	2	3						
	− 5	− 6								
37	2	1	2	4	5	7	49	− 1	6	p. 53
	2	2	4	1						
	− 7	− 7								
41	− 10	− 6	− 7	− 8	− 15	18	4	− 5	4	p. 61
	4	2	2	2						
	3	2								
53	2	3	6	2	7	− 43	47	7	− 2	p. 80
	4	4	2	3						
	− 11	− 9								
61	4	3	2	6	8	42	117	− 5	6	p. 96
	3	3	6	3						
	− 11	− 12								
73	− 16	− 13	− 12	− 14	− 27	− 41	2	3	8	p. 126
	6	5	5	2						
	4	5								
89	− 19	− 18	− 16	− 14	− 33	39	8	− 5	− 8	p. 152
	4	8	5	5						
	6	5								
97	− 22	− 16	− 17	− 18	− 36	91	− 61	− 9	4	p. 167
	8	6	5	5						
	7	5								

TABLE OF THE POWERS OF REUSCHLE'S SELECTED PRIME ROOTS.

	3	5	7	11	13	17	19	23	29	31	37	41	43	47	53	59	61	67	71	73	79	83	89	97	
0	1	1	1	1	1	1	1	1	1	1	1	1	1	1	1	1	1	1	1	1	1	1	1	1	0
1	2	2	5	2	2	3	2	21	2	3	2	6	3	10	2	2	2	2	62	5	3	2	30	10	1
2		4	4	4	4	9	4	4	4	9	4	36	9	6	4	4	4	4	10	25	9	4	10	3	2
3	...	3	6	8	8	10	8	15	8	27	8	11	27	13	8	8	8	8	52	52	27	8	33	30	3
4			2	5	3	13	16	16	16	19	16	25	38	36	16	16	16	16	29	41	2	16	11	9	4
5	...	...	3	10	6	5	13	14	3	26	32	27	28	31	32	32	32	32	23	59	6	32	63	90	5
6				9	12	15	7	18	6	16	27	39	41	28	11	5	3	64	6	3	18	64	21	27	6
7				7	11	11	14	10	12	17	17	29	37	45	22	10	6	61	17	15	54	45	7	76	7
8				3	9	16	9	3	24	20	34	10	25	27	44	20	12	55	60	2	4	7	32	81	8
9	...	...	...	6	5	14	18	17	19	29	31	19	32	35	35	40	24	43	28	10	12	14	70	34	9
10					10	8	17	12	9	25	25	32	10	21	17	21	48	19	32	50	36	28	53	49	10
11	...	...	...	...	7	7	15	22	18	13	13	28	30	22	34	42	35	38	67	31	29	56	77	5	11
12						4	11	2	7	8	26	4	4	32	15	25	9	9	36	9	8	29	85	50	12
13						12	3	19	14	24	15	24	12	38	30	50	18	18	31	45	24	58	58	15	13
14						2	6	8	28	10	30	21	36	4	7	41	36	36	5	6	72	33	49	53	14
15	...	...	...	...	...	6	12	7	27	30	23	3	22	40	14	23	11	5	26	30	58	66	46	45	15
16							5	9	25	28	9	18	23	24	28	46	22	10	50	4	16	49	45	62	16
17	...	...	...	...	...	...	10	5	21	22	18	26	26	5	3	33	44	20	47	20	48	15	15	38	17
18								13	13	4	36	33	35	3	6	7	27	40	3	27	65	30	5	89	18
19								20	26	12	35	34	19	30	12	14	54	13	44	62	37	60	61	17	19
20								6	23	5	33	40	14	18	24	28	47	26	30	18	32	37	50	73	20
21	...	...	...	...	...	...	...	11	17	15	29	35	42	39	48	56	33	52	14	17	17	74	76	51	21
22									5	14	21	5	40	14	43	53	5	37	16	12	51	65	55	25	22
23									10	11	5	30	34	46	33	47	10	7	69	60	74	47	48	56	23
24									20	2	10	16	16	37	13	35	20	14	18	8	64	11	16	75	24
25									11	6	20	14	5	41	26	11	40	28	51	40	34	22	35	71	25
26									22	18	3	2	15	34	52	22	19	56	38	54	23	44	71	31	26
27	...	...	...	...	...	...	...	...	15	23	6	12	2	11	51	44	38	45	13	51	69	5	83	19	27
28										7	12	31	6	16	49	29	15	23	25	36	48	10	87	93	28
29	...	...	...	...	...	...	...	...	...	21	24	22	18	19	45	58	30	46	59	34	68	20	29	57	29
30											11	9	11	2	37	57	60	25	37	24	46	40	69	85	30
31											22	13	33	20	21	55	59	50	22	47	59	80	23	74	31
32											7	37	13	12	42	51	57	33	15	16	19	77	67	61	32
33											14	17	39	26	31	43	53	66	7	7	57	71	52	28	33
34											28	20	31	25	9	27	45	65	8	35	13	59	47	86	34
35	...	...	...	...	...	...	...	...	...	...	19	38	7	15	18	54	29	63	70	29	39	35	75	84	35
36												23	21	9	36	49	58	59	9	72	38	70	25	64	36
37												15	20	43	19	39	55	51	61	68	35	57	38	58	37
38												8	17	7	38	19	49	35	19	48	26	31	72	95	38
39	...	...	...	...	...	...	...	...	...	...	...	7	8	23	23	38	37	3	42	21	78	62	24	77	39
40													24	42	46	17	13	6	48	32	76	41	8	91	40
41	...	...	...	...	...	...	...	...	...	...	...	...	29	44	39	34	26	12	65	14	70	82	62	37	41
42														11	25	9	52	24	45	70	52	81	80	79	42
43														29	50	18	43	28	11	58	77	79	86	14	43
44														8	47	36	25	29	43	71	73	75	88	43	44
45	...	...	...	...	...	...	...	...	...	...	...	...	...	33	41	13	50	58	39	63	61	67	59	42	45
46															29	26	39	49	4	23	25	51	79	32	46
47															5	52	17	31	35	42	75	19	56	29	47
48															10	45	34	62	40	64	67	38	78	96	48
49															20	31	7	57	66	28	43	76	26	87	49
50															40	8	14	47	45	67	50	69	68	94	50

TABLE (*continued*).

	53	59	61	67	71	73	79	83	89	97	
51	27	6	28	27	21	43	71	55	82	67	51
52		12	56	54	24	69	55	27	57	88	52
53		24	51	41	68	53	7	54	19	7	53
54		48	41	15	27	46	21	25	36	70	54
55		37	21	30	41	11	63	50	12	21	55
56		15	42	60	57	55	31	17	4	16	56
57		30	23	53	55	56	14	34	31	63	57
58			46	39	2	61	42	68	40	48	58
59			31	11	53	13	47	53	43	92	59
60				22	20	65	62	23	44	47	60
61				44	33	33	28	46	74	82	61
62				21	58	19	5	9	84	44	62
63				42	46	22	15	18	28	52	63
64				17	12	37	45	36	39	35	64
65				34	34	39	56	72	13	59	65
66					49	49	10	61	34	8	66
67					56	26	30	39	41	80	67
68					64	57	11	78	73	24	68
69					63	66	33	73	54	46	69
70						38	20	63	18	72	70
71						44	60	43	6	41	71
72							22	3	2	22	72
73							26	6	60	26	73
74							40	12	20	66	74
75							41	24	66	78	75
76							44	48	22	4	76
77							53	13	37	40	77
78								26	42	12	78
79								52	14	23	79
80								21	64	36	80
81								42	51	69	81
82									17	11	82
83									65	13	83
84									81	33	84
85									27	39	85
86									9	2	86
87									3	20	87
88										6	88
89										60	89
90										18	90
91										83	91
92										54	92
93										55	93
94										65	94
95										68	95

732.

A THEOREM IN SPHERICAL TRIGONOMETRY.

[From the *Proceedings of the London Mathematical Society*, vol. XI. (1880), pp. 48—50. Read January 8, 1880.]

IN a spherical triangle, where a, b, c are the sides, and A, B, C the opposite angles, we have

$$-\tan\tfrac{1}{2}c\tan\tfrac{1}{2}a\tan\tfrac{1}{2}b\sin(A-B)=\tan\tfrac{1}{2}b\sin A-\tan\tfrac{1}{2}a\sin B,$$

$$\tan\tfrac{1}{2}c\,\{1-\tan\tfrac{1}{2}a\tan\tfrac{1}{2}b\cos(A-B)\}=\tan\tfrac{1}{2}b\cos A+\tan\tfrac{1}{2}a\cos B;$$

which are both included in the form

$$\tan\tfrac{1}{2}a\,(\cos B-i\sin B)=\frac{\tan\tfrac{1}{2}c-\tan\tfrac{1}{2}b\,(\cos A+i\sin A)}{1+\tan\tfrac{1}{2}c\tan\tfrac{1}{2}b\,(\cos A+i\sin A)}.$$

For the first of the two identities: from

$$\cos a=\frac{\cos A+\cos B\cos C}{\sin B\sin C},$$

$$\cos b=\frac{\cos B+\cos A\cos C}{\sin A\sin C},$$

we deduce

$$\begin{aligned}\cos a-\cos b&=\frac{1}{\sin C}\left(\frac{\cos A}{\sin B}-\frac{\cos B}{\sin A}\right)+\frac{\cos C}{\sin C}\left(\frac{\cos B}{\sin B}-\frac{\cos A}{\sin A}\right)\\&=\frac{1}{\sin C}\frac{\tfrac{1}{2}(\sin 2A-\sin 2B)}{\sin A\sin B}+\frac{\cos C\sin(A-B)}{\sin C\sin A\sin B}\\&=\frac{\sin(A-B)}{\sin C\sin A\sin B}\{\cos(A+B)+\cos C\}\\&=\frac{\sin(A-B)}{\sin C}(\cos c-1);\end{aligned}$$

that is,

$$-\sin(A-B)=\frac{\sin C}{1-\cos c}(\cos a-\cos b)$$

$$=\frac{\sin C}{\sin c}\,\frac{\sin c}{1-\cos c}(\cos a-\cos b);$$

or, what is the same thing,

$$-\tan\tfrac{1}{2}c\sin(A-B)=\frac{\sin C}{\sin c}(\cos a-\cos b).$$

Here $\cos a-\cos b$ is $=(1+\cos a)-(1+\cos b)$; substituting for $\frac{\sin C}{\sin c}$ successively $\frac{\sin A}{\sin a}$ and $\frac{\sin B}{\sin b}$, the right-hand side is

$$=\frac{1+\cos a}{\sin a}\sin A-\frac{1+\cos b}{\sin b}\sin B,$$

$$=\cot\tfrac{1}{2}a\sin A-\cot\tfrac{1}{2}b\sin B;$$

whence, multiplying each side by $\tan\tfrac{1}{2}a\tan\tfrac{1}{2}b$, we have the relation in question.

For the second identity which is

$$\tan\tfrac{1}{2}c\,\{1-\tan\tfrac{1}{2}a\tan\tfrac{1}{2}b\cos(A-B)\}=\tan\tfrac{1}{2}b\cos A+\tan\tfrac{1}{2}a\cos B;$$

if on the right-hand side we substitute for $\cos A$, $\cos B$ their values

$$\frac{\cos a-\cos b\cos c}{\sin b\sin c}\quad\text{and}\quad\frac{\cos b-\cos a\cos c}{\sin a\sin c},$$

the right-hand side becomes

$$\frac{1}{\sin c}\left\{\frac{\cos a-\cos b\cos c}{1+\cos b}+\frac{\cos b-\cos a\cos c}{1+\cos a}\right\};$$

whence, multiplying the whole equation by $\sin c\,(1+\cos a)(1+\cos b)$, it becomes

$$(1-\cos c)\{(1+\cos a)(1+\cos b)-\sin a\sin b\cos(A-B)\}$$

$$=(1+\cos a)(\cos a-\cos b\cos c)+(1+\cos b)(\cos b-\cos c\cos a).$$

We have here

$$\cos(A-B)=\cos A\cos B+\sin A\sin B=\frac{(\cos a-\cos b\cos c)(\cos b-\cos c\cos a)+\square}{\sin^2 c\sin a\sin b},$$

by substituting for $\cos A$, $\cos B$ their foregoing values, and for $\sin A$, $\sin B$ their values $\frac{\sqrt{\square}}{\sin b\sin c}$, $\frac{\sqrt{\square}}{\sin a\sin c}$, where

$$\square=1-\cos^2 a-\cos^2 b-\cos^2 c+2\cos a\cos b\cos c.$$

The numerator is

$$\cos a \cos b - \cos c\,(\cos^2 a + \cos^2 b) + \cos a \cos b \cos^2 c$$
$$+ 1 - \cos^2 c - \qquad (\cos^2 a + \cos^2 b) + \cos a \cos b \,.\, 2 \cos c\,;$$

viz. this is

$$= \cos a \cos b\,(1 + \cos c)^2 - (\cos^2 a + \cos^2 b)(1 + \cos c) + 1 - \cos^2 c,$$

having the factor $1 + \cos c$, which is also a factor of $\sin^2 c$, $= 1 - \cos^2 c$, in the denominator. We have, therefore,

$$\cos (A - B) = \frac{\cos a \cos b\,(1 + \cos c) - (\cos^2 a + \cos^2 b) + 1 - \cos c}{(1 - \cos c) \sin a \sin b}\,;$$

and the equation thus is

$$(1 - \cos c)(1 + \cos a)(1 + \cos b) - \{\cos a \cos b\,(1 + \cos c) - (\cos^2 a + \cos^2 b) + 1 - \cos c\}$$
$$= (1 + \cos a)(\cos a - \cos b \cos c) + (1 + \cos b)(\cos b - \cos c \cos a),$$

where each side is in fact

$$= \cos a + \cos^2 a + \cos b + \cos^2 b - \cos c\,(\cos a + \cos b) - 2 \cos a \cos b \cos c\,;$$

and the second identity is thus proved.

733.

ON A FORMULA OF ELIMINATION.

[From the *Proceedings of the London Mathematical Society*, vol. XI. (1880), pp. 139—141. Read June 10, 1880.]

CONSIDER the equations

$$(a, \ldots \between \theta, 1)^n = 0,$$
$$(A, \ldots \between \theta, 1)^m = 0,$$

where $a, \ldots, A, \ldots$ are functions of coordinates. To fix the ideas, suppose that each of these coefficients is a linear function of the four coordinates x, y, z, w. Then, eliminating θ, we obtain $\nabla = 0$, the equation of a surface; and (as is known) this surface has a nodal curve.

It is easy to obtain the equations of the nodal curve in the case where one of the equations, say the second, is a quadric: the process is substantially the same whatever may be the order of the other equation, and I take it to be a cubic; the two equations therefore are

$$(a, b, c, d \between \theta, 1)^3 = 0,$$
$$(A, B, C \between \theta, 1)^2 = 0;$$

giving rise to an equation

$$\nabla, = (a, b, c, d)^2 (A, B, C)^3, = 0.$$

And it is required to perform the elimination so as to put in evidence the nodal line of this surface.

Take θ_1, θ_2 the roots of the second equation, or write

$$(A, B, C \between \theta, 1)^2 = A(\theta - \theta_1)(\theta - \theta_2);$$

that is,

$$\theta_1 + \theta_2 = -\frac{2B}{A}, \quad \theta_1\theta_2 = \frac{C}{A};$$

then, if

$$\Theta_1 = (a, b, c, d \between \theta_1, 1)^3,$$
$$\Theta_2 = (a, b, c, d \between \theta_2, 1)^3,$$

we have

$$\nabla = A^3\Theta_1\Theta_2;$$

viz. on the right-hand side, replacing the symmetrical functions of θ_1, θ_2 by their values in terms of A, B, C, we have the expression of ∇ in its known form

$$\nabla = a^2C^3 + \&c.$$

Form now the expressions

$$\Theta_1 - \Theta_2, \quad \theta_2\Theta_1 - \theta_1\Theta_2, \quad \theta_2^2\Theta_1 - \theta_1^2\Theta_2, \quad \theta_2^3\Theta_1 - \theta_1^3\Theta_2,$$

each divided by $\theta_1 - \theta_2$. These are evidently symmetrical functions of θ_1, θ_2, the values being given by the successive lines of the expression

$$\left(\begin{array}{cccc} 0, & 1, & \theta_1 + \theta_2, & \theta_1^2 + \theta_1\theta_2 + \theta_2^2 \end{array} \between d,\ 3c,\ 3b,\ a\right); \\ \left|\begin{array}{cccc} -1, & 0, & \theta_1\theta_2, & \theta_1\theta_2(\theta_1 + \theta_2) \\ -(\theta_1 + \theta_2), & -\theta_1\theta_2, & 0, & \theta_1^2\theta_2^2 \\ -(\theta_1^2 + \theta_1\theta_2 + \theta_2^2), & -\theta_1\theta_2(\theta_1 + \theta_2), & -\theta_1^2\theta_2^2, & 0 \end{array}\right|$$

and, consequently, these same quantities, each multiplied by A^2, are given by the successive lines of

$$\left(\begin{array}{cccc} 0, & A^2, & -2AB, & -AC + 4B_2 \end{array} \between d,\ 3c,\ 3b,\ a\right). \\ \left|\begin{array}{cccc} -A^2, & 0, & AC, & -2BC \\ 2AB, & -AC, & 0, & C^2 \\ AC - 4B^2, & 2BC, & -C^2, & 0 \end{array}\right|$$

Calling these X, Y, Z, W, that is, writing

$$X = 3A^2c - 6ABb + (-AC + 4B^2)\,a, \ \&c.,$$

then X, Y, Z, W are the values of

$$\Theta_1 - \Theta_2, \quad \theta_2\Theta_1 - \theta_1\Theta_2, \quad \theta_2^2\Theta_1 - \theta_1^2\Theta_2, \quad \theta_2^3\Theta_1 - \theta_1^3\Theta_2,$$

each multiplied by $A^2 \div (\theta_1 - \theta_2)$; and the functions all four of them vanish if only $\Theta_1 = 0$, $\Theta_2 = 0$; or, what is the same thing, the equations $X = 0$, $Y = 0$, $Z = 0$, $W = 0$ constitute only a twofold system.

The functions

$$\left(\begin{array}{ccc} X, & Y, & Z \end{array}\right) \\ \left|\begin{array}{ccc} Y, & Z, & W \end{array}\right|$$

contain each of them the factor $\Theta_1\Theta_2$, that is, ∇; they, in fact, each of them vanish if $\Theta_1=0$, and they also vanish if $\Theta_2=0$; or, by a direct substitution, we have

$$\begin{aligned}
XZ - Y^2 &= \frac{A^4}{(\theta_1-\theta_2)^2} \cdot -(\theta_1-\theta_2)^2\,\Theta_1\Theta_2, && = -A^4\Theta_1\Theta_2,\\
XW - YZ &= \quad ,, \quad -(\theta_1-\theta_2)^2(\theta_1+\theta_2)\,\Theta_1\Theta_2, && = -A^4\Theta_1\Theta_2(\theta_1+\theta_2),\\
YW - Z^2 &= \quad ,, \quad -(\theta_1-\theta_2)^2\,\theta_1\theta_2\Theta_1\Theta_2, && = -A^4\Theta_1\Theta_2\theta_1\theta_2.
\end{aligned}$$

Or, what is the same thing, these are $=-A\nabla$, $2B\nabla$, $-C\nabla$, respectively; thus the first equation is

$$\begin{aligned}
&\{3A^2c - 6ABb + (-AC+4B^2)\,a\}\,\{2ABd - 3ACc + C^2a\}\\
&\qquad\qquad - (-A^2d + 3ACb - 2BCa)^2 = -A\,(A^3d^2 + \&c.),\ = -A\nabla\,;
\end{aligned}$$

and similarly for the other two equations. The nodal curve is thus given by the twofold system $X=0$, $Y=0$, $Z=0$, $W=0$.

The method may be extended to the case where, instead of the quadric equation $(A, B, C \between \theta, 1)^2=0$, we have an equation of any higher order, but the formulæ are less simple.

734.

ON THE KINEMATICS OF A PLANE.

[From the *Quarterly Journal of Pure and Applied Mathematics*, vol. XVI. (1879), pp. 1—8.]

It seems desirable to bring together under this title various questions which have been, or may be, proposed or discussed. We consider two planes in relative motion one upon the other, but, for convenience, they may be distinguished as a moving plane and a fixed plane, the first moving upon the second. Any point of the moving plane traces out on the fixed plane a curve, and any line of the moving plane envelopes on the fixed plane a curve; similarly, any point of the fixed plane traces out on the moving plane a curve, and any line of the fixed plane envelopes on the moving plane a curve. More generally, any curve of the moving plane envelopes on the fixed plane a curve, and any curve of the fixed plane envelopes on the moving plane a curve. There is, moreover, in the moving plane a curve which rolls upon a curve in the fixed plane, and these two curves (a single relative position being given) determine the motion.

Fig. 1.

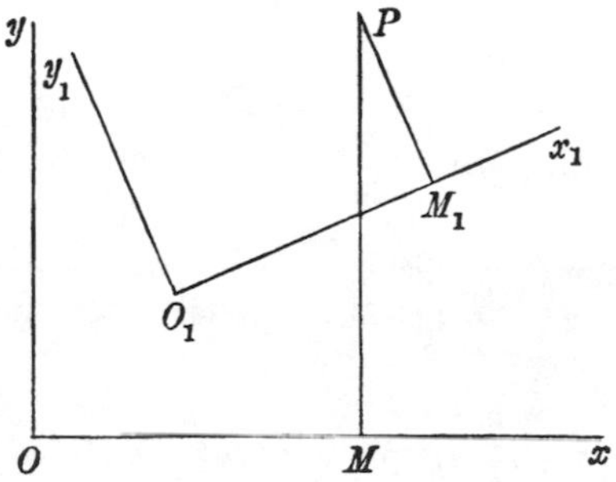

The analytical theory presents no difficulty. Taking in the fixed plane the fixed axes Ox, Oy (fig. 1), and, fixed in the moveable plane so as to move with it, the axes O_1x_1, O_1y_1; then the position of the axes $O_1x_1y_1$ may be determined, say by

α, β, the coordinates of O_1 in regard to Oxy; and by θ, the inclination of O_1x_1 to Ox. And denoting by x, y, x_1, y_1 the coordinates of a point P in regard to the two sets of axes respectively, then

$$x = \alpha + x_1 \cos\theta - y_1 \sin\theta,$$
$$y = \beta + x_1 \sin\theta + y_1 \cos\theta;$$

or, what is the same thing,

$$x_1 = \quad (x-\alpha)\cos\theta + (y-\beta)\sin\theta,$$
$$y_1 = -(x-\alpha)\sin\theta + (y-\beta)\cos\theta;$$

or, as these last equations may be written,

$$x_1 = \alpha_1 + x\cos(-\theta) - y\sin(-\theta),$$
$$y_1 = \beta_1 + x\sin(-\theta) + y\cos(-\theta),$$

where α_1, β_1, $= -\alpha\cos\theta - \beta\sin\theta$, $\alpha\sin\theta - \beta\cos\theta$, are the coordinates of O referred to the axes $O_1x_1y_1$, and $-\theta$ is the inclination of Ox to O_1x_1.

When the motion is given, α, β, θ are given functions of a single variable parameter, say of t*; or, if we please, α, β are given functions of θ.

The velocities of a given point (x, y) are determined by the equations

$$x' = \alpha' \ - (x_1 \sin\theta + y_1\cos\theta)\,\theta',$$
$$y' = \beta' + (x_1\cos\theta - y_1\sin\theta)\,\theta';$$

that is,

$$x' - \alpha' = -(y-\beta)\,\theta',$$
$$y' - \beta' = \quad (x-\alpha)\,\theta';$$

or, as these equations may also be written,

$$-(x'-\alpha')\sin\theta + (y'-\beta')\cos\theta = x_1\theta',$$
$$-(x'-\alpha')\cos\theta - (y'-\beta')\sin\theta = y_1\theta'.$$

Hence if $x' = 0$, $y' = 0$, we have

$$x_1\theta' = \alpha'\sin\theta - \beta'\cos\theta, \text{ or } \quad \alpha' = (y-\beta)\,\theta',$$
$$y_1\theta' = \alpha'\cos\theta + \beta'\sin\theta, \qquad -\beta' = (x-\alpha)\,\theta',$$

which equations determine in terms of t, x_1 and y_1 the coordinates in regard to the axes $O_1x_1y_1$, and x and y the coordinates in regard to the axes Oxy, of I, the centre of instantaneous rotation.

If from the expressions of x_1, y_1 we eliminate t, we obtain an equation between (x_1, y_1), which is that of the rolling curve in the moveable plane; and, similarly, if

* t may be regarded as denoting the time, and then the derived functions of x, y in regard to t will denote velocities; and, to simplify the expression of the theorems, it is convenient to do this.

from the expressions of x, y we eliminate t, we obtain a relation between (x, y), which is that of the rolled-on curve in the fixed plane.

The system may be written

$$x_1 = \frac{\alpha'}{\theta'}\sin\theta - \frac{\beta'}{\theta'}\cos\theta, \quad x = \alpha - \frac{\beta'}{\theta'},$$

$$y_1 = \frac{\alpha'}{\theta'}\cos\theta + \frac{\beta'}{\theta'}\sin\theta, \quad y = \beta + \frac{\alpha'}{\theta'};$$

or, if we take θ as the independent variable,

$$x_1 = \alpha'\sin\theta - \beta'\cos\theta, \quad x = \alpha - \beta',$$

$$y_1 = \alpha'\cos\theta + \beta'\sin\theta, \quad y = \beta + \alpha'.$$

To find the variations of I, we have

$$\begin{aligned} x_1' &= \alpha''\sin\theta - \beta''\cos\theta + \alpha'\cos\theta + \beta'\sin\theta, \quad &= \alpha''\sin\theta - \beta''\cos\theta + y_1, \\ y_1' &= \alpha''\cos\theta + \beta''\sin\theta - \alpha'\sin\theta + \beta'\cos\theta, \quad &= \alpha''\cos\theta + \beta''\sin\theta - x_1, \\ y' &= \beta' + \alpha'', \\ x' &= \alpha' - \beta''. \end{aligned}$$

Hence

$$\begin{aligned} x_1' &= x'\cos\theta + y'\sin\theta, \quad \text{or } x' = x_1'\cos\theta - y_1'\sin\theta, \\ y_1' &= -x'\sin\theta + y'\cos\theta, \quad\quad\;\; y' = x_1'\sin\theta + y_1'\cos\theta, \end{aligned}$$

values which give $x'^2 + y'^2 = x_1'^2 + y_1'^2$, which equation expresses that the motion is in fact a rolling one.

Imagine the two curves, and the initial relative position given; say the two points A, A_1 (fig. 2) were originally in contact, then the arcs AI, A_1I are equal, and, calling each of these s, and X, Y, X_1, Y_1 the coordinates of I in regard to the two

Fig. 2.

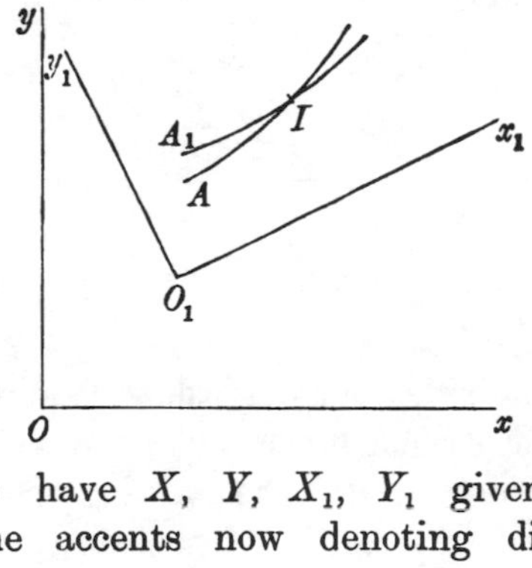

sets of axes respectively, we have X, Y, X_1, Y_1 given functions of s, such that $X'^2 + Y'^2 = 1$, $X_1'^2 + Y_1'^2 = 1$, the accents now denoting differentiation in regard to s. We have, from the figure,

$$\theta = \tan^{-1}\frac{Y'}{X'} - \tan^{-1}\frac{Y_1'}{X_1'};$$

or, what is the same thing,

$$\tan\theta = (Y'X_1' - Y_1'X) \div (X'X_1' + Y'Y_1'),$$

say

$$\sin\theta,\ \cos\theta = Y'X_1' - Y_1'X,\ X'X_1' + Y'Y_1';$$

and then, as before,

$$x = \alpha + x_1\cos\theta - y_1\sin\theta,$$

$$y = \beta + x_1\sin\theta + y_1\cos\theta;$$

or, what is the same thing,

$$x - X = \cos\theta\,(x_1 - X_1) - \sin\theta\,(y_1 - Y_1),$$

$$y - Y = \sin\theta\,(x_1 - X_1) + \cos\theta\,(y_1 - Y_1),$$

where X, Y, X_1, Y_1, and therefore also θ, denote given functions of s. The formulæ will be of a like form if X, Y, X_1, Y_1 are given functions of a parameter t.

A well known but very interesting case is when two points of the moving plane describe right lines on the fixed plane. This may be discussed geometrically as follows: Suppose that we have the points A, C (fig. 3) describing the lines OA_0, OC_0, which meet in O; through A, C, O describe a circle, centre O_1, and with centre

Fig. 3.

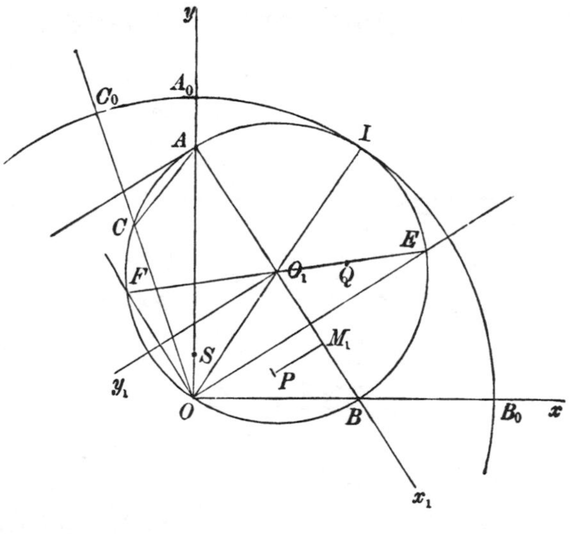

O and radius $= 2OO_1$, describe a circle touching the first circle in a point I; and suppose that A_0, C_0 denote points on the second circle. Then it is at once seen that, considering the first or small circle as belonging to the moving plane, and the second or large circle as belonging to the fixed plane, the motion is in fact the rolling motion of the small upon the large circle; and, moreover, that each point of the small circle describes a right line, which is a diameter of the large circle. In fact, the angle IO_1C at the centre is the double of the angle IOC at the circumference; that is, it is the double of the angle IOC_0; and therefore (the radius of the small circle being half that of the large circle) the arcs IC, IC_0 are equal, so that the rolling motion will carry the point C along the radius OC_0, and will, in like manner, carry the point A along the radius OA_0, or the motion will be as originally assumed. And, in like manner, for any other point B of the small circle the motion will be along the radius OB_0; in particular, taking AB a diameter, the angle A_0OB_0 will be a right angle; and the motion is determined by means of the two points A, B describing respectively the two lines OA_0, OB_0 *at right angles to each other*, viz. there is no loss of generality in assuming that the two fixed lines are at right angles to each other. It thence at once follows, as will presently appear, that each point of the moving plane describes an ellipse (but we have the special case already referred to, each point on the small circle describes a right line, and also the special case, the centre O_1 of the small circle describes a circle). Considering any point Q of the moving plane, let the line QO_1 meet the small circle in the points E, F (or, what is the same thing, let E, F be the extremities of the diameter which passes through Q); then the points E, F describe the lines OE, OF at right angles to each other, and Q is a point on EF or on this line produced; clearly the locus is an ellipse having the lines OE, OF for the directions of its axes, and having the lengths of the semi-axes $= QF$, QE respectively.

Taking the points to be A, B moving along the two lines OB_0, OA_0 at right angles to each other, these lines may be taken for the axes Ox, Oy; the point O_1 for the origin of the coordinates x_1, y_1, the axes O_1x_1 being in the direction O_1B and O_1y_1 at right angles to it; calling the length $AB = 2c$, we have $O_1A = O_1B = c$, and the angle ABO may be called θ (but this angle was previously taken with a contrary sign). We have then for the point P, having in regard to O_1x_1 and O_1y_1 the coordinates (x_1, y_1),

$$\left.\begin{aligned} x &= \alpha + x_1 \cos\theta - y_1 \sin\theta \\ y &= \beta - x_1 \sin\theta - y_1 \cos\theta \end{aligned}\right\},$$

where the sign of y_1 has been changed, and $\alpha = c\cos\theta$, $\beta = c\sin\theta$: the equations thus become

$$x = (c + x_1)\cos\theta - y_1\sin\theta,$$

$$y = (c - x_1)\sin\theta - y_1\cos\theta,$$

where observe that $c + x_1$, $c - x_1$ are the distances M_1A, M_1B respectively. And we have, conversely,

$$x_1 = \quad x\cos\theta - y\sin\theta - c\cos 2\theta,$$

$$y_1 = -x\sin\theta - y\cos\theta + c\sin 2\theta.$$

If, in particular, $y_1 = 0$, then

$$x,\ y = (c + x_1)\cos\theta,\ (c - x_1)\sin\theta;$$

or we have

$$\frac{x^2}{(c + x_1)^2} + \frac{y^2}{(c - x_1)^2} = 1;$$

viz. the curve on the first plane is an ellipse, the semi-axes of which are $\pm(c+x_1)$, $\pm(c-x_1)$, each taken positively; if $x_1^2+y_1^2=c^2$, viz. if P be on the circle having AB for its diameter, then $y_1^2=(c+x_1)(c-x_1)$, and we have

$$y\div x=-(c-x_1)\left(\sin\theta-\frac{y_1}{c-x_1}\cos\theta\right)\div y_1\left(\sin\theta-\frac{c+x_1}{y_1}\cos\theta\right),\ =-(c-x_1)\div y_1,$$

viz. as mentioned above, the curve on the fixed plane is a right line.

In the general case, we have

$$x(c-x_1)+\quad yy_1\quad=(c^2-x_1^2-y_1^2)\cos\theta,$$

$$xy_1+y(c+x_1)=(c^2-x_1^2-y_1^2)\sin\theta,$$

and thence

$$\{x(c-x_1)+yy_1\}^2+\{xy_1+y(c+x_1)\}^2=(c^2-x_1^2-y_1^2)^2;$$

or, what is the same thing,

$$x^2\{(c-x_1)^2+y_1^2\}+4xycy_1+y^2\{(c+x_1)^2+y_1^2\}=(c^2-x_1^2-y_1^2)^2.$$

Considering (x_1, y_1) as given, the curve traced out by P on the fixed plane is of the second order; it would be easy to verify from the equation that it is an ellipse, and to obtain for the position and magnitude of the axes the construction already found geometrically.

The same equation, considering therein (x, y) as constant and (x_1, y_1) as current coordinates, gives the curve traced out on the moving plane; the curve is obviously of the fourth order. Transferring the origin to A, we must in place of x_1 write x_1-c_1; the equation thus becomes

$$x^2\{(x_1-2c)^2+y_1^2\}+4cy_1xy+y^2(x_1^2+y_1^2)=(x_1^2+y_1^2-2cx_1)^2;$$

or, what is the same thing,

$$(x_1^2+y_1^2-2cx_1)^2-(x^2+y^2)(x_1^2+y_1^2)+4cx(xx_1-yy_1)-4c^2x^2=0;$$

and if we suppose herein $x=0$, it becomes

$$(x_1^2+y_1^2-2cx_1)^2-y^2(x_1^2+y_1^2)=0;$$

or, writing $x_1=r_1\cos\theta_1$, $y_1=r_1\sin\theta_1$, where $\theta_1=\text{angle } QAB$, this is

$$(r_1-2c\cos\theta_1)^2-y^2=0,$$

or say it is

$$r_1=2c\cos\theta_1-y,$$

which is the polar equation of the curve described on the moveable plane by the point S, whose coordinates in respect to Ox and Oy are $(0, y)$.

There is no loss of generality in assuming $x=0$. In fact, starting with any point S whatever of the fixed plane, if we draw OS meeting the small circle in A, and through O draw at right angles to this a line meeting the same circle in B, then, as before, the points A and B move along the fixed lines OA_0, OB_0; or as regards the relative motion, taking A, B as fixed points, we have the originally fixed plane now moving in such wise that the two lines OA_0, OB_0 thereof (at right angles to each other) pass always through the points A and B respectively, and the curve is that described by the point S on the line OA; the point O describes the circle on the diameter AB (the small circle), equation $r_1 = 2c\cos\theta_1$; and OQ having a given constant value $=y$, we have for the curve described by the point S the foregoing equation $r_1 = 2c\cos\theta_1 - y$; or writing $y = -f$, that is, taking S on the other side of O at a distance $OS = f$, the equation is $r_1 = 2c\cos\theta_1 + f$; viz. this is a nodal Cartesian or Limaçon, the origin being an acnode or a crunode according as $f >$ or $< 2c$; and if $f = 2c$, then we have the cuspidal curve or cardioid $r_1 = 2c(1+\cos\theta_1)$, $= 4c\cos^2\frac{1}{2}\theta_1$. The general conclusion is that the centre O of the large circle describes on the moving plane a small circle (centre O_1), and that every other point of the fixed plane describes on the moving plane a Limaçon having for its node a point of the small circle, and being, in fact, the curve obtained by measuring off along the radius vector of the small circle from its extremity a constant distance.

Considering in connexion with the point, coordinates (x_1, y_1), (x, y), a second point, coordinates (X_1, Y_1), (X, Y), in regard to the two sets of axes respectively, we have

$$x = (c + x_1)\cos\theta - y_1\sin\theta,\quad X = (c + X_1)\cos\theta - Y_1\sin\theta,$$

$$y = (c - x_1)\sin\theta - y_1\cos\theta,\quad Y = (c - X_1)\sin\theta - Y_1\cos\theta;$$

from the first two equations we have

$$\cos\theta : \sin\theta : 1 = x(c - x_1) + yy_1 : xy_1 + y(c + x_1) : c^2 - x_1^2 - y_1^2;$$

and substituting these values in the second set, we find

$$\begin{aligned} X : Y : 1 & \\ = & \ x\{c^2 + c(X_1 - x_1) - X_1x_1 - Y_1y_1\} + y\{\quad c(y_1 - Y_1) + y_1X_1 - x_1Y_1\} \\ & : x\{\quad c(y_1 - Y_1) - y_1X_1 + x_1Y_1\} + y\{c^2 - c(X_1 - x_1) - X_1x_1 - Y_1y_1\} \\ & : c^2 - x_1^2 - y_1^2; \end{aligned}$$

or the points (x, y), (X, Y), considered as each of them moving on the fixed plane, are homographically related to each other.

To find the curve enveloped on the fixed plane by a given curve of the moving plane, we have only in the equation $f(x_1, y_1) = 0$ of the curve in the moving plane to substitute for x_1, y_1 their values in terms of x, y, θ, and then considering θ as a variable parameter, to find the envelope of the curve represented by this equation. And, similarly, we find the curve enveloped on the moving plane by a given curve of the fixed plane.

Thus, in the particular case of motion above considered, writing, as before,

$$x = (c + x_1)\cos\theta - y_1 \sin\theta,$$

$$y = (c - x_1)\sin\theta - y_1 \cos\theta;$$

or conversely

$$x_1 = x\cos\theta - y\sin\theta - c\cos 2\theta,$$

$$y_1 = -x\sin\theta - y\cos\theta + c\sin 2\theta;$$

the envelope on the moving plane of the line

$$Ax + By + C = 0$$

of the fixed plane is given as the envelope of the line

$$\{A(c + x_1) - By_1\}\cos\theta + \{-A + B(c - x_1)\}\sin\theta + C = 0;$$

viz. this is

$$\{A(c + x_1) - By_1\}^2 + \{Ay_1 - B(c - x_1)\}^2 - C^2 = 0;$$

that is,

$$(A^2 + B^2)(x_1^2 + y_1^2 + c^2) + 2(A^2 - B^2)cx_1 - 4ABcy_1 = 0,$$

a circle.

But the envelope on the fixed plane of the line

$$Ax_1 + By_1 + C = 0$$

of the moving plane is given as the envelope of the line

$$C + (Ax + By)\cos\theta - (Ay + Bx)\sin\theta - AC\cos 2\theta + BC\sin 2\theta = 0,$$

which can be obtained by equating to zero the discriminant of a quartic function, and is apparently a sextic curve.

735.

NOTE ON THE THEORY OF APSIDAL SURFACES.

[From the *Quarterly Journal of Pure and Applied Mathematics*, vol. XVI. (1879), pp. 109—112.]

I OBTAIN in the present Note a system of formulæ which lead very simply to the known theorem, that the apsidals of reciprocal surfaces are reciprocal; or, what is the same thing, that the reciprocal of the apsidal of a given surface is the apsidal of its reciprocal; the surfaces are referred to the same axes, and by the reciprocal is meant the reciprocal surface in regard to a sphere radius unity, having for its centre a determinate point, say the origin; and it is this same point which is used in the construction of the apsidal surfaces. The apsidal of a given surface is constructed as follows; considering the section by any plane through the fixed point, and in this section the apsidal radii from the fixed point (that is, the radii which meet the curve at right angles), then drawing a line through the fixed point at right angles to the plane, and on this line measuring off from the fixed point distances equal to the apsidal radii respectively, the locus of the extremities of these distances is the apsidal surface. We have the surface, its reciprocal, the apsidal of the surface, the apsidal of the reciprocal; and I take

$$(x,\ y,\ z),\ (x',\ y',\ z'),\ (X,\ Y,\ Z),\ (X',\ Y',\ Z')$$

for the coordinates of corresponding points on the four surfaces respectively.

The condition of reciprocity gives $xx' + yy' + zz' - 1 = 0$, and (the equations being $U = 0$, $U' = 0$) x', y', z' proportional to d_xU, d_yU, d_zU, and x, y, z proportional to $d_{x'}U'$, $d_{y'}U'$, $d_{z'}U'$; or, what is the same thing, we must have

$$x'dx + y'dy + z'dz = 0 \text{ and } xdx' + ydy' + zdz' = 0\,;$$

one of these is implied in the other, as appears at once by differentiating the equation $xx' + yy' + zz' - 1 = 0$.

The other two surfaces will therefore be reciprocal if only we have the like relations between the coordinates (X, Y, Z) and (X', Y', Z'); that is, if

$$XX' + YY' + ZZ' - 1 = 0,$$
$$X'dX + Y'dY + Z'dZ = 0,$$
$$XdX' + YdY' + ZdZ' = 0.$$

To find the apsidal surface, we consider an arbitrary section $x\cos\alpha + y\cos\beta + z\cos\gamma = 0$ of the surface $U = 0$, and seek to determine the apsidal radii thereof, that is, the maximum or minimum values of $R^2 = x^2 + y^2 + z^2$ when x, y, z vary subject to these two conditions. Writing x', y', z' to denote functions proportional to d_xU, d_yU, d_zU, we thus have the set of equations

$$x + \lambda x' + \mu\cos\alpha = 0,$$
$$y + \lambda y' + \mu\cos\beta = 0,$$
$$z + \lambda z' + \mu\cos\gamma = 0,$$

where λ, μ are indeterminate coefficients; taking then X, Y, Z as the coordinates of the extremity of the line drawn at right angles to the plane, we have $R^2 = X^2 + Y^2 + Z^2$, and $\cos\alpha, \cos\beta, \cos\gamma = \frac{X}{R}, \frac{Y}{R}, \frac{Z}{R}$; substituting these values in the equation

$$x\cos\alpha + y\cos\beta + z\cos\gamma = 0,$$

we have $Xx + Yy + Zz = 0$, and substituting in the other equations, and instead of λ, μ introducing the new indeterminate coefficients ρ, σ, we obtain

$$X, Y, Z = \rho x + \sigma x', \ \rho y + \sigma y', \ \rho z + \sigma z'.$$

Hence these last equations, together with

$$R^2 = X^2 + Y^2 + Z^2 = x^2 + y^2 + z^2,$$

and

$$Xx + Yy + Zz = 1,$$

contain the solution of the problem. If for convenience we introduce R'^2 to denote $x'^2 + y'^2 + z'^2$, and imagine the absolute values of x', y', z' determined so that $xx' + yy' + zz' = 1$, then substituting for X, Y, Z their values in the equations $X^2 + Y^2 + Z^2 = R^2$ and $Xx + Yy + Zz = 1$, we find

$$R^2 = \rho^2R^2 + 2\rho\sigma + \sigma^2R'^2, \ 0 = \rho R^2 + \sigma,$$

and thence

$$\rho^2 = \frac{1}{R^2R'^2 - 1}, \ \sigma = -\rho R^2,$$

or, finally assuming

$$\rho = \frac{1}{\sqrt{(R^2R'^2 - 1)}},$$

we have

$$X, Y, Z = x - R^2x', \ y - R^2y', \ z - R^2z',$$

each divided by

$$\sqrt{(R^2R'^2 - 1)},$$

where I recall that x', y', z' are proportional to d_xU, d_yU, d_zU, and are such that $xx'+yy'+zz'=1$: they in fact denote

$$d_xU,\ d_yU,\ d_zU, \text{ each divided by } xd_xU+yd_yU+zd_zU;$$

and that R^2 and R'^2 denote $x^2+y^2+z^2$ and $x'^2+y'^2+z'^2$ respectively. The coordinates X, Y, Z of the point of the apsidal surface are thus determined as functions of x, y, z.

For the apsidal of the reciprocal surface, we have in like manner

$$X',\ Y',\ Z' = x'-R'^2x,\ y'-R'^2y,\ z'-R'^2z,$$

each divided by

$$-\sqrt{(R^2R'^2-1)},$$

and then the two sets of values give, not only

$$XX'+YY'+ZZ'=1,$$

as is obvious, but also

$$X'dX+Y'dY+Z'dZ=0, \text{ and } XdX'+YdY'+ZdZ'=0.$$

In fact, writing for a moment ρ, ρ' instead of R^2, R'^2, and $\sqrt{(R^2R'^2-1)}=\sqrt{(\rho\rho'-1)}$, $=\omega$, then

$$\begin{aligned}
&X'dX+Y'dY+Z'dZ\\
&=\frac{x'-x\rho'}{\omega}\,d\,\frac{x-x'\rho}{\omega}+\&\text{c.}\\
&=\frac{x'-x\rho'}{\omega}\left\{\frac{dx-\rho dx'-x'd\rho}{\omega}-\frac{(x-x'\rho)\,d\omega}{\omega^2}\right\}+\&\text{c.}\\
&=\frac{1}{\omega^2}\{\quad x'dx+y'dy+z'dz\\
&\qquad -\ \rho\,(x'dx'+y'dy'+z'dz')\\
&\qquad -\ (x'^2+y'^2+z'^2)\,d\rho\\
&\qquad -\ \rho'(xdx+ydy+zdz)\\
&\qquad +\ \rho\rho'(xdx'+ydy'+zdz')\\
&\qquad +\ \rho'(xx'+yy'+zz')\,d\rho\}\\
&-\frac{d\omega}{\omega^3}\{\quad xx'+yy'+zz'\\
&\qquad -\ \rho\,(x'^2+y'^2+z'^2)\\
&\qquad -\ \rho'(x^2+y^2+z^2)\\
&\qquad +\ \rho\rho'(xx'+yy'+zz')\},
\end{aligned}$$

or, since the terms in { } are

$$0-\rho\,.\,\tfrac{1}{2}d\rho'-\rho'd\rho-\rho'\,.\,\tfrac{1}{2}d\rho+0+\rho'd\rho,\ =-\tfrac{1}{2}(\rho d\rho'+\rho'd\rho),$$

and

$$1-\rho\rho'-\rho\rho'+\rho\rho',\ =1-\rho\rho',\ =-\omega^2,$$

this is

$$=\frac{1}{\omega^2}\{-\tfrac{1}{2}(\rho d\rho'+\rho'd\rho)+\omega d\omega\},\ =0,$$

in virtue of $\omega^2=\rho\rho'-1$. And similarly the other equation $XdX'+YdY'+ZdZ'=0$ might be directly verified.

736.

APPLICATION OF THE NEWTON-FOURIER METHOD TO AN IMAGINARY ROOT OF AN EQUATION.

[From the *Quarterly Journal of Pure and Applied Mathematics*, vol. XVI. (1879), pp. 179—185.]

I CONSIDER only the most simple case, that of a quadric equation $x^2 = n^2$, where n^2 is a given imaginary quantity, having the square roots n, and $-n$; starting from an assumed approximate (imaginary) value $x = a$, we have $(a+h)^2 = n^2$, that is,

$$a^2 + 2ah = n^2, \quad h = -\frac{a^2 - n^2}{2a}, \quad \text{and} \quad a + h = \frac{a^2 + n^2}{2a};$$

that is, the successive values are

$$a_1 = \frac{a^2 + n^2}{2a}, \quad a_2 = \frac{a_1{}^2 + n^2}{2a_1}, \ldots,$$

and the question is, under what conditions do we thus approximate to one determinate root (selected out of the two roots at pleasure), say n, of the given equation.

The nearness of two values is measured by the modulus of their difference; thus a nearer to n, than a_1 is to n, means $\text{mod.}(a-n) < \text{mod.}(a_1 - n)$, and so in other cases; in the course of the approximation $a, a_1, a_2, \ldots$ to n, any step, for instance a to a_1, is regular if a_1 is nearer to n than a is, but otherwise it is irregular; the approximation is regular if all the steps are regular, and if (after one or more irregular steps) all the subsequent steps are regular, then the approximation becomes regular at the step which is the first of the unbroken series of regular steps.

We do by an approximation, which is ultimately regular, obtain the value n, if only the assumed value a is nearer to n than it is to $-n$; or, say, if the condition $\text{mod.}(a-n) < \text{mod.}(a+n)$ is satisfied, and the approximation is regular from the beginning if mod. $(a-n) < \frac{2}{3}$ mod. n, viz. this condition is a sufficient one*; the first step a to a_1 will moreover be regular under a less stringent condition imposed upon a; and it would seem that, without the condition mod. $(a-n) < \frac{2}{3}$ mod. n being satisfied, the subsequent steps will in some cases be also regular; that is, that the last-mentioned condition is not a necessary condition in order to the approximation being regular from the beginning; it is, however, the necessary and sufficient condition, *to be satisfied by the modulus* of $a-n$, in order that the approximation may be regular from the beginning. All this will clearly appear from the geometry.

Fig. 1.

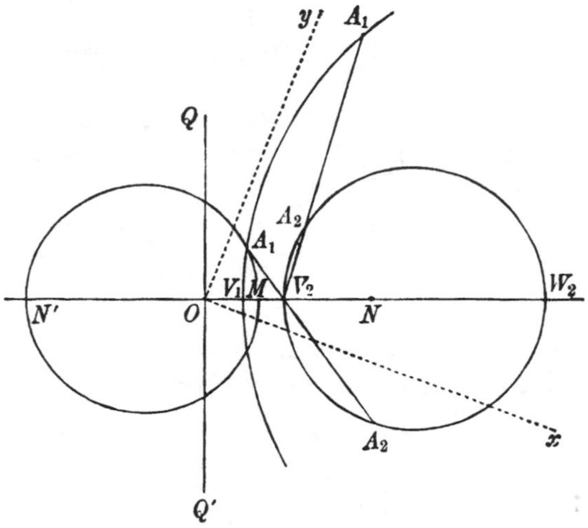

We take N, N' (fig. 1) to represent the values n, $-n$; and similarly A, A_1, &c. to represent the quantities a, a_1, ...; we have then

$$AN = \text{mod.}\,(a-n), \quad A_1N = \text{mod.}\,(a_1-n)\,\ldots,$$

so that the approximation is measured by the approach of the points A, A_1 to N. The line NN' joining the points N, N' passes through, and is bisected at, the origin O; drawing then QQ' through O at right angles to NN' the condition

$$\text{mod.}\,(a-n) < \text{mod.}\,(a+n)$$

means that the point A, which represents the imaginary quantity a, lies on the N-side of QQ', and it will be assumed throughout that this is so. Take now on the line ON, $OM = \frac{1}{3}ON$, and on $N'M$ as diameter, describe a circle, which may be called the "circle of unfitness"; regarding as an area the segment hereof which lies on the N-side of QQ', say this is the "segment of unfitness." It will be shown that if according as A is situate inside, on the boundary of, or outside the segment of

* In the Smith's Prize Examination, Jan. 28, 1879, I gave the theorem under the following form: "If a, n are imaginary quantities, the latter of them given, and the former assumed at pleasure, subject only to the condition mod. $(a-n) < \frac{2}{3}$ mod. n; then if $a_1 = \frac{a^2+n^2}{2a}$, $a_2 = \frac{a_1{}^2+n^2}{2a_1}$, &c., show that the terms a, a_1, a_2,... will converge to the limit n." This is strictly true, but it would have been better to say "will converge regularly."

unfitness, A_1N will be greater than, equal to, or less than AN. It may be added that, if A be within or upon the boundary of the segment of unfitness, then A_1 will be outside it, but this by no means hinders that the next point A_2, or some later point, shall be within the segment of unfitness; and, further, that when A is outside the segment of unfitness, then the next point A_2, or some later point, may very well be within the segment of unfitness; the conclusion is, that A being inside the segment of unfitness, A_1N is less than AN, but that it does *not* thence follow that A_2N is less than A_1N, A_3N than $A_2N, \ldots$; the approximation although regular at the first step, may then, or afterwards, for a step or steps, cease to be regular.

If, however, AN be less than $\frac{2}{3}ON$, that is, if the condition $\text{mod.}(a-n) < \frac{2}{3}\,\text{mod.}\,n$ be satisfied, then the point A lies within the circle centre N and radius NM, and is consequently outside the segment of unfitness; A_1N being less than AN, the point A_1 is *a fortiori* outside the segment of unfitness, and the like for all the subsequent points $A_2, A_3, \ldots$, that is, in this case, the approximation is regular throughout. The circle, centre N, and radius NM, $= \frac{2}{3}\,\text{mod.}\,n$, may be called the "safe circle"; and the conclusion is that, if the point A or any subsequent point be within the safe circle, then every subsequent point will be within the safe circle, and the approximation will be regular.

The successive points $A, A_1, A_2, \ldots$ (or, as it will be convenient to call them, $A_1, A_2, \ldots$) may be obtained each from the preceding one by a simple geometrical construction.

I recall that any circle through the two (imaginary) antipoints of N, N' is a circle having its centre on the indefinite line NN'; it is such that the ratio of the distances of a point thereof from the points N, N' respectively has a certain constant value, viz. for the circles with which we are here alone concerned, those which lie on the N-side of QQ', the centres lie beyond the point N (further away, that is, from O), and the values of the ratio, distance from N to distance from N', are less unity.

Starting then from the given point A_1, for which this ratio $A_1N : A_1N'$ has a given value, suppose $A_1N = kA_1N'$, we describe a first circle (passing of course through A_1) for each point of which this ratio has the value k; let the diameter of this circle be V_1W_1, V_1 being the extremity between O and N, W_1 (not shown in the figure), that beyond N; we then describe a second circle, for which the ratio is $=k^2$; let its diameter be V_2W_2, V_2 being the extremity between O and N (or say between V_1 and N), W_2, that beyond N (or say between N and W_1); the point A_2 lies on this second circle, and is determined as the single intersection of the line V_2A_1 with the second circle. And of course drawing a third circle, for which the ratio is $=k^4$, on the diameter V_3W_3, then A_3 lies on the third circle, and is the intersection with it of the line V_3A_2, and so on; the radii of the successive circles diminish very rapidly, their centres, in like manner, continually approaching the point N; hence, the points $A_1, A_2, A_3, \ldots$, which lie on the several circles respectively approximate, and that very rapidly, to the point O. But by what precedes, if, for instance, the point A_1 be within the segment of unfitness, then also some of the subsequent points may be within the segment of unfitness, and for each point A_p, for which this is the case, the next point A_{p+1} is at a greater distance, so that $NA_{p+1} > NA_p$; it is, however, clear that we always arrive at a point A_q, such that $NA_q < \frac{2}{3}ON$, and so soon as such a point is arrived at the approximation becomes regular.

The point A_2 determined from A_1, as above, is a point such that the subtended angle NA_2N' is = twice the subtended angle NA_1N'; or calling the latter angle ϕ, the former is $= 2\phi$. It is, in fact, this property which gives rise to the construction; for let the values of A_1N, A_1N', regarded as imaginary quantities, be called for a moment

$$\rho_1 (\cos\theta_1 + i\sin\theta_1),\quad \rho_1' (\cos\theta_1' + i\sin\theta_1');$$

and, similarly, those of A_2N, A_2N' be called

$$\rho_2 (\cos\theta_2 + i\sin\theta_2),\quad \rho_2' (\cos\theta_2' + i\sin\theta_2');$$

then these are the values of $a_1 - n$, $a_1 + n$, $a_2 - n$, $a_2 + n$ respectively, or we have

$$\frac{a_1 - n}{a_1 + n} = \frac{\rho_1}{\rho_1'}\{\cos(\theta_1 - \theta_1') + i\sin(\theta_1 - \theta_1')\} = k\ (\cos\ \phi + i\sin\ \phi),$$

$$\frac{a_2 - n}{a_2 + n} = \frac{\rho_2}{\rho_2'}\{\cos(\theta_2 - \theta_2') + i\sin(\theta_2 - \theta_2')\} = k^2 (\cos 2\phi + i\sin 2\phi),$$

that is,

$$\frac{a_2 - n}{a_2 + n} = \left(\frac{a_1 - n}{a_1 + n}\right)^2,$$

which relation between a_2, a_1 is in fact the original relation

$$a_2 = \frac{a_1^2 + n^2}{2a_1};$$

and, conversely, a_1, a_2 being thus connected, then the representative A_2 is obtained from the representative point A_1 by the foregoing geometrical construction.

I give the analytical proofs; we may without loss of generality take, and it is convenient to do so, the axis of x as coinciding with the line ON, and to put also $ON = 1$. We then in place of the original coordinates x, y of any point take the new coordinates k, ϕ which are such that

$$\frac{x + iy - 1}{x + iy + 1} = ke^{i\phi},$$

$$\frac{x - iy - 1}{x - iy + 1} = ke^{-i\phi};$$

equations which may also be written

$$(x-1)^2 + y^2 = k^2 [(x+1)^2 + y^2],$$

$$x^2 + (y - i)^2 = e^{-2i\phi} [x^2 + (y+i)^2];$$

or, what is the same thing,

$$x^2 + y^2 - 1 - 2y\cot\phi = 0,$$

where of course the equation with k shows that k is equal to the ratio of the distances of the point from the points N, N' respectively, and the equation in ϕ, taken in the second form, shows that ϕ is the angle subtended at the point by N, N'.

It is sometimes convenient to write $ke^{i\phi}$, $ke^{-2i\phi} = p$, q respectively; we then have

$$pq = k^2, \quad x + iy = \frac{1+p}{1-p}, \quad x - iy = \frac{1+q}{1-q}.$$

Suppose for a moment that we have (p_1, q_1), (p_2, q_2), (p_3, q_3) as the (p, q) coordinates of any three points, the condition that these three points may lie in a line, is given in the form, determinant $= 0$, where each line of the determinant is of the form

$$\frac{1+p}{1-p}, \quad \frac{1+q}{1-q}, \quad 1,$$

or, what is the same thing, it is

$$1 - pq + p - q, \quad 1 - pq - p + q, \quad 1 + pq - p - q,$$

or, again

$$pq - 1, \quad p - q, \quad p + q - 2,$$

viz. the condition is

$$\begin{vmatrix} p_1q_1 - 1, & p_1 - q_1, & p_1 + q_1 - 2 \\ p_2q_2 - 1, & p_2 - q_2, & p_2 + q_2 - 2 \\ p_3q_3 - 1, & p_3 - q_3, & p_3 + q_3 - 2 \end{vmatrix} = 0.$$

Suppose the (k, ϕ) coordinates of the three points are (l, α), (m, β), (n, γ) respectively; then this equation is

$$\begin{vmatrix} l^2 - 1, & l \sin\alpha, & l \cos\alpha - 1 \\ m^2 - 1, & m \sin\beta, & m \cos\beta - 1 \\ n^2 - 1, & n \sin\gamma, & n \cos\gamma - 1 \end{vmatrix} = 0,$$

viz. it is

$$\begin{vmatrix} l^2 - 1, & l \sin\alpha, & l \cos\alpha \\ m^2 - 1, & m \sin\beta, & m \cos\beta \\ n^2 - 1, & n \sin\gamma, & n \cos\gamma \end{vmatrix} - \begin{vmatrix} l^2 - 1, & l \sin\alpha, & 1 \\ m^2 - 1, & m \sin\beta, & 1 \\ n^2 - 1, & n \sin\gamma, & 1 \end{vmatrix} = 0,$$

or, what is the same thing, it is

$$[(l^2 - 1)\, mn \sin(\beta - \gamma) + (m^2 - 1)\, nl \sin(\gamma - \alpha) + (n^2 - 1)\, lm \sin(\alpha - \beta)]$$
$$+ [(m^2 - n^2)\, l \sin\alpha + (n^2 - l^2)\, m \sin\beta + (l^2 - m^2)\, n \sin\gamma] = 0.$$

If in this equation γ is put $= \pi$, and $\beta = 2\alpha$, so that $\sin(\alpha - \beta) = -\sin\alpha$, the equation will contain only terms in $\sin\alpha$, and $\sin 2\alpha$, viz. it will be

$$[\ (m^2 - n^2)\, l + (m^2 - 1)\, nl - (n^2 - 1)\, lm] \sin\alpha$$
$$+ [-(l^2 - 1)\, mn + m(n^2 - l^2) \quad] \sin 2\alpha = 0,$$

that is,

$$l(m-1)(n+1)(m-n)\sin\alpha + m(m+1)(n - l^2)\sin 2\alpha = 0,$$

or, what is the same thing,

$$(m+1)\sin\alpha\,\{l(n+1)(m-n)+2m(n-l^2)\cos\alpha\}=0,$$

which is satisfied for any values whatever of l, m, n, by a proper value of $\cos\alpha$; and is also satisfied irrespectively of the value of α if only $m=n=l^2$; or, writing k instead of l, say if $l=k$, $m=n=k^2$; that is, writing also ϕ in place of α, the three points

$$(k,\ \phi),\ (k^2,\ 2\phi)\ \text{and}\ (k^2,\ \pi)$$

are in a right line; viz. the point A_1, circle k, subtended angle ϕ; the point A_2, circle k^2, subtended angle 2ϕ; and the point V_2, same circle, subtended angle π; are in a right line.

The equation of the circle of unfitness can be obtained more easily in a different manner; but I have thought it worth while to give the investigation by means of the foregoing (p, q) coordinates.

Suppose that p_1, q_1 refer to the point A_1: then we have

$$(A_1N)^2=(x_1-1)^2+y_1^2=(x_1+iy_1-1)(x_1-iy_1-1),\quad =\left(\frac{1+p_1}{1-p_1}-1\right)\left(\frac{1+q_1}{1-q_1}-1\right),$$

that is,

$$(A_1N)^2=\frac{4p_1q_1}{1-p_1\,.\,1-q_1}.$$

Similarly, if p_2, q_2 refer to the point A_2, then

$$(A_2N)^2=\frac{4p_2q_2}{1-p_2\,.\,1-q_2},\quad =\frac{4p_1^2q_1^2}{1-p_1^2\,.\,1-q_1^2},$$

since $p_2,\ q_2=p_1^2,\ q_1^2$. The two are equal if

$$(1+p_1)(1+q_1)=p_1q_1,$$

that is,

$$p_1+q_1+1=0.$$

Writing for a moment $x_1+iy_1=\xi$, $x_1-iy_1=\eta$, we have

$$p_1,\ q_1=\frac{\xi-1}{\xi+1},\ \frac{\eta-1}{\eta+1},$$

and the equation is

$$\frac{\xi-1}{\xi+1}+\frac{\eta-1}{\eta+1}+1=0,$$

that is,

$$3\xi\eta+\xi+\eta-1=0;$$

or substituting for ξ, η their values, the equation is

$$3(x_1^2+y_1^2)+2x_1-1=0,$$

that is,

$$(x_1+1)(x_1-\tfrac{1}{3})+y_1^2=0,$$

the equation of a circle on the diameter $N'M$, which is, in fact, the before-mentioned circle of unfitness; viz. A_1 being on the circumference of this circle, or say on the boundary of the segment of unfitness, then $A_1N = A_2N$; whence also, according as A_1 is inside or outside the segment, $A_1N < A_2N$ or $> A_2N$.

Suppose A_1 to be on the circle, that is, $p_1 + q_1 + 1 = 0$; it is easy to show that the locus of A_2 is also a circle. We have in fact $(p_1 + q_1)^2 - 1 = 0$, that is,

$$p_2 + q_2 + 2k^2 - 1 = 0,$$

or say

$$\frac{\xi - 1}{\xi + 1} + \frac{\eta - 1}{\eta + 1} + 2k^2 - 1 = 0,$$

viz. this is

$$(2k^2 + 1)\,\xi\eta + (2k^2 - 1)(\xi + \eta) + 2k^2 - 3 = 0,$$

that is,

$$x_2^2 + y_2^2 + \frac{2k^2 - 1}{2k^2 + 1}\,2x_2 + \frac{2k^2 - 3}{2k^2 + 1} = 0,$$

or finally

$$(x_2 + 1)\left(x_2 - \frac{3 - 2k^2}{1 + 2k^2}\right) + y_2^2 = 0.$$

Measuring off from O in the direction of ON, a distance $OS = \dfrac{3 - 2k^2}{1 + 2k^2}$ (always $> \frac{1}{3}$, since $k^2 < 1$), the circle in question is that on the diameter $N'S$; this is a circle touching at N', and containing within it the circle of unfitness; if $k = 1$ (that is, for A_1 on the line QQ') it becomes identical with the circle of unfitness, but except in this limiting case it does not meet the circle of unfitness in any point on the N-side of QQ', that is, A_1 being on the boundary of the segment of unfitness A_2 is never on this boundary; and it thus appears that A_1 being inside the segment, A_2 is always outside the segment.

It is to be further noticed, that we have

$$\frac{(A_1N)^2}{(A_2N)^2} = \frac{1 + p_1 \,.\, 1 + q_1}{p_1 q_1},$$

or

$$\frac{(A_1N)^2}{(A_2N)^2} - 1 = \frac{1 + p_1 + q_1}{k^2} = \frac{3\xi\eta + \xi + \eta - 1}{k^2(\xi\eta + \xi + \eta + 1)}, \quad = \frac{3\,(x_1^2 + y_1^2 + \frac{2}{3}x_1 - \frac{1}{3})}{k^2\,[(x_1 + 1)^2 + y_1^2]},$$

that is,

$$\frac{(A_1N)^2}{(A_2N)^2} - 1 = \frac{3T^2}{k^2(A_1N)^2},$$

where T is the tangential distance of A_1 from the circle of unfitness; there should, it appears to me, be some more elegant formula for the ratio $A_1N \div A_2N$ which determines whether the step is regular or irregular.

It is worth noticing how the conditions

$$\text{mod.}\,(a-n) < \text{mod.}\,(a+n) \quad\text{and}\quad \text{mod.}\,(a-n) < \tfrac{2}{3}\,\text{mod.}\,n,$$

present themselves in the real theory. Making the usual construction by means of the parabola $y=x^2$, the first condition means that the point A must be taken on the N-side of O (fig. 2); the second that, in order to the regularity of the approximation, A must be taken at a distance from $O > \frac{1}{3}ON$; in fact, if (as in the figure) $OA=\frac{1}{3}ON$, then $AN=NA_1$, or the point A_1 is at an equal distance with A from N; and thence, according as OA is greater or less than $\frac{1}{3}ON$, the point A_1 is nearer or further than A to or from N.

Fig. 2.

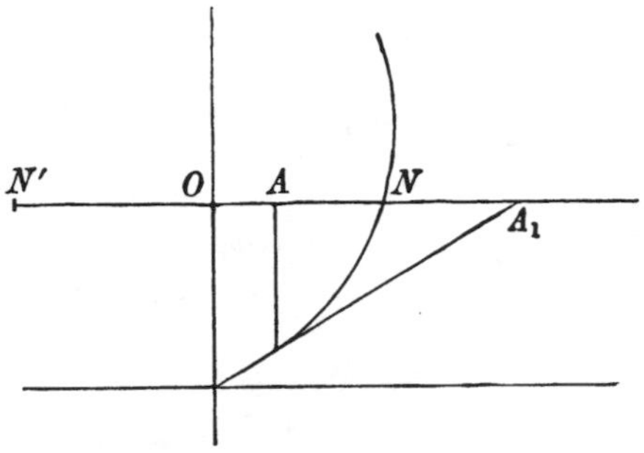

737.

ON A COVARIANT FORMULA.

[From the *Quarterly Journal of Pure and Applied Mathematics*, vol. XVI. (1879), pp. 224—226.]

STARTING from the equation

$$x_1 = x - \frac{fx}{f'x},$$

which presents itself in the Newton-Fourier problem, it is easy to see that, if a be a root of the equation $fx=0$, then

$$x_1 - a, = \frac{(x-a)f'x - fx}{f'x},$$

contains the factor $(x-a)^2$, that is, the equation $(x-x_1)f'x - fx = 0$, considered as an equation in x containing the parameter x_1, will have a twofold root, if x_1 is equal to any root a of the equation $fx=0$; and, consequently, the discriminant in regard to x of the function $(x-x_1)f'x - fx$ will contain the factor fx_1. But if fx be of the order n, then the discriminant is of the order $2n-2$ in x_1, and there is consequently a remaining factor ϕx_1 of the order $n-2$.

The like theorem applies to the homogeneous form

$$(xy_1 - x_1 y)\left(\alpha\frac{d}{dx} + \beta\frac{d}{dy}\right)f(x,\ y) - (\alpha y_1 - \beta x_1)f(x,\ y),$$

which reduces itself to the foregoing on writing $\alpha=1$, $\beta=0$, $y=y_1=1$; or, changing the notation, say to the form

$$(\xi y - \eta x)\left(\alpha\frac{d}{d\xi} + \beta\frac{d}{d\eta}\right)f(\xi,\ \eta) - (\alpha y - \beta x)f(\xi,\ \eta),$$

viz. the discriminant hereof in regard to ξ, η, being a function, homogeneous of the order $2n-2$ in regard to x, y, to α, β, and to the coefficients of $f(\xi, \eta)$, will contain the factor $f(x, y)$, and there will be consequently a remaining factor of the order $n-2$ in (x, y), $2n-2$ in (α, β) and $2n-3$ in the coefficients of $f(\xi, \eta)$.

The most simple case is when $f(\xi, \eta)$ is the quadric function $(a, b, c \between \xi, \eta)^2$. The form here is

$$(\xi y - \eta x)\, 2\,\{(a\alpha + b\beta)\,\xi + (b\alpha + c\beta)\,\eta\} - (\alpha y - \beta x)(a, b, c \between \xi, \eta)^2 = (\mathrm{a}, \mathrm{b}, \mathrm{c} \between \xi, \eta)^2,$$

where the coefficients are

$$\begin{aligned}
\mathrm{a} &= 2y\,(a\alpha + b\beta) - a\,(\alpha y - \beta x), = \quad a\beta x \quad + (a\alpha + 2b\beta)\,y,\\
\mathrm{b} &= y\,(b\alpha + c\beta) - x\,(a\alpha + b\beta) - b\,(\alpha y - \beta x),\\
&\qquad\qquad = -\quad a\alpha x \quad + \quad c\beta y \quad ,\\
\mathrm{c} &= -\,2x\,(b\alpha + c\beta) - c\,(\alpha y - \beta x), = -\,(2b\alpha + c\beta)\,x - \quad c\alpha y \quad ;
\end{aligned}$$

and we then have

$$\begin{aligned}
\mathrm{ac} - \mathrm{b}^2 = &- (2b\alpha\beta + c\beta^2)\,ax^2\\
&- \{2ab\alpha^2 + (2ac + 4b^2)\,\alpha\beta + 2bc\beta^2\}\,xy - (a\alpha^2 + 2b\alpha\beta)\,cy^2\\
&- a\alpha^2 . ax^2 - \{-2ac\alpha\beta\}\,xy - c\beta^2 . cy^2,
\end{aligned}$$

which is

$$= -\,(a\alpha^2 + 2b\alpha\beta + c\beta^2)\,(ax^2 + 2bxy + cy^2).$$

The discriminant is in this case

$$= -\,(a, b, c \between \alpha, \beta)^2 . (a, b, c \between x, y)^2.$$

In the case of the cubic function $(a, b, c, d \between \xi, \eta)^3$, the form is

$$\begin{aligned}
&(\xi y - x\eta)\,\{3\,(a\alpha + b\beta,\ b\alpha + c\beta,\ c\alpha + d\beta \between \xi, \eta)^2\}\\
&\qquad - (\alpha y - \beta x)\,(a, b, c, d \between \xi, \eta)^3 = (\mathrm{a}, \mathrm{b}, \mathrm{c}, \mathrm{d} \between \xi, \eta)^3,
\end{aligned}$$

the values of the coefficients being

$$\begin{aligned}
\mathrm{a} &= \quad a\beta x \qquad\qquad + (2a\alpha + 3b\beta)\,y,\\
\mathrm{b} &= -\,a\alpha x \qquad\qquad + (\ b\alpha + 2c\beta)\,y,\\
\mathrm{c} &= -\,(2b\alpha + \ c\beta)\,x + \qquad d\beta\ y,\\
\mathrm{d} &= -\,(3c\alpha + 2d\beta)\,x - \qquad d\alpha\ y.
\end{aligned}$$

Attending only to the terms in x^2, we have

$$\begin{aligned}
\mathrm{ac} - \mathrm{b}^2 &= -\quad (a\alpha^2 + 2b\alpha\beta + c\beta^2)\,ax^2,\\
\mathrm{ad} - \mathrm{bc} &= -\,2\,(b\alpha^2 + 2c\alpha\beta + d\beta^2)\,ax^2,\\
\mathrm{bd} - \mathrm{c}^2 &= \quad \{(3ac - 4b^2)\,\alpha^2 + (2ad - 4bc)\,\alpha\beta - c^2\beta^2\}\,x^2.
\end{aligned}$$

And hence, in

$$\mathrm{a^2d^2 + 4ac^3 + 4b^3d - 3b^2c^2 - 6abcd}, = \mathrm{(ad - bc)^2 - 4\,(ac - b^2)\,(bd - c^2)},$$

we have the term

$$4ax^3 \,.\, x\,[a\,(b\alpha^2 + 2c\alpha\beta + d\beta^2)^2 + (a\alpha^2 + 2b\alpha\beta + c\beta^2)\,\{(3ac - 4b^2)\,\alpha^2 + (2ad - 4bc)\,\alpha\beta - c^2\beta^2\}];$$

then, forming the analogous term in y^4, and assuming that the whole divides by $(a,\ b,\ c,\ d \between x,\ y)^3$, and also expanding the $\alpha\beta$-functions within the square brackets, we find

$$\text{Discriminant} = 4\,(a,\ b,\ c,\ d \between x,\ y)^3 \text{ multiplied by}$$

$$x \left| \begin{array}{l} 3a^2c - 3ab^2 \\ 2a^2d + 6abc - 8b^3 \\ 6abd + 6ac^2 - 12b^2c \\ 6acd - 6bc^2 \\ ad^2 - c^3 \end{array} \right| \between \alpha,\ \beta)^4 + y \left| \begin{array}{l} a^2d - b^3 \\ 6abd - 6b^2c \\ 6acd + 6b^2d - 12bc^2 \\ 2ad^2 + 6bcd - 8c^3 \\ 3bd^2 - 3c^2d \end{array} \right| \between \alpha,\ \beta)^4.$$

Writing down the Hessian of $(a,\ b,\ c,\ d \between \alpha,\ \beta)^3$,

$$H = (ac - b^2,\ ad - bc,\ bd - c^2 \between \alpha,\ \beta)^2,$$

and the cubicovariant

$$\Phi = \left\{ \begin{array}{l} a^2d - 3abc + 2b^3 \\ abd - 2ac^2 + b^2c \\ -\,acd + 2b^2d - bc^2 \\ -\,ad^2 + 3bcd - 2c^3 \end{array} \right\} (x,\ y)^3,$$

it is easy to see that the coefficient of x is

$$= 3\,(a,\ b,\ c \between \alpha,\ \beta)^2 \,.\, (H - \beta\Phi)\,;$$

hence also that of y is

$$= 3\,(b,\ c,\ d \between \alpha,\ \beta)^2 \,.\, (H + \alpha\Phi)\,,$$

and the final result is that the discriminant $= 4\,(a,\ b,\ c,\ d \between x,\ y)^3$ multiplied by

$$\{3\,(a,\ b,\ c,\ d \between \alpha,\ \beta)^3\,(x,\ y)\,H + (\alpha y - \beta x)\,\Phi\}.$$

It would be interesting to calculate the result for the quartic $(a,\ b,\ c,\ d,\ e \between \xi,\ \eta)^4$.

March 14, 1879.

738.

NOTE ON A HYPERGEOMETRIC SERIES.

[From the *Quarterly Journal of Pure and Applied Mathematics*, vol. XVI. (1879), pp. 268—270.]

IN the memoir on hypergeometric series, Schwarz, "Ueber diejenigen Fälle, &c.," *Crelle*, t. LXXV. (1873), pp. 292—335, the author shows, as part of his general theory, that the equation

$$\frac{d^2y}{dx^2} - \frac{\frac{2}{3} - \frac{7}{6}x}{x\,.\,1-x}\,\frac{dy}{dx} + \frac{\frac{1}{48}}{x\,.\,1-x}\,y = 0,$$

which belongs to the hypergeometric series $F(\frac{1}{4}, -\frac{1}{12}, \frac{2}{3}, x)$, is algebraically integrable, having in fact the two particular integrals

$$y^2 = \surd(\alpha - \alpha^5 x^{\frac{1}{3}}) \pm \surd(-\alpha^5 + \alpha x^{\frac{1}{3}}),$$

where α is a prime sixth root of -1, $\alpha^6 + 1 = 0$, or say $\alpha^4 - \alpha^2 + 1 = 0$ (see p. 326, α being for greater simplicity written instead of δ^2, and the form being somewhat simplified).

It is interesting to verify this directly; writing first $y = \surd(Y)$ and then $x = X^3$, the equation between Y, X is easily found to be

$$Y\frac{d^2Y}{dX^2} - \frac{\frac{3}{2}X^2}{1-X^3}\,Y\frac{dY}{dX} - \frac{1}{2}\left(\frac{dY}{dX}\right)^2 + \frac{\frac{3}{8}X}{1-X^3}\,Y^2 = 0,$$

and the theorem in effect is that that equation has the two particular integrals

$$Y = \surd(P) \pm \surd(Q),$$

P and Q being linear functions of X: in fact,

$$\begin{aligned} P &= \alpha - \alpha^5 X, \\ Q &= -\alpha^5 + \alpha\, X. \end{aligned}$$

Starting say from the equation

$$Y = \surd(P) + \surd(Q),$$

or, as it is convenient to write it,

$$Y = P^{\frac{1}{2}} + Q^{\frac{1}{2}},$$

where P and Q are assumed to be linear functions of X, we have

$$\frac{dY}{dX} = \tfrac{1}{2}P^{-\frac{1}{2}}P' + \tfrac{1}{2}Q^{-\frac{1}{2}}Q',$$

$$\frac{d^2Y}{dX^2} = -\tfrac{1}{4}P^{-\frac{3}{2}}P'^2 - \tfrac{1}{4}Q^{-\frac{3}{2}}Q'^2,$$

and thence

$$Y\frac{d^2Y}{dX^2} = -\tfrac{1}{4}P^{-1}P'^2 - \tfrac{1}{4}Q^{-1}Q'^2 - \tfrac{1}{4}Q^{\frac{1}{2}}P^{-\frac{3}{2}}P'^2 - \tfrac{1}{4}P^{\frac{1}{2}}Q^{-\frac{3}{2}}Q'^2$$

$$Y\frac{dY}{dX} = \tfrac{1}{2}(P' + Q') \qquad + \tfrac{1}{2}P^{-\frac{1}{2}}Q^{\frac{1}{2}}P' + \tfrac{1}{2}P^{\frac{1}{2}}Q^{-\frac{1}{2}}Q',$$

$$\left(\frac{dY}{dX}\right)^2 = \tfrac{1}{4}P^{-1}P'^2 + \tfrac{1}{4}Q^{-1}Q'^2 + \tfrac{1}{2}P^{-\frac{1}{2}}Q^{-\frac{1}{2}}P'Q',$$

where P', Q' are written to denote the derived functions of P, Q respectively.

Substituting these values, the resulting equation contains on the left-hand side a rational part, and a part with the factor $P^{-\frac{3}{2}}Q^{-\frac{3}{2}}$, and it is clear the equation can only be true if these two parts are separately $=0$. We have thus two equations which ought to be verified; viz. after a slight reduction these are found to be

$$\frac{1}{PQ}(QP'^2 + PQ'^2) + \frac{2X^2}{1-X^3}(P' + Q') - \frac{X}{1-X^3}(P + Q) = 0,$$

$$P^2Q'^2 + Q^2P'^2 + PQP'Q' + \frac{3X^2}{1-X^3}PQ(PQ' + P'Q) - \frac{3X}{1-X^3}P^2Q^2 = 0,$$

and it is very interesting to observe the manner in which these equations are, in fact, verified by the foregoing values of P, Q.

We have

$$P + Q = (\alpha - \alpha^5)(1 + X), \quad P' + Q' = \alpha - \alpha^5,$$

and hence

$$2X(P' + Q') - X(P + Q) = -(\alpha - \alpha^5)(1 - X),$$

or, in the first equation, the second part

$$\frac{2X^2}{1-X^3}(P' + Q') - \frac{X}{1-X^3}(P + Q)$$

is

$$= -(\alpha - \alpha^5)\frac{X(1-X)}{1-X^3};$$

viz. this is

$$=\frac{-(\alpha-\alpha^5)X}{1+X+X^2}.$$

We have

$$QP'^2+PQ'^2=\alpha^{10}(-\alpha^5+\alpha X)+\alpha^2(\alpha-\alpha^5X),$$
$$=\alpha^3-\alpha^{15}-(\alpha^7-\alpha^{11})X, \quad =(\alpha-\alpha^5)X;$$

and

$$PQ=-\alpha^6+(\alpha^2+\alpha^{10})X-\alpha^6X^2, \quad =1+X+X^2;$$

hence

$$\frac{1}{PQ}(QP'^2+PQ'^2)=\frac{(\alpha-\alpha^5)X}{1+X+X^2},$$

and the sum of the two parts is $=0$.

Similarly as regards the second equation, the second part

$$\frac{3X^2}{1-X^3}PQ(PQ'+P'Q)-\frac{3X}{1-X^3}P^2Q^2$$

is

$$=\frac{3PQX}{1-X^3}\{(PQ'+P'Q)X-PQ\}.$$

Here $PQ'+P'Q$ is $\alpha(\alpha-\alpha^5X)-\alpha^5(-\alpha^5+\alpha X)$, which is $=1+2X$; and PQ being $=1+X+X^2$, the term in { } is

$$(1+2X)X-(1+X+X^2), \quad =-(1-X)(1+X);$$

hence, outside the { } writing for PQ its value $=1+X+X^2$, the term is

$$=\frac{-3X(1+X+X^2)(1-X)(1+X)}{1-X^3}, \quad =-3X(1+X),$$

which is the value of the second part in question; the first part is

$$(PQ'+QP')^2-PQP'Q', \quad =(1+2X)^2-(1+X+X^2), \quad =3X(1+X);$$

and the sum of the two terms is thus $=0$.

739.

NOTE ON THE OCTAHEDRON FUNCTION.

[From the *Quarterly Journal of Pure and Applied Mathematics*, vol. XVI. (1879), pp. 280, 281.]

A SEXTIC function

$$U = (a,\ b,\ c,\ d,\ e,\ f,\ g \between x,\ y)^6,$$

such that its fourth derivative

$$\begin{aligned}(U,\ U)^4,\ = &\quad\ (ae - 4bd + 3c^2)\, x^4 \\ &+ 2\,(af - 3be + 2cd)\, x^3y \\ &+ \quad (ag - 9ce + 8d^2)\, x^2y^2 \\ &+ 2\,(bg - 3cf + 2de)\, xy^3 \\ &+ \quad (cg - 4df + 3e^2)\, y^4\end{aligned}$$

is identically $=0$, is considered by Dr Klein, and is called by him the octahedron function. Supposing that by a linear transformation the function is made to contain the factors x, y, or what is the same thing assuming $a=0$, $g=0$, then the equations to be satisfied become

$$-4bd + 3c^2 = 0,\quad -3be + 2cd = 0,\quad -9ce + 8d^2 = 0,\quad -3cf + 2de = 0,\quad -4df + 3e^2 = 0,$$

which are all satisfied if only $c = d = e = 0$; and then assuming, as is allowable,

$$b = -f = 1,$$

we have his canonical form $xy\,(x^4 - y^4)$ of the octahedron function.

But the equations may be satisfied in a different manner; viz. the first and last equations give

$$b = \frac{3c^2}{4d},\quad f = \frac{3e^2}{4d},$$

and, substituting these in the remaining equations, they become

$$\frac{c}{4d}(-9ce+8d^2)=0,\quad -9ce+8d^2=0,\quad \frac{e}{4d}(-9ce+8d^2)=0,$$

all satisfied if only $-9ce+8d^2=0$. Assuming $b=f=2$, the values are

$$b,\ c,\ d,\ e,\ f=2,\ 2\sqrt{(2)},\ 3,\ 2\sqrt{(2)},\ 2,$$

and the form is

$$xy\left(x^4+\frac{5}{\sqrt{(2)}}x^3y+5x^2y^2+\frac{5}{\sqrt{(2)}}xy^3+y^4\right),$$

$$=xy\left(x^2+\frac{3}{\sqrt{(2)}}xy+y^2\right)\{x^2+\sqrt{(2)}\,xy+y^2\},$$

$$=xy\left(x+\frac{1+i}{\sqrt{(2)}}y\right)\left(x+\frac{1-i}{\sqrt{(2)}}y\right)\{x+y\sqrt{(2)}\}\left(x+\frac{y}{\sqrt{(2)}}\right).$$

This is, in fact, a linear transformation of the foregoing form $XY(X^4-Y^4)$; for writing

$$X=\left(x+\frac{1+i}{\sqrt{(2)}}y\right),$$

$$Y=\left(x+\frac{1-i}{\sqrt{(2)}}y\right),$$

we have

$$X^2=x^2+(1+i)\sqrt{(2)}\,xy+iy^2,$$
$$Y^2=x^2+(1-i)\sqrt{(2)}\,xy-iy^2;$$

and therefore

$$X^2+Y^2=\quad 2x\{x+\sqrt{(2)}\,y\},$$

$$X^2-Y^2=2i\sqrt{(2)}\,y\left(x+\frac{y}{\sqrt{(2)}}\right),$$

or finally

$$XY(X^4-Y^4)=4i\sqrt{(2)}\,xy\left(x+\frac{1+i}{\sqrt{(2)}}y\right)\left(x+\frac{1-i}{\sqrt{(2)}}y\right)\{x+y\sqrt{(2)}\}\left(x+\frac{y}{\sqrt{(2)}}\right);$$

and the two forms are thus identical.

740.

ON CERTAIN ALGEBRAICAL IDENTITIES.

[From the *Quarterly Journal of Pure and Applied Mathematics*, vol. XVI. (1879), pp. 281, 282.]

If P_0, P_1, P_2 are points on a circle, say the circle $x^2+y^2=1$, then it is possible to find functions of (P_0, P_1) and of (P_1, P_2) respectively, which are really independent of P_1, and consequently functions of only P_0 and P_2: the expression "function of a point or points" being here used to mean algebraical function of the coordinates of the point or points. Thus the functions of (P_0, P_1) and of (P_1, P_2) being $x_0x_1+y_0y_1$, $x_0y_1-x_1y_0$, and $x_1x_2+y_1y_2$, $x_1y_2-x_2y_1$, we have

$$(x_1x_2+y_1y_2)(x_0x_1+y_0y_1)+(x_1y_2-x_2y_1)(x_0y_1-x_1y_0)=x_0x_2+y_0y_2,$$

and another like equation. This depends obviously on the circumstance that the coordinates of a point of the circle are expressible by means of the functions sin, cos, $x=\cos u$, $y=\sin u$; and the identity written down is obtained by expressing the cosine of u_2-u_0, $=(u_2-u_1)+(u_1-u_0)$, in terms of the cosines and sines of u_2-u_1 and u_1-u_0.

Evidently the like property holds good for a curve, such that the coordinates of any point of it can be expressed by means of "additive" functions of a parameter u; where, by an additive function $f(u)$, is meant a function such that $f(u+v)$ is an algebraical function of $f(u)$, $f(v)$; the sine and cosine are each of them an additive function, because

$$\sin(u+v)=\sin u\sqrt{(1-\sin^2 v)}+\sin v\sqrt{(1-\sin^2 u)},$$

and, similarly, for the cosine. But it is convenient to consider pairs or groups $f(u)$, $\phi(u), \ldots$, where $f(u+v)$, $\phi(u+v), \ldots$ are each of them an algebraical (rational) function of $f(u)$, $\phi(u), \ldots, f(v)$, $\phi(v), \ldots$; the sine and cosine are such a group, and so also are the elliptic functions sn, cn, dn; but the H and Θ, or say the ϑ-functions generally, are not additive.

In the case of the elliptic functions, we may consider the quadriquadric curve

$$y^2 = 1 - x^2, \quad z^2 = 1 - k^2x^2,$$

so that the coordinates of a point on the curve are sn u, cn u, dn u. Taking then P_0, P_1, P_2, points on the curve, and (x_0, y_0, z_0), (x_1, y_1, z_1), (x_2, y_2, z_2), the coordinates of these points respectively, we have in the same way, from $u_2 - u_0 = (u_2 - u_1) + (u_1 - u_0)$, three equations, of which the first is

$$\frac{x_2y_0z_0 - x_0y_2z_2}{1 - k^2x_0^2x_2^2} = \frac{\begin{array}{l}(1 - k^2x_1^2x_2^2)(x_2y_1z_1 - x_1y_2z_2)(y_0y_1 + x_0z_0x_1z_1)(z_0z_1 + k^2x_0y_0x_1y_1)\\ + (1 - k^2x_0^2x_1^2)(x_1y_0z_0 - x_0y_1z_1)(y_1y_2 + x_1z_1x_2z_2)(z_1z_2 + k^2x_1y_1x_2y_2)\end{array}}{(1 - k^2x_0^2x_1^2)^2(1 - k^2x_1^2x_2^2)^2 - k^2(x_1y_0z_0 - x_0y_1z_1)^2(x_2y_1z_1 - x_1y_2z_2)^2}.$$

The form of the right-hand side is

$$\frac{A + Bx_1y_1z_1}{C + Dx_1y_1z_1},$$

where A, B, C, D are each of them rational as regards x_1^2; and it is easy to see that the equation can only subsist under the condition that we have separately

$$\frac{x_2y_0z_0 - x_0y_2z_2}{1 - k^2x_0^2x_2^2} = \frac{A}{C} = \frac{B}{D},$$

implying of course the identity $AD - BC = 0$. The values of B and D are found without difficulty; we, in fact, have

$$B = 2k^2(x_2^2 \quad - x_0^2 \quad)(x_1^2y_0z_0y_2z_2 + x_0x_2y_1^2z_1^2),$$

$$D = 2k^2(x_2y_0z_0 + x_0y_2z_2)(x_1^2y_0z_0y_2z_2 + x_0x_2y_1^2z_1^2),$$

so that, comparing the left-hand side with $B \div D$, we have the identity

$$x_2^2y_0^2z_0^2 - x_0^2y_2^2z_2^2 = (x_2^2 - x_0^2)(1 - k^2x_2^2x_0^2),$$

which is right. The comparison with $A \div C$ would be somewhat more difficult to effect.

741.

ON A THEOREM OF ABEL'S RELATING TO A QUINTIC EQUATION.

[From the *Proceedings of the Cambridge Philosophical Society*, vol. III. (1880), pp. 155—159.]

THE theorem in question is given, *Œuvres Complètes*, [Christiania, 1881], t. II., p. 266, as an extract from a letter to Crelle dated 14th March, 1826, as follows:

"Si une équation du cinquième degré dont les coefficients sont *des nombres rationnels* est résoluble algébriquement, on peut donner aux racines la forme suivante:

$$x = c + Aa^{\frac{1}{5}}a_1^{\frac{2}{5}}a_2^{\frac{4}{5}}a_3^{\frac{3}{5}} + A_1a_1^{\frac{1}{5}}a_2^{\frac{2}{5}}a_3^{\frac{4}{5}}a^{\frac{3}{5}} + A_2a_2^{\frac{1}{5}}a_3^{\frac{2}{5}}a^{\frac{4}{5}}a_1^{\frac{3}{5}} + A_3a_3^{\frac{1}{5}}a^{\frac{2}{5}}a_1^{\frac{4}{5}}a_2^{\frac{3}{5}},$$

où

$$a\ = m + n\sqrt{(1+e^2)} + \sqrt{[h\,(1+e^2+\sqrt{(1+e^2)})]},$$
$$a_1 = m - n\sqrt{(1+e^2)} + \sqrt{[h\,(1+e^2-\sqrt{(1+e^2)})]},$$
$$a_2 = m + n\sqrt{(1+e^2)} - \sqrt{[h\,(1+e^2+\sqrt{(1+e^2)})]},$$
$$a_3 = m - n\sqrt{(1+e^2)} - \sqrt{[h\,(1+e^2-\sqrt{(1+e^2)})]},$$
$$A\ = K + K'a + K''a_2 + K'''aa_2,\quad A_1 = K + K'a_1 + K''a_3 + K'''a_1a_3,$$
$$A_2 = K + K'a_2 + K''a + K'''aa_2,\quad A_3 = K + K'a_3 + K''a_1 + K'''a_1a_3.$$

Les quantités c, h, e, m, n, K, K', K'', K''' sont des nombres *rationnels*. Mais de cette manière l'équation $x^5 + ax + b = 0$ n'est pas résoluble tant que a et b sont des quantités quelconques. J'ai trouvé de pareils théorèmes pour les équations du 7ème, 11ème, 13ème, etc. degré."

It is easy to see that x is the root of a quintic equation, the coefficients of which are rational and integral functions of a, a_1, a_2, a_3: these coefficients are not symmetrical functions of a, a_1, a_2, a_3, but they are functions which remain unaltered by the cyclical change a into a_1, a_1 into a_2, a_2 into a_3, a_3 into a. But the coefficients of the quintic equation must be rational functions of c, h, e, m, n, K, K', K'', K''': hence regarding a, a_1, a_2, a_3, as the roots of a quartic equation, the coefficients of this equation being rational functions of m, n, e, h, this equation must be such that every rational function of the roots, unchanged by the aforesaid cyclical change of the roots, shall be rationally expressible in terms of these quantities m, n, e, h: or, what is the same thing, the group of the quartic equation, using the term "group of the equation" in the sense assigned to it by Galois, must be $aa_1a_2a_3$, $a_1a_2a_3a$, $a_2a_3aa_1$, $a_3aa_1a_2$. And conversely, the quartic equation being of this form, x will be the root of a quintic equation, the coefficients whereof are rational and integral functions of c, h, e, m, n, K, K', K'', K'''.

To investigate the form of a quartic equation having the property just referred to, let it be proposed to find γ, γ' functions of e, h, such that $\gamma^2+\gamma'^2$ is a rational function of e, h, but that $\gamma^2-\gamma'^2$, $\gamma\gamma'$ are rational multiples of the same quadric radical $\sqrt{\theta}$. Assume that we have

$$\gamma^2-\gamma'^2=2p\sqrt{\theta},\quad \gamma\gamma'=q\sqrt{\theta};$$

then

$$(\gamma^2+\gamma'^2)^2=4(p^2+q^2)\,\theta;$$

that $\gamma^2+\gamma'^2$ may be rational, we must have $p^2+q^2=\lambda^2\theta$, or say $p^2+q^2=h^2\theta$; hence, $\theta=\frac{p^2}{h^2}+\frac{q^2}{h^2}$ must be a sum of two squares, or, assuming one of these equal to unity and the other of them equal to e^2, say $\theta=1+e^2$, we satisfy the required equation by taking $p=h$, $q=he$: viz. we thus have

$$\gamma^2-\gamma'^2=2h\sqrt{1+e^2},\quad \gamma\gamma'=he\sqrt{1+e^2},\quad \gamma^2+\gamma'^2=2h(1+e^2);$$

and thence also

$$\gamma^2=h(1+e^2+\sqrt{1+e^2}),\quad \gamma'^2=h(1+e^2-\sqrt{1+e^2}),$$

the roots of these expressions, or values of γ, γ', being such that

$$\gamma\gamma'=he\sqrt{1+e^2}.$$

Taking now α rational, $=m$ suppose, and β a rational multiple of

$$\sqrt{1+e^2},\quad =h\sqrt{1+e^2},$$

suppose; it is easy to see that the quartic equation which has for its roots

$$a,\ a_1,\ a_2,\ a_3=\alpha+\beta+\gamma,\ \alpha-\beta+\gamma',\ \alpha+\beta-\gamma,\ \alpha-\beta-\gamma',$$

has the property in question, viz. that every rational function of the roots unchangeable by the cyclical change a into a_1, a_1 into a_2, a_2 into a_3, a_3 into a, is rationally expressible in terms of e, h, m, n.

It will be sufficient to give the proof in the case of a rational and integral function; such a function, unchangeable as aforesaid, is of the form

$$\phi(a,\ a_1,\ a_2,\ a_3)+\phi(a_1,\ a_2,\ a_3,\ a)+\phi(a_2,\ a_3,\ a,\ a_1)+\phi(a_3,\ a,\ a_1,\ a_2);$$

and if $\phi(a, a_1, a_2, a_3)$ contains a term $\alpha^m\beta^n\gamma^p\gamma'^q$, then the other three functions will contain respectively the terms

$$\alpha^m(-\beta)^n\gamma'^p(-\gamma)^q,\quad \alpha^m\beta^n(-\gamma)^p(-\gamma')^q,\quad \alpha^m(-\beta)^n(-\gamma')^p(\gamma)^q;$$

viz. the sum of the four terms is

$$=\alpha^m\beta^n\left[\{1+(-)^{p+q}1\}\gamma^p\gamma'^q+\{(-)^{n+p}1+(-)^{n+q}1\}\gamma^q\gamma'^p\right].$$

This obviously vanishes unless p and q are both even, or both odd; and the cases to be considered are 1°, n even, p and q even; 2°, n odd, p and q even; 3°, n even, p and q odd; 4°, n odd, p and q odd. Writing, for greater distinctness, $2n$ or $2n+1$ for n, according as n is even or odd, and similarly for p and q, the term is, in the four cases respectively,

$$=2\alpha^m\beta^{2n}\,(\gamma^{2p}\gamma'^{2q}+\gamma^{2q}\gamma'^{2p}),$$
$$=2\alpha^m\beta^{2n+1}(\gamma^{2p}\gamma'^{2q}-\gamma^{2q}\gamma'^{2p}),$$
$$=2\alpha^m\beta^{2n}\,(\gamma^{2p+1}\gamma'^{2q+1}-\gamma^{2q+1}\gamma'^{2p+1}),$$
$$=2\alpha^m\beta^{2n+1}(\gamma^{2p+1}\gamma'^{2q+1}+\gamma^{2q+1}\gamma'^{2p+1}).$$

The second, third, and fourth expressions contain the factors

$$\beta(\gamma^2-\gamma'^2),\quad \gamma\gamma'(\gamma^2-\gamma'^2),\quad \beta\gamma\gamma',$$

respectively; and the first expression as it stands, and the other three divested of these factors respectively are rational functions of α, β^2, γ^2, γ'^2, that is, they are rational functions of m, n, e, h. But the omitted factors $\beta(\gamma^2-\gamma'^2)$, $\gamma\gamma'(\gamma^2-\gamma'^2)$, $\beta\gamma\gamma'$, $=2nh(1+e^2)$, $2h^2e(1+e^2)$, $nhe(1+e^2)$ are rational functions of n, h, e; hence each of the original four expressions is a rational function of m, n, h, e; and the entire function

$$\phi(a, a_1, a_2, a_3)+\phi(a_1, a_2, a_3, a)+\phi(a_2, a_3, a, a_1)+\phi(a_3, a, a_1, a_2)$$

is a rational function of m, n, h, e.

Replacing α, β, γ, γ' by their values, the roots of the quartic equation are

$$m+n\sqrt{(1+e^2)}+\sqrt{[h(1+e^2+\sqrt{(1+e^2)})]},$$
$$m-n\sqrt{(1+e^2)}+\sqrt{[h(1+e^2-\sqrt{(1+e^2)})]},$$
$$m+n\sqrt{(1+e^2)}-\sqrt{[h(1+e^2+\sqrt{(1+e^2)})]},$$
$$m-n\sqrt{(1+e^2)}-\sqrt{[h(1+e^2-\sqrt{(1+e^2)})]}.$$

And I stop to remark that taking $m, n, e, h=-\frac{1}{4}, +\frac{1}{4}, 2, -\frac{1}{8}$ respectively, the roots are

$$-\tfrac{1}{4}+\tfrac{1}{4}\sqrt{5}+\sqrt{[-\tfrac{1}{8}(5+\sqrt{5})]},$$
$$-\tfrac{1}{4}-\tfrac{1}{4}\sqrt{5}+\sqrt{[-\tfrac{1}{8}(5-\sqrt{5})]},$$
$$-\tfrac{1}{4}+\tfrac{1}{4}\sqrt{5}-\sqrt{[-\tfrac{1}{8}(5+\sqrt{5})]},$$
$$-\tfrac{1}{4}-\tfrac{1}{4}\sqrt{5}-\sqrt{[-\tfrac{1}{8}(5-\sqrt{5})]},$$

viz. these are the imaginary fifth roots of unity, or roots r, r^2, r^4, r^3 of the quartic equation $x^4+x^3+x^2+x+1=0$; which equation, as is well known, has the group $rr^2r^4r^3$, $r^2r^4r^3r$, $r^4r^3rr^2$, $r^3rr^2r^4$.

Reverting to Abel's expression for x, and writing this for a moment in the form

$$x=c+p+s+r+q,$$

the quintic equation in x is

$$\begin{aligned}0 &= (x-c)^5\\ &+(x-c)^3 . -5\,(pr+qs)\\ &+(x-c)^2 . -5\,(p^2s+q^2p+r^2q+s^2r)\\ &+(x-c)\ \ . -5\,(p^3q+q^3r+r^3s+s^3p)+5\,(p^2r^2+q^2s^2)-5pqrs\\ &+(x-c)^0 . -\ \ (p^5+q^5+r^5+s^5)\\ &\qquad +5\,(p^3rs+q^3sp+r^3pq+s^3qr)\\ &\qquad -5\,(p^2q^2r+q^2r^2s+r^2s^2p+s^2p^2q).\end{aligned}$$

If we substitute herein for p, q, r, s their values, then, altering the order of the terms, the final result is found to be

$$\begin{aligned}0 &= (x-c)^5\\ &+(x-c)^3 . -5\,(AA_2+A_1A_3)\,aa_1a_2a_3\\ &+(x-c)^2 . -5\,(A^2A_1a_2a_3+A_1^2A_2a_3a+A_2^2A_3aa_1+A_3^2Aa_1a_2)\,aa_1a_2a_3\\ &+(x-c)\ \ . -5\,(A^3A_3a_1a_2^2a_3+A_1^3Aa_2a_3^2a+A_2^3A_1a_3a^2a_1+A_3^3A_2aa_1^2a_2)\,aa_1a_2a_3\\ &\qquad +5\,(A^2A_2^2+A_2^2A_3^2-A_1A_2A_3A)\,(aa_1a_2a_3)^2\\ &+(x-c)^0 . -\ \ (A^5a_1a_2^3a_3^2+A_1^5a_2a_3^3a_1^2+A_2^5a_3a^3a_1^2+A_3^5aa_1^3a_2^2)\,aa_1a_2a_3\\ &\qquad +5\,(A^3A_1A_2a_2a_3+A_1^3A_2A_3a_3a+A_2^3A_3Aaa_1+A_3^3AA_1a_1a_2)\,(aa_1a_2a_3)^2\\ &\qquad -5\,(A^2A_3^2A_2a_1a_2+A_1^2A^2A_3a_2a_3+A_2^2A_1^2Aa_3a+A_3^2A_2^2A_1aa_1)\,(aa_1a_2a_3)^2;\end{aligned}$$

viz. considering herein A, A_1, A_2, A_3 as standing for their values

$$K+K'a+K''a_2+K'''aa_2, \text{ \&c.}$$

respectively, each coefficient is a function of a, a_1, a_2, a_3, which is unaltered by the cyclical change of these values and therefore is a rational function of

$$m,\ n,\ e,\ h,\ K,\ K',\ K'',\ K'''.$$

742.

ON THE TRANSFORMATION OF COORDINATES.

[From the *Proceedings of the Cambridge Philosophical Society,* vol. III. (1880), pp. 178—184.]

THE formulæ for the transformation between two sets of oblique coordinates assume a very elegant form when presented in the notation of matrices. I call to mind that a matrix denotes a system of quantities arranged in a square form

$$\left(\begin{array}{ccc} \alpha, & \beta, & \gamma \\ \alpha', & \beta', & \gamma' \\ \alpha'', & \beta'', & \gamma'' \end{array}\right),$$

see my "Memoir on the Theory of Matrices," *Phil. Trans.* t. CXLVIII. (1858), pp. 17—37, [152]; moreover $(\alpha, \beta, \gamma \between x, y, z)$ denotes $\alpha x + \beta y + \gamma z$, and so

$$\left(\begin{array}{ccc} \alpha, & \beta, & \gamma \\ \alpha', & \beta', & \gamma' \\ \alpha'', & \beta'', & \gamma'' \end{array} \between x, y, z\right)$$

denotes

$$(\alpha x + \beta y + \gamma z,\ \alpha' x + \beta' y + \gamma' z,\ \alpha'' x + \beta'' y + \gamma'' z),$$

and again

$$\left(\begin{array}{ccc} \alpha, & \beta, & \gamma \\ \alpha', & \beta', & \gamma' \\ \alpha'', & \beta'', & \gamma'' \end{array} \between x, y, z \between \xi, \eta, \zeta\right) \text{ denotes } \begin{array}{l} \xi(\alpha\, x + \beta\, y + \gamma\, z) \\ +\eta(\alpha' x + \beta' y + \gamma' z) \\ +\zeta(\alpha'' x + \beta'' y + \gamma'' z). \end{array}$$

Consequently

$$\left(\begin{array}{ccc} \alpha, & \beta, & \gamma \\ \alpha', & \beta', & \gamma' \\ \alpha'', & \beta'', & \gamma'' \end{array} \between x, y, z \between \xi, \eta, \zeta\right) = \left(\begin{array}{ccc} \alpha, & \alpha', & \alpha'' \\ \beta, & \beta', & \beta'' \\ \gamma, & \gamma', & \gamma'' \end{array} \between \xi, \eta, \zeta \between x, y, z\right).$$

In the case of a symmetrical matrix

$$\begin{pmatrix} a, & h, & g \\ h, & b, & f \\ g, & f, & c \end{pmatrix},$$

the equal expressions

$$\begin{pmatrix} a, & h, & g \\ h, & b, & f \\ g, & f, & c \end{pmatrix} \between x, y, z \between \xi, \eta, \zeta), = \begin{pmatrix} a, & h, & g \\ h, & b, & f \\ g, & f, & c \end{pmatrix} \between \xi, \eta, \zeta \between x, y, z),$$

are also written

$$(a, b, c, f, g, h \between x, y, z \between \xi, \eta, \zeta), \text{ or } (a, \ldots \between \xi, \eta, \zeta \between x, y, z).$$

In particular, if

$$(\xi, \eta, \zeta) = (x, y, z),$$

then

$$\begin{pmatrix} a, & h, & g \\ h, & b, & f \\ g, & f, & c \end{pmatrix} \between x, y, z)^2 \text{ is written } (a, b, c, f, g, h \between x, y, z)^2.$$

Two matrices are compounded together according to the law

$$\begin{pmatrix} a\,, & b\,, & c \\ a', & b', & c' \\ a'', & b'', & c'' \end{pmatrix} \between \begin{pmatrix} \alpha\,, & \beta\,, & \gamma \\ \alpha', & \beta', & \gamma' \\ \alpha'', & \beta'', & \gamma'' \end{pmatrix} = \begin{array}{r|ccc} & (\alpha, \alpha', \alpha''), & (\beta, \beta', \beta''), & (\gamma, \gamma', \gamma'') \\ \hline (a\,, b\,, c\,) & ,, & ,, & ,, \\ (a', b', c') & ,, & ,, & ,, \\ (a'', b'', c'') & ,, & ,, & ,, \end{array};$$

viz. in the compound matrix, the top-line is

$$(a, b, c \between \alpha, \alpha', \alpha''), \quad (a, b, c \between \beta, \beta', \beta''), \quad (a, b, c \between \gamma, \gamma', \gamma''),$$

and the other two lines are the like functions with (a', b', c'), and (a'', b'', c''), respectively, in the place of (a, b, c).

The inverse matrix is the matrix the terms of which are the minors of the determinant formed out of the original matrix, each minor being divided by this determinant, viz.

$$\begin{pmatrix} \alpha\,, & \beta\,, & \gamma \\ \alpha', & \beta', & \gamma' \\ \alpha'', & \beta'', & \gamma'' \end{pmatrix}^{-1} = \frac{1}{\nabla} \begin{pmatrix} \beta'\gamma'' - \beta''\gamma', & \beta''\gamma - \beta\gamma'', & \beta\gamma' - \beta'\gamma \\ \gamma'\alpha'' - \gamma''\alpha', & \gamma''\alpha - \gamma\alpha'', & \gamma\alpha' - \gamma'\alpha \\ \alpha'\beta'' - \alpha''\beta', & \alpha''\beta - \alpha\beta'', & \alpha\beta' - \alpha'\beta \end{pmatrix},$$

where ∇ is the determinant

$$\begin{vmatrix} \alpha\,, & \beta\,, & \gamma \\ \alpha', & \beta', & \gamma' \\ \alpha'', & \beta'', & \gamma'' \end{vmatrix}.$$

Coming now to the question of transformation, write

$$
\begin{array}{c|ccc:ccc c|c:c}
 & x & y & z & x_1 & y_1 & z_1 & & x \;\; y \;\; z & x_1 \;\; y_1 \;\; z_1 \\
\hline
x & 1 & \nu & \mu & \alpha & \alpha' & \alpha'' & =x & & \\
y & \nu & 1 & \lambda & \beta & \beta' & \beta'' & y & \Omega & W \\
z & \mu & \lambda & 1 & \gamma & \gamma' & \gamma'' & z & & \\
\hdashline
x_1 & \alpha & \beta & \gamma & 1 & \nu_1 & \mu_1 & x_1 & & \\
y_1 & \alpha' & \beta' & \gamma' & \nu_1 & 1 & \lambda_1 & y_1 & V & \Omega_1 \\
z_1 & \alpha'' & \beta'' & \gamma'' & \mu_1 & \lambda_1 & 1 & z_1 & &
\end{array}
$$

viz. the axes of x, y, z are inclined to each other at angles the cosines whereof are λ, μ, ν: those of x_1, y_1, z_1 are inclined to each other at angles the cosines whereof are λ_1, μ_1, ν_1: and the cosines of the inclinations of the two sets of axes to each other are α, β, γ; α', β', γ'; α'', β'', γ'': as is more clearly indicated in the diagram, the top-line showing that cosine-inclinations of x to

$$x,\ y,\ z,\ x_1,\ y_1,\ z_1,$$

are

$$1,\ \nu,\ \mu,\ \alpha,\ \alpha',\ \alpha'',$$

respectively, and the like for the other lines of the diagram. The letters Ω, Ω_1, V, W are used to denote matrices, viz. as appearing by the diagram, these are

$$
\left(\begin{array}{ccc} 1, & \nu, & \mu \\ \nu, & 1, & \lambda \\ \mu, & \lambda, & 1 \end{array}\right),\quad
\left(\begin{array}{ccc} 1, & \nu_1, & \mu_1 \\ \nu_1, & 1, & \lambda_1 \\ \mu_1, & \lambda_1, & 1 \end{array}\right),\quad
\left(\begin{array}{ccc} \alpha, & \beta, & \gamma \\ \alpha', & \beta', & \gamma' \\ \alpha'', & \beta'', & \gamma'' \end{array}\right),\quad
\left(\begin{array}{ccc} \alpha, & \alpha', & \alpha'' \\ \beta, & \beta', & \beta'' \\ \gamma, & \gamma', & \gamma'' \end{array}\right),
$$

respectively.

The coordinates (x, y, z) and (x_1, y_1, z_1) form each set a broken line extending from the origin to the point; hence projecting on the axes of x, y, z and on those of x_1, y_1, z_1 respectively, we have two sets, each of three equations, which may be written

$$(\Omega \between x, y, z) = (W \between x_1, y_1, z_1),$$
$$(V \between x, y, z) = (\Omega_1 \between x_1, y_1, z_1);$$

where of course each set implies the other set.

We have

$$(x, y, z) = (\Omega^{-1} W \between x_1, y_1, z_1), \ = (V^{-1} \Omega_1 \between x_1, y_1, z_1),$$
$$(x_1, y_1, z_1) = (W^{-1} \Omega \between x, y, z), \ = (\Omega_1^{-1} V \between x, y, z),$$

the first giving in two forms (x, y, z) as linear functions of (x_1, y_1, z_1), and the second giving in two forms (x_1, y_1, z_1) as linear functions of (x, y, z); comparing the two forms for each set, we have

$$\Omega^{-1} W = V^{-1} \Omega_1, \quad W^{-1} \Omega = \Omega_1^{-1} V,$$

or, what is the same thing,

$$V\Omega^{-1}W=\Omega_1, \qquad W\Omega_1^{-1}V=\Omega,$$

where in each equation the two sides are matrices which must be equal term by term to each other; but, the matrices being symmetrical, the equation thus gives (not nine but only) six equations. Writing

$$(\text{a, b, c, f, g, h})=(1-\lambda^2,\ 1-\mu^2,\ 1-\nu^2,\ \mu\nu-\lambda,\ \nu\lambda-\mu,\ \lambda\mu-\nu),$$

and

$$K=1-\lambda^2-\mu^2-\nu^2+2\lambda\mu\nu,$$

we have

$$\Omega^{-1}=\frac{1}{K}\begin{pmatrix}\text{a}, & \text{h}, & \text{g}\\ \text{h}, & \text{b}, & \text{f}\\ \text{g}, & \text{f}, & \text{c}\end{pmatrix}.$$

The first equation, written in the form

$$V\begin{pmatrix}\text{a}, & \text{h}, & \text{g}\\ \text{h}, & \text{b}, & \text{f}\\ \text{g}, & \text{f}, & \text{c}\end{pmatrix}W=K\Omega_1,$$

denotes the six equations

$$\begin{aligned}
&(\text{a, b, c, f, g, h})(\alpha\ ,\ \beta\ ,\ \gamma\ )^2 &&=K\ ,\\
&\quad ,, \qquad (\alpha',\ \beta',\ \gamma')^2 &&=K\ ,\\
&\quad ,, \qquad (\alpha'',\ \beta'',\ \gamma'')^2 &&=K\ ,\\
&\quad ,, \qquad (\alpha',\ \beta',\ \gamma')(\alpha'',\ \beta'',\ \gamma'') &&=K\lambda_1,\\
&\quad ,, \qquad (\alpha'',\ \beta'',\ \gamma'')(\alpha\ ,\ \beta\ ,\ \gamma\ ) &&=K\mu_1,\\
&\quad ,, \qquad (\alpha\ ,\ \beta\ ,\ \gamma\ )(\alpha',\ \beta',\ \gamma') &&=K\nu_1.
\end{aligned}$$

And, similarly, writing

$$(\text{a}_1,\ \text{b}_1,\ \text{c}_1,\ \text{f}_1,\ \text{g}_1,\ \text{h}_1)=(1-\lambda_1^2,\ 1-\mu_1^2,\ 1-\nu_1^2,\ \mu_1\nu_1-\lambda_1,\ \nu_1\lambda_1-\mu_1,\ \lambda_1\mu_1-\nu_1),$$

and

$$K_1=1-\lambda_1^2-\mu_1^2-\nu_1^2+2\lambda_1\mu_1\nu_1,$$

then

$$\Omega_1^{-1}=\frac{1}{K_1}\begin{pmatrix}\text{a}_1, & \text{h}_1, & \text{g}_1\\ \text{h}_1, & \text{b}_1, & \text{f}_1\\ \text{g}_1, & \text{f}_1, & \text{c}_1\end{pmatrix};$$

and the second equation, written in the form

$$W\begin{pmatrix}\text{a}_1, & \text{h}_1, & \text{g}_1\\ \text{h}_1, & \text{b}_1, & \text{f}_1\\ \text{g}_1, & \text{f}_1, & \text{c}_1\end{pmatrix}V=K_1\Omega,$$

denotes the six equations

$$
\begin{array}{lll}
(a_1, b_1, c_1, f_1, g_1, h_1 \between \alpha, \alpha', \alpha'')^2 & & = K_1, \\
\quad " \quad (\beta, \beta', \beta'')^2 & & = K_1, \\
\quad " \quad (\gamma, \gamma', \gamma'')^2 & & = K_1, \\
\quad " \quad (\beta, \beta', \beta'' \between \gamma, \gamma', \gamma'') & & = K_1\lambda_1, \\
\quad " \quad (\gamma, \gamma', \gamma'' \between \alpha, \alpha', \alpha'') & & = K_1\mu_1, \\
\quad " \quad (\alpha, \alpha', \alpha'' \between \beta, \beta', \beta'') & & = K_1\nu_1.
\end{array}
$$

The two sets each of six equations are, in fact, equivalent to a single set of six equations, and serve to express the relations between the nine cosines

$$(\alpha, \beta, \gamma, \alpha', \beta', \gamma', \alpha'', \beta'', \gamma''),$$

and the cosines (λ, μ, ν) and $(\lambda_1, \mu_1, \nu_1)$. Observe that the nine cosines are *not* (as in the rectangular transformation) the coefficients of transformation between the two sets of coordinates.

From the original linear relations between the coordinates, multiplying the equations of the first set by x, y, z and adding, and again multiplying the equations of the second set by (x_1, y_1, z_1) and adding, we have

$$(\Omega \between x, y, z)^2 = (W \between x_1, y_1, z_1 \between x, y, z),$$
$$(\Omega_1 \between x_1, y_1, z_1)^2 = (V \between x, y, z \between x_1, y_1, z_1).$$

But

$$(W \between x_1, y_1, z_1 \between x, y, z)$$

and

$$(V \between x, y, z \between x_1, y_1, z_1)$$

denote one and the same function; hence

$$(\Omega \between x, y, z)^2 = (\Omega_1 \between x_1, y_1, z_1)^2,$$

that is,

$$(1, 1, 1, \lambda, \mu, \nu \between x, y, z)^2 = (1, 1, 1, \lambda_1, \mu_1, \nu_1 \between x_1, y_1, z_1)^2,$$

or the linear relations between (x, y, z) and (x_1, y_1, z_1) are such as to transform one of these quadric functions into the other: the two quadrics, in fact, denote the squared distance from the origin expressed in terms of the coordinates (x, y, z) and (x_1, y_1, z_1) respectively.

Since the nine cosines are connected by six equations, there should exist values containing three arbitrary constants, and satisfying these equations identically: but, by what just precedes, it appears that the problem of determining these values is, in fact, that of finding the linear transformation between two given quadric functions: the problem of the linear transformation of a quadric function into itself has an elegant solution; but it would seem that this is not the case for the transformation between two different functions.

The foregoing equation

$$K = (\mathrm{a, b, c, f, g, h} \between \alpha, \beta, \gamma)^2,$$

is a relation between λ, μ, ν, the cosines of the sides of a spherical triangle, and (α, β, γ) the cosines of the distances of a point P from the three vertices: it can be at once verified by means of the relation $A+B+C=2\pi$, and thence

$$1-\cos^2 A-\cos^2 B-\cos^2 C+2\cos A\cos B\cos C=0,$$

which connects the angles A, B, C which the sides subtend at P. Writing a, b, c for λ, μ, ν, and f, g, h for α, β, γ, the relation is

$$\begin{aligned}1-a^2-b^2-c^2+2abc = (1-a^2)f^2+(1-b^2)g^2+(1-c^2)h^2&\\ +2(bc-a)gh+2(ca-b)hf+2(ab-c)fg&,\end{aligned}$$

viz. this is

$$\begin{aligned}1-a^2-b^2-c^2-f^2-g^2-h^2+2abc+2agh+2bhf+2cfg&\\ -a^2f^2-b^2g^2-c^2h^2+2bcgh+2cahf+2abfg=0&;\end{aligned}$$

where (a, b, c, f, g, h) are the cosines of the sides of a spherical quadrangle; (a, b, c), (a, h, g), (h, b, f), (g, f, c) belong respectively to sides forming a triangle, and the remaining sides (f, g, h), (b, c, f), (c, a, g), (a, b, h) are sides meeting in a vertex.

The equation

$$K\nu_1 = (\mathrm{a, b, c, f, g, h} \between \alpha, \beta, \gamma)(\alpha', \beta', \gamma')$$

is a relation between λ, μ, ν, the cosines of the sides of a spherical triangle; α, β, γ, the cosines of the distances of a point P from the three vertices; α', β', γ', the cosines of the distances of a point Q from the three vertices; and ν_1, the cosine of the distance PQ.

Drawing a figure, it is at once seen that

$$\nu_1 = \alpha\alpha' + \sqrt{1-\alpha^2}\sqrt{1-\alpha'^2}\cos(\theta-\theta'),$$

where

$$\cos\theta = \frac{\beta-\alpha\nu}{\sqrt{1-\alpha^2}\sqrt{1-\nu^2}},$$

and therefore

$$\sin\theta = \frac{\sqrt{\nabla}}{\sqrt{1-\alpha^2}\sqrt{1-\nu^2}};$$

also

$$\cos\theta' = \frac{\beta'-\alpha'\nu}{\sqrt{1-\alpha'^2}\sqrt{1-\nu^2}},$$

and therefore

$$\sin\theta' = \frac{\sqrt{\nabla'}}{\sqrt{1-\alpha'^2}\sqrt{1-\nu^2}},$$

the values of ∇, ∇' being

$$\nabla = 1 - \alpha^2 - \beta^2 - \nu^2 + 2\alpha\beta\nu,$$

$$\nabla' = 1 - \alpha'^2 - \beta'^2 - \nu^2 + 2\alpha'\beta'\nu\,;$$

the resulting value of ν_1 is therefore

$$\nu_1 = \alpha\alpha' + \frac{1}{1-\nu^2}\{(\beta - \alpha\nu)(\beta' - \alpha'\nu) + \sqrt{\nabla\nabla'}\}.$$

The equations

$$K = (\mathrm{a, b, c, f, g, h}\between \alpha, \beta, \gamma)^2, \quad K = (\mathrm{a}, \ldots \between \alpha', \beta', \gamma')^2,$$

give

$$(\mathrm{g}\alpha + \mathrm{f}\beta + \mathrm{c}\gamma)^2 = K\nabla,$$

$$(\mathrm{g}\alpha' + \mathrm{f}\beta' + \mathrm{c}\gamma')^2 = K\nabla':$$

and we therefore have

$$(\mathrm{g}\alpha + \mathrm{f}\beta + \mathrm{c}\gamma)(\mathrm{g}\alpha' + \mathrm{f}\beta' + \mathrm{c}\gamma') = K\sqrt{\nabla\nabla'}\,;$$

recollecting that $1 - \nu^2 = \mathrm{c}$, the formula thus is

$$\nu_1 = \alpha\alpha' + \frac{1}{\mathrm{c}}\left\{(\beta - \alpha\nu)(\beta' - \alpha'\nu) + \frac{1}{K}(\mathrm{g}\alpha + \mathrm{f}\beta + \mathrm{c}\gamma)(\mathrm{g}\alpha' + \mathrm{f}\beta' + \mathrm{c}\gamma')\right\},$$

or say,

$$K\nu_1 = K\alpha\alpha' + \frac{1}{\mathrm{c}}\{K(\beta - \alpha\nu)(\beta' - \alpha'\nu) + (\mathrm{g}\alpha + \mathrm{f}\beta)(\mathrm{g}\alpha' + \mathrm{f}\beta')\} + \mathrm{g}(\alpha\gamma' + \alpha'\gamma) + \mathrm{f}(\beta\gamma' + \beta'\gamma) + \mathrm{c}\gamma\gamma'.$$

The sum of the first and second terms is readily found to be

$$= \mathrm{a}\alpha\alpha' + \mathrm{b}\beta\beta' + \mathrm{h}(\alpha\beta' + \alpha'\beta)\,;$$

and the equation thus becomes

$$K\nu_1 = (\mathrm{a, b, c, f, g, h}\between \alpha, \beta, \gamma\between \alpha', \beta', \gamma'),$$

as it should do.

743.

ON THE NEWTON-FOURIER IMAGINARY PROBLEM.

[From the *Proceedings of the Cambridge Philosophical Society*, vol. III. (1880), pp. 231, 232.]

THE Newtonian process of approximation to the root of a numerical equation $f(u)=0$, consists in deriving from an assumed approximate root ξ a new value $\xi_1=\xi-\frac{f(\xi)}{f'(\xi)}$, which should be a closer approximation to the root sought for: taking the coefficients of $f(u)$ to be real, and also the root sought for, and the assumed value ξ, to be each of them real, Fourier investigated the conditions under which ξ_1 is in fact a closer approximation. But the question may be looked at in a more general manner: ξ may be any real or imaginary value, and we have to inquire in what cases the series of derived values

$$\xi_1=\xi-\frac{f(\xi)}{f'(\xi)},\quad \xi_2=\xi_1-\frac{f(\xi_1)}{f'(\xi_1)},\ldots$$

converge to a root, real or imaginary, of the equation $f(u)=0$. Representing as usual the imaginary value ξ, $=x+iy$, by means of the point whose coordinates are x, y, and in like manner ξ_1, $=x_1+iy_1$, &c., then we have a problem relating to an infinite plane; the roots of the equation are represented by points A, B, C,...; the value ξ is represented by an arbitrary point P; and from this by a determinate geometrical construction we obtain the point P_1, and thence in like manner the points P_2, P_3,... which represent the values ξ_1, ξ_2, ξ_3,... respectively. And the problem is to divide the plane into regions, such that, starting with a point P_1 anywhere in one region, we arrive ultimately at the root A; anywhere in another region we arrive ultimately at the root B; and so on for the several roots of the equation. The division into regions is made without difficulty in the case of a quadric equation; but in the next succeeding case, that of a cubic equation, it is anything but obvious what the division is: and the author had not succeeded in finding it.

744.

TABLE OF $\Delta^m 0^n \div \Pi(m)$ UP TO $m=n=20$.

[From the *Transactions of the Cambridge Philosophical Society*, vol. XIII. Part I. (1881), pp. 1—4. Read October 27, 1879.]

THE differences of the powers of zero, $\Delta^m 0^n$, present themselves in the Calculus of Finite Differences, and especially in the applications of Herschel's theorem,

$$f(e^t) = f(1+\Delta)\, e^{t\,.\,0},$$

for the expansion of the function of an exponential. A small Table up to $\Delta^{10}0^{10}$ is given in Herschel's *Examples* (Camb. 1820), and is reproduced in the treatise on Finite Differences (1843) in the *Encyclopædia Metropolitana*. But, as is known, the successive differences $\Delta 0^n$, $\Delta^2 0^n$, $\Delta^3 0^n, \ldots$ are divisible by 1, 1 . 2, 1 . 2 . 3, ... and generally $\Delta^m 0^n$ is divisible by $1\,.\,2\,.\,3 \ldots m, = \Pi(m)$; these quotients are much smaller numbers, and it is therefore desirable to tabulate them rather than the undivided differences $\Delta^m 0^n$: moreover, it is easier to calculate them. A table of the quotients $\Delta^m 0^n \div \Pi(m)$, up to $m=n=12$ is in fact given by Grunert, *Crelle*, t. XXV. (1843), p. 279, but without any explanation in the heading of the meaning of the tabulated numbers $C_n{}^k, = \Delta^n 0^k \div \Pi(n)$, and without using for their determination the convenient formula $C_n{}^{k+1} = nC_n{}^k + C_{n-1}{}^k$ given by Björling in a paper, *Crelle*, t. XXVIII. (1844), p. 284. The formula in question, say

$$\frac{\Delta^m 0^{n+1}}{\Pi(m)} = m\,\frac{\Delta^m 0^n}{\Pi(m)} + \frac{\Delta^{m-1}0^n}{\Pi(m-1)},$$

is given in the second edition (by Moulton) of Boole's *Calculus of Finite Differences*, (London, 1872), p. 28, under the form

$$\Delta^m 0^n = m\,(\Delta^{m-1}0^{n-1} + \Delta^m 0^{n-1}).$$

It occurred to me that it would be desirable to extend the table of the quotients $\Delta^m 0^n \div \Pi(m)$, up to $m=n=20$. The calculation is effected very readily by means of the foregoing theorem, which is used in the following form; viz. any column of the table for instance the fifth, being

$$\begin{array}{lcr} A, \text{ then the following column is} & & A, \\ B, & \ldots & 2B+A, \\ C, & \ldots & 3C+B, \\ D, & \ldots & 4D+C, \\ E, & \ldots & 5E+D, \\ & & +E; \end{array}$$

and then we obtain a good verification by taking the sum of the terms in the new column, and comparing it with the value as calculated from the formula,

$$\text{Sum} = 2A + 3B + 4C + 5D + 6E.$$

Observe that, in the two calculations, we take successive multiples such as $4D$ and $5D$ of each term of the preceding column, and that the verification is thus a safeguard against any error of multiplication or addition.

TABLE, No. 1, OF $\Delta^m 0^n \div \Pi(m)$.

Ind. Δ	0^1	0^2	0^3	0^4	0^5	0^6	0^7	0^8	0^9	0^{10}	0^{11}	0^{12}	0^{13}	0^{14}
1	1	1	1	1	1	1	1	1	1	1	1	1	1	1
2		1	3	7	15	31	63	127	255	511	1 023	2 047	4 095	8 191
3			1	6	25	90	301	966	3 025	9 330	28 501	86 526	261 625	788 970
4				1	10	65	350	1 701	7 770	34 105	145 750	611 501	2 532 530	10 391 745
5					1	15	140	1 050	6 951	42 525	246 730	1 379 400	7 508 501	40 075 035
6						1	21	266	2 646	22 827	179 487	1 323 652	9 321 312	63 436 373
7							1	28	462	5 880	63 987	627 396	5 715 424	49 329 280
8								1	36	750	11 880	159 027	1 899 612	20 912 320
9									1	45	1 155	22 275	359 502	5 135 130
10										1	55	1 705	39 325	752 752
11											1	66	2 431	66 066
12												1	78	3 367
13													1	91
14														1
15														
16														
17														
18														
19														
20														

Ind. Δ	0^{15}	0^{16}	0^{17}	0^{18}	0^{19}	0^{20}	
1	1	1	1	1	1	1	1
2	16 383	32 767	65 535	131 071	262 143	524 287	2
3	2 375 101	7 141 686	21 457 825	64 439 010	193 448 101	580 606 446	3
4	42 355 950	171 798 901	694 337 290	2 798 806 985	11 259 666 950	45 232 115 901	4
5	210 766 920	1 096 190 550	5 652 751 651	28 958 095 545	147 589 284 710	749 206 090 500	5
6	420 693 273	2 734 926 558	17 505 749 898	110 687 251 039	693 081 601 779	4 306 078 895 384	6
7	408 741 333	3 281 882 604	25 708 104 786	197 462 483 400	1 492 924 634 839	11 143 554 045 652	7
8	216 627 840	2 141 764 053	20 415 995 028	189 036 065 010	1 709 751 003 480	15 170 932 662 679	8
9	67 128 490	820 784 250	9 528 822 303	106 175 395 755	1 144 614 626 805	12 011 282 644 725	9
10	12 662 650	193 754 990	2 758 334 150	37 112 163 803	477 297 033 785	5 917 584 964 655	10
11	1 479 478	28 936 908	512 060 978	8 391 004 908	129 413 217 791	1 900 842 429 486	11
12	106 470	2 757 118	62 022 324	1 256 328 866	23 466 951 300	411 016 633 391	12
13	4 550	165 620	4 910 178	125 854 638	2 892 439 160	61 068 660 380	13
14	105	6 020	249 900	8 408 778	243 577 530	6 302 524 580	14
15	1	120	7 820	367 200	13 916 778	452 329 200	15
16		1	136	9 996	527 136	22 350 954	16
17			1	153	12 597	741 285	17
18				1	171	15 675	18
19					1	190	19
20						1	20

Writing down the sloping lines as columns thus:

1 (0)	2 (2)	3 (4)	4 (6)	5 (8)	6 (10)	7 (12)	8 etc. (14) etc.
1							
1	1						
1	3	1					
1	6	7	1				
1	10	25	15	1			
1	15	65	90	31	1		
1	21	140	350	301	63	1	
1	28	266	1 050	1 701	966	127	
1	36	462	2 646	6 951	7 770	3 025	
1	45	750	5 880	22 827	42 525	34 105	
1	55	1 155	11 880	63 987	179 487	246 730	
1	66	1 705	22 275	159 027	627 396	1 323 652	
1	78	2 431	39 325	359 502	1 899 612	5 715 424	
1	91	3 367	66 066	752 752	5 135 130	20 912 320	
1	105	4 550	106 470	1 479 478	12 662 650	67 128 490	
1	120	6 020	165 620	2 757 118	28 936 908	193 754 990	
1	136	7 820	249 900	4 910 178	62 022 324	512 060 978	
1	153	9 996	367 200	8 408 778	125 854 638	1 256 328 866	
1	171	12 597	527 136	13 916 778	243 577 530	2 892 439 160	
1	190	15 675	741 285	22 350 954	452 329 200	6 302 524 580	
20	19	18	17	16	15	14	13 etc.

it appears by inspection that, in the second column the second differences, are constant, in the third column the fourth differences, in the fourth column the sixth differences, and so on, are constant; and we thence deduce the law of the numbers in the successive columns: viz. this can be done up to column 7, in which we have 14 numbers in order to find the 12th differences: but in column 8 we have only 13 numbers, and therefore cannot find the 14th differences. The differences are given in the following

TABLE, No. 2 (*explanation infrà*).

Ind. Δ	1	2	3	4	5	6	7
0	1	1	1	1	1	1	1
1		2	6	14	30	62	126
2		1	12	61	240	841	2 772
3			10	124	890	5 060	25 410
4			3	131	1 830	16 990	127 953
5				70	2 226	35 216	401 436
6				15	1 600	47 062	836 976
7					630	40 796	1 196 532
8					105	21 225	1 182 195
9						10 930	795 718
10						945	349 020
11							90 090
12							10 395

We have, by means of this Table, the general expressions of $\Delta^r 0^r$, $\Delta^{r-1}0^r$, $\Delta^{r-2}0^r, \ldots$ up to $\Delta^{r-6}0^r$, viz. the formulæ are

$$\Delta^r 0^r \div \Pi(r) \quad = 1,$$

$$\Delta^{r-1}0^r \div \Pi(r-1) = 1 + 2\binom{r-2}{1}^1 + 1\binom{r-2}{2}^2,$$

$$\Delta^{r-2}0^r \div \Pi(r-2) = 1 + 6\binom{r-3}{1}^1 + 12\binom{r-3}{2}^2 + 10\binom{r-3}{3}^3 + 3\binom{r-3}{4}^4,$$

&c., &c.,

where the numerical coefficients are the numbers in the successive columns of the table; and where for shortness $\binom{r-m}{k}^k$ is written to denote the binomial coefficient $\frac{[r-m]^k}{[k]^k}$. For instance, $r=10$, we have

$$\Delta^8 0^{10} \div \Pi(8) = 1 + 6.7 + 12.21 + 10.35 + 3.35, \; = 750,$$

agreeing with the principal Table. It will be observed that, in the successive columns of the Table, the last terms are 1, 1, 1.3, 1.3.5, 1.3.5.7, 1.3.5.7.9, and 1.3.5.7.9.11. This is itself a good verification: I further verified the last column by calculating from it the value of $\Delta^{14}0^{20} \div \Pi(14)$, $=6\,302\,524\,580$ as above. The Table shows that we have $\Delta^{r-m}0^r \div \Pi(r-m)$ given as an algebraical rational and integral function of r, of the degree $2m$. But the terms from the top of a column, $\Delta 0^r = 1$, $\Delta^2 0^r \div 1.2 = 2^{r-1} - 1$, &c., are not algebraical functions of r.

22 *October*, 1879.

745.

ON THE SCHWARZIAN DERIVATIVE, AND THE POLYHEDRAL FUNCTIONS.

[From the *Transactions of the Cambridge Philosophical Society*, vol. XIII. Part I. (1881), pp. 5—68. Read March 8, 1880.]

THE quotient s of any two solutions of a linear partial differential equation of the second order, $\frac{d^2y}{dx^2}+p\frac{dy}{dx}+qy=0$, is determined by a differential equation of the third order

$$\frac{\frac{d^3s}{dx^3}}{\frac{ds}{dx}}-\tfrac{3}{2}\left(\frac{\frac{d^2s}{dx^2}}{\frac{ds}{dx}}\right)^2=-\tfrac{1}{2}\left(p^2+2\frac{dp}{dx}-4q\right),$$

where the function on the left-hand is what I call the Schwarzian Derivative; or say this derivative is

$$\{s,\ x\},\ =\frac{s'''}{s'}-\tfrac{3}{2}\left(\frac{s''}{s'}\right)^2,$$

where the accents denote differentiations in regard to the second variable x of the symbol.

Writing in general $(\mathrm{a},\ \mathrm{b},\ \mathrm{c}\ .\!\cdot\!.\between X,\ Y,\ Z)^2$ to denote a quadric function

$$(\mathrm{a},\ \mathrm{b},\ \mathrm{c},\ \tfrac{1}{2}(\mathrm{a}-\mathrm{b}-\mathrm{c}),\ \tfrac{1}{2}(-\mathrm{a}+\mathrm{b}-\mathrm{c}),\ \tfrac{1}{2}(-\mathrm{a}-\mathrm{b}+\mathrm{c})\between X,\ Y,\ Z)^2,$$

then, if the equation of the second order be that of the hypergeometric series, generalised by a homographic transformation upon the variable x, the resulting differential equation of the third order is of the form

$$\{s,\ x\}=(\mathrm{a},\ \mathrm{b},\ \mathrm{c}\ .\!\cdot\!.)\left(\frac{1}{x-a},\ \frac{1}{x-b},\ \frac{1}{x-c}\right)^2;$$

and, presenting themselves in connexion with the algebraically integrable cases of this equation, we have rational and integral functions of s, derived from the polygon, the double pyramid, and the five regular solids. They are called Polyhedral Functions.

The Schwarzian Derivative occurs implicitly in Jacobi's differential equation of the third order for the modulus in the transformation of an elliptic function (*Fund. Nova*, 1829, p. 79, [*Ges. Werke*, t. I., p. 133]) and in Kummer's fundamental equation for the transformation of a hypergeometric series (Kummer, 1836: see list of Memoirs): but it was first explicitly considered and brought into notice in the two Memoirs of Schwarz*, 1869 and 1873. The latter of these, relating to the algebraic integration of the differential equation for the hypergeometric series, is the fundamental Memoir upon the subject, but the theory is in some material points completed in the Memoirs by Klein and Brioschi.

The following list of Memoirs, relating as well to the Polyhedral Functions as to the Schwarzian Derivative, is arranged nearly in chronological order.

Kummer, Ueber die hypergeometrische Reihe $1+\frac{\alpha\,.\,\beta}{1\,.\,\gamma}x+\ldots$ *Crelle*, t. XV. (1836), pp. 39—83 and 127—172.

Schwarz, Ueber einige Abbildungsaufgaben. *Crelle-Borchardt*, t. LXX. (1869), pp. 105—120.

——— Ueber diejenigen Fälle in welchen die *Gaussische* hypergeometrische Reihe eine algebraische Function ihres vierten Elementes darstellt. Do. t. LXXV. (1873), pp. 292—335.

Cayley, Notes on Polyhedra. *Quart. Math. Jour.* t. VII. (1866), pp. 304—316; [375].

——— On the Regular Solids. Do. t. XV. (1878), pp. 127—131; [679].

Fuchs, Ueber diejenigen Differentialgleichungen zweiter Ordnung welche algebraische Integralen besitzen, und eine Anwendung der Invariantentheorie. *Crelle-Borchardt*, t. LXXXI. (1875), pp. 97—142.

Klein, Ueber binäre Formen mit linearen Transformationen in sich selbst. *Math. Ann.* t. IX. (1875), pp. 183—209.

Brioschi, Extrait d'une lettre à M. Klein. *Math. Ann.* t. XI. (1877), pp. 111—114.

Klein, Ueber lineare Differentialgleichungen. *Math. Ann.* t. XI. (1877), pp. 115—118.

Brioschi, La théorie des formes dans l'intégration des équations différentielles linéaires du second ordre. *Math. Ann.* t. XI. (1877), pp. 401—411.

Gordan, Ueber endliche Gruppen linearer Transformationen einer Veränderlichen. *Math. Ann.* t. XII. (1877), pp. 23—46.

——— Binäre Formen mit verschwindenden Covarianten. *Math. Ann.* t. XII. (1877), pp. 147—166.

[* Schwarz, *Ges. Werke*, t. II., p. 351, remarks that the Derivative occurs implicitly in a memoir by Lagrange, "Sur la construction des cartes géographiques," (1779), *Œuvres*, t. IV., p. 651.]

Klein, Ueber lineare Differentialgleichungen. *Math. Ann.* t. XII. (1877), pp. 167—179.

——— Weitere Untersuchungen über das Icosaeder. *Math. Ann.* t. XII. (1877), pp. 503—560.

Cayley, On the Correspondence of Homographies and Rotations. *Math. Ann.* t. XV. (1879), pp. 238—240; [660].

——— On the finite Groups of linear transformations of a Variable. *Math. Ann.* t. XVI. (1880), pp. 260—263, and pp. 439—440; [752].

I propose in the present Memoir to consider the whole theory: and, in particular, to give some additional developments in regard to the Polyhedral Functions.

I remark that Schwarz starts with the foregoing differential equation of the third order

$$\{s,\ x\} = (\mathrm{a},\ \mathrm{b},\ \mathrm{c}\ \therefore)\left(\frac{1}{x-a},\ \frac{1}{x-b},\ \frac{1}{x-c}\right)^2,$$

and he shows (by very refined reasoning founded on the theory of conformable figures, which will be in part reproduced) that this equation is, in fact, algebraically integrable for 16 different sets of values of the coefficients a, b, c. It may I think be taken to be part of his theory, although not very clearly brought out by him, that these integrals are some of them of the form, x = rational function of s; others of the form, rational function of x = rational function of s; the rational functions of s being in fact the same in the last as in the first set of solutions: they are quotients of Polyhedral functions.

But as regards the second set of cases, the solution of these, introducing for convenience a new variable z in place of s, may be made to depend upon the solution in the form, x = rational function of z, of an equation of a somewhat similar form, but involving two quadric functions of x and z respectively, viz. the equation

$$\{x,\ z\} + \left(\frac{dx}{dz}\right)^2 (\mathrm{a},\ \mathrm{b},\ \mathrm{c}\ \therefore)\left(\frac{1}{x-a},\ \frac{1}{x-b},\ \frac{1}{x-c}\right)^2 = (\mathrm{a}_1,\ \mathrm{b}_1,\ \mathrm{c}_1\ \therefore)\left(\frac{1}{z-a_1},\ \frac{1}{z-b_1},\ \frac{1}{z-c_1}\right)^2;$$

and we have the theorem that the solution of this equation depends upon the determination of P, Q, R rational and integral functions of z, containing each of them multiple factors, which are such that $P+Q+R=0$. Using accents to denote differentiation in regard to z, this implies $P'+Q'+R'=0$, and consequently

$$QR' - Q'R = RP' - R'P = PQ' - P'Q.$$

Further, they are such that the equal functions $QR'-Q'R$, $RP'-R'P$, $PQ'-P'Q$ contain only factors which are factors of P, Q or R.

In fact, writing $f,\ g,\ h = b-c,\ c-a,\ a-b$, the required relation between x, z is then expressed in the symmetrical form $f(x-a) : g(x-b) : h(x-c) = P : Q : R$.

The last-mentioned differential equation is considered by Klein and Brioschi: the solutions in 13 cases, or such of them as had not been given by Schwarz, were obtained by Brioschi: and those of the remaining 3 cases, subject to a correction in one of them, were afterwards obtained by Klein.

The first part of the present Memoir relates, say to the foregoing equation

$$\{s,\ x\} = (\mathrm{a},\ \mathrm{b},\ \mathrm{c}\ .\because)\left(\frac{1}{x-a},\ \frac{1}{x-b},\ \frac{1}{x-c}\right)^2,$$

although the other form in $\{x,\ z\}$ may equally well be regarded as the fundamental form.

We consider in the theory:

A. The Derivative $\{s,\ x\}$, meaning as above explained.

B. Quadric functions of any three or more inverts $\dfrac{1}{x-l}$.

C. Rational and integral functions P, Q, R having a sum $=0$, and which are such that $QR'-Q'R$, $=RP'-R'P$, $=PQ'-P'Q$, contains only the factors of P, Q, R.

D. The differential equation of the third order.

E. The Schwarzian theory in regard to conformable figures and the corresponding values of the imaginary variables s and x.

F. Connexion with the differential equation for the hypergeometric series.

The second part of the Memoir relates to the Polyhedral Functions.

The paragraphs of the whole Memoir are numbered consecutively.

PART I.

The Derivative $\{s,\ x\}$. Art. Nos. 1 to 7.

1. If $p = \dfrac{s''}{s'} = \dfrac{d}{dx}\left(\log \dfrac{ds}{dx}\right)$, then $\{s,\ x\} = \dfrac{dp}{dx} - \tfrac{1}{2}p^2$.

2. The derivative $\{s,\ x\}$ may be transformed in regard to either or both of the variables.

Suppose, first, that s is a function of the new variable S, (hence also S is a function of x): using subscript numbers to denote differentiations in regard to S, and the accents as before for differentiations in regard to x, we have

$$s' = S's_1,$$

whence, differentiating the logarithms,

$$\frac{s''}{s'} = S'\frac{s_2}{s_1} + \frac{S''}{S'}.$$

Again differentiating, we have

$$\frac{s'''}{s'} - \left(\frac{s''}{s'}\right)^2 = S'^2\left[\frac{s_3}{s_1} - \left(\frac{s_2}{s_1}\right)^2\right] + S''\frac{s_2}{s_1} + \frac{S'''}{S'} - \left(\frac{S''}{S'}\right)^2.$$

But

$$-\tfrac{1}{2}\left(\frac{s''}{s'}\right)^2 = S'^2\left[-\tfrac{1}{2}\left(\frac{s_2}{s_1}\right)^2\right] - S''\frac{s_2}{s_1} \qquad -\tfrac{1}{2}\left(\frac{S''}{S'}\right)^2,$$

and consequently

$$\frac{s'''}{s'} - \tfrac{3}{2}\left(\frac{s''}{s'}\right)^2 = S'^2\left[\frac{s_3}{s_1} - \tfrac{3}{2}\left(\frac{s_2}{s_1}\right)^2\right] \qquad + \frac{S'''}{S'} - \tfrac{3}{2}\left(\frac{S''}{S'}\right)^2;$$

that is,

$$\{s,\ x\} = \left(\frac{dS}{dx}\right)^2 \{s,\ S\} + \{S,\ x\},$$

the required formula.

In a very similar manner, taking x a function of X, it is shown that

$$\{s,\ x\} = \left(\frac{dX}{dx}\right)^2 (\{s,\ X\} - \{x,\ X\}).$$

3. If in this formula we write S for s, and substitute the resulting value of $\{S,\ x\}$ in the former formula, we have

$$\{s,\ x\} = \left(\frac{dS}{dx}\right)^2 \{s,\ S\} - \left(\frac{dX}{dx}\right)^2 \{x,\ X\} + \left(\frac{dX}{dx}\right)^2 \{S,\ X\},$$

which is the formula for the change of both variables. It, in fact, includes the other two: viz. writing $X = x$, or $S = s$, and observing that $\{s,\ s\} = 0 = \{x,\ x\}$, we have the other two formulæ.

4. By putting in the first formula $X = s$, we obtain

$$\{s,\ x\} = -\left(\frac{ds}{dx}\right)^2 \{x,\ s\},$$

a formula for the interchange of the variables.

5. Writing $S = \dfrac{as+b}{cs+d}$, and using for a moment the accents to denote differentiation in regard to s, we have

$$S' = \frac{ad-bc}{(cs+d)^2}, \quad \frac{S''}{S'} = \frac{-2c}{cs+d},$$

and thence

$$\frac{S'''}{S'} - \left(\frac{S''}{S'}\right)^2 = \frac{2c^2}{(cs+d)^2},$$

$$-\tfrac{1}{2}\left(\frac{S''}{S'}\right)^2 = \frac{-2c^2}{(cs+d)^2}.$$

Consequently $\{S,\ s\} = 0$, whence also $\{s,\ S\} = 0$.

Hence in the first formula $\{S,\ x\} = \{s,\ x\}$, that is,

$$\left\{\frac{as+b}{cs+d},\ x\right\} = \{s,\ x\};$$

viz. we may, in the derivative $\{s,\ x\}$, write for s any homographic function $(as+b) \div (cs+d)$ of s.

6. Again, if $X = \dfrac{\alpha x+\beta}{\gamma x+\delta}$, then from the second formula

$$\{s,\ x\} = \frac{(\alpha\delta-\beta\gamma)^2}{(\gamma x+\delta)^4}\{s,\ X\};$$

that is,

$$\left\{s,\ \frac{\alpha x+\beta}{\gamma x+\delta}\right\} = \frac{(\gamma x+\delta)^4}{(\alpha\delta-\beta\gamma)^2}\{s,\ x\};$$

and here, changing s into $(as+b) \div (cs+d)$, we have finally

$$\left\{\frac{as+b}{cs+d},\ \frac{\alpha x+\beta}{\gamma x+\delta}\right\} = \frac{(\gamma x+\delta)^4}{(\alpha\delta-\beta\gamma)^2}\{s,\ x\},$$

which is the formula for the homographic transformation of the two variables s, x.

7. Let s be a given function of x, the equation $\{S,\ x\} = \{s,\ x\}$ is a differential equation of the third order in S, and by what precedes, its general integral is $S = \dfrac{as+b}{cs+d}$.

The direct process is as follows: we have a first integral $\dfrac{S''}{S'} = \dfrac{s''}{s'} - \dfrac{2cs'}{cs+d}$; a second integral $\log S' = \log s' - 2\log(cs+d) + \text{const.}$, that is, $S' = \dfrac{cAs'}{(cs+d)^2}$; and thence a final integral $S = B - \dfrac{A}{cs+d}$, which is equivalent to the foregoing value of S.

The Quadric Function of three or more Inverts. Art. Nos. 8 to 15.

8. We consider a quadric function of any number of inverts $\dfrac{1}{x-\alpha}$, $\dfrac{1}{x-\beta}, \ldots$, all of them different: it is assumed that the constant term is $=0$, and also that the sum of the coefficients of the linear terms is $=0$. We have therefore square terms $\dfrac{a}{(x-\alpha)^2}$, product terms $\dfrac{h}{x-\alpha\,.\,x-\beta}$, and linear terms $\dfrac{A}{x-\alpha}$, where the sum of the coefficients A is $=0$. Any product term $\dfrac{h}{x-\alpha\,.\,x-\beta}$ is expressible in the form of a difference $\dfrac{h}{\alpha-\beta}\dfrac{1}{x-\alpha} - \dfrac{h}{\alpha-\beta}\dfrac{1}{x-\beta}$ of two linear terms, and (the coefficients of these

being equal), after it is thus expressed, the sum of the coefficients of the linear terms is still $=0$. The function is thus always expressible in the form

$$\frac{a}{(x-\alpha)^2}+\frac{b}{(x-\beta)^2}+\ldots+\frac{A}{x-\alpha}+\frac{B}{x-\beta}+\ldots$$

where the sum $A+B+\ldots$ is $=0$: this may be called the reduced form.

9. Observe that any particular invert $\frac{1}{x-\alpha}$ may disappear altogether from the reduced form: this will be the case if $a=0$, that is, if the original form contains no term in $\frac{1}{(x-\alpha)^2}$, and if also $A=0$. An invert thus disappearing from the reduced form is said to be non-essential: and the inverts which do not disappear are said to be essential. The original form contains in appearance the non-essential inverts, but it is really a quadric function of the essential inverts only.

10. Imagine the original function expressed as a rational fraction, the denominator being the product $(x-\alpha)^2(x-\beta)^2(x-\gamma)^2\ldots$ of the squared factors corresponding to all the inverts (non-essential as well as essential): the numerator will be in general of a degree less by 2 than that of the denominator, but the coefficients of any one or more of the higher powers of x may vanish, and the numerator will then be of a lower degree. But this numerator will for any non-essential invert $\frac{1}{x-\gamma}$ contain the factor $(x-\gamma)^2$, or, dividing the numerator and denominator each by this factor, the difference of the degrees of the numerator and denominator will remain unaltered; that is, the difference will have the same value whether we do or do not attend to the non-essential inverts; or say it will have the same value for the original form and for the reduced form.

11. It is to be remarked that the linear terms $\frac{A}{x-\alpha}+\frac{B}{x-\beta}+\frac{C}{x-\gamma}+\ldots$, where $A+B+C+\ldots=0$, can be (and that in a variety of ways) expressed as a sum of differences $\frac{1}{x-\alpha}-\frac{1}{x-\beta}$, that is, as a sum of product-terms $\frac{1}{x-\alpha\,.\,x-\beta}$. Hence the quadric function can be (and that in a variety of ways) expressed as a homogeneous function $\left(\text{a}, \ldots\middle)\middle(\frac{1}{x-\alpha}, \frac{1}{x-\beta}, \ldots\right)^2$; we must have in the form all the essential inverts, and we need have these only. Supposing that this is so, and that the number of the essential inverts is $=n$, then the number of constants is $=\frac{1}{2}n(n+1)$, whereas the number of constants in the reduced form is only $=2n-1$: hence the coefficients are not determinate; or, what is the same thing, we may have different quadric functions having each of them the same reduced function; these quadric functions, as having the same reduced function, can only differ by multiples of the evanescent expressions

$$\frac{\beta-\gamma}{x-\beta\,.\,x-\gamma}+\frac{\gamma-\alpha}{x-\gamma\,.\,x-\alpha}+\frac{\alpha-\beta}{x-\alpha\,.\,x-\beta}, \text{ \&c.}$$

In particular, if the number of essential inverts is $=3$, then the quadric function is of the form

$$\left(\mathrm{a, b, c, f, g, h}\between \frac{1}{x-\alpha}, \frac{1}{x-\beta}, \frac{1}{x-\gamma}\right)^2,$$

which contains one superfluous constant, and equivalent functions differ only by a multiple of

$$\frac{\beta-\gamma}{x-\beta\,.\,x-\gamma}+\frac{\gamma-\alpha}{x-\gamma\,.\,x-\alpha}+\frac{\alpha-\beta}{x-\alpha\,.\,x-\beta}.$$

12. A quadric function such that the degree of the numerator is less by 4 than that of the denominator is said to be "curtate."

The conditions, in order that the function

$$\left(\mathrm{a, b, c, f, g, h}\between \frac{1}{x-\alpha}, \frac{1}{x-\beta}, \frac{1}{x-\gamma}\right)^2$$

may be curtate, are easily found to be

$$\mathrm{a+b+c+2f+2g+2h}=0,$$
$$\alpha\,(\mathrm{a+h+g})+\beta\,(\mathrm{h+b+f})+\gamma\,(\mathrm{g+f+c})=0\,;$$

and by reason of the superfluous constant we are at liberty to assume a third condition: the three conditions may be taken to be $\mathrm{a+h+g}$, $\mathrm{h+b+f}$, $\mathrm{g+f+c}$ each $=0$; and this being so the values of f, g, h are $=\tfrac{1}{2}(\mathrm{a-b-c})$, $\tfrac{1}{2}(\mathrm{-a+b-c})$, $\tfrac{1}{2}(\mathrm{-a-b+c})$ respectively. Hence the form is

$$\left(\mathrm{a, b, c}, \tfrac{1}{2}(\mathrm{a-b-c}), \tfrac{1}{2}(\mathrm{-a+b-c}), \tfrac{1}{2}(\mathrm{-a-b+c})\between \frac{1}{x-\alpha}, \frac{1}{x-\beta}, \frac{1}{x-\gamma}\right)^2,$$

which, as already mentioned, we denote by

$$\left(\mathrm{a, b, c}\ \therefore\between \frac{1}{x-\alpha}, \frac{1}{x-\beta}, \frac{1}{x-\gamma}\right)^2.$$

We have thus the theorem that a curtate function of any number of inverts, but with only the three essential inverts

$$\frac{1}{x-\alpha}, \frac{1}{x-\beta}, \frac{1}{x-\gamma},$$

is always expressible in the foregoing form

$$\left(\mathrm{a, b, c}\ \therefore\between \frac{1}{x-\alpha}, \frac{1}{x-\beta}, \frac{1}{x-\gamma}\right)^2.$$

13. It may be remarked that the function $(\mathrm{a, b, c}\ \therefore\between X, Y, Z)^2$ is a function of the differences of the variables X, Y, Z; and similarly, in the case of four variables, a function $(\mathrm{a, b, c, d, f, g, h, l, m, n}\between X, Y, Z, W)^2$, for which

$$\mathrm{a+h+g+l},\quad \mathrm{h+b+f+m},\quad \mathrm{g+f+c+n},\quad \mathrm{l+m+n+d},$$

are each $=0$, is a function of the differences of the variables X, Y, Z, W: and so in general. Any such function is said to be "diaphoric": and it is easy to see that, taking for the variables any inverts whatever, a diaphoric function is always curtate.

14. The function

$$\left\{-\frac{\mathrm{a}}{(x-\alpha)^2}-\frac{\mathrm{b}}{(x-\beta)^2}-\frac{\mathrm{c}}{(x-\gamma)^2}-\ldots\right\}$$

$$-\tfrac{1}{2}\left\{\frac{\mathrm{a}}{x-\alpha}+\frac{\mathrm{b}}{x-\beta}+\frac{\mathrm{c}}{x-\gamma}+\ldots\right\}^2,$$

where the coefficients a, b, c, ... satisfy the relation $\mathrm{a}+\mathrm{b}+\mathrm{c}+\ldots=-2$, is diaphoric, and therefore curtate. In fact, forming the sum, coeff. $\frac{1}{(x-\alpha)^2}+\frac{1}{2}$ coeff. $\frac{1}{x-\alpha\,.\,x-\beta}+\ldots$, this is $-\mathrm{a}-\frac{1}{2}\mathrm{a}^2-\frac{1}{2}\mathrm{ab}-\frac{1}{2}\mathrm{ac}-\ldots, =-\frac{1}{2}\mathrm{a}(2+\mathrm{a}+\mathrm{b}+\mathrm{c}+\ldots)$, which is $=0$; and similarly the other conditions are satisfied.

15. The function

$$\left(\mathrm{a},\ \mathrm{b},\ \mathrm{c}\ \therefore\!\right)\!\!\left(\frac{a}{x-\alpha}+\frac{a_1}{x-\alpha_1}+\ldots,\ \ \frac{b}{x-\beta}+\frac{b_1}{x-\beta_1}+\ldots,\ \ \frac{c}{x-\gamma}+\frac{c_1}{x-\gamma_1}+\ldots\right)^2,$$

regarded as a function of the inverts

$$\frac{1}{x-\alpha},\ \frac{1}{x-\alpha_1},\ \ldots,\ \frac{1}{x-\beta},\ \ldots$$

where

$$a+a_1+\ldots=b+b_1+\ldots=c+c_1+\ldots,\ =k \text{ suppose},$$

is diaphoric, and therefore curtate. In fact, the condition in regard to $\frac{1}{x-\alpha}$ is

$$\mathrm{a}\,(a^2+aa_1+aa_2+\ldots)+\tfrac{1}{2}(-\mathrm{a}+\mathrm{b}-\mathrm{c})\,(ab+ab_1+\ldots)+\tfrac{1}{2}(-\mathrm{a}-\mathrm{b}+\mathrm{c})\,(ac+ac_1+\ldots)=0;$$

that is,

$$ak\,\{\mathrm{a}+\tfrac{1}{2}(-\mathrm{a}+\mathrm{b}-\mathrm{c})+\tfrac{1}{2}(-\mathrm{a}-\mathrm{b}+\mathrm{c})\}=0,$$

which is satisfied. And similarly the other conditions are satisfied.

The functions P, Q, R. Art. Nos. 16 to 20.

16. We consider P, Q, R, rational and integral functions of z, such that $P+Q+R=0$: hence, using the accent to denote differentiation in regard to z, we have also $P'+Q'+R'=0$; and therefore $QR'-Q'R=RP'-R'P=PQ'-P'Q, =\Theta$ suppose: and we require to find P, Q, R such that the function Θ contains only the factors of P, Q, R.

17. It is to be observed that, effecting upon a solution P, Q, R any linear substitution $(\alpha z+\beta)\div(\gamma z+\delta)$, and omitting the common denominator, we have a solution; but this is regarded as identical with the original solution. The three functions, if

not originally of the same order, can thus be made to be of the same order; or by taking account of the root $z = \infty$, we may in the original case regard them as being of the same order, and it is convenient so to regard them: say they are taken to be of the same order δ. And there is clearly no loss of generality in taking the three functions to be prime to each other; for any common factor of two of them would divide the third, and might therefore be struck out.

18. We may therefore write

$$P = F\Pi(z-l)^p, \quad Q = G\Pi(z-m)^q, \quad R = H\Pi(z-n)^r,$$

where $(z-l)^p$ is taken to denote the distinct simple or multiple factors of P, and the like as regards Q and R; the factors $z-l$, $z-m$, $z-n$ are thus all of them different. And we have $\delta = \Sigma p, = \Sigma q, = \Sigma r$.

19. It is at once seen that Θ is of the degree $2\delta - 2$, and moreover that it contains the factors $\Pi(z-l)^{p-1}$, $\Pi(z-m)^{q-1}$, $\Pi(z-n)^{r-1}$; hence it contains the factor

$$\Pi(z-l)^{p-1}(z-m)^{q-1}(z-n)^{r-1}.$$

Suppose the number of distinct indices p is $=\sigma_1$, that of distinct indices q is σ_2, and that of distinct indices r is σ_3; then the degree of the factor is $= 3\delta - \sigma_1 - \sigma_2 - \sigma_3$; and if this be $= 2\delta - 2$, then Θ can have no other variable factor: viz. if the numbers $\sigma_1, \sigma_2, \sigma_3$ of the distinct indices p, q, r respectively are such that $\sigma_1 + \sigma_2 + \sigma_3 = \delta + 2$, a relation which is henceforth taken to be satisfied, then we have

$$\Theta = K\Pi(z-l)^{p-1}(z-m)^{q-1}(z-n)^{r-1}.$$

As already in effect remarked, the conclusion extends to the case where P, Q, R are not of the same degree; the equation $P + Q + R = 0$ here implies that two functions, say P, Q, are of the same degree, and the third function R of an inferior degree; but, this being so, we have only to regard R as containing the factor $\left(1 - \dfrac{z}{\infty}\right)^t$ of the degree t proper for raising its degree up to that of P or Q.

20. Solutions are given in the following PQR-Table: in which, where required, the proper factor $\left(1 - \dfrac{z}{\infty}\right)^t$ has been added; the first column headed Ref. No. (Reference Number) will be explained further on. The Annex to the same Table will also be explained.

THE *PQR*-TABLE.

Ref. No.	$P=$	$Q=$	$R=$	$\Theta=$
1	z^n	$-1\left(1-\frac{z}{\infty}\right)^n$	$-z^n+1$	$nz^{n-1}\left(1-\frac{z}{\infty}\right)^{n-1}$
2	$4z^n\left(1-\frac{z}{\infty}\right)^n$	$-(z^n+1)^2$	$(z^n-1)^2$	$-4nz^{n-1}(z^n+1)(z^n-1)\left(1-\frac{z}{\infty}\right)^{n-1}$
3	$(z^4+2\sqrt{-3}z^2+1)^3$	$-12\sqrt{-3}z^2(z^4-1)^2\left(1-\frac{z}{\infty}\right)^2$	$-(z^4-2\sqrt{-3}z^2+1)^3$	$24\sqrt{-3}(z^4+2\sqrt{-3}z^2+1)^2$ $(z^4-2\sqrt{-3}z^2+1)z(z^4-1)\left(1-\frac{z}{\infty}\right)$
4	$(z^8+14z^4+1)^3$	$-(z^{12}-33z^8-33z^4+1)^2$	$-108(z^5-z)^4\left(1-\frac{z}{\infty}\right)^4$	$576(z^8+\ldots)^2(z^{12}-\ldots)(z^5-z)^3\left(1-\frac{z}{\infty}\right)^3$
5	$(z^{20}-228z^{15}+494z^{10}+228z^5+1)^3$	$-(z^{30}-522z^{25}-10005z^{20}$ $+0z^{15}-10005z^{10}+522z^5+1)^2$	$-1728(z^{11}+11z^6-z)^5\left(1-\frac{z}{\infty}\right)^5$	$-8640(z^{20}-\ldots)^2(z^{11}+\ldots)^4\left(1-\frac{z}{\infty}\right)^4$
III, V, VII, VIII	$4z\left(1-\frac{z}{\infty}\right)$	$-(z+1)^2$	$(z-1)^2$	$-4(z+1)(z-1)$
IX	$(z-4)^3$	$-(z-1)(z+8)^2$	$27z^2\left(1-\frac{z}{\infty}\right)$	$27(z-4)^2(z+8)z$
X	$z(z+8)^3$	$-(z^2-20z-8)^2$	$-64(z-1)^3\left(1-\frac{z}{\infty}\right)$	$-64(z+8)^2(z^2-20z\ldots)(z-1)$
XI	$4(z^2-z+1)^3$	$-(2z^3-3z^2-3z+2)^2$	$-27z^2(z-1)^2\left(1-\frac{z}{\infty}\right)^2$	$-108(z^2-z\ldots)^2(2z^3-3z^2\ldots)$ $z(z-1)\left(1-\frac{z}{\infty}\right)$
XII	$z^3(z+5)^2(z+8)$	$-(z^3+9z^2+12z-8)^2$	$-64(3z-1)\left(1-\frac{z}{\infty}\right)^5$	$-960z^3(z+5)(z^3+9z^2\ldots)\left(1-\frac{z}{\infty}\right)^4$
XIII	$(z^2+14z+1)^3$	$-(z^3-33z^2-33z+1)^2$	$-108z(z-1)^4$	$-108(z^2+14z\ldots)^2(z^3-33z^2\ldots)(z-1)^3$
XIV	$(64z+189)(64z^2+133z+49)^3$	$-z(4096z^3+18816z^2+25725z+12005)^2$	$-27\,.\,7^7(z+1)^2\left(1-\frac{z}{\infty}\right)^5$	$-135\,.\,7^7(64z^2+\ldots)^2(4096z^3+\ldots)$ $(z+1)\left(1-\frac{z}{\infty}\right)^4$
XV	$-(5z-27)(125z^3-25z^2-265z-243)^3$	$-(-3125z^5+9375z^4+18750z^3$ $+8750z^2+30750z+19683)^2$	$+1382400000z^3(z+1)^2\left(1-\frac{z}{\infty}\right)^5$	$-1382400000(125z^3-\ldots)^2$ $(-3125z^5+\ldots)z^2(z+1)\left(1-\frac{z}{\infty}\right)^4$

In the second half of the table the functions P, Q, R, except in the lines XII, XIV and XV, were calculated by Brioschi: those for the lines XII and XIV were calculated by Klein, but as regards line XV there would seem to have been some error at the beginning of the calculation, and the values found by him are erroneous.

ANNEX TO THE *PQR*-TABLE.

Ref. No.	(a, b, c)*	Calculation of (a*p*), (b*q*), (c*r*).				Ref. No.	
1	$\tfrac{1}{2}(1-n^{-2}),\ \tfrac{1}{2}(1-n^{-2}),\ 0$	an=0	bn=0	c1=0	$\tfrac{1}{2}(1-n^{-2}),\ \tfrac{1}{2}(1-n^{-2}),\ 0$		Polygon
2	$\tfrac{1}{2}(1-n^{-2}),\ \tfrac{3}{8},\ \tfrac{3}{8}$	an=0	b2=0	c2=0	$\tfrac{1}{2}(1-n^{-2}),\ \tfrac{3}{8},\ \tfrac{3}{8}$	I	Double Pyramid
3	$\tfrac{4}{9},\ \tfrac{3}{8},\ \tfrac{4}{9}$	a3=0	b2=0	c3=0	$\tfrac{4}{9},\ \tfrac{3}{8},\ \tfrac{4}{9}$	II	Tetrahedron
4	$\tfrac{4}{9},\ \tfrac{3}{8},\ \tfrac{15}{32}$	a3=0	b2=0	c4=0	$\tfrac{4}{9},\ \tfrac{3}{8},\ \tfrac{15}{32}$	IV	Cube and Octa-hedron
5	$\tfrac{4}{9},\ \tfrac{3}{8},\ \tfrac{12}{25}$	a3=0	b2=0	c5=0	$\tfrac{4}{9},\ \tfrac{3}{8},\ \tfrac{12}{25}$	VI	Dodecahedron and Icosahedron
III	$\tfrac{4}{9},\ \tfrac{3}{8},\ \tfrac{4}{9}$	a1=$\tfrac{4}{9}$, a1=$\tfrac{4}{9}$	b2=0	c2=$\tfrac{5}{18}$	$\left(\tfrac{4}{9},\ \tfrac{4}{9},\ \tfrac{5}{18} \therefore)(\ \tfrac{1}{z},\ \tfrac{1}{z-\infty},\ \tfrac{1}{z-1}\right)^2$		Tetrahedron
V	$\tfrac{15}{32},\ \tfrac{3}{8},\ \tfrac{4}{9}$	a1=$\tfrac{15}{32}$, a1=$\tfrac{15}{32}$	b2=0	c2=$\tfrac{5}{18}$	$\left(\tfrac{15}{32},\ \tfrac{15}{32},\ \tfrac{5}{18} \therefore)(\ ,,\ ,,\ ,,\right)^2$		Cube and Octa-hedron
VII	$\tfrac{4}{9},\ \tfrac{3}{8},\ \tfrac{12}{25}$	a1=$\tfrac{4}{9}$, a1=$\tfrac{4}{9}$	b2=0	c2=$\tfrac{21}{50}$	$\left(\tfrac{4}{9},\ \tfrac{4}{9},\ \tfrac{21}{50} \therefore)(\ ,,\ ,,\ ,,\right)^2$		Dodecahedron and Icosahedron
VIII	$\tfrac{12}{25},\ \tfrac{3}{8},\ \tfrac{4}{9}$	a1=$\tfrac{12}{25}$, a1=$\tfrac{12}{25}$	b2=0	c2=$\tfrac{5}{18}$	$\left(\tfrac{12}{25},\ \tfrac{12}{25},\ \tfrac{5}{18} \therefore)(\ ,,\ ,,\ ,,\right)^2$		,, ,,
IX	$\tfrac{4}{9},\ \tfrac{3}{8},\ \tfrac{12}{25}$	a3=0	b1=$\tfrac{3}{8}$, b2=0	c1=$\tfrac{12}{25}$, c2=$\tfrac{21}{50}$	$\left(\tfrac{3}{8},\ \tfrac{12}{25},\ \tfrac{21}{50} \therefore)(\ \tfrac{1}{z-1},\ \tfrac{1}{z-\infty},\ \tfrac{1}{z}\right)^2$		,, ,,
X	,, ,, ,,	a1=$\tfrac{4}{9}$, a3=0	b2=0	c1=$\tfrac{12}{25}$, c3=$\tfrac{8}{25}$	$\left(\tfrac{4}{9},\ \tfrac{12}{25},\ \tfrac{8}{25} \therefore)(\ \tfrac{1}{z},\ \tfrac{1}{z-\infty},\ \tfrac{1}{z-1}\right)^2$		,, ,,
XI	,, ,, ,,	a3=0	b2=0	c2=$\tfrac{21}{50}$, c2=$\tfrac{21}{50}$, c2=$\tfrac{21}{50}$	$\left(\tfrac{21}{50},\ \tfrac{21}{50},\ \tfrac{21}{50} \therefore)(\ \tfrac{1}{z},\ \tfrac{1}{z-\infty},\ \tfrac{1}{z-1}\right)^2$		,, ,,
XII	,, ,, ,,	a1=$\tfrac{4}{9}$, a2=$\tfrac{5}{18}$, a3=0	b2=0	c1=$\tfrac{12}{25}$, c5=0	$\left(\tfrac{4}{9},\ \tfrac{5}{18},\ \tfrac{12}{25} \therefore)(\ \tfrac{1}{z+8},\ \tfrac{1}{z+5},\ \tfrac{1}{z-\frac{1}{3}}\right)^2$		,, ,,
XIII	,, ,, ,,	a3=0	b2=0	c1=$\tfrac{12}{25}$, c1=$\tfrac{12}{25}$, c4=$\tfrac{9}{50}$	$\left(\tfrac{12}{25},\ \tfrac{12}{25},\ \tfrac{9}{50} \therefore)(\ \tfrac{1}{z},\ \tfrac{1}{z-\infty},\ \tfrac{1}{z-1}\right)^2$		,, ,,
XIV	,, ,, ,,	a1=$\tfrac{4}{9}$, a3=0	b1=$\tfrac{3}{8}$, b2=0	c2=$\tfrac{21}{50}$, c5=0	$\left(\tfrac{4}{9},\ \tfrac{3}{8},\ \tfrac{21}{50} \therefore)(\ \tfrac{1}{z+\frac{189}{64}},\ \tfrac{1}{z},\ \tfrac{1}{z+1}\right)^2$		,, ,,
XV	,, ,, ,,	a1=$\tfrac{4}{9}$, a3=0	b2=0	c2=$\tfrac{21}{50}$, c3=$\tfrac{8}{25}$, c5=0	$\left(\tfrac{4}{9},\ \tfrac{21}{50},\ \tfrac{8}{25} \therefore)(\ \tfrac{1}{z-\frac{27}{5}},\ \tfrac{1}{z+1},\ \tfrac{1}{z}\right)^2$		,, ,,

* Observe, as regards the (a, b, c) column, that the line III agrees with 3; line V with 4 $\left(\text{only there is a transposition of } \tfrac{4}{9} \text{ and } \tfrac{15}{32}\right)$; lines VII and VIII agree each of them with 5 $\left(\text{only as regards VIII there is a transposition of } \tfrac{4}{9} \text{ and } \tfrac{12}{25}\right)$: the remaining lines IX to XV agree each of them with 5. Observe also as regards the Annex generally, the Roman numbering of the lines 2, 3, 4, 5; the lines after the first have thus the Roman numbers I to XV, corresponding to the Roman numbers used by Schwarz.

The Differential Equations $\{x, z\}$ *and* $\{s, x\}$. Art. Nos. 21 to 45.

21. In reference to what follows, it is convenient to put $P = XP_0$, $P' = X_1P_0$, where P_0 is written for $\Pi(z-l)^{p-1}$, the G.C.M. of P and P'; and X is consequently $= F$ multiplied by the product $\Pi(z-l)$ of the several factors taken each with the index unity; and so for Q and R: viz. we write

$$P,\ Q,\ R\ =\ XP_0,\ \ YQ_0,\ \ ZR_0,$$
$$P',\ Q',\ R' = X_1P_0,\ \ Y_1Q_0,\ \ Z_1R_0,$$

and the foregoing value of Θ then is

$$\Theta = KP_0Q_0R_0.$$

We come now to the investigation of the leading theorem. Take a, b, c arbitrary, f, g, $h = b - c$, $c - a$, $a - b$; P, Q, R functions of z as above; and write

$$f(x-a) : g(x-b) : h(x-c) = P : Q : R,$$

equations, which are consistent with each other and determine x as a rational function of z. Using, as before, the accent to denote differentiation in regard to z, and taking the coefficients (a, b, c) arbitrary, it is required to find the value of

$$\{x, z\} + x'^2 \left(\mathrm{a},\ \mathrm{b},\ \mathrm{c}\ \therefore\!\right)\!\!\left(\frac{1}{x-a},\ \frac{1}{x-b},\ \frac{1}{x-c}\right)^2.$$

22. Calculation of the first term $\{x, z\}$.

We have $x =$ a function $\left(\alpha\dfrac{P}{R} + \beta\right) \div \left(\gamma\dfrac{P}{R} + \delta\right)$, and thence $\{x, z\} = \left\{\dfrac{P}{R},\ z\right\}$, $= \{\xi, z\}$ for a moment; then

$$\xi' = \left(\frac{P}{R}\right)' = \frac{RP' - R'P}{R^2},\ \ = \frac{P_0Q_0R_0}{Z^2R_0^{\,2}},\ \ = \frac{P_0Q_0}{Z^2R_0}.$$

Substituting the values

$$P_0 = \Pi(z-l)^{p-1},\quad Q_0 = \Pi(z-m)^{q-1},\quad R_0 = \Pi(z-n)^{r-1},\quad Z = \Pi(z-n),$$

we have

$$\frac{\xi''}{\xi'} = \Sigma\frac{p-1}{z-l} + \Sigma\frac{q-1}{z-m} - \Sigma\frac{r+1}{z-n},$$

and thence

$$\{x, z\} = \left\{-\Sigma\frac{p-1}{(z-l)^2} - \Sigma\frac{q-1}{(z-m)^2} + \Sigma\frac{r+1}{(z-n)^2}\right\}$$
$$-\tfrac{1}{2}\left\{\ \Sigma\frac{p-1}{z-l} + \Sigma\frac{q-1}{z-m} - \Sigma\ \frac{r+1}{z-n}\right\}^2;$$

or say

$$= \left(-\frac{p-1}{(z-l)^2} - \frac{p_1-1}{(z-l_1)^2} - \dots - \frac{q-1}{(z-m)^2} - \frac{q_1-1}{(z-m_1)^2} - \dots + \frac{r+1}{(z-n)^2} + \frac{r_1+1}{(z-n_1)^2} + \dots\right)$$
$$-\tfrac{1}{2}\left(\ \frac{p-1}{z-l} + \frac{p_1-1}{z-l_1} + \dots + \frac{q-1}{z-m} + \frac{q_1-1}{z-m_1} + \dots - \frac{r+1}{z-n} - \frac{r_1+1}{z-n_1} - \dots\right)^2,$$

where it is to be observed, that

$$\Sigma(p-1)+\Sigma(q-1)-\Sigma(r+1), \quad =\delta-\sigma_1+\delta-\sigma_2-(\delta+\sigma_3)=\delta-\sigma_1-\sigma_2-\sigma_3=-2\,;$$

consequently the function is diaphoric, and therefore curtate.

It is to be remarked that the function, although presenting itself in a form unsymmetric in regard to the factors of P and Q, and of R, is really symmetric as regards the three sets of factors; this is obvious *à priori*, and it will be presently verified.

23. For the calculation of the second term

$$x'^2\left(\text{a, b, c} \therefore \right)\!\!\left(\frac{1}{x-a},\ \frac{1}{x-b},\ \frac{1}{x-c}\right)^2,$$

we have

$$f(x-a),\ g(x-b),\ h(x-c)=\Omega P,\ \Omega Q,\ \Omega R,$$

where Ω is a determinate function of z; hence

$$\frac{x'}{x-a},\ \frac{x'}{x-b},\ \frac{x'}{x-c}=\frac{P'}{P}+\frac{\Omega'}{\Omega},\ \frac{Q'}{Q}+\frac{\Omega'}{\Omega},\ \frac{R'}{R}+\frac{\Omega'}{\Omega}.$$

Then substituting these values, by reason that the function is diaphoric, the terms in $\frac{\Omega'}{\Omega}$ disappear, and we have

$$x'^2\left(\text{a, b, c} \therefore \right)\!\!\left(\frac{1}{x-a},\ \frac{1}{x-b},\ \frac{1}{x-c}\right)^2$$

$$=\left(\text{a, b, c} \therefore \right)\!\!\left(\frac{P'}{P},\ \frac{Q'}{Q},\ \frac{R'}{R}\right)^2,$$

which is

$$=\left(\text{a, b, c} \therefore \right)\!\!\left(\Sigma\frac{p}{z-l},\ \Sigma\frac{q}{z-m},\ \Sigma\frac{r}{z-n}\right)^2.$$

We have $\Sigma p=\Sigma q=\Sigma r$, $=\delta$: and hence by what precedes, this function, considered as a function of the inverts $\frac{1}{z-l}$, &c., is diaphoric, and therefore curtate.

24. We have therefore

$$\{x,\ z\}+x'^2\left(\text{a, b, c} \therefore \right)\!\!\left(\frac{1}{x-a},\ \frac{1}{x-b},\ \frac{1}{x-c}\right)^2=$$

$$\left\{-\Sigma\frac{p-1}{(z-l)^2}-\Sigma\frac{q-1}{(z-m)^2}+\Sigma\frac{r+1}{(z-n)^2}\right\}$$

$$-\tfrac{1}{2}\left(\ \Sigma\frac{p-1}{z-l}+\Sigma\ \frac{q-1}{z-m}\ -\Sigma\ \frac{r+1}{z-n}\right)^2$$

$$+\left(\text{a, b, c} \therefore \right)\!\!\left(\Sigma\frac{p}{z-l},\ \Sigma\frac{q}{z-m},\ \Sigma\frac{r}{z-n}\right)^2,$$

where the whole function on the right-hand side is curtate.

25. We have to bring the function on the right-hand side into the reduced form

$$\frac{a}{(z-\alpha)^2}+\ldots+\frac{A}{z-\alpha}+\ldots$$

for the purpose of getting rid of the non-essential inverts (if any).

We write

$$\Sigma\frac{p-1}{(z-l)^2}=\frac{p-1}{(z-l)^2}+\frac{p_1-1}{(z-l_1)^2}+\ldots$$

$$=\frac{p-1}{(z-l)^2}+\Sigma'\frac{p_1-1}{(z-l_1)^2};$$

viz. $z-l$ here denotes any particular factor, and $z-l_1$ represents any other factor of the same set; and so in other like cases.

26. The whole coefficient of $\dfrac{1}{(z-l)^2}$ is

$$-(p-1)-\tfrac{1}{2}(p-1)^2+\mathrm{a}p^2, \quad =\tfrac{1}{2}(1-p^2)+\mathrm{a}p^2;$$

an expression which, regarded as a function of a and p, is represented by $(\mathrm{a}p)$: the parentheses are used only to avoid ambiguity, and are omitted when p is a number, thus $\mathrm{a}1=\mathrm{a}$, $\mathrm{a}2=-\tfrac{3}{2}+4\mathrm{a}$, and so in other cases.

27. The whole term in $\dfrac{1}{z-l}$ comes from

$$-\frac{p-1}{z-l}\left(\Sigma'\frac{p_1-1}{z-l_1}+\Sigma\frac{q-1}{z-m}-\Sigma\frac{r+1}{z-n}\right)$$

$$+\frac{p}{z-l}\left\{2\mathrm{a}\Sigma'\frac{p_1}{z-l}+(-\mathrm{a}-\mathrm{b}+\mathrm{c})\Sigma\frac{q}{z-m}+(-\mathrm{a}+\mathrm{b}-\mathrm{c})\Sigma\frac{r}{z-n}\right\},$$

viz. each term such as $\dfrac{1}{z-l\,.\,z-l_1}$ is to be replaced by $\dfrac{1}{l-l_1}\left(\dfrac{1}{z-l}-\dfrac{1}{z-l_1}\right)$, giving rise to the term $\dfrac{1}{l-l_1}\dfrac{1}{z-l}$, or contributing the term $\dfrac{1}{l-l_1}$ to the coefficient of $\dfrac{1}{z-l}$. The whole coefficient thus is

$$=-(p-1)\left(\Sigma'\frac{p_1-1}{l-l_1}+\Sigma\frac{q-1}{l-m}-\Sigma\frac{r+1}{l-n}\right)$$

$$+2\mathrm{a}p\Sigma'\frac{p_1}{l-l_1}+p(-\mathrm{a}-\mathrm{b}+\mathrm{c})\Sigma\frac{q}{l-m}+p(-\mathrm{a}+\mathrm{b}-\mathrm{c})\Sigma\frac{r}{l-n}.$$

28. Suppose first that $z-l$ is a multiple factor of P, viz. a factor with an index p greater than 1: then, for $z=l$, we have $Q+R=0$, $Q'+R'=0$, and thence $\dfrac{Q'}{Q}=\dfrac{R'}{R}$, that is, $\Sigma\dfrac{q}{l-m}=\Sigma\dfrac{r}{l-n}$. We have therefore

$$p(-\mathrm{a}-\mathrm{b}+\mathrm{c})\Sigma\frac{q}{l-m}+p(-\mathrm{a}+\mathrm{b}-\mathrm{c})\Sigma\frac{r}{l-n}$$

$$=-\mathrm{a}p\left(\Sigma\frac{q}{l-m}+\Sigma\frac{r}{l-n}\right);$$

moreover, in the top line, the terms $\Sigma\frac{q}{l-m}$ and $-\Sigma\frac{r}{l-n}$ destroy each other. The whole coefficient of $\frac{1}{z-l}$, when $z-l$ is a multiple factor of P, thus is

$$= -(p-1)\left(\Sigma'\frac{p_1-1}{l-l_1} - \Sigma\frac{1}{l-m} - \Sigma\frac{1}{l-n}\right)$$

$$+ 2ap\Sigma'\frac{p_1}{l-l_1} - ap\left(\Sigma\frac{q}{l-m} + \Sigma\frac{r}{l-n}\right),$$

a form which is now symmetrical in regard to the inverts $\frac{1}{l-m}$ and $\frac{1}{l-n}$.

29. The value just obtained must be equal to

$$(1-p^2+2ap^2)\left(\Sigma\frac{\frac{1}{2}q-1}{l-m} + \Sigma\frac{\frac{1}{2}r-1}{l-n} - \Sigma'\frac{1}{l-l_1}\right);$$

viz. comparing the two forms and reducing, they will be identical if only

$$(1-p+2ap)\left\{\Sigma'\frac{p+p_1}{l-l_1} - \Sigma\frac{\frac{1}{2}(1+p)q-p}{l-m} - \Sigma\frac{\frac{1}{2}(1+p)r-p}{l-n}\right\} = 0,$$

and it can be shown that the function inside the { } is in fact $=0$.

30. We have, as before, $\Sigma\frac{q}{l-m} = \Sigma\frac{r}{l-n}$; or writing each of these quantities $=\Phi$, the equation to be verified is

$$\Sigma'\frac{p+p_1}{l-l_1} = (p+1)\Phi - p\Sigma\frac{1}{l-m} - p\Sigma\frac{1}{l-n}.$$

We have

$$\frac{P'}{P} = \frac{p}{z-l} + \Sigma'\frac{p_1}{z-l_1}, \quad = \frac{X_1}{X},$$

that is,

$$\Sigma'\frac{p_1}{l-l_1} = \left[\frac{X_1}{X} - \frac{p}{z-l}\right] \text{ for } z=l,$$

$$= \left[\frac{X_1(z-l)-pX}{X(z-l)}\right].$$

The first derived function of the numerator is $X_1'(z-l)+X_1-pX'$, which for $z=l$ is X_1-pX', which is $=0$; and, for the denominator, it is $X'(z-l)+X$, which is also $=0$. Passing to the second derived functions, we find

$$\Sigma'\frac{p_1}{z-l_1} = \frac{2X_1'-pX''}{2X'}, \quad = \frac{X_1'-\frac{1}{2}pX''}{X'}.$$

From the equation

$$\frac{X'}{X} = \frac{1}{z-l} + \Sigma'\frac{1}{z-l}$$

we find in like manner

$$\Sigma' \frac{1}{l-l_1} = \frac{\frac{1}{2}X''}{X'},$$

and we thence obtain (z being always $=l$)

$$\Sigma' \frac{p+p_1}{z-l_1} = \frac{X_1'}{X'},$$

so that the equation to be verified becomes

$$\frac{X_1'}{X_1} = (p+1)\Phi - p\Sigma\frac{1}{l-m} - p\Sigma\frac{1}{l-n}.$$

31. But from the equation Θ, $=PQ'-P'Q$, $=KP_0Q_0R_0$, we find $XY_1 - X_1Y = KR_0$, and then, differentiating, $XY_1' + X'Y_1 - X_1'Y - X_1Y' = KR_0'$: writing in these equations $z=l$, they become

$$-X_1Y = KR_0,$$

$$X'Y_1 - X_1'Y - X_1Y' = KR_0,$$

so that, dividing the second by the first,

$$-\frac{X'}{X_1}\frac{Y_1}{Y} + \frac{X_1'}{X_1} + \frac{Y'}{Y} = \frac{R_0'}{R_0};$$

or, recollecting that $X_1 = pX'$ and $\frac{Y_1}{Y} = \frac{Q'}{Q}$, we have

$$\frac{X_1'}{X'} = p\left(\frac{R_0'}{R_0} - \frac{Y'}{Y}\right) + \frac{Q'}{Q},$$

that is,

$$\frac{X_1'}{X'} = p\left(\Sigma\frac{r-1}{l-n} - \Sigma\frac{1}{l-m}\right) + \Sigma\frac{q}{l-m},$$

$$= (p+1)\Phi - p\Sigma\frac{1}{l-m} - p\Sigma\frac{1}{l-n},$$

the required relation.

32. The result is that, $z-l$ being a multiple factor of P, the coefficient of the term $\frac{1}{z-l}$ is

$$= (1-p^2+2\mathrm{a}p^2)\left\{\Sigma\frac{\frac{1}{2}q-1}{l-m} + \Sigma\frac{\frac{1}{2}r-1}{l-n} - \Sigma'\frac{1}{l-l_1}\right\},$$

$$= 2(\mathrm{a}p)\left[\tfrac{1}{2}\left(\frac{Q'}{Q} + \frac{R'}{R}\right) - \frac{Y'}{Y} - \frac{Z'}{Z} - \tfrac{1}{2}\frac{X''}{X'}\right].$$

33. In the case where $z-l$ is a simple factor of P we have $p=1$, and the coefficient is

$$= 2\mathrm{a}\Sigma'\frac{p_1}{l-l_1} + (-\mathrm{a}-\mathrm{b}+\mathrm{c})\Sigma\frac{q}{l-m} + (-\mathrm{a}+\mathrm{b}-\mathrm{c})\Sigma\frac{r}{l-n},$$

$$= \mathrm{a}\left(2\Sigma'\frac{p_1}{l-l_1} - \Sigma\frac{q}{l-m} - \Sigma\frac{r}{l-n}\right) - (\mathrm{b}-\mathrm{c})\left(\Sigma\frac{q}{l-m} - \Sigma\frac{r}{l-n}\right).$$

34. Of course the formulæ for the coefficients of $\frac{1}{(z-l)^2}$ and $\frac{1}{z-l}$ give at once, by a mere change of letters, those for the coefficients of $\frac{1}{(z-m)^2}$, $\frac{1}{z-m}$, and $\frac{1}{(z-n)^2}$, $\frac{1}{z-n}$; and the function in question,

$$\{x, z\} + x'^2 \left(\text{a, b, c} \;\therefore\!\!\left)\!\!\!\left(\frac{1}{x-a}, \frac{1}{x-b}, \frac{1}{x-c} \right)^2\right.,$$

is now obtained in the required form

$$= \frac{(\text{a}p)}{(z-l)^2} \cdots + \frac{(\text{b}q)}{(z-m)^2} \cdots + \frac{(\text{c}r)}{(z-n)^2} \cdots + \frac{A}{z-l} \cdots + \frac{B}{z-m} \cdots + \frac{C}{z-n} \cdots$$

where $(\text{a}p)$ denotes $\frac{1}{2}(1-p^2) + \text{a}p^2$, and the like for $(\text{b}q)$ and $(\text{c}r)$; and where, $z-l$ being a multiple factor of P, the coefficient A contains the factor $(\text{a}p)$; and similarly for B and C.

35. Suppose that the coefficients a, b, c are no one of them $=0$; we have a1, $=$a, which does not vanish; that is, $z-l$ being a simple factor of P, the expression contains $\frac{1}{(z-l)^2}$, or the invert $\frac{1}{z-l}$ is essential: and similarly, $z-m$ being a simple factor of Q, or $z-n$ a simple factor of R, the inverts $\frac{1}{z-m}$ and $\frac{1}{z-n}$ are essential. But for $z-l$ a multiple factor of P, the coefficient $(\text{a}p)$ of the term $\frac{1}{(z-l)^2}$ may vanish, viz. this will be the case if $\text{a} = \frac{1}{2}\left(1 - \frac{1}{p^2}\right)$; and, when this is so, the coefficient A of the corresponding term $\frac{1}{z-l}$ also vanishes; that is, $\frac{1}{z-l}$ is a non-essential invert. And similarly for any multiple factor $z-m$ of Q or $z-n$ of R, the invert $\frac{1}{z-m}$ or $\frac{1}{z-n}$ may be non-essential.

36. If P, Q, R contain each of them only multiple factors of the same index, say of the indices p, q, r for the three functions respectively, viz. if the functions are $F(\Pi(z-l))^p$, $G(\Pi(z-m))^q$, $H(\Pi(z-n))^r$, the result contains only the six terms written down: and then, if a, b, c are $= \frac{1}{2}\left(1 - \frac{1}{p^2}\right)$, $\frac{1}{2}\left(1 - \frac{1}{q^2}\right)$, $\frac{1}{2}\left(1 - \frac{1}{r^2}\right)$ respectively the result is $=0$: viz. we then have

$$\{x, z\} + x'^2 \left(\text{a, b, c} \;\therefore\!\!\left)\!\!\!\left(\frac{1}{x-a}, \frac{1}{x-b}, \frac{1}{x-c} \right) = 0,\right.$$

or we in fact have, for the values in question of a, b, c, a solution

$$f(x-a) : g(x-b) : h(x-c) = P : Q : R$$

of this differential equation of the third order.

37. The reasoning applies directly to lines 2, 3, 4, 5 of the *PQR*-Table: and with a slight variation to line 1; viz. here the factors of R $(=-1+z^n)$ are all simple factors, but in virtue of $c=0$ and $a=b$, the corresponding inverts disappear, and, the other inverts also disappearing, the value of the function is $=0$. Hence lines 1, 2, 3, 4, 5 of the *PQR*-Table give each of them a result $=0$, for the values of (a, b, c) appearing by the table itself, and shown explicitly in the corresponding line of the Annex.

Thus line 3 shows that the function x, determined by

$$f(x-a) : g(x-b) : h(x-c) = (z^4+2\sqrt{-3}z^2+1)^3 : -12\sqrt{-3}\,(z^5-z)^2 : -(z^4-2\sqrt{-3}z^2+1)^3,$$

satisfies

$$\{x, z\} + x'^2 \left(\frac{4}{9}, \frac{3}{8}, \frac{4}{9} \therefore \between \frac{1}{x-a}, \frac{1}{x-b}, \frac{1}{x-c}\right)^2 = 0,$$

and so for any other of the five lines.

38. The indices of the factors of P, Q, R may be such that, for proper values of the coefficients a, b, c, there are in all only three essential inverts, say $\frac{1}{z-a_1}$, $\frac{1}{z-b_1}$, $\frac{1}{z-c_1}$, belonging to the three functions P, Q, R respectively, or it may be two, or three, of them to the same function. When this is so, the function of these inverts is, by what precedes, a curtate function, and it is consequently a function

$$\left(a_1, b_1, c_1 \therefore \between \frac{1}{z-a_1}, \frac{1}{z-b_1}, \frac{1}{z-c_1}\right)^2,$$

where a_1, b_1, c_1 are the values of the three which do not vanish in the series of expressions (ap), (bq), (cr).

The remaining lines (III, V, VII, VIII) and IX to XV of the *PQR*-Table give such values of P, Q, R, the values of (a, b, c); and the calculation of the values of (a_1, b_1, c_1) is shown by the corresponding lines of the Annex. And we have thus values of x determined by the equations

$$f(x-a) : g(x-b) : h(x-c) = P : Q : R,$$

and giving

$$\{x, z\} + x'^2 \left(a, b, c \therefore \between \frac{1}{x-a}, \frac{1}{x-b}, \frac{1}{x-c}\right)^2 = \left(a_1, b_1, c_1 \therefore \between \frac{1}{z-a_1}, \frac{1}{z-b_1}, \frac{1}{z-c_1}\right)^2.$$

39. For instance, from line IX we have

$$f(x-a) : g(x-b) : h(x-c) = (z-4)^3 : -(z-1)(z+8)^2 : 27z^2\left(1-\frac{z}{\infty}\right),$$

the values of (a, b, c) are $\frac{4}{9}$, $\frac{3}{8}$, $\frac{12}{25}$; and since P, Q, R contain factors with the exponents 3; 1, 2; and 1, 2 respectively, the coefficients which present themselves on the right-hand side are

a3; b1, b2; c1, c2,

which are

$$=0; \quad \frac{3}{8}, \quad 0; \quad \frac{12}{25}, \quad \frac{21}{50} \text{ respectively.}$$

Hence writing $a_1, b_1, c_1 = \frac{3}{8}, \frac{12}{25}, \frac{21}{50}$, the corresponding inverts are $\frac{1}{z-1}, \frac{1}{z-\infty}, \frac{1}{z}$; and the result is

$$\{x, z\} + x'^2 \left(\frac{4}{9}, \frac{3}{8}, \frac{12}{25} \therefore \frac{1}{x-a}, \frac{1}{x-b}, \frac{1}{x-c}\right)^2 = \left(\frac{3}{8}, \frac{12}{25}, \frac{21}{50} \therefore \frac{1}{z-1}, \frac{1}{z-\infty}, \frac{1}{z}\right)^2.$$

40. It is hardly necessary to remark that an expression

$$\left(a_1, b_1, c_1 \therefore \frac{1}{z-a_1}, \frac{1}{z-b_1}, \frac{1}{z-\infty}\right)^2$$

in fact denotes

$$\frac{a_1}{(z-a_1)^2} + \frac{b_1}{(z-b_1)^2} + \frac{-a_1-b_1+c_1}{(z-a_1)(z-b_1)}.$$

The particular form of the z inverts is immaterial; we could by a general linear transformation upon the z make them to be $\frac{1}{z-a_1}, \frac{1}{z-b_1}, \frac{1}{z-c_1}$ with the (a_1, b_1, c_1) arbitrary; or we can give to the a_1, b_1, c_1 any particular values we please: there would be a propriety in making the inverts to be in every case (as in the foregoing example) $\frac{1}{z}, \frac{1}{z-\infty}, \frac{1}{z-1}$; but the numerical work would be troublesome, and it is not worth while to effect it.

41. The conclusion is that lines (III, V, VII, VIII) and IX to XV of the *PQR*-Table, give, for determinate values of (a, b, c) and (a_1, b_1, c_1), solutions

$$f(x-a) : g(x-b) : h(x-c) = P : Q : R$$

of the equation

$$\{x, z\} + x'^2 \left(a, b, c \therefore \frac{1}{x-a}, \frac{1}{x-b}, \frac{1}{x-c}\right)^2 = \left(a_1, b_1, c_1 \therefore \frac{1}{z-a_1}, \frac{1}{z-b_1}, \frac{1}{z-c_1}\right)^2,$$

where a, b, c, a_1, b_1, c_1 are or can be made arbitrary, but without any real gain of generality herein. This is the Differential Equation $\{x, z\}$.

42. Recurring to the results from the Arabic lines of the *PQR*-Table, but for convenience writing s instead of z, we have

$$f(x-a) : g(x-b) : h(x-c) = P : Q : R,$$

where P, Q, R are now functions of s, a solution of

$$\{x, s\} + \left(\frac{dx}{ds}\right)^2 \left(a, b, c \therefore \frac{1}{x-a}, \frac{1}{x-b}, \frac{1}{x-c}\right)^2_x = 0.$$

But we have

$$\{s, x\} = -\left(\frac{ds}{dx}\right)^2 \{x, s\},$$

and the foregoing is therefore a solution of

$$\{s,\ x\} = \left(\mathrm{a},\ \mathrm{b},\ \mathrm{c}\ \therefore\!\!\between \frac{1}{x-a},\ \frac{1}{x-b},\ \frac{1}{x-c}\right)^2,$$

a differential equation of the third order. This is the Differential Equation $\{s, x\}$.

43. From the Roman lines, if we assume

$$f(x-a) : g(x-b) : h(x-c) = \mathfrak{P} : \mathfrak{Q} : \mathfrak{R},$$

where $\mathfrak{P}$, $\mathfrak{Q}$, $\mathfrak{R}$ are functions of z, not the same functions that P, Q, R are of s, since they belong to a different line of the Table: we have, as before,

$$\{x,\ z\} + \left(\frac{dx}{dz}\right)^2 \left(\mathrm{a},\ \mathrm{b},\ \mathrm{c}\ \therefore\!\!\between \frac{1}{x-a},\ \frac{1}{x-b},\ \frac{1}{x-c}\right)^2 = \left(\mathrm{a}_1,\ \mathrm{b}_1,\ \mathrm{c}_1\ \therefore\!\!\between \frac{1}{z-a_1},\ \frac{1}{z-b_1},\ \frac{1}{z-c_1}\right)^2.$$

44. We may combine any such result with a properly selected result of the preceding system, the two results being such that (a, b, c) have the same values in each of them. (See as to this the foot-note referring to the Annex to the *PQR*-Table.) The last equation then becomes

$$\{x,\ z\} + \left(\frac{dx}{dz}\right)^2 \{s,\ x\} = \left(\mathrm{a}_1,\ \mathrm{b}_1,\ \mathrm{c}_1\ \therefore\!\!\between \frac{1}{z-a_1},\ \frac{1}{z-b_1},\ \frac{1}{z-c_1}\right)^2,$$

or since

$$\{x,\ z\} + \left(\frac{dx}{dz}\right)^2 \{s,\ x\} = \{s,\ z\},$$

this is

$$\{s,\ z\} = \left(\mathrm{a}_1,\ \mathrm{b}_1,\ \mathrm{c}_1\ \therefore\!\!\between \frac{1}{z-a_1},\ \frac{1}{z-b_1},\ \frac{1}{z-c_1}\right)^2,$$

the corresponding relation between s, z being of course obtained by the elimination of x from the two sets of equations

$$f(x-a) : g(x-b) : h(x-c) = P : Q : R, \text{ and } f(x-a) : g(x-b) : h(x-c) = \mathfrak{P} : \mathfrak{Q} : \mathfrak{R};$$

viz. the required relation is

$$P : Q : R = \mathfrak{P} : \mathfrak{Q} : \mathfrak{R},$$

where P, Q, R are functions of s; $\mathfrak{P}$, $\mathfrak{Q}$, $\mathfrak{R}$ functions of z; and, in virtue of

$$P + Q + R = 0, \quad \mathfrak{P} + \mathfrak{Q} + \mathfrak{R} = 0,$$

the relations are equivalent to a single equation between z and s. And writing finally x in place of z, that is, now considering $\mathfrak{P}$, $\mathfrak{Q}$, $\mathfrak{R}$ as functions of x, we have

$$\mathfrak{P} : \mathfrak{Q} : \mathfrak{R} = P : Q : R$$

as a solution of

$$\{s,\ x\} = \left(\mathrm{a}_1,\ \mathrm{b}_1,\ \mathrm{c}_1\ \therefore\!\!\between \frac{1}{x-a_1},\ \frac{1}{x-b_1},\ \frac{1}{x-c_1}\right)^2,$$

a differential equation of the third order of the foregoing form, $\{s, x\}$ = given function of x, but with different values of the coefficients, $(\mathrm{a}_1, \mathrm{b}_1, \mathrm{c}_1)$ instead of (a, b, c).

45. It thus appears that there are in all 16 sets of values of (a, b, c), for which the equation is solved, viz. the 16 sets of values are shown in the right-hand column of the Annex. For greater clearness I exhibit the integral equations as follows:

	Functions of x.		Functions of s.		
1	$f(x-a) : g(x-b) : h(x-c)$	=	$P : Q : R$	(1)	Polygon
I	,,	=	,,	(2)	Double Pyramid
II	,,	=	,,	(3)	Tetrahedron
III	$4x : -(x+1)^2 : (x-1)^2$	=	,,	(3)	,,
IV	$f(x-a) : g(x-b) : h(x-c)$	=	,,	(4)	Cube and Octahedron
V	$(x-1)^2 : -(x+1)^2 : 4x$	=	,,	(4)	,,
VI	$f(x-a) : g(x-b) : h(x-c)$	=	,,	(5)	Dodecahedron and Icosahedron
VII	$4x : -(x+1)^2 : (x-1)^2$	=	,,	(5)	,,
VIII	$(x-1)^2 : -(x+1)^2 : 4x$	=	,,	(5)	,,
IX	$P : Q : R$ (IX)	=	,,	(5)	,,
X	,, (X)	=	,,	(5)	,,
XI	,, (XI)	=	,,	(5)	,,
XII	,, (XII)	=	,,	(5)	,,
XIII	,, (XIII)	=	,,	(5)	,,
XIV	,, (XIV)	=	,,	(5)	,,
XV	,, (XV)	=	,,	(5)	,,

The values of the P, Q, R as functions of x, or of s, are taken out of the PQR-Table: only in the lines III, V, VII, VIII, where P, Q, R are given as

$$=4z, \quad -(z+1)^2, \quad (z-1)^2,$$

and where, as regards V and VIII, there is a transposition of P and R, I have inserted the actual values of the x-functions. (See as to this the foot-note referring to the Annex.)

The Schwarzian Theory. Art. Nos. 46 to 62.

46. Considering the foregoing equation

$$\{s,\ x\} = \left(\mathrm{a}_1,\ \mathrm{b}_1,\ \mathrm{c}_1 \ \therefore\!\!\bigg)\!\!\left(\frac{1}{x-a_1},\ \frac{1}{x-b_1},\ \frac{1}{x-c_1}\right)^2\right.$$

as a particular case of the equation $\{s,\ x\}=$ Rational function of x, $=R(x)$ suppose, then we have in 1, I, II, IV, VI solutions of the form $x=$ Rational function of s.

Consider, in general, a solution of this form, $x = F(s)$ a rational function of s: then s is an irrational function of x, and if s_1, s_2 are any two of its values, $\{s_1, x\} = R(x)$, $\{s_2, x\} = R(x)$; that is, $\{s_2, x\} = \{s_1, x\}$, and therefore (ante, No. 7) $s_2 = \dfrac{as_1 + b}{cs_1 + d}$. And then $x = F(s_2) = F\left(\dfrac{as_1 + b}{cs_1 + d}\right)$, $= F(s_1)$: viz. $F(s)$ is a rational function of s, transformable into itself by the transformation s into $\dfrac{as + b}{cs + d}$: and it is moreover clear that between any two roots s whatever of the equation $x = F(s)$ there exists a homographic relation of the form in question. Further, it is clear that these homographic transformations form a group; and consequently that $F(s)$ is a rational function of s, transformable into itself by the several homographic transformations of a group of such transformations: viz. taking x to be a rational function of s, it is *only* in the case $x = F(s)$, a function of the form in question, that $\{s, x\}$ can be equal to a rational function of x.

47. We may, in any equation between x and s, consider these as imaginary variables $p + qi$ and $u + vi$ respectively; considering then (p, q) and (u, v) as rectangular coordinates of points in different planes, we have a first plane the locus of the points x, and a second plane the locus of the points s: there is between the two planes a correspondence which is in fact the correspondence of conformable figures: to the infinitesimal element dx drawn from a point x of the first figure corresponds an infinitésimal element ds drawn from the corresponding point s of the second figure, these elements being in general connected by an equation of the form $ds = (a + bi)\,dx$, where a and b are functions of x or s; and this signifies that, to obtain the pencil of infinitesimal elements or radii ds proceeding in different directions from the point s, we alter in a determinate ratio the absolute lengths of the infinitesimal elements or radii proceeding from the corresponding point x, and rotate the pencil through a determinate angle: this ratio and angle of rotation, or say, the Auxesis and the Streblosis, being of course variable from point to point. Or, what comes to the same thing, if dx and d_1x be consecutive elements of the path of the point x, and ds, d_1s the corresponding consecutive elements of the path of the point s, then the ratio of the lengths of the elements dx, d_1x is equal to that of the lengths of the elements ds, d_1s; and the mutual inclination of the first pair of elements is equal to that of the second pair of elements. In particular, if at any point the path of x is a curved line without abrupt change of direction, then at the corresponding point the path of s is a curved line without abrupt change of direction. In what precedes, we have the relation at ordinary points; but there may be critical corresponding points (x, s), the relation at a critical point between the corresponding elements dx, ds being of the form $ds = (a + bi)(dx)^\lambda$, ($\lambda$ a positive integer or fraction): here the angle between two elements ds is $= \lambda$ times that between the two elements dx; or, if the path of the point x through the critical point is without abrupt change of direction, say if the angle between the two consecutive elements is the flat angle π, then the angle between the two consecutive elements ds is $= \lambda\pi$: viz. there may be in the path of the point s an abrupt change of direction.

48. I consider the foregoing equation $\{s, x\} = R(x)$, where $R(x)$ is a rational function, and is now taken to be a real function of x: we may assume $s' = i\rho'\theta' e^{i\theta}$, where the accents denote differentiation in regard to x, and where ρ', θ, and therefore also θ', are real functions of x. We have

$$\frac{s''}{s'} = \frac{\rho''}{\rho'} + \frac{\theta''}{\theta'} + i\theta',$$

and thence

$$\begin{aligned} \frac{s'''}{s'} - \left(\frac{s''}{s'}\right)^2 &= \frac{\rho'''}{\rho'} - \left(\frac{\rho''}{\rho'}\right)^2 + \frac{\theta'''}{\theta'} - \left(\frac{\theta''}{\theta'}\right)^2 \qquad\qquad + i\theta'' \\ - \tfrac{1}{2}\left(\frac{s''}{s'}\right)^2 &= \quad - \tfrac{1}{2}\left(\frac{\rho''}{\rho'}\right)^2 \quad - \tfrac{1}{2}\left(\frac{\theta''}{\theta'}\right)^2 + \tfrac{1}{2}\theta'^2 - \frac{\rho''\theta''}{\rho'\theta'} - i\theta'' - i\frac{\rho''\theta'}{\rho'}, \end{aligned}$$

and thence

$$\{s, x\} = \{\rho, x\} + \{\theta, x\} + \tfrac{1}{2}\theta'^2 - \frac{\rho''\theta''}{\rho'\theta'} - i\frac{\rho''\theta'}{\rho'}.$$

Putting this $= R(x)$, and assuming that x is real, we have

$$\{\rho, x\} + \{\theta, x\} + \tfrac{1}{2}\theta'^2 - \frac{\rho''\theta''}{\rho'\theta'} = R(x); \quad 0 = i\frac{\rho''\theta'}{\rho'}.$$

The last equation gives $\rho''\theta' = 0$, that is, $\theta' = 0$, which gives $s' = 0$, and may be disregarded; or else $\rho'' = 0$, therefore ρ', a real constant, $= \gamma$ suppose, and $\{\rho, x\} = 0$: hence for the solution of the equation $\{s, x\} = R(x)$, we have $s' = i\gamma\theta' e^{i\theta}$, θ a real quantity determined by $\{\theta, x\} + \tfrac{1}{2}\theta'^2 = R(x)$: and then, integrating the equation for s', we have $s = \alpha + \beta i + \gamma e^{i\theta}$, α, β, γ real constants.

49. The conclusion is that, if $\{s, x\} = R(x)$, a real function of x, and if x be real, that is, if the point x move along a right line (say the x-line), then $s = \alpha + \beta i + \gamma e^{i\theta}$ (θ, and the constants α, β, γ, being real), that is, the point s moves in a circle, coordinates of the centre α, β, and radius $= \gamma$.

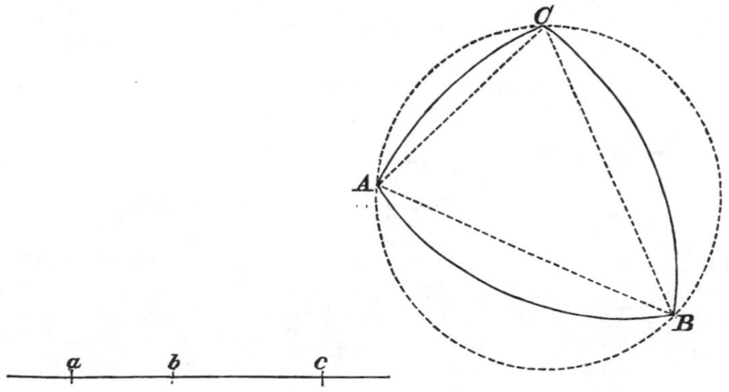

50. Suppose a, b, c are any real values of x representing points a, b, c on the x-line; and A, B, C any given imaginary values of s representing points A, B, C

in the s-plane: since $\{s, x\} = R(x)$ is a differential equation of the third order, the integral contains three arbitrary constants, and we may imagine these so determined that to the values $x = a, b, c$ shall correspond the values $s = A, B, C$ respectively.

If there is not on the x-line any critical point, as the point x moves continuously along this line the point s will move continuously along a circle, which (inasmuch as a, b, c and A, B, C are corresponding points) must be the circle through the three points A, B, C*.

51. If however the points a, b, c are critical points, such that the element ds at the corresponding points A, B, C are equal to multiples of $(dx)^\lambda$, $(dx)^\mu$, $(dx)^\nu$ respectively, then to the flat angles π at a, b, c correspond in the path of s the angles $\lambda\pi$, $\mu\pi$, $\nu\pi$ at the points A, B, C respectively: and, assuming that a, b, c are the only critical points on the x-line, the path of s is made up of the three circular arcs CA, AB, BC meeting at angles $\lambda\pi$, $\mu\pi$, $\nu\pi$ respectively. The arcs are completely determined by these conditions; for supposing the arc BC to make with the chord BC, at the points B and C, the angles f, f, and similarly the arcs CA and AB to make with the corresponding chords the angles g, g and h, h, then the conditions give $\lambda\pi, \mu\pi, \nu\pi = \angle A + g + h, \angle B + h + f, \angle C + f + g$, where the angles referred to are those of the rectilinear triangle ABC: we have thus the values of f, g, h; and the arc BC is the arc on the chord BC meeting it at angles f, f: and the like as regards the arcs CA and AB respectively.

52. The foregoing equation

$$\{s, x\} = \left(\mathrm{a, b, c} \;\therefore\!\right)\!\left(\frac{1}{x-a}, \frac{1}{x-b}, \frac{1}{x-c}\right)^2,$$

where a, b, c have the values $\frac{1}{2}(1-\lambda^2)$, $\frac{1}{2}(1-\mu^2)$, $\frac{1}{2}(1-\nu^2)$, and λ, μ, ν are real and positive, has $x = a, b, c$ for critical points of the kind in question. In fact, writing $x - a = h$, the equation is of the form

$$\{s, h\} = \frac{\frac{1}{2}(1-\lambda^2)}{h^2} + \frac{a_0}{h} + a_1 + a_2 h + \dots,$$

which is satisfied by

$$\frac{d}{dh}\log\frac{ds}{dh} = -\frac{1+\lambda}{h} + b_0 + b_1 h + b_2 h^2 + \dots;$$

we thence obtain an integral of the form

$$s = kh^{-\lambda}(1 + k_1 h + k_2 h^2 + \dots), \quad = k\phi \text{ for shortness.}$$

This is a particular integral, but we have from it the general integral

$$s = \frac{\alpha + \beta k\phi}{\gamma + \delta k\phi}.$$

* Since there is no critical point on the x-line there can be no abrupt change of direction in the path of s, that is, the path of s cannot consist of circular arcs meeting at an angle: but it is in the text further assumed that the path of s cannot consist of different arcs of circle, the one continuing the other without any abrupt change of direction.

If A be the value of s corresponding to $h=0$, then $\beta=\delta A$, and we find

$$s=\frac{\alpha+A\delta k\phi}{\gamma+\delta k\phi}, \quad =\left(A+\frac{\alpha}{\delta k\phi}\right)\left(1+\frac{\gamma}{\delta k\phi}\right)^{-1}, \quad =A+\frac{\alpha-\gamma A}{\delta k}\frac{1}{\phi}+\ldots;$$

viz. reducing $\frac{1}{\phi}$ to its principal term h^λ, and then writing ds, dx for $s-A$, and $h\,(=x-a)$ respectively, we have $ds=K(dx)^\lambda$, or $x=a$ is a critical point with the exponent λ; and similarly $x=b$ and $x=c$ are critical points with the exponents μ and ν respectively.

53. Hence in the equation

$$\{s, x\}=\left(\text{a, b, c } \therefore \!\!\bigg)\!\left(\frac{1}{x-a}, \frac{1}{x-b}, \frac{1}{x-c}\right)^2,$$

as the point x, passing successively through a, b, c, describes the x-line, the point s, passing successively through A, B, C, describes the sides AB, BC, CA of the curvilinear triangle ABC. To points x indefinitely near the x-line correspond points s indefinitely near the boundary AB, BC, CA of the triangle, viz. to points x indefinitely near to and on one side, suppose the upper side, of the x-line, correspond the points s indefinitely near to and within the boundary of the triangle: and in like manner to whole series of the points x on the same upper side of the x-line, correspond the whole series of points s inside the triangle.

54. We have attended so far only to one of the points s which correspond to a given point x, but considering the set of points s which correspond to the same point x, we have in the s-plane entire circles forming by their intersections curvilinear triangles ABC, ABC', &c.; we have thus two systems, say ABC, &c., and ABC', &c., of triangles, such that to a point x on the upper side of the x-line correspond points s, one of them within each of the triangles ABC, &c., and to a point x on the lower side of the x-line correspond points s, one of them within each of the triangles ABC', &c.; and so consequently that, to the two half-planes on opposite sides of the x-line, correspond the two sets of triangles ABC, &c., and ABC', &c., respectively.

55. In order that the relation s and x may be an algebraical one, it is necessary that the two sets of triangles should completely cover, once or a finite number of times, the whole of the s-plane: and this implies that the angles $\lambda\pi$, $\mu\pi$, $\nu\pi$ have certain determinate values; and, in fact, that dividing the surface of a sphere into triangles, each with these angles, the curvilinear triangles ABC, ABC', &c., are the stereographic projections of these triangles. It was by such considerations as these that Schwarz, in the Memoir of 1873, p. 323, obtained the series of values I to XV of λ, μ, ν, giving for a, b, c, $=\frac{1}{2}(1-\lambda^2)$, $\frac{1}{2}(1-\mu^2)$, $\frac{1}{2}(1-\nu^2)$, the series of values mentioned in the Annex of the PQR-Table: and thus showed *à priori* that the equation

$$\{s, x\}=\left(\text{a, b, c } \therefore \!\!\bigg)\!\left(\frac{1}{x-a}, \frac{1}{x-b}, \frac{1}{x-c}\right)^2$$

is algebraically integrable for these values of a, b, c; and only for these values, or for values reducible to them.

56. As an instance, take the double pyramid form: the integral equation is

$$f(x-a) : g(x-b) : h(x-c) = 4s^n : -(s^n-1)^2 : (s^n+1)^2,$$

or say

$$\frac{(c-a)(x-b)}{(a-b)(x-c)} = -\frac{(s^n-1)^2}{(s^n+1)^2};$$

or if, for greater simplicity, we assume $a, b, c = 1, 0, \infty$, this is $x = \frac{(s^n-1)^2}{(s^n+1)^2}$, or say $-(s^n-1) = \sqrt{x}(s^n+1)$, that is, $s^n = \frac{1 \mp \sqrt{x}}{1 \pm \sqrt{x}}$, a solution of the differential equation

$$\{s, x\} = \left(\frac{3}{8},\ \tfrac{1}{2}(1-n^{-2}),\ \frac{3}{8}\ \therefore \left(\frac{1}{x},\ \frac{1}{x-1},\ \frac{1}{x-\infty}\right)^2.$$

In particular, if $n=3$, we have $x = \left(\frac{s^3-1}{s^3+1}\right)^2$ or $s^3 = \frac{1 \mp \sqrt{x}}{1 \pm \sqrt{x}}$, a solution of

$$\{s, x\} = \left(\frac{3}{8},\ \frac{4}{9},\ \frac{3}{8}\ \therefore \left(\frac{1}{x},\ \frac{1}{x-1},\ \frac{1}{x-\infty}\right)^2.$$

57. We have here the spherical surface divided by the equator and three meridians into twelve triangles, each with the angles $\frac{1}{2}\pi$, $\frac{1}{2}\pi$, $\frac{1}{3}\pi$: and then, projecting from the South pole on the plane of the equator, we have the annexed figure of the s-plane,

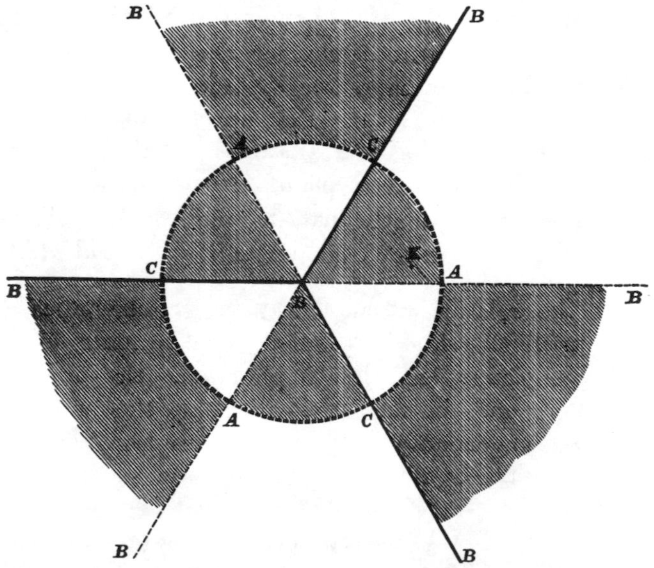

divided into 12 curvilinear triangles, each with these same angles 90°, 90°, 60°; the plane is divided by the shading into two systems, each of 6 triangles. The figure of the x-plane is by the x-line divided into two half-planes, one shaded, the other unshaded; and we have on the line the point c at ∞, a at the origin, and b at the distance unity.

58. Take x real; then, if x is positive and less than 1, s^3 is real and positive, and we have for s the infinite half-lines at the inclinations 0°, 120°, 240°, while if x is positive and greater than 1, s^3 is real and negative, and we have the infinite half-lines at the inclinations 60°, 180°, 300°. If x is real and negative, then s^3 is of the form $\frac{1-ki}{1+ki}$, $=\cos\theta+i\sin\theta$; whence s is of the same form, or the locus of the point s is a circle radius unity. Writing $s^3=\frac{1-\sqrt{x}}{1+\sqrt{x}}$, and supposing that the point x moves along the x-line from b through a to c at $-\infty$, and then from c at $+\infty$ to b, the point s describes the sides BA, AC, CB of the shaded triangle marked K.

59. Suppose that the point x is at k, in the shaded half-plane at an indefinitely small distance from a; say we have $x=-2\kappa^2 i$, (κ small), then taking for $\sqrt{x}$ the value $\kappa(1-i)$, we have $s^3=\frac{1-\kappa(1-i)}{1+\kappa(1-i)}$, $=1-2\kappa(1-i)$ nearly, and hence a value of s is $=1-\frac{2}{3}\kappa+\frac{2}{3}\kappa i$, which belongs to a point K near A, and within the shaded triangle: we have thus, in respect of this value of s, the shaded half of the x-plane corresponding to this shaded triangle. To the same value $x=-2\kappa^2 i$, correspond in all six values of s, giving six points K each lying near a point A within one of the shaded triangles; and hence the shaded half-plane corresponds to the six shaded triangles, and the unshaded half-plane corresponds to the six unshaded triangles.

60. Suppose the equation is

$$\{s,\ x\}=\left(\mathrm{a},\ \mathrm{b},\ \mathrm{c}\ \therefore\right)\!\!\left(\frac{1}{x-a},\ \frac{1}{x-b},\ \frac{1}{x-c}\right)^2,$$

that is,

$$=\frac{-(b-c)(c-a)(a-b)}{x-a\,.\,x-b\,.\,x-c}\left(\frac{\mathrm{a}}{b-c\,.\,x-a}+\frac{\mathrm{b}}{c-a\,.\,x-b}+\frac{\mathrm{c}}{a-b\,.\,x-c}\right),$$

where a, b, c are real, but a, b, c are imaginary. It is to be shown that, if the path of x is the circle passing through the points a, b, c, then the path of s is a circle passing through the corresponding three points.

61. We may find α, β, γ, θ_0, θ_1, θ_2, such that a, b, c are $=\alpha+\beta i+\gamma e^{\theta_0 i}$, $\alpha+\beta i+\gamma e^{\theta_1 i}$, $\alpha+\beta i+\gamma e^{\theta_2 i}$ (this is, in fact, finding α and β the coordinates of the centre, and γ the radius of the circle through the three points a, b, c): we then have $x=\alpha+\beta i+\gamma e^{\theta i}$, θ a variable parameter, the equation which expresses that the point x is situate on the circle in question.

We have $x-a=\gamma(e^{\theta i}-e^{-\theta_0 i})$, $=\gamma e^{\frac{1}{2}(\theta+\theta_0)}\{e^{\frac{1}{2}(\theta-\theta_0)i}-e^{-\frac{1}{2}(\theta-\theta_0)i}\}$; the second factor is $i\sin\frac{1}{2}(\theta-\theta_0)$, $=iP$ suppose, or the equation is $x-a=iP\gamma e^{\frac{1}{2}(\theta+\theta_0)i}$, say

$$x-a=iP\gamma\,\mathrm{expi}\,\tfrac{1}{2}(\theta+\theta_0).$$

Similarly $x-b=iQ\gamma\,\mathrm{expi}\,\frac{1}{2}(\theta+\theta_1)$, and $x-c=iR\gamma\,\mathrm{expi}\,\frac{1}{2}(\theta+\theta_2)$; where P, Q, R denote $\sin\frac{1}{2}(\theta-\theta_0)$, $\sin\frac{1}{2}(\theta-\theta_1)$, $\sin\frac{1}{2}(\theta-\theta_2)$ respectively. In like manner, we have $b-c$, $c-a$, $a-b$, $=iF\gamma\,\mathrm{expi}\,\frac{1}{2}(\theta_1+\theta_2)$, $iG\gamma\,\mathrm{expi}\,\frac{1}{2}(\theta_2+\theta_0)$, $iH\gamma\,\mathrm{expi}\,\frac{1}{2}(\theta_0+\theta_1)$, where F, G, H denote $\sin\frac{1}{2}(\theta_1-\theta_2)$, $\sin\frac{1}{2}(\theta_2-\theta_0)$, $\sin\frac{1}{2}(\theta_0-\theta_1)$ respectively.

We have

$$-\frac{b-c\,.\,c-a\,.\,a-b}{x-a\,.\,x-b\,.\,x-c}=\frac{-FGH}{PQR}\operatorname{expi}\tfrac{1}{2}(\theta_0+\theta_1+\theta_2-3\theta),$$

$$\frac{1}{b-c\,.\,x-a}=\frac{-1}{\gamma^2 PF}\operatorname{expi}-\tfrac{1}{2}(\theta_0+\theta_1+\theta_2+\theta),$$

with the like values for $\dfrac{1}{c-a\,.\,x-b}$ and $\dfrac{1}{a-b\,.\,x-c}$. Hence the right-hand side of the equation is

$$=\frac{FGH}{PQR}\left(\frac{\mathrm{a}}{PF}+\frac{\mathrm{b}}{QG}+\frac{\mathrm{c}}{RH}\right)\operatorname{expi}(-2\theta).$$

62. Considering now the left-hand side of the equation, we have

$$\{s,\ x\}=\frac{1}{\left(\dfrac{dx}{d\theta}\right)^2}(\{s,\ \theta\}-\{x,\ \theta\});$$

substituting for x its value $=\alpha+\beta i+\gamma e^{\theta i}$, this becomes

$$\{s,\ x\}=-\frac{1}{\gamma^2}e^{-2\theta i}(\{s,\ \theta\}-\tfrac{1}{2}),$$

that is,

$$=-\frac{1}{\gamma^2}(\{s,\ \theta\}-\tfrac{1}{2})\operatorname{expi}(-2\theta).$$

Assume $s=L+Mi+Ne^{\Theta i}$, L, M, and N constants; then using the accent to denote differentiation in regard to θ, we find without difficulty $\{s,\ \theta\}=\{\Theta,\ \theta\}+\tfrac{1}{2}\Theta'^2$, and the value of $\{s,\ x\}$ becomes

$$=-\frac{1}{\gamma^2}\left(\{\Theta,\ \theta\}+\tfrac{1}{2}\Theta'^2-\tfrac{1}{2}\right)\operatorname{expi}(-2\theta).$$

Hence, substituting the values of the two sides of the equation, the imaginary factor $\operatorname{expi}(-2\theta)$ divides out, and the equation becomes

$$\{\Theta,\ \theta\}+\tfrac{1}{2}\Theta'^2-\tfrac{1}{2}=-\frac{FGH}{PQR}\left(\frac{\mathrm{a}}{PF}+\frac{\mathrm{b}}{QG}+\frac{\mathrm{c}}{RH}\right),$$

an equation, in which everything is real and which thus determines Θ as a real function of θ: and we have therefore the theorem in question.

Connexion with the differential equation for the hypergeometric series. Art. Nos. 63 to 68.

63. Take p, q given functions of x, and y a function of x determined by the equation

$$\frac{d^2y}{dx^2}+p\frac{dy}{dx}+qy=0;$$

again P, Q given functions of z, and v a function of z determined by the equation

$$\frac{d^2v}{dz^2}+P\frac{dv}{dz}+Qv=0\,;$$

and assume

$$y=wv.$$

Substituting this value of y in the first equation, we obtain for v an equation of the second order (the coefficients of which contain w), and we may make this identical with the second equation; viz. comparing the coefficients of the two equations, we thus have two equations each containing w; and by eliminating w we obtain a differential equation of the third order between z and x. This is, in fact, the basis of Kummer's theory for the transformation of a hypergeometric series: the equation between z, x will be found presently in a different manner.

64. But if with Schwarz, instead of making the equation obtained for v as above identical with the given equation for v, we merely assume that the two equations are consistent, then there is nothing to determine the value of z, which may be regarded as an arbitrary function of x; y and v are then functions of x, and w denotes the quotient $y \div v$ of these two functions, and as such satisfies an equation the form of which will depend on the assumed relation between z and x. In particular, if P and Q denote the same functions of z that p and q are of x; and if we assume $z=x$, P, Q will become $=p$, q respectively: the given equation in v will be

$$\frac{d^2v}{dx^2}+p\frac{dv}{dx}+qv=0\,;$$

and w will thus denote the quotient of any two solutions of the equation

$$\frac{d^2y}{dx^2}+p\frac{dy}{dx}+qy=0\,;$$

viz. writing $X=p^2+2\dfrac{dp}{dx}-4q$, then, by what precedes, the equation for w will be

$$\{w,\ x\}=-\tfrac{1}{2}X.$$

65. Returning now to Kummer's problem, and considering y, v as solutions of the two differential equations respectively, w is a function independent of the particular solutions denoted by these letters: we have $y=wv$, and taking any other two solutions we have $y_1=wv_1$, so that $\dfrac{y}{y_1}=\dfrac{v}{v_1}$; calling each of these equal quantities s, we have s denoting the quotient of two solutions of the equation in y, and also the quotient of two solutions of the equation in v; whence, writing as before $X=p^2+2\dfrac{dp}{dx}-4q$, and similarly $Z=P^2+2\dfrac{dP}{dz}-4Q$, we have

$$\{s,\ x\}=-\tfrac{1}{2}X,\quad \{s,\ z\}=-\tfrac{1}{2}Z,$$

and since in general

$$\{s,\ x\}=\left(\frac{dz}{dx}\right)^2\{s,\ z\}+\{z,\ x\},$$

we obtain

$$\{z,\ x\} = -\tfrac{1}{2}X + \tfrac{1}{2}Z\left(\frac{dz}{dx}\right)^2,$$

as the required equation for the determination of z as a function of x. The process does not give the value of w, but this can be found without difficulty, viz.

$$w^2 = Ce^{\int Pdz - \int pdx} \div \frac{dz}{dx}.$$

If z, x are regarded each of them as a function of the new independent variable θ, then the equation is

$$\{z,\ \theta\} - \tfrac{1}{2}\left(\frac{dz}{d\theta}\right)^2 Z = \{x,\ \theta\} - \tfrac{1}{2}\left(\frac{dx}{d\theta}\right)^2 X.$$

66. Jacobi's differential equation of the third order for the transformed modulus λ, *Fund. Nova*, p. 78, [*Ges. Werke*, t. I, p. 132], is

$$3\,(k'^2\lambda''^2 - \lambda'^2k''^2) - 2k'\lambda'\,(k'\lambda''' - \lambda'k''') + k'^2\lambda'^2\left\{\left(\frac{1+k^2}{k-k^3}\right)^2 k'^2 - \left(\frac{1+\lambda^2}{\lambda-\lambda^3}\right)^2\lambda'^2\right\} = 0,$$

where the accents denote differentiations in regard to an independent variable θ: viz. dividing by $2k'^2\lambda'^2$, this becomes

$$\{k,\ \theta\} + \tfrac{1}{2}k'^2\left(\frac{1+k^2}{k-k^3}\right)^2 = \{\lambda,\ \theta\} + \tfrac{1}{2}\lambda'^2\left(\frac{1+\lambda^2}{\lambda-\lambda^3}\right)^3,$$

which is thus a particular case of Kummer's equation, k, λ corresponding to x, z respectively, and the values of X, Z being

$$X = -\left(\frac{1+k^2}{k-k^3}\right)^2,\quad Z = -\left(\frac{1+\lambda^2}{\lambda-\lambda^3}\right)^3.$$

67. In the case of the hypergeometric series, the two differential equations of the second order are

$$\frac{d^2y}{dx^2} + \frac{\gamma - (\alpha+\beta+1)\,x}{x\,.\,1-x}\,\frac{dy}{dx} - \frac{\alpha\beta y}{x\,.\,1-x} = 0,$$

$$\frac{d^2v}{dz^2} + \frac{\gamma' - (\alpha'+\beta'+1)\,z}{z\,.\,1-z}\,\frac{dv}{dz} - \frac{\alpha'\beta' v}{z\,.\,1-z} = 0.$$

Hence

$$p = \frac{\gamma\,(x + (1-x)) - (\alpha+\beta+1)\,x}{x\,.\,1-x} = \frac{\gamma}{x} + \frac{\gamma-\alpha-\beta-1}{1-x},\quad q = \frac{-\alpha\beta}{x\,.\,1-x},$$

and hence

$$p^2 + 2\frac{dp}{dx} - 4q = \frac{\gamma^2 - 2\gamma}{x^2} + \frac{(\gamma-\alpha-\beta-1)^2 + 2\,(\gamma-\alpha-\beta-1)}{(1-x)^2} + \frac{4\alpha\beta + 2\gamma\,(\gamma-\alpha-\beta-1)}{x\,.\,1-x};$$

viz. writing

$$\begin{aligned} \lambda^2 &= (1-\gamma)^2, & \mathrm{a} &= \tfrac{1}{2}(1-\lambda^2),\\ \mu^2 &= (\alpha-\beta)^2, & \mathrm{b} &= \tfrac{1}{2}(1-\mu^2),\\ \nu^2 &= (\gamma-\alpha-\beta)^2, & \mathrm{c} &= \tfrac{1}{2}(1-\nu^2), \end{aligned}$$

and putting in the formula $x-1, =-(1-x)$, we have

$$-\tfrac{1}{2}\left(p^2+2\frac{dp}{dx}-4q\right)=\frac{\frac{1}{2}(1-\lambda^2)}{x^2}+\frac{\frac{1}{2}(1-\nu^2)}{(x-1)^2}+\frac{\frac{1}{2}(\lambda^2-\mu^2+\nu^2-1)}{x\,.\,x-1},$$

$$=\frac{\mathrm{a}}{x^2}+\frac{\mathrm{c}}{(x-1)^2}+\frac{-\mathrm{a}+\mathrm{b}-\mathrm{c}}{x\,.\,x-1},$$

$$=\left(\mathrm{a},\ \mathrm{b},\ \mathrm{c}\ \therefore\right)\!\left(\frac{1}{x},\ \frac{1}{x-\infty},\ \frac{1}{x-1}\right)^2,$$

with a like formula for $\frac{1}{2}\left(P^2+2\frac{dP}{dz}-4Q\right)$. We then have

$$y=wv,$$

$$w^2=Cx^{-\gamma}(1-x)^{\gamma-\alpha-\beta-1}z^{\gamma'}(1-z)^{-\gamma'+\alpha'+\beta'+1}\frac{dx}{dz},$$

and the differential equation of the third order for the determination of z is

$$\{z,\ x\}+\left(\mathrm{a}_1,\ \mathrm{b}_1,\ \mathrm{c}_1\ \therefore\right)\!\left(\frac{1}{z},\ \frac{1}{z-\infty},\ \frac{1}{z-1}\right)^2\left(\frac{dz}{dx}\right)^2-\left(\mathrm{a},\ \mathrm{b},\ \mathrm{c}\ \therefore\right)\!\left(\frac{1}{x},\ \frac{1}{x-\infty},\ \frac{1}{x-1}\right)^2=0,$$

where a_1, b_1, c_1 are the same functions of α', β', γ' which a, b, c are of α, β, γ. This is, in effect, Kummer's equation for the transformation of the hypergeometric series.

68. And in like manner the Schwarzian equation for the determination of s, the quotient of two solutions, is

$$\{s,\ x\}=\left(\mathrm{a},\ \mathrm{b},\ \mathrm{c}\ \therefore\right)\!\left(\frac{1}{x},\ \frac{1}{x-\infty},\ \frac{1}{x-1}\right)^2.$$

PART II. THE POLYHEDRAL FUNCTIONS.

Origin and Properties. Art. Nos. 69 to 80.

69. The functions in lines 1, ..., 5 of the *PQR*-Table are connected with the geometrical forms:

{1. Polygon or
{2. Double Pyramid *,
3. Tetrahedron,
4. Octahedron and Cube,
5. Dodecahedron and Icosahedron,

(these figures being regarded as situate on a spherical surface), and with the stereographic projections of these figures.

* Prof. Klein regards 1 as belonging to the polygon and 2 to the double pyramid: it seems to me that the fundamental figure, to which 1 and 2 each of them belong, is the polygon.

Consider a spherical surface and upon it any number of points: take at pleasure any point as South Pole, this determines the plane of the equator; and the stereographic projection of any point is the intersection with the plane of the equator of the line joining the point with the South Pole.

To fix the ideas take the radius of the sphere as unity: let the axes of x and y be drawn in the plane of the equator in longitudes 0° and 90° respectively, and the axis of z upwards through the North Pole: the position of a point on the sphere is determined by means of its N.P.D. θ and longitude f: moreover we take X, Y, Z for the coordinates of the point on the surface, and x, y for those of its projection; and we then have

$$X,\ Y,\ Z = \sin\theta\cos f,\ \sin\theta\sin f,\ \cos\theta;$$

$$x = \frac{X}{1+Z} = \tan\tfrac{1}{2}\theta\cos f,\quad y = \frac{Y}{1+Z} = \tan\tfrac{1}{2}\theta\sin f,$$

and conversely,

$$X,\ Y,\ Z = 2x,\ 2y,\ 1 - x^2 - y^2,\ \div (1 + x^2 + y^2).$$

We represent the point (X, Y, Z) on the spherical surface by means of the magnitude $x+iy$, $=\tan\frac{1}{2}\theta(\cos f + i\sin f)$, or say by the linear factor, $s-(x+iy)$: and similarly any system of points on the surface by means of the system of magnitudes $x+iy$, or say by the function $\Pi\{s-(x+iy)\}$, denoting in this manner the product of the linear factors which correspond to the different points respectively.

70. It will presently appear that, if (considering a different stereographic projection, that is, a different position of the South Pole) we take x', y' as the coordinates of the new projection of the point, then $x'+iy'$ is a homographic function

$$a(x+iy)+b \div \{c(x+iy)+d\}$$

of $x+iy$: and consequently that the functions of s, which belong to different projections, are linear transformations one of the other: but at present we consider a single projection.

It may be proper to remark that the figures in question are spherical figures having summits which are points on the spherical surface, edges (or sides) which are arcs of great circle joining two summits, and faces which are portions of the spherical surface: the mid-points of the sides, and the centres of the faces are of course points on the spherical surface.

71. (1), (2). Considering a regular polygon formed by n summits on the equator, the longitude of one of them being 0°, then the stereographic projections correspond with the points themselves, and the values of $x+iy$ are

$$1,\ \cos\frac{2\pi}{n} + i\sin\frac{2\pi}{n},\ \ldots,\ \cos\frac{(n-1)\,2\pi}{n} + i\sin\frac{(n-1)\,2\pi}{n}.$$

The corresponding function of s is $s^n - 1$.

The values of $x+iy$ for the mid-points of the sides are

$$\cos\frac{\pi}{n}+i\sin\frac{\pi}{n},\quad \cos\frac{3\pi}{n}+i\sin\frac{3\pi}{n},\ \ldots,\ \cos\frac{(2n-1)\pi}{n}+i\sin\frac{(2n-1)\pi}{n}.$$

The corresponding function of s is s^n+1.

The North and South Poles, which form with the n points a double pyramid of $n+2$ summits, correspond to the values $s=0$ and $s=\infty$. We have thus

$$s\left(1-\frac{s}{\infty}\right)(s^n-1)$$

as the function corresponding to the double pyramid.

72. (3). Considering for a moment the tetrahedron as a figure with rectilinear edges, this is so placed that two opposite edges are horizontal, and that the vertical planes passing through the centre and these two edges respectively are inclined at angles $\pm 45°$ to the meridian: viz. the upper edge has the longitudes 135°, 315°, and the lower edge the longitudes 45°, 225°. We thus explain the position of the spherical figure.

Corresponding to the summits we have the function $s^4-2i\sqrt{3}\,s^2+1$.

In fact, the equation $s^4-2i\sqrt{3}\,s^2+1=0$ gives $s^2=i(\sqrt{3}\pm 2)$, and hence the values of s are the four values of $x+iy$ shown in the annexed table for the values of X, Y, Z, and $x+iy$ for the summits of the tetrahedron,

long.	X	Y	Z	$x+iy$
45°	$\frac{1}{\sqrt{3}}$	$\frac{1}{\sqrt{3}}$	$-\frac{1}{\sqrt{3}}$	$\frac{1+i}{\sqrt{3}-1}$
135°	$-$	$+$	$+$	$\frac{-1+i}{\sqrt{3}+1}$
225°	$-$	$-$	$-$	$\frac{-1-i}{\sqrt{3}-1}$
315°	$+$	$-$	$+$	$\frac{1+i}{\sqrt{3}+1}$.

Corresponding to the centres of the faces, or summits of the opposite tetrahedron, we have the function $s^4+2i\sqrt{3}\,s^2+1$.

Corresponding to the mid-points of the sides, we have the function

$$s\left(1-\frac{s}{\infty}\right)(s^4-1);$$

viz. the points in question are the North Pole $s=0$, the South Pole $s=\infty$, and the four points $s=\pm 1$, $s=\pm i$ on the equator at longitudes 0°, 90°, 180°, 270° respectively.

73. (4). The octahedron is placed with two of its summits as poles, and the other four summits in the equator at longitudes 0°, 90°, 180°, 270° respectively: the values of s are, as in the last case, 0, ∞, ± 1, $\pm i$, and the function is

$$s\left(1-\frac{s}{\infty}\right)(s^4-1).$$

The function for the centres of the faces, or summits of the cube, is s^8+14s^4+1.

The function for the mid-points of the sides of the octahedron or of the cube is

$$s^{12}-33s^8-33s^4+1.$$

74. (5). The Icosahedron is placed with two of its summits for poles; five summits lying in a small circle above the plane of the equator at longitudes 0°, 72°, 144°, 288°, and the remaining five summits in the corresponding small circle below the equator at longitudes 36°, 108°, 180°, 252° and 324°.

The function for the summits of the Icosahedron is

$$s\left(1-\frac{s}{\infty}\right)(s^{10}+11s^5-1).$$

The function for the centres of the faces of the Icosahedron, or summits of the Dodecahedron, is $s^{20}-228s^{15}+494s^{10}+228s^5-1$.

The function for the mid-points of the sides of the Icosahedron or the Dodecahedron is

$$s^{30}-522s^{25}+10005s^{20}+0s^{15}-10005s^{10}+522s^5+1.$$

I give for the present these results without demonstration.

75. Writing $\frac{x}{y}$ for s so as to obtain homogeneous functions $(*\between x, y)^n$,—it will be recollected that the x, y of these functions have nothing to do with the x, y of the foregoing values $x+iy$—the forms which have thus presented themselves may be denoted as follows:

$$\begin{aligned}
(3):\quad & f3=(1,\ -2i\sqrt{3},\ 1\between x^2,\ y^2)^2, \\
& h3=(1,\ +2i\sqrt{3},\ 1\between x^2,\ y^2)^2, \\
& t3=xy\,(x^4-y^4), \\
(4):\quad & f4=xy\,(x^4-y^4), \\
& h4=(1,\ 14,\ 1\between x^4,\ y^4)^2, \\
& t4=(1,\ -33,\ -33,\ 1\between x^4,\ y^4)^3, \\
(5):\quad & f5=xy\,(1,\ 11,\ -1\between x^5,\ y^5)^2, \\
& h5=(1,\ -228,\ +494,\ +228,\ -1\between x^5,\ y^5)^4, \\
& t5=(1,\ -522,\ 10005,\ 0,\ -10005,\ 522,\ 1\between x^5,\ y^5)^6,
\end{aligned}$$

where observe that $f4$ is the same function as $t3$. In each set of functions f, h, t, we have h and t covariants of f, viz. disregarding numerical factors,

h is the Hessian, or derivative $(f, f)^2$, and t is the derivative (f, h).

76. Since $f4$ is the same function as $t3$, we have of course $f4$, $h4$ and $t4$ themselves covariants of $f3$: but it is convenient to separate the two systems.

77. It is to be observed that $f3$ is a quartic function having its quadrinvariant $(I)=0$; but independently of this, that is, quà quartic function, it has only the covariants $h3$ and $t3$ (the Hessian and the cubicovariant respectively), viz. every other covariant is a rational and integral function of $f3$, $h3$ and $t3$. In particular, $h4$ and $t4$ are rational and integral functions of $f3$, $h3$ and $t3$; but inasmuch as $f3$ and $h3$ are not covariants of $f4$, this is not a property of $h4$ and $t4$ considered as covariants of $f4$, and the relation in question need not be attended to.

78. It has just been stated that $f3$ quà quartic function has (in the sense explained) only the covariants $h3$ and $t3$: $f4$ quà *special* sextic function and $f5$ quà *special* dodecadic function have the like property, viz. $f4$ has only the covariants $h4$ and $t4$; $f5$ only the covariants $h5$ and $t5$. Hence $f3$, $f4$, $f5$ are "Prime-forms" in the sense defined in the paper by Fuchs, of 1875, viz. a Prime-form has no covariant of a lower order than itself, and also no covariant of a higher order which is a power of a form of a lower order.

79. The same functions have also the property that they are functions transformable into themselves by means of a group of linear transformations, and in this point of view they were considered in the nearly contemporaneous paper by Klein, of 1875; it is in this paper shown that the functions so transformable into themselves must be Polyhedral functions as above, the linear transformations in fact corresponding to the rotations whereby the spherical polyhedron can be brought into coincidence with its own original position. This theory will be presently given.

80. It is to be observed that, if U, V are functions $(*\between x, y)^n$ of the same order n, then using the accent to denote differentiation in regard to x, $UV'-U'V$ and (U, V) differ only by a numerical factor: and further that, writing as before $s=\frac{x}{y}$, and in the expression $UV'-U'V$ regarding U, V as functions $(*\between s, 1)^n$, and the accent as denoting differentiation in regard to s, we have $UV'-U'V$ and (U, V) differing by a numerical factor only. We have in the PQR-Table, lines 3, 4, 5, P, Q, R equal to given numerical multiples of h^β, t^γ, f^α, the indices α, β, γ being such as to make these to be functions of the same degree: hence, neglecting numerical multipliers, $PQ'-P'Q$ is equal to a function (h^β, t^γ), which is $=h^{\beta-1}t^{\gamma-1}(h, t)$: and the theorem that $PQ'-P'Q$, $=QR'-Q'R$, $=RP'-R'P$, contains only factors of P, Q, R is in fact the theorem that (h, t), (h, f), and (t, f) are each of them equal to a term or product of f, h, t: which is a result included in the theorem that f has only the covariants h and t. And by this last theorem we know already how from R, assumed to be known, we can derive P and Q: viz. R is a power of f; and we thence have $h=(f, f)^2$ and $t=(h, f)$, equations giving the functions h and t, upon which P and Q depend.

Covariantive Formulæ. Art. Nos. 81 to 84.

81. The various covariantive formulæ will be given with their proper numerical coefficients.

Tetrahedron function. f, h, t stand for the before-mentioned values,

$$f3,\ h3,\ t3\ (P,\ Q,\ R = h^3,\ -12i\sqrt{3}\,.\,t^2,\ -f^3).$$

For $f3$.

$$(a,\ b,\ c,\ d,\ e) = 1,\ 0,\ \frac{-i}{\sqrt{3}},\ 0,\ 1.$$

$$\tfrac{1}{2}(f,\ f)^2 = -96i\sqrt{3}\,.\,h, \qquad \tfrac{1}{2}(h,\ h)^2 = 96i\sqrt{3}\,.\,f, \qquad \tfrac{1}{2}(t,\ t)^2 = -25fh,$$

$$(f,\ h) = \ 32i\sqrt{3}\,.\,t, \qquad (f,\ f)^4 = 576I = 0, \qquad (f,\ h)^4 = 1152J = 1152\,.\,\frac{-4i}{\sqrt{3}},$$

$$(f,\ t) = \ 4\,.\,h^2,$$

$$(h,\ t) = \ 4\,.\,f^2,$$

$$h^3 - f^3 - 12i\sqrt{3}\,t^2 = 0,$$

$$fh = (1,\ 14,\ 1\between x^4,\ y^4)^2\ (=f4).$$

It is convenient to remark that t^2, f^3, h^3 being of the same order we have

$$t^2(f^3,\ h^3) + f^3(h^3,\ t^2) + h^3(t^2,\ f^3) = 0,$$

that is,

$$t^2\,.\,3\,.\,3f^2h^2(f,\ h) + f^3\,.\,3\,.\,2h^2t\,(h,\ t) + h^3\,.\,2\,.\,3tf^2(t,\ f) = 0,$$

an equation which, substituting for $(f,\ h)$, $(h,\ t)$, $(t,\ f)$ their values, reduces itself to the before-mentioned relation $h^3 - f^3 - 12i\sqrt{3}\,t^2 = 0$; and we have thus a verification of the values of $(f,\ h)$, $(h,\ t)$ and $(t,\ f)$. The like remark applies to the other two cases, which follow.

82. Hexahedron function. f, h, t stand for the before-mentioned values

$$f4,\ h4,\ t4\ (P,\ Q,\ R = h^3,\ -t^2,\ -108f^4).$$

For $f4$.

$$(a,\ b,\ c,\ d,\ e,\ f,\ g) = (0,\ \tfrac{1}{4},\ 0,\ 0,\ 0,\ -\tfrac{1}{4},\ 0).$$

$$\tfrac{1}{2}(f,\ f)^2 = -25h, \qquad \tfrac{1}{2}(f,\ f)^4 = 0, \qquad \tfrac{1}{2}(f,\ f)^6 = (720)^2\,.\,\tfrac{3}{8},$$

$$(f,\ h) = -8t, \qquad \tfrac{1}{2}(h,\ h)^2 = 3\,.\,2^6\,.\,7^2\,.\,f^2,$$

$$(f,\ t) = -12h^2, \qquad \tfrac{1}{2}(t,\ t)^2 = 2^4\,.\,3^3\,.\,11^2\,.\,f^2h,$$

$$(h,\ t) = -1728f^3,$$

$$h^3 - t^2 - 108f^4 = 0.$$

83. Dodecahedron function. f, h, t stand for the before-mentioned values

$$f5,\ h5,\ t5\ (P,\ Q,\ R = h^3,\ -t^2,\ -1728f^5).$$

For $f5$.

$$(a,\ b,\ c,\ d,\ e,\ f,\ g,\ h,\ i,\ j,\ k,\ l,\ m) = (0,\ \tfrac{1}{12},\ 0,\ 0,\ 0,\ 0,\ \tfrac{1}{84},\ 0,\ 0,\ 0,\ 0,\ -\tfrac{1}{12},\ 0).$$

$$\begin{array}{lll} \tfrac{1}{2}(f,\ f)^2 = -121h, & \tfrac{1}{2}(f,\ f)^4 = 0, & \tfrac{1}{2}(f,\ f)^6 = \tfrac{1}{2}(924)^2(720)^2 . \tfrac{-5}{84}f^*, \\ \tfrac{1}{2}(f,\ f)^8 = \ \ 0, & \tfrac{1}{2}(f,\ f)^{10} = 0, & \tfrac{1}{2}(f,\ f)^{12} = \tfrac{1}{2}(924)^2(720)^4 . \tfrac{25}{84}{}^*, \\ (f,\ h) = -20t, & \tfrac{1}{2}(h,\ h)^2 = 173280f^3, & \\ (f,\ t) = -30h^2, & \tfrac{1}{2}(t,\ t)^2 = 9082800f^3h, & \\ (h,\ t) = -86400f^5, & & \\ h^3 - t^2 - 1728f^5 = 0. & & \end{array}$$

84. We have

$$t = (x^{10} + y^{10})(1,\ 522,\ -10006,\ -522,\ 1 \between x^5,\ y^5)^4.$$

Write

$$\xi = (x^2 + y^2).(1,\ 2,\ 6,\ -2,\ 1 \between x,\ y)^4,$$

then

$$t = \xi(1,\ -10,\ 45 \between \xi^2,\ f).$$

Or putting

$$p = \frac{\xi}{\sqrt{f}},\quad = \frac{(x^2 + y^2)(1,\ 2,\ 6,\ -2,\ 1 \between x,\ y)^4}{\sqrt{xy}\,(x^{10} + 11x^5y^5 - y^{10})},$$

that is, $\xi = p\sqrt{f}$, then

$$p^5 - 10p^3 + 45p = \frac{t}{\sqrt{f^5}}.\quad \text{(Klein.)}$$

Investigation of the forms $f5$ and $h5$. Art. Nos. 85 and 86.

85. Writing for shortness† $k = \tan\alpha = \dfrac{\sqrt{5} - 1}{2}$, and $g = \cos 36° + i \sin 36°$, then the values of $x + iy$ corresponding to the summits of the Icosahedron are

$$\begin{array}{lllll} 0, & & & & \\ k, & kg^2, & kg^4, & kg^6, & kg^8, \\ k^{-1}g, & k^{-1}g^3, & k^{-1}g^5, & k^{-1}g^7, & k^{-1}g^9, \\ \infty\,; & & & & \end{array}$$

and the function $f5$ is thus

$$= s\left(1 - \frac{s}{\infty}\right)(s^5 - k^5)(s - k^{-5}),$$

* The numerical coefficients $-\frac{5}{84}$ and $\frac{25}{84}$ are Klein's B and A: the latter of them is the ordinary quadrinvariant of a dodecadic function; the former is an invariant linear as regards the coefficients of f, and existing only for the special form f in question: viz. writing for a moment

$$f = \lambda(x^{11}y + 11x^6y^6 - xy^{11}),$$

then $(f,\ f)^6$ contains the factor λ^2, and (f containing the factor λ) the form is

$$\tfrac{1}{2}(f,\ f)^6 = \tfrac{1}{2}(924)^2(720)^2 . -\tfrac{5}{84}\lambda . f,$$

which is linear as regards λ. We have also

$$\tfrac{1}{2}(f,\ f)^{12} = \tfrac{1}{2}(924)^2(720)^4 . \tfrac{25}{84}\lambda^2 :$$

say $A = \frac{25}{84}\lambda^2$, $B = -\frac{5}{84}\lambda$; or $84B^2 = A$. Of course in the case of a general dodecadic function f, we have $(f,\ f)^6$, an irreducible covariant, not breaking up into factors.

† α is the α, γ is the γ, and γ' the $\alpha - \beta$ of the Table, No. 99.

where the product of the last two factors is $s^{10}+(k^{-5}-k^5)\,s^5-1$. We have

$$k^{-5}=\tfrac{1}{32}(80\sqrt{5}+176),\quad =\tfrac{1}{2}(5\sqrt{5}+11),$$
$$k^{5}\;=\tfrac{1}{32}(80\sqrt{5}-176),\quad =\tfrac{1}{2}(5\sqrt{5}-11),$$

and consequently $k^{-5}-k^5=11$; or the function is

$$s\left(1-\frac{s}{\infty}\right)(s^{10}+11s^5-1).$$

86. Similarly, writing for shortness* $l=\tan\tfrac{1}{2}\gamma$, $l'=\tan\tfrac{1}{2}\gamma'$, where

$$\cos^2\gamma=\frac{5+2\sqrt{5}}{15},\quad \sin^2\gamma=\frac{10-2\sqrt{5}}{15};\ \text{and therefore}\ \frac{\cos\gamma}{\sin\gamma}=\frac{3+\sqrt{5}}{4};$$
$$\cos^2\gamma'=\frac{5-2\sqrt{5}}{15},\quad \sin^2\gamma'=\frac{10+2\sqrt{5}}{15};\qquad ,,\qquad \frac{\cos\gamma'}{\sin\gamma'}=\frac{3-\sqrt{5}}{4};$$

and $g=\cos 36°+i\sin 36°$ as before, then the values of $x+iy$ for the summits of the dodecahedron are

$$\begin{array}{lllll} lg, & lg^3, & lg^5, & lg^7, & lg^9, \\ l'g, & l'g^3, & l'g^5, & l'g^7, & l'g^9, \\ l'^{-1}, & l'^{-1}g^2, & l'^{-1}g^4, & l'^{-1}g^6, & l'^{-1}g^8, \\ l^{-1}, & l^{-1}g^2, & l^{-1}g^4, & l^{-1}g^6, & l^{-1}g^8. \end{array}$$

The function $h5$ is therefore

$$=s^{10}+s^5(l^5-l^{-5})+1\,.\,s^{10}+s^5(l'^5-l'^{-5})-1.$$

We have

$$l^{-5}-l^5=\frac{(1+\cos\gamma)^5-(1-\cos\gamma)^5}{\sin^5\gamma}=\frac{2\cos\gamma}{\sin^5\gamma}(5+10\cos^2\gamma+\cos^4\gamma)$$
$$=\frac{2\cos\gamma}{\sin^5\gamma}\cdot\frac{384+64\sqrt{5}}{45}=\frac{128}{45}\,\frac{\cos\gamma}{\sin^5\gamma}(6+\sqrt{5})=114+50\sqrt{5};$$

viz. this last identity depends on

$$\tfrac{32}{45}(3+\sqrt{5})(6+\sqrt{5})=(114+50\sqrt{5})\sin^4\gamma,$$

that is,

$$160(3+\sqrt{5})(6+\sqrt{5})=(114+50\sqrt{5})(120-40\sqrt{5}),$$

or

$$2(3+\sqrt{5})(6+\sqrt{5})=(57+25\sqrt{5})(3-\sqrt{5}),$$

or finally

$$(7+3\sqrt{5})(6+\sqrt{5})=\ 57+25\sqrt{5},$$

which is right.

Similarly

$$l'^{-5}-l'^5=114-50\sqrt{5},$$

and observing that the sum and product of $114+50\sqrt{5}$, $114-50\sqrt{5}$ are $=228$ and 496 respectively, the required function of s is

$$(s^{10}-1)^2-228(s^{15}-s^5)+496s^{10},$$
$$=s^{20}-228s^{15}+494s^{10}+228s^5+1,$$

which is the required value of $h5$.

* α is the α, γ is the γ, and γ' the $\alpha-\beta$ of the Table, No. 99.

Invariantive property of the Stereographic Projection. Art. Nos. 87 to 93.

87. The before-mentioned theorem that the functions derived from two different stereographic projections of the same point are linear transformations one of the other, may be thus stated:

Considering on the surface of a sphere, two fixed points A and B; and determining the position of a point C, first in regard to A by its distance θ and azimuth f, and

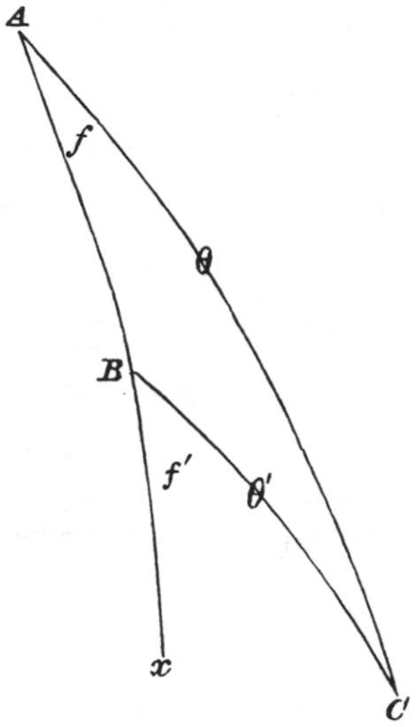

next in regard to B by its distance θ' and azimuth f', the azimuths from the great circle ABx which joins the two points A and B, then we have

$$\tan\tfrac{1}{2}\theta\,(\cos f + i\sin f),\quad\text{and}\quad \tan\tfrac{1}{2}\theta'\,(\cos f' + i\sin f'),$$

homographic functions one of the other: calling them s, s', and putting the distance $AB = c$, the relation between them in fact is

$$s' = \frac{s - \tan\frac{1}{2}c}{1 + s\tan\frac{1}{2}c},$$

or, what is the same thing,

$$\tan\tfrac{1}{2}c\,(1 + ss') = s - s';$$

or, observing that

$$ss' = \tan\tfrac{1}{2}\theta\tan\tfrac{1}{2}\theta'\,\{\cos(f+f') + i\sin(f+f')\},$$

we have the two equations

$$\tan\tfrac{1}{2}c\,\{1 + \tan\tfrac{1}{2}\theta\tan\tfrac{1}{2}\theta'\cos(f+f')\} = \tan\tfrac{1}{2}\theta\cos f - \tan\tfrac{1}{2}\theta'\cos f',$$
$$\tan\tfrac{1}{2}c\,\{\quad\tan\tfrac{1}{2}\theta\tan\tfrac{1}{2}\theta'\sin(f+f')\} = \tan\tfrac{1}{2}\theta\sin f - \tan\tfrac{1}{2}\theta'\sin f'.$$

88. If we denote the angles of the spherical triangle by C, A, B, and the opposite sides by c (as before), a, b, then θ, $\theta' = b$, a; f, $f' = A$, $\pi - B$, whence

$$s,\ s' = \tan\tfrac{1}{2}b\,(\cos A + i\sin A),\quad -\tan\tfrac{1}{2}a\,(\cos B - i\sin B):$$

or we have between the sides a, b, c and angles A, B of a spherical triangle the relations

$$\tan\tfrac{1}{2}c\,\{1 - \tan\tfrac{1}{2}a\tan\tfrac{1}{2}b\cos(A-B)\} = \tan\tfrac{1}{2}b\cos A + \tan\tfrac{1}{2}a\cos B,$$
$$\tan\tfrac{1}{2}c\,\{\quad - \tan\tfrac{1}{2}a\tan\tfrac{1}{2}b\sin(A-B)\} = \tan\tfrac{1}{2}b\sin A - \tan\tfrac{1}{2}a\sin B;$$

equations which may be verified by means of the ordinary formulæ of Spherical Trigonometry.

89. But it is interesting to give the proof with rectangular coordinates.

Taking (X, Y, Z), (X_1, Y_1, Z_1) for the coordinates, referred to two different sets of axes, of a point on the spherical surface: also x, y, x_1, y_1 for the coordinates of the corresponding stereographic projections, we have

$$(X_1, Y_1, Z_1) = \left(\begin{array}{ccc} \alpha, & \beta, & \gamma \\ \alpha', & \beta', & \gamma' \\ \alpha'', & \beta'', & \gamma'' \end{array}\right)(X, Y, Z),$$

$$X : Y : Z : 1 = 2x : 2y : 1 - x^2 - y^2 : 1 + x^2 + y^2,$$
$$X_1 : Y_1 : Z_1 : 1 = 2x_1 : 2y_1 : 1 - x_1^2 - y_1^2 : 1 + x_1^2 + y_1^2,$$

and thence

$$\begin{aligned} x_1 : y_1 : 1 = \; & 2\alpha x + 2\beta y + \gamma\,(1 - x^2 - y^2) \\ : \; & 2\alpha' x + 2\beta' y + \gamma'\,(1 - x^2 - y^2) \\ : \; & 1 + x^2 + y^2 + 2\alpha'' x + 2\beta'' y + \gamma''\,(1 - x^2 - y^2). \end{aligned}$$

90. Introducing z, z_1 for homogeneity, or writing $\frac{x}{z}$, $\frac{y}{z}$ and $\frac{x_1}{z_1}$, $\frac{y_1}{z_1}$ in place of x, y and x_1, y_1, respectively, we have

$$\begin{aligned} x_1 &= 2\alpha x + 2\beta y + \gamma\,(z^2 - x^2 - y^2), & &= (-\gamma, -\gamma, \gamma, \beta, \alpha, 0 \between x, y, z)^2, \\ y_1 &= 2\alpha' x + 2\beta' y + \gamma'\,(z^2 - x^2 - y^2), & &= (-\gamma', -\gamma', \gamma', \beta', \alpha', 0 \between \;\;,,\;\;)^2, \\ z_1 &= z^2 + x^2 + y^2 + 2\alpha'' x + 2\beta'' y + \gamma''\,(z^2 - x^2 - y^2), & &= (1 - \gamma'', 1 - \gamma'', 1 + \gamma'', \beta'', \alpha'', 0 \between \;\;,,\;\;)^2, \end{aligned}$$

and thence without difficulty

$$z_1 = \frac{1}{1 + \gamma''}\{(1 + \gamma'')\,z + (\alpha'' + i\beta'')\,(x - iy)\}\,\{(1 + \gamma'')\,z + (\;\alpha'' - i\beta'')\,(x + iy)\},$$

$$x_1 + iy_1 = \frac{1}{\gamma + \gamma'^2}\{(1 + \gamma'')\,z + (\alpha'' + i\beta'')\,(x - iy)\}\,\{(1 - \gamma'')\,z + (-\alpha'' + i\beta'')\,(x + iy)\},$$

$$x_1 - iy_1 = \frac{1}{\gamma - \gamma'^2}\{(1 - \gamma'')\,z - (\alpha'' + i\beta'')\,(x - iy)\}\,\{(1 + \gamma'')\,z + (\;\alpha'' - i\beta'')\,(x + iy)\},$$

viz. the form is $z_1 : x_1 + iy_1 : x_1 - iy_1 = MN : NL : LM$ (L, M, N linear functions of z, $x + iy$, $x - iy$): showing that the relation between two stereographic projections of the same spherical figure is in fact that of a quadric transformation, the fundamental points in each figure being an arbitrary point and the two circular points at infinity: or, what is the same thing, to any line in the one figure there corresponds a circle in the other figure, which is the "circular relation" of Möbius.

91. The actual values are

$$\frac{x_1 + iy_1}{z_1} = \frac{1 + \gamma''}{\gamma + \gamma' i} \cdot \frac{(1 - \gamma'')\,z - (\alpha'' - i\beta'')\,(x + iy)}{(1 + \gamma'')\,z + (\alpha'' - i\beta'')\,(x + iy)},$$

$$\frac{x_1 - iy_1}{z_1} = \frac{1 + \gamma''}{\gamma - \gamma' i} \cdot \frac{(1 - \gamma'')\,z - (\alpha'' + i\beta'')\,(x - iy)}{(1 + \gamma'')\,z + (\alpha'' + i\beta'')\,(x - iy)},$$

viz. attending only to the former of these, we have $\frac{x_1+iy_1}{z_1}$ a homographic function of $\frac{x+iy}{z}$, which is the before-mentioned theorem.

92. Supposing that the transformation from (X, Y, Z) to (X_1, Y_1, Z_1) is made by a rotation, the coordinates of which are λ, μ, ν: that is, if f, g, h are the inclinations of the resultant axis to the axes of x, y, z respectively, and θ the angle of rotation, putting $\lambda, \mu, \nu = \tan\frac{1}{2}\theta\cos f, \tan\frac{1}{2}\theta\cos g, \tan\frac{1}{2}\theta\cos h$: then the coefficients of transformation are

$$\left(\begin{array}{ccc} \alpha, & \beta, & \gamma \\ \alpha', & \beta', & \gamma' \\ \alpha'', & \beta'', & \gamma'' \end{array}\right) = \left(\begin{array}{ccc} 1+\lambda^2-\mu^2-\nu^2, & 2(\lambda\mu+\nu), & 2(\lambda\nu-\mu) \\ 2(\mu\lambda-\nu), & 1-\lambda^2+\mu^2-\nu^2, & 2(\mu\nu+\lambda) \\ 2(\nu\lambda+\mu), & 2(\mu\nu-\lambda), & 1-\lambda^2-\mu^2+\nu^2 \end{array}\right) \div (1+\lambda^2+\mu^2+\nu^2).$$

Substituting these values, the formulæ become, after an easy reduction,

$$\frac{x_1+iy_1}{z_1} = \frac{-(\nu+i)(x+iy)+(\lambda+i\mu)z}{(\lambda-i\mu)(x+iy)+(\nu-i)z},$$

$$\frac{x_1-iy_1}{z_1} = \frac{-(\nu-i)(x-iy)+(\lambda-i\mu)z}{(\lambda+i\mu)(x-iy)+(\nu+i)z};$$

attending to the former of these, and writing for greater simplicity

$$\frac{x_1+iy_1}{z_1},\ \frac{x+iy}{z} = s_1, s$$

respectively, we have

$$s_1 = \frac{-(\nu+i)s+(\lambda+i\mu)}{(\lambda-i\mu)s+(\nu-i)},$$

or writing this

$$s_1 = \frac{As+B}{Cs+D},$$

then $$A : B : C : D = -\nu-i : \lambda+i\mu : \lambda-i\mu : \nu-i.$$

93. I call to mind that the condition, in order that the homographic transformation $s_1 = (As+B)\div(Cs+D)$ may be periodic of the order n, is

$$(A+D)^2 - 4(AD-BC)\cos^2\frac{m\pi}{n} = 0,$$

m being an integer different from zero and prime to n. In particular, when $n=2$, it is $A+D=0$: $n=3$, it is $A^2+AD+D^2+BC=0$: $n=4$, it is $A^2+D^2+2BC=0$: and $n=5$, it is $(A+D)^2-\frac{1}{2}(3\pm\sqrt{5})(AD-BC)=0$.

Groups of homographic transformations. Art. Nos. 94 and 95.

94. The formulæ just obtained serve to connect the theory of the rotations of a polyhedron with that of the homographic transformations s into $(As+B)\div(Cs+D)$: and, corresponding to the rotations which leave the polyhedron unaltered, we have groups of homographic transformations. We have thus, corresponding to the cases of the tetrahedron, the cube and the octahedron, and the dodecahedron and icosahedron respectively, groups of 12, of 24, and of 60 homographic transformations s into

$(As+B) \div (Cs+D)$. The group of 60 and the group of 24 include each of them as part of itself the group of 12: it is further to be remarked that the group of 12 may be regarded as that of the positive substitutions upon four letters *abcd*, the group of 24 as that of all the substitutions upon the four letters, and the group of 60 as that of the positive substitutions upon five letters *abcde*.

95. I call to mind that a group of functional symbols 1, α, β, ... can always be expressed in the equivalent form 1, $\vartheta\alpha\vartheta^{-1}$, $\vartheta\beta\vartheta^{-1}$, ... where ϑ is any functional symbol whatever: clearly, α, β, ... being homographic transformations, then, ϑ being any homographic transformation whatever, the new symbols $\vartheta\alpha\vartheta^{-1}$, $\vartheta\beta\vartheta^{-1}$, ... will also be homographic transformations; and thus the group of homographic transformations can be expressed in various equivalent forms: these correspond to the different positions of the polyhedron in regard to the axes of coordinates: and there are in fact three cases which it is proper to consider, viz. attending for the moment to the dodecahedron, we may have the axis of z passing through the midpoint of a side, through the centre of a face, or through a summit; that is, in the language presently explained, the cases are 1°, Pole at a point Θ; 2°, Pole at a point A; 3°, Pole at a point B.

The regular Polyhedra. Art. Nos. 96 to 103.

96. We require a theory of the regular Polyhedra considered as systems of points on a sphere. I refer to my two papers [375] and [679]. In the latter paper, I remark that, considering the five regular figures drawn in proper relation to each other on the same spherical surface, the only points which have to be considered are 12 points A, 20 points B, 30 points Θ, and 60 points Φ. Describing these by reference to the dodecahedron, the points A are the centres of the faces, the points B are the summits, the points Θ are the midpoints of the sides, and the points Φ are the midpoints of the diagonals of the faces. Or describing them by reference to the icosahedron, the points A are the summits, the points B are the centres of the faces, the points Θ are the midpoints of the sides: viz. each point Θ is the common midpoint of a side of the dodecahedron and a side of the icosahedron, which there intersect at right angles: and the points Φ are points lying by threes on the faces of the icosahedron, each point Φ of the face being given as the intersection of a perpendicular $A\Theta$ of the face by a line BB joining the centres of two adjacent faces and which intersects $A\Theta$ at right angles.

97. The points Φ are comparatively unimportant, and it is proper in the first instance to attend only to the 12 points A, the 20 points B, and the 30 points Θ: these form 6 pairs of opposite points A, 10 pairs of opposite points B, and 15 pairs of opposite points Θ. Considering the diameters through each pair of opposite points Θ, we have thus a system of 15 axes, which in fact form 5 sets each of 3 rectangular axes: attending to any one of such sets, the diametral plane at right angles to one of the three axes contains of course the other two axes: it contains also two axes each through a pair of opposite points A, and two axes each through a pair of opposite points B. If instead of the plane we consider its intersection with the sphere, we have thus on the sphere 15 circles each containing 4 points Θ,

4 points A and 4 points B. The fifteen circles intersect by fives in the pairs of opposite points A, by threes in the pairs of opposite points B, and by twos in the pairs of opposite points Θ; the mutual inclinations of successive circles at the points A, B, Θ being $=36°$, $60°$ and $90°$ respectively. The whole number 15.14, $=210$, of the intersections of the circles two and two together is thus made up of the 12 points A each counting 10 times, the 20 points B each counting 3 times, and the 30 points Θ each counting once; $210 = 120 + 60 + 30$.

98. The angular magnitudes which present themselves are all obtained from the dodecahedral pentagon, as shown in the annexed figure, in which the angle subtended by a side at the centre is $=72°$, and the angle between two adjacent sides is $=120°$.

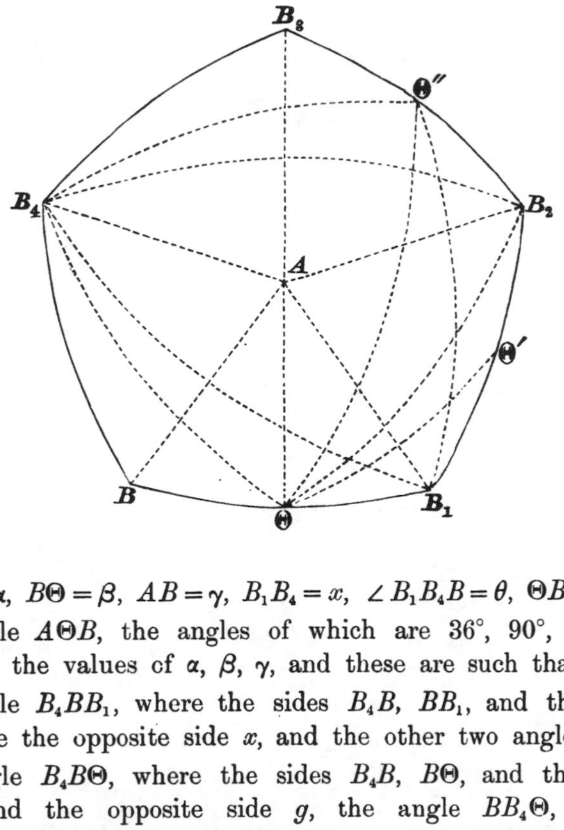

We write $A\Theta=\alpha$, $B\Theta=\beta$, $AB=\gamma$, $B_1B_4=x$, $\angle B_1B_4B=\theta$, $\Theta B_4=g$, $\angle \Theta B_4B=\phi$.

From the triangle $A\Theta B$, the angles of which are $36°$, $90°$, $60°$ and the opposite sides β, γ, α, we find the values of α, β, γ, and these are such that $\alpha+\beta+\gamma=\frac{1}{2}\pi$.

From the triangle B_4BB_1, where the sides B_4B, BB_1, and the included angle are 2β, 2β, $120°$, we have the opposite side x, and the other two angles each $=\theta$.

From the triangle $B_4B\Theta$, where the sides B_4B, $B\Theta$, and the included angle are 2β, β, $120°$, we find the opposite side g, the angle $BB_4\Theta$, $=\phi$, and the angle $B_4\Theta B$, $=45°$.

Hence each of the angles $B_4\Theta B$, $B_2\Theta B_1$, being $=45°$, the angle $B_4\Theta B_2$ is $=90°$: in this triangle the hypothenuse B_2B_4 is $=x$, and each of the other two sides is $=g$: whence we have $\cos x=\cos^2 g$, as is in fact the case, and moreover the values give $x+2g=180°$. Also each of the other angles is found to be $=60°$; that is, we have $\angle B_2B_4\Theta=60°$, or the whole angle at B_4 being $=120°$, the sum of the remaining angles $B_3B_4B_2$ and $BB_4\Theta$ is $=60°$: that is, $\theta+\phi=60°$.

From the triangle $\Theta B_1\Theta'$ where the two sides and the included angle are β, β, $120°$, we find $\Theta\Theta'=36°$.

And from the triangle $\Theta B_4 \Theta''$, where the two sides and the included angle are g, g and $(120^\circ - 2\phi =) 2\theta$, we find $\Theta\Theta'' = 60^\circ$.

99. We thus arrive at the following Table:

			sin	cos
$A\Theta$	α	31° 43′	$\sqrt{\frac{5-\sqrt{5}}{10}}$	$\sqrt{\frac{5+\sqrt{5}}{10}}$
$B\Theta$	β	20° 55′	$\frac{\sqrt{5}-1}{2\sqrt{3}}$	$\frac{\sqrt{5}+1}{2\sqrt{3}}$
AB	γ	37° 22′	$\sqrt{\frac{10-2\sqrt{5}}{15}}$	$\sqrt{\frac{5+2\sqrt{5}}{15}}$
(BB)	x	70° 32′	$\frac{2\sqrt{2}}{3}$	$\frac{1}{3}$
$(B\Theta)$	g	54 44	$\frac{\sqrt{2}}{\sqrt{3}}$	$\frac{1}{\sqrt{3}}$
BBB	θ	37° 46′	$\frac{\sqrt{3}}{2\sqrt{2}}$	$\frac{\sqrt{5}}{2\sqrt{2}}$
$B\Theta B$	ϕ	22 14	$\frac{\sqrt{3}(\sqrt{5}-1)}{4\sqrt{2}}$	$\frac{\sqrt{5}+3}{4\sqrt{2}}$
	2α	63 26	$\frac{2}{\sqrt{5}}$	$\frac{1}{\sqrt{5}}$
	2β	41 50	$\frac{2}{3}$	$\frac{\sqrt{5}}{3}$
	2γ	74 44	$\frac{2(\sqrt{5}+1)}{3\sqrt{5}}$	$\frac{4-\sqrt{5}}{3\sqrt{5}}$
	$\alpha-\beta$		$\sqrt{\frac{5-2\sqrt{5}}{15}}$	$\sqrt{\frac{10+2\sqrt{5}}{15}}$
		18°	$\frac{\sqrt{5}-1}{4}$	$\sqrt{\frac{5+\sqrt{5}}{8}}$
$\Theta\Theta$		36°	$\sqrt{\frac{5-\sqrt{5}}{8}}$	$\frac{\sqrt{5}+1}{4}$

where as above

$$\alpha + \beta + \gamma = 90^\circ,$$
$$x + 2g \quad = 180^\circ,$$
$$\theta + \phi \quad = 60^\circ.$$

100. We now construct three figures of the points A, B, Θ; viz. these are stereographic projections, each showing the Northern hemisphere projected on the plane of the equator by lines drawn to the South Pole: hence, for any pair of opposite points not on the equator, only the point in the Northern hemisphere is shown: but for a pair of opposite points on the equator the two points are each of them shown. In fig. 1 the North Pole is taken to be a point Θ; in fig. 2 it is a point A; and in fig. 3 it is a point B. The position of any point on the sphere is determined by its N.P.D. and its longitude, measured from an arbitrary origin, say from the point E of the centre left-handedly: then, in the three figures, the positions are as follows.

101. Fig. 1. Pole at Θ.

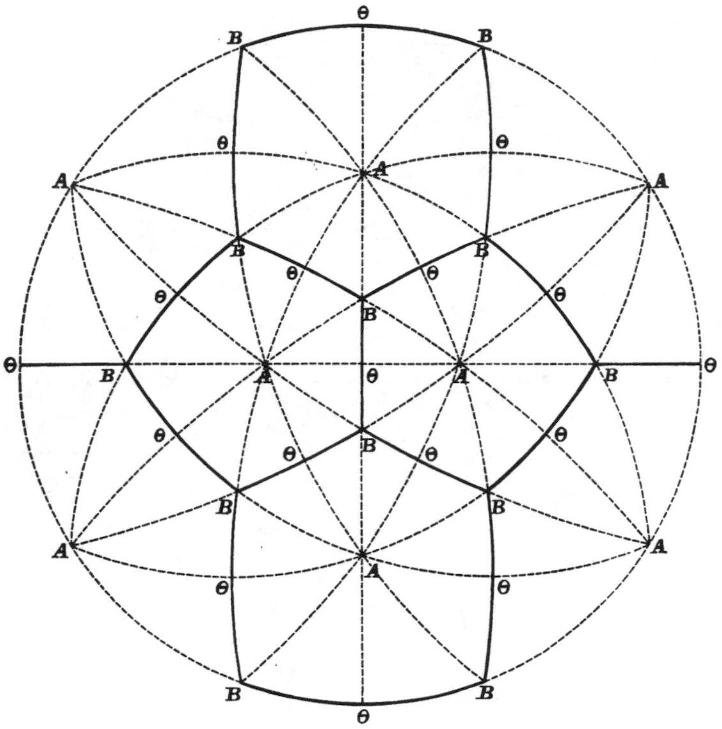

	N.P.D.'s	Longitudes.
2A	$\alpha =$ 31° 43′	0°, 180°
2A	90° − $\alpha =$ 58 17	90 , 270
4A	90	(0 , 180) ± α = 31° 43′
2A	90° + $\alpha =$ 121 43	90 , 270
2A	180° − $\alpha =$ 148 17	0 , 180
2B	$\beta =$ 20° 55′	90°, 270°
4B	$g =$ 54 44	45 , 135, 225, 315
2B	90° − $\beta =$ 69 5	0 , 180
4B	90	(90 , 270) ± β = 20° 55′
2B	90° + $\beta =$ 110 55	0 , 180
4B	180° − $g =$ 125 16	45 , 135, 225, 315
2B	180° − $\beta =$ 159 5	90 , 270
1Θ	0°	—
4Θ	36	(90°, 270°) ± α = 31° 43′
4Θ	60	(0 , 180) ± β = 20 55
4Θ	72	(90 , 270) ± α = 31 43
4Θ	90	0 , 90, 180, 270
4Θ	108	(90 , 270) ± α = 31 43
4Θ	120	(0 , 180) ± β = 20 55
4Θ	144	(90 , 270) ± α = 31 43
1Θ	180	—

102. Fig. 2. Pole at A.

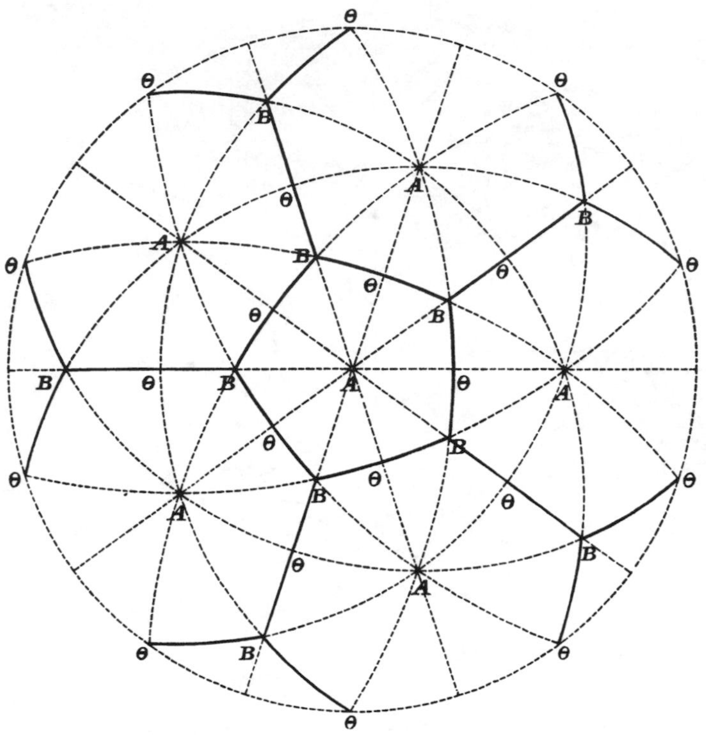

	N.P.D.'s	Longitudes.
A	0	—
$5A$	$2\alpha =$ 63° 26′	0° 72° 144° 216° 288°
$5A$	$180° - 2\alpha =$ 116 34	36 108 180 252 324
A	180	—
$5B$	$\gamma =$ 37 22	36 108 180 252 324
$5B$	$90° - \alpha + \beta =$ 79 12	36 108 180 252 324
$5B$	$90 + \alpha - \beta =$ 100 48	0 72 144 216 288
$5B$	$180 - \gamma =$ 142 38	0 72 144 216 288
5Θ	$\alpha =$ 31 43	0 72 144 216 288
5Θ	$90° - \alpha =$ 58 17	36 108 180 252 324
10Θ	90	(36 108 180 252 324) $\pm$ 18°
5Θ	$90 + \alpha =$ 121 43	0 72 144 216 288
5Θ	$180 - \alpha =$ 144 17	36 108 180 252 324

103. Fig. 3. Pole at B.

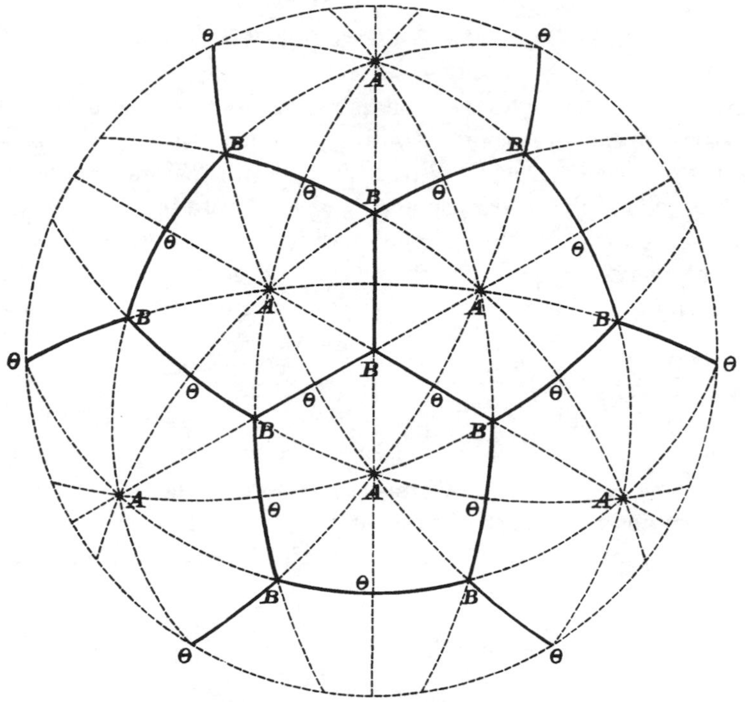

	N.P.D.'s	Longitudes.
3A	$\gamma =$ 37° 22′	30° 150° 270°
3A	$90° - \alpha + \beta =$ 79 12	90 210 330
3A	$90 + \alpha - \beta =$ 100 48	30 150 270
3A	$180 - \gamma =$ 142 38	90 210 330
B	0	—
3B	$2\beta =$ 41 50	90 210 330
6B	$x =$ 70 32	(30 150 270) $\pm \vartheta =$ 37° 46′
6B	$180° - x =$ 109 28	(90 210 330) $\pm \vartheta =$ 37 46
3B	$180 - 2\beta =$ 138 10	30 150 270
B	180	—
3Θ	$\beta =$ 20 55	90 210 330
6Θ	$g =$ 54 44	(90 210 330) $\pm \phi =$ 22° 14′
3Θ	$90° - \beta =$ 69 5	30 150 270
6Θ	90	0 60 120 180° 240° 300°
3Θ	$90 + \beta =$ 110 55	90 210 330
6Θ	$180 - g =$ 125 16	(30 150 270) $\pm \phi =$ 22° 14′
3Θ	$180 - \beta =$ 159 5	30 150 270

The groups of homographic transformations, resumed. Art. Nos. 104 to 117.

104. The axes of rotation for the dodecahedron and the icosahedron are 15 axes each through a pair of opposite points Θ, 6 axes each through a pair of opposite points A, and 10 axes each through a pair of opposite points B; or say 15 Θ-axes, 10 B-axes and 6 A-axes: the corresponding angles of rotation are 180°, 72° and 120°; so that (excluding in each case the original position or that of a rotation 0) we have in respect of each Θ-axis 1 position, in respect of each A-axis 4 positions, and in respect of each B-axis 2 positions; in all, including the original position,

$$1 + 15 + (6 \times 4) + (10 \times 2), \; = 60 \text{ positions},$$

that is, a group of 60 rotations.

To find, in any one of the three forms, the group of homographic transformations, we can in each case obtain from the foregoing tables the values $\cos f$, $\cos g$, $\cos h$ of the cosine-inclination of an axis of rotation to the axes of coordinates, and thence calculate the values of

$$\lambda, \; \mu, \; \nu = \tan \tfrac{1}{2}\vartheta \cos f, \quad \tan \tfrac{1}{2}\vartheta \cos g, \quad \tan \tfrac{1}{2}\vartheta \cos h,$$

and thence the values of

$$A, \; B, \; C, \; D = -\nu - i, \quad \lambda + i\mu, \quad \lambda - i\mu, \quad \nu - i;$$

viz. in the case of a Θ-axis, ϑ is $= 180°$, (so that here $\tan \frac{1}{2}\vartheta = \infty$, or the values of A, B, C, D are $= -\nu$, $\lambda + i\mu$, $\lambda - i\mu$, ν, that is, $-\cos h$, $\cos f + i \cos g$, $\cos f - i \cos g$, $\cos h$); in the case of a B-axis, the values are $\vartheta = 120°$, $240°$, and therefore $\tan \frac{1}{2}\vartheta = \pm \sqrt{3}$; and in the case of an A-axis, they are $\vartheta = 72°$, $144°$, $216°$, $288°$, and therefore

$$\tan \tfrac{1}{2}\vartheta = \pm \frac{\sqrt{10 + 2\sqrt{5}}}{\sqrt{5} - 1}, \quad \pm \frac{\sqrt{10 - 2\sqrt{5}}}{\sqrt{5} + 1}.$$

105. The Θ-form was first given in my paper of 1879, but in obtaining it I used results given in the paper of 1877. As regards the identification with the substitution-symbols, since there is nothing to distinguish *inter se* the letters *a, b, c, d, e,* any transformation A, B, C, D of the fifth order might have been taken for *abcde,* but No. 37 of the group having been taken for this substitution *abcde,* I do not recall in what manner I found that, consistently herewith, the transformation No. 2 $(-1, 0, 0, 1$, that is, s into $-s)$ of the second order could be taken for *ab.cd.* But there is no sub-group of an order divisible by 5; and hence, these two transformations being identified with the two substitutions, the other transformations correspond each of them to a determinate substitution.

106. Homographic Transformations. The group of 60. Pole at Θ.

	$(Ax$	$+B)$ ÷	$(Cx$	$+D)$	
1	1	0	0	1	1
2	-1	0	0	1	*ab . cd*
3	0	1	1	0	*ac . bd*
4	0	1	-1	0	*ad . bc*
5	2	$-3+\sqrt{5}+i(1-\sqrt{5})$	$-3+\sqrt{5}+i(-1+\sqrt{5})$	-2	*bc . de*
6	2	$-3+\sqrt{5}+i(-1+\sqrt{5})$	$-3+\sqrt{5}+i(1-\sqrt{5})$	-2	*ae . bc*
7	2	$3-\sqrt{5}+i(-1+\sqrt{5})$	$3-\sqrt{5}+i(1-\sqrt{5})$	-2	*ad . ce*
8	2	$3-\sqrt{5}+i(1-\sqrt{5})$	$3-\sqrt{5}+i(-1+\sqrt{5})$	-2	*ad . be*
9	2	$-1-\sqrt{5}+i(1-\sqrt{5})$	$-1-\sqrt{5}+i(-1+\sqrt{5})$	-2	*ae . cd*
10	2	$-1-\sqrt{5}+i(-1+\sqrt{5})$	$-1-\sqrt{5}+i(1-\sqrt{5})$	-2	*ab . de*
11	2	$1+\sqrt{5}+i(-1+\sqrt{5})$	$1+\sqrt{5}+i(1-\sqrt{5})$	-2	*be . cd*
12	2	$1+\sqrt{5}+i(1-\sqrt{5})$	$1+\sqrt{5}+i(-1+\sqrt{5})$	-2	*ab . ce*
13	2	$-1-\sqrt{5}+i(-3-\sqrt{5})$	$-1-\sqrt{5}+i(3+\sqrt{5})$	-2	*ac . be*
14	2	$-1-\sqrt{5}+i(3+\sqrt{5})$	$-1-\sqrt{5}+i(-3-\sqrt{5})$	-2	*bd . ce*
15	2	$1+\sqrt{5}+i(3+\sqrt{5})$	$1+\sqrt{5}+i(-3-\sqrt{5})$	-2	*ae . bd*
16	2	$1+\sqrt{5}+i(-3-\sqrt{5})$	$1+\sqrt{5}+i(3+\sqrt{5})$	-2	*ac . de*
17	$-i$	i	1	1	*abc*
18	-1	i	1	i	*acb*
19	1	$-i$	1	i	*adc*
20	$-i$	$-i$	1	-1	*acd*
21	i	i	1	-1	*adb*
22	1	i	1	$-i$	*abd*
23	-1	$-i$	1	$-i$	*bcd*
24	i	$-i$	1	1	*bdc*
25	$-1-\sqrt{5}+i(3+\sqrt{5})$	2	-2	$-1-\sqrt{5}+i(-3-\sqrt{5})$	*aec*
26	$1+\sqrt{5}+i(3+\sqrt{5})$	2	-2	$1+\sqrt{5}+i(-3-\sqrt{5})$	*ace*
27	$1+\sqrt{5}+i(-3-\sqrt{5})$	2	-2	$1+\sqrt{5}+i(3+\sqrt{5})$	*bed*
28	$-1-\sqrt{5}+i(-3-\sqrt{5})$	2	-2	$-1-\sqrt{5}+i(3+\sqrt{5})$	*bde*
29	$-3+\sqrt{5}+i(1-\sqrt{5})$	2	2	$3-\sqrt{5}+i(1-\sqrt{5})$	*bec*
30	$-3+\sqrt{5}+i(-1+\sqrt{5})$	2	2	$3-\sqrt{5}+i(-1+\sqrt{5})$	*bce*
31	$3-\sqrt{5}+i(-1+\sqrt{5})$	2	2	$-3+\sqrt{5}+i(-1+\sqrt{5})$	*aed*
32	$3-\sqrt{5}+i(1-\sqrt{5})$	2	2	$-3+\sqrt{5}+i(1-\sqrt{5})$	*ade*
33	2	$-1-\sqrt{5}+i(-1+\sqrt{5})$	$1+\sqrt{5}+i(-1+\sqrt{5})$	2	*cde*
34	2	$1+\sqrt{5}+i(1-\sqrt{5})$	$-1-\sqrt{5}+i(1-\sqrt{5})$	2	*ced*
35	2	$-1-\sqrt{5}+i(1-\sqrt{5})$	$1+\sqrt{5}+i(1-\sqrt{5})$	2	*aeb*
36	2	$1+\sqrt{5}+i(-1+\sqrt{5})$	$-1-\sqrt{5}+i(-1+\sqrt{5})$	2	*abe*
37	$-1-\sqrt{5}+i(-3-\sqrt{5})$	2	2	$1+\sqrt{5}+i(-3-\sqrt{5})$	*abcde*
38	$-1-\sqrt{5}+i(1-\sqrt{5})$	2	2	$1+\sqrt{5}+i(1-\sqrt{5})$	*acebd*
39	$-1-\sqrt{5}+i(-1+\sqrt{5})$	2	2	$1+\sqrt{5}+i(-1+\sqrt{5})$	*adbec*
40	$-1-\sqrt{5}+i(3+\sqrt{5})$	2	2	$1+\sqrt{5}+i(3+\sqrt{5})$	*aedcb*
41	$1+\sqrt{5}+i(3+\sqrt{5})$	2	2	$-1-\sqrt{5}+i(3+\sqrt{5})$	*adceb*
42	$1+\sqrt{5}+i(-1+\sqrt{5})$	2	2	$-1-\sqrt{5}+i(-1+\sqrt{5})$	*acbde*
43	$1+\sqrt{5}+i(1-\sqrt{5})$	2	2	$-1-\sqrt{5}+i(1-\sqrt{5})$	*aedbc*
44	$1+\sqrt{5}+i(-3-\sqrt{5})$	2	2	$-1-\sqrt{5}+i(-3-\sqrt{5})$	*abecd*
45	$-1-\sqrt{5}+i(-1+\sqrt{5})$	2	-2	$-1-\sqrt{5}+i(1-\sqrt{5})$	*acbed*

46	$-3+\sqrt{5}+i(-1+\sqrt{5})$	2	-2	$-3+\sqrt{5}+i(1-\sqrt{5})$	*abdce*
47	$3-\sqrt{5}+i(-1+\sqrt{5})$	2	-2	$3-\sqrt{5}+i(1-\sqrt{5})$	*aecdb*
48	$1+\sqrt{5}+i(-1+\sqrt{5})$	2	-2	$1+\sqrt{5}+i(1-\sqrt{5})$	*adebc*
49	$1+\sqrt{5}+i(1-\sqrt{5})$	2	-2	$1+\sqrt{5}+i(-1+\sqrt{5})$	*aecbd*
50	$3-\sqrt{5}+i(1-\sqrt{5})$	2	-2	$3-\sqrt{5}+i(-1+\sqrt{5})$	*acdeb*
51	$-3+\sqrt{5}+i(1-\sqrt{5})$	2	-2	$-3+\sqrt{5}+i(-1+\sqrt{5})$	*abedc*
52	$-1-\sqrt{5}+i(1-\sqrt{5})$	2	-2	$-1-\sqrt{5}+i(-1+\sqrt{5})$	*adbce*
53	2	$-3+\sqrt{5}+i(-1+\sqrt{5})$	$3-\sqrt{5}+i(-1+\sqrt{5})$	2	*aebdc*
54	2	$-1-\sqrt{5}+i(3+\sqrt{5})$	$1+\sqrt{5}+i(3+\sqrt{5})$	2	*abced*
55	2	$1+\sqrt{5}+i(-3-\sqrt{5})$	$-1-\sqrt{5}+i(-3-\sqrt{5})$	2	*adecb*
56	2	$3-\sqrt{5}+i(1-\sqrt{5})$	$-3+\sqrt{5}+i(1-\sqrt{5})$	2	*acdbe*
57	2	$-3+\sqrt{5}+i(1-\sqrt{5})$	$3-\sqrt{5}+i(1-\sqrt{5})$	2	*abdec*
58	2	$-1-\sqrt{5}+i(-3-\sqrt{5})$	$1+\sqrt{5}+i(-3-\sqrt{5})$	2	*adcbe*
59	2	$1+\sqrt{5}+i(3+\sqrt{5})$	$-1-\sqrt{5}+i(3+\sqrt{5})$	2	*aebcd*
60	2	$3-\sqrt{5}+i(-1+\sqrt{5})$	$-3+\sqrt{5}+i(-1+\sqrt{5})$	2	*acedb*

107. Taking out of the foregoing group of 60 a group of 12 contained in it, viz. that corresponding to the positive substitutions of the four letters *abcd*, it is easy to see, that there is a transformation $(i, 0, 0, 1)$, that is, s into is, which can be taken for the substitution *adbc*, and also to complete thence the group of 24. And we have thus the following Table.

Groups of 12 and 24. Pole at Θ.

	$(Ax$	$+B)$ $\div$	$(Cx$	$+D)$	
1	1	0	0	1	1
2	-1	0	0	1	*ab . cd*
3	0	1	1	0	*ac . bd*
4	0	1	-1	0	*ad . bc*
5	$-i$	i	1	1	*abc*
6	-1	i	1	i	*acb*
7		$-i$	1	i	*adc*
8	$-i$	$-i$	1	-1	*acd*
9	i	i	1	-1	*adb*
10	1	i	1	$-i$	*abd*
11	-1	$-i$	1	$-i$	*bcd*
12	i	$-i$	1	1	*bdc*
13	i	0	0	1	*adbc*
14	$-i$	0	0	1	*acbd*
15	0	i	1	0	*cd*
16	0	i	-1	0	*ab*
17	1	-1	1	1	*acdb*
18	$-i$	-1	1	i	*bd*
19	i	1	1	i	*abcd*
20	1	1	1	-1	*bc*
21	-1	-1	1	-1	*abdc*
22	i	-1	1	$-i$	*ac*
23	$-i$	1	1	$-i$	*adcb*
24	-1	1	1	1	*ad*

108. The group of 60 was obtained in the A-form by Gordan in his paper. The passage from the Θ-form to the A-form is made as follows: let X, Y, Z be the coordinates of a point when the axes are as in the Θ-form, X_1, Y_1, Z_1 the coordinates of the same point when the axes are as in the A-form: we may write

$$X,\ Y,\ Z = \mathrm{b}X_1 - \mathrm{a}Z_1 : Y_1 : \mathrm{a}X_1 + \mathrm{b}Z_1,$$

where

$$\mathrm{a},\ \mathrm{b} = \sqrt{\frac{5-\sqrt{5}}{10}},\ \sqrt{\frac{5+\sqrt{5}}{10}};$$

then, if the equations of an axis of rotation referred to the first set of coordinates are $X : Y : Z = L : M : N$, those of the same axis referred to the second set of coordinates are

$$\mathrm{b}X_1 + \mathrm{a}Z_1 : Y_1 : -\mathrm{a}X_1 + \mathrm{b}Z_1 = L : M : N;$$

or taking these to be

$$X_1 : Y_1 : Z_1 = L_1 : M_1 : N_1,$$

we may write

$$L_1,\ M_1,\ N_1 = \mathrm{b}L + \mathrm{a}N,\ M,\ -\mathrm{a}L + \mathrm{b}N:$$

these values are such that

$$L_1^2 + M_1^2 + N_1^2 = L^2 + M^2 + N^2,$$

and hence, λ, μ, ν and λ_1, μ_1, ν_1 being the rotations, we may write

$$L,\ M,\ N = \vartheta\lambda,\ \vartheta\mu,\ \vartheta\nu;\quad L_1,\ M_1,\ N_1 = \vartheta\lambda_1,\ \vartheta\mu_1,\ \vartheta\nu_1;$$

where ϑ has the same value in each set of equations. From the equations

$$A : B : C : D = -\nu - i : \lambda + i\mu : \lambda - i\mu : \nu - i,$$

we have

$$B + C : B - C : D - A : D + A = \lambda : i\mu : \nu : -i$$
$$= L : iM : N : -i\vartheta,$$

and similarly

$$B_1 + C_1 : B_1 - C_1 : D_1 - A_1 : D_1 + A_1 = L_1 : iM_1 : N_1 : -i\vartheta.$$

Hence we may write

$$\begin{aligned} B_1 + C_1 &= \mathrm{b}(B + C) + \mathrm{a}(D - A), \\ B_1 - C_1 &= B - C, \\ D_1 - A_1 &= -\mathrm{a}(B + C) + \mathrm{b}(D - A), \\ D_1 + A_1 &= D + A; \end{aligned}$$

or say,

$$\begin{aligned} A_1 &= \mathrm{a}(B + C) - \mathrm{b}(D - A) + (D + A), \\ B_1 &= \mathrm{b}(B + C) + \mathrm{a}(D - A) + (B - C), \\ C_1 &= \mathrm{b}(B + C) + \mathrm{a}(D - A) - (B - C), \\ D_1 &= -\mathrm{a}(B + C) + \mathrm{b}(D - A) + (D + A), \end{aligned}$$

which are the values for a transformation (A_1, B_1, C_1, D_1) in the A-form: of course, as only the ratios are material, the values may be multiplied by any common factor.

109. The results are exhibited in terms of ϵ, an imaginary fifth root of unity: taking $\epsilon = \cos 72^\circ + i \sin 72^\circ$, we have

$$\epsilon,\ \epsilon^4 = \quad \frac{\sqrt{5}-1}{4} \pm i\sqrt{\frac{5+\sqrt{5}}{8}},$$

$$\epsilon^2,\ \epsilon^3 = -\frac{\sqrt{5}+1}{4} \pm i\sqrt{\frac{5-\sqrt{5}}{8}};$$

where the upper signs belong to ϵ, ϵ^2 and the lower to ϵ^4, ϵ^3. It may be remarked that

$$\frac{1}{\mathrm{a}} = \sqrt{\frac{5+\sqrt{5}}{2}},\quad \frac{1}{\mathrm{b}} = \sqrt{\frac{5-\sqrt{5}}{2}},\quad \frac{\mathrm{b}}{\mathrm{a}} = \frac{\sqrt{5}+1}{2},\quad \frac{\mathrm{a}}{\mathrm{b}} = \frac{\sqrt{5}-1}{2}.$$

For instance, we have in the Θ-group $(A, B, C, D) = (-1, 0, 0, 1)$; $ab.cd$: and thence in the A-group $A_1, B_1, C_1, D_1 = (-2\mathrm{b}, 2\mathrm{a}, 2\mathrm{a}, 2\mathrm{b})$; $ab.cd$: or say this is

$$\left(-1, \frac{\mathrm{a}}{\mathrm{b}}, \frac{\mathrm{a}}{\mathrm{b}}, 1\right), = (-1, \epsilon+\epsilon^4, \epsilon+\epsilon^4, 1);$$

which in the Table is given as $(-\epsilon^3, \epsilon^2+\epsilon^4, \epsilon^2+\epsilon^4, \epsilon^3)$; $ab.cd$.

By effecting the passage to the A-group in this manner, we of course obtain the proper substitution corresponding to each transformation: but I found it easier starting from two transformations and the corresponding substitutions, to obtain thence by successive compositions the entire group.

110. Homographic Transformations. The group of 60. Pole at A.

		Θ No.	$(As$	$+B)$	$\div(Cs$	$+D)$	
1		1	1			1	1
2		4	0	-1	1	0	$ad.bc$
3		13	0	$-\epsilon^4$	1	0	$ac.be$
4		9	0	$-\epsilon^3$	1	0	$ae.cd$
5		10	0	$-\epsilon^2$	1	0	$ab.de$
6		14	0	$-\epsilon$	1	0	$bd.ce$
7		6	$\epsilon+\epsilon^2$	ϵ^4	1	$-(\epsilon+\epsilon^3)$	$ae.bc$
8		5	$\epsilon+\epsilon^3$	1	ϵ^4	$-(\epsilon+\epsilon^3)$	$bc.de$
9		16	$\epsilon+\epsilon^3$	ϵ	ϵ^3	$-(\epsilon+\epsilon^3)$	$ac.de$
10		3	$\epsilon+\epsilon^3$	ϵ^2	ϵ^2	$-(\epsilon+\epsilon^3)$	$ac.bd$
11		15	$\epsilon+\epsilon^3$	ϵ^3	ϵ	$-(\epsilon+\epsilon^3)$	$ae.bd$
12		12	-1	$\epsilon+\epsilon^3$	$\epsilon^2+\epsilon^4$	1	$ab.ce$
13		11	$-\epsilon$	ϵ^3+1	$\epsilon^2+\epsilon^4$	ϵ	$be.cd$
14		7	$-\epsilon^2$	$1+\epsilon^2$	$\epsilon^2+\epsilon^4$	ϵ^2	$ad.ce$
15		2	$-\epsilon^3$	$\epsilon^2+\epsilon^4$	$\epsilon^2+\epsilon^4$	ϵ^3	$ab.cd$
16		8	$-\epsilon^4$	$\epsilon^4+\epsilon$	$\epsilon^2+\epsilon^4$	ϵ^4	$ad.be$

17		21	ϵ^3+1	ϵ	1	$-(\epsilon+\epsilon^3)$	*adb*
18		35	ϵ^3+1	ϵ^2	ϵ^4	$-(\epsilon+\epsilon^3)$	*aeb*
19		30	ϵ^3+1	ϵ^3	ϵ^3	$-(\epsilon+\epsilon^3)$	*bce*
20		34	ϵ^3+1	ϵ^4	ϵ^2	$-(\epsilon+\epsilon^3)$	*ced*
21		19	ϵ^3+1	1	ϵ	$-(\epsilon+\epsilon^3)$	*adc*
22		33	$\epsilon+\epsilon^4$	ϵ^2	1	$-(\epsilon+\epsilon^3)$	*cde*
23		20	$\epsilon+\epsilon^4$	ϵ^3	ϵ^4	$-(\epsilon+\epsilon^3)$	*acd*
24		22	$\epsilon+\epsilon^4$	ϵ^4	ϵ^3	$-(\epsilon+\epsilon^3)$	*abd*
25		36	$\epsilon+\epsilon^4$	1	ϵ^2	$-(\epsilon+\epsilon^3)$	*abe*
26		29	$\epsilon+\epsilon^4$	ϵ	ϵ	$-(\epsilon+\epsilon^3)$	*bec*
27		31	$-\epsilon$	$\epsilon^2+\epsilon^4$	$\epsilon^2+\epsilon^4$	1	*aed*
28		17	$-\epsilon^2$	$\epsilon^4+\epsilon$	$\epsilon^2+\epsilon^4$	ϵ	*abc*
29		27	$-\epsilon^3$	$\epsilon+\epsilon^3$	$\epsilon^2+\epsilon^4$	ϵ^2	*bed*
30		25	$-\epsilon^4$	ϵ^3+1	$\epsilon^2+\epsilon^4$	ϵ^3	*aec*
31		23	-1	$1+\epsilon^2$	$\epsilon^2+\epsilon^4$	ϵ^4	*bcd*
32		24	$-\epsilon^4$	$1+\epsilon^2$	$\epsilon^2+\epsilon^4$	1	*bdc*
33		32	-1	$\epsilon^2+\epsilon^4$	$\epsilon^2+\epsilon^4$	ϵ	*ade*
34		18	$-\epsilon$	$\epsilon^4+\epsilon$	$\epsilon^2+\epsilon^4$	ϵ^2	*acb*
35		28	$-\epsilon^2$	$\epsilon+\epsilon^3$	$\epsilon^2+\epsilon^4$	ϵ^3	*bde*
36		26	$-\epsilon^3$	ϵ^3+1	$\epsilon^2+\epsilon^4$	ϵ^4	*ace*
37		44	ϵ	0	0	1	*abecd*
38		43	ϵ^2	0	0	1	*aedbc*
39		42	ϵ^3	0	0	1	*acbde*
40		41	ϵ^4	0	0	1	*adceb*
41		38	$\epsilon^2+\epsilon^4$	1	1	$-(\epsilon+\epsilon^3)$	*acebd*
42		46	$\epsilon^2+\epsilon^4$	ϵ	ϵ^4	$-(\epsilon+\epsilon^3)$	*abdce*
43		58	$\epsilon^2+\epsilon^4$	ϵ^2	ϵ^3	$-(\epsilon+\epsilon^3)$	*adcbe*
44		55	$\epsilon^2+\epsilon^4$	ϵ^3	ϵ^2	$-(\epsilon+\epsilon^3)$	*adecb*
45		50	$\epsilon^2+\epsilon^4$	ϵ^4	ϵ	$-(\epsilon+\epsilon^3)$	*acdeb*
46		51	$1+\epsilon^2$	ϵ^3	1	$-(\epsilon+\epsilon^3)$	*abedc*
47		39	$1+\epsilon^2$	ϵ^4	ϵ^4	$-(\epsilon+\epsilon^3)$	*adbec*
48		47	$1+\epsilon^2$	1	ϵ^3	$-(\epsilon+\epsilon^3)$	*aecdb*
49		59	$1+\epsilon^2$	ϵ	ϵ^2	$-(\epsilon+\epsilon^3)$	*aebcd*
50		54	$1+\epsilon^2$	ϵ^2	ϵ	$-(\epsilon+\epsilon^3)$	*abced*
51		56	$-\epsilon^2$	ϵ^3+1	$\epsilon^2+\epsilon^4$	1	*acdbe*
52		49	$-\epsilon^3$	$1+\epsilon^2$	$\epsilon^2+\epsilon^4$	ϵ	*aecbd*
53		37	$-\epsilon^4$	$\epsilon^2+\epsilon^4$	$\epsilon^2+\epsilon^4$	ϵ^2	*abcde*
54		45	-1	$\epsilon^4+\epsilon$	$\epsilon^2+\epsilon^4$	ϵ^3	*acbed*
55		57	$-\epsilon$	$\epsilon+\epsilon^3$	$\epsilon^2+\epsilon^4$	ϵ^4	*abdec*
56		48	$-\epsilon^3$	$\epsilon^4+\epsilon$	$\epsilon^2+\epsilon^4$	1	*adebc*
57		60	$-\epsilon^4$	$\epsilon+\epsilon^3$	$\epsilon^2+\epsilon^4$	ϵ	*acedb*
58		53	-1	ϵ^3+1	$\epsilon^2+\epsilon^4$	ϵ^2	*aebdc*
59		52	$-\epsilon$	$1+\epsilon^2$	$\epsilon^2+\epsilon^4$	ϵ^3	*adbce*
60		40	$-\epsilon^2$	$\epsilon^2+\epsilon^4$	$\epsilon^2+\epsilon^4$	ϵ^4	*aedcb*

111. Selecting the transformations which correspond to the positive substitutions *abcd*, and completing the group of 24 we have

Homographic Transformations. The groups of 12 and 24. Pole at A.

	$(As$	$+B)$	$\div(Cs$	$+D)$	
1	1	0	0	1	1
2	0	-1	1	0	*ad . bc*
3	$\epsilon+\epsilon^3$	ϵ^2	ϵ^2	$-(\epsilon+\epsilon^3)$	*ac . bd*
4	$-\epsilon^3$	$\epsilon^2+\epsilon^4$	$\epsilon^2+\epsilon^4$	ϵ^3	*ab . cd*
5	$-\epsilon^2$	$\epsilon+\epsilon^4$	$\epsilon^2+\epsilon^4$	ϵ	*abc*
6	$-\epsilon$	$\epsilon+\epsilon^4$	$\epsilon^2+\epsilon^4$	ϵ^2	*acb*
7	$\epsilon+\epsilon^4$	ϵ^3	ϵ^4	$-(\epsilon+\epsilon^3)$	*acd*
8	ϵ^3+1	1	ϵ	$-(\epsilon+\epsilon^3)$	*adc*
9	$\epsilon+\epsilon^4$	ϵ^4	ϵ^3	$-(\epsilon+\epsilon^3)$	*abd*
10	ϵ^3+1	ϵ	1	$-(\epsilon+\epsilon^3)$	*adb*
11	-1	$1+\epsilon^2$	$\epsilon^2+\epsilon^4$	ϵ^4	*bcd*
12	$-\epsilon^4$	$1+\epsilon^2$	$\epsilon^2+\epsilon^4$	1	*bdc*
13	1	$1+2\epsilon^4$	$1+2\epsilon$	-1	*ab*
14	$-\epsilon^2+\epsilon^3$	$1+\epsilon+3\epsilon^4$	$-1-3\epsilon-\epsilon^4$	$\epsilon^2-\epsilon^3$	*cd*
15	$\epsilon^2-\epsilon^4$	$3+\epsilon+\epsilon^3$	$-1-3\epsilon-\epsilon^3$	$-\epsilon^2+\epsilon^4$	*ac*
16	$-1+\epsilon^2$	$-1-\epsilon^2+2\epsilon^4$	$1+\epsilon^2-2\epsilon^3$	$1-\epsilon^2$	*bd*
17	$2+\epsilon^3+2\epsilon^4$	$-2-2\epsilon^2-\epsilon^3$	$2\epsilon+\epsilon^3+2\epsilon^4$	$2\epsilon+2\epsilon^2+\epsilon^3$	*ad*
18	$2+2\epsilon^2+\epsilon^3$	$2+\epsilon^3+2\epsilon^4$	$-2\epsilon-2\epsilon^2-\epsilon^3$	$2\epsilon+\epsilon^3+2\epsilon^4$	*bc*
19	$-2+\epsilon+\epsilon^3$	$-\epsilon+\epsilon^3$	$-\epsilon+\epsilon^3$	$\epsilon+\epsilon^3-2\epsilon^4$	*abcd*
20	1	-1	1	1	*abdc*
21	1	1	-1	1	*acdb*
22	$1+\epsilon+3\epsilon^4$	$\epsilon^2-\epsilon^3$	$\epsilon^2-\epsilon^3$	$1+3\epsilon+\epsilon^4$	*acbd*
23	$1+2\epsilon^4$	-1	-1	$-1-2\epsilon$	*adbc*
24	$3+\epsilon+\epsilon^3$	$-\epsilon^2+\epsilon^4$	$-\epsilon^2+\epsilon^4$	$1+3\epsilon+\epsilon^3$	*adcb*

As an example of the calculation we have $(A, B, C, D)=(0, i, -1, 0)$; *ab*. Hence

$$A_1, B_1, C_1, D_1 = \Big(\mathrm{a}\,(i-1),\ \mathrm{b}\,(i-1)+i+1,\ \mathrm{b}\,(i-1)-(i+1),\ -\mathrm{a}\,(i+1)\Big),$$

$$= \left(1,\ \frac{\mathrm{b}-i}{\mathrm{a}},\ \frac{\mathrm{b}+i}{\mathrm{a}},\ -1\right).$$

The second and third coefficients are

$$\frac{\sqrt{5}+1}{2} - i\sqrt{\frac{5+\sqrt{5}}{2}},\quad \frac{\sqrt{5}+1}{2} + i\sqrt{\frac{5+\sqrt{5}}{2}},$$

which, in virtue of the values of ϵ and ϵ^4, are $=1+2\epsilon^4$ and $1+2\epsilon$ respectively: or the result is as above $(1,\ 1+2\epsilon^4,\ 1+2\epsilon,\ -1)$.

112. In like manner for the passage from the Θ-form to the B-form, if X, Y, Z be the coordinates of a point on the spherical surface in regard to the Θ-axes, X_2, Y_2, Z_2 those of the same point in regard to the B-axes, we may write

$$X : Y : Z = X_2 : \mathrm{b}Y_2 + \mathrm{a}Z_2 : -\mathrm{a}Y_2 + \mathrm{b}Z_2,$$

where

$$\mathrm{a},\ \mathrm{b} = \frac{\sqrt{5}-1}{2\sqrt{3}},\ \frac{\sqrt{5}+1}{2\sqrt{3}}.$$

Hence $X : Y : Z = L : M : N$, being the equations of an axis of rotation in the first set of coordinates, those of the same axis in the second set of coordinates will be

$$X_2 : \mathrm{b}Y_2 + \mathrm{a}Z_2 : -\mathrm{a}Y_2 + \mathrm{b}Z_2 = L : M : N,$$

or calling these

$$X_2 : Y_2 : Z_2 = L_2 : M_2 : N_2,$$

we have

$$L_2,\ M_2,\ N_2 = L : \mathrm{b}M - \mathrm{a}N : \mathrm{a}M + \mathrm{b}N:$$

these values are such that

$$L_2^2 + M_2^2 + N_2^2 = L^2 + M^2 + N^2,$$

or λ, μ, ν, λ_2, μ_2, ν_2 being the rotations, we have

$$L,\ M,\ N = \vartheta\lambda,\ \vartheta\mu,\ \vartheta\nu;\quad L_2,\ M_2,\ N_2 = \vartheta\lambda_2,\ \vartheta\mu_2,\ \vartheta\nu_2,$$

where ϑ has the same value in the two sets of equations. We have thus

$$B + C : B - C : D - A : D + A = L : 2M : N : -i\vartheta,$$

$$B_2 + C_2 : B_2 - C_2 : D_2 - A_2 : D_2 + A_2 = L_2 : 2M_2 : N_2 : -i\vartheta,$$

and hence

$$\begin{aligned} B_2 + C_2 &= B + C, \\ B_2 - C_2 &= \mathrm{b}(B - C) - \mathrm{a}i(D - A), \\ D_2 - A_2 &= -\mathrm{a}i(B - C) + \mathrm{b}(D - A), \\ D_2 + A_2 &= D + A; \end{aligned}$$

and thence

$$\begin{aligned} A_2 &= \mathrm{a}i(B - C) - \mathrm{b}(D - A) + (D + A), \\ B_2 &= \mathrm{b}(B - C) - \mathrm{a}i(D - A) + (B + C), \\ C_2 &= -\mathrm{b}(B - C) + \mathrm{a}i(D - A) + (B + C), \\ D_2 &= -\mathrm{a}i(B - C) + \mathrm{b}(D - A) + (D + A). \end{aligned}$$

113. As an example of the transformation, take

$$(A,\ B,\ C,\ D) = \left(2,\ -3 + \sqrt{5} + i(1 - \sqrt{5}),\ -3 + \sqrt{5} + i(-1 + \sqrt{5}),\ -2\right)\quad [bc\,.\,de]:$$

then

$$B - C,\ B + C,\ D - A,\ D + A = i(1 - \sqrt{5}),\ -3 + \sqrt{5},\ -2,\ 0;$$

and thence

$$A_2 = \frac{1}{2\sqrt{3}}(\ \ 6 - 2\sqrt{5}) + \frac{1}{2\sqrt{3}}(\ \ 2 + 2\sqrt{5}),$$

$$B_2 = \frac{1}{2\sqrt{3}}(-4i) \qquad + \frac{1}{2\sqrt{3}}\Big(2i(1 + \sqrt{5}) + (-3 + \sqrt{5})\Big),$$

$$C_2 = \frac{1}{2\sqrt{3}}(\ \ 4i) \qquad + \frac{1}{2\sqrt{3}}\Big(2i(1 - \sqrt{5}) + (-3 + \sqrt{5})\Big),$$

$$D_2 = \frac{1}{2\sqrt{3}}(-6 + 2\sqrt{5}) + \frac{1}{2\sqrt{3}}(-2 - 2\sqrt{5});$$

viz. multiplying by $2\sqrt{3}$, these are

$$8,\quad i(-6 + 2\sqrt{5}) + 2\sqrt{3}(-3 + \sqrt{5}),\quad i(6 - 2\sqrt{5}) + 2\sqrt{3}(-3 + \sqrt{5}),\quad -8,$$

that is,

$$8,\quad (-6 + 2\sqrt{5})(i + \sqrt{3}),\quad (-6 + 2\sqrt{5})(-i + \sqrt{3}),\quad -8,$$

or since

$$2 + \sqrt{3} = -2i\omega \text{ and } -2 + \sqrt{3} = 2i\omega^2,$$

dividing by 4 these are

$$2,\quad i(3 - \sqrt{5})\,\omega,\quad i(-3 + \sqrt{5})\,\omega^2,\quad -2,$$

as in the table.

114. Homographic Transformations. The group of 60. Pole at *B*.

$$\omega = \tfrac{1}{2}(-1 + i\sqrt{3}).$$

	$(As$	$+B)$	$\div(Cs$	$+D)$	
1	1	0	0	1	1
2	0	1	1	0	*ac . bd*
3	0	ω	1	0	*ae . bd*
4	0	ω^2	1	0	*bd . ce*
5	2	$i(\ \ 3-\sqrt{5})$	$i(\ \ -3+\sqrt{5})$	-2	*ab . cd*
6	2	$i(-3-\sqrt{5})$	$i(\ \ \ 3+\sqrt{5})$	-2	*ad . bc*
7	2	$i(\ \ 3-\sqrt{5})\,\omega$	$i(\ \ -3+\sqrt{5})\,\omega^2$	-2	*bc . de*
8	2	$i(-3-\sqrt{5})\,\omega$	$i(\ \ \ 3+\sqrt{5})\,\omega^2$	-2	*be . cd*
9	2	$i(\ \ 3-\sqrt{5})\,\omega^2$	$i(\ \ -3+\sqrt{5})\,\omega$	-2	*ad . be*
10	2	$i(-3-\sqrt{5})\,\omega^2$	$i(\ \ \ 3+\sqrt{5})\,\omega$	-2	*ab . de*
11	2	$(-\sqrt{3}-i\sqrt{5})\,\omega$	$(-\sqrt{3}+i\sqrt{5})\,\omega^2$	-2	*ab . ce*
12	2	$-\sqrt{3}-i\sqrt{5}$	$-\sqrt{3}+i\sqrt{5}$	-2	*ac . be*
13	2	$(-\sqrt{3}-i\sqrt{5})\,\omega^2$	$(-\sqrt{3}+i\sqrt{5})\,\omega$	-2	*ae . bc*
14	2	$\sqrt{3}-i\sqrt{5}$	$\sqrt{3}+i\sqrt{5}$	-2	*ac . de*
15	2	$(\ \ \sqrt{3}-i\sqrt{5})\,\omega$	$(\ \ \sqrt{3}+i\sqrt{5})\,\omega^2$	-2	*ad . ce*
16	2	$(\ \ \sqrt{3}-i\sqrt{5})\,\omega^2$	$(\ \ \sqrt{3}+i\sqrt{5})\,\omega$	-2	*ae . cd*

17	ω	0	0	1	*ace*
18	ω^2	0	0	1	*aec*
19	$\sqrt{3}-i\sqrt{5}$	2	-2	$\sqrt{3}+i\sqrt{5}$	*bed*
20	$-\sqrt{3}-i\sqrt{5}$	2	-2	$-\sqrt{3}+i\sqrt{5}$	*bde*
21	$-\sqrt{3}-i\sqrt{5}$	$2\omega^2$	-2ω	$-\sqrt{3}+i\sqrt{5}$	*bdc*
22	$\sqrt{3}-i\sqrt{5}$	$2\omega^2$	-2ω	$\sqrt{3}+i\sqrt{5}$	*bcd*
23	$-\sqrt{3}-i\sqrt{5}$	2ω	$-2\omega^2$	$-\sqrt{3}+i\sqrt{5}$	*abd*
24	$\sqrt{3}-i\sqrt{5}$	2ω	$-2\omega^2$	$\sqrt{3}+i\sqrt{5}$	*adb*
25	$2\omega^2$	$-\sqrt{3}-i\sqrt{5}$	$-\sqrt{3}+i\sqrt{5}$	-2ω	*abc*
26	2ω	$-\sqrt{3}-i\sqrt{5}$	$-\sqrt{3}+i\sqrt{5}$	$-2\omega^2$	*acb*
27	$2\omega^2$	$-\sqrt{3}-i\sqrt{5}$	$(-\sqrt{3}+i\sqrt{5})\,\omega^2$	-2	*abe*
28	2	$-\sqrt{3}-i\sqrt{5}$	$(-\sqrt{3}+i\sqrt{5})\,\omega^2$	$-2\omega^2$	*aeb*
29	2ω	$\sqrt{3}-i\sqrt{5}$	$\sqrt{3}+i\sqrt{5}$	$-2\omega^2$	*acd*
30	$2\omega^2$	$\sqrt{3}-i\sqrt{5}$	$\sqrt{3}+i\sqrt{5}$	-2ω	*adc*
31	$2\omega^2$	$\sqrt{3}-i\sqrt{5}$	$(,\ \sqrt{3}+i\sqrt{5})\,\omega^2$	-2	*ade*
32	2	$\sqrt{3}-i\sqrt{5}$	$(\ \sqrt{3}+i\sqrt{5})\,\omega^2$	$-2\omega^2$	*aed*
33	2	$-\sqrt{3}-i\sqrt{5}$	$(-\sqrt{3}+i\sqrt{5})\,\omega$	-2ω	*bce*
34	2ω	$-\sqrt{3}-i\sqrt{5}$	$(-\sqrt{3}+i\sqrt{5})\,\omega$	-2	*bec*
35	2ω	$\sqrt{3}-i\sqrt{5}$	$(\ \sqrt{3}+i\sqrt{5})\,\omega$	-2	*cde*
36	2	$\sqrt{3}-i\sqrt{5}$	$(\ \sqrt{3}+i\sqrt{5})\,\omega$	-2ω	*ced*
37	2	$i(\ 3-\sqrt{5})\,\omega^2$	$i(-3+\sqrt{5})$	$-2\omega^2$	*adceb*
38	$-\sqrt{3}-i\sqrt{5}$	$+2\omega^2$	-2	$(-\sqrt{3}+i\sqrt{5})\,\omega^2$	*acbde*
39	$\sqrt{3}-i\sqrt{5}$	2	-2ω	$(\ \sqrt{3}+i\sqrt{5})\,\omega$	*aedbc*
40	2	$i(\ 3-\sqrt{5})$	$i(-3+\sqrt{5})\,\omega$	-2ω	*abecd*
41	2	$i(\ 3-\sqrt{5})\,\omega$	$i(-3+\sqrt{5})$	-2ω	*aedcb*
42	$-\sqrt{3}-i\sqrt{5}$	2ω	-2	$(-\sqrt{3}+i\sqrt{5})\,\omega$	*adbec*
43	$\sqrt{3}-i\sqrt{5}$	2	$-2\omega^2$	$(\ \sqrt{3}+i\sqrt{5})\,\omega^2$	*acebd*
44	2	$i(\ 3-\sqrt{5})$	$i(-3+\sqrt{5})\,\omega^2$	$-2\omega^2$	*abcde*
45	2	$i(\ 3-\sqrt{5})\,\omega^2$	$i(-3+\sqrt{5})\,\omega^2$	-2ω	*adebc*
46	$\sqrt{3}-i\sqrt{5}$	$2\omega^2$	$-2\omega^2$	$(\ \sqrt{3}+i\sqrt{5})\,\omega$	*aecdb*
47	$-\sqrt{3}-i\sqrt{5}$	2ω	-2ω	$(-\sqrt{3}+i\sqrt{5})\,\omega^2$	*abdce*
48	2	$i(\ 3-\sqrt{5})\,\omega$	$i(-3+\sqrt{5})\,\omega$	$-2\omega^2$	*acbed*
49	2	$i(-3-\sqrt{5})\,\omega$	$i(\ 3+\sqrt{5})\,\omega$	$-2\omega^2$	*acdeb*
50	$\sqrt{3}-i\sqrt{5}$	2ω	-2ω	$(\ \sqrt{3}+i\sqrt{5})\,\omega^2$	*adbce*
51	$-\sqrt{3}-i\sqrt{5}$	$2\omega^2$	$-2\omega^2$	$(-\sqrt{3}+i\sqrt{5})\,\omega$	*aecbd*
52	2	$i(-3-\sqrt{5})\,\omega^2$	$i(\ 3+\sqrt{5})\,\omega^2$	-2ω	*abedc*
53	2	$i(-3-\sqrt{5})\,\omega$	$i(\ 3+\sqrt{5})$	-2ω	*aebcd*
54	$-\sqrt{3}-i\sqrt{5}$	2ω	-2	$(-\sqrt{3}+i\sqrt{5})\,\omega$	*abdec*
55	$\sqrt{3}-i\sqrt{5}$	2	$-2\omega^2$	$(\ \sqrt{3}+i\sqrt{5})\,\omega^2$	*acedb*
56	2	$i(-3-\sqrt{5})$	$i(\ 3+\sqrt{5})\,\omega^2$	$-2\omega^2$	*adcbe*
57	2	$i(-3-\sqrt{5})$	$i(\ 3+\sqrt{5})\,\omega$	-2ω	*adecb*
58	$-\sqrt{3}-i\sqrt{5}$	2	-2ω	$(-\sqrt{3}+i\sqrt{5})\,\omega$	*aebdc*
59	$\sqrt{3}-i\sqrt{5}$	$2\omega^2$	-2	$(\ \sqrt{3}+i\sqrt{5})\,\omega^2$	*acdbe*
60	2	$i(-3-\sqrt{5})\,\omega^2$	$i(\ 3+\sqrt{5})$	$-2\omega^2$	*abced*

115. We hence derive

Homographic Transformations. The groups of 12 and 24. Pole at B.

	$(As$	$+B)$	$\div(Cs$	$+D)$	
1	1	0	0	1	1
2	2	$i(3-\sqrt{5})$	$i(-3+\sqrt{5})$	-2	*ab . cd*
3	0	1	1	0	*ac . bd*
4	2	$i(-3-\sqrt{5})$	$i(3+\sqrt{5})$	-2	*ad . bc*
5	$2\omega^2$	$-\sqrt{3}-i\sqrt{5}$	$-\sqrt{3}+i\sqrt{5}$	-2ω	*abc*
6	2ω	$-\sqrt{3}-i\sqrt{5}$	$\sqrt{3}+i\sqrt{5}$	$-2\omega^2$	*acb*
7	$-\sqrt{3}-i\sqrt{5}$	2ω	$-2\omega^2$	$-\sqrt{3}+i\sqrt{5}$	*abd*
8	$\sqrt{3}-i\sqrt{5}$	2ω	$-2\omega^2$	$\sqrt{3}+i\sqrt{5}$	*adb*
9	2ω	$\sqrt{3}-i\sqrt{5}$	$\sqrt{3}+i\sqrt{5}$	$-2\omega^2$	*acd*
10	$2\omega^2$	$\sqrt{3}-i\sqrt{5}$	$\sqrt{3}+i\sqrt{5}$	-2ω	*adc*
11	$\sqrt{3}-i\sqrt{5}$	$2\omega^2$	-2ω	$\sqrt{3}+i\sqrt{5}$	*bcd*
12	$-\sqrt{3}-i\sqrt{5}$	$2\omega^2$	-2ω	$-\sqrt{3}+i\sqrt{5}$	*bdc*
13	2	$\sqrt{3}(1+\sqrt{5})+(-3-\sqrt{5})$	$\sqrt{3}(1+\sqrt{5})+i(3+\sqrt{5})$	-2	*ab*
14	2	$\sqrt{3}(-1-\sqrt{5})+(-3-\sqrt{5})$	$\sqrt{3}(-1-\sqrt{5})+i(3+\sqrt{5})$	-2	*cd*
15	$\sqrt{5}$	$-i$	i	$-\sqrt{5}$	*ac*
16	1	$i\sqrt{5}$	$-i\sqrt{5}$	-1	*bd*
17	2	$\sqrt{3}(-1+\sqrt{5})+i(3-\sqrt{5})$	$\sqrt{3}(-1+\sqrt{5})+i(-3+\sqrt{5})$	-2	*ad*
18	2	$\sqrt{3}(1-\sqrt{5})+i(3-\sqrt{5})$	$\sqrt{3}(1-\sqrt{5})+i(-3+\sqrt{5})$	-2	*bc*
19	1	i	i	1	*abcd*
20	1	$-i$	$-i$	1	*adcb*
21	$\sqrt{3}(1-\sqrt{5})+i(3+\sqrt{5})$	2	-2	$\sqrt{3}(1-\sqrt{5})+i(-3+\sqrt{5})$	*abdc*
22	$\sqrt{3}(1+\sqrt{5})+i(-3+\sqrt{5})$	2	-2	$\sqrt{3}(1+\sqrt{5})+i(3+\sqrt{5})$	*acbd*
23	$\sqrt{3}(-1+\sqrt{5})+i(3-\sqrt{5})$	2	-2	$\sqrt{3}(-1+\sqrt{5})+i(-3+\sqrt{5})$	*acdb*
24	$\sqrt{3}(-1-\sqrt{5})+i(-3-\sqrt{5})$	2	-2	$\sqrt{3}(-1-\sqrt{5})+i(3+\sqrt{5})$	*adbc*

116. I give also the group of 12, $(abce)$, slightly modifying the form: viz. I write first $\sqrt{3}+i\sqrt{5}=2\sqrt{2k}$, and therefore $\sqrt{3}-i\sqrt{5}=2\sqrt{2}\,.\,\frac{1}{k}$: then for x I write λx, and divide the A and B by λ: the A and B then contain $\frac{k}{\lambda}$, and the C and D contain $\frac{\lambda}{k}$, and assuming $\frac{k}{\lambda}=i$, we have $\frac{\lambda}{k}=-i$. For instance, in the transformation corresponding to abc, the $Ax+B$ and $Cx+D$,

$$=2\omega^2x-(\sqrt{3}+i\sqrt{5}) \text{ and } (-\sqrt{3}+2\sqrt{5})x-2\omega,$$

become first $2\omega^2x-2\sqrt{2k}$, and $-2\sqrt{2}\frac{1}{k}x-2\omega$, and then (omitting also the factor 2) $\omega^2x-\sqrt{2}\frac{k}{\lambda}$ and $-\sqrt{2}\frac{\lambda}{k}x-\omega$, viz. when $\frac{k}{\lambda}=i$, they are $\omega^2x-i\sqrt{2}$ and $x\,.\,i\sqrt{2}-\omega$; that is, the values of A, B, C, D are ω^2, $-i\sqrt{2}$, $i\sqrt{2}$, $-\omega$. The group is

Group of 12. Pole at *B*.

1	0	0	1	1
ω	0	0	1	*ace*
ω^2	0	0	1	*aec*
1	$-i\omega\sqrt{2}$	$i\omega\sqrt{2}$	$-\omega^2$	*abc*
1	$-i\omega^2\sqrt{2}$	$i\omega^2\sqrt{2}$	$-\omega$	*acb*
1	$-i-\omega\sqrt{2}$	$i\sqrt{2}$	$-\omega$	*abe*
1	$-i\sqrt{2}$	$i\omega^2\sqrt{2}$	$-\omega^2$	*aeb*
1	$-i\omega^2\sqrt{2}$	$i\sqrt{2}$	$-\omega^2$	*bec*
1	$-i\sqrt{2}$	$i\omega\sqrt{2}$	$-\omega$	*bce*
1	$-i\omega\sqrt{2}$	$i\omega^2\sqrt{2}$	-1	*ab . ce*
1	$-i\omega^2\sqrt{2}$	$i\omega\sqrt{2}$	-1	*ae . bc*
1	$-i\sqrt{2}$	$i\sqrt{2}$	-1	*ac . be*

117. From the Table of the Groups of 12 and 24, Θ-form, it appears that the group of 12 is

$$x,\ \frac{1}{x},\ -x,\ -\frac{1}{x},\ \frac{i(x-1)}{x+1},\ \frac{-i(x-1)}{x+1},\ \frac{i(x+1)}{x-1},\ \frac{-i(x+1)}{x-1},$$

$$\frac{x+i}{x-i},\ \frac{x-i}{x+i},\ \frac{-(x+i)}{x-i},\ \frac{-(x-i)}{x+i};$$

and if we proceed to form the product of the twelve factors $s-x$, $s-\frac{1}{x}$, $s+x$, &c., we have first the three products

$$s^2-x^2 . s^2-\frac{1}{x^2};\quad s^2+\left(\frac{x-1}{x+1}\right)^2 . s^2+\left(\frac{x+1}{x-1}\right)^2;\quad s^2-\left(\frac{x+i}{x-i}\right)^2 . s^2-\left(\frac{x-i}{x+i}\right)^2$$

$$=s^4+\alpha s^2+1;\qquad s^4+\beta s^2+1;\qquad s^4+\gamma s^2+1;$$

if for shortness

$$\alpha,\ \beta,\ \gamma=-\left(x^2+\frac{1}{x^2}\right),\quad 2\frac{x^4+6x^2+1}{(x^2-1)^2},\quad -2\frac{x^4-6x^2+1}{(x^2+1)^2}.$$

The product of the three quartic functions is

$$=(s^4+1)^3+(s^4+1)^2 s^2(\alpha+\beta+\gamma)+(s^4+1)s^4(\beta\gamma+\gamma\alpha+\alpha\beta)+s^6 . \alpha^2\beta^2\gamma^2;$$

and we have

$$\beta+\gamma=\frac{32x^2(x^4+1)}{(x^4-1)^2},\quad \alpha+\beta+\gamma=\frac{-(x^{12}-33s^8-33s^4+1)}{x^2(x^4-1)^2},$$

$$\beta\gamma=\frac{-4(x^8-34x^4+1)}{(x^4-1)^2},\quad \alpha(\beta+\gamma)=\frac{-32x^2(x^4+1)^2}{x^2(x^4-1)^2},$$

$$\beta\gamma+\gamma\alpha+\alpha\beta,\ =\frac{-36x^2(x^4-1)^2}{x^2(x^4-1)^2},\ =-36,\quad \alpha\beta\gamma=\frac{4(x^{12}-33s^8-33s^4+1)}{x^2(x^4-1)^2}.$$

Hence the product is found to be

$$=(s^{12}-33s^8-33s^4+1)-s^2(s^4-1)^2 . \frac{x^{12}-33x^8-33x^4+1}{x^2(x^4-1)^2},$$

which is

$$= s^2(s^4-1)^2\left\{\frac{s^{12}-33s^8-33s^4+1}{s^2(s^4-1)^2}-\frac{x^{12}-33x^8-33x^4+1}{x^2(x^4-1)^2}\right\}.$$

We thus verify that the twelve transformations x into x, into $\frac{1}{x}$, &c., give each of them a transformation of the function

$$\frac{x^{12}-33x^8-33x^4+1}{x^2(x^4-1)^2}$$

into itself.

The system of 15 *circles.* Art. Nos. 118 to 127.

118. It has been already remarked that we can from the coefficients (A, B, C, D) of the homographic transformation pass back to the position of the axis of rotation: viz. we have

$$A : B : C : D = -\nu - i : \lambda + i\mu : \lambda - i\mu : \nu - i,$$

and thence

$$\lambda : \mu : \nu : 1 = \quad B + C : -i(B - C) \quad : D - A \quad : i(D + A),$$

that is,

$$\lambda,\ \mu,\ \nu = -i(B+C),\ -\ (B-C),\ -i(D-A);\ \div(D+A).$$

The equations of the axis thus are

$$\frac{x}{B+C}=\frac{iy}{B-C}=\frac{z}{D-A},$$

and the equations of the central plane at right angles to the axis are

$$-(B+C)x + i(B-C)y + (A-D)z = 0.$$

119. In particular, we may find the equations of the 15 planes at right angles to the Θ-axes: these are in fact the before-mentioned 15 planes, intersecting the sphere in great circles the projections of which are the circles in the three figures respectively. Taking the equation of the plane to be $Lx + My + Nz = 0$, it is at once seen that the equation of the projecting cone (vertex at the South pole) is

$$N(x^2+y^2+z^2-1) - 2(z+1)(Lx+My+Nz) = 0,$$

and hence, writing $z=0$, we find

$$N(x^2+y^2-1) - 2(Lx+My) = 0$$

for the equation of the circle in the plane figure. We have thus the equations of a system of 15 circles related to each other in the manner before referred to.

120. Taking the Θ-form, the equations of the 15 planes are at once found: and we thence obtain the equations of the 15 circles: viz. writing for shortness

$$\Omega = x^2 + y^2 - 1,$$

the equations are

$$
\begin{array}{rll}
z=0, & (ab\,.\,cd) & \Omega=0, \\
x=0, & (ac\,.\,bd) & x=0, \\
y=0, & (ad\,.\,bc) & y=0, \\
(3-\surd 5)x+(\ \ 1-\surd 5)y+2z=0, & (ae\,.\,bc) & \Omega-[(\ \ 3-\surd 5)x+(\ \ 1-\surd 5)y]=0, \\
(-1-\surd 5)x+(-1+\surd 5)y+2z=0, & (ab\,.\,ce) & \Omega-[(-1-\surd 5)x+(-1+\surd 5)y]=0, \\
(1+\surd 5)x+(\ \ 3+\surd 5)y+2z=0, & (ac\,.\,be) & \Omega-[(\ \ 1+\surd 5)x+(\ \ 3+\surd 5)y]=0; \\
(-3+\surd 5)x+(-1+\surd 5)y+2z=0, & (ad\,.\,be) & \text{and similarly for the other circles.} \\
(1+\surd 5)x+(\ \ 1-\surd 5)y+2z=0, & (ab\,.\,de) & \\
(-1-\surd 5)x+(-3-\surd 5)y+2z=0, & (ae\,.\,bd) & \\
(-3+\surd 5)x+(\ \ 1-\surd 5)y+2z=0, & (ad\,.\,ce) & \\
(1+\surd 5)x+(-1+\surd 5)y+2z=0, & (ae\,.\,cd) & \\
(-1-\surd 5)x+(\ \ 3+\surd 5)y+2z=0, & (ac\,.\,de) & \\
(3-\surd 5)x+(-1+\surd 5)y+2z=0, & (bc\,.\,de) & \\
(-1-\surd 5)x+(\ \ 1-\surd 5)y+2z=0, & (be\,.\,cd) & \\
(1+\surd 5)x+(-3-\surd 5)y+2z=0, & (bd\,.\,ce). &
\end{array}
$$

121. Observe that the arrangement is in sets of 3 planes, or circles, intersecting at right angles. One of the circles is the circle $\Omega, =x^2+y^2-1, =0$ corresponding to the equator, and two of them are the right lines $x=0$ and $y=0$. The equations of the remaining 12 circles may be written in the somewhat different form

$$
\begin{aligned}
&\Omega+(\surd 5-1)\,[y-\tfrac{1}{2}(\surd 5-1)\,x]=0,\\
&\Omega-(\surd 5-1)\,[y-\tfrac{1}{2}(\surd 5+3)\,x]=0,\\
&\Omega-(\surd 5+3)\,[y+\tfrac{1}{2}(\surd 5-1)\,x]=0,\\
&\Omega-(\surd 5-1)\,[y-\tfrac{1}{2}(\surd 5-1)\,x]=0,\\
&\Omega+(\surd 5-1)\,[y-\tfrac{1}{2}(\surd 5+3)\,x]=0,\\
&\Omega+(\surd 5+3)\,[y+\tfrac{1}{2}(\surd 5-1)\,x]=0,\\
&\Omega+(\surd 5-1)\,[y+\tfrac{1}{2}(\surd 5-1)\,x]=0,\\
&\Omega-(\surd 5-1)\,[y+\tfrac{1}{2}(\surd 5+3)\,x]=0,\\
&\Omega-(\surd 5+3)\,[y-\tfrac{1}{2}(\surd 5-1)\,x]=0,\\
&\Omega-(\surd 5-1)\,[y+\tfrac{1}{2}(\surd 5-1)\,x]=0,\\
&\Omega+(\surd 5-1)\,[y+\tfrac{1}{2}(\surd 5+3)\,x]=0,\\
&\Omega+(\surd 5+3)\,[y-\tfrac{1}{2}(\surd 5-1)\,x]=0.
\end{aligned}
$$

It hence appears that 4 and 4 circles have with $\Omega=0$ the common chords $y+\tfrac{1}{2}(\surd 5-1)\,x=0$, $y-\tfrac{1}{2}(\surd 5-1)\,x=0$ respectively: and that 2 and 2 circles have with $\Omega=0$ the common chords $y+\tfrac{1}{2}(\surd 5+3)\,x=0$, $y-\tfrac{1}{2}(\surd 5+3)\,x=0$ respectively.

122. The equations of the 12 circles are, in fact,

$$\Omega \pm (\sqrt{5}-1)\,[y \pm \tfrac{1}{2}(\sqrt{5}-1)\,x] = 0, \quad \Omega \pm (\sqrt{5}+3)\,[y \pm \tfrac{1}{2}(\sqrt{5}-1)\,x] = 0,$$
$$\Omega \pm (\sqrt{5}-1)\,[y \pm \tfrac{1}{2}(\sqrt{5}+3)\,x] = 0:$$

hence the radii are $=\sqrt{5}-1$, 2 and $\sqrt{5}+1$ respectively.

The construction of the 12 circles is as follows. Starting with a circle radius 1.

Lay down the diameters $y \pm \frac{1}{2}(\sqrt{5}-1)\,x = 0$ (*AA* in the figure), and through the extremities of each describe 2 pairs of circles with the radii $\sqrt{5}-1$, $\sqrt{5}+1$ respectively.

Lay down the diameters $y \pm \frac{1}{2}(\sqrt{5}+3)\,x = 0$ (*BB* in the figure), and through the extremities of each describe a pair of circles with the radius 2.

123. For the *A*-form, the equations of the fifteen planes are at once found to be

$$\begin{array}{lllr} & y & = 0, & ad\,.\,bc \\ -x & & +\;(\epsilon+\epsilon^4)\,z = 0, & ac\,.\,bd \\ (\epsilon+\epsilon^4)\,x & & +\;z = 0, & ab\,.\,cd \\ \hline (\epsilon^2-\epsilon^3)\,x & -i\,(\epsilon^2+\epsilon^3)\,y & = 0, & ac\,.\,be \\ -(\epsilon^2+\epsilon^3)\,x & +i\,(\epsilon^2-\epsilon^3)\,y + 2\,(\epsilon+\epsilon^4)\,z & = 0, & ae\,.\,bc \\ -x + i\,(\epsilon^2+\epsilon^4-\epsilon-\epsilon^3)\,y + & 2z & = 0, & ab\,.\,ce \\ \hline (\epsilon-\epsilon^4)\,x & -i\,(\epsilon+\epsilon^4)\,y & = 0, & ab\,.\,de \\ -(\epsilon+\epsilon^4)\,x & +i\,(\epsilon-\epsilon^4)\,y + 2\,(\epsilon+\epsilon^4)\,z & = 0, & ae\,.\,bd \\ +(\epsilon^2+\epsilon^3+2)\,x & -i\,(\epsilon^2-\epsilon^3)\,y + & 2z = 0, & ad\,.\,be \\ \hline (\epsilon-\epsilon^4)\,x & +i\,(\epsilon+\epsilon^4)\,y & = 0, & ae\,.\,cd \\ -(\epsilon+\epsilon^4)\,x & -i\,(\epsilon-\epsilon^4)\,y + 2\,(\epsilon+\epsilon^4)\,z & = 0, & ac\,.\,de \\ (\epsilon^2+\epsilon^3+2)\,x & +i\,(\epsilon^2-\epsilon^3)\,y + & 2z = 0, & ad\,.\,ce \\ \hline (\epsilon^2-\epsilon^3)\,x & +i\,(\epsilon^2+\epsilon^3)\,y & = 0, & bd\,.\,ce \\ -(\epsilon^2+\epsilon^3)\,x & -i\,(\epsilon^2-\epsilon^3)\,y + 2\,(\epsilon+\epsilon^4)\,z & = 0, & bc\,.\,de \\ -x - i\,(\epsilon^2+\epsilon^4-\epsilon-\epsilon^3)\,y + & 2z & = 0, & be\,.\,cd, \end{array}$$

where, as before, the three planes of each set intersect at right angles.

124. Passing to the circles, the first plane of each set gives a right line, and we have thus five of the circles reducing themselves to right lines inclined to the axis of x at angles 0°, 36°, 72°, 108° and 144° respectively.

The remaining 10 circles form 5 pairs, the circles of a pair having different radii, but the two radii being the same for each pair, and so that for the several pairs the common chords with the circle $\Omega = 0$, are the diameters inclined to the axis of x at the angles 18°, 54°, 90°, 126° and 162° respectively. Considering the two circles for which the inclination is 90°, these arise from the planes $-x+(\epsilon+\epsilon^4)\,z=0$, $(\epsilon+\epsilon^4)\,x+z=0$ respectively. The equations of the circles thus are $(\epsilon+\epsilon^4)\,\Omega+2x=0$,

$\Omega - 2(\epsilon + \epsilon^4)x = 0$, or recollecting that $2(\epsilon + \epsilon^4) = \sqrt{5} - 1$ and therefore $\dfrac{2}{\epsilon + \epsilon^4} = \sqrt{5} + 1$, the equations are

$$x^2 + y^2 - (\sqrt{5} - 1)x - 1 = 0, \quad x^2 + y^2 + (\sqrt{5} + 1)x = 0;$$

hence for the first circle the x-coordinate of the centre is $\frac{1}{2}(\sqrt{5} - 1)$ and the radius is $= \frac{1}{2}\sqrt{(10 - 2\sqrt{5})}$; for the second circle the x-coordinate of the centre is $= \frac{1}{2}(\sqrt{5} + 1)$, and the radius $= \frac{1}{2}\sqrt{(10 + 2\sqrt{5})}$. We have thus the construction of these two circles, and consequently the construction of all the 12 circles.

125. For the B-form, after some easy reductions and attending to the relation $\omega - \omega^2 = i\sqrt{3}$, the equations of the 15 planes become

x			$= 0,$	$ac\,.\,bd$
	$(-3 + \sqrt{5})\,y +$		$2z = 0,$	$ab\,.\,cd$
	$(3 + \sqrt{5})\,y +$		$2z = 0,$	$ad\,.\,bc$
$\sqrt{3}x +$	$\sqrt{5}\;y +$		$2z = 0,$	$ac\,.\,be$
$-(1 + \sqrt{5})\sqrt{3}x +$	$(3 - \sqrt{5})\,y +$		$4z = 0,$	$ab\,.\,ce$
$(-1 + \sqrt{5})\sqrt{3}x +$	$(-3 - \sqrt{5})\,y +$		$4z = 0,$	$ae\,.\,bc$
$x +$	$\sqrt{3}\;y$		$= 0,$	$ae\,.\,bd$
$-\sqrt{3}x +$		$y + (3 + \sqrt{5})$	$z = 0,$	$ad\,.\,be$
$\sqrt{3}x -$		$y + (3 - \sqrt{5})$	$z = 0,$	$ab\,.\,de$
$-\sqrt{3}x +$	$\sqrt{5}\;y +$		$2z = 0,$	$ac\,.\,de$
$(1 - \sqrt{5})\sqrt{3}x +$	$(-3 - \sqrt{5})\,y +$		$4z = 0,$	$ad\,.\,ce$
$(1 + \sqrt{5})\sqrt{3}x +$	$(3 - \sqrt{5})\,y +$		$4z = 0,$	$ae\,.\,cd$
$x -$	$\sqrt{3}\;y$		$= 0,$	$bd\,.\,ce$
$\sqrt{3}x +$		$y + (3 + \sqrt{5})$	$z = 0,$	$bc\,.\,de$
$-\sqrt{3}x -$		$y + (3 - \sqrt{5})$	$z = 0,$	$be\,.\,cd.$

126. Of the 15 circles, 3 are the lines $x - y\sqrt{3} = 0$, $x = 0$, $x + y\sqrt{3} = 0$, viz. these are lines at inclinations 30°, 90°, 150° to the axis of x. The equations of the remaining 12 circles are

$$\Omega + (3 - \sqrt{5})\,y = 0,$$
$$\Omega - (3 + \sqrt{5})\,y = 0,$$
$$(3 + \sqrt{5})\,\Omega - 2(y - x\sqrt{3}) = 0,$$
$$(3 - \sqrt{5})\,\Omega + 2(y - x\sqrt{3}) = 0,$$
$$(3 + \sqrt{5})\,\Omega - 2(y + x\sqrt{3}) = 0,$$
$$(3 - \sqrt{5})\,\Omega + 2(y + x\sqrt{3}) = 0,$$

viz. these are pairs of circles having, for their common chords with $\Omega = 0$, the diameters at inclinations 0°, 60°, 120° respectively. And, lastly, we have the circles

$$\begin{array}{l|l} 2\Omega - [(-1+\sqrt{5})\sqrt{3}x - (3+\sqrt{5})y] = 0, & 2\Omega + [(-1+\sqrt{5})\sqrt{3}x + (3+\sqrt{5})y] = 0, \\ \Omega - [\qquad -\sqrt{3}x + \qquad \sqrt{5}\,y] = 0, & \Omega - [\qquad \sqrt{3}x + \qquad \sqrt{5}\,y] = 0, \\ 2\Omega + [(\;1+\sqrt{5})\sqrt{3}x - (3-\sqrt{5})y] = 0, & 2\Omega - [(\;1+\sqrt{5})\sqrt{3}x + (3-\sqrt{5})y] = 0. \end{array}$$

127. The first three of these have, for common chords with $\Omega = 0$, the diameters whose equations are

$$(-1+\sqrt{5})\sqrt{3}x - (3+\sqrt{5})y = 0, \quad -\sqrt{3}x + \sqrt{5}y = 0, \quad (1+\sqrt{5})\sqrt{3}x - (3-\sqrt{5})y = 0:$$

viz. these equations are $y = (-2+\sqrt{5})\,x\sqrt{3}$, $y = \dfrac{\sqrt{3}}{\sqrt{5}}x$, $y = (2+\sqrt{5})\,x\sqrt{3}$. If, as in a foregoing table, $\theta = 37°\,46'$, $\sin\theta = \dfrac{\sqrt{3}}{2\sqrt{2}}$, $\cos\theta = \dfrac{\sqrt{5}}{2\sqrt{2}}$, and therefore $\tan\theta = \dfrac{\sqrt{3}}{\sqrt{5}}$; then the inclinations of these diameters to the axis of x are respectively $60° - \theta$, θ and $120° - \theta$, or say $30° - (\theta - 30°)$, $30° + (\theta - 30°)$ and $90° - (\theta - 30°)$, where $\theta - 30° = 7°\,46'$, i.e. the inclinations are $30° \pm 7°\,46'$ and $90° - 7°\,46'$. And for the other three circles the common chords are the diameters at the same inclinations taken negatively. The geometrical construction of the fifteen circles for the B-case in question is thus not so simple as in the Θ- and A-cases.

The Regular Polyhedra as Solid figures. Art. Nos. 128 to 134.

128. I annex some results relating to the polyhedra considered as solid figures bounded by plane faces; or say results relating to the regular solids: s is in each case taken for the length of the edge of the solid.

	Tetrahedron.	Cube.	Octahedron.	Dodecahedron.	Icosahedron.
Edge	s	s	s	s	s
Rad. of circum. sphere, R	$s\dfrac{\sqrt{3}}{2\sqrt{2}}$	$s\,.\,\tfrac{1}{2}\sqrt{3}$	$s\dfrac{1}{\sqrt{2}}$	$s\dfrac{\sqrt{3}\,(\sqrt{5}+1)}{4}$	$s\sqrt{\dfrac{5+\sqrt{5}}{8}}$
Rad. of inters. sphere, ρ	$s\dfrac{1}{2\sqrt{2}}$	$s\dfrac{1}{\sqrt{2}}$	$s\,.\,\tfrac{1}{2}$	$s\dfrac{3+\sqrt{5}}{4}$	$s\dfrac{1+\sqrt{5}}{4}$
Rad. of inscribed sphere, r	$s\dfrac{1}{2\sqrt{2}\sqrt{3}}$	$s\,.\,\tfrac{1}{2}$	$s\dfrac{1}{\sqrt{2}\sqrt{3}}$	$s\sqrt{\dfrac{25+11\sqrt{5}}{40}}$	$s\dfrac{3+\sqrt{5}}{4\sqrt{3}}$
Rad. of circle circum. to face, R'	$s\,.\,\dfrac{1}{\sqrt{3}}$	$s\dfrac{1}{\sqrt{2}}$	$s\dfrac{1}{\sqrt{3}}$	$s\sqrt{\dfrac{5+\sqrt{5}}{10}}$	$s\dfrac{1}{\sqrt{3}}$
Rad. of circle inscribed to face, r'	$s\,.\,\dfrac{1}{2\sqrt{3}}$	$s\,.\,\tfrac{1}{2}$	$s\dfrac{1}{2\sqrt{3}}$	$s\sqrt{\dfrac{5+2\sqrt{5}}{20}}$	$s\dfrac{1}{2\sqrt{3}}$
Incl. of adjacent faces	$\cos^{-1}\tfrac{1}{3} = 70°\,28'$	90°	$\cos^{-1} - \tfrac{1}{3} = 109°\,32'$		
Incl. of edge to adjacent face	$\cos^{-1}\dfrac{1}{\sqrt{3}} = 54°\,46'$	90°	$\cos^{-1} - \dfrac{1}{\sqrt{3}} \doteq 125°\,44'$		

But we require further data in the cases of the dodecahedron and the icosahedron respectively.

129. For the dodecahedron, taking the edge to be $=s$ as before, then in the pentagonal face

$$\begin{aligned} \text{diagonal, } g \text{ is } &= s\,.\,\tfrac{1}{2}(\sqrt{5}+1), \\ \text{altitude, } k \;\;\text{,,}\;\; &= s\,.\,\tfrac{1}{2}\sqrt{(5+2\sqrt{5})}, \\ \text{segments of do., } e \;\;\text{,,}\;\; &= s\,.\,\tfrac{1}{4}\sqrt{(10-2\sqrt{5})}, \\ f \;\;\text{,,}\;\; &= s\,.\,\tfrac{1}{4}\sqrt{(10+2\sqrt{5})}, \end{aligned}$$

where

$$k = e + f = R' + r'.$$

130. The section through a pair of opposite edges is a hexagon, as shown in the figure, viz. this is constructed by taking the four equal distances $O\Theta$, $=\rho$, $=s\,.\,\tfrac{1}{4}(3+\sqrt{5})$, meeting at right angles in O; then drawing the double ordinates BB, each $=s$, through Θ_1 and Θ_3 respectively, and joining their extremities with Θ_2 and Θ_4: the sides $\Theta_2 B$ and $\Theta_4 B$ are then each $=k$, $=s\,.\,\tfrac{1}{2}\sqrt{(5+2\sqrt{5})}$; and inserting upon them the points A, Φ from the figure of the pentagon, we have several

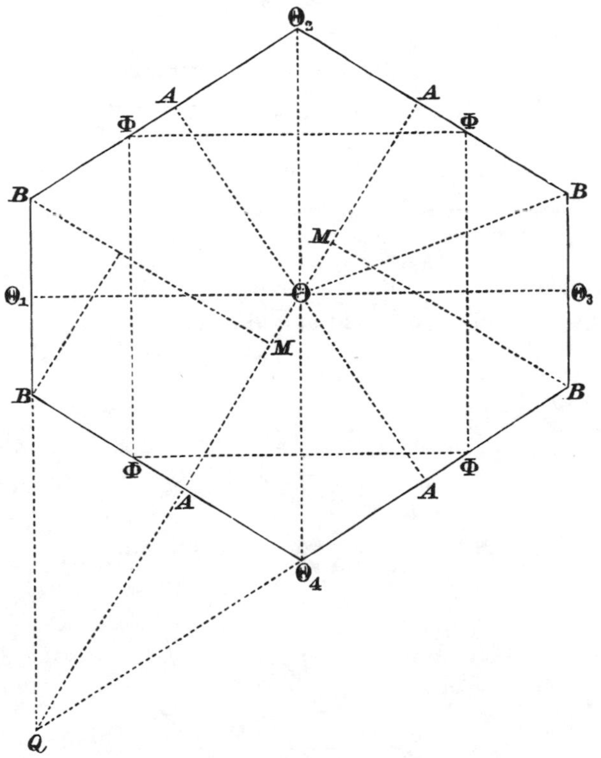

geometrical relations; viz. the line AA cuts the parallel sides $B\Theta_2$, $B\Theta_4$ at right angles, and when produced passes through the intersection of $B\Theta_1$ and $B\Theta_4$: we have OA, OB, $O\Theta = r$, R, ρ respectively: the four points Φ form a square, the side of which is g, $=s\,.\,\tfrac{1}{2}(\sqrt{5}+1)$.

131. We find also

$$AM = s\sqrt{\frac{5+\sqrt{5}}{10}},$$

$$AQ = s\sqrt{\frac{5+2\sqrt{5}}{5}},$$

$$QM = s\sqrt{\frac{25+11\sqrt{5}}{5}}, \quad = r \,.\, 2\sqrt{2},$$

$$OQ = s\sqrt{\frac{25+11\sqrt{5}}{8}}, \quad = r \,.\, \sqrt{5},$$

$$OM = s\sqrt{\frac{5-\sqrt{5}}{40}},$$

$$MB = s\sqrt{\frac{2(5+2\sqrt{5})}{B}}.$$

It may be remarked that in the figure $B\Theta_2$, $B\Theta_4$ are the projections of pentagonal faces, at right angles to the plane of the paper, having their centres at the points A, A, and the perpendicular distance between them $=AA$: the points Q, Q (only one of them shown in the figure) determine the directions of the $5+5$ sides which abut on these pentagonal faces respectively; and the $5+5$ points B which are the other extremities of these sides respectively form two pentagons, centres M, M in the planes MB and MB respectively: the remaining 10 sides of the dodecahedron are the skew decagon obtained by joining in order these 10 points B. We have thus the means of making the perspective delineation of the dodecahedron.

132. The dodecahedron is built up from the cube, by placing on each face a figure of two triangular and two quadrangular faces, the orthogonal projection of which on the face of the cube is as in the figure: the side of the square is g,

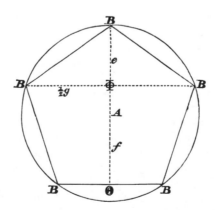

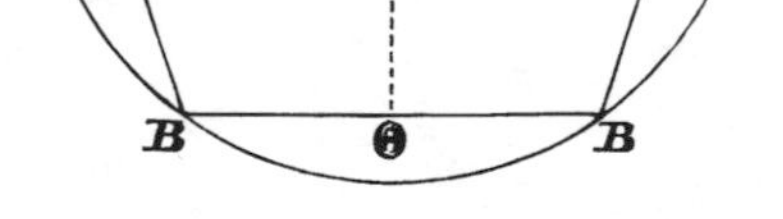

$=s\,.\,\frac{1}{2}(\sqrt{5}+1)$: the slope-breadths of the triangular faces are e, $=s\,.\,\frac{1}{4}\sqrt{(10-2\sqrt{5})}$, and those of the quadrangular faces are f, $=s\,.\,\frac{1}{4}\sqrt{(10+2\sqrt{5})}$; the lines represented by the other lines of the figure are in actual length each $=s$. We have thus a section which is an isosceles triangle, base $=g$, other sides each $=f$; and the square of the altitude is thus $=f^2-\frac{1}{4}g^2=\frac{1}{4}s^2$, or the altitude $=\frac{1}{2}s$; viz. the altitude of the ridge-line BB, above the face of the cube is $=\frac{1}{2}s$, the half-side of the dodecahedron.

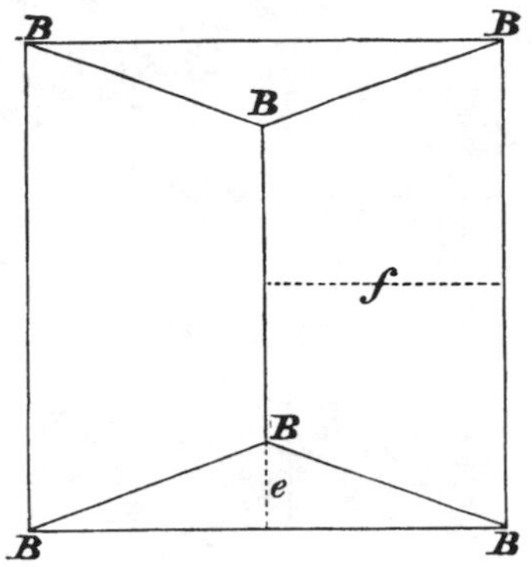

We have in this result the most simple means of forming the perspective delineation of the dodecahedron.

133. For the icosahedron the section through two opposite edges is a hexagon, as shown in the figure (p. 216): to construct it, we take the four distances $O\Theta$ each $=\rho=s.\frac{1}{4}(1+\sqrt{5})$ meeting at right angles; and then the distances $A\Theta_2$, $A\Theta_4$ each $=\frac{1}{2}s$; and complete the hexagon. This gives the sides $A\Theta_1$, $A\Theta_3$ each $=s.\frac{1}{2}\sqrt{3}$, the altitude of the triangular face, side $=s$; and then, taking $\Theta_1 B$ one-third of this, $=s\dfrac{1}{2\sqrt{3}}$, we have OB at right angles to $A\Theta_1$, and OA, OB, $O\Theta=R$, r, ρ respectively.

Moreover, joining $A_1\Theta_2$ and OA_2, we have these lines cutting at right angles in a point M: we find

$$A_1\Theta_3=s.\tfrac{1}{2}\sqrt{(5+2\sqrt{5})},$$

$$M\Theta_3=s\sqrt{\frac{5+2\sqrt{5}}{20}},$$

$$A_1M=s\sqrt{\frac{5+\sqrt{5}}{10}},$$

$$OM=s\sqrt{\frac{5+\sqrt{5}}{40}},\ =\tfrac{1}{2}A_1M,$$

$$A_2M=s\sqrt{\frac{5-\sqrt{5}}{10}}.$$

134. It may be remarked that $A_1\Theta_3$, $A_3\Theta_1$ are the projections of two pentagons in planes perpendicular to that of the paper, their centres being M, M: producing OM, OM to the points A_2, A_4 respectively, we have a pentagonal pyramid, summit A_2, standing on the first pentagon, and an opposite pyramid, summit A_4, standing on

the other pentagon: the 5 + 5 triangular faces of the two pyramids are ten of the faces of the icosahedron, and the remaining ten faces are the triangles each having for its base a side of the one pentagon, and for its vertex a summit of the other

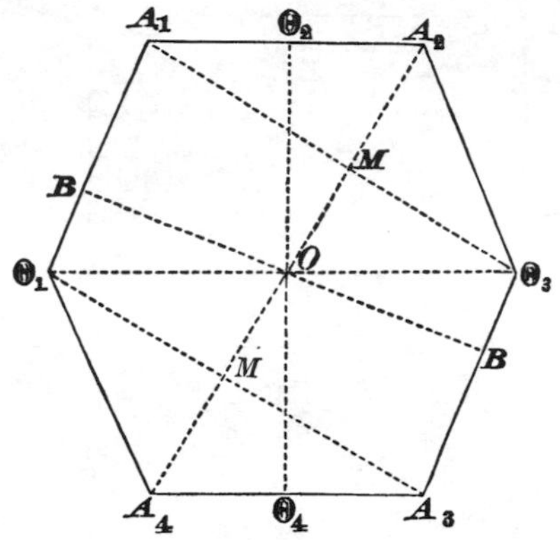

pentagon, viz. the sides are the sides of the skew decagon obtained by joining in order the angular points of the two pentagons. We have thus a convenient method of forming the perspective delineation of the icosahedron.

746.

HIGHER PLANE CURVES.

[From *Salmon's Higher Plane Curves,* (3rd *ed.,* 1879); *see the Preface.*]

ONE chapter and a large number of articles, in the second edition of Salmon's *Higher Plane Curves,* are due to Professor Cayley. Full reference to these is given by Dr Salmon in the preface.

747.

NOTE ON THE DEGENERATE FORMS OF CURVES.

[From *Salmon's Higher Plane Curves*, (3rd *ed.*, 1879), pp. 383—385.]

SOME remarks may be added as to the analytical theory of the degenerate forms of curves. As regards conics, a line-pair can be represented in point-coordinates by an equation of the form $xy = 0$; and reciprocally a point-pair can be represented in line-coordinates by an equation $\xi\eta = 0$, but we have to consider how the point-pair can be represented in point-coordinates: an equation $x^2 = 0$ is no adequate representation of the point-pair, but merely represents (as a two-fold or twice repeated line) the line joining the two points of the point-pair, all traces of the points themselves being lost in this representation: and it is to be noticed, that the conic, or two-fold line $x^2 = 0$, or say $(\alpha x + \beta y + \gamma z)^2 = 0$ is a conic which, analytically, and (in an improper sense) geometrically, satisfies the condition of touching *any line whatever;* whereas the only proper tangents of a point-pair are the lines which pass through one or other of the two points of the point-pair.

The solution arises out of the notion of a point-pair, considered as the limit of a conic, or say as an indefinitely flat conic; we have to consider conics certain of the coefficients whereof are infinitesimals, and which, when the infinitesimal coefficients actually vanish, reduce themselves to two-fold lines; and it is, moreover, necessary to consider the evanescent coefficients as infinitesimals of different orders. Thus consider the conics which pass through two given points, and touch two given lines (four conditions); take $y = 0$, $z = 0$ for the given lines, $x = 0$ for the line joining the given points, and $(x = 0,\ y - \alpha z = 0)$, $(x = 0,\ y - \beta z = 0)$ for the given points; the equation of a conic satisfying the required conditions and containing one arbitrary parameter θ, is

$$x^2 + 2\theta xy + 2\theta\sqrt{(\alpha\beta)}\,xz + \theta^2(y - \alpha z)(y - \beta z) = 0;$$

or, what is the same thing,

$$\{x + \theta y + \theta \sqrt{(\alpha\beta)}\, z\}^2 - \theta^2 (\alpha + \beta)\, yz = 0\,;$$

and this equation, considering therein θ as an infinitesimal, say of the first order, represents the flat conic or point-pair composed of the two given points. Comparing with the general equation

$$(a,\ b,\ c,\ f,\ g,\ h \between x,\ y,\ z)^2 = 0,$$

we have

$$a = 1,\quad b = \theta^2,\quad c = \theta^2\alpha\beta,\quad f = -\tfrac{1}{2}\theta^2 (\alpha + \beta),\quad g = \theta \sqrt{(\alpha\beta)},\quad h = \theta,$$

viz. a being taken to be finite, we have g and h infinitesimals of the first order; b, c, f infinitesimals of the second order; and the four ratios $\sqrt{(b)} : \sqrt{(c)} : \sqrt{(f)} : g : h$ are so determined as to satisfy the prescribed conditions.

Observe that the flat conic, considered as a conic passing through the two given points and touching the two given lines, is represented by a *determinate* equation, viz. considering the condition imposed upon θ (θ = infinitesimal) as a determination of θ, the equation is a completely determinate one; but considering the flat conic merely as a conic passing through the two given points, the equation would contain two arbitrary parameters, determinable if the flat conic was subjected to the condition of touching two given lines, or to any other two conditions.

Generally, we may consider the equation of a curve of the order n; such equation containing certain infinitesimal coefficients and, when these vanish, reducing itself to a composite equation $P^\alpha Q^\beta \ldots = 0$; the equation in its original form represents a curve which may be called the penultimate curve. Consider the tangents from an arbitrary point to the penultimate curve; when this breaks up, the system of tangents reduces itself to (1) the tangents from the fixed point to the several component curves $P = 0$, $Q = 0$, &c. respectively; (2) the lines through the singular points of these same curves respectively; (3) the lines through the points of intersection $P = 0$, $Q = 0$, &c. of each two of the component curves; these points, each reckoned a proper number of times, are called "fixed summits"; (4) the lines from the fixed point to certain determinate points called "free summits" on the several component curves $P = 0$, $Q = 0$, &c. respectively. We have thus a degenerate form of the n-thic curve, which may be regarded as consisting of the component curves, each its proper number of times, and of the foregoing points called summits, and is consequently only inadequately represented by the ultimate equation $P^\alpha Q^\beta \ldots = 0$; the number and distribution of the summits is not arbitrary, but is regulated by laws arising from the consideration of the penultimate curve, and there are of course for any given value of n various forms of degenerate curve, according to the different ultimate forms $P^\alpha Q^\beta \ldots = 0$, and to the number and distribution of the summits on the different component curves. The case of a quartic curve having the ultimate form $x^2y^2 = 0$ has been considered by Cayley, *Comptes Rendus*, t. LXXIV. p. 708 (March, 1872), [515], who states his conclusion as follows:

"there exists a quartic curve the penultimate of $x^2y^2=0$, with nine free summits, three of them on one of the lines (say the line $y=0$), and which are three of the intersections of the quartic by this line (the fourth intersection being indefinitely near to the point $x=0$, $y=0$), six situate at pleasure on the other line $x=0$; and three fixed summits at the intersection of the two lines." Other forms have been considered by Dr Zeuthen, *Comptes Rendus*, t. LXXV. pp. 703 and 950 (September and October, 1872), and some other forms by Zeuthen; the whole question of the degenerate forms of curves is one well deserving further investigation.

The question of the number of cubic curves satisfying given elementary conditions (depending as it does on the consideration of the degenerate forms of these curves) has been solved by Maillard and Zeuthen; that of the number of quartic curves has been solved by Dr Zeuthen.

748.

ON THE BITANGENTS OF A QUARTIC.

[From *Salmon's Higher Plane Curves*, (3rd *ed.*, 1879), pp. 387—389.]

THE equations of the 28 bitangents of a quartic curve were obtained in a very elegant form by Riemann in the paper "Zur Theorie der Abel'schen Functionen für den Fall $p=3$," *Ges. Werke*, Leipzig, 1876, pp. 456—472; and see also Weber's *Theorie der Abel'schen Functionen vom Geschlecht* 3," Berlin, 1876. Riemann connects the several bitangents with the characteristics of the 28 odd functions, thus obtaining for them an algorithm which it is worth while to explain, but they will be given also with the algorithm employed p. 231 *et seq.* of the present work*, which is in fact the more simple one. The characteristic of a triple θ-function is a symbol of the form

$$\begin{matrix} \alpha\,\beta\,\gamma, \\ \alpha'\beta'\gamma', \end{matrix}$$

where each of the letters is $=0$ or 1; there are thus in all 64 such symbols, but they are considered as odd or even according as the sum $\alpha\alpha'+\beta\beta'+\gamma\gamma'$ is odd or even; and the numbers of the odd and even characteristics are 28 and 36 respectively; and, as already mentioned, the 28 odd characteristics correspond to the 28 bitangents respectively.

We have x, y, z trilinear coordinates, α, β, γ, α', β', γ' constants chosen at pleasure, and then α'', β'', γ'' determinate constants, such that the equations

$$x+y+z+\xi+\eta+\zeta=0,$$

$$\alpha x+\beta y+\gamma z+\frac{\xi}{\alpha}+\frac{\eta}{\beta}+\frac{\zeta}{\gamma}=0,$$

$$\alpha' x+\beta' y+\gamma' z+\frac{\xi}{\alpha'}+\frac{\eta}{\beta'}+\frac{\zeta}{\gamma'}=0,$$

$$\alpha'' x+\beta'' y+\gamma'' z+\frac{\xi}{\alpha''}+\frac{\eta}{\beta''}+\frac{\zeta}{\gamma''}=0,$$

[* That is, Salmon's *Higher Plane Curves*.]

are equivalent to three independent equations; this being so, they determine ξ, η, ζ, each of them as a linear function of (x, y, z); and the equations of the bitangents of the curve $\surd(x\xi)+\surd(y\eta)+\surd(z\zeta)=0$ (see Weber, p. 100) are

18	111 111	$x=0$,
28	001 011	$y=0$,
38	011 001	$z=0$,
23	010 010	$\xi=0$,
13	100 110	$\eta=0$,
12	110 100	$\zeta=0$,
48	101 100	$x+y+z=0$,
14	010 011	$\xi+y+z=0$,
58	100 101	$\alpha x+\beta y+\gamma z=0$,
15	011 010	$\frac{\xi}{\alpha}+\beta y+\gamma z=0$,
68	110 010	$\alpha' x+\beta' y+\gamma' z=0$,
16	001 101	$\frac{\xi}{\alpha'}+\beta' y+\gamma' z=0$,
78	010 110	$\alpha'' x+\beta'' y+\gamma'' z=0$,
17	101 001	$\frac{\xi}{\alpha''}+\beta'' y+\gamma'' z=0$,
24	100 111	$x+\eta+z=0$,
34	110 101	$x+y+\zeta=0$,
25	101 110	$\alpha x+\frac{\eta}{\beta}+\gamma z=0$,
35	111 100	$\alpha x+\beta y+\frac{\zeta}{\gamma}=0$,

26	111 001	$\alpha' x + \frac{\eta}{\beta'} \quad + \gamma' z = 0,$
36	101 011	$\alpha' x + \beta' y + \frac{\zeta}{\gamma'} \quad = 0,$
27	011 101	$\alpha'' x + \frac{\eta}{\beta''} \quad + \gamma'' z = 0,$
37	001 111	$\alpha'' x + \beta'' y + \frac{\zeta}{\gamma''} = 0,$
67	100 100	$\frac{x}{1 - \beta\gamma} + \frac{y}{1 - \gamma\alpha} + \frac{z}{1 - \alpha\beta} = 0,$
57	110 011	$\frac{x}{1 - \beta'\gamma'} + \frac{y}{1 - \gamma'\alpha'} + \frac{z}{1 - \alpha'\beta'} = 0,$
56	010 111	$\frac{x}{1 - \beta''\gamma''} + \frac{y}{1 - \gamma''\alpha''} + \frac{z}{1 - \alpha''\beta''} = 0,$
45	001 001	$\frac{\xi}{\alpha (1 - \beta\gamma)} + \frac{\eta}{\beta (1 - \gamma\alpha)} + \frac{\zeta}{\gamma (1 - \alpha\beta)} = 0,$
46	011 110	$\frac{\xi}{\alpha' (1 - \beta'\gamma')} + \frac{\eta}{\beta' (1 - \gamma'\alpha')} + \frac{\zeta}{\gamma' (1 - \alpha'\beta')} = 0,$
47	111 010	$\frac{\xi}{\alpha'' (1 - \beta''\gamma'')} + \frac{\eta}{\beta'' (1 - \gamma''\alpha'')} + \frac{\zeta}{\gamma'' (1 - \alpha''\beta'')} = 0.$

The whole number of ways in which the equation of the curve can be expressed in a form such as $\sqrt{(x\xi)} + \sqrt{(y\eta)} + \sqrt{(z\zeta)} = 0$ is 1260; viz. the three pairs of bitangents entering into the equation of the curve are of one of the types

12.34,	13.24,	14.23	⊠	No. is	70
12.34,	13.24,	56.78	□ | |	"	630
13.23,	14.24,	15.25	◈	"	560
					1260.

It may be remarked that, selecting at pleasure any two pairs out of a system of three pairs, the type is always □ or \|\|\|\|, viz. (see p. 233) the four bitangents are such that their points of contact are situate on a conic.

749.

SOLID GEOMETRY.

[From *Salmon's Treatise on the analytic geometry of three dimensions,* (3rd *ed.*, 1874); *see the Preface.*]

A considerable number of articles in the third edition of Salmon's *Treatise* are due to Professor Cayley. Full reference to these is given by Dr Salmon in the preface.

750.

ON THE THEORY OF RECIPROCAL SURFACES.

[From *Salmon's Treatise on the analytic geometry of three dimensions*, (3rd *ed.*, 1874), pp. 539—550.]

600. In further developing the theory of reciprocal surfaces it has been found necessary to take account of other singularities, some of which are as yet only imperfectly understood. It will be convenient to give the following complete list of the quantities which present themselves:

n, order of the surface.

a, order of the tangent cone drawn from any point to the surface.

δ, number of nodal edges of the cone.

κ, number of its cuspidal edges.

ρ, class of nodal torse.

σ, class of cuspidal torse.

b, order of nodal curve.

k, number of its apparent double points.

f, number of its actual double points.

t, number of its triple points.

j, number of its pinch-points.

q, its class.

c, order of cuspidal curve.

h, number of its apparent double points.

θ, number of its points of an unexplained singularity.

χ, number of its close-points.

ω, number of its off-points.

r, its class.

β, number of intersections of nodal and cuspidal curves, stationary points on cuspidal curve.

γ, number of intersections, stationary points on nodal curve.

i, number of intersections, not stationary points on either curve.

C, number of cnicnodes of surface.

B, number of binodes.

And corresponding reciprocally to these:

n', class of surface.

a', class of section by arbitrary plane.

δ', number of double tangents of section.

κ', number of its inflexions.

ρ', order of node-couple curve.

σ', order of spinode curve.

b', class of node-couple torse.

k', number of its apparent double planes.

f', number of its actual double planes.

t', number of its triple planes.

j', number of its pinch-planes.

q', its order.

c', class of spinode torse.

h', number of its apparent double planes.

θ', number of its planes of a certain unexplained singularity.

χ', number of its close-planes.

ω', number of its off-planes.

r', its order.

β', number of common planes of node-couple and spinode torse, stationary planes of spinode torse.

γ', number of common planes, stationary planes of node-couple torse.

ι', number of common planes, not stationary planes of either torse.

C', number of cnictropes of surface.

B', number of its bitropes.

In all, these are 46 quantities.

601. In part explanation, observe that the definitions of ρ and σ agree with those already given. The nodal torse is the torse enveloped by the tangent planes along the nodal curve; if the nodal curve meets the curve of contact a, then a tangent plane of the nodal torse passes through the arbitrary point, that is, ρ will be the number of these planes which pass through the arbitrary point, viz. the class of the torse. So also the cuspidal torse is the torse enveloped by the tangent planes along the cuspidal curve; and σ will be the number of these tangent planes which pass through the arbitrary point, viz. it will be the class of the torse. Again, as regards ρ' and σ': the node-couple torse is the envelope of the bitangent planes of the surface, and the node-couple curve is the locus of the points of contact of these planes. Similarly, the spinode torse is the envelope of the parabolic planes of the surface, and the spinode curve is the locus of the points of contact of these planes, viz. it is the curve UH of intersection of the surface and its Hessian; the two curves are the reciprocals of the nodal and the cuspidal torses respectively, and the definitions of ρ', σ' correspond to those of ρ and σ.

602. In regard to the nodal curve b, we consider k the number of its apparent double points (excluding actual double points); f the number of its actual double points (each of these is a point of contact of two sheets of the surface, and there is thus at the point a single tangent plane, viz. this is a plane f', and we thus have $f'=f$); t the number of its triple points; and j the number of its pinch-points—these last are not singular points of the nodal curve *per se*, but are singular in regard to the curve as nodal curve of the surface; viz. a pinch-point is a point at which the two tangent planes are coincident. The curve is considered as not having any stationary points other than the points γ, which lie also on the cuspidal curve; and the expression for the class consequently is $q=b^2-b-2k-2f-3\gamma-6t$.

603. In regard to the cuspidal curve c, we consider h the number of its apparent double points; and upon the curve, not singular points in regard to the curve *per se*, but only in regard to it as cuspidal curve of the surface, certain points in number θ, χ, ω respectively. The curve is considered as not having any actual double or other multiple points, and as not having any stationary points except the points β, which lie also on the nodal curve; and the expression for the class consequently is

$$r=c^2-c-2h-3\beta.$$

604. The points γ are points where the cuspidal curve with the two sheets (or say rather half-sheets) belonging to it are intersected by another sheet of the surface; the curve of intersection with such other sheet, belonging to the nodal curve of the surface, has evidently a stationary (cuspidal) point at the point of intersection.

As to the points β, to facilitate the conception, imagine the cuspidal curve to be a semi-cubical parabola, and the nodal curve a right line (not in the plane of the curve) passing through the cusp; then intersecting the two curves by a series of parallel planes, any plane which is, say, above the cusp, meets the parabola in two real points and the line in one real point, and the section of the surface is a curve with two real cusps and a real node; as the plane approaches the cusp, these approach together, and, when the plane passes through the cusp, unite into a singular point in the nature of a triple point (= node + two cusps); and when the plane passes below the cusp, the two cusps of the section become imaginary, and the nodal line changes from crunodal to acnodal.

605. At a point i the nodal curve crosses the cuspidal curve, being on the side away from the two half-sheets of the surface acnodal, and on the side of the two half-sheets crunodal, viz. the two half-sheets intersect each other along this portion of the nodal curve. There is at the point a single tangent plane, which is a plane i'; and we thus have $i = i'$.

606. As already mentioned, a cnicnode C is a point where, instead of a tangent plane, we have a tangent quadri-cone; at a binode B, the quadri-cone degenerates into a pair of planes. A cnictrope C' is a plane touching the surface along a conic; in the case of a bitrope B', the conic degenerates into a flat conic or pair of points.

607. In the original formulæ for $a(n-2)$, $b(n-2)$, $c(n-2)$, we have to write $\kappa - B$ instead of κ, and the formulæ are further modified by reason of the singularities θ and ω. So, in the original formulæ, for $a(n-2)(n-3)$, $b(n-2)(n-3)$, $c(n-2)(n-3)$, we have instead of δ to write $\delta - C - 3\omega$, and to substitute new expressions for $[ab]$, $[ac]$, $[bc]$; viz. these are

$$[ab] = ab - 2\rho - j,$$

$$[ac] = ac - 3\sigma - \chi - \omega,$$

$$[bc] = bc - 3\beta - 2\gamma - i.$$

The whole series of equations thus is

(1) $a' = a.$

(2) $f' = f.$

(3) $i' = i.$

(4) $a = n(n-1) - 2b - 3c.$

(5) $\kappa' = 3n(n-2) - 6b - 8c.$

(6) $\delta' = \tfrac{1}{2}n(n-2)(n^2-9) - (n^2-n-6)(2b+3c) + 2b(b-1) + 6bc + \tfrac{9}{2}c(c-1).$

(7) $a(n-2) = \kappa - B + \rho + 2\sigma + 3\omega.$

(8) $b(n-2) = \rho + 2\beta + 3\gamma + 3t.$

(9) $c(n-2) = 2\sigma + 4\beta + \gamma + \theta + \omega.$

(10) $a(n-2)(n-3) = 2(\delta - C - 3\omega) + 3(ac - 3\sigma - \chi - 3\omega) + 2(ab - 2\rho - j).$

(11) $b(n-2)(n-3) = 4k + (ab - 2\rho - j) + 3(bc - 3\beta - 2\gamma - i).$

(12) $c(n-2)(n-3) = 6h + (ac - 3\sigma - \chi - 3\omega) + 2(bc - 3\beta - 2\gamma - i).$

(13) $q = b^2 - b - 2k - 2f - 3\gamma - 6t.$

(14) $r = c^2 - c - 2h - 3\beta.$

Also, reciprocal to these,

(15) $a' = n'(n'-1) - 2b' - 3c'$.

(16) $\kappa = 3n'(n'-2) - 6b' - 8c'$.

(17) $\delta = \tfrac{1}{2}n'(n'-2)(n'^2-9) - (n'^2 - n' - 6)(2b' + 3c') + 2b'(b'-1) + 6b'c' + \tfrac{9}{2}c'(c'-1)$.

(18) $a'(n'-2) = \kappa' - B' + \rho' + 2\sigma' + 3\omega'$.

(19) $b'(n'-2) = \rho' + 2\beta' + 3\gamma' + 3t'$.

(20) $c'(n'-2) = 2\sigma' + 4\beta' + \gamma' + \theta' + \omega'$.

(21) $a'(n'-2)(n'-3) = 2(\delta' - C' - 3\omega') + 3(a'c' - 3\sigma' - \chi' - 3\omega') + 2(a'b' - 2\rho' - j')$.

(22) $b'(n'-2)(n'-3) = 4k' + (a'b' - 2\rho' - j') + 3(b'c' - 3\beta' - 2\gamma' - \iota')$.

(23) $c'(n'-2)(n'-3) = 6h' + (a'c' - 3\sigma' - \chi' - 3\omega') + 2(b'c' - 3\beta' - 2\gamma' - \iota')$.

(24) $q' = b'^2 - b' - 2k' - 2f' - 3\gamma' - 6t'$.

(25) $r' = c'^2 - c' - 2h' - 3\beta'$,

together with one other independent relation: in all 26 relations between the 46 quantities.

608. The new relation may be presented under several different forms, equivalent to each other in virtue of the foregoing 25 relations; these are

(26) $2(n-1)(n-2)(n-3) - 12(n-3)(b+c) + 6q + 6r + 24t + 42\beta + 30\gamma - \tfrac{3}{2}\theta = \Sigma$,

(27) $26n - 12c - 4C - 10B + \beta - 7j - 8\chi + \tfrac{1}{2}\theta - 4\omega = \Sigma$;

in each of which two equations Σ is used to denote the same function of the accented letters that the left-hand side is of the unaccented letters.

(28)
$$\begin{aligned}\beta' + \tfrac{1}{2}\theta' = {} & 2n(n-2)(11n-24) \\ & + (-66n + 184)b \\ & + (-93n + 252)c \\ & + 22(2\beta + 3\gamma + 3t) \\ & + 27(4\beta + \gamma + \theta) \\ & + \beta + \tfrac{1}{2}\theta \\ & - 24C - 28B - 27j - 38\chi - 73\omega \\ & + 4C' + 10B' + 7j' + 8\chi' - 4\omega'.\end{aligned}$$

Or, reciprocally,

(29)
$$\begin{aligned}\beta + \tfrac{1}{2}\theta = {} & 2n'(n'-2)(11n'-24) \\ & + (-66n' + 184)b' \\ & + (-93n' + 252)c' \\ & + 22(2\beta' + 3\gamma' + 3t') \\ & + 27(4\beta' + \gamma' + \theta') \\ & + \beta' + \tfrac{1}{2}\theta' \\ & - 24C' - 28B' - 27j' - 38\chi' - 73\omega' \\ & + 4C + 10B + 7j + 8\chi - 4\omega.\end{aligned}$$

The equation (26) expresses that the surface and its reciprocal have the same deficiency; viz. the expression for the deficiency is

$$(30)\quad \text{Deficiency} = \tfrac{1}{6}(n-1)(n-2)(n-3) - (n-3)(b+c) + \tfrac{1}{2}(q+r) + 2t + \tfrac{7}{2}\beta + \tfrac{5}{2}\gamma + i - \tfrac{1}{8}\theta,$$

$$= \tfrac{1}{6}(n'-1)(n'-2)(n'-3) - \&c.$$

609. The equation (28) (due to Prof. Cayley) is the correct form of an expression for β', first obtained by him (with some errors in the numerical coefficients) from independent considerations. But it is best obtained by means of the equation (26); and (27) is a relation presenting itself in the investigation. In fact, considering a as standing for its value $n(n-1) - 2b - 3c$, we have from the first 25 equations

6	a	$=\Sigma,$
$+2$	$3n - c - \kappa$	$=\Sigma,$
-2	$a(n-2) - \kappa + B - \rho - 2\sigma - 3\omega$	$=\Sigma,$
-4	$b(n-2) - \rho - 2\beta - 3\gamma - 3t$	$=\Sigma,$
-6	$c(n-2) - 2\sigma - 4\beta - \gamma - \theta - \omega$	$=\Sigma,$
$+2$	$n + \kappa - \sigma - 2C - 4B - 2j - 3\chi - 3\omega$	$=\Sigma,$
-3	$2q - 2\rho + \beta + j$	$=\Sigma,$
-2	$3r + c - 5\sigma - \beta - 4\theta + \chi - \omega$	$=\Sigma;$

multiplying these equations by the numbers set opposite to them respectively, and adding, we find

$$-2n^3 + 12n^2 + 4n + b(12n-36) + c(12n-48)$$
$$-6q - 6r - 4C - 10B - 41\beta - 30\gamma - 24t - 7j - 8\chi + 2\theta - 4\omega = \Sigma,$$

and adding hereto (26) we have the equation (27); and from this (28), or by a like process, (29), is obtained without much difficulty. As to the 8 Σ-equations or symmetries, observe that the first, third, fourth, and fifth are in fact included among the original equations (for an expression which vanishes is in fact $=\Sigma$); we have from them moreover $3n - c = 3a' - \kappa'$, and thence $3n - c - \kappa = 3a' - \kappa - \kappa'$, which is $=\Sigma$, or we have thus the second equation; but the sixth, seventh, and eighth equations have yet to be obtained.

610. The equations (15), (16), (17) give

$$n' = a(a-1) - 2\delta - 3\kappa,$$
$$c' = 3a(a-2) - 6\delta - 8\kappa,$$
$$b' = \tfrac{1}{2}a(a-2)(a^2-9) - (a^2-a-6)(2\delta+3\kappa) + 2\delta(\delta-1) + 6\delta\kappa + \tfrac{9}{2}\kappa(\kappa-1).$$

From (7), (8), (9), we have

$$(a - b - c)(n-2) = \kappa - B - 6\beta - 4\gamma - 3t - \theta + 2\omega,$$
$$(a-2b-3c)(n-2)(n-3) = 2(\delta - C) - 8k - 18h - 6bc + 18\beta + 12\gamma + 6i - 6\omega;$$

substituting these values for κ and δ, and for a its value $= n(n-1) - 2b - 3c$, we obtain the values of n', c', b'; viz. the value of n' is

$$\begin{aligned} n' = n(n-1)^2 - n(7b+12c) + 4b^2 + 8b + 9c^2 + 15c \\ - 8k - 18h + 18\beta + 12\gamma + 12i - 9t \\ - 2C - 3B - 3\theta. \end{aligned}$$

Observe that the effect of a cnicnode C is to reduce the class by 2, and that of a binode B to reduce it by 3.

611. We have

$$(n-2)(n-3) = n^2 - n + (-4n+6) = a + 2b + 3c + (-4n+6);$$

making this substitution in the equations (10), (11), (12), which contain $(n-2)(n-3)$, these become

$$\begin{aligned} a(-4n+6) &= 2(\delta - C) - a^2 - 4\rho - 9\sigma - 2j - 3\chi - 15\omega, \\ b(-4n+6) &= 4k - 2b^2 - 9\beta - 6\gamma - 3i - 2\rho - j, \\ c(-4n+6) &= 6h - 3c^2 - 6\beta - 4\gamma - 2i - 3\sigma - \chi - 3\omega, \end{aligned}$$

which are the foregoing equations (C); adding to each equation four times the corresponding equation with the factor $(n-2)$, these become

$$\begin{aligned} a^2 - 2a &= 2(\delta - C) + 4(\kappa - B) - \sigma - 2j - 3\chi - 3\omega, \\ 2b^2 - 2b &= 4k - \beta + 6\gamma + 12t - 3i + 2\rho - j, \\ 3c^2 - 2c &= 6h + 10\beta + 4\theta - 2i + 5\sigma - \chi + \omega. \end{aligned}$$

Writing in the first of these $a^2 - 2a = n' + 2\delta + 3\kappa - a$, and reducing the other two by means of the values of q, r, the equations become

$$\begin{aligned} n' - a &= -2C - 4B + \kappa - \sigma - 2j - 3\chi - 3\omega, \\ 2q + \beta + 3i + j &= 2\rho, \\ 3r + c + 2i + \chi &= 5\sigma + \beta + 4\theta + \omega, \end{aligned}$$

which give at once the last three of the 8 Σ-equations.

The reciprocal of the first of these is

$$\sigma' = a - n + \kappa' - 2j' - 3\chi' - 2C' - 4B' - 3\omega',$$

viz. writing herein

$$a = n(n-1) - 2b - 3c \text{ and } \kappa' = 3n(n-2) - 6b - 8c,$$

this is

$$\sigma' = 4n(n-2) - 8b - 11c - 2j' - 3\chi' - 2C' - 4B' - 3\omega',$$

giving the order of the spinode curve; viz. for a surface of the order n without singularities, this is $= 4n(n-2)$, the product of the orders of the surface and its Hessian.

612. Instead of obtaining the second and third equations as above, we may to the value of $b(-4n+6)$ add twice the value of $b(n-2)$; and to twice the value of $c(-4n+6)$ add three times the value of $c(n-2)$, thus obtaining equations free from ρ and σ respectively; these equations are

$$b(-2n+2) = 4k - 2b^2 - 5\beta - 3i + 6t - j,$$
$$c(-5n+6) = 12h - 6c^2 - 5\gamma - 4i - 2\chi + 3\theta - 3\omega,$$

equations which, introducing therein the values of q and r, may also be written

$$b(2n-4) \qquad = 2q + 5\beta + 6\gamma + 6t + 3i + j + 4f,$$
$$c(5n-12) + 3\theta = 6r + 18\beta + 5\gamma \qquad + 4i + 2\chi + 3\omega.$$

Considering as given, n the order of the surface; the nodal curve, with its singularities b, k, f, t; the cuspidal curve, with its singularities c, h; and the quantities β, γ, i which relate to the intersections of the nodal and cuspidal curves; the first of the two equations gives j, the number of pinch-points, being singularities of the nodal curve, quoad the surface; and the second equation establishes a relation between θ, χ, ω, the numbers of singular points of the cuspidal curve quoad the surface.

In the case of a nodal curve only, if this be a complete intersection $P=0$, $Q=0$, the equation of the surface is $(A, B, C\between P, Q)^2 = 0$, and the first equation is

$$b(-2n+2) = 4k - 2b^2 + 6t - j;$$

or, assuming $t=0$, say $j = 2(n-1)b - 2b^2 + 4k$, which may be verified; and so in the case of a cuspidal curve only, when this is a complete intersection $P=0$, $Q=0$, the equation of the surface is $(A, B, C\between P, Q)^2 = 0$, where $AC - B^2 = MP + NQ$; and the second equation is

$$c(-5n+6) = 12h - 6c^2 - 2\chi + 3\theta - 3\omega,$$

or, say $2\chi + 3\omega = (5n-6)c - 6c^2 + 12h + 3\theta$, which may also be verified.

613. We may in the first instance out of the 46 quantities consider as given the 14 quantities

$$n : \qquad b, k, f, t : c, h, \theta, \chi \quad : \beta, \gamma, i : C, B,$$

then of the 26 relations, 17 determine the 17 quantities

$$a, \delta, \kappa, \rho, \sigma : j, q \quad : r, \omega$$
$$n' : a', \delta', \kappa' \quad : b', f' \quad : c' \qquad : \qquad i';$$

and there remain the 9 equations

(18), (19), (20), (21), (22), (23), (24), (25), (28),

connecting the 15 quantities

$$\rho', \sigma' : k', t', j', q' : h', \theta', \chi', \omega', r' : \beta', \gamma' : C', B'.$$

Taking then further as given the 5 quantities j', χ', ω', C', B',

equations (18) and (21) give ρ', σ',

equation (19) gives $2\beta' + 3\gamma' + 3t'$,

" (20) " $4\beta' + \gamma' + \theta'$,

" (28) " $\beta' + \frac{1}{2}\theta'$,

so that, taking also t as given, these last three equations determine β', γ', θ'; and finally

equation (22) gives k',

" (23) " h',

" (24) " q',

" (25) " r',

viz. taking as given in all 20 quantities, the remaining 26 will be determined.

614. In the case of the general surface of the order n, without singularities, we have as follow:

$$
\begin{aligned}
n &= n,\\
a &= n(n-1),\\
\delta &= \tfrac{1}{2}n(n-1)(n-2)(n-3),\\
\kappa &= n(n-1)(n-2),\\
n' &= n(n-1)^2,\\
a' &= n(n-1),\\
\delta' &= \tfrac{1}{2}n(n-2)(n^2-9),\\
\kappa' &= 3n(n-2),\\
b' &= \tfrac{1}{2}n(n-1)(n-2)(n^3-n^2+n-12),\\
k' &= \tfrac{1}{8}n(n-2)(n^{10}-6n^9+16n^8-54n^7+164n^6-288n^5\\
&\qquad +547n^4-1058n^3+1068n^2-1214n+1464),\\
t' &= \tfrac{1}{6}n(n-2)(n^7-4n^6+7n^5-45n^4+114n^3-111n^2+548n-960),\\
q' &= n(n-2)(n-3)(n^2+2n-4),\\
\rho' &= n(n-2)(n^3-n^2+n-12),\\
c' &= 4n(n-1)(n-2),\\
h' &= \tfrac{1}{2}n(n-2)(16n^4-64n^3+80n^2-108n+156),\\
r' &= 2n(n-2)(3n-4),\\
\sigma' &= 4n(n-2),\\
\beta' &= 2n(n-2)(11n-24),\\
\gamma' &= 4n(n-2)(n-3)(n^3-3n+16),
\end{aligned}
$$

the remaining quantities vanishing.

615. The question of singularities has been considered under a more general point of view by Zeuthen, in the memoir "Recherche des singularités qui ont rapport à une droite multiple d'une surface," *Math. Annalen*, t. IV. (1871), pp. 1—20. He attributes to the surface:

A number of singular points, viz. points at any one of which the tangents form a cone of the order μ, and class ν, with $y+\eta$ double lines, of which y are tangents to branches of the nodal curve through the point, and $z+\zeta$ stationary lines, whereof z are tangents to branches of the cuspidal curve through the point, and with u double planes and v stationary planes; moreover, these points have only the properties which are the most general in the case of a surface regarded as a locus of points; and Σ denotes a sum extending to all such points. (The foregoing general definition includes the cnicnodes $\mu=\nu=2$, $y=\eta=z=\zeta=u=v=0$, and the binodes $\mu=2$, $\eta=1$, $\nu=y=\&c.=0$.)

And, further, a number of singular planes, viz. planes any one of which touches along a curve of the class μ' and order ν', with $y'+\eta'$ double tangents, of which y' are generating lines of the node-couple torse, $z'+\zeta'$ stationary tangents, of which z' are generating lines of the spinode torse, u' double points and v' cusps; it is, moreover, supposed that these planes have only the properties which are the most general in the case of a surface regarded as an envelope of its tangent planes; and Σ' denotes a sum extending to all such planes. (The definition includes the cnictropes $\mu'=\nu'=2$, $y'=\eta'=z'=\zeta'=u'=v'=0$, and the bitropes $\mu'=2$, $\eta'=1$, $\nu'=y'=\&c.=0$.)

616. This being so, and writing

$$x=\nu+2\eta+3\zeta,\quad x'=\nu'+2\eta'+3\zeta',$$

the equations (7), (8), (9), (10), (11), (12), contain, in respect of the new singularities additional terms, viz. these are

$$\begin{aligned}
a(n-2) &= \ldots+\Sigma\,[x(\mu-2)-\eta-2\zeta],\\
b(n-2) &= \ldots+\Sigma\,[y(\mu-2)],\\
c(n-2) &= \ldots+\Sigma\,[z(\mu-2)],\\
a(n-2)(n-3) &= \ldots+\Sigma\,[x(-4\mu+7)+2\eta+4\zeta],\\
b(n-2)(n-3) &= \ldots+\Sigma\,[y(-4\mu+8)]-\Sigma'(4u'+3v'),\\
c(n-2)(n-3) &= \ldots+\Sigma\,[z(-4\mu+9)]-\Sigma'(2v'),
\end{aligned}$$

and there are of course the reciprocal terms in the reciprocal equations (18), (19), (20), (21), (22), (23). These formulæ are given without demonstration in the memoir just referred to: the principal object of the memoir, as shown by its title, is the consideration not of such singular points and planes, but of the multiple right lines of a surface; and in regard to these, the memoir should be consulted.

751.

NOTE ON RIEMANN'S PAPER "VERSUCH EINER ALLGEMEINEN AUFFASSUNG DER INTEGRATION UND DIFFERENTIATION*."

[From the *Mathematische Annalen*, t. XVI. (1880), pp. 81, 82.]

THE Editors of Riemann's works remark that the paper in question was contained in a MS. of his student time (dated 14 Jan. 1847) and was probably never intended for publication: indeed that he would not in later years have recognised the validity of the principles upon which it is founded. The idea is however a noticeable one: Riemann considers z_{x+h}, a function of $x+h$, expanded in a doubly infinite, necessarily divergent, series of integer or fractional powers of h, according to the law

$$z_{x+h} = \sum_{\nu=-\infty}^{\nu=+\infty} k_\nu \partial^\nu{}_x z \,.\, h^\nu, \tag{2}$$

where the meaning is explained to be that the exponents differ from each other by integer values, in effect, that ν has all the values $\alpha+p$, α a given integer or fractional value, and p any integer number from $-\infty$ to $+\infty$, zero included.

Riemann deduces a theory of fractional differentiation: but without considering the question which has always appeared to me to be the great difficulty in such a theory: what is the real meaning of a complementary function containing an infinity of arbitrary constants? or, in other words, what is the arbitrariness of the complementary function of this nature which presents itself in the theory?

I wish to point out the relation between the paper referred to, and a short paper of my own "On a doubly infinite Series," *Quart. Math. Journ.* t. VI. (1851), pp. 45—47, [102]: this commences with the remark "The following completely paradoxical investigation of the properties of the function Γ (which I have been in possession

* *Werke*, pp. 331—344.

of for some years) may perhaps be found interesting from its connexion with the theories of expansion and divergent series." And I then give the expansion

$$C_n e^x = \Sigma^r [n-r]^r x^{n-1-r},$$

where n is any integer or fractional number whatever, and the summation extends to all positive and negative integer values (zero included) of r. And I remark that, n being an integer, we have $C_n = \Gamma(n)$, and hence that assuming that this is so in general, or writing

$$\Gamma(n) . e^x = \Sigma^r [n-1]^r x^{n-1-r},$$

we have this equation as a definition of $\Gamma(n)$. The point of resemblance of course is that we have a doubly infinite expansion of e^x in a series of integer or fractional powers of x, corresponding to Riemann's like expansion of z_{x+h} in powers of h.

Cambridge, 10 *Sept.* 1879.

752.

ON THE FINITE GROUPS OF LINEAR TRANSFORMATIONS OF A VARIABLE; WITH A CORRECTION.

[From the *Mathematische Annalen*, t. XVI. (1880), pp. 260—263; 439, 440.]

IN the paper "Ueber endliche Gruppen linearer Transformationen einer Veränderlichen," *Math. Ann.* t. XII. (1877), pp. 23—46, Prof. Gordan gave in a very elegant form the groups of 12, 24 and 60 homographic transformations $\frac{ax+b}{cx+d}$. The groups of 12 and 24 are in the like form, the group of 24 thus containing as part of itself the group of 12; but the group of 60 is in a different form, not containing as part of itself the group of 12. It is, I think, desirable to present the group of 60 in the form in which it contains as part of itself Gordan's group of 12: and moreover to identify the group of 60 with the group of the 60 positive permutations of 5 letters: or (writing *abc* for the cyclical permutation *a* into *b*, *b* into *c*, *c* into *a*, and so in other cases) say with the group of the 60 positive permutations 1, *abc*, *ab.cd* and *abcde*.

Any two forms of a group are, it is well known, connected as follows, viz. if 1, α, β, ... are the functional symbols of the one form, then those of the other form are 1, $\vartheta\alpha\vartheta^{-1}$, $\vartheta\beta\vartheta^{-1}$, ... $\left(\text{where in the case in question } \vartheta \text{ is a functional symbol of the like homographic form, } \vartheta x = \frac{Ax+B}{Cx+D}\right)$. But instead of obtaining the new form in this manner, I found it easier to use the values of the rotation-symbol

$$\cos\frac{\pi}{q} + \sin\frac{\pi}{q}(i\cos X + j\cos Y + k\cos Z)$$

for the axes of the icosahedron or dodecahedron, given in my paper "Notes on polyhedra," *Quart. Math. Jour.* t. VII. (1866), pp. 304—316, [375]; viz. if for any axes, λ, μ, ν denote the parameters of rotation $\tan\frac{\pi}{q}\cos X$, $\tan\frac{\pi}{q}\cos Y$, $\tan\frac{\pi}{q}\cos Z$, then, by a formula which is in fact equivalent to that given in my note "On the correspondence of Homographies and Rotations," *Math. Annalen*, t. xv. (1879), pp. 238—240, [660], the corresponding homographic function of x is

$$\frac{(-\nu - i)\,x + \lambda + i\mu}{(\lambda - i\mu)\,x + \nu - i},$$

where i denotes $\sqrt{-1}$ as usual.

The new formulæ for the group of 60, or icosahedron group, of homographic functions $\frac{\alpha x + \beta}{\gamma x + \delta}$ are contained in the following table, where the four columns show the values of the coefficients α, β, γ, δ respectively: and where in the outside column, the substitution is represented as a permutation-symbol on the five letters *abcde*: moreover for shortness Θ is written to denote $\sqrt{5}$.

THE GROUP OF 60.

	α	β	γ	δ	
1	1	0	0	1	1
2	−1	0	0	1	*ab . cd*
3	0	1	1	0	*ac . bd*
4	0	−1	1	0	*ad . bc*
5	2	$-3+\Theta+i(\ \ 1-\Theta)$	$-3+\Theta+i(-1+\Theta)$	−2	*bc . de*
6	2	$-3+\Theta+i(-1+\Theta)$	$-3+\Theta+i(\ \ 1-\Theta)$	−2	*ae . bc*
7	2	$3-\Theta+i(-1+\Theta)$	$3-\Theta+i(\ \ 1-\Theta)$	−2	*ad . ce*
8	2	$3-\Theta+i(\ \ 1-\Theta)$	$3-\Theta+i(-1+\Theta)$	−2	*ad . be*
9	2	$-1-\Theta+i(\ \ 1-\Theta)$	$-1-\Theta+i(-1+\Theta)$	−2	*ae . cd*
10	2	$-1-\Theta+i(-1+\Theta)$	$-1-\Theta+i(\ \ 1-\Theta)$	−2	*ab . de*
11	2	$1+\Theta+i(-1+\Theta)$	$1+\Theta+i(\ \ 1-\Theta)$	−2	*be . cd*
12	2	$1+\Theta+i(\ \ 1-\Theta)$	$1+\Theta+i(-1+\Theta)$	−2	*ab . ce*
13	2	$-1-\Theta+i(-3-\Theta)$	$-1-\Theta+i(\ \ 3+\Theta)$	−2	*ac . be*
14	2	$-1-\Theta+i(\ \ 3+\Theta)$	$-1-\Theta+i(-3-\Theta)$	−2	*bd . ce*
15	2	$1+\Theta+i(\ \ 3+\Theta)$	$1+\Theta+i(-3-\Theta)$	−2	*ae . bd*
16	2	$1+\Theta+i(-3-\Theta)$	$1+\Theta+i(\ \ 3+\Theta)$	−2	*ac . de*
17	$-i$	i	1	1	*abc*
18	−1	i	1	i	*acb*
19	1	$-i$	1	i	*adc*
20	$-i$	$-i$	1	−1	*acd*
21	i	i	1	−1	*adb*
22	1	i	1	$-i$	*abd*
23	−1	$-i$	1	$-i$	*bcd*
24	i	$-i$	1	1	*bdc*

	α	β	γ	δ	
25	$-1-\Theta+i(\ 3+\Theta)$	2	-2	$-1-\Theta+i(-3-\Theta)$	*aec*
26	$1+\Theta+i(\ 3+\Theta)$	2	-2	$1+\Theta+i(-3-\Theta)$	*ace*
27	$1+\Theta+i(-3-\Theta)$	2	-2	$1+\Theta+i(\ 3+\Theta)$	*bed*
28	$-1-\Theta+i(-3-\Theta)$	2	-2	$-1-\Theta+i(\ 3+\Theta)$	*bde*
29	$-3+\Theta+i(\ 1-\Theta)$	2	2	$3-\Theta+i(\ 1-\Theta)$	*bec*
30	$-3+\Theta+i(-1+\Theta)$	2	2	$3-\Theta+i(-1+\Theta)$	*bce*
31	$3-\Theta+i(-1+\Theta)$	2	2	$-3+\Theta+i(-1+\Theta)$	*aed*
32	$3-\Theta+i(\ 1-\Theta)$	2	2	$-3+\Theta+i(\ 1-\Theta)$	*ade*
33	2	$-1-\Theta+i(-1+\Theta)$	$1+\Theta+i(-1+\Theta)$		*cde*
34	2	$1+\Theta+i(\ 1-\Theta)$	$-1-\Theta+i(\ 1-\Theta)$		*ced*
35	2	$-1-\Theta+i(\ 1-\Theta)$	$1+\Theta+i(\ 1-\Theta)$		*aeb*
36	2	$1+\Theta+i(-1+\Theta)$	$-1-\Theta+i(-1+\Theta)$		*abe*
37	$-1-\Theta+i(-3-\Theta)$	2	2	$1+\Theta+i(-3-\Theta)$	*abcde*
38	$-1-\Theta+i(\ 1-\Theta)$	2	2	$1+\Theta+i(\ 1-\Theta)$	*acebd*
39	$-1-\Theta+i(-1+\Theta)$	2	2	$1+\Theta+i(-1+\Theta)$	*adbec*
40	$-1-\Theta+i(\ 3+\Theta)$	2	2	$1+\Theta+i(\ 3+\Theta)$	*aedcb*
41	$1+\Theta+i(\ 3+\Theta)$	2	2	$-1-\Theta+i(\ 3+\Theta)$	*adceb*
42	$1+\Theta+i(\ 1+\Theta)$	2	2	$-1-\Theta+i(-1+\Theta)$	*acbde*
43	$1+\Theta+i(\ 1-\Theta)$	2	2	$-1-\Theta+i(\ 1-\Theta)$	*aedbc*
44	$1+\Theta+i(-3-\Theta)$	2	2	$-1-\Theta+i(-3-\Theta)$	*abecd*
45	$-1-\Theta+i(-1+\Theta)$	2	-2	$-1-\Theta+i(\ 1-\Theta)$	*acbed*
46	$-3+\Theta+i(-1+\Theta)$	2	-2	$-3+\Theta+i(\ 1-\Theta)$	*abdce*
47	$3-\Theta+i(-1+\Theta)$	2	-2	$3-\Theta+i(\ 1-\Theta)$	*aecdb*
48	$1+\Theta+i(-1+\Theta)$	2	-2	$1+\Theta+i(\ 1-\Theta)$	*adebc*
49	$1+\Theta+i(\ 1-\Theta)$	2	-2	$1+\Theta+i(-1+\Theta)$	*aecbd*
50	$3-\Theta+i(\ 1-\Theta)$	2	-2	$3-\Theta+i(-1+\Theta)$	*acdeb*
51	$-3+\Theta+i(\ 1-\Theta)$	2	-2	$-3+\Theta+i(-1+\Theta)$	*abedc*
52	$-1-\Theta+i(\ 1-\Theta)$	2	-2	$-1-\Theta+i(-1+\Theta)$	*adbce*
53	2	$-3+\Theta+i(-1+\Theta)$	$3-\Theta+i(-1+\Theta)$	2	*aebdc*
54	2	$-1-\Theta+i(\ 3+\Theta)$	$1+\Theta+i(\ 3+\Theta)$	2	*abced*
55	2	$1+\Theta+i(-3-\Theta)$	$-1-\Theta+i(-3-\Theta)$	2	*adecb*
56	2	$3-\Theta+i(\ 1-\Theta)$	$-3+\Theta+i(\ 1-\Theta)$	2	*acdbe*
57	2	$-3+\Theta+i(\ 1-\Theta)$	$3-\Theta+i(\ 1-\Theta)$	2	*abdec*
58	2	$-1-\Theta+i(-3-\Theta)$	$1+\Theta+i(-3-\Theta)$	2	*adcbe*
59	2	$1+\Theta+i(\ 3+\Theta)$	$-1-\Theta+i(\ 3+\Theta)$	2	*aebcd*
60	2	$3-\Theta+i(-1+\Theta)$	$-3+\Theta+i(-1+\Theta)$	2	*acedb*

This contains (as one of five groups of 12) the group of the positive permutations of *abcd*; and, completing this into a group of 24, we have

GROUPS OF 12 AND 24.

	α	β	γ	δ	
1	1	0	0	1	1
2	-1	0	0	1	*ab . cd*
3	0	1	1	0	*ac . bd*
4	0	-1	1	0	*ad . bc*
5	$-i$	i	1	1	*abc*
6	-1	i	1	i	*acb*
7	1	$-i$	1	i	*adc*
8	$-i$	$-i$	1	-1	*acd*
9	i	i	1	-1	*adb*
10	1	i	1	$-i$	*abd*
11	-1	$-i$	1	$-i$	*bcd*
12	i	$-i$	1	1	*bdc*
13	i	0	0	1	*adbc*
14	$-i$	0	0	1	*acbd*
15	0	i	1	0	*cd*
16	0	i	-1	0	*ab*
17	1	-1	1	1	*acdb*
18	$-i$	-1	1	i	*bd*
19	i	1	1	i	*abcd*
20	1	1	1	-1	*bc*
21	-1	-1	1	-1	*abdc*
22	i	-1	1	$-i$	*ac*
23	$-i$	1	1	$-i$	*adcb*
24	-1	1	1	1	*ad*

The groups of 60 and 24 thus each of them contain the group of 12,

$$\pm x, \quad \pm\frac{1}{x}, \quad \pm i\frac{1-x}{1+x}, \quad \pm i\frac{1+x}{1-x}, \quad \pm\frac{x+i}{x-i}, \quad \pm\frac{x-i}{x+i}.$$

It may be remarked that, to verify the periodicities of the forms contained in the group of 60, we have as the conditions that

$\dfrac{\alpha x+\beta}{\gamma x+\delta}$ may be periodic of the order 2, $\dfrac{(\alpha+\delta)^2}{\alpha\delta-\beta\gamma}=0$, that is, $\alpha+\delta=0$,

" " " 3, " $=1$,

" " " 5, " $=\frac{1}{2}(3+\sqrt{5})$.

For instance, in the form

$$\frac{[-1-\Theta+i(-3-\Theta)]x+2}{2x+[1+\Theta+i(-3-\Theta)]},$$

we have

$$\alpha\delta = -(1+\Theta)^2-(3+\Theta)^2, \quad = -20-8\Theta, \quad \beta\gamma=4,$$

$$\alpha+\delta=-2i(3+\Theta):$$

and therefore

$$\frac{(\alpha+\delta)^2}{\alpha\delta-\beta\gamma}=\frac{-4(3+\Theta)^2}{-8(3+\Theta)}, \quad =\frac{3+\Theta}{2}=\tfrac{1}{2}(3+\sqrt{5}),$$

as it should be.

Cambridge, 11 *Nov.* 1879.

CORRECTION*, pp. 439, 440.

I erroneously assumed that the symbol *adcb* could be taken as corresponding to the linear transformation ix: but this was obviously wrong, for it gave *bd* as corresponding to the transformation $-ix$, and these are not of the same order, but of the orders 4 and 2 respectively. The proper symbol is *adbc*, as given above, and the remaining eleven symbols are then at once obtained.

Cambridge, 17 *Feb.* 1880.

[* The correction in the Table of the Groups of 12 and 24 has been inserted in the Table as now printed on p. 240; it applies to the second half of the column of symbols on the extreme right-hand.]

753.

ON A THEOREM RELATING TO THE MULTIPLE THETA-FUNCTIONS.

[From the *Mathematische Annalen*, t. XVII. (1880), pp. 115—122.]

I PROPOSE—partly for the sake of the theorem itself, partly for that of the notation which will be employed—to demonstrate the general theorem (3′), p. 4, of Dr Schottky's *Abriss einer Theorie der Abel'schen Functionen von drei Variabeln*, (Leipzig, 1880), which theorem is there presented in the form:

$$e^{-\eta(u_1, \ldots;\, \mu', \nu')}\,\Theta\,(u_1 + 2\overline{\omega}_1', \ldots;\ \mu,\ \nu) = e^{-2\pi i \Sigma \mu_a \nu'_a}\,\Theta\,(u_1, \ldots;\ \mu + \mu',\ \nu + \nu'), \qquad (3')$$

but which I write in the slightly different form

$$\text{exp.}\,[-H(u;\ \mu',\ \nu')]\,.\,\Theta\,(u + 2\varpi';\ \mu,\ \nu) = \text{exp.}\,[-2\pi i\mu\nu']\,.\,\Theta\,(u;\ \mu + \mu',\ \nu + \nu').$$

I remark that the theorem is given in the preliminary paragraphs the contents of which are, as mentioned by the Author, derived from Herr Weierstrass: and that the form of the theta-function is a very general one, depending on the general quadric function

$$G\,(u_1, \ldots, u_\rho;\ n_1, \ldots, n_\rho)$$

of 2ρ variables, ρ being the number of the arguments $u_1, \ldots, u_\rho$ (in fact, the periods are not reduced to the normal form, but are arbitrary); and the characters $\nu_1, \ldots, \nu_\rho$; $\mu_1, \ldots, \mu_\rho$, instead of having each of them the value 0 or 1, have each of them any integer or fractional value whatever. The meaning of the theorem (u denoting a set or row of ρ letters $u_1, \ldots, u_\rho$, and so in other cases), is that the function

$$\Theta\,(u;\ \mu + \mu',\ \nu + \nu')$$

with the new characters $\mu+\mu'$ and $\nu+\nu'$ is, save as to an exponential factor, equal to the function $\Theta(u+2\varpi';\ \mu,\ \nu)$ with the original characters μ, ν, but with the new arguments $u+2\varpi'$.

Notation.

This is in some measure a development of the notation employed in my "Memoir on the Theory of Matrices," *Phil. Trans.* t. CXLVIII. (1858), pp. 17—37, [152] I use certain single letters u, etc. to denote sets or rows each of ρ letters, $u=(u_1, \ldots, u_\rho)$: or if, to fix the ideas $\rho=3$, then $u=(u_1,\ u_2,\ u_3)$, and so in other cases.

But I use certain other letters a, etc. to denote squares or matrices each of ρ^2 letters; thus, if $\rho=3$ as before,

$$a=\left|\begin{matrix} a_{11}, & a_{12}, & a_{13} \\ a_{21}, & a_{22}, & a_{23} \\ a_{31}, & a_{32}, & a_{33} \end{matrix}\right|,$$

and in any such case the transposed matrix is denoted by the same letter enclosed in parentheses

$$(a)=\left|\begin{matrix} a_{11}, & a_{21}, & a_{31} \\ a_{12}, & a_{22}, & a_{32} \\ a_{13}, & a_{23}, & a_{33} \end{matrix}\right|.$$

The sum $u+v$ of the row-letters u, $=(u_1,\ u_2,\ u_3)$ and v, $=(v_1,\ v_2,\ v_3)$ denotes the row $(u_1+v_1,\ u_2+v_2,\ u_3+v_3)$: and in like manner the sum $a+b$ of the two matrices, or square-letters a and b, denotes the matrix

$$\left|\begin{matrix} a_{11}+b_{11}, & a_{12}+b_{12}, & a_{13}+b_{13} \\ a_{21}+b_{21}, & a_{22}+b_{22}, & a_{23}+b_{23} \\ a_{31}+b_{31}, & a_{32}+b_{32}, & a_{33}+b_{33} \end{matrix}\right|;$$

and similarly for a sum of three or more terms.

The product uv, $=(u_1,\ u_2,\ u_3)(v_1,\ v_2,\ v_3)$, of the two row-letters u, v denotes the single term $u_1v_1+u_2v_2+u_3v_3$. We have $uv=vu$.

The product

$$au, =\left|\begin{matrix} a_{11}, & a_{12}, & a_{13} \\ a_{21}, & a_{22}, & a_{23} \\ a_{31}, & a_{32}, & a_{33} \end{matrix}\right|(u_1,\ u_2,\ u_3),$$

of a preceding square-letter a and a succeeding row-letter u, denotes the set or row

$$(a_{11},\ a_{12},\ a_{13})(u_1,\ u_2,\ u_3),\quad (a_{21},\ a_{22},\ a_{23})(u_1,\ u_2,\ u_3),\quad (a_{31},\ a_{32},\ a_{33})(u_1,\ u_2,\ u_3);$$

the notation ua is not employed.

The product

$$auv = \begin{vmatrix} a_{11}, & a_{12}, & a_{13} \\ a_{21}, & a_{22}, & a_{23} \\ a_{31}, & a_{32}, & a_{33} \end{vmatrix} (u_1, u_2, u_3)(v_1, v_2, v_3),$$

of a preceding square-letter a followed by the two row-letters u and v, denotes the single term

$$(a_{11}, a_{12}, a_{13})(u_1, u_2, u_3)v_1 + (a_{21}, a_{22}, a_{23})(u_1, u_2, u_3)v_2 + (a_{31}, a_{32}, a_{33})(u_1, u_2, u_3)v_3.$$

Observe that auv is not in general $= avu$; but it is easy to verify that $auv = (a)vu$; and hence if $(a) = a$, that is, if the matrix a be symmetrical, then $auv = avu$.

A product of two matrices

$$ab, = \begin{vmatrix} a_{11}, & a_{12}, & a_{13} \\ a_{21}, & a_{22}, & a_{23} \\ a_{31}, & a_{32}, & a_{33} \end{vmatrix} . \begin{vmatrix} b_{11}, & b_{12}, & b_{13} \\ b_{21}, & b_{22}, & b_{23} \\ b_{31}, & b_{32}, & b_{33} \end{vmatrix},$$

denotes a matrix

	(b_{11}, b_{21}, b_{31}),	(b_{12}, b_{22}, b_{32}),	(b_{13}, b_{23}, b_{33})
(a_{11}, a_{12}, a_{13})	,,	,,	,,
(a_{21}, a_{22}, a_{23})	,,	,,	,,
(a_{31}, a_{32}, a_{33})	,,	,,	,,

viz. the top-line of the compound matrix is

$$(a_{11}, a_{12}, a_{13})(b_{11}, b_{21}, b_{31}), \quad (a_{11}, a_{12}, a_{13})(b_{12}, b_{22}, b_{32}), \quad (a_{11}, a_{12}, a_{13})(b_{13}, b_{23}, b_{33}),$$

and so for the other lines: or expressing this in words, we say that any *line* of the compound matrix is obtained by compounding the corresponding *line* of the first or further component matrix with the several columns of the second or nearer component matrix.

Clearly ab is not in general $= ba$. We may easily verify that $(ab) = (b)(a)$, that is, the transposed matrix (ab) is that obtained by the composition of the transposed matrix (b) as first or further matrix, with the transposed matrix (a) as second or nearer matrix. Even if a and b are each symmetrical, we do not in general have $ab = ba$, but only $(ab) = ba$, or what is the same thing, $ab = (ba)$.

In a symbol such as $abuv$, we first combine a, b into a single matrix ab, and then regard the expression as a combination such as auv: the expression denotes therefore a single term. The theory might be explained in greater detail; but the mode of working with row- and square-letters will be readily understood from what precedes.

In all that follows, u, μ, ν, μ', ν', n, ϖ', ζ are row-letters; a, b, h, ω, ω', η, η' are square-letters: a and b are symmetrical, viz. $a = (a)$, $b = (b)$.

And I write

$$\begin{aligned}(*)(u,\ \nu)^2, &= (a,\ h,\ b)(u,\ \nu)^2 \\ &= au^2 + 2hu\nu + b\nu^2 \\ &= \begin{vmatrix} a_{11}, & a_{12}, & a_{13} \\ a_{21}, & a_{22}, & a_{23} \\ a_{31}, & a_{32}, & a_{33} \end{vmatrix} (u_1,\ u_2,\ u_3)^2 \\ &+ 2\begin{vmatrix} h_{11}, & h_{12}, & h_{13} \\ h_{21}, & h_{22}, & h_{23} \\ h_{31}, & h_{32}, & h_{33} \end{vmatrix} (u_1,\ u_2,\ u_3)(\nu_1,\ \nu_2,\ \nu_3) \\ &+ \begin{vmatrix} b_{11}, & b_{12}, & b_{13} \\ b_{21}, & b_{22}, & b_{23} \\ b_{31}, & b_{32}, & b_{33} \end{vmatrix} (\nu_1,\ \nu_2,\ \nu_3)^2\end{aligned}$$

to denote the general quadric function of the 2ρ letters u, ν, with

$$\tfrac{1}{2}\rho(\rho+1) + \rho^2 + \tfrac{1}{2}\rho(\rho+1),\ = \rho(2\rho+1)$$

coefficients. It is assumed that the determinant formed with the $\frac{1}{2}\rho(\rho+1)$ coefficients b is negative: this is the necessary and sufficient condition for the convergence of the series.

Definition of $\Theta(u\,;\ \mu,\ \nu)$.

$\Theta(u\,;\ \mu,\ \nu)$, the general theta-function with ρ arguments u, and 2ρ characters μ, ν, is the sum of a ρ-tuple series of exponentials

$$\Theta(u\,;\ \mu,\ \nu) = \Sigma\ \text{exp.}\,[(*)(u,\ n+\nu)^2 + 2\pi i\mu(n+\nu)],$$

where each of the letters n, $=(n_1, \ldots, n_\rho)$, has all integer values (zero included) from $-\infty$ to $+\infty$.

The general theorem in regard to $\Theta(u\,;\ \mu,\ \nu)$.

This is

$$\text{exp.}\,[-H(u\,;\ \mu',\ \nu')]\,.\,\Theta(u+2\varpi'\,;\ \mu,\ \nu) = \text{exp.}\,[-2\pi i\mu\nu']\,.\,\Theta(u\,;\ \mu+\mu',\ \nu+\nu'),$$

establishing a relation between the function $\Theta(u\,;\ \mu+\mu',\ \nu+\nu')$, with arbitrary character-increments μ', ν', and the function $\Theta(u+2\varpi'\,;\ \mu,\ \nu)$ with the original characters, but with new arguments $u+2\varpi'$. Also $H(u\,;\ \mu',\ \nu')$ denotes a function, linear as regards the arguments u, but quadric as regards μ' and ν'; $-2\pi i\mu\nu'$ is a single term depending only on μ and ν'; and the theorem thus is that the two functions differ only by an exponential factor. The relations between the constants will be obtained in the course of the investigation.

Demonstration.

The truth of the theorem depends on the equality of corresponding exponentials on the two sides of the equation: viz. substituting for the theta-functions their values, and comparing the exponents or arguments of the exponentials: writing also for convenience

$$G(u+2\varpi',\ n+\nu),$$

to denote the quadric function $(*)(u+2\varpi',\ n+\nu)^2$; we ought to have

$$-H(u;\ \mu',\ \nu')+G(u+2\varpi',\ n+\nu)+2\pi i\mu(n+\nu)$$
$$=-2\pi i\mu\nu'+G(u,\ n+\nu+\nu')+2\pi i(\mu+\mu')(n+\nu+\nu'),$$

or say

$$H(u;\ \mu',\ \nu')=G(u+2\varpi',\ n+\nu)-G(u,\ n+\nu+\nu')-2\pi i(n+\nu+\nu')\mu'.$$

In this equation, if true at all, the terms containing n must destroy each other; assuming that they do so, the equation becomes

$$H(u;\ \mu',\ \nu')=G(u+2\varpi',\ \nu)-G(u,\ \nu+\nu')-2\pi i(\nu+\nu')\mu'.$$

Consider first the terms in n: the right-hand side is

$$\begin{aligned}=\ & a(u+2\varpi')^2+2h(u+2\varpi')(n+\nu)+b(n+\nu)^2\\ & -au^2\qquad -2hu(n+\nu+\nu')-b(n+\nu+\nu')^2-2\pi in\mu';\end{aligned}$$

and the terms herein which contain n thus are

$$\begin{aligned}& 2h(u+2\varpi')n+bn^2+2bn\nu\\ & -2hun\qquad -bn^2-2bn(\nu+\nu')-2\pi in\mu',\\ & =4h\varpi'n\qquad -2bn\nu'\qquad -2\pi in\mu',\end{aligned}$$

which, b being symmetrical, may be written

$$=2(2h\varpi'-b\nu'-\pi i\mu')n,$$

and these terms will vanish if, and only if

$$2h\varpi'-b\nu'-\pi i\mu'=0,$$

a system of ρ equations connecting $\varpi',\ \mu',\ \nu'$.

Assuming them to be satisfied, the remaining relation,

$$H(u;\ \mu',\ \nu')=\ G(u+2\varpi',\ \nu)-G(u,\ \nu+\nu')-2\pi i(\nu+\nu')\mu',$$

becomes

$$\begin{aligned}H(u;\ \mu',\ \nu')=\ & a(u+2\varpi')^2+2h(u+2\varpi')n+b\nu^2\\ & -au^2\qquad -2hu(\nu+\nu')-b(\nu+\nu')^2-2\pi i(\nu+\nu')\mu'.\end{aligned}$$

Here, a and b being symmetrical, we have

$$a(u+2\varpi')^2=au^2+4a\varpi'u+4a\varpi'^2,\quad b(\nu+\nu')^2=b\nu^2+2b\nu'\nu+b\nu'^2,$$

and the value therefore is

$$= 4a(\varpi' u + \varpi'^2) + 2h(2\varpi' v - uv') - b(2v' v + v'^2) - 2\pi i (v + v')\mu'.$$

On the right-hand side, putting the term in h under the form

$$-2h(u + \varpi')v' + 2h\varpi'(2v + v'), \quad = -2(h)v'(u + \varpi') + 2h\varpi'(2v + v'),$$

and the last term under the form

$$-\pi i\mu'(2v + v') - \pi i\mu' v',$$

the equation becomes

$$\begin{aligned} H(u;\ \mu',\ v') = &\ (4a\varpi' - 2(h)v')(u + \varpi') - \pi i\mu' v' \\ &+ (2h\varpi' - bv' - \pi i\mu')(2v + v'), \end{aligned}$$

where the second line vanishes in virtue of the foregoing equation

$$2h\varpi' - bv' - \pi i\mu' = 0\,;$$

the equation thus is

$$H(u;\ \mu',\ v') = (4a\varpi' - 2(h)v')(u + \varpi') - \pi i\mu' v',$$

which equation, regarding therein ϖ' as a linear function of μ' and v', shows that $H(u;\ \mu',\ v')$ is a function linear as regards u (and containing this only through $u + \varpi'$), but quadric as regards μ', v'.

Introducing the new row-letter ζ', we may write

$$H(u;\ \mu',\ v') = 2\zeta'(u + \varpi') - \pi i\mu' v',$$

viz. the expression on the right-hand side is here assumed as the value of the function

$$H(u;\ \mu',\ v'), \quad = G(u + 2\varpi',\ v) - G(u,\ v + v') - 2\pi i(v + v')\mu'\,;$$

and the theorem then is

$$\exp.[-H(u;\ \mu',\ v')]\,.\,\Theta(u + 2\varpi';\ \mu,\ v) = \exp.[-2\pi i\mu v']\,.\,\Theta(u;\ \mu + \mu',\ v + v'),$$

where, by what precedes,

$$\begin{aligned} 2h\varpi' - \ bv' \ - \pi i\mu' &= 0, \\ 2a\varpi' - (h)v' - \zeta' \ &= 0, \end{aligned}$$

2ρ equations for determining the 2ρ functions ϖ', ζ' as linear functions of μ', v': which equations depend on the $\rho(2\rho + 1)$ constants a, b, h.

Suppose that the resulting values of ϖ' and ζ' are

$$\begin{aligned} \varpi' &= \omega\mu' + \omega' v', \\ \zeta' &= \eta\mu' + \eta' v', \end{aligned}$$

where ω, ω', η, η' are square-letters; then, regarding a, b, h as arbitrary, the $4\rho^2$ new constants ω, ω', η, η' cannot be all of them arbitrary, but must be connected by $4\rho^2 - \rho(2\rho + 1)$, $= \rho(2\rho - 1)$ equations.

We may regard ω, ω', η, η' as satisfying these $\rho(2\rho-1)$ equations, but as being otherwise arbitrary; the foregoing equations then are

$$2h\varpi' - b\nu' - \pi i\mu' = 0,$$
$$2a\varpi' - (h)\nu' - \zeta' = 0,$$
$$\varpi' = \omega\mu' + \omega'\nu',$$
$$\zeta' = \eta\mu' + \eta'\nu',$$

which lead to the equations connecting a, b, h with ω, ω', η, η'.

The first and second equations, substituting for ϖ' and ζ' their values, become

$$(2h\omega - \pi i)\,\mu' + (2h\omega' - b)\,\nu' = 0,$$
$$(2a\omega - \eta\ )\,\mu' + (2a\omega' - (h) - \eta')\,\nu' = 0,$$

or μ', ν' being arbitrary, we thus obtain the $4\rho^2$ equations

$$2a\omega - \eta = 0,$$
$$2h\omega - \pi i = 0,$$
$$2a\omega' - \eta' - (h) = 0,$$
$$2h\omega' - b = 0,$$

which are the equations in question. It is to be observed that πi is, like the other symbols, a matrix, viz. it is regarded as containing the matrix unity; or, what is the same thing, it denotes

$$\pi i \left| \begin{array}{lll} 1, & 0, & 0, \ldots \\ 0, & 1, & 0, \\ \vdots & & \end{array} \right| .$$

We can eliminate a, b, h from these equations and thus obtain the $\rho(2\rho-1)$ equations before referred to, which connect the $4\rho^2$ constants ω, ω', η, η'. I give, but without a complete explanation, the steps of the elimination.

The equation $2a\omega - \eta = 0$, may be written in the form

$$2\,(a\omega) - (\eta) = 0,$$

that is,

$$2\,(\omega)\,(a) - (\eta) = 0,$$

or since $(a) = a$, this is

$$2\,(\omega)\,a - (\eta) = 0\,;$$

from the original form, and the new form respectively, we find

$$2\,(\omega)\,a\omega - (\omega)\,\eta = 0, \quad 2\,(\omega)\,a\,(\omega) - (\eta)\,\omega = 0\,;$$

and comparing these

$$(\omega)\,\eta - (\eta)\,\omega = 0, \quad \text{(first result)}.$$

The equation $2a\omega' - \eta' - (h) = 0$, or say $(h) = -\eta' + 2a\omega'$, may be written in the form

$$h = -\,(\eta') + 2\,(a\omega'),$$

that is, since $a = (a)$,

$$h = -(\eta') + 2(\omega')a;$$

and we thence deduce

$$h\omega = -(\eta')\omega + 2(\omega')a\omega.$$

But from the equation $2a\omega - \eta = 0$, we have $2(\omega')a\omega - (\omega')\eta = 0$, and the equation thus becomes $h\omega = -(\eta')\omega + (\omega')\eta$; which, in virtue of $2h\omega - \pi i = 0$, becomes

$$\tfrac{1}{2}\pi i = -(\eta')\omega + (\omega')\eta, \quad \text{(second result)}.$$

From the equation above obtained, $h = -(\eta') + 2(\omega')a$, we have

$$h\omega' = -(\eta')\omega' + 2(\omega')a\omega';$$

in virtue of $2h\omega' - b = 0$, this becomes $-2(\eta')\omega' + 4(\omega')a\omega' = b$; an equation which may also be written $-2((\eta')\omega') + 4((\omega')a\omega') = (b)$, or, what is the same thing, $-2(\omega')\eta' + 4(\omega')(a)\omega' = (b)$; or since $(a) = a$ and $(b) = b$, this is

$$-2(\omega')\eta' + 4(\omega')a\omega' = b:$$

and comparing with the original equation

$$-2(\eta')\omega' + 4(\omega')a\omega' = b,$$

we obtain

$$(\omega')\eta' - (\eta')\omega' = 0, \quad \text{(third result)}.$$

We have thus the three systems

$$\begin{array}{lll} (\omega)\eta - (\eta)\omega = 0, & \tfrac{1}{2}\rho(\rho-1) & \text{equations,} \\ (\omega')\eta - (\eta')\omega = \tfrac{1}{2}\pi i, & \rho^2 & \quad " \\ (\omega')\eta' - (\eta')\omega' = 0, & \tfrac{1}{2}\rho(\rho-1) & \quad " \end{array}$$

in all $\rho(2\rho-1)$ equations. As to these systems, observe that $(\omega)\eta$, $(\eta)\omega$, etc., are all of them matrices of ρ^2 terms; each of the three systems denotes therefore in the first instance ρ^2 equations, viz. the equations obtained by equating to zero the several terms of such a matrix: but in the first system each diagonal term so equated to zero gives the identity $0 = 0$; and equating to zero the terms which are symmetrical in regard to the diagonal we obtain twice over, in the forms $P = 0$, and $-P = 0$, one and the same equation; the number of equations is thus diminished from ρ^2 to $\tfrac{1}{2}\rho(\rho-1)$; and similarly in the third system the number of equations is $= \tfrac{1}{2}\rho(\rho-1)$: but for the second system the number of equations is really $= \rho^2$. It is hardly necessary to remark that in this second system $\tfrac{1}{2}\pi i$ is as before regarded as a matrix.

The foregoing three systems of equations are in fact the equations (6) p. 4 of Dr Schottky's work.

Cambridge, 12 *July*, 1880.

754.

ON THE CONNEXION OF CERTAIN FORMULÆ IN ELLIPTIC FUNCTIONS.

[From the *Messenger of Mathematics*, vol. IX. (1880), pp. 23—25.]

IN reference to a like question in the theory of the double ϑ-functions, it is interesting to show that (if not completely, at least very nearly) the single formula

$$\Pi(u,\ a) = u\frac{\Theta' a}{\Theta a} + \tfrac{1}{2}\log\frac{\Theta(u-a)}{\Theta(u+a)},$$

that is,

$$\int_0 \frac{k^2 \operatorname{sn} a \operatorname{cn} a \operatorname{dn} a \operatorname{sn}^2 u\, du}{1 - k^2 \operatorname{sn}^2 a \operatorname{sn}^2 u} = u\frac{\Theta' a}{\Theta a} + \tfrac{1}{2}\log\frac{\Theta(u-a)}{\Theta(u+a)},$$

leads not only to the relation

$$\log \Theta u = \tfrac{1}{2}\log\frac{2k'K}{\pi} + \tfrac{1}{2}\left(1 - \frac{E}{K}\right)u^2 - k^2\int_0 du \int_0 du \operatorname{sn}^2 u,$$

between the functions Θ, sn, but also to the addition-equation for the function sn.

Writing in the equation a indefinitely small, and assuming only that $\operatorname{sn} a$, $\operatorname{cn} a$, $\operatorname{dn} a$ then become a, 1, 1, respectively, the equation is

$$k^2 a \int_0 \operatorname{sn}^2 u\, du = u\frac{a\Theta'' 0}{\Theta 0} + \tfrac{1}{2}\log\frac{\Theta u - a\Theta' u}{\Theta u + a\Theta' u},$$

$$= ua\frac{\Theta'' 0}{\Theta 0} - a\frac{\Theta' u}{\Theta u},$$

that is,

$$\frac{\Theta' u}{\Theta u} = u\frac{\Theta'' 0}{\Theta 0} - k^2\int_0 du \operatorname{sn}^2 u,$$

or, integrating from $u = 0$, this is

$$\log \Theta u = C + \tfrac{1}{2}u^2\frac{\Theta'' 0}{\Theta 0} - k^2\int_0 du \int_0 du \operatorname{sn}^2 u,$$

which, except as regards the determination of the constants, is the required equation for $\log \Theta u$.

Next, differentiating twice the equation for $\Pi(u, a)$, and once the equation obtained for $\frac{\Theta' u}{\Theta u}$, we have

$$k^2 \operatorname{sn} a \operatorname{cn} a \operatorname{dn} a \frac{d}{du}\left(\frac{\operatorname{sn}^2 u}{1 - k^2 \operatorname{sn}^2 a \operatorname{sn}^2 u}\right) = \tfrac{1}{2}\frac{\Theta''\Theta - \Theta'^2}{\Theta^2}(u-a) - \tfrac{1}{2}\frac{\Theta''\Theta - \Theta'^2}{\Theta^2}(u+a),$$

and

$$\frac{\Theta''\Theta - \Theta'^2}{\Theta^2} u = \frac{\Theta'' 0}{\Theta 0} - k^2 \operatorname{sn}^2 u,$$

where, for shortness, $\frac{\Theta''\Theta - \Theta'^2}{\Theta^2} u$ is written to denote $\frac{\Theta'' u \,\Theta u - (\Theta' u)^2}{\Theta^2 u}$, and the like in the first equation; the right-hand side of the first equation therefore is

$$-\tfrac{1}{2}k^2 \{\operatorname{sn}^2(u-a) - \operatorname{sn}^2(u+a)\},$$

or the equation becomes

$$2 \operatorname{sn} a \operatorname{cn} a \operatorname{dn} a \frac{d}{du} \frac{\operatorname{sn}^2 u}{1 - k^2 \operatorname{sn}^2 u \operatorname{sn}^2 a} = \operatorname{sn}^2(u+a) - \operatorname{sn}^2(u-a),$$

that is,

$$\frac{4 \operatorname{sn} u \operatorname{sn}' u \operatorname{sn} a \operatorname{cn} a \operatorname{dn} a}{(1 - k^2 \operatorname{sn}^2 u \operatorname{sn}^2 a)^2} = \operatorname{sn}^2(u+a) - \operatorname{sn}^2(u-a).$$

The numerator on the left-hand side must be a symmetrical function of u, a, and hence (even if the value of $\operatorname{sn}' u$ were unknown) it would appear that $\operatorname{sn}' u$ must be a mere constant multiple of $\operatorname{cn} u \operatorname{dn} u$; assuming, however, the actual value, $\operatorname{sn}' u = \operatorname{cn} u \operatorname{dn} u$, the formula is

$$\frac{4 \operatorname{sn} u \operatorname{cn} u \operatorname{dn} u \operatorname{sn} a \operatorname{cn} a \operatorname{dn} a}{(1 - k^2 \operatorname{sn}^2 u \operatorname{sn}^2 a)^2}$$
$$= \operatorname{sn}^2(u+a) - \operatorname{sn}^2(u-a)$$
$$= \{\operatorname{sn}(u+a) + \operatorname{sn}(u-a)\}\{\operatorname{sn}(u+a) - \operatorname{sn}(u-a)\}.$$

The factor $\{\operatorname{sn}(u+a) + \operatorname{sn}(u-a)\}$ becomes $= 2 \operatorname{sn} u$ for $a = 0$, and this suggests that the factor $\operatorname{sn} u$ on the left-hand side is a factor of $\{\operatorname{sn}(u+a) + \operatorname{sn}(u-a)\}$. That $\operatorname{cn} u$ is *not* a factor hereof would follow from the properties of the period K; viz. for $u = K$, $\operatorname{cn} u = 0$, but $\{\operatorname{sn}(u+a) + \operatorname{sn}(u-a)\}$, $= 2 \operatorname{sn}(K+a)$ is not $= 0$; and, similarly, that $\operatorname{dn} u$ is *not* a factor from the properties of the period iK; hence, $\operatorname{cn} u$, $\operatorname{dn} u$ belong to the other factor $\{\operatorname{sn}(u+a) - \operatorname{sn}(u-a)\}$, and by symmetry $\operatorname{cn} a$, $\operatorname{dn} a$ belong to the first-mentioned factor. And we are thus led to assume

$$\operatorname{sn}(u+a) + \operatorname{sn}(u-a) = 2M \operatorname{sn} u \operatorname{cn} a \operatorname{dn} a,$$
$$\operatorname{sn}(u+a) - \operatorname{sn}(u-a) = 2M' \operatorname{sn} a \operatorname{cn} u \operatorname{dn} u,$$

where

$$\text{denom.} = 1 - k^2 \operatorname{sn}^2 a \operatorname{sn}^2 u,$$

and $MM' = 1$. Some further investigation is wanting to show that M and M' are constants, but assuming that they are so and each $= 1$, the formulæ give at once the ordinary expression for $\operatorname{sn}(u+a)$; that is, we have the addition-equation for the function sn.

755.

ON THE MATRIX $\begin{pmatrix} a, & b \\ c, & d \end{pmatrix}$, AND IN CONNEXION THEREWITH THE FUNCTION $\dfrac{ax+b}{cx+d}$.

[From the *Messenger of Mathematics*, vol. IX. (1880), pp. 104—109.]

IN the preceding paper, [due to Prof. W. W. Johnson,] the theory of the symbolic powers and roots of the function $\dfrac{ax+b}{cx+d}$ is developed in a complete and satisfactory manner; the results in the main agreeing with those obtained in the original memoir, Babbage, "On Trigonometrical Series," *Memoirs of the Analytical Society* (1813), Note I. pp. 47—50, and which are to some extent reproduced in my "Memoir on the Theory of Matrices," *Phil. Trans.*, t. CXLVIII. (1858), pp. 17—37, [152]. I had recently occasion to reconsider the question, and have obtained for the nth function $\phi^n x$, where $\phi x = \dfrac{ax+b}{cx+d}$, a form which, although substantially identical with Babbage's, is a more compact and convenient one; viz. taking λ to be determined by the quadric equation

$$\frac{(\lambda+1)^2}{\lambda} = \frac{(a+d)^2}{ad-bc},$$

the form is

$$\phi^n(x) = \frac{(\lambda^{n+1}-1)(ax+b)+(\lambda^n-\lambda)(-dx+b)}{(\lambda^{n+1}-1)(cx+d)+(\lambda^n-\lambda)(\ \ cx\ -a)}.$$

The question is, in effect, that of the determination of the nth power of the matrix $\begin{pmatrix} a, & b \\ c, & d \end{pmatrix}$; viz. in the notation of matrices

$$(x_1,\ y_1) = \begin{pmatrix} a, & b \\ c, & d \end{pmatrix}(x,\ y),$$

means the two equations $x_1 = ax + by$, $y_1 = cx + dy$; and then if x_2, y_2 are derived in like manner from x_1, y_1, that is, if $x_2 = ax_1 + by_1$, $y_2 = cx_1 + dy_1$, and so on, x_n, y_n will be linear functions of x, y; say we have $x_n = a_n x + b_n y$, $y_n = c_n x + d_n y$: and the nth power of $\begin{pmatrix} a, & b \\ c, & d \end{pmatrix}$ is, in fact, the matrix $\begin{pmatrix} a_n, & b_n \\ c_n, & d_n \end{pmatrix}$.

In particular, we have

$$\begin{pmatrix} a, & b \\ c, & d \end{pmatrix}^2, = \begin{pmatrix} a_2, & b_2 \\ c_2, & d_2 \end{pmatrix}, = \begin{pmatrix} a^2+bc, & b(a+d) \\ c(a+d), & d^2+bc \end{pmatrix},$$

and hence the identity

$$\begin{pmatrix} a, & b \\ c, & d \end{pmatrix}^2 - (a+d)\begin{pmatrix} a, & b \\ c, & d \end{pmatrix} + (ad-bc)\begin{pmatrix} 1, & 0 \\ 0, & 1 \end{pmatrix} = 0;$$

viz. this means that the matrix

$$\begin{pmatrix} a_2-(a+d)a+ad-bc, & b_2-(a+d)b \\ c_2-(a+d)c\quad, & d_2-(a+d)d+ad-bc \end{pmatrix} = \begin{pmatrix} 0, & 0 \\ 0, & 0 \end{pmatrix},$$

or, what is the same thing, that each term of the left-hand matrix is $=0$; which is at once verified by substituting for a_2, b_2, c_2, d_2 their foregoing values.

The explanation just given will make the notation intelligible and show in a general way how a matrix may be worked in like manner with a single quantity: the theory is more fully developed in my Memoir above referred to. I proceed with the solution in the algorithm of matrices. Writing for shortness $M = \begin{pmatrix} a, & b \\ c, & d \end{pmatrix}$, the identity is

$$M^2 - (a+d)M + (ad-bc) = 0,$$

the matrix $\begin{pmatrix} 1, & 0 \\ 0, & 1 \end{pmatrix}$ being in the theory regarded as $=1$; viz. M is determined by a quadric equation; and we have consequently $M^n =$ a linear function of M. Writing this in the form

$$M^n - AM + B = 0,$$

the unknown coefficients A, B can be at once obtained in terms of α, β, the roots of the equation

$$u^2 - (a+d)u + ad - bc = 0,$$

viz. we have

$$\alpha^n - A\alpha + B = 0,$$

$$\beta^n - A\beta + B = 0;$$

or more simply from these equations, and the equation for M^n, eliminating α, β, we have

$$\begin{vmatrix} M^n, & M, & 1 \\ \alpha^n, & \alpha, & 1 \\ \beta^n, & \beta, & 1 \end{vmatrix} = 0;$$

that is,

$$M^n(\alpha-\beta)-M(\alpha^n-\beta^n)+\alpha\beta(\alpha^{n-1}-\beta^{n-1})=0.$$

But instead of α, β, it is convenient to introduce the ratio λ of the two roots, say we have $\alpha=\lambda\beta$; we thence find

$$(\lambda+1)\beta=a+d,$$
$$\lambda\beta^2=ad-bc,$$

giving

$$\frac{(\lambda+1)^2}{\lambda}=\frac{(a+d)^2}{ad-bc}$$

for the determination of λ, and then

$$\beta=\frac{a+d}{\lambda+1},$$
$$\alpha=\frac{(a+d)\lambda}{\lambda+1}.$$

The equation thus becomes

$$M^n(\lambda-1)\beta-M(\lambda^n-1)\beta^n+(\lambda^n-\lambda)\beta^{n+1}=0,$$

or we have

$$M^n=\frac{\beta^{n-1}}{\lambda-1}\{(\lambda^n-1)M-(\lambda^n-\lambda)\beta\}.$$

It is convenient to multiply the numerator and denominator by $\lambda+1$, viz. we thus have

$$M^n=\frac{\beta^{n-1}}{\lambda^2-1}[(\lambda^{n+1}-1)M+(\lambda^n-\lambda)\{M-(\lambda+1)\beta\}].$$

The exterior factor is here

$$=\frac{1}{\lambda^2-1}\left(\frac{a+d}{\lambda+1}\right)^{n-1},$$

moreover $(\lambda+1)\beta$ is $=a+d$: hence

$$M=\left(\begin{array}{|cc|} a, & b \\ c, & d \end{array}\right),$$

and

$$M-(\lambda+1)\beta=\left(\begin{array}{|cc|} a, & b \\ c, & d \end{array}\right)-\left(\begin{array}{|cc|} a+d, & 0 \\ 0\;, & a+d \end{array}\right),=\left(\begin{array}{|cc|} -d, & b \\ c\;, & -a \end{array}\right);$$

the formula thus is

$$M^n=\frac{1}{\lambda^2-1}\left(\frac{a+d}{\lambda+1}\right)^{n-1}\left\{(\lambda^{n+1}-1)\left(\begin{array}{|cc|} a, & b \\ c, & d \end{array}\right)+(\lambda^n-\lambda)\left(\begin{array}{|cc|} -d, & b \\ c\;, & -a \end{array}\right)\right\},$$

viz. we have thus the values of the several terms of the nth matrix

$$M^n=\left(\begin{array}{|cc|} a_n, & b_n \\ c_n, & d_n \end{array}\right);$$

and, if instead of these we consider the combinations $a_n x + b_n$ and $c_n x + d_n$, we then obtain

$$a_n x + b_n = \frac{1}{\lambda^2 - 1}\left(\frac{a+d}{\lambda+1}\right)^{n-1}\{(\lambda^{n+1}-1)(ax+b)+(\lambda^n-\lambda)(-dx+b)\},$$

$$c_n x + d_n = \quad " \quad\quad " \quad\quad \{(\lambda^{n+1}-1)(cx+d)+(\lambda^n-\lambda)(\ \ cx-a)\};$$

and in dividing the first of these by the second, the exterior factor disappears.

It is to be remarked that, if $n=0$, the formulæ become as they should do $a_0 x + b_0 = x$, $c_0 x + d_0 = 1$; and if $n=1$, they become $a_1 x + b_1 = ax + b$, $c_1 x + d_1 = cx + d$.

If $\lambda^m - 1 = 0$, where m, the least exponent for which this equation is satisfied, is for the moment taken to be greater than 2, the terms in { } are

$$(\lambda-1)(ax+b)+(1-\lambda)(-dx+b),$$

and

$$(\lambda-1)(cx+d)+(1-\lambda)(\ \ cx-a);$$

viz. these are $(\lambda-1)(a+d)x$, and $(\lambda-1)(a+d)$, or if for $(\lambda-1)(a+d)$ we write $(\lambda^2-1)\dfrac{a+d}{\lambda+1}$, the formulæ become for $n=m$

$$a_m x + b_m = \left(\frac{a+d}{\lambda+1}\right)^m x,$$

$$c_m x + d_m = \left(\frac{a+d}{\lambda+1}\right)^m;$$

viz. we have here

$$\frac{a_m x + b_m}{c_m x + d_m} = x,$$

or the function is periodic of the mth order. Writing for shortness $\vartheta = \dfrac{s\pi}{n}$, s being any integer not $=0$, and prime to n, we have $\lambda = \cos 2\vartheta + i \sin 2\vartheta$, hence

$$1+\lambda = 2\cos\vartheta(\cos\vartheta + i\sin\vartheta),$$

or $\dfrac{(1+\lambda)^2}{\lambda} = 4\cos^2\vartheta$; consequently, in order to the function being periodic of the nth order, the relation between the coefficients is

$$4\cos^2\frac{s\pi}{n} = \frac{(a+d)^2}{ad-bc}.$$

The formula extends to the case $m=2$, viz. $\cos\frac{1}{2}(s\pi)=0$, or the condition is $a+d=0$. But here $\lambda+1=0$, and the case requires to be separately verified. Recurring to the original expression for M^2, we see that, for $a+d=0$, this becomes

$$\begin{vmatrix} a^2+bc, & 0 \\ 0\ \ , & d^2+bc \end{vmatrix}, \; = (a^2+bc)\begin{vmatrix} 1, & 0 \\ 0, & 1 \end{vmatrix};$$

that is,

$$\frac{a_2 x + b_2}{c_2 x + d_2} = x,$$

or the result is thus verified.

But the case $m=1$ is a very remarkable one; we have here $\lambda=1$, and the relation between the coefficients is thus $(a+d)^2=4(ad-bc)$, or what is the same thing $(a-d)^2+4bc=0$. And then determining the values for $\lambda=1$ of the vanishing fractions which enter into the formulæ, we find

$$a_n x+b_n=\frac{1}{2^n}(a+d)^{n-1}\{(n+1)(ax+b)+(n-1)(-dx+b)\},$$

$$c_n x+d_n=\frac{1}{2^n}(a+d)^{n-1}\{(n+1)(cx+d)+(n-1)(\ \ cx-a)\},$$

or as these may also be written

$$a_n x+b_n=\frac{1}{2^n}(a+d)^{n-1}\{x[n(a-d)+(a+d)]+2nb\},$$

$$c_n x+d_n=\frac{1}{2^n}(a+d)^{n-1}\{x\,.\,2nc+[-n(a-d)+a+d]\},$$

which for $n=0$, become as they should do $a_0x+b_0=x$, $c_0x+d_0=1$, and for $n=1$ they become $a_1x+b_1=ax+b$, $c_1x+d_1=cx+d$. We thus do *not* have $\dfrac{a_1x+b_1}{c_1x+d_1}=x$, and the function is *not* periodic of any order. This remarkable case is noticed by Mr Moulton in his edition (2nd edition, 1872) of Boole's *Finite Differences*.

If to satisfy the given relation $(a-d)^2+4bc=0$, we write $2b=k(a-d)$, $2c=-\dfrac{1}{k}(a-d)$, then the function of x is

$$\frac{ax+\frac{1}{2}k(a-d)}{-\frac{1}{2}k^{-1}(a-d)x+d},$$

and the formulæ for the nth function are

$$a_n x+b_n=\frac{1}{2^n}(a+d)^{n-1}\{(a+d)x+n(a-d)(x+k)\},$$

$$c_n x+d_n=\frac{1}{2^n}(a+d)^{n-1}\left\{(a+d)\quad -n(a-d)\left(\frac{x}{k}+1\right)\right\};$$

which may be verified successively for the different values of n.

Reverting to the general case, suppose $n=\infty$, and let u be the value of $\phi^\infty(x)$. Supposing that the modulus of λ is not $=1$, we have λ^n indefinitely large or indefinitely small. In the former case, we obtain

$$u=\frac{\lambda(ax+b)+(-dx+b)}{\lambda(cx+d)+(\ \ cx-a)},\quad =\frac{(\lambda a-d)x+b(\lambda+1)}{c(\lambda+1)x+\lambda d-a};$$

which, observing that the equation in λ may be written

$$\frac{\lambda a-d}{c(\lambda+1)}=\frac{b(\lambda+1)}{\lambda d-a},$$

is independent of x, and equal to either of these equal quantities; and if from these two values of u we eliminate λ, we obtain for u the quadric equation

$$cu^2 - (a-d)\,u - b = 0,$$

that is,

$$u = \frac{au+b}{cu+d},$$

as is, in fact, obvious from the consideration that n being indefinitely large the nth and $(n+1)$th functions must be equal to each other. In the latter case, as λ^n is indefinitely small, we have the like formulæ, and we obtain for u the same quadric equation: the two values of u are however not the same, but (as is easily shown) their product is $= -b \div c$; u is therefore the other root of the quadric equation. Hence, as n increases, the function $\phi^n x$ continually approximates to one or the other of the roots of this quadric equation. The equation has equal roots if $(a-d)^2 + 4bc = 0$, which is the relation existing in the above-mentioned special case; and here $u = \frac{1}{2c}(a-d), = \frac{-2b}{a-d}$, which result is also given by the formulæ of the special case on writing therein $n = \infty$.

756.

A GEOMETRICAL CONSTRUCTION RELATING TO IMAGINARY QUANTITIES.

[From the *Messenger of Mathematics,* vol. X. (1881), pp. 1—3.]

LET A, B, C be given imaginary quantities, and let it be required to construct the roots of the quadric equation

$$\frac{1}{X-A}+\frac{1}{X-B}+\frac{1}{X-C}=0.$$

The equation is

$$(X-B)(X-C)+(X-C)(X-A)+(X-A)(X-B)=0,$$

that is,

$$3X^2-2(A+B+C)X+BC+CA+AB=0,$$

and we have therefore

$$3X-(A+B+C)=\pm\sqrt{\{(A+B+C)^2-3(BC+CA+AB)\}},$$
$$=\pm\sqrt{\{A^2+B^2+C^2-BC-CA-AB\}};$$

or as this may be written

$$X=\tfrac{1}{3}(A+B+C)\pm\sqrt{\{\tfrac{1}{3}(A+B\omega+C\omega^2)\,.\,\tfrac{1}{3}(A+B\omega^2+C\omega)\}},$$

where ω is an imaginary cube root of unity,

$$=\cos 120^\circ+i\sin 120^\circ \text{ suppose.}$$

Taking an arbitrary point O as the origin, let the imaginary quantity A, $=\alpha+\alpha' i$ suppose, be represented by the point A, coordinates α and α'; and in like manner the imaginary quantities B and C by the points B and C respectively.

Then $B\omega$, $B\omega^2$ are represented by points B_1, B_2, obtained by rotating the point B about the origin through angles of 120° and 240° respectively; $C\omega^2$, $C\omega$ are represented by points C_1, C_2 obtained by rotating the point C about the origin through angles of 240° and 480° (=120°) respectively: and

$$\tfrac{1}{3}(A+B+C),\quad \tfrac{1}{3}(A+B\omega+C\omega^2),\quad \tfrac{1}{3}(A+B\omega^2+C\omega)$$

are represented by the points G, G_1, G_2 which are the C.G.'s of the triangles ABC, AB_1C_1, AB_2C_2 respectively. The formula therefore is

$$X=OG\pm\sqrt{(OG_1 . OG_2)},$$

where, if a, a′ are the coordinates of G, then OG is written to denote the imaginary quantity $\mathrm{a}+\mathrm{a}'i$; and the like as regards OG_1, OG_2. Taking $\sqrt{(OG_1 . OG_2)}=OH$, we then have H a point such, that the distance OH from the origin is = geometric mean of the distances OG_1, OG_2, and that the radial direction* of the distance OH bisects the radial directions of the distances OG_1, OG_2 respectively. Finally, measuring off from G in the radial direction OH, and in the opposite radial direction, the distances GX', GX'' each $= OH$; we have the two points X', X'' representing the two roots X.

The construction is somewhat simplified if we take for the origin the point G; for then $OG=0$, and we have $X=\pm\sqrt{(GG_1 . GG_2)}$, so that the points X', X'' are in fact the point H, and the opposite point in regard to G.

The theory of the more general equation

$$\frac{p}{X-A}+\frac{q}{X-B}+\frac{r}{X-C}=0,$$

(p, q, r real) is somewhat similar, but the construction is less simple; we have

$$(p+q+r)X^2-\{(q+r)A+(r+p)B+(p+q)C\}X+pBC+qCA+rAB=0.$$

Writing herein $q+r$, $r+p$, $p+q=l$, m, n, the equation becomes

$$(l+m+n)X^2-2(lA+mB+nC)X+(-l+m+n)BC+(l-m+n)CA+(l+m-n)AB=0,$$

that is,

$$\{(l+m+n)X-lA-mB-nC\}^2$$
$$=(lA+mB+nC)^2+\{l^2-(m+n)^2\}BC+\{m^2-(n+l)^2\}CA+\{n^2-(l+m)^2\}AB.$$

Here the right-hand side is

$$=l^2A^2+m^2B^2+n^2C^2+(l^2-m^2-n^2)BC+(-l^2+m^2-n^2)CA+(-l^2-m^2+n^2)AB,$$

which is

$$=-l^2(C-A)(A-B)-m^2(A-B)(B-C)-n^2(C-A)(A-B),$$

and consequently is a product of two linear factors; these, in fact, are

$$\frac{1}{l}\{l^2A+\tfrac{1}{2}(-l^2-m^2+n^2\pm\sqrt{\Delta})B+\tfrac{1}{2}(-l^2+m^2-n^2\mp\sqrt{\Delta})C\},$$

* Radial direction is, I think, a convenient expression for the direction of a line considered as drawn as a radius of a circle from the centre, and not as a diameter in two opposite radial directions.

where

$$\Delta = l^4 + m^4 + n^4 - 2m^2n^2 - 2n^2l^2 - 2l^2m^2.$$

It is to be observed that Δ, $= (l^2 - m^2 - n^2)^2 - 4m^2n^2$, is negative; hence, calling the factors $fA + gB + hC$, $f'A + g'B + h'C$ respectively, the coefficients f, g, h, and f', g', h' are imaginary; moreover $f + g + h = 0$, $f' + g' + h' = 0$.

The values of X thus are

$$(l + m + n) X = lA + mB + nC \pm \sqrt{\{(fA + gB + hC)(f'A + g'B + h'C)\}},$$

and then passing to the geometrical representation, we have $\dfrac{lA + mB + nC}{l + m + n}$ represented by the point which is the C.G. of weights l, m, n at the points A, B, C respectively; on account of the imaginary values of the coefficients the construction is not immediately applicable to the factors

$$fA + gB + hC, \quad f'A + g'B + h'C;$$

but a construction, such as was used for the factors

$$A + \omega B + \omega^2 C, \quad A + \omega^2 B + \omega C,$$

might be found without difficulty.

757.

ON A SMITH'S PRIZE QUESTION, RELATING TO POTENTIALS.

[From the *Messenger of Mathematics*, vol. XI. (1882), pp. 15—18.]

A SPHERICAL shell is divided by a plane into two segments A and B, one of them so small that it may be regarded as a plane disk: trace the curves which exhibit the potentials of the two segments and of the whole shell respectively, in regard to a point P moving along the axis of symmetry of the two segments.

Criticise the following argument:

The potential of the segment A in regard to a point P, coordinates (x, y, z), is one and the same function of (x, y, z) whatever be the position of P; similarly the potential of the segment B in regard to the same point P is one and the same function of (x, y, z) whatever be the position of P: hence the potential of the whole shell in regard to the point P is one and the same function of (x, y, z) whatever be the position of P.

The question is taken from my memoir "On Prepotentials," *Phil. Trans.* vol. 165 (1875), pp. 675—774, [607]; and the figure of the curves is given p. 689*. There is no difficulty in tracing them by means of the expression for the potential of a plane circular disk in regard to a point on its axis of symmetry: it was in order that they might be so traced, that one of the segments was taken to be small; but I had overlooked the circumstance that the formula for the disk is in fact only a particular case of a similar and equally simple formula for the spherical segment: viz. (as was found in one of the papers) the potential of a spherical segment in regard to a point on the axis is $=\dfrac{2\pi a}{\rho}(\rho_1 \sim \rho_2)$, where ρ, ρ_1, ρ_2 are the distances of the attracted point from the centre of the sphere and from the centre and the circumference respectively of the segment. The segments might therefore just as well have been any two segments whatever, or (to take the most symmetrical case) they might have been hemispheres.

As to the argument: the assertion in regard to the potential of the segment

[* This Collection, vol. IX. p. 333.]

A is based upon the consideration of this segment *alone*; and, on the ground that we can without crossing the segment pass from any one position of P to any other position of P, it is inferred that the potential is one and the same function of the coordinates, whatever be the position of P: it is therefore unassailable by any considerations in relation to the non-existent segment B. Similarly the assertion in regard to the potential of the segment B is based upon the consideration of this segment alone, and it is unassailable upon any considerations in regard to the non-existent segment A: the potential of the whole sphere is certainly the sum of the potentials of the segments A and B: it is therefore altogether off the purpose to object that in the case of the whole sphere we cannot pass from a point outside the sphere to a point inside the sphere without crossing one or other of the segments A and B. I consider that the two assertions are each of them true, and that the conclusion is a legitimate one, but it is true only in the sense in which $a+x+\sqrt{[(a-x)^2]}$ is one and the same function of x whatever be the value of x: this is so, if $\sqrt{[(a-x)^2]}$ denotes indifferently or successively the two functions $\pm(a-x)$: but if, a and x being real, $\sqrt{[(a-x)^2]}$ is taken to mean the positive value, then the function $a+x+\sqrt{[(a-x)^2]}$ is $=2a$ or $=2x$ according as $a-x$ is positive or negative.

Fig. 1.

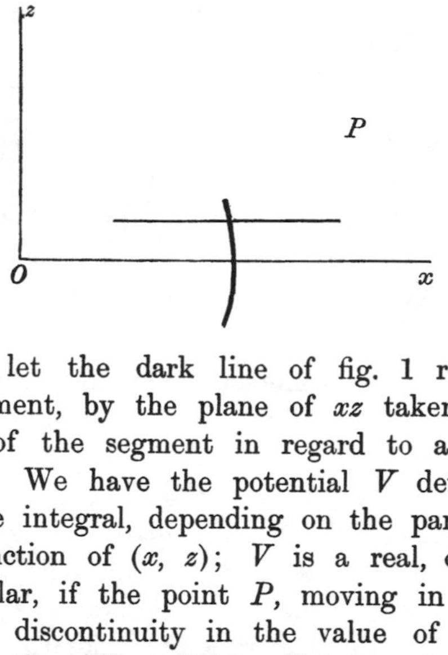

In further illustration, let the dark line of fig. 1 represent the intersection of an unclosed surface, or segment, by the plane of xz taken to be that of the paper, and consider the potential of the segment in regard to a point P in the plane of the paper, coordinates x, z. We have the potential V defined as a function of x, z by an equation $V=$ a definite integral, depending on the parameters x, z, and being in general a transcendental function of (x, z); V is a real, one-valued, finite, continuous function of x, z: in particular, if the point P, moving in any manner, traverses the dark line, there is not any discontinuity in the value of V. There is however in this case a discontinuity in the differential coefficients of V: if to fix the ideas we imagine P moving parallel to the axis of x, so that z is taken to be constant and V a function of x only, then when the path of P crosses the black line there is in general an abrupt change of value in $\frac{dV}{dx}$. Taking V as a coordinate y at right angles to the plane of the paper, a section by any plane parallel to that of xy is (when the trace of the plane upon that of xz does not meet the dark line) a continuous curve; but when the trace meets the dark line, then for this value of x there is an abrupt change of direction in the section.

If (as may very well happen in particular cases) V is algebraically determinable, then, *quâ* one-valued function of (x, z), V is not any root y at pleasure of an algebraical equation $\phi(x, y, z) = 0$, but it is for any given values of (x, z), some one determinate root y_1 of this equation: and we thus see how in this case the before-mentioned discontinuity in the value of $\frac{dV}{dx}$ must arise: viz. when the trace of the plane meets the dark line the section is a curve having a double point; and, for the positions of P on the two sides of the dark line, we have V the ordinate belonging to different branches of the curve of section. If the path of P passes through an extremity of the dark line, then the curve of section will, instead of a double point, have in general a cusp; and when the path of P does not cross the dark line, then the curve of section is a continuous line without singularity. It may be added that the surface $\phi(x, y, z) = 0$ must have a nodal line which as to a certain finite portion thereof is crunodal, giving the before-mentioned double points of the sections, but as to the residue thereof is acnodal or isolated.

It may happen that (the surface being algebraical) any particular section thereof, instead of being a single curve having a double point as above, breaks up into two distinct curves, so that for the two positions of P, we have V the ordinate of two distinct curves: and this is what really happens in the case of P a point on the axis of a circular disk or a spherical segment: thus in the case of the disk, taking c for the radius, and x for the distance from the centre of the disk, the formula is $V = 2\pi\{\sqrt{(c^2 + x^2)} \pm x\}$; or writing $V \div 2\pi = y$, the section is made up of the two distinct hyperbolas $y(y - 2x) = c^2$, and $y(y + 2x) = c^2$.

It may be remarked that in each case, it is only for P on the axis that the potential is algebraical.

In the case of the hemispheres, drawing OM a radius at right angles to the axis, the formula for the potential of an axial point P is of the form

$$V = \frac{2\pi \,.\, OM}{OP}(PM \sim PA),$$

or writing $V = 2\pi y$ we have for the hemisphere A, the curve (1) or (2) according as $(x - a)$ is positive or negative; and for the hemisphere B the curve (3) or (4) according as $x + a$ is positive or negative; viz. the equations are

$$(1)\quad y = \frac{a}{x}\{\sqrt{(a^2 + x^2)} - (x - a)\},$$

$$(2)\quad y = \frac{a}{x}\{\sqrt{(a^2 + x^2)} + (x - a)\},$$

$$(3)\quad y = \frac{a}{x}\{\sqrt{(a^2 + x^2)} - (x + a)\},$$

$$(4)\quad y = \frac{a}{x}\{\sqrt{(a^2 + x^2)} + (x + a)\},$$

being four cubic curves. The whole curve (1) is shown in fig. 2, and the others are

Fig. 2.

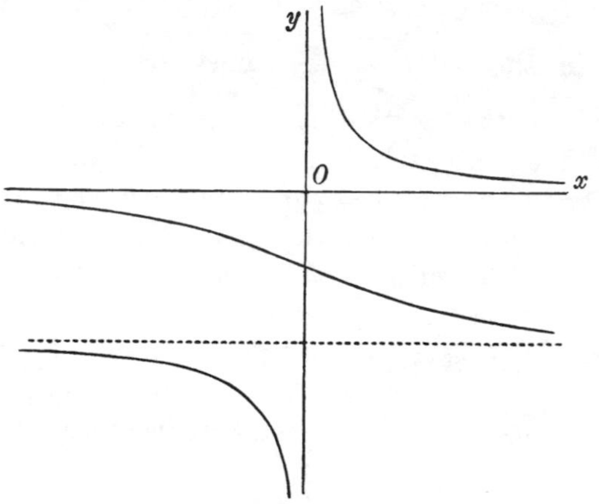

equal or opposite curves: the rationalised equation of (1) is in fact

$$x = \frac{2a^2(y+a)}{(y+a)^2 - a^2},$$

and by writing $-a$ for a, and in each equation $-x$ for x, we have the rational equations of the other three curves.

But, drawing only the required portions of the curves, we have fig. 3 exhibiting

Fig. 3.

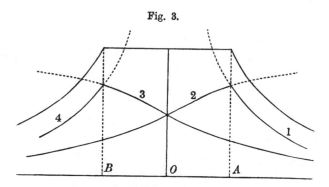

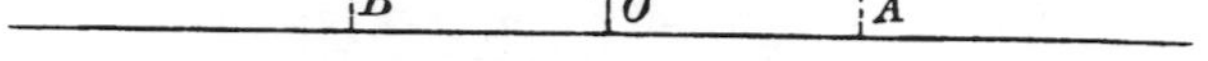

the potentials of the two hemispheres A and B; and also the discontinuous potential of the whole shell, the ordinate for this last being the sum of the ordinates for the two hemispheres respectively.

758.

SOLUTION OF A SENATE-HOUSE PROBLEM.

[From the *Messenger of Mathematics*, vol. XI. (1882), pp. 23—25.]

PROVE that, if $a+b+c=0$ and $x+y+z=0$, then

$$\begin{aligned}&4(ax+by+cz)^3\\&-3(ax+by+cz)(a^2+b^2+c^2)(x^2+y^2+z^2)\\&-2(b-c)(c-a)(a-b)(y-z)(z-x)(x-y)\\&-54abcxyz=0.\end{aligned}$$

I do not know the origin of this identity, nor do I see any very simple way of proving it: that which seems the most straightforward way is to transform the third line, which, omitting the factor -2, is

$$\begin{vmatrix}1, & 1, & 1\\ a, & b, & c\\ a^2, & b^2, & c^2\end{vmatrix} \cdot \begin{vmatrix}1, & 1, & 1\\ x, & y, & z\\ x^2, & y^2, & z^2\end{vmatrix},$$

$$= \begin{vmatrix}3, & a+b+c, & a^2+b^2+c^2\\ x+y+z, & ax+by+cz, & a^2x+b^2y+c^2z\\ x^2+y^2+z^2, & ax^2+by^2+cz^2, & a^2x^2+b^2y^2+c^2z^2\end{vmatrix};$$

and therefore when $a+b+c=0$ and $x+y+z=0$, is

$$\begin{aligned}=\;&3(ax+by+cz)(a^2x^2+b^2y^2+c^2z^2)\\&-3(a^2x+b^2y+c^2z)(ax^2+by^2+cz^2)\\&-\;(ax+by+cz)(a^2+b^2+c^2)(x^2+y^2+z^2);\end{aligned}$$

or, as this may be written,

$$\begin{aligned} = \;& 6\,(ax + by + cz)\,(a^2x^2 + b^2y^2 + c^2z^2) \\ - \;& (ax + by + cz)\,(a^2 + b^2 + c^2)\,(x^2 + y^2 + z^2) \\ -\;& 3\,(ax + by + cz)\,(a^2x^2 + b^2y^2 + c^2z^2) \\ -\;& 3\,(a^2x + b^2y + c^2z)\,(ax^2 + by^2 + cz^2). \end{aligned}$$

Here the third and fourth lines, omitting the factor -3, are

$$2\,(a^3x^3 + b^3y^3 + c^3z^3) + (ab^2 + a^2b)\,(xy^2 + x^2y) + (ac^2 + a^2c)\,(xz^2 + x^2z) + (bc^2 + b^2c)\,(yz^2 + y^2z),$$

where, in virtue of the two relations, each of the last three product-terms is $= abcxyz$, and the whole is thus

$$\begin{aligned} = \;& 2\,(a^3x^3 + b^3y^3 + c^3z^3) \\ +\;& 3abcxyz. \end{aligned}$$

The product of the two determinants is thus

$$\begin{aligned} = \;& 6\,(ax + by + cz)\,(a^2x^2 + b^2y^2 + c^2z^2) \\ -\;& (ax + by + cz)\,(a^2 + b^2 + c^2)\,(x^2 + y^2 + z^2) \\ -\;& 6\,(a^3x^3 + b^3y^3 + c^3z^3) \\ -\;& 9\; abcxyz\,; \end{aligned}$$

and this being so the identity to be verified is

$$\begin{aligned} & 4\;\;(ax + by + cz)^3 \\ +(-3 + 2 =) - 1\;\; & (ax + by + cz)\,(a^2 + b^2 + c^2)\,(x^2 + y^2 + z^2) \\ & - 12\,(ax + by + cz)\,(a^2x^2 + b^2y^2 + c^2z^2) \\ & + 12\,(a^3x^3 + b^3y^3 + c^3z^3) \\ +(18 - 54 =) - 36abcxyz \;\; & = 0. \end{aligned}$$

We have here the terms

$$\begin{aligned} & 12\,(a^3x^3 + b^3y^3 + c^3z^3 - 3abcxyz), \\ = \;& 12\,(ax + by + cz)\,(a^2x^2 + b^2y^2 + c^2z^2 - bcyz - cazx - abxy), \end{aligned}$$

so that the left-hand side is now divisible by $ax + by + cz$, and throwing out this factor the equation becomes

$$\begin{aligned} & 4\;\;(ax + by + cz)^2 \\ -\;& (a^2 + b^2 + c^2)\,(x^2 + y^2 + z^2) \\ -\;& 12\,(a^2x^2 + b^2y^2 + c^2z^2) \\ +\;& 12\,(a^2x^2 + b^2y^2 + c^2z^2 - bcyz - cazx - abxy) = 0\,; \end{aligned}$$

or, as this may be written,

$$4(a^2x^2+b^2y^2+c^2z^2-bcyz-cazx-abxy)$$
$$-\quad (a^2+b^2+c^2)(x^2+y^2+z^2)=0,$$

which under the assumed relations $a+b+c=0$, $x+y+z=0$ may be verified without difficulty. It may be remarked that we have identically

$$8(a^2x^2+b^2y^2+c^2z^2-bcyz-cazx-abxy)$$
$$-2(a^2+b^2+c^2)(x^2+y^2+z^2)$$
$$=(x+y+z)\left\{\begin{array}{l} x(\ \ 3a^2-\ b^2-\ c^2+2bc-2ca-2ab) \\ +y(-\ a^2+3b^2-\ c^2-2bc+2ca-2ab) \\ +z(-\ a^2-\ b^2+3c^2-2bc-2ca+2ab) \end{array}\right\}$$
$$+(a+b+c)\left\{\begin{array}{l} a(\ \ 3x^2-\ y^2-\ z^2+2yz-2zx-2xy) \\ +b(-\ x^2+3y^2-\ z^2-2yz+2zx-2xy) \\ +c(-\ x^2-\ y^2+3z^2-2yz-2zx+2xy) \end{array}\right\},$$

which is a more complete form of the last-mentioned theorem.

759.

ILLUSTRATION OF A THEOREM IN THE THEORY OF EQUATIONS.

[From the *Messenger of Mathematics*, vol. XI. (1882), pp. 111—113.]

THE knowledge of the value of an unsymmetrical function of the roots of a numerical equation adds something to what is given by the equation itself; but it may or may not add anything to what is given by the equation itself in regard to each root separately. If, for instance, α, β, γ being the roots of a cubic equation, it is known that $\alpha^2\beta+\beta^2\gamma+\gamma^2\alpha=$ a given value k, then α, β, γ must denote the roots, taken not in any order whatever, nor yet in a uniquely determinate order, but with a certain restriction as to order, viz. if the roots in a certain order are a, b, c, these roots being such that $a^2b+b^2c+c^2a=k$, then clearly the relation in question $\alpha^2\beta+\beta^2\gamma+\gamma^2\alpha=k$, will be satisfied if $\alpha, \beta, \gamma=a, b, c$, or $=b, c, a$, or $=c, a, b$ (but not if $\alpha, \beta, \gamma=b, a, c$, or $=$ either of the remaining two arrangements); the relation thus allows α to be $=a$, or $=b$, or $=c$; that is, α is $=$ any one at pleasure of the roots of the cubic equation, and it is thus determined by the cubic equation, and not by any inferior equation; but α being known, the other two roots β and γ will be uniquely, and therefore rationally, determined.

It is worth while to see how the result works out; suppose, for greater simplicity, the cubic equation is $x^3-7x+6=0$ having roots $(1, 2, -3)$, and that the given relation is $\alpha^2\beta+\beta^2\gamma+\gamma^2\alpha=-1$, then the cubic equation gives

$$\alpha+\beta+\gamma=0, \quad \alpha\beta+\alpha\gamma+\beta\gamma=-7, \quad \alpha\beta\gamma=-6,$$

and we have, besides, the relation in question

$$\alpha^2\beta+\beta^2\gamma+\gamma^2\alpha=-1;$$

eliminating γ we have

$$\alpha^2+\alpha\beta+\beta^2=7, \quad \alpha\beta(\alpha+\beta)=6, \quad \alpha^3+3\alpha^2\beta-\beta^3+1=0;$$

or, as it is convenient to write these equations,

$$\beta^2+\ \ \alpha\beta+\alpha^2-7=0,$$

$$\beta^2+\ \ \alpha\beta-\frac{6}{\alpha}\qquad=0,$$

$$\beta^3-3\alpha^2\beta-\alpha^3-1=0.$$

If from these equations we eliminate β, we obtain two equations in α, which it might be supposed would determine α uniquely; but, by what precedes, α is any root at pleasure of the cubic equation and can thus be determined only by the cubic equation itself, and it follows that any equation obtained by the elimination of β must contain as a factor the cubic function $\alpha^3 - 7\alpha + 6$, and be thus of the form $M(\alpha^3 - 7\alpha + 6) = 0$, where M is a function of α; one result of the elimination is $\alpha^3 - 7\alpha + 6 = 0$, and every other result is of the form just referred to, $M(\alpha^3 - 7\alpha + 6) = 0$; hence we have definitely $\alpha^3 - 7\alpha + 6 = 0$, viz. the roots of the equation $M = 0$ do not apply to the question.

In verification, observe that the first and second equations give $\alpha^2 - 7 = \frac{6}{\alpha}$, that is, $\alpha^2 - 6\alpha + 7 = 0$. To eliminate β from the first and third equations we first find

$$\alpha\beta^2 + (4\alpha^2 - 7)\ \beta + \alpha^3 + 1 = 0,$$

or say

$$\beta^2 + \left(4\alpha - \frac{7}{\alpha}\right)\beta + \alpha^2 + \frac{1}{\alpha} = 0,$$

and combining herewith the first equation

$$\beta^2 + \alpha\beta + \alpha^2 - 7 = 0,$$

we obtain

$$\beta\left(3\alpha - \frac{7}{\alpha}\right) + 7 + \frac{1}{\alpha} = 0,$$

that is,

$$\beta = \frac{7\alpha + 1}{-3\alpha^2 + 7};$$

substituting in the first equation,

$$\begin{aligned} &(7\alpha + 1)^3 \\ &+ \alpha(7\alpha + 1)(-3\alpha^2 + 7) \\ &+ \quad (\alpha^2 - 7)(-3\alpha^2 + 7)^2 = 0, \end{aligned}$$

that is,

$$\begin{array}{rrrrrrr} & & & & 49 & 14 & 1 \\ & & -\ 21 & -3 & +\ 49 & +\ 7 & \\ 9 & 0 & -105 & & +343 & & -343 \\ \hline 9 & 0 & -126 & -3 & +441 & +21 & -342, \end{array}$$

or, dividing by 3,

$$3\alpha^6 - 42\alpha^4 - \alpha^3 + 147\alpha^2 + 7\alpha - 114 = 0,$$

which, in fact, is

$$(\alpha^3 - 7\alpha + 6)(3\alpha^3 - 21\alpha - 19) = 0,$$

of the form in question $M(\alpha^3 - 7\alpha + 6) = 0$. Thus α has any one at pleasure of the three values 1, 2, -3, but α being known we have $\beta = \frac{7\alpha + 1}{-3\alpha^2 + 7}$, and thence

$$\gamma = -\alpha + \frac{-7\alpha - 1}{-3\alpha^2 + 7}, \quad = \frac{3\alpha^3 - 14\alpha - 1}{-3\alpha^2 + 7};$$

in particular, as $\alpha = 1$, then $\beta = 2$ and $\gamma = -3$.

760.

REDUCTION OF $\int \frac{dx}{(1-x^3)^{\frac{2}{3}}}$ TO ELLIPTIC INTEGRALS.

[From the *Messenger of Mathematics*, vol. XI. (1882), pp. 142, 143.]

WRITING s, c, d for the sn, cn, and dn of u to a modulus k, which will be determined, and denoting by θ a constant which will also be determined, the formula of reduction is

$$x = \frac{-1+\theta scd}{1+\theta scd}.$$

To find from this the value of y, $=\sqrt[3]{(1-x^3)}$, putting for shortness $X = \theta scd$, the formula is $x = \frac{-1+X}{1+X}$, and we thence have

$$y^3, = 1 - x^3, = \frac{2(1+3X^2)}{(1+X)^3},$$

where

$$1 + 3X^2 = 1 + 3\theta^2 s^2 (1-s^2)(1-k^2 s^2),$$

$$= 1 + 3\theta^2 s^2 - 3\theta^2 (1+k^2) s^4 + 3\theta^2 k^2 s^6,$$

may be put equal to $(1+\theta^2 s^2)^3$, that is,

$$= 1 + 3\theta^2 s^2 + 3\theta^4 s^4 \qquad + \theta^6 s^6;$$

viz. this will be the case if

$$3\theta^4 = -3\theta^2 (1+k^2), \quad \theta^6 = 3\theta^2 k^2;$$

that is,

$$\theta^2 = -1 - k^2, \qquad \theta^4 = 3k^2;$$

these give

$$k^4 - k^2 + 1 = 0;$$

that is, $k^2 = \omega$, if $\omega = -\frac{1}{2} + \frac{1}{2}i\sqrt{3}$, an imaginary cube root of unity; and then

$$\theta^2 = -1 + \omega, \quad = \omega^2(\omega^2 - \omega), \quad = -i\omega^2\sqrt{3};$$

that is,

$$\theta = \pm \frac{(1-\sqrt{3}) - i(1+\sqrt{3})}{2\sqrt{2}} \sqrt[4]{3},$$

as may be verified by squaring.

Hence finally, θ and k denoting the values just obtained,

$$x = \frac{-1+\theta scd}{1+\theta scd},$$

$$y = \sqrt[3]{(1-x^3)} = \frac{\sqrt[3]{2}\,(1+\theta^2 s^2)}{1+\theta scd};$$

or, writing as before, $X = \theta scd$, we have

$$dx = \frac{2dX}{(1+X)^2}, \quad y^2 = \frac{2^{\frac{2}{3}}(1+\theta^2 s^2)^2}{(1+X)^2};$$

whence

$$\frac{dx}{(1-x^3)^{\frac{2}{3}}}, \quad = \frac{dx}{y^2}, \quad = \frac{2^{\frac{1}{3}} dX}{(1+\theta^2 s^2)^2},$$

and then

$$dX = \theta\,\{1 - 2(1+k^2)\,s^2 + 3k^2 s^4\}\,du, \quad = \theta\,(1+\theta^2 s^2)^2\,du;$$

that is,

$$\frac{dx}{(1-x^3)^{\frac{2}{3}}} = 2^{\frac{1}{3}}\theta\,.\,du;$$

or say

$$\int \frac{dx}{(1-x^3)^{\frac{2}{3}}} = 2^{\frac{1}{3}}\theta u,$$

the required formula.

761.

ON THE THEOREM OF THE FINITE NUMBER OF THE COVARIANTS OF A BINARY QUANTIC.

[From the *Quarterly Journal of Pure and Applied Mathematics,* vol. XVII. (1881), pp. 137—147.]

GORDAN'S proof, the only one hitherto given, is based upon the theory of derivatives (Uebereinanderschiebungen). It is shown that the irreducible covariants of the binary quantic f are included in the series

$$(f, f)^2, (f, f)^4, \dots (f, h), (f, h)^2, \dots$$

of the derivatives of the quantic upon itself or upon some other covariant, and that the number of the irreducible covariants thus obtained is finite. And not only so, but for the quintic and the sextic the complete systems were formed, and the numbers shown to be $=23$ and 26 respectively.

It would seem that there ought to be a more simple proof based upon the consideration of the fundamental covariants: for the cubic $(a, b, c, d \between x, y)^3$, these are the cubic itself $(a, \dots \between x, y)^3$, the Hessian $(ac-b^2, \dots \between x, y)^2$, and the cubicovariant $(a^2d-3abc+2b^3, \dots \between x, y)^3$; and so in general for the quantic $(a, \dots \between x, y)^n$, we have a series of fundamental covariants the leading coefficients whereof are the seminvariants

$$a,\ ac-b^2,\ a^2d-3abc+2b^3,\ a^3e-4a^2bd+6ab^2c-3b^4, \ \&c.$$

It is known that every covariant can be expressed as a *rational* function of these, or more precisely that every covariant multiplied by a positive integral power of the quantic itself can be expressed as a *rational* and *integral* function of the fundamental covariants, and we may for the covariants substitute their leading coefficients, or say the seminvariants; hence, every seminvariant is a *rational* function of the fundamental seminvariants, and more precisely, every seminvariant multiplied by a positive integral power of the first coefficient a is a *rational* function of the fundamental seminvariants. Thus, in the case of the cubic, we have the discriminant ∇,

$$= a^2d^2 - 6abcd + 4ac^3 + 4b^3d - 3b^2c^2,$$

obtained from

$$a,\ ac - b^2,\ a^2d - 3abc + 2b^3,$$

by the formula

$$a^2\nabla = (a^2d - 3abc + 2b^3)^2 + 4(ac - b^2)^3,$$

and it is easily shown that this invariant ∇ is the only new covariant thus obtainable, and that every other covariant is thus a *rational and integral* function of the irreducible covariants, the leading coefficients of which are

$$a,\ ac - b^2,\ a^2d - 3abc + 2b^3,$$

and ∇. It appears a truism, and it might be thought that it would be, if not easy, at least practicable, to show for a quantic of any given finite order n, that we can in this manner, as *rational* functions of the $n-1$ seminvariants, obtain only a *finite* number of new seminvariants, so that all the seminvariants would be expressible as *rational and integral* functions of a finite number of seminvariants; and, consequently, all the covariants be expressible as rational and integral functions of a finite number of irreducible covariants. But the large number, 23, of the covariants of the quintic is enough to show that the proof, even if it could be carried out, would involve algebraical operations of great complexity.

The theory may be considered from a different point of view, in connexion with the root-form $a(x - \alpha y)(x - \beta y) \ldots$, or say $(x - \alpha)(x - \beta) \ldots$ of the quantic; we have here what may be called the monomial form of covariant, viz. the general monomial form is

$$(\alpha - \beta)^m (\alpha - \gamma)^n (\beta - \gamma)^p \ldots (x - \alpha)^q (x - \beta)^r \ldots,$$

where in all the factors (whether $\alpha - \beta$ or $x - \alpha$) which contain α, in all the factors which contain $\beta, \ldots$, and so for each root in succession, the sum of the indices has one and the same value, $= \theta$ suppose. Thus, for the cubic

$$(x - \alpha)(x - \beta)(x - \gamma),$$

we have the monomial covariants

$$(\alpha - \beta)(\alpha - \gamma)(\beta - \gamma),$$

$$(\beta - \gamma)(x - \alpha),\ (\alpha - \gamma)(x - \beta),\ (\alpha - \beta)(x - \gamma),$$

$$(x - \alpha)(x - \beta)(x - \gamma);$$

and so for the quartic

$$(x - \alpha)(x - \beta)(x - \gamma)(x - \delta),$$

we have the monomial invariants

$$(\alpha - \beta)(\gamma - \delta),\ (\alpha - \gamma)(\beta - \delta),\ (\alpha - \delta)(\beta - \gamma).$$

Observe that the monomial form is considered as essential; a syzygetic function of two or more monomials is not a monomial, and we are thus in no wise concerned with identities such as

$$(\beta-\gamma)(x-\alpha)-(\alpha-\gamma)(x-\beta)+(\alpha-\beta)(x-\gamma)=0,$$

or

$$(\alpha-\delta)(\beta-\gamma)-(\beta-\delta)(\alpha-\gamma)+(\gamma-\delta)(\alpha-\beta)=0\,;$$

notwithstanding these syzygies respectively,

$$(\beta-\gamma)(x-\alpha),\;\; (\alpha-\gamma)(x-\beta),\text{ and } (\alpha-\beta)(x-\gamma)$$

are regarded as independent covariants of the cubic, and

$$(\alpha-\delta)(\beta-\gamma),\;\; (\beta-\delta)(\alpha-\gamma),\text{ and } (\gamma-\delta)(\alpha-\beta),$$

as independent invariants of the quartic.

It is only when a monomial covariant is equal to a power or product of simple or other powers of lower monomial covariants that it is regarded as a function of these lower monomial covariants and therefore as not irreducible. Thus

$$(\alpha-\beta)(\alpha-\gamma)(\beta-\delta)(\gamma-\delta)=(\alpha-\beta)(\gamma-\delta)\,.\,(\alpha-\gamma)(\beta-\delta),$$

is a reducible monomial covariant, expressible in terms of the lower irreducible monomial covariants

$$(\alpha-\beta)(\gamma-\delta)\text{ and }(\alpha-\gamma)(\beta-\delta).$$

The theorem of the finite number of the irreducible monomial covariants (as just explained) of the root-quantic is a question of the same kind as, but entirely distinct from, that of the finite number of the covariants of the quantic in the ordinary form; and there are thus the two questions; (*A*), that of the finite number of the irreducible monomial covariants of the root-quantic; and (*C*), that of the finite number of the irreducible covariants of the ordinary quantic.

But we can pass from (*A*) to (*C*) by means of a lemma (*B*), which I have not proved, but which seems highly probable, and which I enunciate as follows: (*B*) The infinite system of terms *X*, rational and integral functions of a finite set of letters (a, b, c, ...) which remain unaltered by all the substitutions of a certain group *G*(a, b, c, ...) of substitutions upon these letters, includes always a finite set of terms *P* such that every term *X* whatever is a rational and integral function of these terms *P*.

In explanation of this lemma, observe that, if *G*(a, b, c, ...) denotes the entire group of substitutions upon these letters, so that the functions which remain unaltered by the substitutions of the group are in fact the symmetrical functions of (a, b, c, ...), then the theorem is "The infinite system of rational and integral symmetrical functions of (a, b, c, ...) includes always a finite set of terms *P* such that every such rational and integral symmetrical function is a rational and integral function of the terms *P*, viz. the terms *P* are here the several symmetrical functions

$$\mathrm{a}+\mathrm{b}+\mathrm{c}+\ldots,\;\; \mathrm{ab}+\mathrm{ac}+\mathrm{bc}+\ldots,\;\; \mathrm{abc}+\ldots,\;\; \&\mathrm{c}.\text{"};$$

and so, if $G(\mathrm{a, b, c}, \ldots)$ is the group of all the positive substitutions, then we have the like theorem for the rational and integral two-valued functions of (a, b, c, ...), viz. the terms P are here the two-valued function $(\mathrm{a}-\mathrm{b})(\mathrm{a}-\mathrm{c})(\mathrm{b}-\mathrm{c})\ldots$, and the symmetrical functions

$$\mathrm{a}+\mathrm{b}+\mathrm{c}+\ldots,\ \mathrm{ab}+\mathrm{ac}+\mathrm{bc}+\ldots,\ \mathrm{abc}+\ldots,\ \&c.,$$

as before.

I return to the theorem (A), but instead of the covariants of a root-quantic of any order, I consider first the invariants of a root-quantic of any even order. The general form is

$$(\alpha-\beta)^m(\alpha-\gamma)^n(\beta-\gamma)^p\ldots,$$

where in all the factors which contain α, in all the factors which contain β, and so for each root in succession, the sum of the indices has one and the same value $=\theta$. Writing 12 for the index of $\alpha-\beta$, 13 for that of $\alpha-\gamma$, and so in other cases, then assuming always $12=21$, $13=31$, &c., the indices, taken each twice, form the square

0	12	13	...
21	0	23	...
31	32	0	
⋮			

the order of which, or number of its rows or columns, is equal to the order of the quantic; the terms of the dexter diagonal are each $=0$, and the square is symmetrical in regard to this dexter diagonal. Moreover, the square is such, that the sum of the terms in each row (or column) has one and the same value $=\theta$; and conversely, every such square, say R_θ, represents an invariant.

Thus, for the quartic $(x-\alpha)(x-\beta)(x-\gamma)(x-\delta)$, the square R_θ is a square of four rows (or columns) representing the invariant

$$\begin{aligned}(\alpha-\beta)^{12}(\alpha-\gamma)^{13}(\alpha-\delta)^{14},&\\ (\beta-\gamma)^{23}(\beta-\delta)^{24},&\\ (\gamma-\delta)^{34},&\end{aligned}$$

in which

$$\begin{aligned}12+13+14&=\theta,\\ 21\qquad+23+24&=\theta,\\ 31+32\qquad+34&=\theta,\\ 41+42+43\qquad&=\theta.\end{aligned}$$

There are three squares R_1, viz. these are the squares

	1		
1			
			1
		1	

,

		1	
			1
1			
	1		

,

			1
		1	
	1		
1			

representing the before-mentioned invariants $(\alpha-\beta)(\gamma-\delta)$, $(\alpha-\gamma)(\beta-\delta)$, $(\alpha-\delta)(\beta-\gamma)$ respectively: say these are a, b, c, and every other invariant is a *rational and integral* function of these; in fact, the θ-equations give easily $12=34$, $13=24$, $14=23$, so that the general form of the invariant is $=\mathrm{a}^{12}\mathrm{b}^{13}\mathrm{c}^{14}$, where 12, 13, 14 are each of them a positive integer number (which may be $=0$). Or, what is the same thing, the square R_θ $(\theta=12+13+14)$ is a sum

$$=12\,.\,R_1+13\,.\,R_1'+14\,.\,R_1'',$$

with positive integer coefficients 12, 13, 14, say for shortness it is a *sum* of squares R_1. And so any like expression with a negative coefficient or coefficients may, for shortness, be called a difference of squares R_1.

Observe that, in general, two squares R_θ, R_ϕ are added together by adding their corresponding terms, the result being a square $R_{\theta+\phi}$; similarly, if each term of R_ϕ be less than or at most equal to the corresponding term of R_θ, then (but not otherwise) the square R_ϕ may be subtracted from R_θ, giving a square $R_{\theta-\phi}$.

In the case of the sextic

$$(x-\alpha)(x-\beta)(x-\gamma)(x-\delta)(x-\epsilon)(x-\zeta),$$

there are fifteen squares R_1, which may be represented as follows:

12.34.56	x_1
12.35.46	y_1
12.36.45	z_1
13.24.56	x_2
13.25.46	y_2
13.26.45	z_2
14.23.56	x_3
14.25.36	y_3
14.26.35	z_3
15.23.46	x_4
15.24.36	y_4
15.26.34	z_4
16.23.45	x_5
16.24.35	y_5
16.25.34	z_5;

viz. $12.34.56$ here represents the square R_1, for which the terms 12, 34, 56 (and of course the symmetrical terms 21, 43, 65) are each $=1$, the other terms all vanishing; or, what is the same thing, it represents the invariant $(\alpha-\beta)^{12}(\gamma-\delta)^{34}(\epsilon-\zeta)^{56}$. But it is not true that *every* square R_θ is a sum of squares R_1; this is not the case, for the square R_2,

$$=12.13.23.45.46.56,$$

representing the invariant

$$(\alpha-\beta)^{12}(\alpha-\gamma)^{13}(\beta-\gamma)^{23}(\delta-\epsilon)^{45}(\delta-\zeta)^{46}(\epsilon-\zeta)^{56},$$

is not a sum of squares R_1.

But the square last referred to is a *difference* of squares R_1: it is in fact

$$=12.36.45+13.25.46+14.23.56-14.25.36,$$

or, what is the same thing, the corresponding invariant is the product of the invariants $12.36.45$, $13.25.46$, $14.23.56$, divided by the invariant $14.25.36$; viz. it is a rational function of invariants R_1.

It is required to show, first, that *every* square R_θ is a difference of squares R_1; and thence, secondly, that it is a sum of a finite number of squares R_k (being, in fact, squares R_1 and R_2).

For the first theorem we equate the general expression of R_θ with the assumed value

$$x_1.12.34.56+y_1.12.35.46+z_1.12.36.45+\ldots+z_5.16.25.34.$$

We thus obtain

fifteen equations	satisfied by
$12=y_1+x_1+z_1$	$x_1=34-26+r+s-t,$
$13=x_2+y_2+z_2$	$x_2=13-25+p-r+t,$
$14=x_3+y_3+z_3$	$x_3=14-p-s\ ,$
$15=x_4+y_4+z_4$	$x_4=15-26-36+p+q+r+s\ ,$
$16=x_5+y_5+z_5$	$x_5=45-q-r\ ,$
$23=x_3+x_4+x_5$	$y_1=12-34+26-q-r-s+t,$
$24=x_2+y_4+y_5$	$y_2=25-p\ ,$
$25=y_2+y_3+z_5$	$y_3=p\ ,$
$26=z_2+z_3+z_4$	$y_4=36-p-q\ ,$
$34=x_1+z_4+z_5$	$y_5=16-45+q+r-t,$
$35=y_1+y_5+z_3$	$z_1=q\ ,$
$36=y_3+y_4+z_1$	$z_2=r\ ,$
$45=x_5+z_1+z_2$	$z_3=s\ ,$
$46=x_4+y_1+y_2$	$z_4=26-r-s\ ,$
$56=x_1+x_2+x_3$	$z_5=t,$

connecting $x_1, y_1, z_1, \ldots, z_5$ with the terms 12, 13, etc. of R_θ (or indices of the corresponding invariants). The fifteen equations are not independent, for regarding them as giving the values of 12, 13, ... in terms of the $x_1, y_1, z_1, \ldots, z_5$, these values satisfy identically the relations which ought to be satisfied by the terms 12, 13, etc., viz. the equations obtained by the elimination of θ from the equations

$$\begin{array}{l} 12 + 13 + 14 + 15 + 16 = \theta, \\ 12 \qquad + 23 + 24 + 25 + 26 = \theta, \\ \vdots \\ 16 + 26 + 36 + 46 + 56 \qquad = \theta. \end{array}$$

The equations are thus insufficient to determine the values of $x_1, y_1, z_1, \ldots, z_5$, and the general values given by the equations will contain five indeterminate quantities which are taken to be p, q, r, s, t (these being in fact the values of y_3, z_1, z_2, z_3, z_5 respectively), and we then have the equations all of them satisfied by the above-mentioned values containing these indeterminate quantities; taking them to be positive or negative integers, then $x_1, y_1, z_1, \ldots, z_5$, will be all of them integers; but by what precedes, it appears that they cannot all of them be made to be *positive* integers, so that we have consequently R_θ,

$$= x_1 . 12 . 34 . 56 + y_1 . 12 . 35 . 46 + z_1 . 12 . 36 . 45 + \ldots + z_5 . 16 . 25 . 34,$$

equal in general to a difference of squares R_1.

Suppose in such difference of squares R_1 we have any term, say $-12.34.56$, occurring with the coefficient -1. Since the expression represents a square R_θ, we must have among the positive terms, $12.35.46$ or $12.36.45$ to render possible the subtraction of the 12; $15.26.34$ or $16.25.34$ to render possible the subtraction of the 34; and $13.24.56$ or $14.23.56$ to render possible the subtraction of the 56; that is, the expression must contain one of the eight combinations

$$\begin{array}{l} 12.35.46 + 15.26.34 + 13.24.56 - 12.34.56, \\ 12.35.46 + 15.26.34 + 14.23.56 - 12.34.56, \\ 12.35.46 + 16.25.34 + 13.24.56 - 12.34.56, \\ 12.35.46 + 16.25.34 + 14.23.56 - 12.34.56, \\ 12.36.45 + 15.26.34 + 13.24.56 - 12.34.56, \\ 12.36.45 + 15.26.34 + 14.23.56 - 12.34.56, \\ 12.36.45 + 16.25.34 + 13.24.56 - 12.34.56, \\ 12.36.45 + 16.25.34 + 14.23.56 - 12.34.56. \end{array}$$

The first of these is $35.46.15.26.13.24$, viz. it is $13.15.35.24.26.46$ which is a square R_2 (of the form mentioned above); the second is $35.46.15.26.14.23$, which is $15.23.46 + 14.26.35$, a sum of squares R_1; and similarly each of the other combinations is either a square R_2 or a sum of squares R_1. We have thus got rid of the negative term $-12.34.56$, and in like manner if the negative term had been

$$-m.12.34.56, = -12.34.56 - 12.34.56 - \&c.,$$

or, whatever the negative terms may be, we get rid one by one of each negative term; and thus ultimately express R_θ as a sum of squares R_1 and R_2. Or, what is the same thing, the invariant R_θ originally expressed as a *rational* function of invariants R_1, is finally expressed as a *rational and integral* function of invariants R_1 and R_2.

Similarly for a root-quantic of any even order n, we have the general square R_θ expressed, first as a difference of squares R_1, and then as a sum of squares R_1, R_2, or it may be higher squares R_3, &c., but certainly as a sum of a *finite* number of squares R_k. For a root-quantic of any odd order n, the investigation would be of a somewhat different form, since here there are no squares R_1, but the lowest squares are squares R_2 of a form such as $12.23.34.45.15$; but the general conclusion would still follow that every square R_θ is a sum of a *finite* number of squares R_k. And a like reasoning would apply to covariants instead of invariants: viz. the reasoning (although for simplicity it has been given for a very particular and special case) does, I think, really establish the theorem (A) in its generality, viz. the theorem that for a root-quantic of any given finite order, the number of irreducible monomial covariants is finite.

From any monomial covariant of the root-quantic, by taking the sum of the forms belonging to the different roots, so as to obtain a symmetrical function of the roots, that is, a rational and integral function of the coefficients, we obtain a covariant of the quartic in its ordinary form $(a, \ldots \between x, y)^n$. Consider for a moment the before-mentioned case of the invariants of the root-quartic

$$(x-\alpha y)(x-\beta y)(x-\gamma y)(x-\delta y),$$

now put

$$=\frac{1}{a}(a,\ b,\ c,\ d,\ e \between x,\ y)^4;$$

and to make the reasoning clearer, take a, b, c, f, g, h $=(\alpha-\delta)(\beta-\gamma)$, $(\beta-\delta)(\gamma-\alpha)$, $(\gamma-\delta)(\alpha-\beta)$, $(\alpha-\delta)(\gamma-\beta)$, $(\beta-\delta)(\alpha-\gamma)$, $(\gamma-\delta)(\beta-\alpha)$ respectively, these being, with the signs $\pm$, the before-mentioned three monomial invariants. In the root-theory, every monomial invariant is a rational and integral function of a, b, c, f, g, h. Every invariant of $(a, \ldots \between x, y)^4$, *quà* rational and integral function of the coefficients, is, when expressed in terms of the roots, a rational and integral function of the roots, and then *quà* invariant is a sum of monomial invariants, and as such a rational and integral function of a, b, c, f, g, h. But every such rational and integral function of a, b, c, f, g, h is not a symmetrical function of α, β, γ, δ, and consequently not in the present theory an invariant of $(a, \ldots \between x, y)^4$; the invariants are those rational and integral functions of a, b, c, f, g, h which are symmetrical functions of $(\alpha, \beta, \gamma, \delta)$, that is, which remain unaltered by every substitution whatever upon the roots $(\alpha, \beta, \gamma, \delta)$. Now each such substitution gives a substitution upon a, b, c, f, g, h, and the 24 substitutions upon α, β, γ, δ give a group of 6, $=\frac{1}{4}.24$ substitutions upon (a, b, c, f, g, h); the invariants are thus the rational and integral functions of (a, b, c, f, g, h) which are unaltered by each of the substitutions of a certain group G(a, b, c, d, e, f) of 6 substitutions. Theorem (B) asserts that, among the terms in question, that is, among such rational and integral functions of (a, b, c, f, g, h), we have a finite number of terms P, such that every one of the terms is a rational and integral function of the terms P; and recollecting that $\mathrm{a}+\mathrm{b}+\mathrm{c}=0$, these terms P are in fact two terms $\mathrm{bc}+\mathrm{ca}+\mathrm{ab}$ and $(\mathrm{b}-\mathrm{c})(\mathrm{c}-\mathrm{a})(\mathrm{a}-\mathrm{b})$; the conclusion being, that the invariants of the quartic $(a, b, c, d, e \between x, y)^4$ are all of them rational and integral functions of the last-mentioned two functions, that is, of

$$I, = ae - 4bd + 3c^2, \text{ and } J, = ace - ad^2 - b^2e + 2bcd - c^3.$$

As regards the group G(a, b, c, f, g, h) of 6 substitutions upon a, b, c, f, g, h, observe that the 24 substitutions of $(\alpha, \beta, \gamma, \delta)$ operating upon a, b, c, f, g, h give 6 substitutions taken each four times; for instance, the substitutions 1, $\alpha\beta.\gamma\delta$, $\alpha\gamma.\beta\delta$, $\alpha\delta.\beta\gamma$ leave each of them a, b, c, f, g, h unaltered, that is, they each give the substitution 1. And we thus find for the group G(a, b, c, f, g, h) the 6 substitutions

1,

abc . fgh,

acb . fhg,

af . bh . cg,

ah . bg . cf,

ag . bf . ch.

For the functions of a, b, c, f, g, h, which remain unaltered by the substitution of this group, observe that we have f, g, h = − a, − b, − c; so that any function of the six letters may be represented as a function of a, b, c. An odd symmetrical function, for instance abc, does *not* remain unaltered, for it is by any one of the last three substitutions changed into fgh, that is, into −abc; on the other hand, the two-valued function $(\mathrm{b}-\mathrm{c})(\mathrm{c}-\mathrm{a})(\mathrm{a}-\mathrm{b})$ does remain unaltered: the functions which remain unaltered are therefore the even symmetrical functions of a, b, c (that is, the symmetric functions $\mathrm{a}^2+\mathrm{b}^2+\mathrm{c}^2$, or $\mathrm{ab}+\mathrm{ac}+\mathrm{bc}$, &c., which are of an even order in a, b, c conjointly), and the same even functions multiplied by $(\mathrm{b}-\mathrm{c})(\mathrm{c}-\mathrm{a})(\mathrm{a}-\mathrm{b})$; and having regard to the relation $\mathrm{a}+\mathrm{b}+\mathrm{c}=0$, all these can be expressed as already mentioned as rational and integral functions of $\mathrm{bc}+\mathrm{ca}+\mathrm{ab}$ and $(\mathrm{b}-\mathrm{c})(\mathrm{c}-\mathrm{a})(\mathrm{a}-\mathrm{b})$.

The proof applies to the general case of the theorem (C), viz. taking the theorem (A) to be proved, and putting the root-quantic

$$(x-\alpha y)(x-\beta y)\ldots = \frac{1}{a}(a, \ldots \between x, y)^n,$$

then we have a, b, c, d, ... a system of monomial covariants of the root-quantic; and all the covariants of $(a, \ldots \between x, y)$ are rational and integral functions of (a, b, c, d, ...) which remain unaltered by the substitutions of a certain group G(a, b, c, d, ...); hence, assuming the theorem (B), they are rational and integral functions of a finite number of irreducible covariants. And the demonstration thus depends upon that of the theorem (B).

762.

ON SCHUBERT'S METHOD FOR THE CONTACTS OF A LINE WITH A SURFACE.

[From the *Quarterly Journal of Pure and Applied Mathematics*, vol. XVII. (1881), pp. 244—258.]

I WISH to reproduce in part § 33, "Coincidenz von Schnittpunkten einer Geraden mit einer Fläche" of Schubert's very interesting work *Calcul der abzählende Geometrie*, Leipzig, 1879, explaining in the first instance (but not altogether in the manner or from the point of view of the author) the general principles of the theory.

We have to do with *conditions* relating to a *subject;* the subject is a geometrical form or entity of any kind depending upon a certain number of constants; and the condition is onefold, twofold, &c., according as it imposes a onefold, twofold, &c., relation upon these constants. The number of constants is the Postulandum of the subject, and the manifoldness of the condition is called also its Postulation. A condition is incomplete when its postulation is less than postulandum of subject, complete when its postulation is equal to postulandum of subject; two or more incomplete conditions, making up a complete condition, are supplementary to each other. The case where the postulation exceeds the postulandum, or say that of a more than complete condition, is not in general considered; it may however sometimes present itself. For instance, the subject may be a line with n points upon it; the number of constants is here $=n+4$. A condition that the line shall meet a given line, or that a certain one of the n points shall lie on a given plane, is a onefold condition; the condition that such point shall lie upon a given line is a twofold condition; and so in other cases.

Conditions are denoted by letters, and simultaneous conditions by a product; for instance, the subject is a line carrying the n points $1, 2, \ldots, n$; g is the condition that the line meets a given line; p_1 the condition that the point 1 lies on a given plane; then gp_1 is the twofold condition that the line meets a given line and that the point 1 lies on a given plane; p_1^2 is the twofold condition that the point 1 lies on each of two given planes (in fact, on their line of intersection). The letters p, g, e are used as the initials of Punkt, Gerade, Ebene.

The letter or combination of letters denoting an incomplete condition, or, say, the incomplete condition itself, has no numerical value; but for a complete condition there exists a definite number of subjects satisfying the condition, and the condition is regarded as having this number as its value. A more than complete condition has the value 0.

Conditions of the same postulation may be connected by the sign +; for instance,

subject a line,

g_e the condition that it lies in a given plane,

g_p the condition that it passes through a given point,

then $g_e + g_p$ is the condition that the line shall either lie in the given plane or else pass through the given point.

I abstain from attempting any definition in regard to the sign −.

Conditions of the same postulation may be connected by an equation or equations; for instance,

subject a point,

p the condition that the point shall lie in a given plane,

p_g the condition that the point shall lie in a given line,

then $p^2 = p_g$.

This equation has (so far) no numerical signification; it has the logical signification that the condition that a point shall lie on each of two given planes is equivalent to the condition that the point shall lie on a given line.

Second example. Subject a line,

g the condition that the line meets a given line,

g_e the condition that it lies in a given plane,

g_p the condition that it passes through a given point,

then $g^2 = g_e + g_p$.

This equation has (so far) no numerical signification, and I regard it as having no logical signification. Schubert, however, gives it a logical signification by means of his "Princip der speciellen Lage" (Principle of Special Situation), viz. the condition of the line meeting each of two given lines is, in the particular case where the two given lines meet, equivalent to the condition, that the line shall either lie in the plane of the two given lines or else pass through their point of intersection.

Third example. Subject a line bearing upon it the points 1 and 2,

ϵ the condition of the coincidence of the two points,

p „ that the point 1 shall lie on a given plane,

q „ „ 2 „ „

g „ that the line shall meet a given line,

then $\epsilon = p + q - g$.

This equation has (so far) no numerical signification, and it does not appear to have any logical signification. In fact, in the actual form of the equation we have a sign $-$ which has not had given to it any logical interpretation; and if we write the equation in the form $\epsilon + g = p + q$, there seems to be no logical signification in the assertion, the condition that either the points shall coincide, or else the line meet a given line, is equivalent to the condition that either the first point, or else the second point, shall lie in a given plane.

Any equation connecting complete conditions is a numerical equation; and to render a condition complete, we have only to join to it a supplementary condition X of the proper postulation. Thus, in the last example the postulandum is $=6$; ϵ, p, q, g are onefold conditions, and joining to each of them one and the same fivefold condition X, we have $X\epsilon = Xp + Xq - Xg$. And, taking X to be an arbitrary fivefold condition, the original equation $\epsilon = p + q - g$ has in fact the meaning

$$X\epsilon = Xp + Xq - Xg.$$

For instance, the fivefold condition X may be that the line shall belong to a given regulus (scroll or developable surface), and that the points 1, 2 upon the line shall be the intersections of the line with given surfaces S_1, S_2 respectively. The subject is the line of the given regulus with its two points; and the meaning of the equation is that the number of subjects with two coincident points is equal to the number of subjects with the point 1 on a given plane, *plus* the number of subjects with the point 2 on a given plane, *minus* the number of subjects for which the line meets a given line. Although for the moment concerned only with the meaning of the theorem, not with its truth, I stop to show *à posteriori* that the theorem is in fact true: take k for the order of the regulus; m_1, m_2 for the orders of the surfaces S_1, S_2 respectively; then it is to be shown that $X\epsilon$, Xp, Xq, Xg are each $=km_1m_2$ (values which satisfy the equation). First $X\epsilon$: the points 1 and 2 here coincide at a point of the curve of the order m_1m_2, which is the intersection of S_1 and S_2; the regulus meets this curve in km_1m_2 points, and through each of these we have a line of the regulus having upon it the two coincident points; that is, $X\epsilon = km_1m_2$. Next Xp: the point 1 is here on the plane curve of the order m_1, which is the intersection of S_1 with the corresponding given plane; the regulus meets this plane curve in km_1 points; through each of these we have a line of the regulus intersecting S_2 in m_2 points, any one of which may be taken for the point 2; that is, the number of subjects is Xp, $=km_1 . m_2$. Then Xq: in precisely the same manner we have $Xq = km_2 . m_1$. Lastly Xg: the given line meets the regulus in k points, and through each of these there is a line of the regulus meeting S_1 in m_1 points, any one of which may be taken for the point 1, and meeting S_2 in m_2 points, any one of which may be taken for the point 2; the number of the subjects Xg is thus $Xg, = k . m_1 . m_2$.

The general theorem $X\epsilon = Xp + Xq - Xg$ is proved by means of Chasles' theorem of united points as follows: the subject is a line, or say, for convenience, an *axis* ξ, bearing upon it the two points 1 and 2; we consider in conjunction therefore a given line λ, and through this draw the planes P_1, P_2 passing through the points 1 and 2 respectively.

Suppose that when 2 lies in a given plane there are α' positions of the axis, and on each of these β' positions of the point 1; and, similarly, that when 1 lies on a given plane there are α positions of the axis, and on each of these β positions of the point 2; then, 1 lying in a given plane, the number of subjects is $\alpha\beta$, or we have $Xp = \alpha\beta$; and, similarly, $Xq = \alpha'\beta'$. Take now for the point P_1 an arbitrary plane through λ; then, 1 lying on this plane, the number of the points 2 is $= \alpha\beta$, or, since each of these determines with λ a position of the plane P_2, the number of these planes is $= \alpha\beta$, that is, it is $= Xp$; and, similarly, taking P_2 an arbitrary plane through λ, the number of the planes P_1 is $\alpha'\beta'$, that is, it is $= Xq$; viz. the two planes P_1, P_2 through the line λ have an (Xp, Xq) correspondence; hence, by Chasles' theorem, the number of united planes is $= Xp + Xq$.

But we have a united plane, 1°, if the points 1 and 2 coincide, that is, if the condition $X\epsilon$ be satisfied, and the number of these united planes is $X\epsilon$; 2°, if the axis ξ meet the arbitrary line λ, that is, if the condition Xg be satisfied, and the number of these united planes is $= Xg$; hence the whole number is $= X\epsilon + Xg$; or we have $Xp + Xq = X\epsilon + Xg$, that is, $X\epsilon = Xp + Xq - Xg$, which is the theorem in question.

The conclusion is that the equation $\epsilon = p + q - g$, which in this, its original form, has neither a numerical nor a logical signification, is to be understood as meaning the numerical equation $X\epsilon = Xp + Xq - Xg$, the truth of which numerical equation has just been proved. Or we may, without explicit introduction of the condition X, understand the equation $\epsilon = p + q - g$ as a numerical equation as follows, viz. taking for the subject a line with two points, *which line and points are regarded as satisfying a given fivefold condition*, then

ϵ is the (additional onefold) condition				that the two points shall coincide,
p	"	"	"	that the point 1 shall lie in given plane,
q	"	"	"	that the point 2 shall lie in given plane,
g	"	"	"	that line shall meet given line.

The conditions ϵ, p, q, g are thus in effect complete conditions, having values which may be connected by an equation; there, in fact, exists between them the relation

$$\epsilon = p + q - g.$$

The like remarks would apply to the before-mentioned equation (subject a point) $p^2 = p_g$: either adding to it a onefold condition X, and so taking it in the form $Xp^2 = Xp_g$, or understanding it in its original form $p^2 = p_g$ as belonging to a point which satisfies already a onefold condition, the equation is true as a numerical equation; and this in fact follows at once from its truth as a logical equation. But observe the difference: the equation in question $p^2 = p_g$ has, the equation $\epsilon = p + q - g$ has not, a logical signification.

I regard as the fundamental notion of the theory the existence of equations between conditions such as the foregoing equation $\epsilon = p + q - g$; equations which in their original form have not (of necessity) any logical signification, and have not any numerical signification; but which, when we adjoin to them a supplementary condition X of the proper postulation, become numerical equations, which are true, independently of the form of the supplementary condition X and whatever this condition may be. And this being so, it seems to follow at once that such equations may be treated and worked with as ordinary algebraical equations. For instance, let M be any condition of less postulation than X: then if from the equation $\epsilon = p + q - g$, assumed to be true, we deduce $M\epsilon = Mp + Mq - Mg$, this (like the original equation $\epsilon = p + q - g$) is in its actual form an equation without logical or numerical signification; but if we adjoin to it a supplementary condition K, such that postulation of K + do. of M = do. of X (or, what is the same thing, that the condition KM shall be supplementary to the several conditions contained in the original equation $\epsilon = p + q - g$), then the equation in question, $M\epsilon = Mp + Mq - Mg$, is to be interpreted as meaning

$$KM\epsilon = KMp + KMq - KMg,$$

that is,

$$X\epsilon = Xp + Xq - Xg,$$

which is numerically true. We thus see that the original equation $\epsilon = p + q - g$ implies the new equation

$$M\epsilon = Mp + Mq - Mg,$$

which is its algebraical consequence. And if we regard, for instance, $A + B$ as the condition that either the condition A shall be satisfied or else the condition B shall be satisfied, then $A + B$ is a condition, and as such we have

$$(A + B)\,\epsilon = (A + B)\,p + (A + B)\,q - (A + B)\,\epsilon.$$

It is going a step further to say that if we have, for instance, an equation $M = A + B - C$ between conditions M, A, B, C, then that, instead of

$$M\epsilon = Mp + Mq - M\epsilon,$$

we may write

$$(A + B - C)\,\epsilon = (A + B - C)\,p + (A + B - C)\,q - (A + B - C)\,\epsilon\,;$$

this is, in fact, treating $A + B - C$ as being to all intents and purposes a condition such as M, or an alternative condition $A + B$. It is, in fact, assumed that the step is permissible; and we thus make such deductions as

$$(\epsilon + p + q - g)(\epsilon - p - q + g) = 0\,;$$

that is,

$$\epsilon^2 - (p+q-g)^2 = 0,$$

or

$$\epsilon^2 = (p+q-g)^2, \; = p^2 + 2pq + q^2 - 2pg - 2qg + g^2;$$

viz. this is an equation such as the original equation $\epsilon = p+q-g$, acquiring a numerical signification when we adjoin to it a supplementary condition X of the proper postulation.

The section above referred to deals with the question to determine the number of lines which satisfy the several relations of contact in regard to a given surface F of the order n, without point-singularities, that is, the surface represented by the *general* equation $(*\between x, y, z, w)^n = 0$.

The chief results are contained in the following table, the notation of which will be explained:

1. $\epsilon_2 g_s \quad = n(n-1)$,
2. $\epsilon_2 b_2 g_e \quad = n$,
3. $\epsilon_3 g_e \quad = 3n(n-2)$,
4. $\epsilon_3 g_p \quad = n(n-1)(n-2)$,
5. $\epsilon_3 b_3^2 \quad = 2n$,
6. $\epsilon_{22} g_e \quad = \tfrac{1}{2}n(n-2)(n-3)(n+3)$,
7. $\epsilon_{22} g_p \quad = \tfrac{1}{2}n(n-1)(n-2)(n-3)$,
8. $\epsilon_{22} b_2^2 \quad = n(n-3)(n+2)$,
9. $\epsilon_{22} b_2 c_2 \quad = n(n^3 - 2n^2 + 2n - 6)$,
10. $\epsilon_4 g \quad = 2n(n-3)(3n-2)$,
11. $\epsilon_4 b_4 \quad = n(11n-24)$,
12. $\epsilon_{32} g \quad = n(n-3)(n-4)(n^2+6n-4)$,
13. $\epsilon_{32} b_3 \quad = n(n-4)(3n^2+5n-24)$,
14. $\epsilon_{32} b_2 \quad = n(n-2)(n-4)(n^2+2n+12)$,
15. $\epsilon_{222} g \quad = \tfrac{1}{3}n(n-3)(n-4)(n-5)(n^2+3n-2)$,
16. $\epsilon_{222} b_2 \quad = \tfrac{1}{2}n(n-2)(n-4)(n-5)(n^2+5n+12)$,
17. $\epsilon_5 \quad = 5n(n-4)(7n-12)$,
18. $\epsilon_{42} \quad = 2n(n-4)(n-5)(n+6)(3n-5)$,
19. $\epsilon_{33} \quad = \tfrac{1}{2}n(n-4)(n-5)(n^3+3n^2+29n-60)$,
20. $\epsilon_{322} \quad = \tfrac{1}{2}n(n-4)(n-5)(n-6)(n^3+9n^2+20n-60)$,
21. $\epsilon_{2222} \quad = \tfrac{1}{12}n(n-4)(n-5)(n-6)(n-7)(n^3+6n^2+7n-30)$,
22. $\epsilon_{222} b_1 \quad = \tfrac{1}{3}n(n-4)(n-5)(n-6)(n^3+3n^2-2n-12)$,
23. $\epsilon_3 b_1^2 \quad = (n-3)(n^2+2)$,
24. $\epsilon_2 b_1 c_1 d_1 = n^2(n-4)(2n^2-3n-3)$.

In the foregoing formulæ the suffixes of the ϵ refer to the contacts, viz. ϵ_2 denotes a 2-pointic intersection, ϵ_{32} a 3-pointic and a 2-pointic intersection. The letters b, c, d refer to the points of contact or intersection, thus $\epsilon_{32}b_3$, b_3 is the point of 3-pointic intersection; $\epsilon_{222}b_1$, b_1 is one of the points of simple intersection; b_1 is also the condition that the point in question lies on a given plane; g, g_s, g_e, g_p have their ordinary signification explained a little further on. Thus (15) $\epsilon_{222}g$ denotes the number of triple tangents which can be drawn to meet a given line; or, what is the same thing, it is the order of the regulus formed by the triple tangents.

The following are elementary formulæ used in the investigation of the foregoing results.

Subject a line having upon it a point,

			Postul.
p	the condition that	point is in a given plane	1
p_g	"	" line	2
g	"	line meets a given line	1
g_e	"	" is in a given plane	2
g_p	"	" passes through a given point	2
g_s	"	" lies in a given plane and passes through a given point of that plane	3
G	"	" coincides with a given line	4

We have (p. 22 *et seq.*)

	Postul.
$p_g = p^2$	2 (*logical*)
$p_g = p^2 + g_e$	2
$g^2 = g_e + g_p$	2
$g_s = gg_e$	3 (*logical*)
$g_s = gg_p$	3 (*logical*)
$pg_p = p^3 + g_s$	3 (*demons. infrà*)
$p^4 = 0$	4
$g_eg_p = 0$	4
$g_e^2 = G$	4
$g_p^2 = G$	4
$p^3g = p^2g_e$	4 (*demons. infrà*)
$pg_s = p^2g_p$	4 "
$pg_s = p^2g_e + G$	4 "
$p^3g_e = 0$	5 "
$p^3g_p = pG$	5 "
$p^2g_s = pG$	5 "

$pg_p = p^3 + g_s$; we have $0 = g_e + p^2 - pg$, $0 = g_e + g_p - g^2$, and thence

$$\begin{aligned} 0 = &\quad (p+g)\{g_e + p(p-g)\} - p\{g_e + g_p - g^2\} \\ = &\quad pg_e + gg_e + p^3 - pg^2 \\ &\quad - pg_e - pg_p \qquad + pg^2 \\ = &\quad gg_e - pg_p + p^3 \\ = &\quad g_s - pg_p + p^3. \end{aligned}$$

$p^3g = p^2g_e$: from $pg = p^2 + g_e$, we have $p^2g = p^4 + p^2g_e = p^2g_e$, since $p^4 = 0$,

$pg_s = p^2g_e + G$ „ $g_s = gg_e$ „ $pg_e = pgg_e = (p^2 + g_e)\,g_e = p^2g_e + G$,

$pg_s = p^2g_p$ „ $g_s = gg_p$ „ $pg_s = pgg_p = (p^2 + g_e)\,g_p = p^2g_p$, since $g_eg_p = 0$;

and in a similar manner we prove the last three equations.

For the demonstration of the formulæ of the table we take the subject to be a line bearing upon it the points 1, 2, ..., n, which are its intersections with a given surface of the order n. The symbols p_1, p_2, ... refer to these points respectively; thus, p_1 is the condition that the point 1 may lie on a given plane; and then, writing

$$\begin{aligned} \epsilon\ &= p_1 + p_2 - g, \\ \epsilon'\ &= p_1 + p_3 - g, \\ \epsilon'' &= p_3 + p_4 - g, \end{aligned}$$

it appears that ϵ will denote the condition of the coincidence of the points 1 and 2; ϵ' that of the points 1 and 3, &c. Hence also, $\epsilon\epsilon'$ will denote the twofold condition of the coincidence of the points 1, 2, 3; and so in other cases. But, according to the notation above explained, ϵ is also denoted by ϵ_2, $\epsilon\epsilon'$ by ϵ_3, $\epsilon\epsilon''$ by ϵ_{22}, &c.

We thus have

$$\begin{aligned} \epsilon_2 &= p_1 + p_2 - g, \\ \epsilon_3 &= (p_1 + p_2 - g)(p_1 + p_3 - g), \\ 2\epsilon_{22} &= (p_1 + p_2 - g)(p_3 + p_4 - g), \\ \epsilon_4 &= (p_1 + p_2 - g)(p_1 + p_3 - g)(p_1 + p_4 - g), \\ \epsilon_{32} &= (p_1 + p_2 - g)(p_1 + p_3 - g)(p_4 + p_5 - g), \\ 6\epsilon_{222} &= (p_1 + p_2 - g)(p_3 + p_4 - g)(p_5 + p_6 - g), \\ \epsilon_5 &= (p_1 + p_2 - g)(p_1 + p_3 - g)(p_1 + p_4 - g)(p_1 + p_5 - g), \\ \epsilon_{42} &= (p_1 + p_2 - g)(p_1 + p_3 - g)(p_1 + p_4 - g)(p_5 + p_6 - g), \\ 2\epsilon_{33} &= (p_1 + p_2 - g)(p_1 + p_3 - g)(p_4 + p_5 - g)(p_4 + p_6 - g), \\ 2\epsilon_{322} &= (p_1 + p_2 - g)(p_1 + p_3 - g)(p_4 + p_5 - g)(p_6 + p_7 - g), \\ 24\epsilon_{2222} &= (p_1 + p_2 - g)(p_3 + p_4 - g)(p_5 + p_6 - g)(p_7 + p_8 - g). \end{aligned}$$

We can now, by a mere analytical process of development and reduction, express each of the foregoing values as a linear function of

$$p_1^2p_2^2,\ p_1^2p_2p_3,\ p_1p_2p_3p_4,\ \text{and}\ G.$$

(Schubert says, as a linear function of these four symbols and $p_1p_2g_e$; but in fact $p_1p_2g_e$ is $=p_1^2p_2^2$.)

Observe, first, that we may, p. 287, in all the general equations instead of p write p_1, p_2, &c.; and, further, that any symbol containing for instance p_1^3 is $=0$. For the symbols now belong to the intersections of the line with a given surface; p_1^3 is the condition that a certain one of these intersections shall lie in three given planes, that is, that it shall coincide with a given arbitrary point; this cannot be the case, for the arbitrary point is not on the surface F; and therefore $p_1^3=0$.

We thus have $p_1g=p_1^2+g_e$, thence $p_1^2g=p_1^3+p_1g_e$, that is, $p_1^2g=p_1g_e$; and thence further $p_1^3g=p_1^2g_e$, that is, $p_1^2g_e=0$.

Again, from $p_2g=p_2^2+g_e$, $p_1g=p_1^2+g_e$, we have

$$p_1^2(p_2^2+g_e)=p_1p_2(p_1^2+g_e),$$

which, in virtue of $p_1^2g_e=0$ and $p_1^3p_2=0$, becomes

$$p_1^2p_2^2=p_1p_2g_e.$$

As a simple instance of the reductions, take

$$\epsilon_2g_s,\ =(p_1+p_2-g)\,g_s.$$

Here

$$p_1g_s,\ =p_2g_s,\ =p_1^2g_e+G,\ =G,\ \text{since}\ p_1^2g_e=0;$$

and

$$gg_s=g^2g_e=(g_e+g_p)\,g_e=g_e^2+g_eg_p=G,\ \text{since}\ g_e^2=0,\ g_eg_p=G;$$

whence the value is

$$\epsilon_2g_s=G+G-G,\ =G.$$

As a more complicated example, take

$$\epsilon_5,\ =(p_1+p_2-g)(p_1+p_3-g)(p_1+p_4-g)(p_1+p_5-g).$$

Observe that, after the multiplication is effected we may, in any way we please, interchange the suffixes, $p_1^2p_3p_4=p_1^2p_2p_3$, $p_3^2p_4^2=p_1^2p_2^2$, &c.; the suffixes serve only to distinguish from each other symbols in the same product (thus p_1^4 is different from $p_1p_2p_3p_4$), but there is nothing to distinguish one point of intersection from another. Thus the foregoing expression containing the terms $(p_2+p_3+p_4+p_5)(p_1-g)^3$, these may be combined into the single term $4p_2(p_1-g)^3$; expanding in powers of p_1-g and reducing in this manner, the value of ϵ_5 is, in fact, found to be

$$=(p_1-g)^4+4p_2(p_1-g)^3+6p_2p_3(p_1-g)^2+4p_2p_3p_4(p_1-g)+p_1p_2p_3p_4.$$

Developing this in powers of g, omitting the terms containing p_2^3 which vanish, and further reducing, the value is

$$6p_1^2p_2p_3+5p_1p_2p_3p_4+g(-12p_1^2p_2-16p_1p_2p_3)+g^2(6p_1^2+18p_1p_2)-8p_1g^3+g^4.$$

We have

$$g^4 = 2G,\ p_1g^3 = p_1g_s = p_1^2g_e + G,\ = G.$$

Next for the terms in g^2, from $p_1g = p_1^2 + g_e$ we have

$$p_1^2g = \qquad p_1g_e,$$

$$p_1p_2g = p_1^2p_2 \ + p_1g_e,$$

and thence

$$p_1^2g^2 = \qquad p_1g_s;$$

$$p_1p_2g^2 = p_1^2p_2g + p_1g_s,$$

or, since $p_1g_s = G$ as before, the whole term is $= 18p_1^2p_2g + 24G$. The terms in g thus become $= g(6p_1^2p_2 - 16p_1p_2p_3)$, and from the same equation $p_1g = p_1^2 + g_e$ we find

$$p_1^2p_2g = p_1^2p_2^2 \text{ and } p_1p_2p_3g = p_1^2p_2p_3 + p_1^2p_2^2.$$

The value is thus finally found to be

$$= -10p_1^2p_2^2 - 10p_1^2p_2p_3 + 5p_1p_2p_3p_4 + 10G.$$

The whole series of like results is

			$p_1^2p_2^2$	$p_1^2p_2p_3$	$p_1p_2p_3p_4$	G
1.	ϵ_2	g_s				+ 1
2.	,,	b_2g_e	+ 1			− 1
3.	ϵ_3	g_e	3			− 3
4.	,,	g_p				+ 1
5.	,,	b_3^2	− 2	+ 1		+ 1
6.	$2\epsilon_{22}$	g_e	+ 4			− 3
7.	2 ,,	g_p				+ 1
8.	,,	b_2^2	− 3	+ 2		+ 1
9.	ϵ_2	b_2c_2	− 2		+ 1	+ 1
10.	ϵ_4	g	− 2	+ 4		− 2
11.	,,	b_4	− 6		+ 1	+ 4
12.	ϵ_{32}	g	− 3	+ 6		− 2
13.	,,	b_3	− 7	− 1	+ 2	+ 4
14.	,,	b_2	− 6	− 3	+ 3	+ 4
15.	$6\epsilon_{222}$	g	− 4	+ 8		− 2
16.	2 ,,	b_2	− 7	− 4	+ 4	+ 4
17.	ϵ_5		− 10	− 10	+ 5	+ 10
18.	ϵ_{42}		− 10	− 16	+ 8	+ 10
19.	$2\epsilon_{33}$		− 9	− 18	+ 9	+ 10
20.	$2\epsilon_{322}$		− 9	− 24	+ 12	+ 10
21.	$24\epsilon_{2222}$		− 8	− 32	+ 16	+ 10
22.	$6\epsilon_{222}$	b_1	− 6	− 12	+ 8	+ 4
23.	$n\epsilon_3$	b_1^2	− 3	+ 3		+ 1
24.	ϵ_2	$b_1c_1d_1$	− 1	− 1	+ 2	

But in these formulæ $p_1^2p_2^2$, $p_1^2p_2p_3$, $p_1p_2p_3p_4$, G have numerical values which are different according to the number of points of intersection presenting themselves in the several formulæ; viz. this number being called i, we have for the formulæ in

	ϵ_2	ϵ_3	ϵ_{22}	ϵ_4	ϵ_{32}	ϵ_{222}	ϵ_5	ϵ_{42}	ϵ_{33}	ϵ_{322}	ϵ_{2222}	$\epsilon_{222}b_1$	$\epsilon_3 b_1^2$	$\epsilon_2 b_1 c_1 d_1$
$i=$	2	3	4	4	5	6	5	6	6	7	8	7	4	5,

and the values of the symbols are

$$\begin{aligned} p_1^2p_2^2 &= n^2(n-2)(n-3) \quad\ldots(n-i+1), \\ p_1^2p_2p_3 &= n^2(n-1)(n-3) \quad\ldots(n-i+1), \\ p_1p_2p_3p_4 &= n^2(2n^2-6n+3)(n-4)\ldots(n-i+1), \\ G &= n\,(n-1)(n-2) \quad\ldots(n-i+1). \end{aligned}$$

Thus, suppose $i=4$, the subject is a line bearing the points 1, 2, 3, 4, which are intersections of the line with the surface F; we have then G as the condition in order that this line (or, say, the line of the subject) may coincide with a given line, which given line intersects the surface in n points; any four of these (their order being attended to) may be regarded as being the points 1, 2, 3, 4; or there are $n(n-1)(n-2)(n-3)$ subjects satisfying the prescribed condition (that the line of the subject may coincide with the given line). Hence here $G=n(n-1)(n-2)(n-3)$; and so in general $G=n(n-1)(n-2)\ldots(n-i+1)$.

Next, for $p_1^2p_2^2$. Here p_1^2 is the condition that the point 1 shall lie in each of two given planes, that is, in a given line, say L_1; and, similarly, p_2^2 is the condition that 2 may lie in a given line L_2. We take any one of the n intersections of L_1 with F for the point 1, and any one of the n intersections of L_2 with F for the point 2; this determines the line of the subject, but the $i-2$ points 3, 4, ..., i are then any $i-2$ of the remaining $n-2$ intersections of this line with F; that is, $p_1^2p_2^2=n^2(n-2)(n-3)\ldots(n-i+1)$ as above.

Again, for $p_1^2p_2p_3$. Here p_1^2 is the condition that 1 shall lie in a given line L_1; we therefore take for 1 any one of the n intersections of L_1 with F; p_2 is the condition that 2 may lie in a given plane P_2, it lies therefore in the curve of intersection of P_2 with F; and, similarly, 3 lies in the curve of intersection of a plane P_3 with F; the two planes intersect in a line meeting F in n points σ, and the two cones, vertex 1, which stand upon the plane curves respectively, intersect in the n lines joining 1 with the n points σ, and in n^2-n other lines. The line of the subject is then any one of these n^2-n lines, or, since the vertex is any one of n points, the line is any one of $n(n^2-n)$, $=n^2(n-1)$ lines; the remaining points 4, 5, ..., i are any $i-3$ of the remaining $n-3$ intersections of the line with F; hence the formula

$$p_1^2p_2p_3=n^2(n-1)(n-3)(n-4)\ldots(n-i+1).$$

For $p_1p_2p_3p_4$. We have here 1, 2, 3, 4 lying in given plane sections of the surface F, and we have consequently to find the number of lines which can be drawn to meet each of these four sections. Observing that any two of the sections meet in the n

intersections with F of the line of intersection of their planes, the order of the scroll generated by the lines which meet three of the sections is $2n^3-3n^2$; this scroll meets the fourth section in $n(2n^3-3n^2)$, $=2n^4-3n^3$ points; or we have this number of lines meeting each of the four sections. But among these are included $3n^2(n-1)$ lines which have to be rejected, viz. the sections 1 and 4 meet in n points, each of which is the vertex of cones through the sections 1 and 2 respectively; these cones meet in n lines, which are to be disregarded, and in n^2-n other lines, and we have thus $n(n^2-n)$, $=n^2(n-1)$ lines; and similarly from the intersections of 2 and 4, and from the intersections of 3 and 4, $n^2(n-1)$ and $n^2(n-1)$ lines, in all $3n^2(n-1)$ lines. Hence the number of lines meeting the four sections is

$$2n^4-3n^3-3n^3+3n^2, \quad =2n^4-6n^3+3n^2;$$

taking any one of these for the line of the subject, the remaining points 5, 6, ..., i are any $i-4$ of the remaining $n-4$ intersections, or we have the required formula

$$p_1p_2p_3p_4=n^2(2n^2-6n+3)(n-4)\ldots(n-i+1).$$

The four numbers $p_1^2p_2^2$, $p_1^2p_2p_3$, $p_1p_2p_3p_4$, G for any line of the table being now known, we can at once calculate the required values $\epsilon_2 g_8$, &c., as the case may be; for instance,

$$\begin{aligned} i=5, \quad \epsilon_5 = &-10p_1^2p_2^2 && = -10n^2(n-2)(n-3)(n-4) \\ &-10p_1^2p_2p_3 && \quad -10n^2(n-1)(n-3)(n-4) \\ &+\ 5p_1p_2p_3p_4 && \quad +\ 5n^2(2n^2-6n+3)(n-4) \\ &+10G && \quad +10n\ (n-1)(n-2)(n-3)(n-4) \\ & && = \quad 5n\ (n-4)(7n-12). \end{aligned}$$

In fact, throwing out $n(n-4)$, the remaining terms give

$$\begin{array}{l} -10n^3+50n^2-\ \ 60n \\ -10n^3+40n^2-\ \ 30n \\ +10n^3-30n^2+\ \ 15n \\ +10n^3-60n^2+110n-60 \\ \hline \qquad\qquad 35n-60, \ =5(7n-12). \end{array}$$

And we obtain in like manner the other formulæ of the table.

The remainder of § 33 contains investigations of less systematically connected theorems, and I quote the results only.

25. If on the surface F_n there is a curve order r, then of the tangent planes of F_n along this curve there pass $r(n-1)$ through an arbitrary point of space; *aliter*, class of torse is $=r(n-1)$.

In particular, for curve of 4-pointic contact, $r=n(11n-24)$, class of torse is

$$=n(n-1)(11n-24).$$

No. of tangent planes through line, or class of surface, $=n(n-1)^2$.

26. $\epsilon_3 b_3 g = \epsilon_3 b_3{}^2 + \epsilon_3 g_e = 2n + 3n(n-2), = n(3n-4)$.

$\epsilon_3 b_3 g$, $= n(3n-4)$, is the order of curve of contact of the 3-pointic (chief) tangents which meet a given line.

Parabolic tangents are coincident chief tangents.

No. of 4-pointic parabolic tangents $= 2n(n-2)(11n-24)$.

27. Order of parabolic curve $= 4n(n-2)$.

Order of regulus formed by parabolic tangents

$$= 2n(n-2)(3n-4).$$

The parabolic curve and curve of contacts of an ϵ_4 tangent meet in

$$4n(n-2)(11n-24)$$

points, i.e., they *touch* in $2n(n-2)(11n-24)$ points.

28. Umbilici. No. is $= 2n(5n^2 - 14n + 11)$.

29. No. of points at which the chief tangents being distinct are each of them 4-pointic, or, what is the same thing, No. of actual double points of curve ϵ_4,

$$= 5n(7n^2 - 28n + 30),$$

$n = 3$, No. is $15(63 - 84 + 30)$, $= 135$, viz. this is the number of points of intersection of two of the 27 lines; or, what is the same thing, the number of triple tangent planes is $= 45$.

30. No. of parabolic tangents which have besides a 2-pointic contact is

$$= 2n(n-2)(n-4)(3n^2 + 5n - 24).$$

31. No. of double tangent planes such that line through points of contact is at one of these points 3-pointic

$$= n(n-2)(n-4)(n^3 + 3n^2 + 13n - 48).$$

32. No. of points where one chief tangent is 4-pointic, the other 3-pointic and (at another point of the surface) 2-pointic is

$$= n(n-4)(27n^3 - 13n^2 - 264n + 396).$$

33. No. of points where chief tangents being distinct are each of them at another point of the surface 2-pointic is

$$= n(n-4)(4n^5 - 4n^4 - 95n^3 + 99n^2 + 544n - 840).$$

34. The curve of contacts b_3 of an ϵ_{32} tangent has with the parabolic curve 2-pointic intersections only, and these are at the points for which the chief tangent is (at another point of the surface) 2-pointic.

35. The curve of contacts b_3 of an ϵ_{32} tangent has, with the curve of contacts of an ϵ_4 tangent, 2-pointic intersections at the contacts of an ϵ_5 tangent; and has also simple intersections with the same curve, 1° at the contacts b_4 of an ϵ_{42} tangent, 2° at the points where the chief tangents are ϵ_4 and ϵ_{32}.

763.

ON THE THEOREMS OF THE 2, 4, 8, AND 16 SQUARES.

[From the *Quarterly Journal of Pure and Applied Mathematics,* vol. XVII. (1881), pp. 258—276.]

A SUM of 2 squares multiplied by a sum of 2 squares is a sum of 2 squares; a sum of 4 squares multiplied by a sum of 4 squares is a sum of 4 squares; a sum of 8 squares multiplied by a sum of 8 squares is a sum of 8 squares; but a sum of 16 squares multiplied by a sum of 16 squares is *not* a sum of 16 squares. These theorems were considered in the paper, Young, "On an extension of a theorem of Euler, with a determination of the limit beyond which it fails," *Trans. R. I. A.*, t. XXI. (1848), pp. 311—341; and the later history of the question is given in the paper by Mr S. Roberts, "On the Impossibility of the general Extension of Euler's Theorem &c.," *Quart. Math. Jour.* t. XVI. (1879), pp. 159—170; as regards the 16-question, it has been throughout assumed that there is only one type of synthematic arrangement (what this means will appear presently); but as regards this type, it is, I think, well shown that the signs cannot be determined. It will appear in the sequel, that there are in fact four types (the last three of them possibly equivalent) of synthematic arrangement; and for a complete proof, it is necessary to show in regard to each of these types that the signs cannot be determined. The existence of the four types has not (so far as I am aware) been hitherto noticed; and it hence follows, that no complete proof of the non-existence of the 16-square theorem has hitherto been given.

For the 2 squares the theorem is of course

$$(x_1^2+x_2^2)(y_1^2+y_2^2)=(x_1y_1+x_2y_2)^2+(x_1y_2-x_2y_1)^2.$$

For the 4 squares (for which the nature of the theorem is better seen) it is

$$\begin{aligned}(x_1^2+x_2^2+x_3^2+x_4^2)(y_1^2+y_2^2+y_3^2+y_4^2)= \quad &(x_1y_1+x_2y_2+x_3y_3+x_4y_4)^2\\ &+(x_1y_2-x_2y_1+x_3y_4-x_4y_3)^2\\ &+(x_1y_3-x_3y_1-x_2y_4+x_4y_2)^2\\ &+(x_1y_4-x_4y_1+x_2y_3-x_3y_2)^2;\end{aligned}$$

or, as this may be written,

$$(x_1^2 + x_2^2 + x_3^2 + x_4^2)(y_1^2 + y_2^2 + y_3^2 + y_4^2) - (x_1y_1 + x_2y_2 + x_3y_3 + x_4y_4)^2$$
$$= (12 + 34)^2$$
$$+ (13 - 24)^2$$
$$+ (14 + 23)^2;$$

where 12 is used to denote $x_1y_2 - x_2y_1$, &c., and the truth of the theorem depends on the identity $12.34 - 13.24 + 14.23 = 0$. Clearly, the first step for forming the equation is to arrange the duads in a synthematic form

12.34
13.24
14.23,

and then to determine the signs: such an arrangement exists in the case of 8, and the signs can be determined; it exists also in the case of 16, but the signs *cannot* be determined to satisfy all the necessary relations.

In the case of 8, we have the synthematic arrangement

12.34.56.78
13.24.57.68
14.23.58.67
15.26.37.48
16.25.38.47
17.28.35.46
18.27.36.45,

being the *only type* of synthematic arrangement. This is, in fact, important as regards the 16-question, and it will appear that the case is so; but in the 8-question, starting from this arrangement, we have to show that there exists an equation which, for convenience, I write as follows:

$$(x_1^2 + \ldots + x_8^2)(y_1^2 + \ldots + y_8^2) - (x_1y_1 + \ldots + x_8y_8)^2$$
$$= (12 + 34 + 56 + 78)^2$$
$$+ (13 + 24 + 57 + 68)^2$$
$$+ (14 + 23 + 58 + 67)^2$$
$$+ (15 + 26 + 37 + 48)^2$$
$$+ (16 + 25 + 38 + 47)^2$$
$$+ (17 + 28 + 35 + 46)^2$$
$$+ (18 + 27 + 36 + 45)^2,$$

but in which it is to be understood that each duad is affected by a factor ± 1 which is to be determined; say the factor of 12 is ϵ_{12}, that of 34, ϵ_{34}; and so in other cases. It is however assumed that ϵ_{12}, ϵ_{34}, ϵ_{56}, ϵ_{78}; ϵ_{13}, ϵ_{14}, ϵ_{15}, ϵ_{16}, ϵ_{17}, ϵ_{18} are each $=+1$.

We have then on the right-hand side triads of terms such as, 2 into

$$\epsilon_{12}\epsilon_{34}\, 12\,.\,34 + \epsilon_{13}\epsilon_{24}\, 13\,.\,24 + \epsilon_{14}\epsilon_{23}\, 14\,.\,23,$$

which triad ought to vanish identically, as reducing itself to a multiple of

$$12\,.\,34 - \;\; 13\,.\,24 + 14\,.\,23;$$

viz. we ought to have

$$\epsilon_{12}\epsilon_{34} = - \;\; \epsilon_{13}\epsilon_{24} \;\; = \;\; \epsilon_{14}\epsilon_{23};$$

or, *using now and henceforward when occasion requires*, 12, 34, &c. to denote ϵ_{12}, ϵ_{34}, &c. respectively, we have

$$12\,.\,34 = +k,$$

$$13\,.\,24 = -k,$$

$$14\,.\,23 = +k,$$

where $k, = \pm 1$, has to be determined (in the actual case we have $12 = +1$, $34 = +1$, $13 = 1$, $14 = 1$; and therefore the first equation gives $k = 1$, and the other two then give $24 = -1$, $23 = +1$).

We have in this way triads of values corresponding to the different tetrads

1234
1256
1278
1357
1368
1458
1467
2358
2367
2457
2468
3456
3478
5678,

which can be formed with the several lines of the formula. Thus we have from the first line 1234, 1256, 1278; then from the second line (not 1324 which in the form 1234 has been taken already) 1357, 1368, ...; and finally from the last line 5678.

We might consider each line as giving 6 tetrads, but the tetrads would then be obtained 3 times over; the number of tetrads is thus $6 \times 7 \div 3, = 14$ as above. And observe, that the systems of values for the coefficients $\epsilon = \pm 1$ are obtained directly from the tetrads, without the employment of any other formula.

We thus obtain the system of signs as follows:

12	$+1$	
13	$+1$	
14	$+1$	
15	$+1$	
16	$+1$	
17	$+1$	
18	$+1$	
23	$+1$	
24	-1	
25	$+1$	
26	-1	
27	$+1$	
28	-1	
34	$+1$	
35	a	$-\theta$
36	b	θ
37	$-a$	θ
38	$-b$	$-\theta$
45	c	θ
46	d	θ
47	$-d$	$-\theta$
48	$-c$	$-\theta$
56	$+1$	
57	a	$-\theta$
58	c	θ
67	d	θ
68	b	θ
78	$+1$	

viz. the original assumptions $12 = +1$, &c., and the tetrads 1234, 1256, 1278 give all the signs ± 1 up to $34 = +1$; from the tetrad 1357 we have

$$\begin{array}{llll} 13.57 & + & 1 & a, \\ 15.37 & - & 1 & a, \\ 17.35 & + & 1 & a, \end{array}$$

that is, $35 = a$, $37 = -a$, $57 = a$, where a, $= \pm 1$, is still undetermined; and similarly, the tetrads 1368, 1458, 1467 give the remaining signs b, c, d. The tetrad 2358 then gives

$$\begin{array}{llll} 23.58 & + & 1 & c, \\ 25.38 & - & 1 & -b, \\ 28.35 & + & -1 & a, \end{array}$$

that is, $-a = b = c$; and similarly the tetrads 2367, 2457, 2468 give $-a = b = d$, $-a = c = d$, $b = c = d$ respectively; the four tetrads thus give $-a = b = c = d$, say each of these $= \theta$. But retaining for the moment a, b, c, d, the tetrad 3456 then gives

$$\begin{array}{llll} 34.56 & + & 1 & 1, \\ 35.46 & - & a & d, \\ 36.45 & + & b & c, \end{array}$$

that is, $1 = -ad = bc$, and similarly the last two tetrads 3478 and 5678 give $1 = -ac = bd$ and $1 = -ab = cd$ respectively; substituting the values in terms of θ, the several equations give only $\theta^2 = 1$, that is, $\theta = \pm 1$ at pleasure; and the series of signs for the 8-formula, containing this one arbitrary sign $\theta = \pm 1$, is thus determined.

Passing to the case of 16, we have in like manner to form a synthematic arrangement of the numbers 1, 2, ..., 16 in 15 lines, each containing the 16 numbers in 8 duads (no duad twice repeated), and this containing all the 120 duads. And, using for the moment letters instead of numbers, the necessary condition is, that $ab.cd$ occurring in one line, $ac.bd$ must occur in another line, and $ad.bc$ in a third line. Observe that as well the order of the letters in a duad as the order of the duads is thus far immaterial; so that a line containing $bd.ca$ may be considered as containing $ac.bd$.

Considering any such combination $ab.cd$, the line which contains it may be taken to be the first line; and the line which contains $ac.bd$ may be taken to be the second line. And then writing 1, 2, 3, 4 in place of a, b, c, d respectively, the first line will contain 12.34, and the second line will contain 13.24. Let e be any other symbol occurring in the first line, say in the duad ef, and in the second line say in the duad eg; then g must occur in the first line in some duad gh, or the first line will contain $ef.gh$, and then the second line as containing eg will contain also *fh*, that is, it will contain *eg*.*fh*. And then writing 5, 6, 7, 8 in place of *e*, *f*, *g*, *h* respectively, the first line will contain 56.78 and the second line will contain 57.68. And continuing the like reasoning, it appears that the first line and the second line may be taken to be

1 2. 3 4. 5 6. 7 8. 9 10. 11 12. 13 14. 15 16,

and

1 3. 2 4. 5 7. 6 8. 9 11. 10 12. 13 15. 14 16,

respectively. There will then be a line containing 1 4 which may be taken for the third line, a line containing 1 5 which may be taken for the fourth line, and so on; viz. the successive lines may be taken to begin with 1 2, 1 3, 1 4, ..., 1 16 respectively.

Proceeding to form the synthematic arrangement, and starting with the first and second lines and first column as above, it appears that in each of the remaining lines there are three duads which occur of necessity, and putting these in the second, third, and fourth places (the order of the duads in any line being immaterial), it is seen that the second, third, and fourth columns can be filled up in one, and only one way; see the annexed first-half:

First-half common to all.

1	2	3	4	5	6	7	8
1	3	2	4	5	7	6	8
1	4	2	3	5	8	6	7
1	5	2	6	3	7	4	8
1	6	2	5	3	8	4	7
1	7	2	8	3	5	4	6
1	8	2	7	3	6	4	5
1	9	2	10	3	11	4	12
1	10	2	9	3	12	4	11
1	11	2	12	3	9	4	10
1	12	2	11	3	10	4	9
1	13	2	14	3	15	4	16
1	14	2	13	3	16	4	15
1	15	2	16	3	13	4	14
1	16	2	15	3	14	4	13

Four forms of second-half.

I.

9	10	11	12	13	14	15	16
9	11	10	12	13	15	14	16
9	12	10	11	13	16	14	15
9	13	10	14	11	15	12	16
9	14	10	13	11	16	12	15
9	15	10	16	11	13	12	14
9	16	10	15	11	14	12	13
5	13	6	14	7	15	8	16
5	14	6	13	7	16	8	15
5	15	6	16	7	13	8	14
5	16	6	15	7	14	8	13
5	9	6	10	7	11	8	12
5	10	6	9	7	12	8	11
5	11	6	12	7	9	8	10
5	12	6	11	7	10	8	9

II.

9	10	11	12	13	14	15	16
9	11	10	12	13	15	14	16
9	12	10	11	13	16	14	15
9	14	10	13	11	16	12	15
9	13	10	14	11	15	12	16
9	16	10	15	11	14	12	13
9	15	10	16	11	13	12	14
5	14	6	13	7	16	8	15
5	13	6	14	7	15	8	16
5	16	6	15	7	14	8	13
5	15	6	16	7	13	8	14
5	10	6	9	7	12	8	11
5	9	6	10	7	11	8	12
5	12	6	11	7	10	8	9
5	11	6	12	7	9	8	10

III.

9	10	11	12	13	14	15	16
9	11	10	12	13	15	14	16
9	12	10	11	13	16	14	15
9	15	10	16	11	13	12	14
9	16	10	15	11	14	12	13
9	13	10	14	11	15	12	16
9	14	10	13	11	16	12	15
5	15	6	16	7	13	8	14
5	16	6	15	7	14	8	13
5	13	6	14	7	15	8	16
5	14	6	13	7	16	8	15
5	11	6	12	7	9	8	10
5	12	6	11	7	10	8	9
5	9	6	10	7	11	8	12
5	10	6	9	7	12	8	11

IV.

9	10	11	12	13	14	15	16
9	11	10	12	13	15	14	16
9	12	10	11	13	16	14	15
9	16	10	15	11	14	12	13
9	15	10	16	11	13	12	14
9	14	10	13	11	16	12	15
9	13	10	14	11	15	12	16
5	16	6	15	7	14	8	13
5	15	6	16	7	13	8	14
5	14	6	13	7	16	8	15
5	13	6	14	7	15	8	16
5	12	6	11	7	10	8	9
5	11	6	12	7	9	8	10
5	10	6	9	7	12	8	11
5	9	6	10	7	11	8	12

And it is to be noticed that in this first-half the upper part, or first seven lines, give in fact the synthematic arrangement for the 8-question; so that (as remarked above) in this 8-question there is but one form of synthematic arrangement.

Proceeding to fill up the remaining columns, the duad 59 cannot be placed in any line which contains a 5 or a 9; that is, it must be placed in some one of the last 4 lines; and placing it successively in each of these, it appears that the columns can be filled up in one, and only one, way; we have thus the above "four forms of second-half," each of which, taken in conjunction with the common first-half, gives a synthematic arrangement of the 16 numbers.

Each of these synthematic arrangements may be converted into a square, the first line of which is formed with the numbers 1 to 16 in order, and the other fifteen lines of which are derived from the fifteen lines of the synthematic arrangement respectively: thus the line

1 2. 3 4. 5 6. 7 8. 9 10. 11 12. 13 14. 15 16

gives the second line of

1 2. 3 4. 5 6. 7 8. 9 10. 11 12. 13 14. 15 16,

2 1. 4 3. 6 5. 8 7. 10 9. 12 11. 14 13. 16 15,

and so in other cases. And conversely, by comparing with the first line of the square each of the other fifteen lines respectively, we have the fifteen lines of the synthematic arrangement; we thus obtain the four squares presently given. These squares are not required in the sequel, but they serve to put in a clearer light the construction of the synthematic arrangements; by converting in like manner into a square the formula p. 332 of Young's paper, it appears that his arrangement is in fact the first of the foregoing four arrangements. The squares are

I.

1	2	3	4	5	6	7	8	9	10	11	12	13	14	15	16
2	1	4	3	6	5	8	7	10	9	12	11	14	13	16	15
3	4	1	2	7	8	5	6	11	12	9	10	15	16	13	14
4	3	2	1	8	7	6	5	12	11	10	9	16	15	14	13
5	6	7	8	1	2	3	4	13	14	15	16	9	10	11	12
6	5	8	7	2	1	4	3	14	13	16	15	10	9	12	11
7	8	5	6	3	4	1	2	15	16	13	14	11	12	9	10
8	7	6	5	4	3	2	1	16	15	14	13	12	11	10	9
9	10	11	12	13	14	15	16	1	2	3	4	5	6	7	8
10	9	12	11	14	13	16	15	2	1	4	3	6	5	8	7
11	12	9	10	15	16	13	14	3	4	1	2	7	8	5	6
12	11	10	9	16	15	14	13	4	3	2	1	8	7	6	5
13	14	15	16	9	10	11	12	5	6	7	8	1	2	3	4
14	13	16	15	10	9	12	11	6	5	8	7	2	1	4	3
15	16	13	14	11	12	9	10	7	8	5	6	3	4	1	2
16	15	14	13	12	11	10	9	8	7	6	5	4	3	2	1

II.

1	2	3	4	5	6	7	8	9	10	11	12	13	14	15	16
2	1	4	3	6	5	8	7	10	9	12	11	14	13	16	15
3	4	1	2	7	8	5	6	11	12	9	10	15	16	13	14
4	3	2	1	8	7	6	5	12	11	10	9	16	15	14	13
5	6	7	8	1	2	3	4	14	13	16	15	10	9	12	11
6	5	8	7	2	1	4	3	13	14	15	16	9	10	11	12
7	8	5	6	3	4	1	2	16	15	14	13	12	11	10	9
8	7	6	5	4	3	2	1	15	16	13	14	11	12	9	10
9	10	11	12	14	13	16	15	1	2	3	4	6	5	8	7
10	9	12	11	13	14	15	16	2	1	4	3	5	6	7	8
11	12	9	10	16	15	14	13	3	4	1	2	8	7	6	5
12	11	10	9	15	16	13	14	4	3	2	1	7	8	5	6
13	14	15	16	10	9	12	11	6	5	8	7	1	2	3	4
14	13	16	15	9	10	11	12	5	6	7	8	2	1	4	3
15	16	13	14	12	11	10	9	8	7	6	5	3	4	1	2
16	15	14	13	11	12	9	10	7	8	5	6	4	3	2	1

III.

1	2	3	4	5	6	7	8	9	10	11	12	13	14	15	16
2	1	4	3	6	5	8	7	10	9	12	11	14	13	16	15
3	4	1	2	7	8	5	6	11	12	9	10	15	16	13	14
4	3	2	1	8	7	6	5	12	11	10	9	16	15	14	13
5	6	7	8	1	2	3	4	15	16	13	14	11	12	9	10
6	5	8	7	2	1	4	3	16	15	14	13	12	11	10	9
7	8	5	6	3	4	1	2	13	14	15	16	9	10	11	12
8	7	6	5	4	3	2	1	14	13	16	15	10	9	12	11
9	10	11	12	15	16	13	14	1	2	3	4	7	8	5	6
10	9	12	11	16	15	14	13	2	1	4	3	8	7	6	5
11	12	9	10	13	14	15	16	3	4	1	2	5	6	7	8
12	11	10	9	14	13	16	15	4	3	2	1	6	5	8	7
13	14	15	16	11	12	9	10	7	8	5	6	1	2	3	4
14	13	16	15	12	11	10	9	8	7	6	5	2	1	4	3
15	16	13	14	9	10	11	12	5	6	7	8	3	4	1	2
16	15	14	13	10	9	12	11	6	5	8	7	4	3	2	1

IV.

1	2	3	4	5	6	7	8	9	10	11	12	13	14	15	16
2	1	4	3	6	5	8	7	10	9	12	11	14	13	16	15
3	4	1	2	7	8	5	6	11	12	9	10	15	16	13	14
4	3	2	1	8	7	6	5	12	11	10	9	16	15	14	13
5	6	7	8	1	2	3	4	16	15	14	13	12	11	10	9
6	5	8	7	2	1	4	3	15	16	13	14	11	12	9	10
7	8	5	6	3	4	1	2	14	13	16	15	10	9	12	11
8	7	6	5	4	3	2	1	13	14	15	16	9	10	11	12
9	10	11	12	16	15	14	13	1	2	3	4	8	7	6	5
10	9	12	11	15	16	13	14	2	1	4	3	7	8	5	6
11	12	9	10	14	13	16	15	3	4	1	2	6	5	8	7
12	11	10	9	13	14	15	16	4	3	2	1	5	6	7	8
13	14	15	16	12	11	10	9	8	7	6	5	1	2	3	4
14	13	16	15	11	12	9	10	7	8	5	6	2	1	4	3
15	16	13	14	10	9	12	11	6	5	8	7	3	4	1	2
16	15	14	13	9	10	11	12	5	6	7	8	4	3	2	1

The foregoing investigation of the synthematic arrangements is exhaustive: it thereby appears that there are at most four types, viz. that every synthematic arrangement is of the type of one or other of the four arrangements above written down. The real nature of these is perhaps more clearly seen by means of the corresponding squares; and it will be observed, that there is in the first square a repetition of parts without transposition, which does not occur in the other three squares; this seems to suggest, that while the first square (and therefore the first synthematic arrangement) is really of a distinct type, the other three squares (or synthematic arrangements) *may possibly* belong to one and the same type. If this were so, it would be sufficient to prove the 16-theorem (viz. the non-existence of the 16-square formula) for the first and for any one of the other three synthematic arrangements; but I provisionally assume that the four types are really distinct, and propose therefore to prove the theorem for each of the four arrangements separately.

The process is the same as for the 8-theorem; we require the tetrads 1234, &c., contained in the synthematic arrangements. In any one of these, each line gives $\frac{1}{2}8.7, =28$ tetrads, and the 15 lines give therefore $15.28, =420$ tetrads: but we thus obtain each tetrad 3 times, or the number of the tetrads is $420 \div 3, =140$.

For the four arrangements respectively, these are as follows: the word "same" means same as in column I.

I.				II.				III.				IV.			
1	2	3	4	same				same				same			
		5	6												
		7	8												
		9	10												
		11	12												
		13	14												
		15	16												
1	3	5	7												
		6	8												
		9	11												
		10	12												
		13	15												
		14	16												
1	4	5	8												
		6	7												
		9	12												
		10	11												
		13	16												
		14	15												
1	5	9	13	1	5	9	14	1	5	9	15	1	5	9	16
		10	14			10	13			10	16			10	15
		11	15			11	16			11	13			11	14
		12	16			12	15			12	14			12	13
1	6	9	14	1	6	9	13	1	6	9	16	1	6	9	15
		10	13			10	14			10	15			10	16
		11	16			11	15			11	14			11	13
		12	15			12	16			12	13			12	14
1	7	9	15	1	7	9	16	1	7	9	13	1	7	9	14
		10	16			10	15			10	14			10	13
		11	13			11	14			11	15			11	16
		12	14			12	13			12	16			12	15
1	8	9	16	1	8	9	15	1	8	9	14	1	8	9	13
		10	15			10	16			10	13			10	14
		11	14			11	13			11	16			11	15
		12	13			12	14			12	15			12	16

I.	II.	III.	IV.
2 3 5 8 6 7 9 12 10 11 13 16 14 15	same	same	same
2 4 5 7 6 8 9 11 10 12 13 15 14 16	same	same	same
2 5 9 14 10 13 11 16 12 15	2 5 9 13 10 14 11 15 12 16	2 5 9 16 10 15 11 14 12 13	2 5 9 15 10 16 11 13 12 14
2 6 9 13 10 14 11 15 12 16	2 6 9 14 10 13 11 16 12 15	2 6 9 15 10 16 11 13 12 14	2 6 9 16 10 15 11 14 12 13
2 7 9 16 10 15 11 14 12 13	2 7 9 15 10 16 11 13 12 14	2 7 9 14 10 13 11 16 12 15	2 7 9 13 10 14 11 15 12 16
2 8 9 15 10 16 11 13 12 14	2 8 9 16 10 15 11 14 12 13	2 8 9 13 10 14 11 15 12 16	2 8 9 14 10 13 11 16 12 15
3 4 5 6 7 8 9 10 11 12 13 14 15 16	same	same	same

I.				II.				III.				IV.			
3	5	9	15	3	5	9	16	3	5	9	13	3	5	9	14
		10	16			10	15			10	14			10	13
		11	13			11	14			11	15			11	16
		12	14			12	13			12	16			12	15
3	6	9	16	3	6	9	15	3	6	9	14	3	6	9	13
		10	15			10	16			10	13			10	14
		11	14			11	13			11	16			11	15
		12	13			12	14			12	15			12	16
3	7	9	13	3	7	9	14	3	7	9	15	3	7	9	16
		10	14			10	13			10	16			10	15
		11	15			11	16			11	13			11	14
		12	16			12	15			12	14			12	13
3	8	9	14	3	8	9	13	3	8	9	16	3	8	9	15
		10	13			10	14			10	15			10	16
		11	16			11	15			11	14			11	13
		12	15			12	16			12	13			12	14
4	5	9	16	4	5	9	15	4	5	9	14	4	5	9	13
		10	15			10	16			10	13			10	14
		11	14			11	13			11	16			11	15
		12	13			12	14			12	15			12	16
4	6	9	15	4	6	9	16	4	6	9	13	4	6	9	14
		10	16			10	15			10	14			10	13
		11	13			11	14			11	15			11	16
		12	14			12	13			12	16			12	15
4	7	9	14	4	7	9	13	4	7	9	16	4	7	9	15
		10	13			10	14			10	15			10	16
		11	16			11	15			11	14			11	13
		12	15			12	16			12	13			12	14
4	8	9	13	4	8	9	14	4	8	9	15	4	8	9	16
		10	14			10	13			10	16			10	15
		11	15			11	16			11	13			11	14
		12	16			12	15			12	14			12	13
5	6	7	8	same				same				same			
		9	10												
		11	12												
		13	14												
		15	16												

I.				II.	III.	IV.
5	7	9	11	same	same	same
		10	12			
		13	15			
		14	16			
5	8	9	12			
		10	11			
		13	16			
		14	15			
6	7	9	12			
		10	11			
		13	16			
		14	15			
6	8	9	11			
		10	12			
		13	15			
		14	16			
7	8	9	10			
		11	12			
		13	14			
		15	16			
9	10	11	12			
		13	14			
		15	16			
9	11	13	15			
		14	16			
9	12	13	16			
		14	15			
10	11	13	16			
		14	15			
10	12	13	15			
		14	16			
11	12	13	14			
		15	16			
13	14	15	16			

As regards the signs, observe that the first line may always be written

$$ab + cd + ef + \&c.,$$

with the signs all of them +; and then writing $a, b, c, \ldots = 1, 2, 3, \ldots, 16$ respectively, the first line will be

$$1\ 2 + 3\ 4 + 5\ 6 + 7\ 8 + 9\ 10 + 11\ 12 + 13\ 14 + 15\ 16,$$

with the signs all of them +; that is, we may assume ϵ_{12}, ϵ_{34}, &c., or say

$$1\ 2,\ 3\ 4,\ 5\ 6,\ 7\ 8,\ 9\ 10,\ 11\ 12,\ 13\ 14,\ 15\ 16,$$

all of them $= +1$. And in the other lines, the signs of all the terms of any line may be reversed at pleasure, that is, we may assume ϵ_{13}, ϵ_{14}, &c., or say 1 3, 1 4, 1 5, 1 6, 1 7, 1 8, 1 9, 1 10, 1 11, 1 12, 1 13, 1 14, 1 15, 1 16, all of them $= +1$.

Making these assumptions, then for any one of the synthematic arrangements the several tetrads give as before relations between the signs; among these are included the results already obtained for the 8-question, and taking as before

$$-a = b = c = d = \theta,$$

we have the signs of the several terms belonging to the 8-question given as $= \pm 1$ or $\pm \theta$ as before. The remaining tetrads up to 1 8 12 13 then serve to express all the remaining signs in terms of the as yet undetermined signs $e, f, g, h, i, j, k, l, m, n, o, p, q, r, s, t, u, v, w, x, y, z, \alpha, \beta$, for instance

$$1\ \ 3\,.\,9\ 11 +\ \ 1\ \ e,$$

$$1\ \ 9\,.\,3\ 11 -\ \ 1\ \ e,$$

$$1\ 11\,.\,3\ \ 9 +\ \ 1\ \ e,$$

that is, $3\ 9 = e$, $3\ 11 = -e$, $9\ 11 = e$; and then the tetrads up to 2 8 9 15 serve to express these signs in terms of the undetermined signs $\lambda, \mu, \nu, \rho, \sigma, \tau$; for instance

$$2\ \ 3\,.\,9\ 12 +\ \ 1\ \ i,$$

$$2\ \ 9\,.\,3\ 12 -\ \ 1 - f,$$

$$2\ 12\,.\,3\ \ 9 + -1\ \ e,$$

that is, $-e = f = i$; and in like manner 2 3 10 11, 2 4 9 11 and 2 4 10 12 give respectively $-e = f = j$, $-e = i = j$, $f = i = j$; that is, we have $-e = f = i = j$, $= \lambda$ suppose. And in this way we have, for each of the four synthematic arrangements the signs of all the terms expressed in terms of the undetermined signs $\theta, \lambda, \mu, \nu, \rho, \sigma, \tau$, as shown in the following table; where observe that the results apply to the four synthematic arrangements *separately*, viz. the e, f, g, &c., and the $\theta, \lambda, \mu, \nu, \rho, \sigma, \tau$ in each column are altogether independent of the like symbols in the other three columns.

Signs for the four synthematic arrangements:

I.				II.	III.	IV.
1	2	+ 1		same	same	same
	3	+ 1				
	4	+ 1				
	5	+ 1				
	6	+ 1				
	7	+ 1				
	8	+ 1				
	9	+ 1				
	10	+ 1				
	11	+ 1				
	12	+ 1				
	13	+ 1				
	14	+ 1				
	15	+ 1				
	16	+ 1				
2	3	+ 1		same	same	same
	4	− 1				
	5	+ 1				
	6	− 1				
	7	+ 1				
	8	− 1				
	9	+ 1				
	10	− 1				
	11	+ 1				
	12	− 1				
	13	+ 1				
	14	− 1				
	15	+ 1				
	16	− 1				
3	4	+ 1		same	same	same
	5	$-\theta$				
	6	θ				
	7	θ				
	8	$-\theta$				
	9	e	$-\lambda$			
	10	f	λ			
	11	$-e$	λ			
	12	$-f$	$-\lambda$			
	13	g	$-\mu$			
	14	h	μ			
	15	$-g$	μ			
	16	$-h$	$-\mu$			

		I.		II.		III.		IV.	
4	5	θ		same		same		same	
	6	θ							
	7	$-\theta$							
	8	$-\theta$							
	9	i	λ						
	10	j	λ						
	11	$-j$	$-\lambda$						
	12	$-i$	$-\lambda$						
	13	k	μ						
	14	l	μ						
	15	$-l$	$-\mu$						
	16	$-k$	$-\mu$						
5	6	$+1$		$+1$		$+1$		1	
	7	$-\theta$		$-\theta$		$-\theta$		$-\theta$	
	8	θ		θ		θ		θ	
	9	m	$-\nu$	m	ν	m	$-\nu$	m	ν
	10	n	ν	n	ν	n	ν	n	ν
	11	o	$-\rho$	o	ρ	o	$-\rho$	o	ρ
	12	p	ρ	p	ρ	p	ρ	p	ρ
	13	$-m$	ν	$-n$	$-\nu$	$-o$	ρ	$-p$	$-\rho$
	14	$-n$	$-\nu$	$-m$	$-\nu$	$-p$	$-\rho$	$-o$	$-\rho$
	15	$-o$	ρ	$-p$	$-\rho$	$-m$	ν	$-n$	$-\nu$
	16	$-p$	$-\rho$	$-o$	$-\rho$	$-n$	$-\nu$	$-m$	$-\nu$
6	7	θ		θ		θ		θ	
	8	θ		θ		θ		θ	
	9	q	ν	q	ν	q	ν	q	ν
	10	r	ν	r	$-\nu$	r	ν	r	$-\nu$
	11	s	ρ	s	ρ	s	ρ	s	ρ
	12	t	ρ	t	$-\rho$	t	ρ	t	$-\rho$
	13	$-r$	$-\nu$	$-q$	$-\nu$	$-t$	$-\rho$	$-s$	$-\rho$
	14	$-q$	$-\nu$	$-r$	ν	$-s$	$-\rho$	$-t$	ρ
	15	$-t$	$-\rho$	$-s$	$-\rho$	$-r$	$-\nu$	$-q$	$-\nu$
	16	$-s$	$-\rho$	$-t$	ρ	$-q$	$-\nu$	$-r$	ν
7	8	$+1$		$+1$		$+1$		$+1$	
	9	u	$-\sigma$	u	$-\sigma$	u	$-\sigma$	u	σ
	10	v	σ	v	σ	v	σ	v	σ
	11	w	$-\tau$	w	τ	w	$-\tau$	w	τ
	12	x	τ	x	τ	x	τ	x	τ
	13	$-w$	τ	$-x$	$-\tau$	$-u$	σ	$-v$	$-\sigma$
	14	$-x$	$-\tau$	$-w$	$-\tau$	$-v$	$-\sigma$	$-u$	$-\sigma$
	15	$-u$	σ	$-v$	$-\sigma$	$-w$	τ	$-x$	$-\tau$
	16	$-v$	$-\sigma$	$-u$	$-\sigma$	$-x$	$-\tau$	$-w$	$-\tau$

		I.		II.		III.		IV.	
8	9	y	σ	y	σ	y	σ	y	σ
	10	z	σ	z	$-\sigma$	z	σ	z	$-\sigma$
	11	a	τ	a	τ	a	τ	a	τ
	12	β	τ	β	$-\tau$	β	τ	β	$-\tau$
	13	$-\beta$	$-\tau$	$-a$	$-\tau$	$-z$	$-\sigma$	$-y$	$-\sigma$
	14	$-a$	$-\tau$	$-\beta$	τ	$-y$	$-\sigma$	$-z$	σ
	15	$-z$	$-\sigma$	$-y$	$-\sigma$	$-\beta$	$-\tau$	$-a$	$-\tau$
	16	$-y$	$-\sigma$	$-z$	σ	$-a$	$-\tau$	$-\beta$	τ
9	10	$+1$		$+1$		$+1$		$+1$	
	11	e	$-\lambda$	e	$-\lambda$	e	$-\lambda$	e	$-\lambda$
	12	i	λ	i	λ	i	λ	i	λ
	13	m	$-\nu$	q	ν	u	$-\sigma$	y	σ
	14	q	ν	m	ν	y	σ	u	σ
	15	u	$-\sigma$	y	σ	m	$-\nu$	q	ν
	16	y	σ	u	σ	q	ν	m	ν
10	11	j	λ	j	λ	j	λ	j	λ
	12	f	λ	f	λ	f	λ	f	λ
	13	r	ν	n	ν	z	σ	v	σ
	14	n	ν	r	$-\nu$	v	σ	z	$-\sigma$
	15	z	σ	v	σ	r	ν	n	ν
	16	v	σ	z	$-\sigma$	n	ν	r	$-\nu$
11	12	$+1$		$+1$		1		$+1$	
	13	w	$-\tau$	a	τ	o	$-\rho$	s	ρ
	14	a	τ	w	τ	s	ρ	o	ρ
	15	o	$-\rho$	s	ρ	w	$-\tau$	a	τ
	16	s	ρ	o	ρ	a	τ	w	τ
12	13	β	τ	x	τ	t	ρ	p	ρ
	14	x	τ	β	$-\tau$	p	ρ	t	$-\rho$
	15	t	ρ	p	ρ	β	τ	x	τ
	16	p	ρ	t	$-\rho$	x	τ	β	$-\tau$
13	14	$+1$		$+1$		$+1$		$+1$	
	15	g	$-\mu$	g	$-\mu$	g	$-\mu$	g	$-\mu$
	16	k	μ	k	μ	k	μ	k	μ
14	15	l	μ	l	μ	l	μ	l	μ
	16	h	μ	h	μ	h	μ	h	μ
15	16	$+1$		$+1$		$+1$		$+1$	

We have now for the four arrangements respectively, by means of hitherto unused tetrads, the following determinations of sign: these being in each case inconsistent with each other.

First arrangement.

3	5	9	15	$+-\theta.-\sigma$	that is,
3	9	5	15	$--\lambda.\ \rho$	$\theta\sigma = \lambda\rho = -\mu\nu$,
3	15	5	9	$+\ \mu.-\nu$	
3	5	10	16	$+-\theta.\ \sigma$	
3	10	5	16	$-\ \lambda.-\rho$	$-\theta\sigma = \lambda\rho = -\mu\nu$,
3	16	5	10	$+-\mu.\ \nu$	
3	5	11	13	$+-\theta.-\tau$	
3	11	5	13	$-\ \lambda.\ \nu$	$\theta\tau = -\lambda\nu = \mu\rho$,
3	13	5	11	$+-\mu.-\rho$	
3	5	12	14	$+-\theta.\ \tau$	
3	12	5	14	$--\lambda.-\nu$	$-\theta\tau = -\lambda\nu = \mu\rho$.
3	14	5	12	$+\ \mu.\ \rho$	

Second arrangement.

3	5	9	16	$+-\theta.\ \sigma$	
3	9	5	16	$--\lambda.-\rho$	$\theta\sigma = \lambda\rho = \mu\nu$,
3	16	5	9	$+-\mu.\ \nu$	
3	5	10	15	$+-\theta.\ \sigma$	
3	10	5	15	$-\ \lambda.-\rho$	$-\theta\sigma = \lambda\rho = \mu\nu$,
3	15	5	10	$+\ \mu.\ \nu$	
3	5	11	14	$+-\theta.\ \tau$	
3	11	5	14	$-\ \lambda.-\nu$	$-\theta\tau = \lambda\nu = \mu\rho$,
3	14	5	11	$+\ \mu.\ \rho$	
3	5	12	13	$+-\theta.\ \tau$	
3	12	5	13	$--\lambda.-\nu$	$\theta\tau = \lambda\nu = \mu\rho$.
3	13	5	12	$+-\mu.\ \rho$	

Third arrangement.

3	5	9	13	$+-\theta.-\sigma$	
3	9	5	13	$--\lambda.\ \rho$	$\theta\sigma=-\lambda\rho=\ \mu\nu,$
3	13	5	9	$+-\mu.-\nu$	
3	5	10	14	$+-\theta.\ \sigma$	
3	10	5	14	$-\ \lambda.-\rho$	$\theta\sigma=-\lambda\rho=-\mu\nu,$
3	14	5	10	$+\ \mu.\ \nu$	
3	5	11	15	$+-\theta.-\tau$	
3	11	5	15	$-\ \lambda.\ \nu$	$\theta\tau=-\lambda\nu=-\mu\rho,$
3	15	5	11	$+\ \mu.-\rho$	
3	5	12	16	$+-\theta.\ \tau$	
3	12	5	16	$--\lambda.-\nu$	$\theta\tau=-\lambda\nu=\ \mu\rho.$
3	16	5	12	$+-\mu.\ \rho$	

Fourth arrangement.

3	5	9	14	$+-\theta.\ \sigma$	
3	9	5	14	$--\lambda.-\rho$	$\theta\sigma=\ \lambda\rho=-\mu\nu,$
3	14	5	9	$+\ \mu.\ \nu$	
3	5	10	13	$+-\theta.\ \sigma$	
3	10	5	13	$-\ \lambda.-\rho$	$\theta\sigma=-\lambda\rho=\ \mu\nu,$
3	13	5	10	$+-\mu.\ \nu$	
3	5	11	16	$+-\theta.\ \tau$	
3	11	5	16	$-\ \lambda.-\nu$	$\theta\tau=-\lambda\nu=\ \mu\rho,$
3	16	5	11	$+-\mu.\ \rho$	
3	5	12	15	$+-\theta.-\tau$	
3	12	5	15	$--\lambda.-\nu$	$\theta\tau=-\lambda\nu=-\mu\rho.$
3	15	5	12	$+-\mu.\ \rho$	

And it hence finally appears, that we cannot, in any one of the four arrangements, determine the signs so as to give rise to a 16-square theorem; that is, the product of a sum of 16 squares into a sum of 16 squares cannot be made equal to a sum of 16 squares.

764.

THE BINOMIAL EQUATION $x^p-1=0$: QUINQUISECTION.

[From the *Proceedings of the London Mathematical Society*, vol. XII. (1881), pp. 15, 16. Read December 9, 1880.]

THE theory should be precisely analogous to those for the trisection and quartisection (see my paper, "The Binomial Equation $x^p-1=0$, Trisection and Quartisection," *Proceedings of the London Mathematical Society*, vol. XI. (1879), pp. 4—17, [731]), only I have not been able to carry it so far. We have in the present case five periods X, Y, Z, W, T, the actual expressions for which, $X=\eta^1+\ldots$, $Y=\eta^g+\ldots$, etc., with Reuschle's selected prime root g, can be (for the primes $5n+1$ under 100) at once written down by means of the table given, pp. 16, 17, of that paper; [see this volume, pp. 95, 96]. The relations between the periods are of the form

	X	Y	Z	W	T
$X^2=$	a	b	c	d	e
$XY=$	f	g	h	i	j
$XZ=$	k	l	m	n	o;

that is, we have

$$X^2=(a,\ b,\ c,\ d,\ e\between X,\ Y,\ Z,\ W,\ T),$$

and thence, by cyclical permutations,

$$Y^2=(e,\ a,\ b,\ c,\ d\between \quad ,, \quad), \text{ etc.};$$

viz. from the value of X^2 we have those of Y^2, Z^2, W^2, T^2; from the value of XY those of YZ, ZW, WT, TX; and from the value of XZ those of YW, ZT, WX, TY.

From the equation $X + Y + Z + W + T = -1$, multiplying by X and then substituting for X^2, XY, &c., their values, we obtain

$$\begin{aligned}
-a &= 1 + f + k + m + g,\\
-b &= g + l + n + h,\\
-c &= h + m + o + i,\\
-d &= i + n + k + j,\\
-e &= j + o + l + f,
\end{aligned}$$

which determine (a, b, c, d, e) in terms of (f, g, h, i, j) and (k, l, m, n, o). It is, moreover, easy to prove that

$$\begin{aligned}
f + g + h + i + j &= \tfrac{1}{5}(p-1),\\
k + l + m + n + o &= \tfrac{1}{5}(p-1),
\end{aligned}$$

whence also

$$a + b + c + d + e = -1 - \tfrac{4}{5}(p-1).$$

We obtain other relations between the coefficients by considering the two triple products XYZ and XYW: these are all that need be considered, since the other triple products are deducible from them by cyclical permutations. From the first of these we have

$$X.YZ = Y.XZ = Z.XY,$$

and from the second

$$X.YW = Y.XW = W.XY;$$

and if we herein substitute for YZ, XZ, &c., their values, and then in the resulting equations for X^2, XY, &c., their values as linear functions of X, Y, Z, W, T, we obtain in all $5.2.2 = 20$ quadric relations between the 15 coefficients; or if we substitute for (a, b, c, d, e) their foregoing values, in all 20 relations between the 10 coefficients (f, g, h, i, j) and (k, l, m, n, o). These are at most equivalent to 8 independent equations, since we have, besides, the sums $f+g+h+i+j$ and $k+l+m+n+o$ each $= \frac{1}{5}(p-1)$; but I have not succeeded in finding the connexions between them, or even in ascertaining to how many independent equations they are equivalent.

For any given prime $p = 5n + 1$, the values of the coefficients, and also the coefficients of the quintic equation for the periods, could of course be calculated directly from the expressions of the periods; but for the primes under 100, that is, for the values 11, 31, 41, 61, 71, they are at once obtained from Reuschle. We have thus the two Tables, the former giving the coefficients $a, b, \ldots, n, o$, and the latter the coefficients of the quintic equations.

TABLE 1.

p	a f k	b g l	c h m	d i n	e j o
11	− 2	− 1	− 2	− 2	− 2
	1	0	0	1	0
	0	0	0	1	1
31	− 4	− 6	− 6	− 4	− 5
	0	1	2	1	2
	0	2	2	1	1
41	− 8	− 5	− 6	− 6	− 8
	3	0	2	1	2
	2	2	2	1	1
61	− 10	− 9	− 12	− 8	− 10
	3	2	2	3	2
	0	2	4	3	3
71	− 14	− 10	− 12	− 9	− 12
	4	2	3	2	3
	2	3	5	2	2

TABLE 2 OF THE QUINTIC EQUATIONS.

COEFFICIENTS OF

p	η^5	η^4	η^3	η^2	η^1	1	= 0.
11	1	1	− 4	− 3	+ 3	+ 1	
31	1	1	− 12	− 2	+ 1	+ 5	
41	1	1	− 16	+ 5	+ 21	− 9	
61	1	1	− 24	− 17	+ 41	− 23	
71	1	1	− 28	+ 37	+ 25	+ 1	

765.

ON THE FLEXURE AND EQUILIBRIUM OF A SKEW SURFACE.

[From the *Proceedings of the London Mathematical Society*, vol. XII. (1881), pp. 103—108. Read March 10, 1881.]

THE skew surface is taken to be such that the strip between two consecutive generating lines is rigid, and that the flexure takes place by the rotation of the strips about the generating lines successively. The theory of the flexure is well known, but I am not aware that the theory of the equilibrium of such a surface, when acted upon by any given forces, has been considered; it is, however, a question which presents itself naturally in connexion with those relating to other continuous bodies treated of in the *Mécanique Analytique*, and forms a good example of the principles made use of.

To begin with the mechanical theory: we may regard the forces as acting on the generating lines regarded as material lines; and if for an element of mass dm, coordinates (x, y, z) of a particular generating line G, the forces parallel to the axes are X', Y', Z', then the corresponding term in the equation of equilibrium is

$$S(X'\delta x + Y'\delta y + Z'\delta z)\,dm\,;$$

and observing that there are (as will afterwards appear) *five* geometrical conditions, which I represent by $U_1 = 0$, $U_2 = 0$, ..., $U_5 = 0$, the equation of equilibrium is

$$S\{(X'\delta x + Y'\delta y + Z'\delta z)\,dm + T_1\delta U_1 + T_2\delta U_2 + T_3\delta U_3 + T_4\delta U_4 + T_5\delta U_5\} = 0,$$

where T_1, T_2, ..., T_5 are the indeterminate multipliers, representing colligation-forces which correspond to the five geometrical conditions respectively.

Taking (ξ, η, ζ) for the coordinates of a particular point P on the generating line; p, q, r for the cos-inclinations of the line (whence $U_1 = p^2 + q^2 + r^2 - 1 = 0$ is one of the geometrical relations), and ρ for the distance of dm from P, we have

$$x,\ y,\ z = \xi + \rho p,\ \eta + \rho q,\ \zeta + \rho r,$$

$$\delta x,\ \delta y,\ \delta z = \delta\xi + \rho\delta p,\ \delta\eta + \rho\delta q,\ \delta\zeta + \rho\delta r.$$

The summation S extends first to the different points of the generating line, and then to the different generating lines; applying it first to the particular generating line, we write

$$\begin{array}{cccccc} SX'dm, & SY'dm, & SZ'dm, & SX'\rho dm, & SY'\rho dm, & SZ'\rho dm \\ = X, & Y, & Z, & L, & M, & N, \end{array}$$

where X, Y, Z are the whole forces, and L, M, N the whole moments about the point P, for the generating line G; retaining the same summatory symbol S, as now referring to the different generating lines, the equation becomes

$$S\{X\delta\xi + Y\delta\eta + Z\delta\zeta + L\delta p + M\delta q + N\delta r + T_1\delta U_1 + \ldots + T_5\delta U_5\} = 0.$$

We have now to consider the geometrical theory of the flexure. Taking on the skew surface an arbitrary curve cutting each generating line G in a point P, coordinates (ξ, η, ζ), and taking σ for the distance along the curve of the point P from a fixed point of the curve; also p, q, r, as before, for the cos-inclinations of the generating line G, then when the surface is in a determinate state, ξ, η, ζ, p, q, r are given functions of σ; but these functions vary with the flexure of the surface, with, however, certain relations unaffected by the flexure; and the problem is to find first these relations. As already mentioned, one of them is $p^2 + q^2 + r^2 - 1 = 0$.

Taking P' as the consecutive point on the curve, so that the direction of the element PP' is that of the tangent PT at P, it is convenient to write l, m, n for the cosine-inclinations of the tangent; we have, it is clear,

$$l, m, n = \frac{d\xi}{d\sigma}, \frac{d\eta}{d\sigma}, \frac{d\zeta}{d\sigma}; \quad l^2 + m^2 + n^2 - 1 = 0.$$

The conditions in order to the rigidity of the strip, are that the angles GPP', $G'P'P$ $(= 180° - G'P'T)$, and the inclination $G'P'$ to GP, shall have given values,

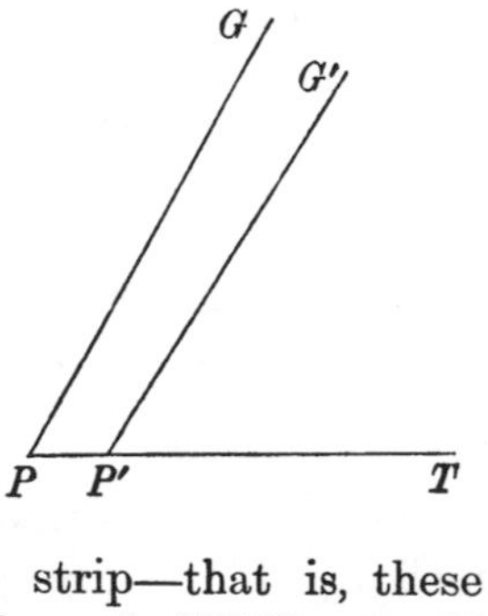

variable it may be from strip to strip—that is, these values must be given functions of σ. Taking $\angle GPT = I$, the value of $G'P'T$ can differ only infinitesimally from that of GPT, and we take it to be $G'P'T = I - \Omega d\sigma$; also the inclination GP to $G'P'$ is an infinitesimal, $= \Theta d\sigma$: we have I, Ω, Θ given functions of σ. It is to be remarked that these conditions imply, inclination of $G'P'$ to tangent plane GPT at P has a given value $\Lambda d\sigma$; in fact, if through P we draw a line $P\gamma$ parallel to $P'G'$, then, if P is regarded as the centre of a sphere which meets PG, $P\gamma$, PT in the points g, g', t respectively, we have a spherical triangle $gg't$, the sides of which are $I - \Omega d\sigma$, I, and $\Theta d\sigma$, and of which the perpendicular $g'm$ is $= \Lambda d\sigma$; we have thus an infinitesimal right-angled triangle, the base and altitude of which are $\Omega d\sigma$, $\Lambda d\sigma$,

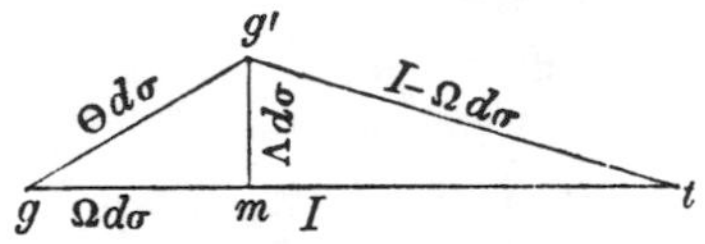

and the hypothenuse is $\Theta d\sigma$; whence $\Theta^2 = \Omega^2 + \Lambda^2$. In the case of the developable surface $\Lambda = 0$ and $\Theta = \Omega$. It may be remarked that, when the curve on the skew surface is the line of striction, we have $\Omega = 0$; in fact, taking P to be on the line of striction, the line

$$\frac{X-\xi}{qr'-q'r} = \frac{Y-\eta}{rp'-r'p} = \frac{Z-\zeta}{pq'-p'q},$$

through (ξ, η, ζ) at right angles to the two generating lines, meets the consecutive generating line $X, Y, Z = \xi' + \rho p', \eta' + \rho q', \zeta' + \rho r'$; and the condition that this may be so is easily found to be $\Omega = 0$.

Take, for a moment, p', q', r' for the cos-inclinations of the consecutive generating line $P'G'$; we have

$$lp + mq + nr = \cos I,$$
$$lp' + mq' + nr' = \cos (I - \Omega d\sigma),$$
$$pp' + qq' + rr' = \cos \Theta d\sigma;$$

and then writing $p', q', r' = p + dp, q + dq, r + dr$, and observing that the equation $p'^2 + q'^2 + r'^2 = 1$ gives

$$pdp + qdq + rdr = -\tfrac{1}{2}(dp^2 + dq^2 + dr^2),$$

these equations and the before-mentioned two equations become

$$(U_1)\quad p^2 + q^2 + r^2 - 1 = 0,$$
$$(U_2)\quad l^2 + m^2 + n^2 - 1 = 0,$$
$$(U_3)\quad lp + mq + nr - \cos I = 0,$$
$$(U_4)\quad ldp + mdq + ndr - \Omega \sin I d\sigma = 0,$$
$$(U_5)\quad dp^2 + dq^2 + dr^2 - \Theta^2 d\sigma^2 = 0,$$

which equations, considering therein l, m, n as standing for their values $\frac{d\xi}{d\sigma}, \frac{d\eta}{d\sigma}, \frac{d\zeta}{d\sigma}$, are the geometrical relations which connect the six variables $\xi, \eta, \zeta, p, q, r$, considered as functions of σ. And in these equations I, Ω, Θ denote given functions of σ, invariable by any flexure of the surface.

To complete the geometrical theory, it is to be observed that we can by flexure bring the generating lines of the surface to be parallel to those of any given cone

$C(p, q, r) = 0$, where $C(p, q, r)$ denotes a homogeneous function of (p, q, r). Hence, joining to the foregoing five equations this new equation

$$C(p, q, r) = 0,$$

these six equations determine $\xi, \eta, \zeta, p, q, r$ as functions of σ. To make the solution completely determinate, we have only to assume for the point P, which corresponds, say, to the value $\sigma = 0$, a position in space at pleasure, and to take the corresponding generating line PG parallel to a generating line, at pleasure, of the cone.

As an example, writing γ to denote an arbitrary constant angle, if the invariable conditions are

$$I = \gamma, \quad \Theta = \sin\gamma, \quad \Omega = 0,$$

then the five equations are

$$\begin{aligned} p^2 + q^2 + r^2 - 1 &= 0, \\ l^2 + m^2 + n^2 - 1 &= 0, \\ lp + mq + nr - \cos\gamma &= 0, \\ dp^2 + dq^2 + dr^2 - \sin^2\gamma\, d\sigma^2 &= 0, \\ ldp + mdq + ndr &= 0. \end{aligned}$$

We assume *first*

$$C(p, q, r) = p^2 + q^2 - r^2 \tan^2\gamma, \ = 0;$$

and *secondly*

$$C(p, q, r) = r, \ = 0.$$

Then, in the former case, we find the solution

$$\begin{aligned} p, q, r &= -\sin\gamma\sin\sigma, \ \sin\gamma\cos\sigma, \ \cos\gamma; \\ \xi, \eta, \zeta &= \cos\sigma, \ \sin\sigma, \ 0; \end{aligned}$$

giving

$$x, y, z = \cos\sigma - \rho\sin\gamma\sin\sigma, \ \sin\sigma + \rho\sin\gamma\cos\sigma, \ \cos\gamma;$$

and consequently

$$x^2 + y^2 - z^2\tan^2\gamma = 0,$$

the hyperboloid of revolution. And, in the latter case,

$$\begin{aligned} p, q, r &= \cos(\sigma\sin\gamma), \ \sin(\sigma\sin\gamma), \ 0, \\ \xi, \eta, \zeta &= \cot\gamma\sin(\sigma\sin\gamma), \ -\cot\gamma\cos(\sigma\sin\gamma), \ \sigma\sin\gamma, \end{aligned}$$

that is,

$$x, y = \cot\gamma\sin z + \rho\cos z, \ -\cot\gamma\cos z + \rho\sin z,$$

whence

$$x\sin z - y\cos z = \cot\gamma,$$

a skew helicoid generated by horizontal tangents of the cylinder $x^2 + y^2 = \cot^2\gamma$. This is a known deformation of the hyperboloid.

Returning now to the mechanical problem, we have to consider the terms

$$\begin{aligned} S\ .\ & T_1\delta\tfrac{1}{2}(p^2+q^2+r^2-1) \\ &+T_2\delta\tfrac{1}{2}(l^2+m^2+n^2-1) \\ &+T_3\delta(lp+mq+nr-\cos I) \\ &+T_4\delta\left(l\frac{dp}{d\sigma}+m\frac{dq}{d\sigma}+n\frac{dr}{d\sigma}\Omega\sin I\right) \\ &+T_5\delta\tfrac{1}{4}\left\{\left(\frac{dp}{d\sigma}\right)^2+\left(\frac{dq}{d\sigma}\right)^2+\left(\frac{dr}{d\sigma}\right)^2-\Theta^2\right\}. \end{aligned}$$

The first term gives, under the sign S,

$$T_1(p\delta p+q\delta q+r\delta r). \qquad (*)$$

The second term gives, in the first instance,

$$\frac{T_2}{d\sigma}(ld\delta\xi+md\delta\eta+nd\delta\zeta);$$

or, since in general

$$S\Omega d\delta\xi=\Omega''\delta\xi''-\Omega'\delta\xi'+S(-d\Omega\,.\,\delta\xi),$$

then, attending only to the terms under the sign S, these are

$$=-\frac{d}{d\sigma}T_2l\,.\,\delta\xi-\frac{d}{d\sigma}T_2m\,.\,\delta\eta-\frac{d}{d\sigma}T_2n\,.\,\delta\zeta. \qquad (*)$$

The third term gives

$$\begin{aligned} &T_3(l\delta p+m\delta q+n\delta r) \qquad (*) \\ +\ &T_3(p\delta l+q\delta m+r\delta n), \end{aligned}$$

where the second line,

$$=\frac{T_3}{d\sigma}(pd\delta\xi+qd\delta\eta+rd\delta\zeta),$$

attending only to the terms under the sign S, gives

$$-\frac{d}{d\sigma}T_3p\,.\,\delta\xi-\frac{d}{d\sigma}T_3q\,.\,\delta\eta-\frac{d}{d\sigma}T_3r\,.\,\delta\zeta. \qquad (*)$$

The fourth term gives

$$\begin{aligned} &T_4\left(\frac{dp}{d\sigma}\delta l+\frac{dq}{d\sigma}\delta m+\frac{dr}{d\sigma}\delta n\right) \\ +\ &T_4\left(\frac{l}{d\sigma}d\delta p+\frac{m}{d\sigma}d\delta q+\frac{n}{d\sigma}d\delta r\right), \end{aligned}$$

where the first line, written under the form

$$\frac{T_4}{d\sigma}\left(\frac{dp}{d\sigma}d\delta\xi+\frac{dq}{d\sigma}d\delta\eta+\frac{dr}{d\sigma}d\delta\zeta\right),$$

and attending only to the terms under the sign S, gives

$$-\frac{d}{d\sigma}\left(T_4\frac{dp}{d\sigma}\right).\,\delta\xi-\frac{d}{d\sigma}\left(T_4\frac{dq}{d\sigma}\right).\,\delta\eta-\frac{d}{d\sigma}\left(T_4\frac{dr}{d\sigma}\right).\,\delta\zeta; \qquad (*)$$

and the second line, attending in like manner only to the terms under the sign S, gives

$$-\frac{d}{d\sigma}T_4 l\,.\,\delta p - \frac{d}{d\sigma}T_4 m\,.\,\delta q - \frac{d}{d\sigma}T_4 n\,.\,\delta r. \qquad (*)$$

The fifth term, written under the form

$$\frac{T_5}{d\sigma}\left(\frac{dp}{d\sigma}d\delta p + \frac{dq}{d\sigma}d\delta q + \frac{dr}{d\sigma}d\delta r\right), \qquad (*)$$

and attending only to the terms under the sign S, gives

$$-\frac{d}{d\sigma}T_5\frac{dp}{d\sigma}\,.\,\delta p - \frac{d}{d\sigma}T_5\frac{dq}{d\sigma}\,.\,\delta q - \frac{d}{d\sigma}T_5\frac{dr}{d\sigma}\,.\,\delta r; \qquad (*)$$

where in each case I have marked with an asterisk the lines which present themselves in the final result.

Hence, joining to the foregoing the force-terms

$$X\delta\xi + Y\delta\eta + Z\delta\zeta + L\delta p + M\delta q + N\delta r, \qquad (*)$$

and equating to zero the coefficients of $\delta\xi$, $\delta\eta$, $\delta\zeta$, δp, δq, δr respectively, we have

$$\left\{\begin{array}{lllll}
0 = X & -\frac{d}{d\sigma}T_2 l & -\frac{d}{d\sigma}T_3 p & -\frac{d}{d\sigma}T_4\frac{dp}{d\sigma}, & \\
0 = Y & -\frac{d}{d\sigma}T_2 m & -\frac{d}{d\sigma}T_3 q & -\frac{d}{d\sigma}T_4\frac{dq}{d\sigma}, & \\
0 = Z & -\frac{d}{d\sigma}T_2 n & -\frac{d}{d\sigma}T_3 r & -\frac{d}{d\sigma}T_4\frac{dr}{d\sigma}, & \\
0 = L + T_1 p & & + \quad T_3 l & -\frac{d}{d\sigma}T_4 l & -\frac{d}{d\sigma}T_5\frac{dp}{d\sigma}, \\
0 = M + T_1 q & & + \quad T_3 m & -\frac{d}{d\sigma}T_4 m & -\frac{d}{d\sigma}T_5\frac{dq}{d\sigma}, \\
0 = N + T_1 r & & + \quad T_3 n & -\frac{d}{d\sigma}T_4 n & -\frac{d}{d\sigma}T_5\frac{dr}{d\sigma},
\end{array}\right.$$

where it will be recollected that l, m, n stand for $\frac{d\xi}{d\sigma}$, $\frac{d\eta}{d\sigma}$, $\frac{d\zeta}{d\sigma}$, the variables being ξ, η, ζ, p, q, r, and σ. The elimination of T_1, T_2, ..., T_5 from the six equations should lead to a relation between ξ, η, ζ, p, q, r, which, with the foregoing five relations, would determine the six variables ξ, η, ζ, p, q, r in terms of σ.

In particular, the forces and moments X, Y, Z, L, M, N may all of them vanish; assuming that T_1, T_2, ..., T_5 do not all of them vanish, we still have the sixth relation, which (with the foregoing five relations) determines ξ, η, ζ, p, q, r in terms of σ; and it is to be remarked that the problem in question, of the figure of equilibrium of the skew surface not acted upon by any forces, is analogous to that of the geodesic line in space; only whilst here the solution is, curve a straight line, the solution for the case of the skew surface depends upon equations of a complex enough form; in the case of the developable surface, the required figure is of course the plane.

766.

ON THE GEODESIC CURVATURE OF A CURVE ON A SURFACE.

[From the *Proceedings of the London Mathematical Society*, vol. XII. (1881), pp. 110—117. Read April 14, 1881.]

THERE is contained in Liouville's Note II. to his edition of Monge's *Application de l'Analyse à la Géométrie* (Paris, 1850), see pp. 574 and 575, the following formula,

$$\frac{1}{\rho} = -\frac{di}{ds} + \frac{1}{2G\sqrt{E}}\frac{dG}{du}\cos i - \frac{1}{2E\sqrt{G}}\frac{dE}{dv}\sin i,$$

$$= -\frac{di}{ds} + \frac{\cos i}{\rho_2} + \frac{\sin i}{\rho_1},$$

which gives the radius of geodesic curvature of a curve upon a surface when the position of a point on the surface is defined by the parameters u, v, belonging to a system of orthotomic curves; or, what is the same thing, such that

$$ds^2 = Edu^2 + Gdv^2.$$

Writing with Gauss p, q instead of u, v, I propose to obtain the corresponding formula in the general case where the parameters p, q are such that

$$ds^2 = Edp^2 + 2Fdp\,dq + Gdq^2.$$

I call to mind that, if PQ, PQ' are equal infinitesimal arcs on the given curve and on its tangent geodesic, then the radius of geodesic curvature ρ is, by definition, a length ρ such that $2\rho \,.\, QQ' = \overline{PQ}^2$. More generally, if the curves on the surface are any two curves which touch each other, then ρ as thus determined is the radius of relative curvature of the two curves.

The notation is that of the Memoir, "Disquisitiones generales circa superficies curvas" (1827), Gauss, *Werke*, t. III.; see also my paper "On geodesic lines, in particular those of a quadric surface," *Proc. Lond. Math. Society*, t. IV. (1872), pp. 191—211, [508]; and Salmon's *Solid Geometry*, 3rd ed., 1874, pp. 251 *et seq.* The coordinates (x, y, z) of a point on the surface are taken to be functions of two independent parameters p, q; and we then write

$$dx + \tfrac{1}{2}d^2x = a\,dp + a'dq + \tfrac{1}{2}\,(\alpha\,dp^2 + 2\alpha'\,dp\,dq + \alpha''\,dq^2),$$

$$dy + \tfrac{1}{2}d^2y = b\,dp + b'dq + \tfrac{1}{2}\,(\beta\,dp^2 + 2\beta'\,dp\,dq + \beta''\,dq^2),$$

$$dz + \tfrac{1}{2}d^2z = c\,dp + c'dq + \tfrac{1}{2}\,(\gamma\,dp^2 + 2\gamma'\,dp\,dq + \gamma''\,dq^2):$$

$$E,\ F,\ G = a^2+b^2+c^2,\quad aa'+bb'+cc',\quad a'^2+b'^2+c'^2;\quad V^2 = EG-F^2;$$

and therefore

$$ds^2 = Edp^2 + 2Fdp\,dq + Gdq^2,$$

where E, F, G are regarded as given functions of p and q.

To determine a curve on the surface, we establish a relation between the two parameters p, q, or, what is the same thing, take p, q to be functions of a single parameter θ; and we write as usual p', p'', q', etc., to denote the differential coefficients of p, q, etc., in regard to θ; we write also E_1, E_2, etc., to denote the differential coefficients $\dfrac{dE}{dp}$, $\dfrac{dE}{dq}$, etc. In the first instance, θ is taken to be an arbitrary parameter, but we afterwards take it to be the length s of the curve from a fixed point thereof.

First formula for the radius of relative curvature.

Consider any two curves touching at the point P, coordinates (x, y, z) which are regarded as given functions of (p, q); where (p, q) are for the one curve given functions, and for the other curve other given functions, of θ.

The coordinates of a consecutive point for the one curve are then

$$x + dx + \tfrac{1}{2}d^2x,\quad y + dy + \tfrac{1}{2}d^2y,\quad z + dz + \tfrac{1}{2}d^2z,$$

where

$$dp = p'd\theta + \tfrac{1}{2}p''d\theta^2,\quad dq = q'd\theta + \tfrac{1}{2}q''d\theta^2;$$

hence these coordinates are

$$x + (ap' + a'q')\,d\theta + \tfrac{1}{2}\,(\alpha p'^2 + 2\alpha'p'q' + \alpha''q'^2)\,d\theta^2 + \tfrac{1}{2}\,(ap'' + a'q'')\,d\theta^2,$$

$$\vdots$$

and for the other curve they are in like manner

$$x + (ap' + a'q')\,d\theta + \tfrac{1}{2}\,(\alpha p'^2 + 2\alpha'p'q' + \alpha''q'^2)\,d\theta^2 + \tfrac{1}{2}\,(aP'' + a'Q'')\,d\theta^2,$$

$$\vdots$$

the only difference being in the terms which contain the second differential coefficients, p'', q'' for the first curve, and P'', Q'' for the second curve. Hence the differences of the coordinates are

$$\tfrac{1}{2}\{a(p''-P'')+a'(q''-Q'')\}\,d\theta^2,\quad \tfrac{1}{2}\{b(p''-P'')+b'(q''-Q'')\}\,d\theta^2,$$
$$\tfrac{1}{2}\{c(p''-P'')+c'(q''-Q'')\}\,d\theta^2,$$

and consequently the distance QQ' of the two consecutive points Q, Q' is

$$=\tfrac{1}{2}\sqrt{(E,\ F,\ G\between p''-P'',\ q''-Q'')^2}\,d\theta^2.$$

The squared arc $\overline{PQ}^2$ is

$$=(E,\ F,\ G\between p',\ q')^2\,d\theta^2;$$

and hence, if as before $2\rho\,.\,QQ'=\overline{PQ}^2$, that is, $\dfrac{1}{\rho}=2QQ'\div\overline{PQ}^2$, then

$$\frac{1}{\rho}=\frac{\sqrt{(E,\ F,\ G\between p''-P'',\ q''-Q'')^2}}{(E,\ F,\ G\between p',\ q')^2},$$

the required formula for ρ.

Second formula for the radius of relative curvature.

We now take the variable θ to be the length s of the curve measured from a fixed point thereof, so that p', p'', etc. denote $\dfrac{dp}{ds}$, $\dfrac{d^2p}{ds^2}$, etc. We have therefore

$$1=(E,\ F,\ G\between p',\ q')^2,$$

and the formula becomes

$$\frac{1}{\rho}=\sqrt{(E,\ F,\ G\between p''-P'',\ q''-Q'')^2}.$$

But, differentiating the above equation as regards the curve, we find

$$0=2(E,\ F,\ G\between p',\ q'\between p'',\ q'')+(\dot{E},\ \dot{F},\ \dot{G}\between p',\ q')^2,$$

where $\dot{E}$, $\dot{F}$, $\dot{G}$ are used to denote the complete differential coefficients $E_1p'+E_2q'$, etc. And similarly, differentiating in regard to the tangent geodesic, we obtain

$$0=2(E,\ F,\ G\between p',\ q'\between P'',\ Q'')+(\dot{E},\ \dot{F},\ \dot{G}\between p',\ q')^2;$$

and hence, taking the difference of the two equations,

$$0=(E,\ F,\ G\between p',\ q'\between p''-P'',\ q''-Q'').$$

Hence, in the equation for $\dfrac{1}{\rho}$, the function under the radical sign may be written

$$(E,\ F,\ G\between p',\ q')^2\,.\,(E,\ F,\ G\between p''-P'',\ q''-Q'')^2-\{(E,\ F,\ G\between p',\ q'\between p''-P'',\ q''-Q'')\}^2,$$

which is identically

$$= (EG - F^2)\{p'(q'' - Q'') - q'(p'' - P'')\}^2.$$

Hence, extracting the square root, and for $\sqrt{EG - F^2}$ writing V, we have

$$\frac{1}{\rho} = V\{p'(q'' - Q'') - q'(p'' - P'')\},$$

or say

$$\frac{1}{\rho} = V(p'q'' - q'p'') - V(p'Q'' - q'P''),$$

which is the new formula for the radius of relative curvature.

Formula for the radius of geodesic curvature.

In the paper "On Geodesic Lines, etc.," p. 195, [vol. VIII. of this Collection, p. 160], writing $EG - F^2 = V^2$, and P'', Q'' in place of p'', q'', the differential equation of the geodesic line is obtained in the form

$$\begin{aligned}&(Ep' + Fq')\{(2F_1 - E_2)p'^2 + 2G_1p'q' + G_2q'^2\}\\ &-(Fp' + Gq')\{E_1p'^2 + 2E_2p'q' + (2F_2 - G_1)q'^2\}\\ &+2V^2(p'Q'' - q'P'') = 0;\end{aligned}$$

or, denoting by Ω the first two lines of this equation, we have

$$V(p'Q'' - q'P'') = -\frac{\tfrac{1}{2}}{V}\Omega.$$

The foregoing equation gives therefore, for the radius of geodesic curvature,

$$\frac{1}{\rho} = V(p'q'' - p''q') + \frac{\tfrac{1}{2}}{V}\Omega,$$

which is an expression depending only upon p', q', the first differential coefficients (common to the curve and geodesic), and on p'', q'', the second differential coefficients belonging to the curve.

Observe that Ω is a cubic function of p', q': we have

$$\Omega = (\mathfrak{A}, \mathfrak{B}, \mathfrak{C}, \mathfrak{D} \between p', q')^3,$$

the values of the coefficients being

$$\begin{aligned}\mathfrak{A} &= 2EF_1 - EE_2 - FE_1,\\ \mathfrak{B} &= 2EG_1 + 2FF_1 - 3FE_2 - GE_1,\\ \mathfrak{C} &= EG_2 + 3FG_1 - 2FF_2 - 2GE_2,\\ \mathfrak{D} &= FG_2 - 2GF_2 + GG_1.\end{aligned}$$

The Special Curves, $p=$ constant and $q=$ constant.

Consider the curve $p=\text{const.}$ For this curve $p'=0$, $p''=0$; therefore also $Gq'^2=1$, and, if R be the radius of geodesic curvature, then

$$\frac{1}{R}=\frac{\tfrac{1}{2}}{V}\mathfrak{D}q'^3,\quad =\frac{\tfrac{1}{2}}{V}\frac{\mathfrak{D}}{G\sqrt{G}}.$$

Similarly for the curve $q=\text{const.}$ Here $q'=0$, $q''=0$; therefore $Ep'^2=1$, and, if S be the radius of geodesic curvature, then

$$\frac{1}{S}=\frac{\tfrac{1}{2}}{V}\mathfrak{A}p'^3,\quad =\frac{\tfrac{1}{2}}{V}\frac{\mathfrak{A}}{E\sqrt{E}}.$$

These values of R and S are interesting for their own sakes, and they will be introduced into the expression for the radius of geodesic curvature ρ of the general curve.

Transformed Formula for the Radius of Geodesic Curvature.

From the values of $\frac{1}{R}$, $\frac{1}{S}$, we have

$$\frac{1}{\rho}-\frac{q'\sqrt{G}}{R}-\frac{p'\sqrt{E}}{S}=V(p'q''-p''q')+\frac{\tfrac{1}{2}}{V}\left\{\Omega-\frac{\mathfrak{A}}{E}p'-\frac{\mathfrak{D}}{G}q'\right\},$$

where the term in $\{\ \}$ is

$$=\mathfrak{A}p'^3-\frac{\mathfrak{A}}{E}p'+\mathfrak{B}p'^2q'+\mathfrak{C}p'q'^2+\mathfrak{D}q'^3-\frac{\mathfrak{D}}{G}q'.$$

The terms in $\mathfrak{A}$ are

$$=-\frac{\mathfrak{A}}{E}p'(1-Ep'^2),\quad =-\frac{\mathfrak{A}}{E}p'(2Fp'q'+Gq'^2),$$

and those in $\mathfrak{D}$ are

$$=-\frac{\mathfrak{D}}{G}q'(1-Gq'^2),\quad =-\frac{\mathfrak{D}}{G}q'(Ep'^2+2Fp'q').$$

Hence the whole expression contains the factor $p'q'$, and is, in fact,

$$=p'q'\left\{p'\left(\mathfrak{B}-\frac{2\mathfrak{A}F}{E}-\frac{\mathfrak{D}E}{G}\right)+q'\left(\mathfrak{B}-\frac{\mathfrak{A}G}{E}-\frac{2\mathfrak{D}F}{G}\right)\right\};$$

or substituting for $\mathfrak{A}$, $\mathfrak{B}$, $\mathfrak{C}$, $\mathfrak{D}$ their values, this is

$$=p'q'\left\{p'\left(-GE_1+EG_1+\frac{2F^2E_1}{E}-2FF_1-FE_2+2EF_2-\frac{EFG_2}{G}\right)\right.$$
$$\left.+q'\left(-GE_2+EG_2-\frac{2F^2G_2}{G}+2FF_2+FG_1-2GF_1+\frac{GFE_1}{E}\right)\right\},$$

say this is

$$= p'q'(Lp' + Mq');$$

and the formula thus is

$$\frac{1}{\rho} - \frac{q'\sqrt{G}}{R} - \frac{p'\sqrt{E}}{S} = V(p'q'' - p''q') + \frac{\tfrac{1}{2}}{V} p'q'(Lp' + Mq').$$

Taking ϕ, θ to be the inclination of the curve to the curves $q = \text{const.}$, $p = \text{const.}$, respectively, and $\omega\,(= \phi + \theta)$ the inclination of these two curves to each other, then

$$\cos\phi = \frac{Fp' + Gq'}{\sqrt{G}}, \quad \cos\theta = \frac{Ep' + Fq'}{\sqrt{E}}, \quad \cos\omega = \frac{F}{\sqrt{EG}},$$

$$\sin\phi = \frac{Vp'}{\sqrt{G}}, \quad \sin\theta = \frac{Vq'}{\sqrt{E}}, \quad \sin\omega = \frac{V}{\sqrt{EG}};$$

hence $\dfrac{\sin\phi}{\sin\omega} = p'\sqrt{E}$, $\dfrac{\sin\theta}{\sin\omega} = q'\sqrt{G}$, and the formula may also be written

$$\frac{1}{\rho} - \frac{\sin\theta}{\sin\omega}\frac{1}{R} - \frac{\sin\phi}{\sin\omega}\frac{1}{S} = V(p'q'' - p''q') + \frac{\tfrac{1}{2}}{V} p'q'(Lp' + Mq').$$

The Orthotomic Case $F = 0$, *or* $ds^2 = Edp^2 + Gdq^2$.

The formula becomes in this case much more simple. We have

$$1 = Ep'^2 + Gq'^2, \quad V = \sqrt{EG}, \quad \omega = 90^\circ, \quad \sin\theta = \cos\phi;$$

and the term $Lp' + Mq'$ becomes $= E\dot{G} - \dot{E}G$, if, as before, $\dot{E}$, $\dot{G}$ denote the complete differential coefficients $E_1p' + E_2q'$ and $G_1p' + G_2q'$. The formula then is

$$\frac{1}{\rho} - \frac{\cos\phi}{R} - \frac{\sin\phi}{S} = V(p'q'' - p''q') + \frac{\tfrac{1}{2}}{V}(E\dot{G} - \dot{E}G),$$

where the values $\dfrac{1}{R}$ and $\dfrac{1}{S}$ are now $= \dfrac{\tfrac{1}{2}G_1}{G\sqrt{E}}$ and $\dfrac{-\tfrac{1}{2}E_2}{E\sqrt{G}}$, respectively. But we have moreover $\phi = \tan^{-1}\dfrac{p'\sqrt{E}}{q'\sqrt{G}}$, and thence

$$\phi' = q'\sqrt{G}\left(p''\sqrt{E} + \frac{\tfrac{1}{2}p'\dot{E}}{\sqrt{E}}\right) - p'\sqrt{E}\left(q''\sqrt{G} + \frac{\tfrac{1}{2}q'\dot{G}}{\sqrt{G}}\right),$$

$$= -V(p'q'' - p''q') - \frac{\tfrac{1}{2}}{V} p'q'(E\dot{G} - \dot{E}G);$$

or the formula finally is

$$\frac{1}{\rho} - \frac{\cos\phi}{R} - \frac{\sin\phi}{S} + \phi' = 0,$$

which is Liouville's formula referred to at the beginning of the present paper. It will be recollected that ϕ' is the differential coefficient $\dfrac{d\phi}{ds}$ with respect to the arc s of the curve.

ADDITION.—Since the foregoing paper was written, I have succeeded in obtaining a like interpretation of the term

$$V(p'q''-p''q')+\frac{\tfrac{1}{2}}{V}p'q'(Lp'+Mq'),$$

which belongs to the general case. I find that these terms are, in fact, $=-\dot{\phi}+\omega_1 p'$; or, what is the same thing (since $\omega=\phi+\theta$ and therefore $\omega_1 p'+\omega_2 q'=\dot{\phi}+\dot{\theta}$), are $=\dot{\theta}-\omega_2 q'$. It will be recollected that ϕ is the inclination of the curve to the curve $q=c$, which passes through a given point of the curve, $\dot{\phi}$ is the variation of ϕ corresponding to the passage to the consecutive point of the curve, viz., $\phi+\dot{\phi}ds$ is the inclination at this consecutive point to the curve $q=c+dc$, which passes through the consecutive point; ω is the inclination to each other of the curves $p=b$, $q=c$, which pass through the given point of the curve, ω_1 the variation corresponding to the passage along the curve $q=c$, viz., $\omega+\omega_1 ds$ is the inclination to each other of the curves $p=b+db$, $q=c$; and the like as regards $\dot{\theta}$ and ω_2.

For the demonstration, we have, as above,

$$\phi=\tan^{-1}\frac{Vp'}{Fp'+Gq'},\quad \omega=\tan^{-1}\frac{V}{F},$$

where

$$V=\sqrt{EG-F^2};$$

and moreover $Ep'^2+2Fp'q'+Gq'^2=1$. In virtue of this last equation,

$$V^2p'^2+(Fp'+Gq')^2=G;$$

and we have

$$\dot{\phi}=-V(p'q''-p''q')+\frac{1}{G}\square,$$

where

$$\square=(Fp'+Gq')\,p'\dot{V}-Vp'(\dot{F}p'+\dot{G}q');$$

or, since $V^2=EG-F^2$, and thence $2V\dot{V}=G\dot{E}-2F\dot{F}+E\dot{G}$, we have

$$\square=\frac{\tfrac{1}{2}p'}{V}\{(Fp'+Gq')(G\dot{E}-2F\dot{F}+E\dot{G})-2(EG-F^2)(\dot{F}p'+\dot{G}q')\}.$$

Substituting herein for $\dot{E}$, $\dot{F}$, $\dot{G}$ their values $E_1p'+E_2q'$, $F_1p'+F_2q'$, $G_1p'+G_2q'$, the term in { } becomes

$$=Ip'^2+Jp'q'+Kq'^2,$$

where

$$I=FGE_1-2EGF_1+EFG_1,$$
$$J=G^2E_1-2FGF_1+(-EG+2F^2)\,G_1+FGE_2-2EGF_2+EFG_2,$$
$$K=G^2E_2-2FGF_2+(-EG+2F^2)\,G_2.$$

But from the equation $\omega=\tan^{-1}\frac{V}{F}$, differentiating in regard to p, we obtain

$$\omega_1=\frac{\tfrac{1}{2}}{EGV}(FG\dot{E}-2EG\dot{F}+EF\dot{G})=\frac{\tfrac{1}{2}}{EGV}I;$$

or, for p writing

$$p'(Ep'^2+2Fp'q'+Gq'^2),\quad =Ep'\left(p'^2+2\frac{F}{E}p'q'+\frac{G}{E}q'^2\right),$$

we have

$$\dot{\phi}-\omega_1 p'=-V(p'q''-p''q')+\frac{\tfrac{1}{2}p'}{GV}(Ip'^2+Jp'q'+Kq'^2)$$

$$-\frac{\tfrac{1}{2}p'}{GV}I\left(p'^2+2\frac{F}{E}p'q'+\frac{G}{E}q'^2\right).$$

The terms in p'^3 destroy each other, and the form thus is

$$\dot{\phi}-\omega_1 p'=-V(p'q''-p''q')-\frac{\tfrac{1}{2}}{V}p'q'(Lp'+Mq'),$$

where

$$L=-\frac{J}{G}+\frac{2IF}{GE},$$

$$M=-\frac{K}{G}+\frac{I}{E};$$

and, upon substituting herein for I, J, K their values, we find

$$L=-GE_1+EG_1+\frac{2F^2E_1}{F}-2FF_1-FE_2+2EF_2-\frac{EFG_2}{G},$$

$$M=-GE_2+EG_2-\frac{2F^2G_2}{G}+2FF_2+FG_1-2GF_1+\frac{GFE_1}{E};$$

viz., these are the values denoted above by the same letters L, M. The final result thus is

$$\frac{1}{\rho}-\frac{q'\sqrt{G}}{R}-\frac{p'\sqrt{E}}{S}=-\dot{\phi}+\omega_1 p',$$

$$=\quad \dot{\theta}-\omega_2 q',$$

where the meanings of the symbols have been already explained. A formula substantially equivalent to this, but in a different (and scarcely properly explained) notation, is given, Aoust, "Théorie des coordonnées curvilignes quelconques," *Annali di Matem.*, t. II. (1868), pp. 39—64; and I was, in fact, led thereby to the foregoing further investigation.

As to the definition of the radius of geodesic curvature, I remark that, for a curve on a given surface, if PQ be an infinitesimal arc of the curve, then if from Q we let fall the perpendicular QM on the tangent plane at P (the point M being thus a point on the tangent PT of the curve), and if from M, in the tangent plane and at right angles to the tangent, we draw MN to meet the osculating plane of the curve in N, then MN is in fact equal to the infinitesimal arc QQ' mentioned near the beginning of the present paper, and the radius of geodesic curvature ρ is thus a length such that $2\rho\,.\,MN=\overline{PQ}^2$.

767.

ON THE GAUSSIAN THEORY OF SURFACES.

[From the *Proceedings of the London Mathematical Society*, vol. XII. (1881), pp. 187—192. Read June 9, 1881.]

IN the Memoir, Bour, "Théorie de la déformation des surfaces" (*Jour. de l'Éc. Polyt.*, Cah. 39 (1862), pp. 1—148), the author, working with the form $ds^2 = dv^2 + g^2du^2$ as a special case of Gauss's formula $ds^2 = Edp^2 + 2Fdpdq + Gdq^2$, obtains (p. 29) the following equations which he calls *fundamental*:—

$$[\text{IV.}] \ldots\ldots \begin{cases} \dfrac{1}{g}\dfrac{dg_1}{dv} = T^2 - HH_1, \\ \dfrac{dT}{du} + \dfrac{d\,.\,Hg}{dv} - H_1 g_1 = 0, \\ \dfrac{d\,.\,Tg^2}{dv} + g\dfrac{dH_1}{du} \qquad = 0, \end{cases}$$

where g_1 is written to denote $\dfrac{dg}{dv}$, and where (see p. 26)

H is the curvature of the normal section containing the tangent to the curve $v = \text{constant}$,

H_1 is the curvature of the normal section at right angles to the preceding, containing the tangent to the (geodesic) curve $u = \text{constant}$,

T is the torsion of the same geodesic curve;

or, what is the same thing (see p. 25), the quadric equation for the determination of the principal radii of curvature at the point of the surface is

$$\left(\frac{1}{\rho} - H\right)\left(\frac{1}{\rho} - H_1\right) - T^2 = 0.$$

Writing for greater convenience K in place of the suffixed letter H_1, also V instead of g, so that the differential formula is $ds^2 = dv^2 + V^2du^2$, the equations become

$$\left\{\begin{aligned}&\frac{1}{V}\frac{d^2V}{dv^2} = T - HK,\\ &\frac{dT}{du} + \frac{d\,.\,HV}{dv} - K\frac{dV}{dv} = 0,\\ &\frac{d\,.\,TV^2}{dv} + V\frac{dK}{du} \qquad = 0;\end{aligned}\right.$$

or, if we use the suffix 1 to denote differentiation in regard to v, and the suffix 2 to denote differentiation in regard to u, then the equations are

$$\frac{V_{11}}{V} = T^2 - HK,$$

$$T_2 + (HV)_1 - KV_1 = 0,$$

$$(TV^2)_1 + K_2V = 0,$$

or, what is the same thing,

$$\left\{\begin{aligned}&V_{11} = V(T^2 - HK),\\ &T_2 + H_1V + (H - K)\,V_1 = 0,\\ &T_1V + 2TV_1 + K_2 \qquad = 0.\end{aligned}\right.$$

I wish to show how these formulæ connect themselves with formulæ belonging to the general form $ds^2 = Edp^2 + 2Fdpdq + Gdq^2$. These involve not only Gauss's coefficients E, F, G, but also the coefficients E', F', G' belonging to the inflexional tangents; and, for convenience, I quote the system of definitions, Salmon's *Geometry of Three Dimensions*, 3rd ed., 1874, p. 251, viz.

$$dx,\ dy,\ dz = adp + a'dq,\quad bdp + b'dq,\quad cdp + c'dq;$$

$$d^2x = \alpha dp^2 + 2\alpha' dpdq + \alpha'' dq^2,$$

$$d^2y = \beta dp^2 + 2\beta' dpdq + \beta'' dq^2,$$

$$d^2z = \gamma dp^2 + 2\gamma' dpdq + \gamma'' dq^2;$$

$$A,\ B,\ C = bc' - b'c,\quad ca' - c'a,\quad ab' - a'b;\quad V^2 = EG - F^2;$$

$$E' = A\alpha + B\beta + C\gamma,\quad F' = A\alpha' + B\beta' + C\gamma',\quad G' = A\alpha'' + B\beta'' + C\gamma'',$$

so that E', F', G' are, in fact, the determinants

$$\begin{vmatrix} a, & b, & c \\ a', & b', & c' \\ \alpha, & \beta, & \gamma \end{vmatrix},\quad \begin{vmatrix} a, & b, & c \\ a', & b', & c' \\ \alpha', & \beta', & \gamma' \end{vmatrix},\quad \begin{vmatrix} a, & b, & c \\ a', & b', & c' \\ \alpha'', & \beta'', & \gamma'' \end{vmatrix}.$$

The equation for the determination of the principal radii of curvature is

$$(E'\rho - EV)(G'\rho - GV) - (F'\rho - FV)^2 = 0,$$

which, in the particular case $F=0$ (and therefore $V^2=EG$), becomes

$$(E'\rho - EV)(G'\rho - GV) - F'^2\rho^2 = 0,$$

or, as this may be written,

$$\left(\frac{1}{\rho} - \frac{E'}{EV}\right)\left(\frac{1}{\rho} - \frac{G'}{GV}\right) - \frac{F'^2}{EGV^2} = 0,$$

an equation which corresponds with Bour's form

$$\left(\frac{1}{\rho} - K\right)\left(\frac{1}{\rho} - H\right) - T^2 = 0,$$

and becomes identical with it, if

$$E' = EVK, \quad G' = GVH, \quad F' = -V^2T.$$

But, making p, q correspond to Bour's variables, p to v, and q to u, it is necessary to show that the foregoing values (and not the interchanged values $E' = GVH$, $G' = EVK$) are the correct ones. We have, Salmon, p. 254,

$$\left\| \begin{matrix} dq, & \rho E' - VE, & \rho F' - VF \\ -dp, & \rho F' - VF, & \rho G' - VG \end{matrix} \right\| = 0;$$

or, putting herein $F=0$, the equations may be written

$$\frac{dq}{-dp} = \frac{E'}{F'}\left(1 - \frac{VE}{\rho E'}\right) = \frac{F'}{G'} \div \left(1 - \frac{VG}{\rho G'}\right);$$

or, we see that to $dq=0$ corresponds the value $\frac{1}{\rho} = \frac{E'}{EV}$, and to $dp=0$ the value $\frac{1}{\rho} = \frac{G'}{GV}$. Hence the former of these values of $\frac{1}{\rho}$ corresponds to Bour's $du=0$, that is, to his $\frac{1}{\rho} = K$; and the latter to Bour's $dv=0$, that is, to his $\frac{1}{\rho} = H$; or the values are, as stated,

$$E' = EVK, \quad G' = GVH.$$

The formula $ds^2 = Edp^2 + 2Fdpdq + Gdq^2$ agrees with Bour's $ds^2 = dv^2 + g^2du^2$, if $p=u$, $q=v$, $E=1$, $F=0$, $G=g^2$. With these values, $V^2 = EG - F^2 = g^2$, or say $g = V$, and Bour's equation is, as it was before written, $ds^2 = dv^2 + V^2du^2$. And we have to find the three equations which, putting therein $p=u$, $q=v$, $E=1$, $F=0$, $G=V^2$, $E' = VK$, $F' = -V^2T$, $G' = V^3H$, reduce themselves to Bour's equations.

The first of these is nothing else than the equation for the measure of curvature, viz. Salmon, p. 262 (but, using the suffixes 1 and 2 to denote differentiation in regard to p and q respectively), this is

$$\begin{aligned} 4(E'G' - F'^2) = {} & E(E_2G_2 - 2F_1G_2 + G_1^2) \\ & + F(E_1G_2 - E_2G_1 - 2E_2F_1 + 4F_1F_2 - 2F_1G_1) \\ & + G(E_1G_1 - 2E_1F_2 + E_2^2) \\ & - 2(EG - F^2)(E_{22} - 2F_{12} + G_{11}). \end{aligned}$$

In fact, writing herein $E=1$, $F=0$, and therefore the differential coefficients of E and F each $=0$, the equation becomes

$$4(E'G'-F'^2)=G_1^2-2GG_{11},$$

which is

$$4V^4(HK-T^2)=(2VV_1)^2-2V^2(2V_1^2+2VV_{11}),\ =-4V^3V_{11};$$

or finally it is

$$V_{11}=V(T^2-HK).$$

The other two of Bour's equations are derived from equations which give respectively the values of $E_2'-F_1'$ and $F_2'-G_1'$; viz. starting from the equations

$$E'=A\alpha\ +B\beta\ +C\gamma\ ,$$
$$F'=A\alpha'\ +B\beta'\ +C\gamma'\ ,$$
$$G'=A\alpha''+B\beta''+C\gamma'',$$

we see at once that E_2' and F_1' contain, E_2' the terms $A\alpha_2+B\beta_2+C\gamma_2$, and F_1' the terms $A\alpha_1'+B\beta_1'+C\gamma_1'$, which are equal to each other ($\alpha_2=\alpha_1'$ since α and α' are the differential coefficients x_{11}, x_{12} of x, and so $\beta_2=\beta_1'$ and $\gamma_2=\gamma_1'$). Hence

$$E_2'-F_1'=A_2\alpha+B_2\beta+C_2\gamma-A_1\alpha'-B_1\beta'-C_1\gamma';$$

and similarly

$$F_2'-G_1'=A_2\alpha'+B_2\beta'+C_2\gamma'-A_1\alpha''-B_1\beta''-C_1\gamma''.$$

Here, from the values of A, B, C, we have

$$A=bc'-cb';\quad A_1=\beta c'-\gamma b'+b\gamma'-c\beta';\quad A_2=\beta'c'-\gamma'b'+b\gamma''-c\beta'';$$
$$B=ca'-ac';\quad B_1=\gamma a'-\alpha c'+c\alpha'-a\gamma';\quad B_2=\gamma'a'-\alpha'c'+c\alpha''-a\gamma'';$$
$$C=ab'-ba';\quad C_1=\alpha b'-\beta a'+a\beta'-b\alpha';\quad C_2=\alpha'b'-\beta'a'+a\beta''-b\alpha'';$$

and, substituting, we find

$$E_2'-F_1'=\ 2a'\alpha\alpha'+a\alpha''\alpha,$$
$$F_2'-G_1'=-2a\alpha'\alpha''-a'\alpha''\alpha,$$

if, for shortness, $a'\alpha\alpha'$ denotes the determinant

$$\begin{vmatrix} a', & \alpha, & \alpha' \\ b', & \beta, & \beta' \\ c', & \gamma, & \gamma' \end{vmatrix},$$

and so for the other like symbols. Observe that, with

$$\begin{vmatrix} a, & a', & \alpha, & \alpha', & \alpha'' \\ b, & b', & \beta, & \beta', & \beta'' \\ c, & c', & \gamma, & \gamma', & \gamma'' \end{vmatrix},$$

we have in all 10 determinants, viz. these are $aa'\alpha$, $=E'$; $aa'\alpha'$, $=F'$; $aa'\alpha''$, $=G'$; $\alpha\alpha'\alpha''$; and the six determinants $a\alpha\alpha'$, $a\alpha'\alpha''$, $a\alpha''\alpha$; $a'\alpha\alpha'$, $a'\alpha'\alpha''$, $a'\alpha''\alpha$. The foregoing expressions of $E_2'-F_1'$ and $F_2'-G_1'$ respectively, substituting therein for the determinants $a'\alpha\alpha'$, $a\alpha''\alpha$, $a\alpha'\alpha''$, $a'\alpha''\alpha$ their values as about to be obtained, are the required two equations. We have

$$\begin{array}{ll}
aa\ +bb\ +cc\ =E, & aa'\ +bb'\ +cc'\ =F,\\
a'a\ +b'b\ +c'c\ =F, & a'a'\ +b'b'\ +c'c'\ =G,\\
\alpha a\ +\beta b\ +\gamma c\ =\tfrac{1}{2}E_1, & \alpha a'\ +\beta b'\ +\gamma c'\ =F_1-\tfrac{1}{2}E_2,\\
\alpha' a\ +\beta' b\ +\gamma' c\ =\tfrac{1}{2}E_2, & \alpha' a'\ +\beta' b'\ +\gamma' c'\ =\tfrac{1}{2}G_1,\\
\alpha'' a+\beta'' b+\gamma'' c=F_2-\tfrac{1}{2}G_1, & \alpha'' a'+\beta'' b'+\gamma'' c'=\tfrac{1}{2}G_2;
\end{array}$$

and if from the first five equations, regarded as equations linear in (a, b, c), we eliminate these quantities, and from the second five equations, regarded as linear in (a', b', c'), we eliminate these quantities, we obtain two sets each of five equations,

$$\left\|\begin{array}{ccccc}
a, & a', & \alpha, & \alpha', & \alpha''\\
b, & b', & \beta, & \beta', & \beta''\\
c, & c', & \gamma, & \gamma', & \gamma''\\
E, & F, & \tfrac{1}{2}E_1, & \tfrac{1}{2}E_2, & F_2-\tfrac{1}{2}G_1
\end{array}\right\|=0,\ \text{and}\ \left\|\begin{array}{ccccc}
a, & a', & \alpha, & \alpha', & \alpha''\\
b, & b', & \beta, & \beta', & \beta''\\
c, & c', & \gamma, & \gamma', & \gamma''\\
F, & G, & F_1-\tfrac{1}{2}E_2, & \tfrac{1}{2}G_1, & \tfrac{1}{2}G_2
\end{array}\right\|=0.$$

These may be written,

$$\begin{array}{l}
F\alpha\,\alpha'\,\alpha''-\tfrac{1}{2}E_1 a'\alpha'\,\alpha''-\tfrac{1}{2}E_2 a'\alpha''\alpha-(F_2-\tfrac{1}{2}G_1)\,a'\alpha\alpha'=0,\\
-E\alpha\,\alpha'\,\alpha''+\tfrac{1}{2}E_1 a\,\alpha'\,\alpha''+\tfrac{1}{2}E_2 a\alpha''\alpha+(F_2-\tfrac{1}{2}G_1)\,a\alpha\alpha'=0,\\
Ea'\alpha'\,\alpha''-\ F\,a\,\alpha'\,\alpha''+\tfrac{1}{2}E_2G'-(F_2-\tfrac{1}{2}G_1)\,F'=0,\\
Ea'\alpha''\alpha\ -\ F\,a\,\alpha''\alpha\ -\tfrac{1}{2}E_1G'+(F_2-\tfrac{1}{2}G_1)\,E'=0,\\
Ea'\alpha\,\alpha'-\ F\,a\,\alpha\,\alpha'+\tfrac{1}{2}E_1F'-\tfrac{1}{2}E_2E'=0;
\end{array}$$

and

$$\begin{array}{l}
G\alpha\,\alpha'\,\alpha''-(F_1-\tfrac{1}{2}E_2)\,a'\alpha'\,\alpha''-\tfrac{1}{2}G_1 a'\alpha''\alpha-\tfrac{1}{2}G_2 a'\alpha\alpha'=0,\\
-F\alpha\,\alpha'\,\alpha''+(F_1-\tfrac{1}{2}E_2)\,a\,\alpha'\,\alpha''+\tfrac{1}{2}G_1 a\alpha''\alpha+\tfrac{1}{2}G_2 a\alpha\alpha'=0,\\
Fa'\alpha'\,\alpha''-\ G\ a\,\alpha'\,\alpha''+\tfrac{1}{2}G_1G-\tfrac{1}{2}G_2F'=0,\\
Fa'\alpha''\alpha\ -\ G\ a\,\alpha''\alpha\ -(F_1-\tfrac{1}{2}E_2)\,G'+\tfrac{1}{2}G_2E'=0,\\
Fa'\alpha\,\alpha'\ -\ G\ a\,\alpha\,\alpha'+(F_1-\tfrac{1}{2}E_2)\,F'-\tfrac{1}{2}G_1E'=0.
\end{array}$$

Attending in each set only to the third, fourth, and fifth equations, and combining these in pairs, we obtain

$$\begin{array}{l}
V^2a\,\alpha'\,\alpha''+(\ \tfrac{1}{2}FG_1-FF_2+\tfrac{1}{2}EG_2)\,F'+(-\tfrac{1}{2}EG_1+\tfrac{1}{2}FE_2)\,G'=0,\\
V^2a'\alpha'\,\alpha''+(\ \tfrac{1}{2}GG_1-GF_2+\tfrac{1}{2}FG_2)\,F'+(-\tfrac{1}{2}FG_1+\tfrac{1}{2}GE_2)\,G'=0;\\[6pt]
V^2a\,\alpha''\alpha\ +(-\tfrac{1}{2}FE_1+EF_1-\tfrac{1}{2}EE_2)\,G'+(-\tfrac{1}{2}FG_1+FF_2-\tfrac{1}{2}EG_2)\,E'=0,\\
V^2a'\alpha''\alpha\ +(-\tfrac{1}{2}GE_1+FF_1-\tfrac{1}{2}FE_2)\,G'+(-\tfrac{1}{2}GG_1+GF_2-\tfrac{1}{2}FG_2)\,E'=0;\\[6pt]
V^2a\,\alpha\,\alpha'\ +(\ \tfrac{1}{2}EG_1-\tfrac{1}{2}FE_2)\,E'+(\ \tfrac{1}{2}FE_1-EF_1+\tfrac{1}{2}EE_2)\,F'=0,\\
V^2a'\alpha\,\alpha'\ +(\ \tfrac{1}{2}FG_1-\tfrac{1}{2}GE_2)\,E'+(\ \tfrac{1}{2}GE_1-FF_1+\tfrac{1}{2}FE_2)\,F'=0.
\end{array}$$

We thus obtain

$$E_2' - F_1' = \frac{2}{V^2}\{(\qquad -\tfrac{1}{2}FG_1 + \tfrac{1}{2}GE_2)\,E' + (-\tfrac{1}{2}GE_1 + FF_1 \ -\tfrac{1}{2}FF_2)\,F'\}$$

$$+\frac{1}{V^2}\{(\ \tfrac{1}{2}FE_1 - \ EF_1 + \tfrac{1}{2}EE_2)\,G' + (\ \tfrac{1}{2}FG_1 - FF_2 \ +\tfrac{1}{2}EG_2)\,E'\},$$

$$F_2' - G_1' = \frac{2}{V^2}\{(\ \tfrac{1}{2}FG_1 - \ FF_2 + \tfrac{1}{2}EG_2)\,F' + (\qquad -\tfrac{1}{2}EG_1 + \tfrac{1}{2}FE_2)\,G'\}$$

$$+\frac{1}{V^2}\{(-\tfrac{1}{2}GE_1 + \ FF_1 - \tfrac{1}{2}FE_2)\,G' + (-\tfrac{1}{2}GG_1 + GF_2 \ -\tfrac{1}{2}FG_2)\,E'\};$$

or, finally,

$$E_2' - F_1' = \frac{1}{V^2}\{(-\tfrac{1}{2}FG_1 + GE_2 - FF_2 + \tfrac{1}{2}EG_2)\,E'$$

$$+(-GE_1 + 2FF_1 - FF_2)\,F' + (\tfrac{1}{2}FE_1 - EF_1 + \tfrac{1}{2}EE_2)\,G'\},$$

$$F_2' - G_1' = \frac{1}{V^2}\{(-\tfrac{1}{2}GG_1 + GF_2 - \tfrac{1}{2}FG_2)\,E'$$

$$+(FG_1 - 2FF_2 + EG_2)\,F' + (-\tfrac{1}{2}GE_1 + FF_1 - EG_1 + \tfrac{1}{2}FE_2)\,G'\},$$

which are the required formulæ; and which may, I think, be regarded as new formulæ in the Gaussian theory of surfaces.

Writing herein as before, the first of these becomes

$$(VK)_2 + (V^2T)_1 = \frac{1}{V^2}\{\tfrac{1}{2}(V^2)_2\,VK\},\ = V_2K,$$

that is,

$$V_2K + VK_2 + V^2T_1 + 2VV_1T = V_2K;$$

or finally

$$VT_1 + 2TV_1 + K_2 = 0,$$

which is Bour's third equation. And the second equation becomes

$$-(V^2T)_2 - (V^3H)_1 = \frac{1}{V^2}\{-\tfrac{1}{2}V^2(V^2)_1\,VK + (V^2)_2(-V^2T) - (V^2)_1\,V^3H\},$$

$$= -V^2V_1K - 2VV_2T - 2V^2V_1H,$$

that is,

$$-V^2T_2 - 2VV_2T - V^3H_1 - 3V^2V_1H = -V^2V_1K - 2VV_2T - 2V^2V_1H;$$

or finally

$$T_2 + VH_1 + (H-K)\,V_1 = 0,$$

which is Bour's second equation.

768.

NOTE ON LANDEN'S THEOREM.

[From the *Proceedings of the London Mathematical Society*, vol. XIII. (1882), pp. 47, 48. Read November 10, 1881.]

LANDEN'S theorem, as given in the paper "An Investigation of a General Theorem for finding the length of any Arc of any Conic Hyperbola by means of two Elliptic Arcs, with some other new and useful Theorems deduced therefrom," *Phil. Trans.*, t. LXV. (1775), pp. 283—289, is, as appears by the title, a theorem for finding the length of a hyperbolic arc in terms of the length of two elliptic arcs; this theorem being obtained by means of the following differential identity, viz., if

$$t = gx\sqrt{\frac{m^2 - x^2}{m^2 - gx^2}},$$

where

$$g = \frac{m^2 - n^2}{n^2},$$

then

$$\sqrt{\frac{m^2 - gx^2}{m^2 - x^2}}\,dx = \left\{\tfrac{1}{2} + \tfrac{1}{4}\sqrt{\frac{(m-n)^2 - t^2}{(m+n)^2 - t^2}} + \tfrac{1}{4}\sqrt{\frac{(m+n)^2 - t^2}{(m-n)^2 - t^2}}\right\} dt,$$

(this is exactly Landen's form, except that he of course writes $\dot{x}$, $\dot{t}$ in place of dx, dt respectively): viz., integrating each side, and interpreting geometrically in a very ingenious and elegant manner the three integrals which present themselves, he arrives at his theorem for the hyperbolic arc; but with this I am not now concerned.

Writing for greater convenience $m = 1$, $n = k'$, and therefore $g = k^2$, if as usual $k^2 + k'^2 = 1$, the transformation is

$$t = k^2 x\sqrt{\frac{1 - x^2}{1 - k^2x^2}},$$

leading to

$$\sqrt{\frac{1-k^2x^2}{1-x^2}}\,dx=\left\{\tfrac{1}{2}+\tfrac{1}{4}\sqrt{\frac{(1-k')^2-t^2}{(1+k')^2-t^2}}+\sqrt{\frac{(1+k')^2-t^2}{(1-k')^2-t^2}}\right\}dt.$$

The form in which the transformation is usually employed (see my *Elliptic Functions*, pp. 177, 178) is

$$y=(1+k')\,x\sqrt{\frac{1-x^2}{1-k^2x^2}},$$

leading to

$$\frac{(1+k')\,dx}{\sqrt{1-x^2\,.\,1-k^2x^2}}=\frac{dy}{\sqrt{1-y^2\,.\,1-\lambda^2y^2}},$$

where

$$\lambda=\frac{1-k'}{1+k'}.$$

If, to identify the two forms, we write $y=\dfrac{t}{1-k'}$ and in the last equation introduce t in place of y, the last equation becomes

$$\frac{dx}{\sqrt{1-x^2\,.\,1-k^2x^2}}=\frac{dt}{\sqrt{\{(1-k')^2-t^2\}\,\{(1+k')^2-t^2\}}}.$$

Comparing with Landen's form, in order that the two may be identical, we must have

$$1-k^2x^2=\left\{\tfrac{1}{2}+\tfrac{1}{4}\sqrt{\frac{(1-k')^2-t^2}{(1+k')^2-t^2}}+\tfrac{1}{4}\sqrt{\frac{(1+k')^2-t^2}{(1-k')^2-t^2}}\right\}$$

$$\times\sqrt{(1-k')^2-t^2}\,\sqrt{(1+k')^2-t^2},$$

viz., this is

$$1-k^2x^2=\tfrac{1}{4}\left\{\sqrt{(1-k')^2-t^2}+\sqrt{(1+k')^2-t^2}\right\}^2,$$

that is,

$$1-k^2x^2=\tfrac{1}{2}\left[1+k'^2-t^2+\sqrt{\{(1-k')^2-t^2\}\,\{(1+k')^2-t^2\}}\right],$$

where the function under the radical sign is

$$(1-k'^2)^2-2\,(1+k'^2)\,t^2+t^4\ (=T\ \text{suppose});$$

and this must consequently be a form of the original integral equation

$$t=k^2x\sqrt{\frac{1-x^2}{1-k^2x^2}}.$$

In fact, squaring and solving in regard to x^2 with the assumed sign of the radical, we have

$$x^2=\frac{k^2+t^2-\sqrt{T}}{2k^2},$$

corresponding to an equation given by Landen. And we thence have

$$1 - k^2x^2 = \tfrac{1}{2}\{2 - k^2 - t^2 + \sqrt{T}\}, \quad = \tfrac{1}{2}\{1 + k'^2 - t^2 + \sqrt{T}\},$$

which is the required expression for $1 - k^2x^2$.

The trigonometrical form $\sin(2\phi' - \phi) = c \sin\phi$ of the relation between y and x does not occur in Landen; it is employed by Legendre, I believe, in an early paper, *Mém. de l'Acad. de Paris*, 1786, and in the *Exercices*, 1811, and also in the *Traité des Fonctions Elliptiques*, 1825, and by means of it he obtains an expression for the arc of a hyperbola in terms of two elliptic functions, $E(c, \phi)$, $E(c', \phi')$, showing that the arc of the hyperbola is expressible by means of two elliptic arcs,—this, he observes, "est le beau théorème dont Landen a enrichi la géométrie." We have, then (1828), Jacobi's proof, by two fixed circles, of the addition-theorem (see my *Elliptic Functions*, p. 28), and the application of this (p. 30) to Landen's theorem is also due to Jacobi, see the "Extrait d'une lettre adressée à M. Hermite," *Crelle*, t. XXXII. (1846), pp. 176—181; the connection of the demonstrations, by regarding the point, which is alone necessary for Landen's theorem as the limit of the smaller circle in the figure for the addition-theorem is due to Durège (see his *Theorie der elliptischen Functionen*, Leipzig, 1861, pp. 168, *et seq.*).

769.

ON A FORMULA RELATING TO ELLIPTIC INTEGRALS OF THE THIRD KIND.

[From the *Proceedings of the London Mathematical Society*, vol. XIII. (1882), pp. 175, 176. Presented May 11, 1882.]

THE formula for the differentiation of the integral of the third kind

$$\Pi = \int_0 \frac{d\phi}{(1 + n\sin^2\phi)\,\Delta}$$

in regard to the parameter n, see my *Elliptic Functions*, Nos. 174 *et seq.*, may be presented under a very elegant form, by writing therein

$$\sin^2\phi = x = \text{sn}^2 u, \quad \sin\phi\cos\phi\,\Delta = y = \text{sn}\, u\, \text{cn}\, u\, \text{dn}\, u,$$

and thus connecting the formula with the cubic curve

$$y^2 = x(1-x)(1-k^2x).$$

The parameter must, of course, be put under a corresponding form, say $n = -\dfrac{1}{a}$, where $a = \text{sn}^2\theta$, $b = \text{sn}\,\theta\,\text{cn}\,\theta\,\text{dn}\,\theta$, and therefore (a, b) are the coordinates of the point corresponding to the argument θ. The steps of the substitution may be effected without difficulty, but it will be convenient to give at once the final result and then verify it directly. The result is

$$\frac{d}{d\theta}\frac{b}{a-x} - \frac{d}{du}\frac{y}{x-a} = k^2(a-x).$$

We, in fact, have

$$\frac{dx}{du} = 2\,\text{sn}\, u\,\text{cn}\, u\,\text{dn}\, u = 2y,$$

and thence

$$y\frac{dx}{du}=2y^2,$$

that is,

$$y\frac{dx}{du}=2x\,[1-(1+k^2)\,x+k^2x^2].$$

Also

$$\frac{dy}{du}=\text{cn}^2 u\,\text{dn}^2 u-\text{sn}^2 u\,\text{dn}^2 u-k^2\,\text{sn}^2 u\,\text{cn}^2 u$$

$$=1-2\,(1+k^2)\,x+3k^2x^2,$$

and hence

$$\frac{d}{du}\frac{y}{x-a}=\frac{1}{(a-x)^2}\left\{(x-a)\frac{dy}{du}-y\frac{dx}{du}\right\}$$

$$=\frac{1}{(a-x)^2}\{-x-a+2\,(1+k^2)\,ax+k^2x^3-3k^2ax^2\}.$$

Interchanging the letters, we have

$$\frac{d}{d\theta}\frac{b}{a-x}=\frac{1}{(a-x)^2}\{-x-a+2\,(1+k^2)\,ax+k^2a^3-3k^2a^2x\},$$

and hence, subtracting,

$$\frac{d}{d\theta}\frac{b}{a-x}-\frac{d}{du}\frac{y}{x-a}=\frac{1}{(a-x)^2}\{k^2a^3-3k^2a^2x+3k^2ax^2-k^2x^3\}$$

$$=\frac{1}{(a-x)^2}k^2\,(a-x)^3$$

$$=k^2\,(a-x),$$

which is the required result.

770.

ON THE 34 CONCOMITANTS OF THE TERNARY CUBIC.

[From the *American Journal of Mathematics*, vol. IV. (1881), pp. 1—15.]

I HAVE, (by aid of Gundelfinger's formulæ, afterwards referred to), calculated, and I give in the present paper, the expressions of the 34 concomitants of the canonical ternary cubic $ax^3+by^3+cz^3+6lxyz$, or, what is the same thing, the 34 covariants of this cubic and the adjoint linear function $\xi x+\eta y+\zeta z$: this is the chief object of the paper. I prefix a list of memoirs, with short remarks upon some of them; and, after a few observations, proceed to the expressions for the 34 concomitants; and, in conclusion, exhibit the process of calculation of these concomitants other than such of them as are taken to be known forms. I insert a supplemental table of 6 derived forms.

The list of memoirs (not by any means a complete one) is as follows:

HESSE, Ueber die Elimination der Variabeln aus drei algebraischen Gleichungen vom zweiten Grade mit zwei Variabeln: *Crelle*, t. XXVIII. (1844), pp. 68—96. Although purporting to relate to a different subject, this is in fact the earliest, and a very important, memoir in regard to the general ternary cubic; and in it is established the canonical form, as Hesse writes it, $y_1^3+y_2^3+y_3^3+6\pi y_1y_2y_3$.

ARONHOLD, Zur Theorie der homogenen Functionen dritten Grades von drei Variabeln: *Crelle*, t. XXXIX. (1850), pp. 140—159.

CAYLEY, A Third Memoir on Quantics: *Phil. Trans.*, t. CXLVI. (1856), pp. 627—647; [144].

ARONHOLD, Theorie der homogenen Functionen dritten Grades von drei Variabeln: *Crelle*, t. LV. (1858), pp. 97—191.

SALMON, *Lessons Introductory to the Modern Higher Algebra:* 8°, Dublin, 1859.

CAYLEY, A Seventh Memoir on Quantics: *Phil. Trans.*, t. CLI. (1861), pp. 277—292; [269].

BRIOSCHI, Sur la théorie des formes cubiques à trois indéterminées: *Comptes Rendus*, t. LVI. (1863), pp. 304—307.

HERMITE, Extrait d'une lettre à M. Brioschi: *Crelle*, t. LXIII. (1864), pp. 30—32, followed by a note by Brioschi, pp. 32—33.

The skew covariant of the ninth order, which is $y^3 - z^3 . z^3 - x^3 . x^3 - y^3$ for the canonical form $x^3 + y^3 + z^3 + 6lxyz$, and the corresponding contravariant $\eta^3 - \zeta^3 . \zeta^3 - \xi^3 . \xi^3 - \eta^3$, alluded to p. 116 of Salmon's *Lessons*, were obtained, the covariant by Brioschi and the contravariant by Hermite, in the last-mentioned papers.

CLEBSCH and GORDAN, Ueber die Theorie der ternären cubischen Formen: *Math. Annalen*, t. I. (1869), pp. 56—89.

The establishment of the complete system of the 34 covariants, contravariants and *Zwischenformen*, or, as I have here called them, the 34 concomitants, was first effected by Gordan in the next following memoir:

GORDAN, Ueber die ternären Formen dritten Grades: *Math. Annalen*, t. I. (1869), pp. 90—128.

And the theory is further considered:

GUNDELFINGER, Zur Theorie der ternären cubischen Formen: *Math. Annalen*, t. VI. (1871), pp. 144—163. The author speaks of the 34 forms as being "theils mit den von Gordan gewählten identisch, theils möglichst einfache Combinationen derselben." They are, in fact, the 34 forms given in the present paper for the canonical form of the cubic, and the meaning of the adopted combinations of Gordan's forms will presently clearly appear.

There is an advantage in using the form $ax^3 + by^3 + cz^3 + 6lxyz$ rather than the Hessian form $x^3 + y^3 + z^3 + 6lxyz$, employed in my Third and Seventh Memoirs on Quantics: for the form $ax^3 + by^3 + cz^3 + 6lxyz$ is what the general cubic

$$(a, b, c, f, g, h, i, j, k, l)(x, y, z)^3$$

becomes by no other change than the reduction to zero of certain of its coefficients; and thus any concomitant of the canonical form consists of terms which are leading terms of the same concomitant of the general form.

The concomitants are functions of the coefficients $(a, b, \ldots, l)$, of (ξ, η, ζ), and of (x, y, z): the dimensions in regard to the three sets respectively may be distinguished as the degree, class, and order; and we have thus to consider the deg-class-order of a concomitant.

Two or more concomitants of the same deg-class-order may be linearly combined together: viz., the linear combination is the sum of the concomitants each multiplied by a mere number. The question thus arises as to the selection of a representative concomitant. As already mentioned, I follow Gundelfinger, viz., my 34 concomitants of the canonical form correspond each to each (with only the difference of a numerical factor of the entire concomitant) to his 34 concomitants of the general form. The principle underlying the selection would, in regard to the general form, have to be explained altogether differently; but this principle exhibits itself in a very remarkable manner in regard to the canonical form $ax^3 + by^3 + cz^3 + 6lxyz$.

Each concomitant of the general form is an indecomposable function, not breaking up into rational factors; but this is not of necessity the case in regard to a canonical form: only a concomitant which *does* break up must be regarded as indecomposable, no factor of such concomitant being rejected, or separated. So far from it, there is, in regard to the canonical form in question, a frequent occurrence of $abc+8l^3$ or a power thereof, either as a factor of a unique concomitant, or when there are two or more concomitants of the same deg-class-order, then as a factor of a properly selected linear combination of such concomitants: and the principle referred to is, in fact, that of the selection of such combination for the representative concomitant; or (in other words) the representative concomitant is taken so as to contain as a factor the highest power that may be of $abc+8l^3$. As to the signification of this expression $abc+8l^3$, I call to mind that the discriminant of the form is $abc(abc+8l^3)^3$.

As to numerical factors: my principle has been, and is, to throw out any common numerical divisor of all the terms: thus I write $S=-abcl+l^4$, instead of Aronhold's $S=-4abcl+4l^4$. There is also the question of nomenclature: I retain that of my Seventh Memoir on Quantics, except that I use single letters H, P, &c., instead of the same letters with U, thus HU, PU, &c.; in particular, I use U, H, P, Q instead of Aronhold's f, Δ, S_f, T_f. It is thus at all events necessary to make some change in Gundelfinger's letters; and there is moreover a laxity in his use of accented letters; his B, B', B'', B''', and so in other cases E, E', E'', &c., are used to denote functions derived in a determinate manner each from the preceding one (by the δ-process explained further on); whereas his L, L'; M, M'; N, N' are functions having to each other an altogether different relation; also three of his functions are not denoted by any letters at all. Under the circumstances, I retain only a few of his letters; use the accent where it denotes the δ-process; and introduce barred letters $\bar{J}$, $\bar{K}$, &c., to denote a different correspondence with the unbarred letters J, K, &c. But I attach also to each concomitant a numerical symbol showing its deg-class-order, thus: 541 (degree $=5$, class $=4$, order $=1$) or 1290, (there is no ambiguity in the two-digit numbers 10, 11, 12 which present themselves in the system of the 34 symbols); and it seems to me very desirable that the significations of these deg-class-order symbols should be considered as permanent and unalterable. Thus, in writing $S=400=-abcl+l^4$, I wish the 400 to be regarded as denoting its expressed value $-abcl+l^4$: if the same letter S is to be used in Aronhold's sense to denote $-4abcl+4l^4$, this would be completely expressed by the new definition $S=4.400$, the meaning of the symbol 400 being explained by reference to the present memoir, or by the actual quotation $400=-abcl+l^4$.

I proceed at once to the table: for shortness, I omit, in general, terms which can be derived from an expressed term by mere cyclical interchanges of the letters (a, b, c), (ξ, η, ζ), (x, y, z).

Table of the 34 Covariants of the Canonical Cubic $ax^3+by^3+cz^3+6lxyz$ and the linear form $\xi x+\eta y+\zeta z$.

First Part, 10 Forms. Class = Order.

Current No.

1 $S = 400 = -abcl + l^4.$

2 $T = 600 = a^2b^2c^2 - 20abcl^3 - 8l^6.$

3 $\Lambda = 011 = \xi x + \eta y + \zeta z.$

4 $\Theta = 222 = x^2[-l^2\xi^2 - 2al\eta\zeta]\ldots$
$+ yz[bc\xi^2 + 2l^2\eta\zeta]\ldots.$

5 $\Theta' = 422 = x^2[l(abc+2l^3)\xi^2 + a(abc-4l^3)\eta\zeta]\ldots$
$+ yz[6bcl^2\xi^2 - 2l(abc+2l^3)\eta\zeta]\ldots.$

6 $\Theta'' = 622 = x^2[-(abc+2l^3)^2\xi^2 + 12al^2(abc+2l^3)\eta\zeta]\ldots$
$+ yz[36bcl^4\xi^2 + 2(abc+2l^3)^2\eta\zeta]\ldots.$

7 $B = 333 = x^3[a^2(c\eta^3 - b\zeta^3)]\ldots$
$+ y^2z[(abc+8l^3)\eta^2\zeta + 12bl^2\zeta^2\xi + 6bcl\xi^2\eta]\ldots$
$+ yz^2[-(abc+8l^3)\eta\zeta^2 - 6bcl\zeta\xi^2 - 12cl^2\xi\eta^2]\ldots.$

8 $B' = 533 = x^3[3a^2l^2(c\eta^3 - b\zeta^3)]\ldots$
$+ y^2z[-l^2(abc+8l^3)\eta^2\zeta + 4bl(-abc+l^3)\zeta^2\xi - bc(abc-10l^3)\xi^2\eta]\ldots$
$+ yz^2[l^2(abc+8l^3)\eta\zeta^2 + bc(abc-10l^3)\zeta\xi^2 - 4cl(-abc+l^3)\xi\eta^2]\ldots.$

9 $B'' = 733 = x^3[9a^2l^4(c\eta^3 - b\zeta^3)]\ldots$
$+ y^2z[l(abc+8l^3)(2abc+l^3)\eta^2\zeta$
$+ b(abc+2l^3)(abc-10l^3)\zeta^2\xi + 6bcl^2(-abc+l^3)\xi^2\eta]\ldots$
$+ yz^2[-l(abc+8l^3)(2abc+l^3)\eta\zeta^2$
$- 6bcl^2(-abc+l^3)\zeta\xi^2 - c(abc+2l^3)(abc-10l^3)\xi\eta^2]\ldots.$

10 $B''' = 933 = x^3[27a^2l^6(c\eta^3 - b\zeta^3)]\ldots$
$+ y^2z[-(abc+8l^3)(abc-l^3)^2\eta^2\zeta + 9bl^2(abc+2l^3)^2\zeta^2\xi$
$- 27bcl^4(abc+2l^3)\xi^2\eta]\ldots$
$+ yz^2[(abc+8l^3)(abc-l^3)^2\eta\zeta^2 + 27bcl^4(abc+2l^3)\zeta\xi^2$
$- 9cl^2(abc+2l^3)^2\xi\eta^2]\ldots.$

Second Part, (4 + 4 =) 8 forms. Class = 0, and Order = 0.

Class = 0.

11 $U = 103 = ax^3 + by^3 + cz^3 + 6lxyz.$

12 $H = 303 = l^2(ax^3+by^3+cz^3) - (abc+2l^3)xyz.$

13 $\Psi = 806 = (abc+8l^3)^2\{a^2x^6 + b^2y^6 + c^2z^6 - 10(bcy^3z^3 + caz^3x^3 + abx^3y^3)\}.$

14 $\Omega = 1209 = (abc+8l^3)^3\{by^3 - cz^3 \,.\, cz^3 - ax^3 \,.\, ax^3 - by^3\}.$

Current No. Order = 0.

15 $P = 330 = -l(bc\xi^3 + ca\eta^3 + ab\zeta^3) + (-abc + 4l^3)\xi\eta\zeta.$

16 $Q = 530 = (abc - 10l^3)(bc\xi^3 + ca\eta^3 + ab\zeta^3) - 6l^2(5abc + 4l^3)\xi\eta\zeta.$

17 $F = 460 = b^2c^2\xi^6 + c^2a^2\eta^6 + a^2b^2\zeta^6 - 2(abc + 16l^3)(a\eta^3\zeta^3 + b\zeta^3\xi^3 + c\xi^3\eta^3)$
$- 24l^2(bc\xi^3 + ca\eta^3 + ab\zeta^3)\xi\eta\zeta - 24l(abc + 2l^3)\xi^2\eta^2\zeta^2.$

18 $\Pi = 1290 = (abc + 8l^3)^3\{c\eta^3 - b\zeta^3 \,.\, a\zeta^3 - c\xi^3 \,.\, b\xi^3 - a\eta^3\}.$

Third Part, $(8 + 8 =)$ 16 forms. Class less or greater than Order.

Class less than Order.

19 $J = 414 = (abc + 8l^3)\{\xi x(by^3 - cz^3) + \eta y(cz^3 - ax^3) + \zeta z(ax^3 - by^3)\}.$

20 $K = 514 = (abc + 8l^3)\{\xi[alx^4 - 2blxy^3 - 2clxz^3 + 3bcy^2z^2]\ldots\}.$

21 $K' = 714 = (abc + 8l^3)\{\xi[(abc + 2l^3)(ax^4 - 2bxy^3 - 2cxz^3) - 18bcl^2y^2z^2]\ldots\}.$

22 $E = 625 = (abc + 8l^3)\{\xi^2(by^3 - cz^3)[2l^2x^2 + bcyz]\ldots$
$+ \eta\zeta(by^3 - cz^3)[4alx^2 + 2l^2yz]\ldots\}.$

23 $E' = 825 = (abc + 8l^3)\{\xi^2(by^3 - cz^3)[l(abc + 2l^3)x^2 - 3bcl^2yz]\ldots$
$+ \eta\zeta(by^3 - cz^3)[a(abc - 4l^3)x^2 + l(abc + 2l^3)yz]\ldots\}.$

24 $E'' = 1025 = (abc + 8l^3)\{\xi^2(by^3 - cz^3)[(abc + 2l^3)^2x^2 + 18bcl^4yz]\ldots$
$+ \eta\zeta(by^3 - cz^3)[-12al^2(abc + 2l^3)x^2 + (abc + 2l^3)^2yz]\ldots\}.$

25 $M = 917 = (abc + 8l^3)^2\{\xi(by^3 - cz^3)[5alx^4 - blxy^3 - clxz^3 - 3bcy^2z^2]\ldots\}.$

26 $M' = 1117 = (abc + 8l^3)^2\{\xi(by^3 - cz^3)[(abc + 2l^3)(5ax^4 - bxy^3 - cxz^3)$
$+ 18bcl^2y^2z^2]\ldots\}.$

Order less than Class.

27 $\bar{J} = 841 = (abc + 8l^3)^2\{x\xi a(c\eta^3 - b\zeta^3) + y\eta b(a\zeta^3 - c\xi^3) + z\zeta c(b\xi^3 - a\eta^3)\}.$

28 $\bar{K} = 541 = (abc + 8l^3)\{x[bc\xi^4 - 2ca\xi\eta^3 - 2ab\xi\zeta^3 - 6al\eta^2\zeta^2]\ldots\}.$

29 $\bar{K'} = 741 = (abc + 8l^3)\{x[l^2(bc\xi^4 - 2ca\xi\eta^3 - 2ab\xi\zeta^3) + a(abc + 2l^3)\eta^2\zeta^2]\ldots\}.$

30 $\bar{E} = 652 = (abc + 8l^3)\{x^2(c\eta^3 - b\zeta^3)[2al\xi^2 + a^2\eta\zeta]\ldots$
$+ yz(c\eta^3 - b\zeta^3)[4l^2\xi^2 + 2al\eta\zeta]\ldots\}.$

31 $\bar{E'} = 852 = (abc + 8l^3)\{x^2(c\eta^3 - b\zeta^3)[a(abc - 4l^3)\xi^2 - 6a^2l^2\eta\zeta]\ldots$
$+ yz(c\eta^3 - b\zeta^3)[4l(abc + 2l^3)\xi^2 + a(abc - 4l^3)\eta\zeta]\ldots\}.$

32 $\bar{E''} = 1052 = (abc + 8l^3)\{x^2(c\eta^3 - b\zeta^3)[-3al^2(abc + 2l^3)\xi^2 + 9a^2l^4\eta\zeta]\ldots$
$+ yz(c\eta^3 - b\zeta^3)[(abc + 2l^3)^2\xi^2 - 3al^2(abc - 4l^3)\eta\zeta]\ldots\}.$

33 $\bar{M} = 771 = (abc + 8l^3)\{x(c\eta^3 - b\zeta^3)[(abc - 8l^3)\xi^4 - a^2c\xi\eta^3 - a^2b\xi\zeta^3$
$- 12al^2\xi^2\eta\zeta - 6a^2l\eta^2\zeta^2]\ldots\}.$

34 $\bar{M'} = 971 = (abc + 8l^3)\{x(c\eta^3 - b\zeta^3)[l^2(7abc + 8l^3)\xi^4 - 3a^2cl^2\xi\eta^3 - 3a^2bl^2\xi\zeta^3$
$+ 4al(abc - l^3)\xi^2\eta\zeta + a^2(abc - 10l^3)\eta^2\zeta^2]\ldots\}.$

To this may be joined the following Supplemental Table of certain Derived Forms:

Current No.		
35	$R = 1200 =$	$64S^3 - T^2 = -abc\,(abc + 8l^3)^3.$
36	$C = 703 =$	$-TU + 24SH = (abc + 8l^3)\,\{(-abc + 4l^3)(ax^3 + by^3 + cz^3) + 18abclxyz\}.$
37	$D = 903 =$	$8S^2U - 3TH = (abc + 8l^3)\,\{l^2(5abc + 4l^3)(ax^3 + by^3 + cz^3) + 3abc\,(abc - 10l^3)\,xyz\}.$
38	$Y = 930 =$	$3TP - 4SQ = (abc + 8l^3)^2\,\{l\,(bc\xi^3 + ca\eta^3 + ab\zeta^3) - 3abc\xi\eta\zeta\}.$
39	$Z = 1130 =$	$-48S^2P + TQ = (abc + 8l^3)^2\,\{(abc + 2l^3)(bc\xi^3 + ca\eta^3 + ab\zeta^3) + 18abcl^2\xi\eta\zeta\}.$
40	$\Phi = 1660 =$	$12\,(abc + 8l^3)^3F - 288STP^2 + 768S^2PQ - 8TQ^2 = (abc + 8l^3)^4\,\{b^2c^2\xi^6 + c^2a^2\eta^6 + a^2b^2\zeta^6 - 10abc\,(a\eta^3\zeta^3 + b\zeta^3\xi^3 + c\xi^3\eta^3)\},$

viz. these are derived forms characterized by having a power of $abc + 8l^3$ as a factor: R is the discriminant; C, D, Y, Z occur in Aronhold, and in my Seventh memoir on Quantics [269]: Φ in Clebsch and Gordan's memoir of 1869.

I regard as known forms Λ, U, H, P, Q, S, T, F, that is, the eight forms 3, 11, 12, 15, 16, 1, 2, 17; the remaining 26 forms are expressed in terms of these by formulæ involving notations which will be explained, viz. we have

13 $\Psi = 3\,(\mathrm{bc}' + \mathrm{b}'\mathrm{c} - 2\mathrm{ff}', \ldots, \mathrm{gh}' + \mathrm{g}'\mathrm{h} - \mathrm{af}' - \mathrm{a}'\mathrm{f}, \ldots \between X, Y, Z \between X', Y', Z') + TU^2.$

14 $\Omega = \frac{1}{18}\,\mathrm{Jac}\,(U, H, \Psi).$

18 $\Pi = -\frac{1}{36}\,[\mathrm{Jac}]\,(P, Q, F).$

4 $\Theta = (\mathrm{bc} - \mathrm{f}^2, \ldots, \mathrm{gh} - \mathrm{af}, \ldots \between \xi, \eta, \zeta)^2.$

5 $\Theta' = \frac{1}{2}\,\delta\Theta.$

6 $\Theta'' = \frac{1}{2}\,\delta^2\Theta.$

7 $B = -\frac{1}{3}\,\mathrm{Jac}\,(U, \Theta, \Lambda).$

8 $B' = \frac{1}{6}\,\delta B.$

9 $B'' = \frac{1}{24}\,\delta^2 B.$

10 $B''' = \frac{1}{48}\,\delta^3 B.$

19 $J = -\frac{1}{3}\,\mathrm{Jac}\,(U, H, \Lambda).$

27 $\bar{J} = \frac{1}{3}\,[\mathrm{Jac}]\,(P, Q, \Lambda).$

20 $K = -\frac{3}{2}\,\{\partial_\xi\Theta\partial_x H + \partial_\eta\Theta\partial_y H + \partial_\zeta\Theta\partial_z H\} - SU\Lambda.$

21 $K' = -(\delta)\,K.$

28 $\bar{K} = 3\,\{\partial_x\Theta\partial_\xi P + \partial_y\Theta\partial_\eta P + \partial_z\Theta\partial_\zeta P\} + Q\Lambda.$

29 $\bar{K}' = \frac{1}{6}\,(\delta)\,\bar{K}.$

22 $E = -\frac{1}{18}\,\mathrm{Jac}\,(K, U, \Lambda).$

23 $E' = -\frac{1}{4}\,(\delta)\,E.$

24 $E'' = \ \tfrac{1}{4}(\delta^2)E.$

30 $\bar{E} = -\tfrac{1}{9}\text{Jac}(\bar{K},\ U,\ \Lambda).$

31 $\bar{E}' = -\tfrac{1}{2}(\delta)\bar{E}.$

32 $\bar{E}'' = -\tfrac{1}{8}(\delta^2)\bar{E}.$

25 $M = \ \tfrac{1}{36}\text{Jac}(U,\ \Psi,\ \Lambda).$

26 $M' = -(\delta)M.$

33 $\bar{M} = -\tfrac{1}{6}[\text{Jac}](P,\ F,\ \Lambda).$

34 $\bar{M}' = \ \tfrac{1}{6}(\delta)\bar{M}.$

In explanation of the notations, observe that

$$U = \quad ax^3 + by^3 + cz^3 + 6lxyz,$$

$$H = l^2(ax^3 + by^3 + cz^3) - (abc + 2l^3)\,xyz.$$

Hence, writing

$$6H = \quad a'x^3 + b'y^3 + c'z^3 + 6l'xyz,$$

we have

$$a',\ b',\ c',\ l' = 6al^2,\ 6bl^2,\ 6cl^2,\ -(abc + 2l^3).$$

And this being so, we write

$$X,\ Y,\ Z \qquad = ax^2 + 2lyz,\ by^2 + 2lzx,\ cz^2 + 2lxy,$$

$$\text{a, b, c, f, g, h} = ax,\ by,\ cz,\ lx,\ ly,\ lz,$$

for $\tfrac{1}{3}$ of the first differential coefficients, and $\tfrac{1}{6}$ of the second differential coefficients of U; and in like manner

$$X',\ Y',\ Z' \qquad = a'x^2 + 2l'yz,\ b'y^2 + 2l'zx,\ c'z^2 + 2l'xy,$$

$$\text{a}',\ \text{b}',\ \text{c}',\ \text{f}',\ \text{g}',\ \text{h}' = a'x,\ b'y,\ c'z,\ l'x,\ l'y,\ l'z,$$

for $\tfrac{1}{3}$ of the first differential coefficients, and $\tfrac{1}{6}$ of the second differential coefficients of $6H$.

Jac is written to denote the Jacobian, viz.:

$$\text{Jac}(U,\ H,\ \Psi) = \begin{vmatrix} \partial_x U, & \partial_y U, & \partial_z U \\ \partial_x H, & \partial_y H, & \partial_z H \\ \partial_x \Psi, & \partial_y \Psi, & \partial_z \Psi \end{vmatrix},$$

and in like manner [Jac] to denote the Jacobian, when the differentiations are in regard to $(\xi,\ \eta,\ \zeta)$ instead of $(x,\ y,\ z)$: δ is the symbol of the δ-process, or substitution of the coefficients $(a',\ b',\ c',\ l')$ in place of $(a,\ b,\ c,\ l)$; in fact,

$$\delta = a'\partial_a + b'\partial_b + c'\partial_c + l'\partial_l:$$

δ, δ^2, &c., each operate directly on a function of $(a,\ b,\ c,\ l)$, the $(a',\ b',\ c',\ l')$ of the symbol δ being in the first instance regarded as constants, and being replaced ultimately by their values; for instance,

$$\delta abc = a'bc + ab'c + abc',\ \ \delta^2 abc = 2\,(ab'c' + a'bc' + a'b'c),\ \ \delta^3 abc = 6a'b'c'.$$

In several of the formulæ, instead of δ or δ^2, the symbol used is (δ) or (δ^2); in these cases, the function operated upon contains the factor $(abc+8l^3)$ or $(abc+8l^3)^2$, and is of the form $(abc+8l^3)(aU+bV+cW)$ or $(abc+8l^3)^2\ (a^2U+abV+\&c.)$: the meaning is, that the δ or δ^2 is supposed to operate through the $(abc+8l^3)a$, or $(abc+8l^3)^2a^2$, &c., as if this were a constant, upon the U, V, &c., only; thus: $(\delta).(abc+8l^3)(aU+bV+cW)$ is used to denote $(abc+8l^3)(a\delta U+b\delta V+c\delta W)$. As to this, observe that, operating with δ instead of (δ), there would be the additional terms $U\delta(abc+8l^3)a+\&c.$; we have in this case

$$\delta(abc+8l^3)a, =a(2a'bc+ab'c+abc'+24l^2l')+8l^3a',$$
$$=24a^2bcl^2-24al^2(abc+2l^3)+48al^5, =0;$$

or the rejected terms in fact vanish. For $(\delta^2).(abc+8l^3)(aU+bV+cW)$, operating with δ^2, we should have, in like manner, terms $U\delta^2(abc+8l^3)a$, &c.; here

$$\delta^2(abc+8l^3)a=a'^2bc+2aba'c'+2aca'b'+a^2b'c'+24l^2a'l'+24all'^2,$$

which is found to be $=-24a(abc+8l^3)(-abcl+l^4)$, that is, $=-24S(abc+8l^3)a$; and the terms in question are thus $=-24S(abc+8l^3)(aU+bV+cW)$, viz.

$$(abc+8l^3)(aU+bV+cW)$$

being a covariant, this is also a covariant; that is, in using (δ^2) instead of δ^2, we in fact reject certain covariant terms; or say, for instance, δ^2E being a covariant, then $(\delta^2)E$ is also a covariant, but a different covariant. The calculation with (δ) or (δ^2) is more simple than it would have been with δ or δ^2. See *post*, the calculations of K, $\overline{K}'$, &c.

I give for each of the 26 covariants a calculation showing how at least a single term of the final result is arrived at, and, in the several cases for which there is a power of $abc+8l^3$ as a factor, showing how this factor presents itself.

Calculations for the 26 Covariants.

13. $$\Psi=3(\mathrm{bc}'+\mathrm{b}'\mathrm{c}-2\mathrm{ff}', \ldots, \mathrm{gh}'+\mathrm{g}'\mathrm{h}-\mathrm{af}'-\mathrm{a}'\mathrm{f}, \ldots \between X, Y, Z \between X', Y', Z')+TU^2,$$
$$=3((bc'+b'c)yz-2ll'x^2, \ldots, 2ll'yz-(al'+a'l)x^2, \ldots \between ax^2+2lyz, \ldots \between a'x^2+2l'yz, \ldots)$$
$$+T(a^2x^6+\ldots).$$

The whole coefficient of x^6 is

$$-6ll'aa'+Ta^2, =36a^2l^3(abc+2l^3)+Ta^2,$$

viz. the coefficient of a^2x^6 is

$$=36l^3(abc+2l^3)+a^2b^2c^2-20abcl^3-8l^6$$
$$=a^2b^2c^2+16abcl^3+64l^6$$
$$=(abc+8l^3)^2.$$

14. $$\Omega=\tfrac{1}{18}\operatorname{Jac}(U, H, \Psi), =\tfrac{1}{2}\begin{vmatrix} X, & X', & \tfrac{1}{6}\partial_x\Psi \\ Y, & Y', & \tfrac{1}{6}\partial_y\Psi \\ Z, & Z', & \tfrac{1}{6}\partial_z\Psi \end{vmatrix}.$$

Here

$$\begin{aligned} YZ' - Y'Z &= (by^2 + 2lzx)(c'z^2 + 2l'xy) - (cz^2 + 2lxy)(b'z^2 + 2l'xy) \\ &= (bc' - b'c)\, y^2z^2 + (2bl' - b'l)\, xy^3 - 2(cl' - c'l)\, xz^3 \\ &= -2(abc + 8l^3)\, x\,(by^3 - cz^3); \\ \tfrac{1}{2} \cdot \tfrac{1}{6} \partial_x \Psi &= \tfrac{1}{2}(a^2x^5 - 5abx^2y^3 - 5acx^2z^3). \end{aligned}$$

Hence the whole is

$$\begin{aligned} &= -(abc + 8l^3)\{a^2x^6(by^3 - cz^3) + b^2y^6(cz^3 - ax^3) + c^2z^6(ax^3 - by^3)\}, \\ &= \quad (abc + 8l^3)(by^3 - cz^3)(cz^3 - ax^3)(ax^3 - by^3). \end{aligned}$$

18. $$\Pi = -\tfrac{1}{36}[\text{Jac}](P,\ Q,\ F) = -\tfrac{1}{36}\begin{vmatrix} \partial_\xi P, & \partial_\xi Q, & \partial_\xi F \\ \partial_\eta P, & \partial_\eta Q, & \partial_\eta F \\ \partial_\zeta P, & \partial_\zeta Q, & \partial_\zeta F \end{vmatrix},$$

viz. if, in this calculation, we write

$$\begin{aligned} 6P &= \mathrm{a}\,\xi^3 + \mathrm{b}\,\eta^3 + \mathrm{c}\,\zeta^3 + 6\mathrm{l}\,\xi\eta\zeta, \quad \textit{i.e.}\ \mathrm{a},\ \mathrm{b},\ \mathrm{c},\ \mathrm{l} = -6lbc,\ -6lca,\ -6lab,\ -abc + 4l^3, \\ Q &= \mathrm{a}'\xi^3 + \mathrm{b}'\eta^3 + \mathrm{c}'\zeta^3 + 6\mathrm{l}'\xi\eta\zeta, \quad ,,\ \mathrm{a}',\ \mathrm{b}',\ \mathrm{c}',\ \mathrm{l}' = (abc - 10l^3)(bc,\ ca,\ ab),\ -l^2(5abc + 4l^3), \end{aligned}$$

then

$$\Pi = -\tfrac{1}{4}\begin{vmatrix} \mathrm{a}\xi^2 + 2\mathrm{l}\eta\zeta, & \mathrm{a}'\xi^2 + 2\mathrm{l}'\eta\zeta, & \tfrac{1}{6}\partial_\xi F \\ \mathrm{b}\eta^2 + 2\mathrm{l}\zeta\xi, & \mathrm{b}'\eta^2 + 2\mathrm{l}'\zeta\xi, & \tfrac{1}{6}\partial_\eta F \\ \mathrm{c}\zeta^2 + 2\mathrm{l}\xi\eta, & \mathrm{c}'\zeta^2 + 2\mathrm{l}'\xi\eta, & \tfrac{1}{6}\partial_\zeta F \end{vmatrix}.$$

Here

$$\begin{aligned} &(\mathrm{b}\eta^2 + 2\mathrm{l}\zeta\xi)(\mathrm{c}'\zeta^2 + 2\mathrm{l}'\xi\eta) - (\mathrm{b}'\eta^2 + 2\mathrm{l}'\zeta\xi)(\mathrm{c}\zeta^2 + 2\mathrm{l}\xi\eta) \\ &= (\mathrm{bc}' - \mathrm{b}'\mathrm{c})\,\eta^2\zeta^2 + 2(\mathrm{bl}' - \mathrm{b}'\mathrm{l})\,\xi\eta^3 - 2(\mathrm{cl}' - \mathrm{c}'\mathrm{l})\,\xi\zeta^3, \end{aligned}$$

or since

$$\begin{aligned} \mathrm{bc}' - \mathrm{b}'\mathrm{c} &= 0, \\ \mathrm{bl}' - \mathrm{b}'\mathrm{l} &= -6lca \,.\, -l^2(5abc + 4l^3) - (abc - 10l^3)\,ca\,(-abc + 4l^3) \\ &= ca\,\{6l^3(5abc + 4l^3) + (abc - 4l^3)(abc - 10l^3)\} \\ &= ca\,(abc + 8l^3)^2, \end{aligned}$$

and the like for $\mathrm{cl}' - \mathrm{c}'\mathrm{l}$, the expression is

$$= 2(abc + 8l^3)^2(ca\eta^3 - ab\zeta^3)\,\xi;$$

and the whole is thus

$$\begin{aligned} &= -\tfrac{1}{2}(abc + 8l^3)^2\{(ca\eta^3 - ab\zeta^3)\,\xi \,.\, \tfrac{1}{6}\partial_\xi F + \ldots\} \\ &= -\tfrac{1}{2}(abc + 8l^3)^2\{(ca\eta^3 - ab\zeta^3)[b^2c^2\xi^6 - (abc + 16l^3)(b\zeta^3\xi^3 + c\xi^3\eta^3) + \&\text{c.}] \\ &\qquad + (ab\zeta^3 - bc\xi^3)[c^2a^2\eta^6 - (abc + 16l^3)(c\xi^3\eta^3 + a\eta^3\zeta^3) + \&\text{c.}] \\ &\qquad + (bc\xi^3 - ca\eta^3)[a^2b^2\zeta^6 - (abc + 16l^3)(a\eta^3\zeta^3 + b\zeta^3\xi^3) + \&\text{c.}]\}. \end{aligned}$$

Here the coefficient of $\xi^6\eta^3$, inside the $\{\}$, is

$$ab^2c^3 + bc^2(abc + 16l^3),\ = 2bc^2(abc + 8l^3),$$

and consequently the whole is

$$= -(abc + 8l^3)^3 (bc^2\xi^6\eta^3 - \ldots),$$
$$= \quad (abc + 8l^3)^3 \{(c\eta^3 - b\zeta^3)(a\zeta^3 - c\xi^3)(b\xi^3 - a\eta^3)\}.$$
$$\Theta = (\text{bc} - \text{f}^2, \ldots, \text{gh} - \text{af}, \ldots \between \xi, \eta, \zeta)^2$$
$$= (bcyz - l^2x^2)\,\xi^2 + \ldots + 2\,(l^2yz - alx^2)\,\eta\zeta + \ldots$$

which are the terms of the final result

$$\Theta = x^2\,[-l^2\xi^2 - 2al\eta\zeta] + yz\,[bc\xi^2 + 2l^2\eta\zeta].$$

5 and 6. The δ-process applied to the terms of Θ just written down gives

$$\Theta' = \tfrac{1}{2}\,\delta\Theta \;= x^2\,[-ll'\xi^2 - (al' + a'l)\,\eta\zeta] + yz\,[\tfrac{1}{2}(bc' + b'c)\,\xi^2 + 2ll'\eta\zeta],$$
$$\Theta'' = \tfrac{1}{2}\,\delta^2\Theta = x^2\,[-l'^2\xi^2 - 2a'l'\eta\zeta] + yz\,[b'c'\xi^2 + 2l'^2\eta\zeta];$$

substituting for a', b', c', l' their values, we have the corresponding terms of Θ' and Θ'' respectively.

7. $$B = -\tfrac{1}{3}\,\text{Jac}\,(U,\ \Theta,\ \Lambda), \ = -\begin{vmatrix} X, & \partial_x\Theta, & \xi \\ Y, & \partial_y\Theta, & \eta \\ Z, & \partial_z\Theta, & \zeta \end{vmatrix}.$$

A term is $X\,(\eta\partial_z\Theta - \zeta\partial_y\Theta)$, and if, in this calculation, we write

$$\Theta = (A,\ B,\ C,\ F,\ G,\ H \between x,\ y,\ z)^2,\ \textit{i.e.}\ A = -l^2\xi^2 - 2al\eta\zeta,\ \ldots,\ F = \tfrac{1}{2}\,bc\xi^2 + l^2\eta\zeta,$$

then the term is

$$= (ax^2 + 2lyz)\,\{x\,.\,2\,(G\eta - H\zeta) + y\,.\,2\,(F\eta - B\zeta) + z\,.\,2\,(C\eta - F\zeta)\}.$$

Here

$$2\,(G\eta - H\zeta) = \eta\,(ca\eta^2 + l^2\zeta\xi) - \zeta\,(ab\zeta^2 + l^2\xi\eta),\ = a\,(c\eta^3 - b\zeta^3),$$

and hence the whole term in x^3 is $= a^2x^3\,(c\eta^3 - b\zeta^3)$.

8, 9, 10. The coefficient of $x^3\eta^3$ in B is a^2c, and hence in δB, δ^2B, δ^3B the coefficients of this term are $2a'ac + a^2c'$, $2a'^2c + 4aa'c'$, $6a'^2c'$, whence in

$$B',\ B'',\ B''' = \tfrac{1}{6}\,\delta B,\ \tfrac{1}{24}\,\delta^2B,\ \tfrac{1}{48}\,\delta^3B \text{ respectively,}$$

the coefficients are

$$\tfrac{1}{6}\,(a^2c' + 2aa'c),\quad \tfrac{1}{12}\,(a'^2c + 2aa'c'),\quad \tfrac{1}{8}\,a'^2c',$$
$$= \qquad 3l^2a^2c, \qquad\qquad 9l^4a^2c, \qquad\qquad 27l^6a^2c \text{ respectively.}$$

19 $$J = -\tfrac{1}{3}\,\text{Jac}\,(U,\ H,\ \Lambda) = -\tfrac{1}{2}\begin{vmatrix} X, & X', & \xi \\ Y, & Y', & \eta \\ Z, & Z', & \zeta \end{vmatrix};$$

a term is $-\tfrac{1}{2}\,(YZ' - Y'Z)\,\xi$, where, as in a previous calculation,

$$YZ' - Y'Z = -2\,(abc + 8l^3)\,x\,(by^3 - cz^3).$$

Hence, the whole is

$$= (abc + 8l^3)\{\xi x(by^3 - cz^3) + \eta y(cz^3 - ax^3) + \zeta z(ax^3 - by^3)\}.$$

27.
$$\bar{J} = \tfrac{1}{3}[\text{Jac}](P, Q, \Lambda) = \tfrac{1}{2}\begin{vmatrix} \mathrm{a}\xi^2 + 2\mathrm{l}\eta\zeta, & \mathrm{a}'\xi^2 + 2\mathrm{l}'\eta\zeta, & x \\ \mathrm{b}\eta^2 + 2\mathrm{l}\zeta\xi, & \mathrm{b}'\eta^2 + 2\mathrm{l}'\zeta\xi, & y \\ \mathrm{c}\zeta^2 + 2\mathrm{l}\xi\eta, & \mathrm{c}'\zeta^2 + 2\mathrm{l}'\xi\eta, & z \end{vmatrix},$$

if, as in a previous calculation

$$6P = \mathrm{a}\xi^3 + \mathrm{b}\eta^3 + \mathrm{c}\zeta^3 + 6\mathrm{l}\xi\eta\zeta, \quad Q = \mathrm{a}'\xi^3 + \mathrm{b}'\eta^3 + \mathrm{c}'\zeta^3 + 6\mathrm{l}'\xi\eta\zeta.$$

Here, as before,

$$(\mathrm{b}\eta^2 + 2\mathrm{l}\zeta\xi)(\mathrm{c}'\zeta^2 + 2\mathrm{l}'\xi\eta) - (\mathrm{b}'\eta^2 + 2\mathrm{l}'\zeta\xi)(\mathrm{c}\zeta^2 + 2\mathrm{l}\xi\eta) = 2(abc + 8l^3)^2(ca\eta^3 - ab\zeta^3)\,\xi.$$

Hence, the whole is

$$= (abc + 8l^3)^2\{x\xi a(c\eta^3 - b\zeta^3) + y\eta b(a\zeta^3 - c\xi^3) + z\zeta c(b\xi^3 - a\eta^3)\}.$$

20.
$$K = -\tfrac{3}{2}(\partial_\xi\Theta\partial_x H + \partial_\eta\Theta\partial_y H + \partial_\zeta\Theta\partial_z H) - SU\Lambda,$$

which, H being

$$= \tfrac{1}{6}(a'x^3 + b'y^3 + c'z^3 + 6l'xyz),$$

and putting

$$\Theta = (A, B, C, F, G, H \between \xi, \eta, \zeta)^2, \quad A = -l^2x^2 + bcyz, \ldots, F = -alx^2 + l^2yz, \ldots,$$

is

$$\begin{aligned} = -\tfrac{3}{2}\{&(a'x^2 + 2l'yz)(A\xi + H\eta + G\zeta) - (-abcl + l^4)\,U(\xi x + \eta y + \zeta z) \\ &+ (b'y^2 + 2l'zx)(H\xi + B\eta + F\zeta) \\ &+ (c'z^2 + 2l'xy)(G\xi + F\eta + C\zeta)\}. \end{aligned}$$

The whole coefficient of ξ is thus

$$\begin{aligned} &= -\tfrac{3}{2}\{(a'x^2 + 2l'yz)A + (b'y^2 + 2l'zx)H + (c'z^2 + 2l'xy)G\} - (-abcl + l^4)\,Ux \\ &= -\tfrac{3}{2}\{(a'x^2 + 2l'yz)(-l^2x^2 + bcyz) + (b'y^2 + 2l'zx)(-clz^2 + l^2xy) \\ &\quad + (c'z^2 + 2l'xy)(-bly^2 + l^2zx)\} - (-abcl + l^4)\{ax^4 + bxy^3 + cxz^3 + 6lx^2yz\}, \end{aligned}$$

and herein the coefficient of x^4 is

$$= \tfrac{3}{2}a'l^2 - al(-abc + l^3), \; = 9al^4 - al(-abc + l^3), \; = (abc + 8l^3)\,al;$$

viz. we have thus the term $(abc + 8l^3)\,\xi\,.\,alx^4$ of the final result.

21. $K' = -(\delta)K$, where K is of the form $(abc + 8l^3)(aU + bV + cW)$; operating with (δ), we obtain $(abc + 8l^3)(a\delta U + b\delta V + c\delta W)$. Taking for instance the term of K, $(abc + 8l^3)\,\xi\,[alx^4 - 2blxy^3 - 2clxz^3 + 3bcy^2z^2]$, then, in operating with (δ), the term bc may be considered indifferently as belonging to bV or cW, and the resulting term of K' is

$$\begin{aligned} K' = -(\delta)K &= -(abc + 8l^3)\,\xi\,[al'x^4 - 2bl'xy^3 - 2cl'xz^3 + 3bc'y^2z^2], \\ &= \quad (abc + 8l^3)\,\xi\,[(abc + 2l^3)(ax^4 - 2bxy^3 - 2cxz^3) - 18bcl^2y^2z^2]. \end{aligned}$$

28. $$\overline{K} = 3\,\{\partial_x\Theta\partial_\xi P + \partial_y\Theta\partial_\eta P + \partial_z\Theta\partial_\zeta P\} + Q\Lambda\,;\ \text{viz. writing}$$

$$\Theta = (A,\ B,\ C,\ F,\ G,\ H \between x,\ y,\ z)^2,\quad A = -\,l^2\xi^2 - 2al\eta\zeta,\ \ldots\ F = \tfrac{1}{2}bc\xi^2 + l^2\eta\zeta,\ \ldots,$$

then this is

$$\begin{aligned} = 3\,\{&[-\,3bcl\xi^2 + (-\,abc + 4l^3)\,\eta\zeta]\,2\,(Ax + Hy + Gz)\\ &+ [-\,3cal\eta^2 + (-\,abc + 4l^3)\,\zeta\xi]\,2\,(Hx + By + Fz)\\ &+ [-\,3abl\zeta^2 + (-\,abc + 4l^3)\,\xi\eta]\,2\,(Gx + Fy + Cz)\}\\ &+ \{(abc - 10l^3)\,(bc\xi^3 + ca\eta^3 + ab\zeta^3) - 6l^2\,(5abc + 4l^3)\,\xi\eta\zeta\}\,(\xi x + \eta y + \zeta z).\end{aligned}$$

The whole coefficient of x is thus

$$\begin{aligned} = 3\,\{&[-\,3bcl\xi^2 + (-\,abc + 4l^3)\,\eta\zeta]\,(-\,2l^2\xi^2 - 4al\eta\zeta)\\ &+ [-\,3cal\eta^2 + (-\,abc + 4l^3)\,\zeta\xi]\,(ab\zeta^2 + l^2\xi\eta)\\ &+ [-\,3abl\zeta^2 + (-\,abc + 4l^3)\,\xi\eta]\,(ac\eta^2 + l^2\zeta\xi)\}\\ &+ \{(abc - 10l^3)\,(bc\xi^4 + ca\xi\eta^3 + ab\xi\zeta^3) - 6l^2\,(5abc + 4l^3)\,\xi^2\eta\zeta\}\,;\end{aligned}$$

herein the coefficient of ξ^4 is $18bcl^3 + (abc - 10l^3)\,bc$, $= (abc + 8l^3)\,bc$, giving, in the final result, the term $(abc + 8l^3)\,\xi\,.\,bcx^4$.

29. $$\overline{K}' = \tfrac{1}{6}(\delta)\,\overline{K}.$$

Here $\overline{K}$ is of the form $(abc + 8l^3)\,(aU + bV + cW)$, and we have

$$\overline{K}' = \tfrac{1}{6}\,(abc + 8l^3)\,(a\delta U + b\delta V + c\delta W).$$

A term of $aU + bV + cW$ is $x\,[bc\xi^4 - 2ca\xi\eta^3 - 2ab\xi\zeta^3 - 6al\eta^2\zeta^2]$, where $bc\xi^4$ may be considered as belonging indifferently to bV or cW; and so for the other terms. The resulting term in $\tfrac{1}{6}(a\delta U + b\delta V + c\delta W)$ is thus

$$\tfrac{1}{6}\,x\,[bc'\xi^4 - 2ca'\xi\eta^3 - 2ab'\xi\zeta^3 - 6al'\eta^2\zeta^2],$$

which is

$$= x\,[l^2\,(bc\xi^4 - 2ca\xi\eta^3 - 2ab\xi\zeta^3) + a\,(abc + 2l^3)\,\eta^2\zeta^2],$$

and we have thus a term of $\overline{K}'$.

22. $$E = -\,\tfrac{1}{18}\,\text{Jac}\,(K,\ U,\ \Lambda):$$

K contains the factor $abc + 8l^3$, and if, omitting this factor, the value of K is called $A\xi + B\eta + C\zeta$, then we have

$$\begin{aligned} E = -\,\tfrac{1}{6}\,\{&(\xi\partial_x A + \eta\partial_x B + \zeta\partial_x C)\,(Y\zeta - Z\eta) + (\xi\partial_y A + \eta\partial_y B + \zeta\partial_y C)\,(Z\xi - X\zeta)\\ &+ (\xi\partial_z A + \eta\partial_z B + \zeta\partial_z C)\,(X\eta - Y\xi)\},\end{aligned}$$

and the term herein in ξ^2 is $-\,\tfrac{1}{6}\,\xi^2\,(Z\partial_y A - Y\partial_z A)$, where A is

$$= alx^4 - 2blxy^3 - 2clxz^3 + 3bcy^2z^2\,;$$

viz. the coefficient of ξ^2 is

$$\begin{aligned}&= -\tfrac{1}{8}\{(cz^2+2lxy)(-6blxy^2+6bcyz^2)-(by^2+2lzx)(-6clxz^2+6bcy^2z)\}\\ &= b^2cy^4z - bc^2yz^4 + 2bl^2x^2y^3 - 2cl^2x^2z^3\\ &= (2l^2x^2+bcyz)(by^3-cz^3).\end{aligned}$$

Hence, restoring the omitted factor $(abc+8l^3)$, we have in E the term

$$(abc+8l^3)\,\xi^2(by^3-cz^3)\,[2l^2x^2+bcyz].$$

23, 24. $$E' = -\tfrac{1}{4}(\delta)\,E,\; E'' = \tfrac{1}{4}(\delta^2)\,E:$$

E is of the form $(abc+8l^3)(aU+bV+cW)$, and, as before, in a term such as

$$(abc+8l^3)\,\xi^2(by^3-cz^3)(2l^2x^2+bcyz),$$

we operate with δ or δ^2 only on the factor $2l^2x^2+bcyz$; and in E' and E'' respectively, operating upon this factor, we obtain

$$-\tfrac{1}{4}\{4ll'x^2+(bc'+b'c)\,yz\},\text{ and } \tfrac{1}{4}\{4l'^2x^2+2b'c'yz\},$$

viz. we thus obtain in E' the term

$$(abc+8l^3)\,\xi^2(by^3-cz^3)\,[l\,(abc+2l^3)\,x^2-3bcl^2yz],$$

and in E'' the term

$$(abc+8l^3)\,\xi^2(by^3-cz^3)\,[(abc+2l^3)^2x^2+18bcl^4yz].$$

30. $$\overline{E} = -\tfrac{1}{9}\,\mathrm{Jac}\,(\overline{K},\, U,\, \Lambda),\; = -\tfrac{1}{3}\begin{vmatrix}\partial_x\overline{K}, & X, & \xi\\ \partial_y\overline{K}, & Y, & \eta\\ \partial_z\overline{K}, & Z, & \zeta\end{vmatrix},$$

and, if omitting in $\overline{K}$ the factor $abc+8l^3$, we write $\overline{K}=Ax+By+Cz$, where

$$A = bc\xi^4 - 2ca\xi\eta^3 - 2ab\xi\zeta^3 - 6al\eta^2\zeta^2,\text{ this is } = -\tfrac{1}{3}\begin{vmatrix}A, & X, & \xi\\ B, & Y, & \eta\\ C, & Z, & \zeta\end{vmatrix},$$

which contains the term

$$\begin{aligned}\tfrac{1}{3}X(B\zeta - C\eta),\; &= \tfrac{1}{3}(ax^2+2lyz)\,\{\zeta\,(ca\eta^4-2ab\eta\zeta^3-2bc\eta\xi^3-6bl\zeta^2\xi^2)\\ &\qquad -\eta\,(ab\zeta^4-2bc\zeta\xi^3-2ca\zeta\eta^3-6cl\xi^2\eta^2)\},\\ &= (ax^2+2lyz)(c\eta^3-b\zeta^3)(2l\xi^2+a\eta\zeta).\end{aligned}$$

Hence, restoring the factor $abc+8l^3$, we have the terms

$$\overline{E} = (abc+8l^3)\,\{x^2(c\eta^3-b\zeta^3)\,[2al\xi^2+a^2\eta\zeta]+yz\,(c\eta^3-b\zeta^3)\,[4l^2\xi^2+2al\eta\zeta]\}.$$

31 and 32. $$\overline{E}' = -\tfrac{1}{2}(\delta)\,\overline{E},\quad \overline{E}'' = -\tfrac{1}{8}(\delta^2)\,\overline{E}:$$

$\overline{E}$ is of the form $(abc+8l^3)(aU+bV+cW)$, and we operate with δ and δ^2 on the factors $2al\xi^2+a^2\eta\zeta$, &c.; viz.

$$\delta\,(2al\xi^2+a^2\eta\zeta) = 2\,(al'+a'l)\,\xi^2+2aa'\eta\zeta,\quad \delta^2(2al\xi^2+a^2\eta\zeta) = 4a'l'\xi^2+2a'^2\eta\zeta,$$

and we thus obtain in $\bar{E}'$ the term

$$(abc+8l^3)\,x^2\,(c\eta^3-b\zeta^3)\,[a\,(abc-4l^3)\,\xi^2-6a^2l^2\eta\zeta],$$

and in $\bar{E}''$ the term

$$(abc+8l^3)\,x^2\,(c\eta^3-b\zeta^3)\,[-3al^2\,(abc+2l^3)\,\xi^2+9a^2l^4\eta\zeta].$$

25. $M=\frac{1}{36}\,\text{Jac}\,(U,\ \Psi,\ \Lambda)$: this, omitting the factor $(abc+8l^3)^2$ of Ψ, is

$$=\tfrac{1}{2}\begin{vmatrix} ax^2+2lyz, & ax^2\,(ax^3-5by^3-5cz^3), & \xi \\ by^2+2lzx, & by^2\,(by^3-5cz^3-5ax^3), & \eta \\ cz^2+2lxy, & cz^2\,(cz^3-5ax^3-5by^3), & \zeta \end{vmatrix};$$

the coefficient of ξ herein is

$$=\tfrac{1}{2}\,\{(bcy^2z^2+2clxz^3)\,(cz^3-5ax^3-5by^3)-(bcy^2z^2+2blxy^3)\,(by^3-5cz^3-5ax^3)\},$$

$$=\tfrac{1}{2}\,\{bcy^2z^2\,(-6by^3+6cz^3)+2lx\,[-b^2y^6+c^2z^6+5ax^3\,(by^3-cz^3)]\},$$

$$=\ \ (by^3-cz^3)\,[5alx^4-blxy^3-clxz^3-3bcy^2z^2].$$

Hence, restoring the factor $(abc+8l^3)^2$, we have the term

$$(abc+8l^3)^2\,.\,\xi\,(by^3-cz^3)\,[5alx^4-blxy^3-clxz^3-3bcy^2z^2].$$

26. $M'=-(\delta)\,M$. Here M is of the form $(abc+8l^3)^2\,(a^2U+\&\text{c.})$; and the δ operates through the $(abc+8l^3)^2\,a^2$, &c.; we, in fact, have in M' the term

$$-(abc+8l^3)^2\,.\,\xi\,(by^3-cz^3)\,[5al'x^4-bl'xy^3-cl'xz^3-3bc'y^2z^2],$$

which is

$$=(abc+8l^3)^2\,.\,\xi\,(by^3-cz^3)\,[(abc+2l^3)\,(5ax^4-bxy^3-cxz^3)+18bcl^2y^2z^2].$$

33. $$\bar{M}=-\tfrac{1}{6}\,[\text{Jac}]\,(P,\ F,\ \Lambda),\ =-\tfrac{1}{6}\begin{vmatrix} -3lbc\xi^2+(-abc+4l^3)\,\eta\zeta, & \partial_\xi F, & x \\ -3lca\eta^2+(-abc+4l^3)\,\zeta\xi, & \partial_\eta F, & y \\ -3lab\zeta^2+(-abc+4l^3)\,\xi\eta, & \partial_\zeta F, & z \end{vmatrix},$$

and the whole coefficient of x is thus

$$=\tfrac{1}{6}\,\{[3lca\eta^2+(abc-4l^3)\,\zeta\xi]\,\partial_\zeta F-[3lab\zeta^2+(abc-4l^3)\,\xi\eta]\,\partial_\eta F\},$$

or substituting for $\frac{1}{6}\,\partial_\zeta F$, $\frac{1}{6}\,\partial_\eta F$ their values, this is

$$=\{3lca\eta^2+(abc-4l^3)\,\zeta\xi\}\,[a^2b^2\zeta^5-(abc+16l^3)\,(b\zeta^2\xi^3+a\zeta^2\eta^3)$$
$$-4l^2\,(bc\xi^4\eta+ca\xi\eta^4+4ab\xi\eta\zeta^3)-8l\,(abc+2l^3)\,\xi^2\eta^2\zeta]$$

$$-\{3lab\zeta^2+(abc-4l^3)\,\xi\eta\}\,[a^2c^2\eta^5-(abc+16l^3)\,(a\eta^2\zeta^3+c\eta^2\xi^3)$$
$$-4l^2\,(bc\zeta\xi^4+4ca\xi\eta^3\zeta+ab\xi\zeta^4)-8l\,(abc+2l^3)\,\xi^2\eta\zeta^2].$$

Collecting, first, the terms independent of $abc-4l^3$, and, next, those which contain $abc-4l^3$, each set contains the factor $c\eta^3-b\zeta^3$, and the whole is $=c\eta^3-b\zeta^3$ multiplied by

$$-3la^3bc\eta^2\zeta^2-3a^2l(abc+8l^3)\eta^2\zeta^2-12l^3(abc\xi^4+a^2c\xi\eta^3+a^2b\xi\zeta^3)-24al^2(abc+2l^3)\xi^2\eta\zeta$$
$$+(abc-4l^3)\{a^2c\xi\eta^3+a^2b\xi\zeta^3-(abc+16l^3)\xi^4+12al^2\xi^2\eta\zeta\};$$

and here collecting the terms in ξ^4, $\xi(c\eta^3+b\zeta^3)$, $\xi^2\eta\zeta$, and $\eta^2\zeta^2$, each of these contains the factor $abc+8l^3$, and, finally, the term of $\overline{M}$ is

$$=(abc+8l^3)(c\eta^3-b\zeta^3)[(abc-8l^3)\xi^4-a^2c\xi\eta^3-a^2b\xi\zeta^3-12al^2\xi^2\eta\zeta-6a^2l\eta^2\zeta^2]\,x.$$

34. $$\overline{M}'=\tfrac{1}{6}(\delta)\,\overline{M}.$$

Here M is of the form $(abc+8l^3)(aU+bV+cW)$; and, operating with δ through the $(abc+8l^3)\,a$, &c., we obtain in $\overline{M}'$ the term

$$\tfrac{1}{6}(abc+8l^3)\,x\,(c\eta^3-b\zeta^3)\,[(a'bc+ab'c+abc'-24l^2l')\,\xi^4+\&c.],$$

where

$$a'bc+ab'c+abc'-24l^2l'=18abcl^2+24l^2(abc+2l^3),\ =6l^2(7abc+8l^3),$$

and the term thus is

$$=(abc+8l^3)\,x\,(c\eta^3-b\zeta^3)\,[(7abc+8l^3)\,l^2\xi^4+\ldots].$$

This concludes the series of calculations.

Cambridge, England, 17 *May,* 1881.

771.

SPECIMEN OF A LITERAL TABLE FOR BINARY QUANTICS, OTHERWISE A PARTITION TABLE.

[From the *American Journal of Mathematics*, vol. IV. (1881), pp. 248—255.]

THE Table, commencing 1; b; c, b^2; d, bc, b^3; ..., is in fact a Partition Table, viz. considering the letters b, c, d, ... as denoting 1, 2, 3, ... respectively, it is 1°; 1; 2, 11; 3, 12, 111; ... a table of the partitions of the numbers 0, 1, 2, 3, ..., expressed however in the literal form, in order to its giving the literal terms which enter into the coefficients of any covariant of a binary quantic. The table ought to have been made and published many years ago, before the calculation of the covariants of the quintic; and the present publication of it is, in some measure, an anachronism: but I in fact felt the need of it in some calculations in regard to the sextic; and I think the table may be found useful on other occasions. I have contented myself with calculating the table up to $s=18$, that is, so as to include in it all the partitions of 18: it would, I think, be desirable to extend it further, say to $z=26$; or even beyond this point, but perhaps without introducing any new letters, (that is, so as to give for the higher numbers only the partitions with a largest part not exceeding 26): the question of the space which such a table would occupy will be considered presently.

As to the employment of the table, observe that, in applying it to the case of a quantic $(a, b, c, d \between x, y)^3$, the terms containing the letters e, f, etc., posterior to the last coefficient d of the quantic are to be disregarded; and that the terms are to be rendered homogeneous by the introduction of the proper power of the first coefficient a, rejecting any term for which the exponent of a would be negative (or what is the same thing, any term of too high a degree in the coefficients b, c, d);

thus, for the cubicovariant, where the coefficients are of the degree 3, and of the weights 3, 4, 5, 6 respectively, from the portion of the table

d	e	f	g
bc	bd	be	bf
b^3	c^2	cd	ce
	b^2c	b^2d	d^2
	b^4	bc^2	b^2e
		b^3c	bcd
		b^5	c^3
			b^3d
			etc.

we at once copy out the terms

a^2d	abd	acd	ad^2
abc	ac^2	b^2d	bcd
b^3	b^2c	bc^2	c^3

which compose the coefficients in question.

As regards the formation of the table, this is at once effected, and the successive terms are obtained *currente calamo*, by Arbogast's rule of the last and the last but one: observing that each term is to be regarded as containing implicitly a power of a, so that operating on any term such as b^4, the operation on the last letter gives b^3c, and that on the last but one letter gives b^5. There is little risk of error except in the accidental omission of a term; but of course any one omission would occasion the omission of all the subsequent terms derivable from the omitted term, and would so be fatal: to remove this source of error, observe that for the successive numbers 0, 1, 2, 3, etc., the number of partitions should be

0	1	2	3	4	5	6	7	8	9	10	11	12	13	14	15	16	17	18 ...
1	1	2	3	5	7	11	15	22	30	42	56	77	101	135	176	231	297	385 ...

and we can thus, for each partible number successively, verify that the right number of partitions has been obtained.

But as the number of partitions becomes large, a further control is convenient, and even necessary—say we have the 176 partitions of 15, we have by the rule to derive thence the 231 partitions of 16, and it is not until the whole of this derivation is gone through, that we could by counting the number of the new terms ascertain that the right number of 231 terms has been obtained. To break up the verification, it is convenient to know that for the partitions of 16 into 1 part, 2 parts, 3 parts, 4 parts, etc., the numbers of partitions are 1, 8, 21, 34, etc., respectively: we can then as soon as the derivations giving the partitions into 1 part, 2 parts, 3 parts, etc., respectively, have been performed, verify that the right numbers 1, 8, 21, 34, etc., of terms have been obtained. The numbers are contained in the following table, each column of which is calculated from the preceding columns according to a rule which is easily obtained, and which is itself verified by the condition that the sums of the numbers in the several columns give the before mentioned series of numbers 1, 1, 2, 3, 5, 7, etc.

No. of Parts.	Partible Number. 0	1	2	3	4	5	6	7	8	9	10	11	12	13	14	15	16	17	18
1	1	1	1	1	1	1	1	1	1	1	1	1	1	1	1	1	1	1	1
2			1	1	2	2	3	3	4	4	5	5	6	6	7	7	8	8	9
3				1	1	2	3	4	5	7	8	10	12	14	16	19	21	24	27
4					1	1	2	3	5	6	9	11	15	18	23	27	34	39	47
5						1	1	2	3	5	7	10	13	18	23	30	37	47	57
6							1	1	2	3	5	7	11	14	20	26	35	44	58
7								1	1	2	3	5	7	11	15	21	28	38	49
8									1	1	2	3	5	7	11	15	22	29	40
9										1	1	2	3	5	7	11	15	22	30
10											1	1	2	3	5	7	11	15	22
11												1	1	2	3	5	7	11	15
12													1	1	2	3	5	7	11
13														1	1	2	3	5	7
14															1	1	2	3	5
15																1	1	2	3
16																	1	1	2
17																		1	1
18																			1
	1	1	2	3	5	7	11	15	22	30	42	56	77	101	135	176	231	297	385

The practical rule for the construction of the table thus is:—On a sheet of paper ruled in squares, and which is read as a continuous column from the bottom of one column to the top of the next column, form the terms by Arbogast's method as already explained; writing down in pencil a batch of terms, and counting them to see that the right number has been obtained, then, at the same time verifying the derivations, mark these over in ink; and so on with another batch of terms, until the whole number of the partitions of any particular number is obtained.

The foregoing series 1, 1, 2, 3, ..., 385, for the number of the partitions of the successive numbers 0, 1, 2, 3, ..., 18 is carried by Euler up to the number of partitions of 59, = 831820, see the paper "De Partitione Numerorum," *Op. Arith. Coll.* I., bottom line of the table pp. 97—101: the continuation from the number 385 and for the partible numbers 19 to 30 is as follows:

19	20	21	22	23	24	25	26	27	28	29	30
490	627	792	1002	1255	1575	1958	2436	3010	3718	4565	5604;

the whole number of terms 1, 1, ..., 5604 amounts to 28629, which at the rate of 500 to a page would occupy somewhat under 60 pages; or, at the rate here employed of 369 to a page, somewhat under 78 pages.

THE PARTITION TABLE, 0 TO 18.

0.3	4.5	6.7	7.8	8.9	9	9.10	10	10.11
0 1 1 **1** 1 b **2** 2 c b^2 **3** 3 d bc b^3	**4** 5 e bd c^2 b^2c b^4 **5** 7 f be cd b^2d bc^2 b^3c b^5	**6** 11 g bf ce d^2 b^2e bcd c^3 b^3d b^2c^2 b^4c b^6 **7** 15 h bg	cf de b^2f bce bd^2 c^2d b^3e b^2cd bc^3 b^4d b^3c^2 b^5c b^7 **8** 22 i bh cg df e^2	b^2g bcf bde c^2e cd^2 b^3f b^2ce b^2d^2 bc^2d c^4 b^4e b^3cd b^2c^3 b^5d b^4c^2 b^6c b^8 **9** 30 j	bi ch dg ef b^2h bcg bdf be^2 c^2f cde d^3 b^3g b^2cf b^2de bc^2e bcd^2 c^3d b^4f b^3ce b^3d^2 b^2c^2d	bc^4 b^5e b^4cd b^3c^3 b^6d b^5c^2 b^7c b^9 **10** 42 k bj ci dh eg f^2 b^2i bch	bdg bef c^2g cdf ce^2 d^2e b^3h b^2cg b^2df b^2e^2 bc^2f $bcde$ bd^3 c^3e c^2d^2 b^4g b^3cf b^3de b^2c^2e b^2cd^2 bc^3d	c^5 b^5f b^4ce b^4d^2 b^3c^2d b^2c^4 b^6e b^5cd b^4c^3 b^7d b^6c^2 b^8c b^{10} **11** 56 l bk cj

THE PARTITION TABLE, 0 TO 18 (*continued*).

11	11.12	12	12.13	13	13.14	14	14	14
di	b^6f	$bcef$	b^6d^2	$bceg$	bc^3d^2	gi	b^4k	c^5e
eh	b^5ce	bd^2f	b^5c^2d	bcf^2	c^5d	h^2	b^3cj	c^4d^2
fg	b^5d^2	bde^2	b^4c^4	bd^2g	b^6h	b^2m	b^3di	b^6i
b^2j	b^4c^2d	c^3g	b^8e	$bdef$	b^5cg	bcl	b^3eh	b^5ch
bci	b^3c^4	c^2df	b^7cd	be^3	b^5df	bdk	b^3fg	b^5dg
bdh	b^7e	c^2e^2	b^6c^3	c^3h	b^5e^2	bej	b^2c^2i	b^5ef
beg	b^6cd	cd^2e	b^9d	c^2dg	b^4c^2f	bfi	b^2cdh	b^4c^2g
bf^2	b^5c^3	d^4	b^8c^2	c^2ef	b^4cde	bgh	b^2ceg	b^4cdf
c^2h	b^8d	b^4i	$b^{10}c$	cd^2f	b^4d^3	c^2k	b^2cf^2	b^4ce^2
cdg	b^7c^2	b^3ch	b^{12}	cde^2	b^3c^3e	cdj	b^2d^2g	b^4d^2e
cef	b^9c	b^3dg	**13**	d^3e	$b^3c^2d^2$	cei	b^2def	b^3c^3f
d^2f	b^{11}	b^3ef	101	b^4j	b^2c^4d	cfh	b^2e^3	b^3c^2de
de^2	**12**	b^2c^2g		b^3ci	bc^6	cg^2	bc^3h	b^3cd^3
b^3i	77	b^2cdf	n	b^3dh	b^7g	d^2i	bc^2dg	b^2c^4e
b^2ch		b^2ce^2	bm	b^3eg	b^6cf	deh	bc^2ef	$b^2c^3d^2$
b^2dg	m	b^2d^2e	cl	b^3f^2	b^6de	dfg	bcd^2f	bc^5d
b^2ef	bl	bc^3f	dk	b^2c^2h	b^5c^2e	e^2g	$bcde^2$	c^7
bc^2g	ck	bc^2de	ej	b^2cdg	b^5cd^2	ef^2	bd^3e	b^7h
$bcdf$	dj	bcd^3	fi	b^2cef	b^4c^3d	b^3l	c^4g	b^6cg
bce^2	ei	c^4e	gh	b^2d^2f	b^3c^5	b^2ck	c^3df	b^6df
bd^2e	fh	c^3d^2	b^2l	b^2de^2	b^8f	b^2dj	c^3e^2	b^6e^2
c^3f	g^2	b^5h	bck	bc^3g	b^7ce	b^2ei	c^2d^2e	b^5c^2f
c^2de	b^2k	b^4cg	bdj	bc^2df	b^7d^2	b^2fh	cd^4	b^5cde
cd^3	bcj	b^4df	bei	bc^2e^2	b^6c^2d	b^2g^2	b^5j	b^5d^3
b^4h	bdi	b^4e^2	bfh	bcd^2e	b^5c^4	bc^2j	b^4ci	b^4c^3e
b^3cg	beh	b^3c^2f	bg^2	bd^4	b^9e	$bcdi$	b^4dh	$b^4c^2d^2$
b^3df	bfg	b^3cde	c^2j	c^4f	b^8cd	$bceh$	b^4eg	b^3c^4d
b^3e^2	c^2i	b^3d^3	cdi	c^3de	b^7c^3	$bcfg$	b^4f^2	b^2c^6
b^2c^2f	cdh	b^2ce^2	ceh	c^2d^3	$b^{10}d$	bd^2h	b^3c^2h	b^8g
b^2cde	ceg	$b^2c^2d^2$	cfg	b^5i	b^9c^2	$bdeg$	b^3cdg	b^7cf
b^2d^3	cf^2	bc^4d	d^2h	b^4ch	$b^{11}c$	bdf^2	b^3cef	b^7de
bc^3e	d^2g	c^6	deg	b^4dg	b^{13}	be^2f	b^3d^2f	b^6c^2e
bc^2d^2	def	b^6g	df^2	b^4ef	**14**	c^3i	b^3de^2	b^6cd^2
c^4d	e^3	b^5cf	e^2f	b^3c^2g	135	c^2dh	b^2c^3g	b^5c^3d
b^5g	b^3j	b^5de	b^3k	b^3cdf		c^2eg	b^2c^2df	b^4c^5
b^4cf	b^2ci	b^4c^2e	b^2cj	b^3ce^2	o	c^2f^2	$b^2c^2e^2$	b^9f
b^4de	b^2dh	b^4cd^2	b^2di	b^3d^2e	bn	cd^2g	b^2cd^2e	b^8ce
b^3c^2e	b^2eg	b^3c^3d	b^2eh	b^2c^3f	cm	$cdef$	b^2d^4	b^8d^2
b^3cd^2	b^2f^2	b^2c^5	b^2fg	b^2c^2de	dl	ce^3	bc^4f	b^7c^2d
b^2c^3d	bc^2h	b^7f	bc^2i	b^2cd^3	ek	d^3f	bc^3de	b^6c^4
bc^5	$bcdg$	b^6ce	$bcdh$	bc^4e	fj	d^2e^2	bc^2d^3	$b^{10}e$

THE PARTITION TABLE, 0 TO 18 (*continued*).

14.15	15	15	15	15.16	16	16	16	16
b^9cd	b^2gh	$bcdef$	b^5f^2	b^4c^4d	cdl	cef^2	cd^2e^2	b^5fg
b^8c^3	bc^2k	bce^3	b^4c^2h	b^3c^6	cek	d^3h	d^4e	b^4c^2i
$b^{11}d$	$bcdj$	bd^3f	b^4cdg	b^9g	cfj	d^2eg	b^5l	b^4cdh
$b^{10}c^2$	$bcei$	bd^2e^2	b^4cef	b^8cf	cgi	d^2f^2	b^4ck	b^4ceg
$b^{12}c$	$bcfh$	c^4h	b^4d^2f	b^8de	ch^2	de^2f	b^4dj	b^4cf^2
b^{14}	bcg^2	c^3dg	b^4de^2	b^7c^2e	d^2k	e^4	b^4ei	b^4d^2g
15	bd^2i	c^3ef	b^3c^3g	b^7cd^2	dej	b^4m	b^4fh	b^4def
176	$bdeh$	c^2d^2f	b^3c^2df	b^6c^3d	dfi	b^3cl	b^4g^2	b^4e^3
	$bdfg$	c^2de^2	$b^3c^2e^2$	b^5c^5	dgh	b^3dk	b^3c^2j	b^3c^3h
p	be^2g	cd^3e	b^3cd^2e	$b^{10}f$	e^2i	b^3ej	b^3cdi	b^3c^2dg
bo	bef^2	d^5	b^3d^4	b^9ce	efh	b^3fi	b^3ceh	b^3c^2ef
cn	c^3j	b^5k	b^2c^4f	b^9d^2	eg^2	b^3gh	b^3cfg	b^3cd^2f
dm	c^2di	b^4cj	b^2c^3de	b^8c^2d	f^2g	b^2c^2k	b^3d^2h	b^3cde^2
el	c^2eh	b^4di	$b^2c^2d^3$	b^7c^4	b^3n	b^2cdj	b^3deg	b^3d^3e
fk	c^2fg	b^4eh	bc^5e	$b^{11}e$	b^2cm	b^2cei	b^3df^2	b^2c^4g
gj	cd^2h	b^4fg	bc^4d^2	$b^{10}cd$	b^2dl	b^2cfh	b^3e^2f	b^2c^3df
hi	$cdeg$	b^3c^2i	c^6d	b^9c^3	b^2ek	b^2cg^2	b^2c^3i	$b^2c^3e^2$
b^2n	cdf^2	b^3cdh	b^7i	$b^{12}d$	b^2fj	b^2d^2i	b^2c^2dh	$b^2c^2d^2e$
bcm	ce^2f	b^3ceg	b^6ch	$b^{11}c^2$	b^2gi	b^2deh	b^2c^2eg	b^2cd^4
bdl	d^3g	b^3cf^2	b^6dg	$b^{13}c$	b^2h^2	b^2dfg	$b^2c^2f^2$	bc^5f
bek	d^2ef	b^3d^2g	b^6ef	b^{15}	bc^2l	b^2e^2g	b^2cd^2g	bc^4de
bfj	de^3	b^3def	b^5c^2g	**16**	$bcdk$	b^2ef^2	b^2cdef	bc^3d^3
bgi	b^4l	b^3e^3	b^5cdf	231	$bcej$	bc^3j	b^2ce^3	c^6e
bh^2	b^3ck	b^2c^3h	b^5ce^2		$bcfi$	bc^2di	b^2d^3f	c^5d^2
c^2l	b^3dj	b^2c^2dg	b^5d^2e	q	$bcgh$	bc^2eh	$b^2d^2e^2$	b^7j
cdk	b^3ei	b^2c^2ef	b^4c^3f	bp	bd^2j	bc^2fg	bc^4h	b^6ci
cej	b^3fh	b^2cd^2f	b^4c^2de	co	$bdei$	bcd^2h	bc^3dg	b^6dh
cfi	b^3g^2	b^2cde^2	b^4cd^3	dn	$bdfh$	$bcdeg$	bc^3ef	b^6eg
cgh	b^2c^2j	b^2d^3e	b^3c^4e	em	bdg^2	$bcdf^2$	bc^2d^2f	b^6f^2
d^2j	b^2cdi	bc^4g	$b^3c^3d^2$	fl	be^2h	bce^2f	bc^2de^2	b^5c^2h
dei	b^2ceh	bc^3df	b^2c^5d	gk	$befg$	bd^3g	bcd^3e	b^5cdg
dfh	b^2cfg	bc^3e^2	bc^7	hj	bf^3	bd^2ef	bd^5	b^5cef
dg^2	b^2d^2h	bc^2d^2e	b^8h	i^2	c^3k	bde^3	c^5g	b^5d^2f
e^2h	b^2dey	bcd^4	b^7cg	b^2o	c^2dj	c^4i	c^4df	b^5de^2
efg	b^2df^2	c^5f	b^7df	bcn	c^2ei	c^3dh	c^4e^2	b^4c^3g
f^3	b^2e^2f	c^4de	b^7e^2	bdm	c^2fh	c^3eg	c^3d^2e	b^4c^2df
b^3m	bc^3i	c^3d^3	b^6c^2f	bel	c^2g^2	c^3f^2	c^2d^4	$b^4c^2e^2$
b^2cl	bc^2dh	b^6j	b^6cde	bfk	cd^2i	c^2d^2g	b^6k	b^4cd^2e
b^2dk	bc^2eg	b^5ci	b^6d^3	bgj	$cdeh$	c^2def	b^5cj	b^4d^4
b^2ej	bc^2f^2	b^5dh	b^5c^3e	bhi	$cdfg$	c^2e^3	b^5di	b^3c^4f
b^2fi	bcd^2g	b^5eg	$b^5c^2d^2$	c^2m	ce^2g	cd^3f	b^5eh	b^3c^3de

THE PARTITION TABLE, 0 TO 18 (*continued*).

16	16.17	17	17	17	17	17	17	17.18
$b^3c^2d^3$	b^9c^2d	efi	def^2	c^2deg	bc^2e^3	bc^5g	b^7f^2	$b^7c^2d^2$
b^2c^5e	b^8c^4	egh	e^3f	c^2df^2	bcd^3f	bc^4df	b^6c^2h	b^6c^4d
$b^2c^4d^2$	$b^{12}e$	f^2h	b^4n	c^2e^2f	bcd^2e^2	bc^4e^2	b^6cdg	b^5c^6
bc^6d	$b^{11}cd$	fg^2	b^3cm	cd^3g	bd^4e	bc^3d^2e	b^6cef	$b^{11}g$
c^8	$b^{10}c^3$	b^3o	b^3dl	cd^2ef	c^5h	bc^2d^4	b^6d^2f	$b^{10}cf$
b^8i	$b^{13}d$	b^2cn	b^3ek	cde^3	c^4dg	c^6f	b^6de^2	$b^{10}de$
b^7ch	$b^{12}c^2$	b^2dm	b^3fj	d^4f	c^4ef	c^5de	b^5c^3g	b^9c^2e
b^7dg	$b^{14}c$	b^2el	b^3gi	d^3e^2	c^3d^2f	c^4d^3	b^5c^2df	b^9cd^2
b^7ef	b^{16}	b^2fk	b^3h^2	b^5m	c^3de^2	b^7k	$b^5c^2e^2$	b^8c^3d
b^6c^2g	**17**	b^2gj	b^2c^2l	b^4cl	c^2d^3e	b^6cj	b^5cd^2e	b^7c^5
b^6cdf	297	b^2hi	b^2cdk	b^4dk	cd^5	b^6di	b^5d^4	$b^{12}f$
b^6ce^2		bc^2m	b^2cej	b^4ej	b^6l	b^6eh	b^4c^4f	$b^{11}ce$
b^6d^2e	r	$bcdl$	b^2cfi	b^4fi	b^5ck	b^6fg	b^4c^3de	$b^{11}d^2$
b^5c^3f	bq	$bcek$	b^2cgh	b^4gh	b^5dj	b^5c^2i	$b^4c^2d^3$	$b^{10}c^2d$
b^5c^2de	cp	$bcfj$	b^2d^2j	b^3c^2k	b^5ei	b^5cdh	b^3c^5e	b^9c^4
b^5cd^3	do	$bcgi$	b^2dei	b^3cdj	b^5fh	b^5ceg	$b^3c^4d^2$	$b^{13}e$
b^4c^4e	en	bch^2	b^2dfh	b^3cei	b^5g^2	b^5cf^2	b^2c^6d	$b^{12}cd$
$b^4c^3d^2$	fm	bd^2k	b^2dg^2	b^3cfh	b^4c^2j	b^5d^2g	bc^8	$b^{11}c^3$
b^3c^5d	gl	$bdej$	b^2e^2h	b^3cg^2	b^4cdi	b^5def	b^9i	$b^{14}d$
b^2c^7	hk	$bdfi$	b^2efg	b^3d^2i	b^4ceh	b^5e^3	b^8ch	$b^{13}c^2$
b^9h	ij	$bdgh$	b^2f^3	b^3deh	b^4cfg	b^4c^3h	b^8dg	$b^{15}c$
b^8cg	b^2p	be^2i	bc^3k	b^3dfg	b^4d^2h	b^4c^2dg	b^8ef	b^{17}
b^8df	bco	$befh$	bc^2dj	b^3e^2g	b^4deg	b^4c^2ef	b^7c^2g	**18**
b^8e^2	bdn	beg^2	bc^2ei	b^3ef^2	b^4df^2	b^4cd^2f	b^7cdf	385
b^7c^2f	bem	bf^2g	bc^2fh	b^2c^3j	b^4e^2f	b^4cde^2	b^7ce^2	
b^7cde	bfl	c^3l	bc^2g^2	b^2c^2di	b^3c^3i	b^4d^3e	b^7d^2e	s
b^7d^3	bgk	c^2dk	bcd^2i	b^2c^2eh	b^3c^2dh	b^3c^4g	b^6c^3f	br
b^6c^3e	bhj	c^2ej	$bcdeh$	b^2c^2fg	b^3c^2eg	b^3c^3df	b^6c^2de	cq
$b^6c^2d^2$	bi^2	c^2fi	$bcdfg$	b^2cd^2h	$b^3c^2f^2$	$b^3c^3e^2$	b^6cd^3	dp
b^5c^4d	c^2n	c^2gh	bce^2g	b^2cdeg	b^3cd^2g	$b^3c^2d^2e$	b^5c^4e	eo
b^4c^6	cdm	cd^2j	$bcef^2$	b^2cdf^2	b^3cdef	b^3cd^4	$b^5c^3d^2$	fn
$b^{10}g$	cel	$cdei$	bd^3h	b^2ce^2f	b^3ce^3	b^2c^5f	b^4c^5d	gm
b^9cf	cfk	$cdfh$	bd^2eg	b^2d^3g	b^3d^3f	b^2c^4de	b^3c^7	hl
b^9de	cgj	cdg^2	bd^2f^2	b^2d^2ef	$b^3d^2e^2$	$b^2c^3d^3$	$b^{10}h$	ik
b^8c^2e	chi	ce^2h	bde^2f	b^2de^3	b^2c^4h	bc^6e	b^9cg	j^2
b^8cd^2	d^2l	$cefg$	be^4	bc^4i	b^2c^3dg	bc^5d^2	b^9df	b^2q
b^7c^3d	dek	cf^3	c^4j	bc^3dh	b^2c^3ef	c^7d	b^9e^2	bcp
b^6c^5	dfj	d^3i	c^3di	bc^3eg	$b^2c^2d^2f$	b^8j	b^8c^2f	bdo
$b^{11}f$	dgi	d^2eh	c^3eh	bc^3f^2	$b^2c^2de^2$	b^7ci	b^8cde	ben
$b^{10}ce$	dh^2	d^2fg	c^3fg	bc^2d^2g	b^2cd^3e	b^7dh	b^8d^3	bfm
$b^{10}d^2$	e^2j	de^2g	c^2d^2h	bc^2def	b^2d^5	b^7eg	b^7c^3e	bgl

THE PARTITION TABLE, 0 TO 18 (*continued*).

18	18	18	18	18	18	18	18	18
bhk	$befi$	b^2dej	d^3ef	bc^2deg	b^3cdf^2	b^4c^3i	b^5cd^2f	b^8c^2g
bij	$begh$	b^2dfi	d^2e^3	bc^2df^2	b^3ce^2f	b^4c^2dh	b^5cde^2	b^8cdf
c^2o	bf^2h	b^2dgh	b^5n	bc^2e^2f	b^3d^3g	b^4c^2eg	b^5d^3e	b^8ce^2
cdn	bfg^2	b^2e^2i	b^4cm	bcd^3g	b^3d^2ef	$b^4c^2f^2$	b^4c^4g	b^8d^2e
cem	c^3m	b^2efh	b^4dl	bcd^2ef	b^3de^3	b^4cd^2g	b^4c^3df	b^7c^3f
cfl	c^2dl	b^2eg^2	b^4ek	$bcde^3$	b^2c^4i	b^4cdef	$b^4c^3e^2$	b^7c^2de
cgk	c^2ek	b^2f^2g	b^4fj	bd^4f	b^2c^3dh	b^4ce^3	$b^4c^2d^2e$	b^7cd^3
chj	c^2fj	bc^3l	b^4gi	bd^3e^2	b^2c^3eg	b^4d^3f	b^4cd^4	b^6c^4e
ci^2	c^2gi	bc^2dk	b^4h^2	c^5i	$b^2c^3f^2$	$b^4d^2e^2$	b^3c^5f	$b^6c^3d^2$
d^2m	c^2h^2	bc^2ej	b^3c^2l	c^4dh	$b^2c^2d^2g$	b^3c^4h	b^3c^4de	b^5c^5d
del	cd^2k	bc^2fi	b^3cdk	c^4eg	b^2c^2def	b^3c^3dg	$b^3c^3d^3$	b^4c^7
dfk	$cdej$	bc^2gh	b^3cej	c^4f^2	$b^2c^2e^3$	b^3c^3ef	b^2c^6e	$b^{11}h$
dgj	$cdfi$	bcd^2j	b^3cfi	c^3d^2g	b^2cd^3f	$b^3c^2d^2f$	$b^2c^5d^2$	$b^{10}cg$
dhi	$cdgh$	$bcdei$	b^3cgh	c^3def	$b^2cd^2e^2$	$b^3c^2de^2$	bc^7d	$b^{10}df$
e^2k	ce^2i	$bcdfh$	b^3d^2j	c^3e^3	b^2d^4e	b^3cd^3e	c^9	$b^{10}e^2$
efj	$cefh$	$bcdg^2$	b^3dei	c^2d^3f	bc^5h	b^3d^5	b^9j	b^9c^2f
egi	ceg^2	bce^2h	b^3dfh	$c^2d^2e^2$	bc^4dg	b^2c^5g	b^8ci	b^9cde
eh^2	cf^2g	$bcefg$	b^3dg^2	cd^4e	bc^4ef	b^2c^4df	b^8dh	b^9d^3
f^2i	d^3j	bcf^3	b^3e^2h	d^6	bc^3d^2f	$b^2c^4e^2$	b^8eg	b^8c^3e
fgh	d^2ei	bd^3i	b^3efg	b^6m	bc^3de^2	$b^2c^3d^2e$	b^8f^2	$b^8c^2d^2$
g^3	d^2fh	bd^2eh	b^3f^3	b^5cl	bc^2d^3e	$b^2c^2d^4$	b^7c^2h	b^7c^4d
b^3p	d^2g^2	bd^2fg	b^2c^3k	b^5dk	bcd^5	bc^6f	b^7cdg	b^6c^6
b^2co	de^2h	bde^2g	b^2c^2dj	b^5ej	c^6g	bc^5de	b^7cef	$b^{12}g$
b^2dn	$defg$	$bdef^2$	b^2c^2ei	b^5fi	c^5df	bc^4d^3	b^7d^2f	$b^{11}cf$
b^2em	df^3	be^3f	b^2c^2fh	b^5gh	c^5e^2	c^7e	b^7de^2	$b^{11}de$
b^2fl	e^3g	c^4k	$b^2c^2g^2$	b^4c^2k	c^4d^2e	c^6d^2	b^6c^3g	$b^{10}c^2e$
b^2gk	e^2f^2	c^3dj	b^2cd^2i	b^4cdj	c^3d^4	b^8k	b^6c^2df	$b^{10}cd^2$
b^2hj	b^4o	c^3ei	b^2cdeh	b^4cei	b^7l	b^7cj	$b^6c^2e^2$	b^9c^3d
b^2i^2	b^3cn	c^3fh	b^2cdfg	b^4cfh	b^6ck	b^7di	b^6cd^2e	b^8c^5
bc^2n	b^3dm	c^3g^2	b^2ce^2g	b^4cg^2	b^6dj	b^7eh	b^6d^4	$b^{13}f$
$bcdm$	b^3el	c^2d^2i	b^2cef^2	b^4d^2i	b^6ei	b^7fg	b^5c^4f	$b^{12}ce$
$bcel$	b^3fk	c^2deh	b^2d^3h	b^4deh	b^6fh	b^6c^2i	b^5c^3de	$b^{12}d^2$
$bcfk$	b^3gj	c^2dfg	b^2d^2eg	b^4dfg	b^6g^2	b^6cdh	$b^5c^2d^3$	$b^{11}c^2d$
$bcgj$	b^3hi	c^2e^2g	$b^2d^2f^2$	b^4e^2g	b^5c^2j	b^6ceg	b^4c^5e	$b^{10}c^4$
$bchi$	b^2c^2m	c^2ef^2	b^2de^2f	b^4ef^2	b^5cdi	b^6cf^2	$b^4c^4d^2$	$b^{14}e$
bd^2l	b^2cdl	cd^3h	b^2e^4	b^3c^3j	b^5ceh	b^6d^2g	b^3c^6d	$b^{13}cd$
$bdek$	b^2cek	cd^2eg	bc^4j	b^3c^2di	b^5cfg	b^6def	b^2c^8	$b^{12}c^3$
$bdfj$	b^2cfj	cd^2f^2	bc^3di	b^3c^2eh	b^5d^2h	b^6e^3	$b^{10}i$	$b^{15}d$
$bdgi$	b^2cgi	cde^2f	bc^3eh	b^3c^2fg	b^5deg	b^5c^3h	b^9ch	$b^{14}c^2$
bdh^2	b^2ch^2	ce^4	bc^3fg	b^3cd^2h	b^5df^2	b^5c^2dg	b^9dg	$b^{16}c$
be^2j	b^2d^2k	d^4g	bc^2d^2h	b^3cdeg	b^5e^2f	b^5c^2ef	b^9ef	b^{18}

772.

ON THE ANALYTICAL FORMS CALLED TREES.

[From the *American Journal of Mathematics*, vol. IV. (1881), pp. 266—268.]

IN a tree of N knots, selecting any knot at pleasure as a root, the tree may be regarded as springing from this root, and it is then called a root-tree. The same tree thus presents itself in various forms as a root-tree; and if we consider the different root-trees with N knots, these are not all of them distinct trees. We have thus the two questions, to find the number of root-trees with N knots; and, to find the number of distinct trees with N knots.

I have in my paper "On the Theory of the Analytical Forms called Trees," *Phil. Mag.*, t. XIII. (1857), pp. 172—176, [203] given the solution of the first question; viz. if ϕ_N denotes the number of the root-trees with N knots, then the successive numbers ϕ_1, ϕ_2, ϕ_3, etc., are given by the formula

$$\phi_1 + x\phi_2 + x^2\phi_3 + \ldots = (1-x)^{-\phi_1}(1-x^2)^{-\phi_2}(1-x^3)^{-\phi_3}\ldots,$$

viz. we thus find

suffix of ϕ	1	2	3	4	5	6	7	8	9	10	11	12	13
$\phi =$	1	1	2	4	9	20	48	115	286	719	1842	4766	12486.

And I have, in the paper "On the analytical forms called Trees, with application to the theory of chemical combinations," *Brit. Assoc. Report*, 1875, pp. 257—305, [610] also shown how by the consideration of the centre or bicentre "of length" we can obtain formulæ for the number of central and bicentral trees, that is, for the number of distinct trees, with N knots: the numerical result obtained for the total number of distinct trees with N knots is given as follows:

No. of Knots	1	2	3	4	5	6	7	8	9	10	11	12	13
No. of Central Trees	1	0	1	1	2	3	7	12	27	55	127	284	682
„ Bicentral „	0	1	0	1	1	3	4	11	20	51	108	267	619
Total	1	1	1	2	3	6	11	23	47	106	235	551	1301.

But a more simple solution is obtained by the consideration of the centre or bicentre "of number." A tree of an odd number N of knots has a centre of number, and a tree of an even number N of knots has a centre or else a bicentre of number. To explain this notion (due to M. Camille Jordan) we consider the branches which proceed from any knot, and (excluding always this knot itself) we count the number of the knots upon the several branches; say these numbers are α, β, γ, δ, ϵ, etc., where of course $\alpha+\beta+\gamma+\delta+\epsilon+\text{etc.}=N-1$. If N is even we may have, say $\alpha=\frac{1}{2}N$; and then $\beta+\gamma+\delta+\epsilon+\text{etc.}=\frac{1}{2}N-1$, viz. α is larger by unity than the sum of the remaining numbers: the branch with α knots, or the number α, is said to be "merely dominant." If N be odd, we cannot of course have $\alpha=\frac{1}{2}N$, but we may have $\alpha>\frac{1}{2}N$; here α exceeds by 2 at least the sum of the other numbers; and the branch with α knots, or the number α, is said to be "predominant." In every other case, viz. in the case where each number α is less than $\frac{1}{2}N$, (and where consequently the largest number α does not exceed the sum of the remaining numbers), the several branches, or the numbers α, β, γ, etc., are said to be subequal. And we have the theorem. First, when N is odd, there is always one knot (and only one knot) for which the branches are subequal: such knot is called the centre of number. Secondly, when N is even, either there is one knot (and only one knot) for which the branches are subequal: and such knot is then called the centre of number; or else there is no such knot, but there are two adjacent knots (and no other knot) each having a merely-dominant branch: such two knots are called the bicentre of number, and each of them separately is a half-centre.

Considering now the trees with N knots as springing from a centre or a bicentre of number, and writing ψ_N for the whole number of distinct trees with N knots, we readily obtain these in terms of the foregoing numbers ϕ_1, ϕ_2, ϕ_3, etc., viz. we have

$$\psi_1 = 1,$$

$$\psi_2 = \tfrac{1}{2}\phi_1(\phi_1+1),$$

$$\psi_3 = \text{coeff. } x^2 \text{ in } (1-x)^{-\phi_1},$$

$$\psi_4 = \tfrac{1}{2}\phi_2(\phi_2+1) + \text{coeff. } x^3 \text{ in } (1-x)^{-\phi_1},$$

$$\psi_5 = \text{coeff. } x^4 \text{ in } (1-x)^{-\phi_1}(1-x^2)^{-\phi_2},$$

$$\psi_6 = \tfrac{1}{2}\phi_3(\phi_3+1) + \text{coeff. } x^5 \text{ in } (1-x)^{-\phi_1}(1-x^2)^{-\phi_2},$$

$$\psi_7 = \text{coeff. } x^6 \text{ in } (1-x)^{-\phi_1}(1-x^2)^{-\phi_2}(1-x^3)^{-\phi_3},$$

and so on, the law being obvious. And the formulæ are at once seen to be true. Thus for $N=6$, the formula is

$$\psi_6 = \tfrac{1}{2}\phi_3(\phi_3+1) + \tfrac{1}{2}\phi_2(\phi_2+1)\,.\,\phi_1 + \phi_2\,.\,\tfrac{1}{6}\phi_1(\phi_1+1)(\phi_1+2)$$
$$+\tfrac{1}{120}\phi_1(\phi_1+1)(\phi_1+2)(\phi_1+3)(\phi_1+4).$$

We have ϕ_3 root-trees with 3 knots, and by simply joining together any two of them, treating the two roots as a bicentre, we have all the bicentral trees with 6 knots: this accounts for the term $\tfrac{1}{2}\phi_3(\phi_3+1)$. Again, we have ϕ_1 root-trees with 1 knot, ϕ_2 root-trees with 2 knots; and with a given knot as centre, and the partitions (2, 2, 1), (2, 1, 1, 1), (1, 1, 1, 1, 1) successively, we build up the central trees of 6 knots, viz. 1° we take as branches any two ϕ_2's and any one ϕ_1; 2° any one ϕ_2 and any three ϕ_1's; 3° any five ϕ_1's; the partitions in question being all the partitions of 5 with no part greater than 2, that is, all the partitions with sub-equal parts. We easily obtain

suffix of ψ	1	2	3	4	5	6	7	8	9	10	11	12	13
$\psi=$	1	1	1	2	3	6	11	23	47	106	235	551	1301

agreeing with the results obtained by the much more complicated formulæ of the paper of 1875.

773.

ON THE 8-SQUARE IMAGINARIES.

[From the *American Journal of Mathematics*, vol. IV. (1881), pp. 293—296.]

I WRITE throughout **0** to denote positive unity, and uniting with it the seven imaginaries **1**, ..., **7**, form an octavic system **0**, **1**, **2**, **3**, **4**, **5**, **6**, **7**, the laws of combination being

$$\mathbf{0}^2 = \mathbf{0},\quad \mathbf{1}^2 = \mathbf{2}^2 = \mathbf{3}^2 = \mathbf{4}^2 = \mathbf{5}^2 = \mathbf{6}^2 = \mathbf{7}^2 = -\mathbf{0},$$

$$\mathbf{123} = \epsilon_1,\qquad \mathbf{145} = \epsilon_2,\qquad \mathbf{167} = \epsilon_3,$$

$$\mathbf{246} = \epsilon_4,\qquad \mathbf{257} = \epsilon_5,$$

$$\mathbf{347} = \epsilon_6,\qquad \mathbf{356} = \epsilon_7,$$

where $\epsilon = \pm$, viz. each ϵ has a determinate value $+$ or $-$ as the case may be; and where the formula, $\mathbf{123} = \epsilon_1$, denotes the six equations

$$\mathbf{23} = \epsilon_1\mathbf{1},\qquad \mathbf{31} = \epsilon_1\mathbf{2},\qquad \mathbf{12} = \epsilon_1\mathbf{3},$$

$$\mathbf{32} = -\epsilon_1\mathbf{1},\qquad \mathbf{13} = -\epsilon_1\mathbf{2},\qquad \mathbf{21} = -\epsilon_1\mathbf{3},$$

and so for the other formulæ. The multiplication table of the eight symbols thus is

	0	1	2	3	4	5	6	7
0	0	1	2	3	4	5	6	7
1	1	$-$ 0	ϵ_1 3	$-\epsilon_1$ 2	ϵ_2 5	$-\epsilon_2$ 4	ϵ_3 7	$-\epsilon_3$ 6
2	2	$-\epsilon_1$ 3	$-$ 0	ϵ_1 1	ϵ_4 6	ϵ_5 7	$-\epsilon_4$ 4	$-\epsilon_5$ 5
3	3	ϵ_1 2	$-\epsilon_1$ 1	$-$ 0	ϵ_6 7	ϵ_7 6	$-\epsilon_7$ 5	$-\epsilon_6$ 4
4	4	$-\epsilon_2$ 5	$-\epsilon_4$ 6	$-\epsilon_6$ 7	$-$ 0	ϵ_2 1	ϵ_4 2	ϵ_6 3
5	5	ϵ_2 4	$-\epsilon_5$ 7	$-\epsilon_7$ 6	$-\epsilon_2$ 1	$-$ 0	ϵ_7 3	ϵ_5 2
6	6	$-\epsilon_3$ 7	ϵ_4 4	ϵ_7 5	$-\epsilon_4$ 2	$-\epsilon_7$ 3	$-$ 0	ϵ_3 1
7	7	ϵ_3 6	ϵ_5 5	ϵ_6 4	$-\epsilon_6$ 3	$-\epsilon_5$ 2	$-\epsilon_3$ 1	$-$ 0

Hence if 0, 1, 2, 3, 4, 5, 6, 7 and 0′, 1′, 2′, 3′, 4′, 5′, 6′, 7′ denote ordinary algebraical magnitudes, and we form the product

$$(0\mathbf{0}+1\mathbf{1}+2\mathbf{2}+3\mathbf{3}+4\mathbf{4}+5\mathbf{5}+6\mathbf{6}+7\mathbf{7})(0'\mathbf{0}+1'\mathbf{1}+2'\mathbf{2}+3'\mathbf{3}+4'\mathbf{4}+5'\mathbf{5}+6'\mathbf{6}+7'\mathbf{7}),$$

this is at once found to be =

$$\begin{array}{ll}
(00'-11'-22'-33'-44'-55'-66'-77') & \mathbf{0} \\
+\,(01'+0'1+\epsilon_1 23+\epsilon_2 45+\epsilon_3 67 &)\,\mathbf{1} \\
+\,(02'+0'2+\epsilon_1 31+\epsilon_4 46+\epsilon_5 57 &)\,\mathbf{2} \\
+\,(03'+0'3+\epsilon_1 12+\epsilon_6 47+\epsilon_7 56 &)\,\mathbf{3} \\
+\,(04'+0'4+\epsilon_1 51+\epsilon_4 62+\epsilon_6 73 &)\,\mathbf{4} \\
+\,(05'+0'5+\epsilon_2 14+\epsilon_5 72+\epsilon_7 63 &)\,\mathbf{5} \\
+\,(06'+0'6+\epsilon_3 71+\epsilon_4 24+\epsilon_7 35 &)\,\mathbf{6} \\
+\,(07'+0'7+\epsilon_3 16+\epsilon_5 25+\epsilon_6 34 &)\,\mathbf{7},
\end{array}$$

where 12 is written to denote $12'-1'2$, and so in other cases.

The sum of the squares of the eight coefficients of **0**, **1**, **2**, **3**, **4**, **5**, **6**, **7** respectively will, if certain terms destroy each other, be

$$=(0^2+1^2+2^2+3^2+4^2+5^2+6^2+7^2)(0'^2+1'^2+2'^2+3'^2+4'^2+5'^2+6'^2+7'^2);$$

viz. the sum of the squares contains the several terms

$$\begin{array}{lllllll}
\epsilon_1\epsilon_2 23.45, & \epsilon_1\epsilon_3 23.67, & \epsilon_1\epsilon_4 31.46, & \epsilon_1\epsilon_5 31.57, & \epsilon_1\epsilon_6 12.47, & \epsilon_1\epsilon_7 12.56, & \epsilon_2\epsilon_3 45.67, \\
\epsilon_4\epsilon_7 24.35, & \epsilon_4\epsilon_6 62.73, & \epsilon_2\epsilon_7 14.63, & \epsilon_2\epsilon_6 51.73, & \epsilon_2\epsilon_5 14.72, & \epsilon_2\epsilon_4 51.62, & \epsilon_4\epsilon_5 46.57, \\
\epsilon_5\epsilon_6 25.34, & \epsilon_5\epsilon_7 72.63, & \epsilon_3\epsilon_6 16.34, & \epsilon_3\epsilon_7 71.35, & \epsilon_3\epsilon_4 71.24, & \epsilon_3\epsilon_5 16.25, & \epsilon_6\epsilon_7 47.56,
\end{array}$$

and observing that $\mathbf{21}=-\mathbf{12}$, etc., and that we have identically

$$\mathbf{23}.\mathbf{45}+\mathbf{24}.\mathbf{53}+\mathbf{25}.\mathbf{34}=\text{zero, etc.},$$

then the three terms of each column will vanish, provided a proper relation exists between the ϵ's: viz. the conditions which we thus obtain are

$$\begin{array}{l}
\epsilon_1\epsilon_2=-\epsilon_4\epsilon_7=\epsilon_5\epsilon_6, \\
\epsilon_1\epsilon_3=-\epsilon_4\epsilon_6=\epsilon_5\epsilon_7, \\
\epsilon_1\epsilon_4=-\epsilon_3\epsilon_6=-\epsilon_2\epsilon_7, \\
\epsilon_1\epsilon_5=\epsilon_3\epsilon_7=\epsilon_2\epsilon_6, \\
\epsilon_1\epsilon_6=\epsilon_2\epsilon_5=-\epsilon_3\epsilon_4, \\
\epsilon_1\epsilon_7=-\epsilon_2\epsilon_4=\epsilon_3\epsilon_5, \\
\epsilon_2\epsilon_3=-\epsilon_4\epsilon_5=\epsilon_6\epsilon_7.
\end{array}$$

We may without loss of generality assume $\epsilon_1=\epsilon_2=\epsilon_3=+$; the equations then become

$$\begin{array}{l}
+=-\epsilon_4\epsilon_7=\epsilon_5\epsilon_6, \\
+=-\epsilon_4\epsilon_6=\epsilon_5\epsilon_7, \\
+=-\epsilon_4\epsilon_5=\epsilon_6\epsilon_7, \\
\epsilon_4=-\epsilon_6=-\epsilon_7, \\
\epsilon_5=\epsilon_7=\epsilon_6, \\
\epsilon_6=\epsilon_5=-\epsilon_4, \\
\epsilon_7=-\epsilon_4=\epsilon_5;
\end{array}$$

and writing $\theta = \pm$ at pleasure, these are all satisfied if $-\epsilon_4 = \epsilon_5 = \epsilon_6 = \epsilon_7 = \theta$. The terms written down all disappear, and the sum of the squares of the eight coefficients thus becomes equal to the product of two sums each of them of eight squares, viz. this is the case if $\epsilon_1 = \epsilon_2 = \epsilon_3 = +$, $-\epsilon_4 = \epsilon_5 = \epsilon_6 = \epsilon_7 = \theta$, θ being $= \pm$ at pleasure: the resulting system of imaginaries may be said to be an 8-square system.

We may inquire whether the system is associative; for this purpose, supposing in the first instance that the ϵ's remain arbitrary, we form the complete system of the values of the triplets **12 . 3**, **1 . 23**, etc., (read the top line **12 . 3** $= -\epsilon_1$**0**, **1 . 23** $= -\epsilon_1$**0**, the next line **12 . 4** $= \epsilon_1\epsilon_6$**7**, **1 . 24** $= \epsilon_3\epsilon_4$**7**, and so in other cases):

12 . 3 =	**1 . 23** =	$-\epsilon_1$	$-\epsilon_1$	**0**
12 . 4 =	**1 . 24** =	$\epsilon_1\epsilon_6$	$\epsilon_3\epsilon_4$	**7**
12 . 5 =	**1 . 25** =	$\epsilon_1\epsilon_7$	$-\epsilon_3\epsilon_5$	**6**
12 . 6 =	**1 . 26** =	$-\epsilon_1\epsilon_7$	$-\epsilon_2\epsilon_4$	**5**
12 . 7 =	**1 . 27** =	$-\epsilon_1\epsilon_6$	$\epsilon_2\epsilon_5$	**4**
13 . 4 =	**1 . 34** =	$-\epsilon_1\epsilon_4$	$-\epsilon_3\epsilon_6$	**6**
13 . 5 =	**1 . 35** =	$-\epsilon_1\epsilon_5$	. $\epsilon_3\epsilon_7$	**7**
13 . 6 =	**1 . 36** =	$\epsilon_1\epsilon_4$	$\epsilon_2\epsilon_7$	**4**
13 . 7 =	**1 . 37** =	$\epsilon_1\epsilon_5$	$-\epsilon_2\epsilon_6$	**5**
14 . 5 =	**1 . 45** =	$-\epsilon_2$	$-\epsilon_2$	**0**
14 . 6 =	**1 . 46** =	$\epsilon_2\epsilon_7$	$\epsilon_1\epsilon_4$	**3**
14 . 7 =	**1 . 47** =	$\epsilon_2\epsilon_5$	$-\epsilon_1\epsilon_6$	**2**
15 . 6 =	**1 . 56** =	$-\epsilon_2\epsilon_4$	$-\epsilon_1\epsilon_7$	**2**
15 . 7 =	**1 . 57** =	$-\epsilon_2\epsilon_6$	$\epsilon_1\epsilon_5$	**3**
16 . 7 =	**1 . 67** =	$-\epsilon_3$	$-\epsilon_3$	**0**
23 . 4 =	**2 . 34** =	$\epsilon_1\epsilon_2$	$-\epsilon_5\epsilon_6$	**5**
23 . 5 =	**2 . 35** =	$-\epsilon_1\epsilon_2$	$-\epsilon_4\epsilon_7$	**4**
23 . 6 =	**2 . 36** =	$\epsilon_1\epsilon_3$	$-\epsilon_5\epsilon_7$	**7**
23 . 7 =	**2 . 37** =	$-\epsilon_1\epsilon_3$	$-\epsilon_4\epsilon_6$	**6**
24 . 5 =	**2 . 45** =	$-\epsilon_4\epsilon_7$	$-\epsilon_1\epsilon_2$	**3**
24 . 6 =	**2 . 46** =	$-\epsilon_4$	$-\epsilon_4$	**0**
24 . 7 =	**2 . 47** =	$\epsilon_3\epsilon_4$	$\epsilon_1\epsilon_6$	**1**
25 . 6 =	**2 . 56** =	$-\epsilon_3\epsilon_5$	$\epsilon_1\epsilon_7$	**1**
25 . 7 =	**2 . 57** =	$-\epsilon_5$	$-\epsilon_5$	**0**
26 . 7 =	**2 . 67** =	$-\epsilon_4\epsilon_6$	$-\epsilon_1\epsilon_3$	**3**
34 . 5 =	**3 . 45** =	$-\epsilon_5\epsilon_6$	$\epsilon_1\epsilon_2$	**2**
34 . 6 =	**3 . 46** =	$-\epsilon_3\epsilon_6$	$-\epsilon_1\epsilon_4$	**1**
34 . 7 =	**3 . 47** =	$-\epsilon_6$	$-\epsilon_6$	**0**
35 . 6 =	**3 . 56** =	$-\epsilon_7$	$-\epsilon_7$	**0**
35 . 7 =	**3 . 57** =	$\epsilon_3\epsilon_7$	$-\epsilon_1\epsilon_5$	**1**
36 . 7 =	**3 . 67** =	$-\epsilon_5\epsilon_7$	$\epsilon_1\epsilon_3$	**2**
45 . 6 =	**4 . 56** =	$\epsilon_2\epsilon_3$	$-\epsilon_6\epsilon_7$	**7**
45 . 7 =	**4 . 57** =	$-\epsilon_2\epsilon_3$	$-\epsilon_4\epsilon_5$	**6**
46 . 7 =	**4 . 67** =	$-\epsilon_4\epsilon_5$	$-\epsilon_2\epsilon_3$	**5**
56 . 7 =	**5 . 67** =	$-\epsilon_6\epsilon_7$	$\epsilon_2\epsilon_3$	**4.**

Write as before $\epsilon_1 = \epsilon_2 = \epsilon_3 = +$; then, disregarding the lines (such as the first line) which contain the symbol **0**, and writing down only the signs as given in the third and fourth columns, these are

ϵ_6	ϵ_4
ϵ_7	$-\epsilon_5$
$-\epsilon_7$	$-\epsilon_4$
$-\epsilon_6$	ϵ_5
$-\epsilon_4$	$-\epsilon_6$
$-\epsilon_5$	ϵ_7
ϵ_4	ϵ_7
ϵ_5	$-\epsilon_6$
ϵ_7	ϵ_4
ϵ_5	$-\epsilon_6$
$-\epsilon_4$	$-\epsilon_7$
$-\epsilon_6$	ϵ_5
$+$	$-\epsilon_5\epsilon_6$
$-$	$-\epsilon_4\epsilon_7$
$+$	$-\epsilon_5\epsilon_7$
$-$	$-\epsilon_4\epsilon_6$
$-\epsilon_4\epsilon_7$	$-$
ϵ_4	ϵ_6
$-\epsilon_5$	ϵ_7
$-\epsilon_4\epsilon_6$	$-$
$-\epsilon_5\epsilon_6$	$+$
$-\epsilon_6$	$-\epsilon_4$
ϵ_7	$-\epsilon_5$
$-\epsilon_5\epsilon_7$	$+$
$+$	$-\epsilon_6\epsilon_7$
$-$	$-\epsilon_4\epsilon_5$
$-\epsilon_4\epsilon_5$	$-$
$-\epsilon_6\epsilon_7$	$+$

We hence see at once that the pairs of signs in the two columns respectively cannot be made identical: to make them so, we should have $\epsilon_6 = \epsilon_4$, $\epsilon_7 = -\epsilon_5$, $\epsilon_7 = \epsilon_4$, that is, $\epsilon_4 = \epsilon_6 = \epsilon_7 = -\epsilon_5$, which is inconsistent with the last equation of the system $-\epsilon_6\epsilon_7 = +$. Hence the imaginaries **1**, **2**, **3**, **4**, **5**, **6**, **7**, as defined by the original conditions, are not in any case associative.

If we have $\epsilon_1 = \epsilon_2 = \epsilon_3 = +$ and also $-\epsilon_4 = \epsilon_5 = \epsilon_6 = \epsilon_7 = \theta$, that is, if the imaginaries belong to the 8-square formula, then it is at once seen that each pair consists of two opposite signs; that is, for the several triads **123**, **145**, **167**, **246**, **257**, **347**, **356** used for the definition of the imaginaries, the associative property holds good, **12** . **3** = **1** . **23**, etc.; but for each of the remaining twenty-eight triads, *the two terms are equal but of opposite signs*, viz. **12** . **4** = − **1** . **24**, etc.; so that the product **124** of any such three symbols has no determinate meaning.

Baltimore, March 5th, 1882.

774.

TABLES FOR THE BINARY SEXTIC.

The Leading Coefficients of the First 18 of the 26 Covariants.

[From the *American Journal of Mathematics*, vol. IV. (1881), pp. 379—384.]

Including the sextic itself, the number of covariants of the binary sextic is $=26$, as shown in the table p. 296 of Clebsch's *Theorie der binären algebraischen Formen*, Leipzig, 1872; viz. this is

	Order						
Deg.	0	2	4	6	8	10	12
1				f			
2	A		i		H		
3		l		p	(f, i)		T
4	B		$(f, l)_2$	(f, l)		(H, i)	
5		$(i, l)_2$	(i, l)		(H, l)		
6	A_{u}			(p, l) $((f, i), l)_2$			
7		$(f, l^2)_4$	$(f, l^2)_3$				
8		$(i, l^2)_3$					
9			$((f, i), l^2)_4$				
10	$(f, l^3)_6$	$(f, l^3)_5$					
12		$((f, i), l^3)_6$					
15	$((f, i), l^4)_8$						

Or, using the capital letters $A, B, \ldots, Z$ to denote the 26 covariants in the same order, the table is

	0	2	4	6	8	10	12
1				A			
2	B		C		D		
3		E		F	$G, = (A, C)^1$		H
4	I		$J, = (A, E)^2$	$K, = (A, E)^1$		$L, = (D, C)^1$	
5		$M, = (C, E)^2$	$N, = (C, E)^1$		$O, = (D, E)^1$		
6	P			$Q, = (F, E)^1$ $R, = (G, E)^2$			
7		$S, = (A, E^2)^4$	$T, = (A, E^2)^3$				
8		$U, = (C, E^2)^3$					
9			$V, = (G, E^2)^4$				
10	$W, = (A, E^3)^6$	$X, = (A, E^3)^5$					
12		$Y, = (G, E^3)^6$					
15	$Z, = (G, E^4)^8$						

A is the sextic. P is Salmon's C, p. 204.
B is Salmon's A, p. 202. W ,, ,, D, p. 207.
I ,, ,, B, p. 203. Z ,, ,, E, p. 253.
The references are to Salmon's *Higher Algebra*, 2nd Ed., 1866.

In the present short paper I give the leading coefficients of the first 18 covariants, A to R (some of these are of course known values, but it is convenient to include them): for the next four covariants S, T, U, V, the leading coefficients depend upon the coefficients of A, C, G and E^2, viz. writing

$$\begin{aligned} A &= (a,\ b,\ c,\ d,\ e,\ f,\ g \between x,\ y)^6, \\ E^2 &= (\alpha,\ \tfrac{1}{4}\beta,\ \tfrac{1}{6}\gamma,\ \tfrac{1}{4}\delta,\ \epsilon \between x,\ y)^4, \\ C &= (\alpha',\ \tfrac{1}{4}\beta',\ \tfrac{1}{6}\gamma',\ \tfrac{1}{4}\delta',\ \epsilon' \between x,\ y)^4, \\ G &= (\alpha'',\ \tfrac{1}{8}\beta'',\ \tfrac{1}{28}\gamma'',\ \tfrac{1}{56}\delta'',\ \tfrac{1}{70}\epsilon'',\ \ldots \between x,\ y)^8, \end{aligned}$$

we have

$$\begin{aligned} &S, \text{ Coeff. } x^2 = a\epsilon - b\delta + c\gamma - d\beta + e\alpha, \\ &T, \quad ,, \quad x^4 = a\delta - 2b\gamma + 3c\beta - 4d\alpha, \\ &U, \quad ,, \quad x^2 = 2\alpha'\delta - \beta'\gamma + \gamma'\beta - 2\delta'\alpha, \\ &V, \quad ,, \quad x^4 = 280\alpha''\epsilon - 35\beta''\delta + 10\gamma''\gamma - 20\delta''\beta + 24\epsilon''\alpha. \end{aligned}$$

Similarly the invariant W and the leading coefficients of X, Y depend on the coefficients of A, G and E^3; and the invariant Z depends on the coefficients of G and E^4.

But these two invariants W and Z have been already calculated; viz. as already mentioned, W is Salmon's invariant D, and Z his invariant E, given each of them in the second edition of his *Higher Algebra* (but not reproduced in the third edition): on account of the great length of these expressions, it has been thought that it was not expedient to give them here.

For the reason appearing above, I have added the expressions for the remaining coefficients of C, E, G.

$A,\ x^6$	$B,\ x^0$	$C,\ x^4$	$D,\ x^8$	$E,\ x^2$	$F,\ x^6$	$G,\ x^8$	$H,\ x^{12}$
$a + 1$	$a\,g + 1$	$a\,e + 1$	$a\,c + 1$	$a\,cg + 1$	$a\,ce + 1$	$a^2f + 1$	$a^2d + 1$
	$a^0bf - 6$	$a^0bd - 4$	$a^0b^2 - 1$	$df - 3$	$d^2 - 1$	$a\,be - 5$	$a\,bc - 3$
	$ce + 15$	$c^2 + 3$		$e^2 + 2$	$a^0b^2e - 1$	$cd + 2$	$a^0b^3 + 2$
	$d^2 - 10$			$a^0b^2g - 1$	$bcd + 2$	$a^0b^2d + 8$	
				$bcf + 3$	$c^3 - 1$	$bc^2 - 6$	
				$bde - 1$			
				$c^2e - 3$			
				$cd^2 + 2$			

$I,\ x^0$	$J,\ x^4$	$K,\ x^6$	$L,\ x^{10}$	$M,\ x^2$	$N,\ x^4$	$O,\ x^8$
$a\,ceg + 1$	$a^2f^2 + 1$	$a^2dg + 1$	$a^2cf + 1$	$a^2cg^2 + 1$	$a^2cfg - 1$	$a^2cdg\ \ 0$
$cf^2 - 1$	$a\,bef - 10$	$ef - 1$	$de - 1$	$dfg - 6$	$deg + 1$	$cef - 1$
$d^2g - 1$	$cdf + 4$	$a\,bcg - 3$	$a\,b^2f - 1$	$e^2g + 8$	$df^2 + 3$	$d^2f + 3$
$def + 2$	$ce^2 + 16$	$bdf - 2$	$bce - 2$	$ef^2 - 3$	$e^2f - 3$	$de^2 - 2$
$e^3 - 1$	$d^2e - 12$	$be^2 + 5$	$bd^2 + 4$	$a\,b^2g^2 - 1$	$a\,b^2fg + 1$	$a\,b^2dg\ \ 0$
$a^0b^2eg - 1$	$a^0b^2df + 16$	$c^2f + 9$	$c^2d - 1$	$bcfg + 6$	$bceg + 2$	$b^2ef + 1$
$b^2f^2 + 1$	$b^2e^2 + 9$	$cde - 17$	$a^0b^3e + 3$	$bdeg - 34$	$bcf^2 - 3$	$bc^2g\ \ 0$
$bcdg + 2$	$bc^2f - 12$	$d^3 + 8$	$b^2cd - 6$	$bdf^2 + 48$	$bd^2g - 4$	$bcdf - 14$
$bcef - 2$	$bcde - 76$	$a^0b^3g + 2$	$bc^3 + 3$	$be^2f - 18$	$bdef - 12$	$bce^2 + 11$
$bd^2f - 2$	$bd^3 + 48$	$b^2cf - 6$		$c^2eg + 18$	$be^3 + 15$	$bd^2e + 1$
$bde^2 + 2$	$c^3e + 48$	$b^2de + 2$		$c^2f^2 - 45$	$c^2dg + 1$	$c^3f + 9$
$c^3g - 1$	$c^2d^2 - 32$	$bc^2e + 6$		$cd^2g + 4$	$c^2ef + 9$	$c^2de - 14$
$c^2df + 2$		$bcd^2 - 4$		$cdef + 78$	$cd^2f + 4$	$cd^3 + 6$
$c^2e^2 + 1$				$ce^3 - 36$	$cde^2 - 21$	$a^0b^3cg\ \ 0$
$cd^2e - 3$				$d^3f - 48$	$d^3e + 8$	$b^3df + 8$
$d^4 + 1$				$d^2e^2 + 28$	$a^0b^3eg - 3$	$b^3e^2 - 9$
				$a^0b^2ceg\ \ 0$	$b^2cdg + 6$	$b^2c^2f - 6$
				$b^2d^2g + 64$	$b^2cef + 9$	$b^2cde + 16$
				$b^2def - 144$	$b^2d^2f + 32$	$b^2d^3 - 8$
				$b^2e^3 + 81$	$b^2de^2 - 39$	$bc^3e - 3$
				$bc^2dg - 96$	$bc^3g - 3$	$bc^2d^2 + 2$
				$bc^2ef + 108$	$bc^2df - 66$	
				$bcd^2f + 96$	$bc^2e^2 + 18$	
				$bcde^2 - 126$	$bcd^2e + 76$	
				$bd^3e + 16$	$bd^4 - 32$	
				$c^4g + 36$	$c^4f + 27$	
				$c^3df - 72$	$c^3de - 45$	
				$c^3e^2 - 27$	$c^2d^3 + 20$	
				$c^2d^2e + 96$		
				$cd^4 - 32$		

P, x^0	
$a^2d^2g^2$	+ 1
$defg$	− 6
df^3	+ 4
e^3g	+ 4
e^2f^2	− 3
$a\,bcdg^2$	− 6
$bcefg$	+ 18
bcf^3	− 12
bd^2fg	+ 12
bde^2g	− 18
be^3f	+ 6
c^3g^2	+ 4
c^2e^2g	− 24
c^2dfg	− 18
c^2ef^2	+ 30
cd^2eg	+ 54
cd^2f^2	− 12
cde^2f	− 42
ce^4	+ 12
d^4g	− 20
d^3ef	+ 24
d^2e^3	− 8
$a^0b^3dg^2$	+ 4
b^3efg	− 12
b^3f^3	+ 8
$b^2c^2g^2$	− 3
b^2ce^2g	+ 30
b^2cef^2	− 24
b^2d^2eg	− 12
$b^2d^2f^2$	− 24
b^2de^2f	+ 60
b^2e^4	− 27
bc^3fg	+ 6
bc^2deg	− 42
bc^2df^2	+ 60
bc^2e^2f	− 30
bcd^3g	+ 24
bcd^2ef	− 84
$bcde^3$	+ 66
bd^4f	+ 24
bd^3e^2	− 24
c^4eg	+ 12
c^4f^2	− 27
c^3d^2g	− 8
c^3def	+ 66
c^3e^3	− 8
c^2d^3f	− 24
$c^2d^2e^2$	− 39
cd^4e	+ 36
d^6	− 8

	Q, x^6	R, x^6
a^3dg^2		− 1
efg		+ 9
f^3		− 8
a^2bcg^2		+ 3
$bdfg$		− 24
be^2g		− 45
bef^2		+ 66
c^2fg	− 2	+ 3
$cdeg$	+ 5	+ 48
cdf^2	+ 6	− 12
ce^2f	− 7	− 51
d^3g	− 3	− 16
d^2ef	− 3	+ 36
de^3	+ 4	− 8
$a\,b^3g^2$	0	− 2
b^2cfg	+ 4	+ 12
b^2deg	− 5	+ 192
b^2df^2	− 6	− 48
b^2e^2f	+ 7	− 144
bc^2eg	− 5	− 159
bc^2f^2	− 6	+ 18
bcd^2g	+ 7	− 48
$bcdef$	− 16	+ 24
bce^3	+ 23	+ 279
bd^3f	+ 30	− 48
bd^2e^2	− 33	− 84
c^3dg	− 1	+ 42
c^3ef	+ 36	+ 153
c^2d^2f	− 37	− 36
c^2de^2	− 53	− 399
cd^3e	+ 79	+ 312
d^5	− 24	− 64
a^0b^4fg	− 2	0
b^3ceg	+ 5	0
b^3cf^2	+ 6	0
b^3d^2g	+ 2	− 224
b^3def	+ 22	+ 144
b^3e^3	− 27	+ 54
b^2c^2dg	− 8	+ 336
b^2c^2ef	− 39	− 108
b^2cd^2f	− 50	+ 384
b^2cde^2	+ 107	− 684
b^2d^3e	− 22	+ 144
bc^4g	+ 3	− 126
bc^3df	+ 84	− 648
bc^3e^2	− 21	+ 432
bc^2d^2e	− 102	+ 564
bcd^4	+ 44	− 288
c^5f	− 27	+ 270
c^4de	+ 45	− 450
c^3d^3	− 20	+ 200

Remaining Coefficients of C, E, G.

C

x^3y	
af	$+2$
be	-6
cd	$+4$
x^2y^2	
ag	$+1$
ce	-9
d^2	$+8$
xy^3	
bg	$+2$
cf	-6
de	$+4$
y^4	
cg	$+1$
df	-4
e^2	$+3$

E

xy	
adg	$+1$
aef	-1
bcg	-1
bdf	-8
be^2	$+9$
c^2f	$+9$
cde	-17
d^3	$+8$
y^2	
aeg	$+1$
af^2	-1
bdg	-3
bef	$+3$
c^2g	$+2$
cdf	-1
ce^2	-3
d^2e	$+2$

G

x^7y	
a^2g	$+1$
abf	$+2$
ace	-19
ad^2	$+8$
b^2e	-6
bcd	$+44$
c^3	-30
x^6y^2	
abg	$+7$
acf	-14
ade	-14
b^2f	0
bce	-21
bd^2	$+112$
c^2d	-70
x^5y^3	
acg	$+7$
adf	-28
ae^2	-14
b^2g	$+14$
bcf	-42
bde	$+168$
c^2e	-105
x^4y^4	
adg	0
aef	-35
bcg	$+35$
bdf	0
be^2	$+105$
c^2f	-105

G

x^3y^5	
aeg	-7
af^2	-14
bdg	$+28$
bef	$+42$
c^2g	$+14$
cdf	-168
ce^2	$+105$
x^2y^6	
afg	-7
beg	$+14$
bf^2	0
cdg	$+14$
cef	$+21$
d^2f	-112
de^2	$+70$
xy^7	
ag^2	-1
bfg	-2
ceg	$+19$
cf^2	$+6$
d^2g	-8
def	-44
e^3	$+30$
y^8	
bg^2	-1
cfg	$+5$
deg	-2
df^2	-8
e^2f	$+6$

Note.—**In the tables on this page, a has been treated like the other letters; on the preceding pages, the powers of a have been suppressed except in the first of every series of terms containing a common power of a.**

The final result is that we have the values of the invariants B, I, P, W, Z and the leading coefficients of the covariants A, C, D, E, F, G, H, J, K, L, M, N, O, Q, R: also the means of calculating the leading coefficients of the remaining covariants S, T, U, V, X, Y.

775.

TABLES OF COVARIANTS OF THE BINARY SEXTIC.

[*Written in* 1894: *now first published.*]

THE binary sextic has in all (including the sextic itself and the invariants) 26 covariants which I have represented by the capital letters $A, B, C, \ldots, Z$. The leading coefficients of the covariants A to R (of course for an invariant this means the invariant itself) are given in my paper "Tables for the binary sextic," *Amer. Math. Jour.* vol. IV. (1881), pp. 379—384, [774]; the two invariants Z and W (Salmon's invariants D and E) had been already calculated. But I did not in my values of the leading coefficients, nor did Salmon in his values of the two invariants, insert the literal terms with zero coefficients: as remarked in my paper [143] "Tables of the covariants M to W of the binary quintic," it is very desirable to have in every case the complete series of literal terms, and I have accordingly in the expressions of the covariants A to R obtained for the leading coefficients, and in the expressions obtained from Salmon for the invariants W and Z, inserted in each case the complete series of literal terms.

I give a list of the 26 covariants nearly in the form of that given in the latter paper [143] for the covariants of the quintic, only instead of a separate column of deg-weights I insert these in the body of the symbol; thus

$$C = (3,\ 3,\ 4,\ 3,\ 3)^2\ 4 \text{ to } 8\ (x,\ y)^4,$$

the 5 coefficients of the quartic function contain respectively 3, 3, 4, 3, 3 terms (some of them it may be with zero coefficients), are of the degree 2, and of the weights 4, 5, 6, 7, 8 respectively.

The list is as follows:

$A = (1,\ 1,\ 1,\ 1,\ 1,\ 1,\ 1)^1\ 0 \text{ to } 6\ (x,\ y)^6,$

$B = (4)^2\ 6\ (x,\ y)^0,$ Invt.,

$C = (3,\ 3,\ 4,\ 3,\ 3)^2\ 4 \text{ to } 8\ (x,\ y)^4,$

$D = (2,\ 2,\ 3,\ 3,\ 4,\ 3,\ 3,\ 2,\ 2)^2\ 2 \text{ to } 10\ (x,\ y)^8,$

$E = (8,\ 8,\ 8)^3\ 8 \text{ to } 10\ (x,\ y)^2,$

$F = (7,\ 7,\ 8,\ 8,\ 8,\ 7,\ 7)^3\ 6 \text{ to } 12\ (x,\ y)^6,$

$G = (5,\ 7,\ 7,\ 8,\ 8,\ 8,\ 7,\ 7,\ 5)^3\ 5 \text{ to } 13\ (x,\ y)^8,$

$H = (3,\ 4,\ 5,\ 7,\ 7,\ 8,\ 8,\ 8,\ 7,\ 7,\ 5,\ 4,\ 3)^3\ 3 \text{ to } 15\ (x,\ y)^{12},$

$I = (18)^4\ 12\ (x,\ y)^0,$ Invt.,

$J = (16,\ 16,\ 18,\ 16,\ 16)^4\ 10 \text{ to } 14\ (x,\ y)^4,$

$K = (14,\ 16,\ 16,\ 18,\ 16,\ 16,\ 14)^4\ 9 \text{ to } 15\ (x,\ y)^6,$

$L = (10,\ 13,\ 14,\ 16,\ 16,\ 18,\ 16,\ 16,\ 14,\ 13,\ 10)^4\ 7 \text{ to } 17\ (x,\ y)^{10},$

$M = (32,\ 32,\ 32)^5\ 14 \text{ to } 16\ (x,\ y)^2,$

$N = (30,\ 32,\ 32,\ 32,\ 30)^5\ 13 \text{ to } 17\ (x,\ y)^4,$

$O = (25,\ 29,\ 30,\ 32,\ 32,\ 32,\ 30,\ 29,\ 25)^5\ 11 \text{ to } 19\ (x,\ y)^8,$

$P = (58)^6\ 18\ (x,\ y)^0,$ Invt.,

$Q = (51,\ 55,\ 55,\ 58,\ 55,\ 55,\ 51)^6\ 15 \text{ to } 21\ (x,\ y)^6,$

$R = (51,\ 55,\ 55,\ 58,\ 55,\ 55,\ 51)^6\ 15 \text{ to } 21\ (x,\ y)^6,$

$S = (94,\ 94,\ 94)^7\ 20 \text{ to } 22\ (x,\ y)^2,$

$T = (90,\ 94,\ 94,\ 94,\ 90)^7\ 19 \text{ to } 23\ (x,\ y)^4,$

$U = (147,\ 151,\ 147)^8\ 23 \text{ to } 25\ (x,\ y)^2,$

$V = (221,\ 227,\ 227,\ 227,\ 221)^9\ 25 \text{ to } 29\ (x,\ y)^4,$

$W = (338)^{10}\ 30\ (x,\ y)^0,$ Invt.,

$X = (332,\ 338,\ 332)^{10}\ 29 \text{ to } 31\ (x,\ y)^2,$

$Y = (668,\ 676,\ 668)^{12}\ 35 \text{ to } 37\ (x,\ y)^2,$

$Z = (1636)^{15}\ 45\ (x,\ y)^0,$ Invt.

$$A = (\ \mathfrak{X}\!\!\; x,\ y)^6$$

x^6	x^5y	x^4y^2	x^3y^3	x^2y^4	xy^5	y^6
$a + 1$	$b + 6$	$c + 15$	$d + 20$	$e + 15$	$f + 6$	$g + 1$

$$B = (\ \mathfrak{X}\!\!\; x,\ y)^0, \text{ Invt.}$$

ag	$+\ \ 1$
bf	$-\ \ 6$
ce	$+\ 15$
d^2	$-\ 10$
	$\pm\ 16$

$C = (\quad)(x, y)^4$

x^4	x^3y	x^2y^2	xy^3	y^4
$ae + 1$	$af + 2$	$ag + 1$	$bg + 2$	$cg + 1$
$bd - 4$	$be - 6$	bf ...	$cf - 6$	$df - 4$
$c^2 + 3$	$cd + 4$	$ce - 9$	$de + 4$	$e^2 + 3$
		$d^2 + 8$		
± 4	± 6	± 9	± 6	± 4

$D = (\quad)(x, y)^8$

x^8	x^7y	x^6y^2	x^5y^3	x^4y^4	x^3y^5	x^2y^6	xy^7	y^8
$ac + 1$	$ad + 4$	$ae + 6$	$af + 4$	$ag + 1$	$bg + 4$	$cg + 6$	$dg + 4$	$eg + 1$
$b^2 - 1$	$bc - 4$	$bd + 4$	$be + 16$	$bf + 14$	$cf + 16$	$df + 4$	$ef - 4$	$f^2 - 1$
		$c^2 - 10$	$cd - 20$	$ce + 5$	$de - 20$	$e^2 - 10$		
				$d^2 - 20$				
± 1	± 4	± 10	± 20	± 20	± 20	± 10	± 4	± 1

$E = (\quad)(x, y)^2$

x^2	xy	y^2
$a\,cg + 1$	$a\,dg + 1$	$a\,eg + 1$
$df - 3$	$ef - 1$	$f^2 - 1$
$e^2 + 2$	$a^0bcg - 1$	$a^0bdg - 3$
$a^0b^2g - 1$	$bdf - 8$	$bef + 3$
$bcf + 3$	$be^2 + 9$	$c^2g + 2$
$bde - 1$	$c^2f + 9$	$cdf - 1$
$c^2e - 3$	$cde - 17$	$ce^2 - 3$
$cd^2 + 2$	$d^3 + 8$	$d^3 + 2$
± 3	± 1	± 1
± 5	± 26	± 7

$F = (\quad)(x, y)^6$

x^6	x^5y	x^4y^2	x^3y^3	x^2y^4	xy^5	y^6
a^2g ...	$a\,bg$ ...	$a\,cg + 1$	$a\,dg + 2$	$a\,eg + 1$	$a\,fg$ ...	$a\,g^2$ ...
$a\,bf$ ...	$cf + 2$	$df + 2$	$ef - 2$	$f^2 - 1$	$a^0beg + 2$	a^0bfg ...
$ce + 1$	$de - 2$	$e^2 - 3$	$a^0bcg - 2$	$a^0bdg + 2$	$bf^2 - 2$	$ceg + 1$
$d^2 - 1$	$a^0b^2f - 2$	$a^0b^2g - 1$	$bdf + 4$	$bef - 2$	$cdg - 2$	$cf^2 - 1$
$a^0b^2e - 1$	$bce + 2$	$bcf - 2$	$be^2 - 2$	$c^2g - 3$	$cef + 2$	$d^2g - 1$
$bcd + 2$	$bd^2 + 2$	$bde + 4$	$c^2f - 2$	$cdf + 4$	$d^2f + 2$	$def + 2$
$c^3 - 1$	$c^2d - 2$	$c^2e + 2$	$cde + 6$	$ce^2 + 2$	$de^2 - 2$	$e^3 - 1$
		$d^2 - 3$	$d^3 - 4$	$d^2e - 3$		
± 1	± 2	± 3	± 2	± 1	± 6	± 3
± 2	± 4	± 6	± 10	± 8		

$G = (\ \between\!x,\ y)^8$

x^8	x^7y	x^6y^2	x^5y^3	x^4y^4	x^3y^5	x^2y^6	xy^7	y^8
$a^2f\ +\ 1$	$a^2g\ +\ 1$	$a\,bg\ +\ 7$	$a\,cg\ +\ 7$	$a\,dg\ \ldots$	$a\,eg\ -\ 7$	$a\,fg\ -\ 7$	$a\,g^2\ -\ 1$	$a^0bg^2-\ 1$
$a\,be\ -\ 5$	$bf\ +\ 2$	$cf\ -\ 14$	$df\ -\ 28$	$ef\ -\ 35$	$f^2\ -\ 14$	$a^0beg+\ 14$	$a^0bfg-\ 2$	$cfg+\ 5$
$cd\ +\ 2$	$ce\ -19$	$de\ -\ 14$	$e^2\ -\ 14$	$a^0bcg+\ 35$	$a^0bdg+\ 28$	$bf^2\ \ldots$	$ceg+19$	$deg-\ 2$
$a^0b^2d+\ 8$	$d^2\ +\ 8$	$a^0b^2f\ \ldots$	$a^0b^2g\ +\ 14$	$bdf\ \ldots$	$bef\ +\ 42$	$cdg+\ 14$	$cf^2+\ 6$	$df^2-\ 8$
$bc^2-\ 6$	$a^0b^2e\ -\ 6$	$bce-\ 21$	$bcf-\ 42$	$be^2\ -105$	$c^2g\ +\ 14$	$cef+\ 21$	$d^2g-\ 8$	$e^2f+\ 6$
	$bcd+44$	bd^2+112	$bde+168$	$c^2f\ +105$	$cdf-168$	d^2f-112	$def-44$	
	$c^3\ -30$	$c^2d-\ 70$	$c^2e\ -105$	$cde\ \ldots$	$ce^2\ +105$	$de^2+\ 70$	$e^3\ +30$	
			$d^2\ \ldots$	$d^3\ \ldots$	$d^2e\ \ldots$			
± 11	± 55	± 119	± 189	± 140	± 189	± 119	± 55	± 11

$H = (\ \between\!x,\ y)^{12}$

x^{12}	$x^{11}y$	$x^{10}y^2$	x^9y^3	x^8y^4	x^7y^5	x^6y^6	x^5y^7	x^4y^8	x^3y^9	x^2y^{10}	xy^{11}	y^{12}
a^2d+1	$a^2e\ +\ 3$	$a^2f\ +\ 3$	$a^2g\ +\ 1$	$a\,bg\ +\ 9$	$a\,cg\ +\ 12$	$a\,dg\ \ldots$	$a\,eg\ -\ 12$	$a\,fg\ -\ 9$	$a\,g^2\ -\ 1$	$a^0bg^2-\ 3$	$a^0cg^2\ -\ 3$	a^0dg^2-1
$abc\ -3$	$a\,bd\ \ldots$	$a\,be\ +18$	$a\,bf+\ 24$	$cf\ +\ 15$	$df\ -\ 48$	$ef\ -\ 84$	$f^2\ -\ 24$	$a^0beg-\ 15$	$a^0bfg-\ 24$	$cfg-18$	$dfg\ \ldots$	$efg+3$
$b^3\ +2$	$c^2\ -15$	$cd\ -60$	$ce\ -\ 30$	$de\ -150$	$e^2\ -\ 90$	$a^0bcg+\ 84$	$a^0bdg+\ 48$	$bf^2-\ 66$	$ceg\ +\ 30$	$deg+60$	$e^2g\ +15$	$f^3\ -2$
	a^0b^2c+12	a^0b^2d+24	$d^2-\ 80$	$a^0b^2f\ +\ 66$	$a^0b^2g\ +\ 24$	$bdf\ ..$	$bef-192$	$cdg+150$	$cf^2\ -\ 60$	df^2-24	ef^2-12	
		bc^2+15	$a^0b^2e+\ 60$	$bce+105$	$bcf+192$	$be^2\ -210$	$c^2g\ +\ 90$	$cef\ -105$	$d^2g+\ 80$	e^2f-15		
			$bcd\ \ldots$	bd^2-120	$bde-240$	$c^2f\ +210$	$cdf+240$	$d^2f\ +120$	$def\ \ldots$			
			$c^3\ +\ 25$	$c^2d\ +\ 75$	$c^2e\ +150$	$cde\ \ldots$	de^2-150	$de^2-\ 75$	$e^3\ -\ 25$			
					$cd^2\ \ldots$	$d^3\ \ldots$	$d^2e\ \ldots$					
± 3	± 15	± 60	± 110	± 270	± 378	± 294	± 378	± 270	± 110	± 60	± 15	± 3

$I = (\ ☊x,\ y)^0$, Invt.

a^2g^2 …
$a\,bfg$ …
ceg + 1
cf^2 − 1
d^2g − 1
def + 2
e^3 − 1
a^0b^2eg − 1
b^2f^2 + 1
$bcdg$ + 2
$bcef$ − 2
bd^2f − 2
bde^2 + 2
c^3g − 1
c^2df + 2
c^2e^2 + 1
cd^2e − 3
d^4 + 1
± 3
± 9

$J = (\ ☊x,\ y)^4$

x^4	x^3y	x^2y^2	xy^3	y^4
a^2eg …	a^2fg + 2	a^2g^2 + 1	$a\,bg^2$ + 2	$a\,cg^2$ …
f^2 + 1	$a\,beg$ − 10	$a\,bfg$ − 6	cfg − 10	dfg …
$a\,bdg$ …	bf^2 − 8	ceg − 6	deg + 4	e^2g …
bef − 10	cdg + 4	cf^2 + 6	df^2 + 16	ef^2 …
c^2g …	cef + 26	d^2g + 4	e^2f − 12	$a^0b^2g^2$ + 1
cdf + 4	d^2f − 8	def + 12	a^0b^2fg − 8	$bcfg$ − 10
ce^2 + 16	de^2 − 8	e^3 − 12	$bceg$ + 26	$bdeg$ + 4
d^2e − 12	a^0b^2dg + 16	a^0b^2eg + 6	bcf^2 + 24	bdf^2 + 16
a^0b^2cg …	b^2ef + 24	b^2f^2 …	bd^2g − 8	be^2f − 12
b^2df + 16	bc^2g − 12	$bcdg$ + 12	$bdef$ − 64	c^2eg + 16
b^2e^2 + 9	$bcdf$ − 64	$bcef$ + 18	be^3 + 36	c^2f^2 + 9
bc^2f − 12	bce^2 − 42	bd^2f − 96	c^2dg − 8	cd^2g − 12
$bcde$ − 76	bd^2e + 56	bde^2 + 60	c^2ef − 42	$cdef$ − 76
bd^3 + 48	c^3f + 36	c^3g − 12	cd^2f + 56	ce^3 + 48
c^3e + 48	c^2de + 4	c^2df + 60	cde^2 + 4	d^3f + 48
c^2d^2 − 32	cd^3 − 16	c^2e^2 − 99	d^3e − 16	d^2e^2 − 32
		cd^2e + 84		
		d^4 − 32		
± 142	± 168	± 263	± 168	± 142

$K = (\ ☊x,\ y)^6$

x^6	x^5y	x^4y^2	x^3y^3	x^2y^4	xy^5	y^6
a^2dg + 1	a^2eg + 2	a^2fg …	a^2g^2 …	$a\,bg^2$ …	$a\,cg^2$ − 2	$a\,dg^2$ − 1
ef − 1	f^2 − 2	$a\,beg$ + 10	$a\,bfg$ …	cfg − 10	dfg + 2	efg + 3
$a\,bcg$ − 3	$a\,bdg$ − 2	bf^2 − 10	ceg …	deg + 15	e^2g + 6	f^3 − 2
bdf − 2	bef + 2	cdg − 15	cf^2 − 20	df^2 + 10	ef^2 − 6	a^0bcg^2 + 1
be^2 + 5	c^2g − 6	cef − 5	d^2g …	e^2f − 15	$a^0b^2g^2$ + 2	$bdfg$ + 2
c^2f + 9	cdf + 28	d^2f + 60	def + 60	a^0b^2fg + 10	$bcfg$ − 2	be^2g − 9
cde − 17	ce^2 − 26	de^2 − 40	e^3 − 40	$bceg$ + 5	$bdeg$ − 28	bef^2 + 6
d^3 + 8	d^2e + 4	a^0b^2dg − 10	a^0b^2eg + 20	bcf^2 − 30	bdf^2 + 32	c^2fg − 5
a^0b^3g + 2	a^0b^2cg + 6	b^2ef + 30	b^2f^2 …	bd^2g − 60	be^2f − 6	$cdeg$ + 17
b^2cf − 6	b^2df − 32	bc^2g + 15	$bcdg$ − 60	$bdef$ + 110	c^2eg + 26	cdf^2 − 2
b^2de + 2	b^2e^2 + 36	$bcdf$ − 110	$bcef$ …	be^3 − 45	c^2f^2 − 36	ce^2f − 6
bc^2e + 6	bc^2f + 6	bce^2 + 15	bd^2f …	c^2dg + 40	cd^2g − 4	d^3g − 8
bcd^2 − 4	$bcde$ − 58	bd^2e + 40	bde^2 + 20	c^2ef − 15	$cdef$ + 58	d^2ef + 4
c^3d …	bd^3 + 32	c^3f + 45	c^3g + 40	cd^2f − 40	ce^3 − 30	de^3 …
	c^3e + 30	c^2de − 25	c^2df − 20	cde^2 + 25	d^3f − 32	
	c^2d^2 − 20	cd^3 …	c^2e^2 …	d^3e …	d^2e^2 + 20	
			cd^2e …			
			d^4 …			
± 33	± 146	± 215	± 140	± 215	± 146	± 33

$L = (\ \mathfrak{L} \between x, y)^{10}$

x^{10}	x^9y	x^8y^2	x^7y^3	x^6y^4	x^5y^5	x^4y^6	x^3y^7	x^2y^8	xy^9	y^{10}
a^2bg ...	$a^2cg\ +1$	$a^2dg\ +3$	$a^2eg\ +2$	a^2fg ...	a^2g^2 ...	$a\,bg^2$ ...	$a\,cg^2\ -2$	$a\,dg^2\ -3$	$a\,eg^2\ -1$	$a\,fg^2$ ...
$cf\ +1$	$df\ +2$	$ef\ -3$	$f^2\ -2$	$a\,beg\ +14$	$a\,bfg$ ...	$cfg\ -14$	$dfg\ -16$	efg ...	$f^2g\ +1$	a^0beg^2-1
$de\ -1$	$e^2\ -3$	$a\,bcg$ ...	$a\,bdg+16$	$bf^2\ -14$	ceg ...	deg ...	$e^2g\ +6$	$f^3\ +3$	a^0bdg^2-2	bf^2g+1
$a\,b^2f\ -1$	$a\,b^2g\ -1$	$bdf\ +12$	$bef\ -16$	cdg ...	$cf^2\ -42$	$df^2\ -28$	$ef^2\ +12$	a^0bcg^2+3	$befg\ +2$	cdg^2+1
$bce\ -2$	$bcf\ -2$	$be^2\ -12$	$c^2g\ -6$	$cef\ -70$	d^2g ...	$e^2f\ +42$	$a^0b^2g^2+2$	$bdfg-12$	bf^3 ...	$cefg\ +2$
$bd^2\ +4$	$bde\ +4$	$c^2f\ -18$	$cdf\ -32$	d^2f ...	def ...	a^0b^2fg+14	$bcfg+16$	be^2g+18	$c^2g^2\ +3$	$cf^3\ -3$
$c^2d\ -1$	$c^2e\ -13$	$cde\ -6$	$ce^2\ -26$	$de^2\ +70$	$e^3\ +42$	$bceg\ +70$	$bdeg+32$	bef^2-9	$cdfg-4$	d^2fg-4
$a^0b^3e\ +3$	$cd^2\ +12$	$d^3\ +24$	$d^2e\ +64$	a^0b^2dg+28	a^0b^2eg+42	bcf^2-42	bdf^2-64	$c^2fg\ +12$	$ce^2g\ +13$	de^2g+1
b^2cd-6	a^0b^3f ...	$a^0b^3g\ -3$	a^0b^2cg-12	b^2ef+42	b^2f^2 ...	bd^2g ...	be^2f+12	$cdeg+6$	cef^2-15	def^2+6
$bc^3\ +3$	b^2ce+15	b^2cf+9	b^2df+64	bc^2g-42	$bcdg$ ...	$bdef-56$	c^2eg+26	cdf^2-42	d^2eg-12	$e^3f\ -3$
	b^2d^2 ...	b^2de+42	$b^2e^2\ +18$	$bcdf+56$	$bcef$ ...	be^3 ...	c^2f^2-18	ce^2f+9	d^2f^2 ...	
	bc^2d-30	bc^2e-9	bc^2f-12	bce^2-42	bd^2f ...	c^2dg-70	cd^2g-64	$d^3g\ -24$	de^2f+30	
	$c^4\ +15$	bcd^2-84	$bcde-64$	bd^2e-112	bde^2-84	c^2ef+42	$cdef+64$	d^2ef+84	$e^4\ -15$	
		$c^3d\ +45$	$bd^3\ -64$	c^3f ...	$c^3g\ -42$	cd^2f+112	$ce^3\ -30$	$de^3\ -45$		
			$c^3e\ +30$	$c^2de\ +70$	c^2df+84	cde^2-70	$d^3f\ +64$			
			c^2d^2+40	cd^3 ...	c^2e^2 ...	d^3e ...	d^2e^2-40			
					cd^2e ...					
					d^4 ...					

$$M = (\ \between x,\ y)^2$$

x^2		xy		y^2	
a^2cg^2	+ 1	a^2dg^2	− 2	a^2eg^2	+ 1
dfg	− 6	efg	+ 8	f^2g	− 1
e^2g	+ 8	f^3	− 6	$a\ bdg^2$	− 6
ef^2	− 3	$a\ bcg^2$	+ 8	$befg$	+ 6
$a\ b^2g^2$	− 1	$bdfg$	− 20	bf^3	...
$bcfg$	+ 6	be^2g	− 24	c^2g^2	+ 8
$bdeg$	− 34	bef^2	+ 36	$cdfg$	− 34
bdf^2	+ 48	c^2fg	− 24	ce^2g	+ 18
be^2f	− 18	$cdeg$	+ 76	cef^2	...
c^2eg	+ 18	cdf^2	+ 36	d^2eg	+ 4
c^2f^2	− 45	ce^2f	− 72	d^2f^2	+ 64
cd^2g	+ 4	d^3g	− 32	de^2f	− 96
$cdef$	+ 78	d^2ef	− 8	e^4	+ 36
ce^3	− 36	de^3	+ 24	$a^0b^2cg^2$	− 3
d^3f	− 48	$a^0b^3g^2$	− 6	b^2dfg	+ 48
d^2e^2	+ 28	b^2cfg	+ 36	b^2e^2g	− 45
a^0b^3fg	...	b^2deg	+ 36	b^2ef^2	...
b^2ceg	...	b^2df^2	...	bc^2fg	− 18
b^2cf^2	...	b^2e^2f	− 54	$bcdeg$	+ 78
b^2d^2g	+ 64	bc^2eg	− 72	$bcdf^2$	− 144
b^2def	− 144	bc^2f^2	− 54	bce^2f	+ 108
b^2e^3	+ 81	bcd^2g	− 8	bd^3g	− 48
bc^2dg	− 96	$bcdef$	− 36	bd^2ef	+ 96
bc^2ef	+ 108	bce^3	+ 216	bde^3	− 72
bcd^2f	+ 96	bd^3f	+ 128	c^3eg	− 36
$bcde^2$	− 126	bd^2e^2	− 192	c^3f^2	+ 81
bd^3e	+ 16	c^3dg	+ 24	c^2d^2g	+ 28
c^4g	+ 36	c^3ef	+ 216	c^2def	− 126
c^3df	− 72	c^2d^2f	− 192	c^2e^3	− 27
c^3e^2	− 27	c^2de^2	− 378	cd^3f	+ 16
c^2d^2e	+ 96	cd^3e	+ 464	cd^2e^2	+ 96
cd^4	− 32	d^5	− 128	d^4e	− 32
	± 9		± 8		± 1
	182		180		± 136
	497		1120		± 551
	± 688		± 1308		± 688

$$N = (\quad \between x, y)^4$$

x^4		x^3y		x^2y^2		xy^3		y^4	
a^2bg^2	...	a^2cg^2	-1	a^2dg^2	...	a^2eg^2	$+1$	a^2fg^2	...
cfg	-1	dfg	$+4$	efg	$+3$	f^2g	-1	$a\,beg^2$	$+1$
deg	$+1$	e^2g	...	f^3	-3	$a\,bdg^2$	-4	bf^2g	-1
df^2	$+3$	ef^2	-3	$a\,bcg^2$	-3	$befg$	$+4$	cdg^2	-1
e^2f	-3	$a\,b^2g^2$	$+1$	$bdfg$	...	bf^3	...	$cefg$	-2
$a\,b^2fg$	$+1$	$bcfg$	-4	be^2g	-15	c^2g^2	...	cf^3	$+3$
$bceg$	$+2$	$bdeg$	-16	bef^2	$+18$	$cdfg$	$+16$	d^2fg	$+4$
bcf^2	-3	bdf^2	...	c^2fg	$+15$	ce^2g	-22	de^2g	-1
bd^2g	-4	be^2f	$+18$	$cdeg$	...	cef^2	$+6$	def^2	-6
$bdef$	-12	c^2eg	$+22$	cdf^2	-36	d^2eg	$+8$	e^3f	$+3$
be^3	$+15$	c^2f^2	$+3$	ce^2f	$+9$	d^2f^2	-32	$a^0b^2dg^2$	-3
c^2dg	$+1$	cd^2g	-8	d^3g	...	de^2f	$+36$	b^2efg	$+3$
c^2ef	$+9$	$cdef$	-48	d^2ef	$+24$	e^4	-12	b^2f^3	...
cd^2f	$+4$	ce^3	$+12$	de^3	-12	$a^0b^2cg^2$	$+3$	bc^2g^2	$+3$
cde^2	-21	d^3f	$+32$	$a^0b^3g^2$	$+3$	b^2dfg	...	$bcdfg$	$+12$
d^3e	$+8$	d^2e^2	-12	b^2cfg	-18	b^2e^2g	-3	bce^2g	-9
a^0b^3eg	-3	a^0b^3fg	...	b^2deg	$+36$	b^2ef^2	...	$bcef^2$	-9
b^3f^2	...	b^2ceg	-6	b^2df^2	...	bc^2fg	-18	bd^2eg	-4
b^2cdg	$+6$	b^2cf^2	...	b^2e^2f	-27	$bcdeg$	$+48$	bd^2f^2	-32
b^2cef	$+9$	b^2d^2g	$+32$	bc^2eg	-9	$bcdf^2$	...	bde^2f	$+66$
b^2d^2f	$+32$	b^2def	...	bc^2f^2	$+27$	bce^2f	-18	be^4	-27
b^2de^2	-39	b^2e^3	-27	bcd^2g	-24	bd^3g	-32	c^3fg	-15
bc^3g	-3	bc^2dg	-36	$bcdef$	...	bd^2ef	$+32$	c^2deg	$+21$
bc^2df	-66	bc^2ef	$+18$	bce^3	$+27$	bde^3	-12	c^2df^2	$+39$
bc^2e^2	$+18$	bcd^2f	-32	bd^3f	...	c^3eg	-12	c^2e^2f	-18
bcd^2e	$+76$	$bcde^2$	$+84$	bd^2e^2	-12	c^3f^2	$+27$	cd^3g	-8
bd^4	-32	bd^3e	-32	c^3dg	$+12$	c^2d^2g	$+12$	cd^2ef	-76
c^4f	$+27$	c^4g	$+12$	c^3ef	-27	c^2def	-84	cde^3	$+45$
c^3de	-45	c^3df	$+12$	c^2d^2f	$+12$	c^2e^3	$+45$	d^4f	$+32$
c^2d^3	$+20$	c^3e^2	-45	c^2de^2	...	cd^3f	$+32$	d^3e^2	-20
		c^2d^2e	$+20$	cd^3e	...	cd^2e^2	-20		
		cd^4	...	d^5	...	d^4e	...		

$O = (\mathfrak{X}x, y)^8$

x^8	x^7y	x^6y^2	x^5y^3	x^4y^4	x^3y^5	x^2y^6	xy^7	y^8
a^3fg ...	a^3g^2 ...	a^2bg ...	a^2cg^2 − 1	a^2dg^2 ...	a^2eg^2 + 1	a^2fg^2 ...	a^2g^3 ...	$a\,bg^3$ ...
a^2beg ...	a^2bfg ...	cfg − 3	dfg + 4	efg + 5	f^2g − 1	$a\,beg^2$ + 3	$a\,bfg^2$ ...	cfg^2 ...
cdg ...	ceg − 1	deg + 10	e^2g + 7	f^3 − 5	$a\,bdg^2$ − 4	bf^2g − 3	ceg^2 + 1	deg^2 ...
bf^2 ...	cf^2 − 2	df^2 + 2	ef^2 − 10	$a\,bcg^2$ − 5	$befg$ + 18	cdg^2 − 10	cf^2g − 1	df^2g ...
cef − 1	d^2g + 3	e^2f − 9	$a\,b^2g^2$ + 1	$bdfg$ ...	bf^3 − 14	$cefg$ + 22	d^2g^2 − 3	e^2fg ...
d^2f + 3	def + 6	$a\,b^2fg$ + 3	$bcfg$ − 18	be^2g + 10	c^2g^2 − 7	cf^3 − 12	$defg$ + 14	ef^3 ...
de^2 − 2	e^3 − 6	$bceg$ − 22	$bdeg$ − 16	bef^2 − 5	$cdfg$ + 16	d^2fg + 19	df^3 − 8	$a^0b^2fg^2$ ...
$a\,b^3dg$ ...	$a\,b^2eg$ + 1	bcf^2 − 2	bdf^2 + 28	c^2fg − 10	ce^2g − 1	de^2g − 24	e^3g − 9	$bceg^2$ + 1
b^2ef + 1	b^2f^2 + 2	bd^2g − 19	be^2f + 4	$cdeg$ ...	cef^2 − 1	def^2 − 4	e^2f^2 + 6	bcf^2g − 1
bc^2g ...	$bcdg$ − 14	$bdef$ + 34	c^2eg + 1	cdf^2 + 80	d^2eg − 13	e^3f + 9	$a^0b^2eg^2$ + 2	bd^2g^2 − 3
$bcdf$ − 14	$bcef$ ...	be^3 + 3	c^2f^2 + 24	ce^2f − 55	d^2f^2 + 38	$a^0b^2dg^2$ − 2	b^2f^2g − 2	$bdefg$ + 14
bce^2 + 11	bd^2f − 18	c^2dg + 24	cd^2g + 13	d^3g ...	de^2f − 62	b^2efg + 2	$bcdg^2$ − 6	bdf^3 − 8
bd^2e + 1	bde^2 + 26	c^2ef + 4	$cdef$ − 6	d^2ef − 65	e^4 + 30	b^2f^3 ...	$bcefg$ ...	be^3g − 9
c^3f + 9	c^3g + 9	cd^2f − 58	ce^3 − 37	de^3 + 50	$a^0b^2cg^2$ + 10	bc^2g^2 + 9	bcf^3 + 6	be^2f^2 + 6
c^2de − 14	c^2df + 10	cde^2 − 42	d^3f − 52	$a^0b^3g^2$ + 5	b^2dfg − 28	$bcdfg$ − 34	bd^2fg + 18	c^2dg^2 + 2
cd^3 + 6	c^2e^2 + 13	d^3e + 38	d^2e^2 + 58	b^2cfg + 5	b^2e^2g − 24	bce^2g − 4	bde^2g − 10	c^2efg − 11
a^0b^3cg ...	cd^2e − 53	a^0b^3eg + 12	a^0b^3fg + 14	b^2deg − 80	b^2ef^2 + 42	$bcef^2$ + 57	$bdef^2$ − 20	c^2f^3 + 9
b^3df + 8	d^4 + 24	b^3f^2 ...	b^2ceg + 1	b^2df^2 ...	bc^2fg − 4	bd^2eg + 58	be^3f + 12	cd^2fg − 1
b^3e^2 − 9	a^0b^3dg + 8	b^2cdg + 4	b^2cf^2 − 42	b^2e^2f + 60	$bcdeg$ + 6	bd^2f^2 + 16	c^3g^2 + 6	cde^2g + 14
b^2c^2f − 6	b^3ef − 6	b^2cef − 57	b^2d^2g − 38	bc^2eg + 55	$bcdf^2$ ...	bde^2f − 110	c^2dfg − 26	$cdef^2$ − 16
b^2cde + 16	b^2c^2g − 6	b^2d^2f − 16	b^2def ...	bc^2f^2 − 60	bce^2f − 18	be^4 + 45	c^2e^2g − 13	ce^3f + 3
b^2d^3 − 8	b^2cdf + 20	b^2de^2 + 30	b^2e^3 + 36	bcd^2g + 55	bd^3g + 52	c^3fg − 3	c^2ef^2 + 21	d^3eg − 6
bc^3e − 3	b^2ce^2 − 21	bc^3g − 9	bc^2dg + 62	$bcdef$ ...	bd^2ef − 66	c^2deg + 42	cd^2eg + 53	d^3f^2 + 8
bc^2d^2 + 2	b^2d^2e − 2	bc^2df + 110	bc^2ef + 18	bce^3 − 60	bde^3 + 30	c^2df^2 − 30	cd^2f^2 + 2	d^2e^2f − 2
c^4d ...	bc^3f − 12	bc^2e^2 + 12	bcd^2f + 66	bd^3f ...	c^3eg + 37	c^2e^2f − 12	cde^2f − 52	de^4 ...
	bc^2de + 52	bcd^2e − 87	$bcde^2$ − 126	bd^2e^2 + 60	c^3f^2 − 66	cd^3g − 38	ce^4 + 15	
	bcd^3 − 28	bd^4 + 16	bd^3e + 24	c^3dg − 50	c^2d^2g − 58	cd^2ef + 87	d^4g − 24	
	c^4e − 15	c^4f + 45	c^4g − 30	c^3ef + 60	c^2def + 126	cde^3 − 40	d^3ef + 28	
	c^3d^2 + 10	c^3de + 40	c^3df − 30	c^2d^2f − 15	c^2e^3 − 60	d^4f − 16	d^2e^3 − 10	
		c^2d^3 − 10	c^3e^2 + 60	c^2de^2 ...	cd^3f − 24	d^3e^2 + 10		
			c^2d^2e − 15	cd^3e ...	cd^2e^2 + 15			
			cd^4 ...	d^5 ...	d^4e ...			

$P = (\ \between x, y)^0$, Invt.

$a^3b^0g^3$	...
a^2bfg^2	...
$a^2b^0ceg^2$	...
cfg^2	...
d^2g^2	+ 1
$defg$	− 6
df^3	+ 4
e^3g	+ 4
e^2f^2	− 3
$a\,b^2eg^2$	...
f^2g	...
$a\,b\,cdg^2$	− 6
$cefg$	+ 18
cf^3	− 12
d^2fg	+ 12
de^2g	− 18
def^2	...
e^3f	+ 6
$b^0c^3g^2$	+ 4
c^2dfg	− 18
c^2e^2g	− 24
c^2ef^2	+ 30
cd^2eg	+ 54
cd^2f^2	− 12
cd^2ef	− 42
ce^4	+ 12
d^4g	− 20
d^3ef	+ 24
d^2e^3	− 8

$a^0b^3dg^2$	+ 4
efg	− 12
f^3	+ 8
$b^2c^2g^2$	− 3
$cdfg$	...
ce^2g	+ 30
cef^2	− 24
d^2eg	− 12
d^2f^2	− 24
de^2f	+ 60
e^4	− 27
$b\,c^3fg$	+ 6
c^2deg	− 42
c^2df^2	+ 60
c^2e^2f	− 30
cd^3g	+ 24
cd^2ef	− 84
cde^3	+ 66
d^4f	+ 24
d^3e^2	− 24
b^0c^4eg	+ 12
c^4f^2	− 27
c^3d^2g	− 8
c^3def	+ 66
c^3e^3	− 8
c^2d^3f	− 24
$c^2d^2e^2$	− 39
cd^4e	+ 36
d^6	− 8

$Q = (\ \natural x, y)^6$

51	55	55	58	55	55	51
$a^3b^0dg^2$ ...	$a^3b^0eg^2$ ...	$a^3b^0fg^2$ ...	$a^3b^0g^3$ ...	$a^2b\,g^3$ ...	$a^2b^0cg^3$ ...	$a^2b^0dg^3$ ...
efg ...	f^2g ...	$a^2b\,eg^2$ ...	$a^2b\,fg^2$ ...	b^0cfg^2 ...	dfg^2 ...	efg^2 ...
f^3 ...	$a^2b\,dg^2$ ...	f^2g ...	b^0ceg^2 ...	deg^2 + 5	e^2g^2 + 2	f^3g ...
$a^2b\,cg^2$ ...	efg ...	b^0cdg^2 − 5	cf^2g ...	df^2g − 5	ef^2g − 4	$a\,b\,cg^3$ ...
dfg ...	f^3 ...	$cefg$ + 15	d^2g^2 ...	e^2fg − 5	f^4 + 2	dfg^2 ...
e^2g ...	$a^2b^0c^2g^2$ − 2	cf^3 − 10	$defg$ + 20	ef^3 + 5	$a\,b^2g^3$ ...	e^2g^2 + 2
ef^2 ...	$cdfg$ + 6	d^2fg + 20	df^3 − 20	$a\,b^2fg^2$ ...	$b\,cfg^2$ ...	ef^2g − 4
b^0c^2fg − 2	ce^2g + 8	de^2g − 25	e^3g − 20	$b\,ceg^2$ − 15	deg^2 − 6	f^4 + 2
$cdeg$ + 5	cef^2 − 10	def^2 − 10	e^2f^2 + 20	cf^2g + 15	df^2g + 6	$b^0c^2fg^2$ ...
cdf^2 + 6	d^2eg − 10	e^3f + 15	$a\,b^2eg^2$ ...	d^2g^2 − 20	e^2fg + 6	$cdeg^2$ − 5
ce^2f − 7	d^2f^2 + 18	$a\,b^2dg^2$ + 5	f^2g ...	$defg$ + 90	ef^3 − 6	cdf^2g + 5
d^3g − 3	de^2f − 22	efg − 15	$b\,cdg^2$ − 20	df^3 − 50	$b^0c^2eg^2$ − 8	ce^2fg + 5
d^2ef − 3	e^4 + 12	f^3 + 10	$cefg$ ...	e^3g − 35	c^2fg + 8	cef^3 − 5
de^3 + 4	$a\,b^2cg^2$ + 4	$b\,c^2g^2$ + 5	cf^3 + 20	e^2f^2 − 15	cd^2g^2 + 10	d^3g^2 + 3
$a\,b^3g^2$ ...	dfg − 6	$cdfg$ − 90	d^2fg ...	$b^0c^2dg^2$ + 25	$cdefg$ + 4	d^2efg − 7
b^2cfg + 4	e^2g − 8	ce^2g + 40	de^2g + 40	c^2efg − 40	cdf^3 − 24	d^2f^3 − 2
deg − 5	ef^2 + 10	cef^2 + 40	def^2 − 20	c^2f^3 + 15	ce^3g − 8	de^3g + 1
df^2 − 6	$a\,b\,c^2fg$ − 6	d^2eg + 50	e^3f − 20	cd^2fg − 50	ce^2f^2 + 18	de^2f^2 + 8
e^2f + 7	$cdeg$ − 4	d^2f^2 + 10	$b^0c^3g^2$ + 20	cde^2g + 45	d^3fg − 22	e^4f − 3
$b\,c^2eg$ − 5	cdf^2 − 68	de^2f − 40	c^2dfg − 40	$cdef^2$ + 5	d^2e^2g + 14	$a^0b^3g^3$ ...
c^2f^2 − 6	ce^2f + 76	e^4 − 15	c^2e^2g ...	ce^3f ...	d^2ef^2 + 42	b^2cfg^2 ...
cd^2g + 7	d^3g + 22	b^0c^3fg + 35	c^2ef^2 − 20	d^3eg − 5	de^3f − 46	deg^2 − 6
$cdef$ − 16	d^2ef + 38	c^2deg − 45	cd^2eg ...	d^3f^2 + 50	e^5 + 12	df^2g + 6
ce^3 + 23	de^3 − 58	c^2df^2 ...	cd^2f^2 + 60	d^2e^2f − 65	$a^0b^3fg^2$ ...	e^2fg + 6
d^3f + 30	b^0c^3eg + 8	c^2e^2f − 65	cde^2f − 20	de^4 + 20	b^2ceg^2 + 10	ef^3 − 6
d^2e^2 − 33	c^3f^2 + 42	cd^3g + 5	ce^4 + 20	$a^0b^3eg^2$ + 10	cf^2g − 10	$b\,c^2eg^2$ + 7
b^0c^3dg − 1	c^2d^2g − 14	cd^2ef + 65	d^4g ...	f^2g − 10	d^2g^2 − 18	c^2f^2g − 7
c^3ef + 36	c^2def − 82	cde^3 − 45	d^3ef − 20	b^2cdg^2 + 10	$defg$ + 68	cd^2g^2 + 3
c^2d^2f − 37	c^2e^3 − 44	d^4f − 20	d^2e^3 ...	$cefg$ − 40	df^3 − 32	$cdefg$ + 16
c^2de^2 − 53	cd^3f + 12	d^3e^2 − 20	$a^0b^3dg^2$ + 20	cf^3 + 30	e^3g − 42	cdf^3 − 22

$Q = (\ \mathfrak{Q}x,\ y)^6$ (*continued*).

51	55	55	58	55	55	51
$a\,b^0cd^3e$ + 79	$a\,b^0cd^2e^2$ + 122	$a^0b^3cg^2$ − 5	a^0b^3efg − 20	$a^0b^2d^2fg$ − 10	$a^0b^2e^2f^2$ + 24	$a^0b\,ce^3g$ − 86
d^5 − 24	d^4e − 44	dfg + 50	f^3 ...	de^2g ...	$b\,c^2dg^2$ + 22	ce^2f^2 + 39
a^0b^4fg − 2	$a^0b^4g^2$ − 2	e^2g − 15	$b^2c^2g^2$ − 20	def^2 + 10	c^2efg − 76	d^3fg − 30
b^3ceg + 5	b^3cfg + 6	ef^2 − 30	$cdfg$ + 20	e^3f ...	c^2f^3 + 54	d^2e^2g + 37
cf^2 + 6	deg + 24	b^2c^2fg − 15	ce^2g + 20	$b\,c^3g^2$ − 15	cd^2fg − 38	d^2ef^2 + 50
d^2g + 2	df^2 + 32	$cdeg$ − 5	cef^2 ...	c^2dfg + 40	cde^2g + 82	de^3f − 84
def + 22	e^2f − 54	cdf^2 − 10	d^2eg − 60	c^2e^2g + 65	$cdef^2$ − 50	e^5 + 27
e^3 − 27	b^2c^2eg − 18	ce^2f + 60	d^2f^2 ...	c^2ef^2 + 60	ce^3f + 6	$b^0c^3dg^2$ − 4
b^2c^2dg − 8	c^2f^2 − 24	d^3g − 50	de^2f + 40	cd^2eg − 65	d^3eg − 12	c^3efg − 23
c^2ef − 39	cd^2g − 42	d^2ef − 10	e^4 ...	cd^2f^2 + 10	d^3f^2 + 64	c^3f^3 + 27
cd^2f − 50	$cdef$ + 50	de^3 + 30	$b\,c^3fg$ + 20	cde^2f + 10	d^2e^2f − 82	c^2d^2fg + 33
cde^2 + 107	ce^3 + 54	$b\,c^3eg$ ...	c^2deg + 20	ce^4 ...	de^4 + 30	c^2de^2g + 53
d^3e − 22	d^3f − 64	c^3f^2 ...	c^2df^2 − 40	d^4g + 20	$b^0c^4g^2$ − 12	c^2def^2 − 107
$b\,c^4g$ + 3	d^2e^2 + 32	c^2d^2g + 65	c^2e^2f ...	d^3ef − 10	c^3dfg + 58	c^2e^3f + 21
c^3df + 84	$b\,c^3dg$ + 46	c^2def − 10	cd^3g + 20	d^2e^3 + 5	c^3e^2g + 44	cd^3eg − 79
c^3e^2 − 21	c^3ef − 6	c^2e^3 − 30	cd^2ef ...	b^0c^4fg + 15	c^3ef^2 − 54	cd^3f^2 + 22
c^2d^2e − 102	c^2d^2f + 82	cd^3f + 10	cde^3 − 40	c^3deg + 45	c^2d^2eg − 122	cd^2e^2f + 102
cd^4 + 44	c^2de^2 − 112	cd^2e^2 − 75	d^4f ...	c^3df^2 − 30	$c^2d^2f^2$ − 32	cde^4 − 45
b^0c^5f − 27	cd^3e − 34	d^4e + 40	d^3e^2 + 20	c^3e^2f + 30	c^2de^2f + 112	d^5g + 24
c^4de + 45	d^5 + 32	b^0c^4dg − 20	b^0c^4eg − 20	c^2d^3g + 20	c^2e^4 − 30	d^4ef − 44
c^3d^3 − 20	b^0c^5g − 12	c^4ef ...	c^4f^2 ...	c^2d^2ef + 75	cd^4g + 44	d^3e^3 + 20
	c^4df − 30	c^3d^2f − 5	c^3d^2g ...	c^2de^3 − 50	cd^3ef + 34	
	c^4e^2 + 30	c^3de^2 + 50	c^3def + 40	cd^4f − 40	cd^2e^3 − 30	
	c^3d^2e + 30	c^2d^3e − 25	c^3e^3 ...	cd^3e^2 + 25	d^5f − 32	
	c^2d^4 − 20	cd^5 ...	c^2d^3f − 20	d^5e ...	d^4e^2 + 20	
			$c^2d^2e^2$ ...			
			cd^4e ...			
			d^6 ...			

776.

ON THE JACOBIAN SEXTIC EQUATION.

[From the *Quarterly Journal of Pure and Applied Mathematics*, vol. XVIII. (1882), pp. 52—65.]

THE Jacobian sextic equation has been discussed under the form

$$(z-a)^6 - 4a(z-a)^5 + 10b(z-a)^3 - 4c(z-a) + 5b^2 - 4ac = 0,$$

(see references at end of paper), but the connexion of this form with the general sextic equation has not, so far as I am aware, been considered. And although this is probably known, I do not find it to have been explicitly stated that the group of the equation is the positive half-group, or group of the 60 positive substitutions out of the 120 substitutions, which leave unaltered Serret's 6-valued function of six letters.

Invariantive Property of the Jacobian Sextic.

Taking $z-a$ as the variable, and comparing the equation with the general sextic equation

$$(\mathrm{a, b, c, d, e, f, g})(z-a,\ 1)^6 = 0,$$

we have

$$\begin{array}{ccccccc} \mathrm{a}, & \mathrm{b}, & \mathrm{c}, & \mathrm{d}, & \mathrm{e}, & \mathrm{f}, & \mathrm{g} \\ =1, & -\tfrac{2}{3}a, & 0, & \tfrac{1}{2}b, & 0, & -\tfrac{2}{3}c, & 5b^2-4ac; \end{array}$$

the Jacobian equation is thus an equation

$$(\mathrm{a, b, c, d, e, f, g})(x,\ y)^6 = 0,$$

for which $\mathrm{c}=0$, $\mathrm{e}=0$, $\mathrm{ag}+9\mathrm{bf}-20\mathrm{d}^2=0$; but of course any equation, which can be by a linear transformation upon the variables brought into this form, may be regarded as a Jacobian equation.

Hence, using henceforward the small italic in place of the small roman letters, the Jacobian sextic may be regarded as an equation

$$(a, b, c, d, e, f, g)(x,\ y)^6 = 0,$$

linearly transformable into the form

$$(a, b, 0, d, 0, f, g)(x,\ y)^6 = 0,$$

where $ag + 9bf - 20d^2 = 0$. It is to be shown, that this implies a single relation between the four invariants A, B, C, and Δ of the sextic function.

I call to mind that the general sextic has five invariants A, B, C, D, E of the orders 2, 4, 6, 10, 15 respectively; the last of them E is not independent, but its square is equal to a rational and integral function of A, B, C, D; and instead of D, we consider the discriminant Δ which is an invariant of the same order 10. The values of A, B, C are given, Table Nos. 31, 34, and 35 of my Third Memoir on Quantics, *Phil. Trans.*, vol. CXLVI. (1856), pp. 627—647, [144]; those of D, Δ, E were obtained by Dr Salmon, see his *Higher Algebra*, second ed. 1866, where the values of A, B, C, D, Δ, E are all given; only those of A, B, C, Δ are reproduced in the third edition, 1876.

It may be remarked, that for the general form we have $A = ag - 6bf + 15ce - 10d^2$, and that B is the determinant

$$\begin{vmatrix} a, & b, & c, & d \\ b, & c, & d, & e \\ c, & d, & e, & f \\ d, & e, & f, & g \end{vmatrix} :$$

C and Δ are complicated forms, the latter of them containing 246 terms. But writing $c = 0$, $e = 0$, there is a great reduction; we have

$A =$	$B =$	$C =$	$D =$
ag + 1	ad^2g − 1	$a^2d^2g^2$ + 1	a^5g^5 + 1
bf − 6	b^2f^2 + 1	,, df^3 + 4	a^4bfg^4 − 30
d^2 − 10	bd^2f − 2	$a\,bd^2fg$ + 12	,, d^2g^2 − 300
	d^4 + 1	,, d^4g − 20	,, df^2g − 2500
		$a^0b^3dg^2$ + 4	,, f^6 − 3125
		,, b^3f^3 + 8	$a^3b^2f^2g^3$ − 15
		,, $b^2d^2f^2$ − 24	,, bd^2fg^2 − 4800
		,, bd^4f + 24	,, bdf^4g − 7500
		,, d^6 − 8	,, d^4g^3 + 30000
			,, d^3f^3g + 50000
			$a^2b^3dg^4$ − 2500
			,, $b^3f^3g^2$ − 410
			,, $b^2d^2f^2g^2$ − 171300
			,, b^2df^5 − 240000
			,, bd^4fg^2 + 780000
			,, bd^3f^4 + 1200000
			,, d^6g^2 − 1000000
			,, d^5f^2 − 1600000
			$a\,b^4dfg^3$ − 7500
			,, b^4f^4g − 11520
			,, $b^3d^3g^3$ + 50000
			,, $b^3d^2f^3g$ + 83200
			$a^0b^6g^4$ − 3125
			,, $b^5df^2g^2$ − 240000
			,, b^5f^5 − 331776
			,, $b^4d^3fg^2$ + 1200000
			,, $b^4d^2f^4$ + 1843200
			,, $b^3d^5g^2$ − 1600000
			,, $b^3d^4f^3$ − 2560000

It is clear that these are all functions of ag, bf, d^2 and $a^2f^3+b^3g^2$, say of α, β, δ and ϕ. In fact, A and B are functions of α, β, δ; C contains two terms, coefficient 4, which are $= 4\sqrt{\delta}\,.\,\phi$; Δ contains two terms

$$-3125\,(a^4f^6+b^6g^4),$$

which are $=-3125\,(\phi^2-2\alpha^2\beta^3)$; and also several pairs of terms, each which pair contains the factor ϕ. We thus have

$A=$	$B=$	$C=$	$\Delta=$
$\alpha\ +\ 1$	$\alpha\delta\ -\ 1$	$\alpha^2\delta\ \ +\ 1$	$\alpha^5\ \ +\ \ 1$
$\beta\ -\ 6$	$\beta^2\ +\ 1$	$\alpha\beta\delta\ \ +12$	$\alpha^4\beta\ \ -\ \ 30$
$\delta\ -\ 10$	$\beta\delta\ -\ 2$	$\alpha\delta^2\ \ -20$	$\alpha^4\delta\ \ -\ \ 300$
	$\delta^2\ +\ 1$	$\beta^3\ \ +\ 8$	$\alpha^2\beta^3\ \ +\ \ 5840\ (=6250-410)$
		$\beta^2\delta\ \ -24$	$\alpha^3\beta^3\ \ -\ \ 15$
		$\beta\delta^2\ \ +24$	$\alpha^3\beta\delta\ \ -\ \ 4800$
		$\delta^3\ \ -\ 8$	$\alpha^3\delta^2\ \ +\ \ 3000$
		$\phi\sqrt{\delta}\ +\ 4$	$\alpha^2\beta^2\delta\ \ -\ \ 171300$
			$\alpha^2\beta\delta^2\ \ +\ \ 780000$
			$\alpha^2\delta^3\ \ -\ \ 1000000$
			$\alpha\beta^4\ \ -\ \ 11520$
			$\alpha\beta^3\delta\ \ +\ \ 83200$
			$\beta^5\ \ -\ \ 331776$
			$\beta^4\delta\ \ +\ \ 1843200$
			$\beta^3\delta^2\ \ -\ \ 2560000$
			$\phi\sqrt{\delta}\,.\,\alpha^3\ -\ \ 2500$
			,, $\alpha\beta\ -\ \ 7500$
			,, $\alpha\delta\ +\ \ 50000$
			,, $\beta^2\ -\ \ 240000$
			,, $\beta\delta\ +\ \ 1200000$
			,, $\delta^2\ -\ \ 1600000$
			$\phi^2.\ \ -\ \ 3125$

We have *ante*, the relation $\alpha+9\beta-20\delta=0$, and using this to eliminate α, we have A, B, C, Δ as functions of β, δ, ϕ (that is, of bf, d^2 and $a^2f^3+b^3g^2$). Effecting the substitution, we find the values of A, B, C without difficulty. As regards the value of Δ, this is

$$=-3125\phi^2+2\phi K\sqrt{\delta}+\text{terms without }\phi,$$

where

$$\begin{aligned}2K = &- 2500\,(81\beta^2 - 360\beta\delta + 400\delta^2)\\ &- 7500\,(-9\beta^2 + 20\beta\delta)\\ &+ 50000\,(- 9\beta\delta + 20\delta^2)\\ &- 240000\,(\beta^2)\\ &+1200000\,(\beta\delta)\\ &-1600000\,(\delta^2),\end{aligned}$$

or, reducing and dividing by 2,

$$K = -\,3125\,(60\beta^2 - 240\beta\delta + 256\delta^2).$$

The calculation of the terms without ϕ is much more laborious, but they come out

$$= -\,3125\,(60\beta^2 - 240\beta\delta + 256\delta^2)^2\,\delta.$$

Hence the value of Δ is

$$\begin{aligned}\Delta = -\,3125\,\{\,&\phi^2\\ &+ 2\phi\,(60\beta^2 - 240\beta\delta + 256\delta^2)\sqrt{\delta}\\ &+ (60\beta^2 - 240\beta\delta + 256\delta^2)^2\ \delta\},\end{aligned}$$

say this is

$$\Delta = -\,3125h^2,$$

where

$$h = \phi + (60\beta^2 - 240\beta\delta + 256\delta^2)\sqrt{\delta},$$

that is,

$$= a^2f^3 + b^3g^2 + (60b^2df^2 - 240bd^3f + 256d^5).$$

The values of A, B, C, and the foregoing value of h then are

$A =$	$B =$	$C =$	$h =$
β − 15	β^2 + 1	β^3 + 8	$\beta^2\sqrt{\delta}$ + 60
δ + 10	$\beta\delta$ + 7	$\beta^2\delta$ + 51	$\beta\delta\sqrt{\delta}$ − 240
	δ^2 − 19	$\beta\delta^2$ + 84	$\delta^2\sqrt{\delta}$ + 256
		δ^3 − 8	ϕ + 1
		$\phi\sqrt{\delta}$ + 4	

We may, if we please, regard β, δ, ϕ as irrational invariants of the sextic, viz. A, B, C being rational and integral functions of β, δ, ϕ, we have conversely β, δ, ϕ irrational functions of A, B, C; and then the equation for h, say

$$\frac{1}{25\sqrt{5}}\sqrt{(-\Delta)} = \phi + \sqrt{\delta}\,(60\beta^2 - 240\beta\delta + 256\delta^2)$$

is the invariantive relation which characterises the Jacobian sextic.

The expression for Δ in terms of A, B, C, D is

$$\Delta = A^5 - 375A^3B - 625A^2C + 3125D,$$

and it was in the foregoing investigation proper to use Δ in place of D. But I annex the value of D for the case in question $b=0$, $f=0$; and also its value in terms of α, β, δ, ϕ. These are

$D=$		$D=$	
a^4f^6	$-\ 1$	ϕ^2	$-\ 1$
a^3bdf^4g	$-\ 12$	$\alpha^2\beta^3$	$+\ 2$
,, d^4g^3	$+\ 5$	$\alpha\beta^4$	$-\ 12$
$a^2b^2d^2f^2g^2$	$-\ 90$	β^5	$-\ 72$
,, b^2df^5	$-\ 48$	$\delta\,\alpha^2\beta^2$	$-\ 90$
,, bd^4fg^2	$+\ 246$	,, $\alpha\beta^3$	$+\ 168$
,, bd^3f^4	$+\ 480$	,, β^4	$+\ 552$
,, d^6g^2	$-\ 258$	$\delta^2\alpha^3$	$+\ 5$
,, d^5f^3	$-\ 432$	,, $\alpha^2\beta$	$+\ 246$
$a\,b^4dfg^3$	$-\ 12$	,, $\alpha\beta^2$	$+\ 240$
,, b^4f^4g	$-\ 12$	,, β^3	$-\ 976$
,, $b^3d^2f^3g$	$+\ 168$	$\delta^3\alpha^2$	$-\ 258$
,, $b^2d^4f^2g$	$+\ 240$	,, $\alpha\beta$	$-\ 168$
,, bd^6fg	$-\ 168$	,, β^2	$+\ 336$
,, d^8g	$-\ 228$	$\delta^4\alpha$	$-\ 228$
$a^0b^6g^4$	$-\ 1$	,, β	$-\ 408$
,, $b^5df^2g^2$	$-\ 48$	δ^5	$-\ 240$
,, b^5f^5	$-\ 72$	$\phi\sqrt{\delta}\,\alpha\beta$	$-\ 12$
,, $b^4d^3fg^2$	$+\ 480$	,, β^2	$-\ 48$
,, $b^4d^2f^3$	$+\ 552$	$\phi\delta\sqrt{\delta}\,\beta$	$-\ 480$
,, $b^3d^5g^2$	$-\ 432$	$\phi\delta^2\sqrt{\delta}$	$-\ 432$
,, $b^3d^4f^3$	$-\ 976$		
,, $b^2d^6f^2$	$+\ 336$		
,, bd^8f	$+\ 408$		
,, d^{10}	$-\ 248$		

The Group of the Jacobian Sextic.

The solution of the Jacobian sextic equation depends upon that of a quintic; in fact, calling the roots z_∞, z_0, z_1, z_2, z_3, z_4, then there exists a quintic having the roots

$$\sqrt{(z_\infty - z_0 \,.\, z_2 - z_3 \,.\, z_4 - z_1)},$$
$$\sqrt{(z_\infty - z_1 \,.\, z_3 - z_4 \,.\, z_0 - z_2)},$$
$$\sqrt{(z_\infty - z_2 \,.\, z_4 - z_0 \,.\, z_1 - z_3)},$$
$$\sqrt{(z_\infty - z_3 \,.\, z_0 - z_1 \,.\, z_2 - z_4)},$$
$$\sqrt{(z_\infty - z_4 \,.\, z_1 - z_2 \,.\, z_3 - z_0)},$$

the coefficients of which are rational functions of the coefficients a, b, d, f, g, and of the fourth root of the discriminant, i.e., $\sqrt{h}$. But the meaning of this has not, so far as I am aware, been noticed. Passing to the quintic whose roots are the squares of the foregoing values, i.e., $z_\infty - z_0 . z_2 - z_3 . z_4 - z_1$, &c., the coefficients are here rational functions of a, b, d, f, g and h; that is, they are rational functions of a, b, d, f, g. The symmetrical functions of these roots $z_\infty - z_0 . z_2 - z_3 . z_4 - z_1$, &c., are thus rational functions of the coefficients of the sextic; each such rational function is a 12-valued function of $z_\infty, z_0, z_1, z_2, z_3, z_4$, invariable by all the substitutions of a group of 60 substitutions; and therefore also every like 12-valued function of the roots $z_\infty, z_0, z_1, z_2, z_3, z_4$ is invariable by the substitutions of this group of 60; or, in other words, this group of 60 is the group of the Jacobian sextic equation.

I write for convenience, in this section only,

$$z_\infty, z_0, z_1, z_2, z_3, z_4 = f, a, b, c, d, e;$$

and writing further ab for shortness instead of $a-b$, &c., (so that of course $ba = -ab$), and putting $B, C, D, E, F = -ab.cd.ef,\ -ac.bf.de,\ ad.bc.ef,\ ae.bd.cf,\ af.be.cd$, then the five functions are B, C, D, E, F, and the group of 60 which leaves unaltered every symmetrical function of these functions is made up of the substitutions

1.				1
ab . ce,	*ab . df*,	*ce . df*,		15
ac . bf,	*ac . de*,	*bf . de*,		
ad . bc,	*ad . ef*,	*bc . ef*,		
ae . bd,	*ae . cf*,	*bd . cf*,		
af . be,	*af . cd*,	*be . cd*.		
abcde,	*acebd*,	*adbec*,	*aedbc*,	24
afbce,	*abefc*,	*acfeb*,	*aecbf*,	
abdef,	*adfbe*,	*aebfd*,	*afedb*,	
afced,	*acdfe*,	*aefdc*,	*adecf*,	
afdbc,	*adcfb*,	*abfcd*,	*acbdf*,	
bdcef,	*bcfde*,	*bedfc*,	*bfecd*.	
abc . dfe,	*acb . def*,			20
abd . cfe,	*adb . cef*,			
abe . cfd,	*aeb . cdf*,			
abf . ced,	*afb . cde*,			
acd . bef,	*adc . bfe*,			
ace . bfd,	*aec . bdf*,			
acf . bed,	*afc . bde*,			
ade . bfc,	*aed . bcf*,			
adf . bce,	*afd . bec*,			
aef . bcd,	*afe . bdc*,			60

where the symbols, *ab*, *abcde*, *abc*, &c. denote cyclical substitutions. It is easy to verify that each of these substitutions does in fact merely permute B, C, D, E, F; thus

$$\begin{array}{lccccc} & B & C & D & E & F \\ abcde \text{ on } & -ab\,.\,ce\,.\,df, & -ac\,.\,bf\,.\,de, & ad\,.\,bc\,.\,ef, & ae\,.\,bd\,.\,cf, & af\,.\,be\,.\,cd \\ & =-bc\,.\,da\,.\,ef, & -bd\,.\,cf\,.\,ea, & be\,.\,cd\,.\,af, & ba\,.\,ce\,.\,df, & bf\,.\,ca\,.\,de \\ & =+ad\,.\,bc\,.\,ef, & ae\,.\,bd\,.\,cf, & af\,.\,be\,.\,cd, & -ab\,.\,ce\,.\,df, & -ac\,.\,bf\,.\,de \\ & = \quad D & E & F & B & C, \end{array}$$

which (expressed as a cyclical substitution) is $= BDFCE$, and so in other cases.

We may to the foregoing 60 substitutions join the 60 other substitutions:

cdef,	*cfed*,	30
bdfe,	*befd*,	
becf,	*bfce*,	
bcdf,	*bfdc*,	
bced,	*bdec*,	
aedf,	*afde*,	
acef,	*afec*,	
acfd,	*adfc*,	
adce,	*aecd*,	
abfe,	*aefb*,	
adbf,	*afbd*,	
abed,	*adeb*,	
abcf,	*afcb*,	
acbe,	*aebc*,	
abdc,	*acdb*.	
ab . cd . ef,	*ab . cf . de*,	10
ac . bd . ef,	*ac . be . df*,	
ad . be . cf,	*ad . bf . ce*,	
ae . bc . df,	*ae . bf . cd*,	
af . bc . de,	*af . bd . ce*.	
abcefd,	*adfecb*,	20
abfdec,	*acedfb*,	
abecdf,	*afdceb*,	
abdfce,	*aecfdb*,	
acfbde,	*aedbfc*,	
acbfed,	*adefbc*,	
acdebf,	*afbedc*,	
adbcfe,	*aefcbd*,	
adcbef,	*afebcd*,	—
aebdcf,	*afcdbe*,	60

each of which changes B, C, D, E, F into a permutation of $-B$, $-C$, $-D$, $-E$, $-F$.

The 60 and 60 substitutions form together a group of 120 substitutions, which leave unaltered any even symmetrical function of B, C, D, E, F, or say any symmetrical function of B^2, C^2, D^2, E^2, F^2; such a function is thus a 6-valued function of a, b, c, d, e, f, viz. it is Serret's 6-valued function of 6 letters.

Transformation of the Jacobian Sextic into the Resolvent Sextic of a special quintic equation.

Starting from the Jacobian Sextic Equation

$$(a,\ b,\ 0,\ d,\ 0,\ f,\ g \between z,\ 1)^6 = 0,$$

$ag + 9bf - 20d^2 = 0$, I effect upon it the Tschirnhausen transformation

$$X = -az^3 - 6bz^2 - 10d;$$

which, it may be remarked, is a particular case of the Tschirnhausen-Hermite form

$$\begin{aligned} X\ (az+b)\,B + (az^2+6bz+5c)\,C + (az^3+6bz^2+15cz+10d)\,D \\ + (az^4+6bz^3+15cz^2+20dz+10e)\,E + (az^5+6bz^4+15cz^3+20dz^2+15ez+5f)\,F. \end{aligned}$$

Writing for convenience $Y = X + 10d$, $Z = X - 10d$, this is

$$\begin{array}{rrrrrrl} . & . & az^3 + & 6bz^2 - & . & +Y & = 0, \end{array}$$

and we thence have

$$\begin{array}{rrrrrrl} . & az^4 + & 6bz^3 & . & +\ Yz & . & = 0, \\ az^5 + & 6bz^4 & . & +\ Yz^2 & . & . & = 0, \\ & - Zz^3 & . & -6fz & -g & & = 0, \\ - Zz^4 & . & -6fz^2 - & gz & . & & = 0, \\ -Zz^5 & . & -6fz^3 - & gz^2 & . & . & = 0, \end{array}$$

or, eliminating, the resulting equation is

$$\begin{vmatrix} . & . & a, & 6b, & . & Y \\ . & a, & 6b, & . & Y, & . \\ a, & 6b, & . & Y, & . & . \\ . & . & Z, & ., & 6f, & g \\ . & Z, & . & 6f, & g, & . \\ Z, & . & 6f, & g, & . & . \end{vmatrix} = 0.$$

The developed form is most easily obtained by expanding the determinant in the form

$$123\ .\ \overline{456} - 456\ .\ \overline{123},\ \&c.,$$

where the terms 123, &c., belong to the matrix

$$\begin{vmatrix} . & . & a, & 6b, & . & Y \\ . & a, & 6b, & . & Y, & \\ a, & 6b, & . & Y, & & \end{vmatrix},$$

and those of $\overline{123}$, &c., to the matrix

$$\left|\begin{array}{cccccc} . & . & Z, & . & 6f, & g \\ . & Z, & . & 6f, & g, & . \\ Z, & . & 6f, & g, & . & . \end{array}\right| .$$

The several terms are

$123 \,.\, \overline{456}$	$+-a^3$	$.\,-g^3$
$-124 \,.\, 356$	$--6a^2b$	$.\,-6fg^2$
$+125 \,.\, 346$	$+\ \ 0$	$.\,-36f^2g$
$-126 \,.\, 345$	$--a^2Y$	$.\ \ g^2Z-216f^3$
$+134 \,.\, 256$	$+-36ab^2$	$.\ \ 0$
$-135 \,.\, 246$	$-\ \ a^2Y$	$.\,-g^2Z$
$+136 \,.\, 245$	$+-6ab\ Y$	$.\,-6fgZ$
$+145 \,.\, 236$	$+\ \ 6ab\ Y$	$.\,-6fgZ$
$-146 \,.\, 235$	$-\ \ 0$	$.\,-36f^2Z$
$+156 \,.\, 234$	$+-a\ Y^2$	$.\,-gZ^2$
$-234 \,.\, 156$	$--a^2Y-216b^3$	$.\ \ g^2Z$
$+235 \,.\, 146$	$+\ \ 6ab\ Y$	$.\ \ 6fgZ$
$-236 \,.\, 145$	$--36b^2Y$	$.\ \ 36f^2Z$
$-245 \,.\, 136$	$-\ \ 36b^2Y$	$.\ \ 0$
$+246 \,.\, 135$	$+\ \ aY^2$	$.\ \ g\ Z^2$
$-256 \,.\, 134$	$--6b\ Y^2$	$.\ \ 6fZ^2$
$+345 \,.\, 126$	$+-a\ Y^2$	$.\,-g\ Z^2$
$-346 \,.\, 125$	$-\ \ 6b\ Y^2$	$.\,-6fZ^2$
$+356 \,.\, 124$	$+\ \ 0$	$.\ \ 0$
$-456 \,.\, 123$	$--Y^3$	$.\ \ Z^3.$

Hence, collecting and reducing, the equation is

$$\begin{aligned} 0 = \ & Y^3Z^3 \,. \\ & + Y^2Z^2 \,.\ (3ag+72bf) \\ & + YZ \ .\ (3a^2g^2+36agbf+1296b^2f^2) \\ & + Y \ .\ 216a^2f^3 \\ & + Z \ .\ -216b^3g^2 \\ & + \qquad a^3g^3-36a^2g^2bf, \end{aligned}$$

where Y, Z denote $X+10d$, $X-10d$ respectively, and consequently $YZ=X^2-100d^2$. Hence, writing as before α, β, δ, ϕ to denote ag, bf, d^2 and $a^2f^3+b^3g^2$ respectively, the result finally is

$$\left(\begin{array}{c|c|c|c|c|c|c} 1 & 0 & \alpha + 3 & 0 & \alpha^2 + 3 & a^2f^3 + 216 & \phi\sqrt{\delta} + 2160 \\ & & \beta + 72 & & \alpha\beta + 36 & b^3g^2 - 216 & \alpha^3 + 1 \\ & & \delta - 300 & & \alpha\delta - 600 & & \alpha^2\beta - 36 \\ & & & & \beta^2 + 1296 & & \alpha^2\delta - 30 \\ & & & & \beta\delta - 14400 & & \alpha\beta\delta - 360 \\ & & & & \delta^2 + 30000 & & \alpha\delta^2 + 30000 \\ & & & & & & \beta^2\delta + 12960 \\ & & & & & & \beta\delta^2 + 720000 \\ & & & & & & \delta^3 - 1000000 \end{array}\between X,\ 1)^6=0,$$

where observe that the coefficient of the term in X is $216\,(a^2f^3-b^3g^2)$, $=216\sqrt{(\phi^2-4\alpha^2\beta^3)}$. We have as before $ag+9bf-20d^2=0$, that is, $\alpha+9\beta-20\delta=0$; and using this equation to eliminate α, also in the constant term writing its value for ϕ in terms of h,

$$\phi = h + (-60\beta^2+240\beta\delta-256\delta^2)\sqrt{\delta},$$

the new equation is

$$\left(\begin{array}{c|c|c|c|c|c|c} 1 & 0 & -5\times\overbrace{\beta - 9} & 0 & 5\times\overbrace{\beta^2 - 243} & -216\sqrt{\Lambda} & 5\times\overbrace{h\sqrt{\delta} + 432} \\ & & \delta + 48 & & \beta\delta - 1872 & & \beta^3 - 729 \\ & & & & \delta^2 + 3840 & & \beta^2\delta + 4184 \\ & & & & & & \beta\delta^2 - 11520 \\ & & & & & & \delta^3 + 8292 \end{array}\between X,\ 1)^6=0,$$

where

$$\Lambda = \{h+(-60\beta^2+240\beta\delta-256\delta^2)\sqrt{\delta}\}^2 - 4\,(-9\beta+20\delta)^2\beta^3$$

$$= h^2 + 2h\sqrt{\delta}\,.\left|\begin{array}{l} \beta^2 - 60 \\ \beta\delta + 240 \\ \delta^2 - 256 \end{array}\right| - 4\,(\beta-4\delta)^3\,(9\beta-16\delta)^2.$$

It is to be shown that this Tschirnhausen-transformation of the Jacobian sextic is, in fact, the resolvent sextic of the quintic equation

$$(\mathrm{a},\ 0,\ \mathrm{c},\ 0,\ \mathrm{e},\ \mathrm{f}\between x,\ 1)^5=0,$$

where

$$\mathrm{a}=1,\quad \mathrm{c}=2d,\quad \mathrm{e}=-9bf+36d^2,\quad \mathrm{f}^2=216h.$$

I consider the general quintic $(a, b, c, d, e, f \between x, 1)^5 = 0$; taking the roots to be x_1, x_2, x_3, x_4, x_5, and writing

$$\phi_1 = 12345 - 24135,$$
$$\phi_2 = 13425 - 32145,$$
$$\phi_3 = 14235 - 43125,$$
$$\phi_4 = 21435 - 13245,$$
$$\phi_5 = 31245 - 14325,$$
$$\phi_6 = 41325 - 12435,$$

where 12345 is used to denote the function

$$= (x_1x_2 + x_2x_3 + x_3x_4 + x_4x_5 + x_5x_1)\sqrt{(20)},$$

(this numerical factor $\sqrt{(20)}$ being inserted for greater convenience), then the equation whose roots are $\phi_1, \phi_2, \phi_3, \phi_4, \phi_5, \phi_6$, which equation may be regarded as the resolvent sextic of the given quintic equation, is

$a^6 \times$		$-5a^4 \times$	$5a^2 \times$	$-\sqrt{(\square)}\,.\,a^2$	$+5$
1	0	ae	$-2a^2df$	$+1$	$+1a^3cf^3$
		$-4bd$	$+3a^2e^2$		$-2a^2def$
		$+3c^2$	&c.		+ &c.

$$(\ldots \between \phi, 1)^6 = 0,$$

$\square = a^4f^4 +$ &c., the discriminant of the quintic: see p. 274* of my paper "On a new auxiliary equation in the theory of equations of the fifth order," *Phil. Trans.* t. CLI. (1861), pp. 263—276, [268].

I now write $b = 0$, $d = 0$, but, to avoid confusion again, write roman instead of italic letters, viz. I consider the resolvent sextic of the quintic equation

$$(\mathrm{a}, 0, \mathrm{c}, 0, \mathrm{e}, \mathrm{f} \between x, 1)^5.$$

Many of the terms thus vanish, and the equation assumes the form

$\mathrm{a}^6 \times$		$-5\mathrm{a}^4$	$5\mathrm{a}^2$	$-\mathrm{a}^2\sqrt{\square}$	$+5$
1	0	$\mathrm{ae} + 1$	$\mathrm{a^2e^2} + 3$	$+1$	$\mathrm{a^3cf^2} + 1$
		$\mathrm{c}^2 + 3$	$\mathrm{ac^2e} - 2$		$\mathrm{a^3e^3} + 1$
			$\mathrm{c}^4 + 15$		$\mathrm{a^2c^2e^2} - 11$
					$\mathrm{ac^4e} + 35$
					$\mathrm{c}^6 - 25$

$$(\ldots \between X, 1)^6 = 0,$$

and then if, as before,

$$\mathrm{a} = 1,\ \mathrm{c} = 2d,\quad \mathrm{e} = -9bf + 36d^2,\ \mathrm{f}^2 = 216h,$$

or say

$$\mathrm{a} = 1,\ \mathrm{c} = 2\sqrt{\delta},\ \mathrm{e} = -9\beta + 36\delta,\ \mathrm{f}^2 = 216h,$$

* [This Collection, vol. IV., p. 321.]

this becomes identical with the foregoing Tschirnhausen-transformation equation; thus

$$\text{ae} + 3\text{c}^2 = -9\beta + 36\delta + 12\delta, \; = \beta - 9$$
$$\delta + 48;$$

and similarly

$$3\text{a}^2\text{e}^2 - 2\text{ac}^2\text{e} + 15\text{c}^4 = \beta^2 + 243,$$
$$\beta\delta - 1872,$$
$$\delta^2 + 3840.$$

So for the constant term, $+1\text{a}^3\text{cf}^2$ gives the term $432h\sqrt{\delta}$, and $+1\text{a}^3\text{e}^3$, &c., give the remaining terms $-729\beta^3$, &c. of the value in question.

It only remains to verify the equality of the coefficients of X,

$$216\sqrt{\Lambda} = \sqrt{\square} \text{ or } 46656\Lambda = \square.$$

Here $\square$, the discriminant of the quintic $(\text{a}, 0, \text{c}, 0, \text{e}, \text{f} \between x, 1)^5$, from the general form (see my Second Memoir on Quantics, [141], or Salmon's *Higher Algebra*, third edition, p. 209) putting therein $\text{b} = 0$, $\text{d} = 0$, is

$$\begin{aligned}\square = \;& \text{a}^4\text{f}^4 \quad + 1,\\ & \text{a}^3\text{ce}^2\text{f}^2 + 160,\\ & \text{a}^3\text{e}^5 \quad + 256,\\ & \text{a}^2\text{c}^3\text{ef}^2 - 1440,\\ & \text{a}^2\text{c}^2\text{e}^4 \; - 2560,\\ & \text{ac}^5\text{f}^2 \; + 3456,\\ & \text{ac}^4\text{e}^3 \; + 6400,\end{aligned}$$

and writing for a, c, e, f their values 1, $2\sqrt{\delta}$, $9(-\beta + 4\delta)$, $216h$, the value becomes

$$\begin{aligned}\square = \;& (216)^2 \,.\, h^2 \;.\\ & + 432h\sqrt{\delta} \;.\; 12960\,(\beta - 4\delta)^2\\ & \qquad\qquad + 34560\,(\beta - 4\delta)\,\delta\\ & \qquad\qquad + 55296\;\delta^2\\ & - 256 \;.\; 9^5 \,.\, (\beta - 4\delta)^5\\ & - 10240 \;.\; 9^4 \,.\, (\beta - 4\delta)^4\,\delta\\ & - 102400 \;.\; 9^3 \,.\, (\beta - 4\delta)^3\,\delta^2.\end{aligned}$$

The whole divides by $(216)^2$, and we thus obtain

$$\begin{aligned}\Lambda = h^2 + 2h\sqrt{\delta} \,.\; & 60\;(-\beta + 4\delta)^2 \,.+ (-\beta - 4\delta)^3 \,.\; && 324\,(\beta - 4\delta)^2\\ & - 240\,(-\beta + 4\delta)\,\delta && + 1440\,(\beta - 4\delta)\,\delta\\ & + 256\;\;\delta^2 && + 1600\;\delta^2,\end{aligned}$$

which is, in fact, equal to the foregoing value of Λ.

The conclusion is that, starting from the Jacobian sextic

$$(a,\ b,\ 0,\ d,\ 0,\ f,\ g \between z,\ 1)^6 = 0,$$

where $ag + 9bf - 20d^2 = 0$, and effecting upon it the Tschirnhausen-transformation

$$X = -az^3 - 6bz^2 - 10d,$$

so as to obtain from it a sextic equation in X, this sextic equation in X is the resolvent sextic of the quintic equation

$$(1,\ 0,\ \mathrm{c},\ 0,\ \mathrm{e},\ \mathrm{f} \between x,\ 1)^5 = 0,$$

where

$$\mathrm{c} = 2d,\quad \mathrm{e} = -9bf + 36d^2,\quad \mathrm{f} = \sqrt{(216h)},$$

and, Δ being the discriminant of the Jacobian sextic, then

$$h = \frac{1}{5^2\sqrt{5}}\sqrt{(-\Delta)},\ = a^2f^3 + b^3g^2 + 60b^2df^2 - 240bd^3f + 256d^5.$$

As to the subject of the present paper, see in particular Brioschi, "Ueber die Auflösung der Gleichungen vom fünften Grade," *Math. Annalen*, t. XIII. (1878), pp. 109—160, and the third Appendix to his translation of my *Elliptic Functions*, Milan, 1880, each containing references to the earlier papers.

777.

A SOLVABLE CASE OF THE QUINTIC EQUATION.

[From the *Quarterly Journal of Pure and Applied Mathematics*, vol. XVIII. (1882), pp. 154—157.]

THE roots of the general quintic equation

$$(a, b, c, d, e, f \between x, 1)^5 = 0$$

may be taken to be

$$\begin{array}{lllll} -\dfrac{b}{a} + & B + & C + & D + & E \\ - \;,, + \omega^4 & ,, + \omega^3 & ,, + \omega^2 & ,, + \omega & ,, \\ - \;,, + \omega^3 & ,, + \omega & ,, + \omega^4 & ,, + \omega^2 & ,, \\ - \;,, + \omega^2 & ,, + \omega^4 & ,, + \omega & ,, + \omega^3 & ,, \\ - \;,, + \omega & ,, + \omega^2 & ,, + \omega^3 & ,, + \omega^4 & ,,\, , \end{array}$$

where ω is an imaginary fifth root of unity; and if one of the four functions B, C, D, E is $=0$, say if $E=0$ (this implies of course a single relation between the coefficients), then the equation is solvable.

Writing $x = \xi - \dfrac{b}{a}$, we have

$$(a, b, c, d, e, f)\left(\xi - \frac{b}{a}, 1\right)^5 = (a', 0, c', d', e', f' \between \xi, 1)^5,$$

where

$$\begin{aligned} a' &= a, \\ ac' &= ac - b^2, \\ a^2d' &= a^2d - 3abc + 2b^3, \\ a^3e' &= a^3e - 4a^2bd + 6ab^2c - 3b^4, \\ a^4f' &= a^4f - 5a^3be + 10ab^2d - 10ab^2c + 4b^5, \end{aligned}$$

and the roots of the new equation

$$(a', 0, c', d', e', f' \between \xi, 1)^5 = 0$$

have the above-mentioned values, omitting therefrom the terms $-\frac{b}{a}$; we find without difficulty

$$2\frac{c'}{a'} = -BE - CD,$$

$$2\frac{d'}{a'} = -B^2D - BC^2 - CE^2 - D^2E,$$

$$\frac{e'}{a'} = -B^3C - B^2E^2 + BCDE + BD^3 + C^3E + C^2D^2 - DE^2,$$

$$\frac{f'}{a'} = -B^5 + 5B^3DE - 5B^2C^2E - 5B^2CD^2 + 5BC^3D + 5BCE^3$$
$$- 5BD^2E^2 - C^5 + 5CD^3E - 5CD^2E^2 - D^5 - E^5,$$

and hence, when $E = 0$, we have

$$2\frac{c'}{a'} = -CD,$$

$$2\frac{d'}{a'} = -B^2D - BC^2,$$

$$\frac{e'}{a'} = -B^3C - BD^3 - C^2D^2,$$

$$\frac{f'}{a'} = -B^5 - 5B^2CD^2 + 5BC^3D - C^5 - D^5,$$

or, as these may be written,

$$-2\frac{c'}{a'} = CD,$$

$$-2\frac{d'}{a'} = B^2D + BC^2,$$

$$-\frac{e'}{a'} - 4\frac{c'^2}{a'^2} = B^3C - BD^3,$$

$$-\frac{f'}{a'} = B^5 + C^5 + D^5 - 10\frac{c'}{a'}(B^2D - BC^2),$$

equations which imply a single relation between the coefficients a', c', d', e', f'. Supposing this satisfied, we may attend only to the first three equations; or, writing for convenience,

$$\gamma = -2\frac{c'}{a'}, \quad = -\frac{2}{a^2}(ac - b^2),$$

$$\delta = -2\frac{d'}{a'}, \quad = -\frac{2}{a^3}(a^2d - 3abc + 2b^3),$$

$$\theta = -\frac{e'}{a'} - 4\frac{c'^2}{a'^2}, \quad = -\frac{1}{a^4}\{a^2(ae - 4bd + 3c^2) + (ac - b^2)^2\},$$

the equations are

$$\gamma = CD,$$
$$\delta = B(BD + C^2),$$
$$\theta = B(D^3 - B^2C).$$

The first equation gives $C=\dfrac{\gamma}{D}$, and substituting this value in the other two equations, we have

$$B^2D^3+B\gamma^4\ -\delta D^2=0,$$
$$B^3\gamma\ +BD^4+\theta D\ =0.$$

Eliminating B, the result is obtained in the form Det. $=0$, where in the last column of the determinant each term is divisible by D; and omitting this factor, the result is

$$\left|\begin{array}{ccccc} & & D^3, & \gamma^2, & -\delta D \\ & D^3, & \gamma^2, & -\delta D^2, & \\ D^3, & \gamma^2, & -\delta D^2, & & \\ & \gamma, & 0, & -D^4, & \theta \\ \gamma, & 0, & -D^4, & \theta D, & \end{array}\right|=0.$$

If, in order to develope the determinant, we consider it as a sum of products, each first factor being a minor composed out of columns 1 and 2, and the second factor being the complementary minor composed out of columns 3, 4, 5 (the several products being of course taken each with its proper sign), the expansion presents itself in the form

$$\begin{aligned} &D^3\gamma\,(-\theta\delta\gamma^2D^2+\delta^2D^7),\\ &-D^6\,(-\theta\gamma^2D^4+\delta D^9-\theta^2D^4)\\ &-\gamma D^3\,.\,-\delta D^2\,(\delta D^5-\theta\gamma^2)\\ &+\gamma^3\,(\gamma^2\delta D^5-\theta\delta D^5-\theta\gamma^4)\\ &-\gamma^2\delta^3D^5. \end{aligned}$$

Hence, collecting, and changing the sign of the whole expression, we obtain

$$\delta D^{15}-(2\gamma\delta^2+\gamma^2\theta+\theta^2)\,D^{10}+(-\gamma^3\delta+3\gamma\delta\theta+\delta^3)\,\gamma^2D^5+\gamma^7\theta=0,$$

a cubic equation for D^5. We have then as above $C=\dfrac{\gamma}{D}$, and B is given rationally as the common root of the foregoing quadric and cubic equations satisfied by B.

Substituting for γ, δ, θ their values in terms of the original coefficients, the equation for D^5 becomes

$$\begin{aligned} &2\,(a^2d-3abc+2b^3)\,(aD)^{15}\\ &+\left\{\begin{array}{l} a^4\,(ae-4bd+3c^2)^2\\ +\,a^2\,(ac-b^2)^2\,(ae-4bd+3c^2)\\ -\,16\,(ac-b^2)\,(a^2d-3abc+2b^3)^2 \end{array}\right\}(aD)^{10}\\ &+4\,(ac-b^2)^2\left\{\begin{array}{l} 28\ \ (ac-b^2)^3\,(a^2d-3abc+2b^3)\\ +\,12a^2\,(ac-b^2)\,(a^2d-3abc+2b^3)\,(ae-4bd+3c^2)\\ +\ 8\ \ (a^2d-3abc+2b^3)^3 \end{array}\right\}(aD)^5\\ &-128\,(ac-b^2)^7\,\{a^2\,(ae-4bd+3c^2)+(ac^2-b^2)^2\}=0, \end{aligned}$$

and the solution of the given quintic equation thus ultimately depends upon that of this cubic equation.

778.

[ADDITION TO MR HUDSON'S PAPER "ON EQUAL ROOTS OF EQUATIONS."]

[From the *Quarterly Journal of Pure and Applied Mathematics*, vol. XVIII. (1882), pp. 226—229.]

It seems desirable to present in a more developed form some of the results of the foregoing paper.

Thus, if the equation $(a_0, a_1, \ldots, a_n \between x, 1)^n = 0$ of the order n has $n-v$ equal roots, where v is not $> \tfrac{1}{2}n - 1$, then we have $\psi(r, v+1, m) = 0$, where m has any one of the values $0, 1, \ldots, n-2v-2$, and r any one of the values

$$2v+2,\ 2v+3,\ \ldots,\ n-m.$$

The signification is

$$\begin{aligned}
\psi(r, v+1, m) = \quad & r && . \quad 1 && . \frac{1}{[r]^{v+2}} && a_m \quad a_{r+m} \\
& -(r-2) && . \frac{v+1}{1} && . \frac{1}{[r-1]^{v+2}} && a_{m+1} \quad a_{r+m-1} \\
& +(r-4) && . \frac{v+1 \,.\, v+2}{1\,.\,2} && . \frac{1}{[r-2]^{v+2}} && a_{m+2} \quad a_{r+m-2} \\
& \ \ \vdots \\
& +(-)^s (r-2s) && . \frac{[v+1]^s}{[s]^s} && . \frac{1}{[r-s]^{v+2}} && a_{m+s} \quad a_{r+m-s} \\
& \ \ \vdots \\
& +(-)^{v+1}(r-2v-2) \,. && && . \frac{1}{[r-v-1]^{v+2}} && a_{m+v+1}\, a_{r+m-v-1}.
\end{aligned}$$

Thus, when $v = 0$, the condition is

$$\left.\begin{array}{l} r \qquad \dfrac{1}{r\,.\,r-1} \quad a_m \quad a_{r+m} \\ -(r-2)\,\dfrac{1}{r-1\,.\,r-2}\, a_{m+1} a_{r+m-1} \end{array}\right\} = 0,$$

that is,

$$a_m a_{r+m} - a_{m+1} a_{r+m-1} = 0,$$

satisfied when the equation has all its roots equal.

The values of m are $0, 1, 2, \ldots, n-2$, and those of r are $2v+2, 2v+3, \ldots, n-m$; in particular, if $m=0$, the values of r are $2, 3, \ldots, n$, and the corresponding conditions are

$$\begin{aligned} a_0 a_2 - a_1^2 \quad &= 0, \\ a_0 a_3 - a_1 a_2 \quad &= 0, \\ \vdots \\ a_0 a_n - a_1 a_{n-1} &= 0, \end{aligned}$$

and so for the different values of m up to the final value $n-2$, for which $r=2$, and the condition is

$$a_{n-2} a_n - a^2_{n-1} = 0;$$

we have thus, it is clear, the whole series of conditions included in

$$\left\| \begin{matrix} a_0, & a_1, & a_2, & \ldots, & a_{n-2}, & a_{n-1} \\ a_1, & a_2, & a_3, & \ldots, & a_{n-1}, & a_n \end{matrix} \right\| = 0,$$

which are obviously satisfied in the case in question of the roots being all equal.

Again, when $v=1$, the condition for $n-1$ equal roots is

$$\left.\begin{aligned} r \qquad .\,1\,.\frac{1}{r\,.\,r-1\,.\,r-2}\quad & a_m \; a_{r+m} \\ -(r-2)\,.\,2\,.\frac{1}{r-1\,.\,r-2\,.\,r-3} & a_{m+1} a_{r+m-1} \\ +(r-4)\,.\,1\,.\frac{1}{r-2\,.\,r-3\,.\,r-4} & a_{m+2} a_{r+m-2} \end{aligned}\right\} = 0,$$

that is,

$$\frac{a_m a_{r+m}}{r-1\,.\,r-2} - \frac{2a_{m+1} a_{r+m-1}}{r-1\,.\,r-3} + \frac{a_{m+2} a_{r+m-2}}{r-2\,.\,r-3} = 0;$$

or, what is the same thing,

$$(r-3)\, a_m a_{r+m} - 2\,(r-2)\, a_{m+1} a_{r+m-1} + (r-1)\, a_{m+2} a_{r+m-2} = 0,$$

where $n=4$ at least, and m, r have the values

$m=$	$0, 1, 2, \ldots, n-4$
$r=$	$4,\ 4,\qquad\qquad 4$
	$5,\ 5$
	$\vdots\ \ \vdots$
	$\vdots\ \ n-1$
	n

thus, when $n=4$, the only values are $m=0$, $r=4$, and the condition is

$$a_0 a_4 - 4a_1 a_3 + 3a_2^2 = 0.$$

Similarly, when $v=2$, the condition for $n-2$ equal roots is found to be

$$\frac{a_m a_{r+m}}{r-1 \,.\, r-2 \,.\, r-3} - \frac{3a_{m+1}a_{r+m-1}}{r-1 \,.\, r-3 \,.\, r-4} + \frac{3a_{m+2}a_{r+m-2}}{r-2 \,.\, r-3 \,.\, r-5} - \frac{a_{m+3}a_{r+m-3}}{r-3 \,.\, r-4 \,.\, r-5} = 0\,;$$

or, what is the same thing,

$$\begin{aligned} & r-4 \,.\, r-5 \,.\, a_m \; a_{r+m} \\ & -3 \,.\, r-2 \,.\, r-5 \,.\, a_{m+1}a_{r+m-1} \\ & +3 \,.\, r-1 \,.\, r-4 \,.\, a_{m+2}a_{r+m-2} \\ & - \quad . \, r-1 \,.\, r-2 \,.\, a_{m+3}a_{r+m-3} = 0, \end{aligned}$$

where $n=6$ at least, and m, r have the values

$m=$	$0,$	$1,$	$\ldots,$	$n-6$
$r=$	$6,$	$6,$		6
	$7,$	7		
	$\vdots$	$\vdots$		
	$\vdots$	$n-1$		
	n			

Observe that the sum of the coefficients is $=0$, viz.

$$(r-4)(r-5)-3(r-2)(r-5)+3(r-1)(r-4)-(r-1)(r-2)=0,$$

this should obviously be the case, since the conditions for $n-2$ equal roots must be satisfied when the roots are all of them equal; and the property serves as a verification.

It is to be remarked that the equation $\psi(r,\ v+1,\ m)=0$ does not in all cases give all the conditions for the existence of $n-v$ equal roots in an equation of the order n; thus when $n=3$ and $v=1$, we cannot by means of it obtain the condition that a cubic equation may have 2 equal roots. The problem really considered is that of the determination of those *quadric* functions of the coefficients which vanish in the case of $n-v$ equal roots; and in the case in question ($n=3$, $v=1$) there is no quadric function which vanishes, but the condition depends on a cubic function.

The question of the quadric functions which vanish in the case of $n-v$ equal roots, and to a small extent that of the *cubic* functions which thus vanish, is considered in Dr Salmon's "Note on the conditions that an equation may have equal roots," *Camb. and Dublin Math. Jour.*, t. v. (1850), pp. 159—165, and in particular the equation there obtained p. 161 is the equation $\psi(0,\ v+1,\ n)=0$.

779.

[NOTE ON MR JEFFERY'S PAPER "ON CERTAIN QUARTIC CURVES WHICH HAVE A CUSP AT INFINITY, WHEREAT THE LINE AT INFINITY IS A TANGENT."]

[From the *Proceedings of the London Mathematical Society*, vol. XIV. (1883), p. 85.]

THE assumed form $\kappa\alpha^3\beta = u_2$, or, as this is afterwards written,

$$2\kappa x^3 y = ax^2 + 2bxy + cy^2 + 2ex + 2dy + \lambda,$$

is, I think, introduced without a proper explanation. Say, the form is $x^3y = z^2(*\between x, y, z)^2$, it ought to be shown how for a cuspidal quartic we arrive at this form; viz. taking the cusp to be at the point $(x=0,\ z=0)$, $z=0$ for the tangent at the cusp, and $x=0$ an arbitrary line through the cusp; then the line $z=0$ besides intersects the curve in a single point, and, if $y=0$ is taken as the tangent at that point, the equation of the curve must, it can be seen, be of the form

$$(x^3 + \theta x^2 z)\, y = z^2(a,\ b,\ c,\ f,\ g,\ h\between x,\ y,\ z)^2.$$

The conic $(a, b, c, f, g, h\between x, y, z)^2 = 0$ touches the quartic at each of the two intersections of the quartic with the arbitrary line $x=0$; and we cannot, so long as the line remains arbitrary, find a conic which shall osculate the quartic at the two points in question; but, for the particular line $x + \frac{1}{3}\theta z = 0$, there exists such a conic, viz. writing x instead of $x + \frac{1}{3}\theta z$, the form is $x^3y = z^2(a', b', c', f', g', h'\between x, y, z)^2$, and the new conic $(a', \ldots\between x, y, z)^2 = 0$ has the property in question. This is the adopted form, and it thus appears that in it the line $x=0$ is a determinate line, viz. the line passing through the cusp and the two points of osculation of the osculating conic. It thus appears that in the assumed form the lines $x=0$, $y=0$, $z=0$ are determinate lines.

780.

[ADDITION TO MR HAMMOND'S PAPER "NOTE ON AN EXCEPTIONAL CASE IN WHICH THE FUNDAMENTAL POSTULATE OF PROFESSOR SYLVESTER'S THEORY OF TAMISAGE FAILS."]

[From the *Proceedings of the London Mathematical Society*, vol. XIV. (1883), pp. 88—91. Read Dec. 14, 1882.]

THE extreme importance of Mr Hammond's result, as regards the entire subject of Covariants, leads me to reproduce his investigation in the notation of my Memoirs on Quantics, and with a somewhat different arrangement of the formulæ. For the binary seventhic

$$(a,\ b,\ c,\ d,\ e,\ f,\ g,\ h \between x,\ y)^7,$$

the four composite seminvariants of the deg-weight 5 . 11 (sources of covariants of the deg-order 5 . 13) are

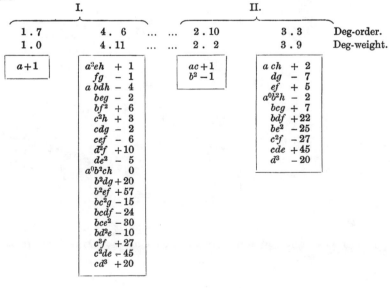

I.			II.		
1 . 7	4 . 6	... ...	2 . 10	3 . 3	Deg-order.
1 . 0	4 . 11	... ...	2 . 2	3 . 9	Deg-weight.
a +1	a^2eh + 1		ac +1	$a\,ch$ + 2	
	fg − 1		b^2 −1	dg − 7	
	$a\,bdh$ − 4			ef + 5	
	beg − 2			a^0b^2h − 2	
	bf^2 + 6			bcg + 7	
	c^2h + 3			bdf +22	
	cdg − 2			be^2 −25	
	cef − 6			c^2f −27	
	d^2f +10			cde +45	
	de^2 − 5			d^3 −20	
	a^0b^2ch 0				
	b^2dg +20				
	b^2ef +57				
	bc^2g −15				
	$bcdf$ −24				
	bce^2 −30				
	bd^2e −10				
	c^3f +27				
	c^2de −45				
	cd^3 +20				

III.			IV.		
2 . 6	3 . 7	......	2 . 2	3 . 11	Deg-order.
2 . 4	3 . 7	......	2 . 6	3 . 5	Deg-weight.

ae	+1
bd	−4
c^2	+3

a^2h	+ 1
$a\,bg$	− 7
cf	+ 9
de	− 5
a^0b^2f	+12
bce	−30
bd^2	+20

ag	+ 1
bf	− 6
ce	+15
d^2	−10

a^2f	+1
$a\,be$	−5
cd	+2
a^0b^2d	+8
bc^2	−6

and it is here at once obvious that there exists a syzygy of the form I. = III. − IV.; in fact, if in III. and IV. we write $a = 0$, then the values are each

$$= -2b(4bd - 3c^2)(6bf - 15ce + 10d^2);$$

hence III. − IV. must divide by a, the quotient being a seminvariant of the deg-weight 4 . 11, which can only be a numerical multiple of the second factor of I., and is in fact = this second factor, that is, we have the syzygy I. = III. − IV.

Working out the values of the four products, and joining to them the expression for the irreducible seminvariant of the same deg-weight 5 . 11 (O, x^8 of my tables [774] for the binary sextic), we have the table:

5 . 10
a^3dh
eg
f
a^2bch
bdg
bef
c^2g
cdf
ce^2
d^2e
$a\,b^3h$
b^2cg
b^2df
b^2e^2
bc^2f
$bcde$
bd^3
c^3e
c^2d^2
a^0b^4g
b^3cf
b^3de
b^2c^2e
b^2cd^2
bc^3d
c^5

5 . 11	O	I.	III.	IV.	II.
a^3eh		+ 1	+ 1		
fg		− 1		+ 1	
a^2bdh		− 4	− 4		
beg		− 2	− 7	− 5	
bf^2		+ 6		− 6	
c^2h		+ 3	+ 3		+ 2
cdg		− 2		+ 2	− 7
cef	− 1	− 6	+ 9	+ 15	+ 5
d^2f	+ 3	+10		− 10	
de^2	− 2	− 5	− 5		
$a\,b^2ch$					− 4
b^2dg		+20	+ 28	+ 8	+ 7
b^2ef	+ 1	+57	+ 12	− 45	− 5
bc^2g		−15	− 21	− 6	+ 7
$bcdf$	−14	−24	− 36	− 12	+22
bce^2	+11	−30	− 30		−25
bd^2e	+ 1	−10	+ 40	+ 50	
c^3f	+ 9	+27	+ 27		−27
c^2de	−14	−45	− 15	+ 30	+45
cd^3	+ 6	+20		− 20	−20
a^0b^4h					− 2
b^3cg					− 7
b^3df	+ 8		− 48	− 48	−22
b^3e^2	− 9				+25
b^2c^2f	− 6		+ 36	+ 36	+27
b^2cde	+16		+120	+120	−45
b^2d^3	− 8		− 80	− 80	+20
bc^3e	− 3		− 90	− 90	
bc^2d^2	+ 2		+ 60	+ 60	
c^4d					

I have prefixed to the table the literal terms of the deg-weight 5 . 10; for the deg-weights 5 . 11 and 5 . 10, the numbers of terms are = 30 and 26 respectively; and it is the difference of these 30 − 26, = 4, which gives the number of asyzygetic seminvariants of the deg-weight 5 . 11.

781.

ON THE AUTOMORPHIC TRANSFORMATION OF THE BINARY CUBIC FUNCTION.

[From the *Proceedings of the London Mathematical Society*, vol. XIV. (1883), pp. 103—108. Read Jan. 11, 1883.]

I CONSIDER the cubic equation $(a, b, c, d \between x, 1)^3 = 0$. It is shown (Serret, *Cours d'Algèbre supérieure*, 4th ed., Paris, 1879, t. II. pp. 466—471) how, given one root of the equation, the other two roots can be each of them expressed rationally in terms of this root and of the square root of the discriminant; viz. making the proper changes of notation, and writing

$$A,\ B,\ C = ac - b^2,\ ad - bc,\ bd - c^2,\ \lambda = \sqrt{-\tfrac{1}{3}},$$

$$\Omega = B^2 - 4AC,\ = a^2d^2 + 4ac^3 + 4b^3d - 3b^2c^2 - 6abcd,$$

$$\alpha = \frac{-\lambda\sqrt{\Omega} + B}{2\lambda\sqrt{\Omega}},\quad \beta = \frac{+2C}{2\lambda\sqrt{\Omega}},\quad \gamma = \frac{-2A}{2\lambda\sqrt{\Omega}},\quad \delta = \frac{-\lambda\sqrt{\Omega} - B}{2\lambda\sqrt{\Omega}},$$

(values which give $\alpha + \delta = -1$, $\alpha\delta - \beta\gamma = -1$, and therefore also

$$\alpha^2 + \alpha\delta + \delta^2 + \beta\gamma = 0,$$

which is the condition in order that the function $\phi x,\ = \dfrac{\alpha x + \beta}{\gamma x + \delta}$, may be periodic of the third order, $\phi^3 x = x$), then, u being a root of the equation, say $(a, b, c, d \between u, 1)^3 = 0$, the other two roots are

$$\phi u,\ = \frac{\alpha u + \beta}{\gamma u + \delta},$$

and

$$\phi^2 u,\ = \phi^{-1} u,\ = \frac{(\alpha^2 + \beta\gamma)\, u + \beta\,(\alpha + \delta)}{\gamma\,(\alpha + \delta)\, u + \delta^2 + \beta\gamma},\ = \frac{\delta u - \beta}{-\gamma u + \alpha},$$

where observe that, by the change of $\sqrt{\Omega}$ into $-\sqrt{\Omega}$, $\alpha, \beta, \gamma, \delta$ become $\delta, -\beta, -\gamma, \alpha$; so that the last-mentioned value $\phi^{-1}u$ is, in fact, the value obtained from ϕu by the mere change of sign of the radical.

It is to be observed that, if we have between two roots u, v of the equation

$$(a, b, c, d \between x, 1)^3 = 0,$$

a relation $v = \dfrac{\alpha u + \beta}{\gamma u + \delta}$, where α, β, γ, δ have given values, this implies in the first place a relation between a, b, c, d (and the given values of α, β, γ, δ), and it implies moreover that u, and consequently also v, are not any roots whatever, but two determinate roots of the equation; viz. u, v will be each of them expressible rationally in terms of a, b, c, d and α, β, γ, δ. And if, in order that (a, b, c, d) may remain arbitrary, we consider α, β, γ, δ as given quantities satisfying the relation which exists between these quantities and (a, b, c, d), then in general we still have u, v determinate roots of the cubic equation. But in the foregoing solution u is any root whatever of the cubic equation.

To examine how this is, starting from the equations

$$(a, b, c, d \between u, 1)^3 = 0, \quad (a, b, c, d \between v, 1)^3 = 0, \quad v = \frac{\alpha u + \beta}{\gamma u + \delta},$$

we have

$$a(u^3 - v^3) + 3b(u^2 - v^2) + 3c(u - v) = 0,$$

and therefore

$$a(u^2 + uv + v^2) + 3b(u + v) + 3c = 0,$$

that is,

$$av^2 + (au + 3b)v + au^2 + 3bu + 3c = 0;$$

or, writing herein for v its value,

$$a(\alpha u + \beta)^2 + (\alpha u + \beta)(\gamma u + \delta)(au + 3b) + (\gamma u + \delta)^2 (au^2 + 3bu + 3) = 0;$$

that is,

$$\begin{aligned} a(\alpha u + \beta)^2 + {} & \alpha\gamma\ (au^3 + 3bu^2) + \gamma^2 (au^4 + 3bu^3 + 3cu^2) \\ + {} & (\alpha\delta + \beta\gamma)(au^2 + 3bu) + 2\gamma\delta(au^3 + 3bu^2 + 3cu) \\ + {} & \beta\delta\ (au + 3b) + \delta^2 (au^2 + 3bu + c) = 0; \end{aligned}$$

or, reducing by the equation $au^3 + 3bu^2 + 3cu + d = 0$, this is

$$\begin{aligned} a(\alpha u + \beta)^2 + {} & \alpha\gamma\ (-3cu - d) + \gamma^2 (-du) \\ + {} & (\alpha\delta + \beta\gamma)(au^2 + 3bu) + 2\gamma\delta(-d) \\ + {} & \beta\delta\ (au + 3b) + \delta^2 (au^2 + 3bu + 3c) = 0, \end{aligned}$$

and, collecting the terms, this is

$$\begin{aligned} & u^2 a(\alpha^2 + \alpha\delta + \delta^2 + \beta\gamma) \\ & + u\{a(2\alpha\beta + \beta\delta) + 3b(\alpha\delta + \delta^2 + \beta\gamma) - 3c\alpha\gamma - d\gamma^2\} \\ & + \quad a\beta^2 + 3b\beta\delta + 3c\delta^2 + d(-\alpha\gamma - 2\gamma\delta) = 0. \end{aligned}$$

We can, from this equation, and the equation $au^3 + 3bu^2 + 3cu + d = 0$, eliminate u, thus obtaining a relation between a, b, c, d, α, β, γ, δ; and, this relation being satisfied, the two equations then determine u rationally in terms of these quantities.

We may without loss of generality assume $\alpha\delta - \beta\gamma = 1$; and, this being so, if we then further assume $\alpha + \delta = -1$, then we have

$$\alpha^2 + \alpha\delta + \delta^2 + \beta\gamma = 0,$$

which is, as above appearing, the condition for $\phi^3 x = 0$. The foregoing equation in u thus becomes

$$u\,\{a\beta(\alpha - 1) - 3b\alpha^2 - 3c\alpha\gamma - d\gamma^2\}$$
$$+ \{a\beta^2 + 3b\beta\delta + 3c\delta^2 - d\gamma(\delta - 1)\} = 0;$$

a linear equation giving (in a simplified form) the like results to those given by the quadric equation; viz. substituting in the cubic equation the value of u given by the linear equation, we have a relation between $a, b, c, d, \alpha, \beta, \gamma, \delta$; and, this relation being satisfied, u has the determinate value given by the linear equation.

The only way in which u can cease to have this determinate value, and so be capable of being any root whatever of the cubic equation, is when the linear equation becomes $0 = 0$; viz. if

$$a\beta(\alpha - 1) - 3b\alpha^2 - 3c\alpha\gamma - d\gamma^2 = 0,$$
$$a\beta^2 + 3b\beta\delta + 3c\delta^2 - d\gamma(\delta - 1) = 0,$$

equations which are, in fact, satisfied by the foregoing values of $\alpha, \beta, \gamma, \delta$, as may be verified without difficulty.

It is to be remarked that if, instead of the root u and the equation $v = \dfrac{\alpha u + \beta}{\gamma u + \delta}$, we consider the root v and the equation $u = \dfrac{\delta v - \beta}{-\gamma v + \alpha}$; then, instead of $\alpha, \beta, \gamma, \delta$, we have $\delta, -\beta, -\gamma, \alpha$, and the corresponding equations are

$$a\beta(\delta - 1) + 3b\delta^2 + 3c\gamma\delta + d\gamma^2 = 0,$$
$$a\beta^2 + 3b\alpha\beta + 3c\alpha^2 + d\gamma(\alpha - 1) = 0,$$

equations which are also satisfied by the foregoing values of $\alpha, \beta, \gamma, \delta$. And the four equations, together with $\alpha\delta - \beta\gamma = 1$ and $\alpha + \delta = 1$, are more than sufficient to determine the foregoing values of $\alpha, \beta, \gamma, \delta$.

But we further verify without difficulty that the foregoing values of $\alpha, \beta, \gamma, \delta$ give identically

$$(a, b, c, d \between \alpha x + \beta y, \gamma x + \delta y)^3 = (a, b, c, d \between x, y)^3;$$

or the formulæ lead to an automorphic transformation of the binary cubic $(a, b, c, d \between x, y)^3$. And conversely, starting from the notion of the automorphic transformation of the binary cubic, we ought to be able to obtain the foregoing formulæ.

For greater convenience, I write the equation of transformation in the form

$$(a, b, c, d \between \alpha x + \beta y, \gamma x + \delta y)^3 = -\theta\,(a, b, c, d \between x, y)^3;$$

the equations to be satisfied by α, β, γ, δ, θ then are

$$\begin{aligned}
&a\alpha^3 + b\,.\,3\alpha^2\gamma + c\,.\,3\alpha\gamma^2 + d\gamma^3 = -a\theta,\\
&a\alpha^2\beta + b(\alpha^2\delta + 2\alpha\beta\gamma) + c(2\alpha\gamma\delta + \beta\gamma^2) + d\gamma^2\delta = -b\theta,\\
&a\alpha\beta^2 + b(2\alpha\beta\delta + \beta^2\gamma) + c(\alpha\delta^2 + 2\beta\gamma\delta) + d\gamma\delta^2 = -c\theta,\\
&a\beta^3 + b\,.\,3\beta^2\delta + c\,.\,3\beta\delta^2 + d\delta^3 = -d\theta.
\end{aligned}$$

Writing $\alpha\delta - \beta\gamma = \nabla$, and as before Ω for the discriminant, the theory of invariants gives $\Omega\nabla^6 = \Omega\theta^4$. We are considering the case of the general cubic function $(a, b, c, d \between x, y)^3$, for which Ω is not $=0$; and we have therefore $\nabla^6 - \theta^4 = 0$, or, what is the same thing, we may write $\nabla = q^2$, $\theta = q^3$, where q is arbitrary.

It is to be shown that $\alpha + \delta$ is $= q$ or $-2q$, the latter value giving the trivial solution $\alpha x + \beta y$, $\gamma x + \delta y = (x, y)$. The proper solution thus corresponds to $\nabla = q^2$, $\alpha + \delta = q$, that is,

$$(\alpha + \delta)^2 - (\alpha\delta - \beta\gamma) = 0, \text{ or } \alpha^2 + \delta^2 + \alpha\delta + \beta\gamma = 0,$$

the condition for the periodic function $\phi^3 x - x = 0$.

For this purpose, from the foregoing equations eliminating a, b, c, d, we have

$$\left|\begin{array}{cccc}
\alpha^3 + \theta, & 3\alpha^2\gamma\ , & 3\alpha\gamma^2\ , & \gamma^3\\
\alpha^2\beta\ , & \alpha^2\delta + 2\alpha\beta\gamma + \theta, & 2\alpha\gamma\delta + \beta\gamma^2\ , & \gamma^2\delta\\
\alpha\beta^2\ , & 2\alpha\beta\delta + \beta^2\gamma\ , & \alpha\delta^2 + 2\beta\gamma\delta + \theta, & \gamma\delta^2\\
\beta^3\ , & 3\beta^2\delta\ , & 3\beta\delta^2\ , & \delta^3 + \theta
\end{array}\right| = 0;$$

an equation which may be written

$$\square + \theta(123 + 234 + 341 + 412) + \theta^2(12 + 23 + 34 + 41 + 13 + 42) + \theta^3(1 + 2 + 3 + 4) + \theta^4 = 0,$$

where 123, &c., are the first diagonal minors, 12, &c., the second diagonal minors, 1, &c., the third diagonal minors, or diagonal terms of the foregoing determinant, writing therein $\theta = 0$. We find without difficulty

$$\begin{aligned}
1,\ 2,\ 3,\ 4 &= \alpha^3,\ \alpha^2\delta + 2\alpha\beta\gamma,\ \alpha\delta^2 + 2\beta\gamma\delta,\ \delta^3,\\
12,\ 13,\ 14,\ 23,\ 24,\ 34 &= \{\alpha^4,\ \alpha^2(\alpha\delta + 3\beta\gamma),\ (\alpha^2\delta^2 + \alpha\delta\beta\gamma + \beta^2\gamma^2),\\
&\qquad (\alpha^2\delta^2 + \alpha\delta\beta\gamma + \beta^2\gamma^2),\ \delta^2(\alpha\delta + 3\beta\gamma),\ \delta^4\}\,\nabla,\\
123,\ 124,\ 134,\ 234 &= \{\alpha^3,\ \alpha\,(\alpha\delta + 2\beta\gamma),\ \delta\,(\alpha\delta + 2\beta\gamma),\ \delta^3\}\,\nabla^3,\\
\square &= \nabla^6,
\end{aligned}$$

and the equation thus is

$$\begin{aligned}
&\theta^4\\
&+ \theta^3[\alpha^3 + \alpha^2\delta + \alpha\delta^2 + \delta^3 + 2\alpha\beta\gamma + 2\beta\gamma\delta]\\
&+ \theta^2\,\nabla\,[\alpha^4 + \alpha^3\delta + \alpha\delta^3 + \delta^4 + 3\alpha^2\beta\gamma + 3\beta\gamma\delta^2 + 2\alpha^2\delta^2 + 2\alpha\delta\beta\gamma + 2\beta^2\gamma^2]\\
&+ \theta\,\nabla^3[\alpha^3 + \delta^3 + (\alpha + \delta)(\alpha\delta + 2\beta\gamma)]\\
&+ \quad \nabla^6 = 0.
\end{aligned}$$

Putting herein $\alpha+\delta=m$, $\alpha\delta=n$, $\beta\gamma=n-\nabla$, it is found that n disappears altogether from the equation; viz. the resulting form is

$$\theta^4+\theta^3 m(m^2-2\nabla)+\theta^2\nabla(m^4-3\nabla m^2+2\nabla^2)+\theta\nabla^3 m(m^2-2\nabla)+\nabla^6=0,$$

or, what is the same thing,

$$m^4.\theta^2\nabla+m^3(\theta^3+\theta\nabla^3)-3\theta^2\nabla^2m^2-2(\theta\nabla^4+\theta^3\nabla)m+\theta^4+2\theta^2\nabla^3+\nabla^6=0.$$

Putting for θ, ∇, their values q^3, q^2, the equation divides by q^8, and omitting this factor it becomes

$$m^4+2m^3q-3m^2q^2-4mq^3+4q^4=0;$$

viz. this is

$$\{(m-q)(m+2q)\}^2=0,$$

or we have

$$m=q \text{ or } -2q; \text{ that is, } \alpha+\delta=q, \text{ or } \alpha+\delta=-2q.$$

Writing, as before, A, B, $C=ac-b^2$, $ad-bc$, $bd-c^2$, we deduce from the foregoing equations

$$\begin{aligned}\theta^2A&=\nabla^2[A\alpha^2+B.\alpha\gamma+C\gamma^2],\\ \theta^2B&=\nabla^2[A.2\alpha\beta+B(\alpha\delta+\beta\gamma)+C.2\gamma\delta],\\ \theta^2C&=\nabla^2[A\beta^2+B.\beta\delta+C\delta^2],\end{aligned}$$

which are, in fact, the equations for the automorphic transformation of the Hessian $(A, B, C\between x, y)^2$. And, writing herein θ, $\nabla=q^3$, q^2, the equations become

$$\begin{aligned}&A(\alpha^2-q^2)+B\alpha\gamma+C.\gamma^2=0,\\ &A\,2\alpha\beta+B(\alpha\delta+\beta\gamma-q^2)+C.2\gamma\delta=0,\\ &A\beta^2+B\beta\delta+C(\delta^2-q^2)=0.\end{aligned}$$

From the first and second of these we have

$$A:B:C=\gamma^2(\nabla+q^2):-2\gamma\alpha\nabla+2\gamma\delta q^2:\alpha^2\nabla-(\alpha^2+\alpha\delta+\beta\gamma)q^2+q^4;$$

or, writing herein for ∇, q^4 the values q^2, $q^2(\alpha\delta-\beta\gamma)$, the three expressions divide by $2\gamma q^2$, and we have

$$A:B:C=\gamma:\delta-\alpha:-\beta.$$

Combining these values in the first place with the equation $\alpha+\delta=-2q$, we may write

$$\alpha, \beta, \gamma, \delta=-q-pB, -2pC, 2pA, -q+pB,$$

where p is to be determined. Substituting in the last of the three equations, we have

$$A.4p^2C^2-2pBC(-q+pB)+C(-2pqB+p^2B^2)=0,$$

that is,

$$p^2C(4AC-B^2), =-p^2.C\Omega, =0,$$

and the form $(a, b, c, d\between x, y)^3$ being arbitrary, neither C nor Ω is $=0$; whence $p=0$, and the values are α, β, γ, $\delta=-q$, 0, 0, $-q$, that is,

$$(a, b, c, d\between -qx, -qy)^3=-q^3.(a, b, c, d\between x, y)^3,$$

a trivial result.

But, combining the same values with $\alpha+\delta=q$, we have

$$\alpha,\ \beta,\ \gamma,\ \delta = \tfrac{1}{2}q - pB,\ -2pC,\ 2pA,\ \tfrac{1}{2}q + pB;$$

and then, substituting in the third equation, we have

$$A\ .\ 4p^2C^2 - 2pBC\,(\tfrac{1}{2}q + pB) + C\,(-\tfrac{3}{4}q^2 + pqB + p^2B^2) = 0,$$

that is,

$$C\,\{(4AC - B^2)\,p^2 - \tfrac{3}{4}q^2\} = 0,$$

or, omitting the factor C, and introducing the foregoing notation $\lambda^2 = -\tfrac{1}{3}$, this is

$$4\Omega\lambda^2p^2 - q^2 = 0, \text{ or say } p = \frac{1}{2\lambda\sqrt{\Omega}}\,q.$$

For the unimodular substitution $\alpha\delta - \beta\gamma = 1$, we must have $q^2 = 1$: but, the transformation being

$$(a,\ b,\ c,\ d \between \alpha x + \beta y,\ \gamma x + \delta y)^3 = -q^3\ .\ (a,\ b,\ c,\ d \between x,\ y)^3,$$

to make this strictly automorphic, we must have $q^3 = -1$, and the two conditions are satisfied by $q = -1$; we then have $p = -\dfrac{1}{2\lambda\sqrt{\Omega}}$; and the coefficients are

$$\alpha,\ \beta,\ \gamma,\ \delta,\ = \frac{-\lambda\sqrt{\Omega} + B}{2\lambda\sqrt{\Omega}},\ \frac{2C}{2\lambda\sqrt{\Omega}},\ \frac{-2A}{2\lambda\sqrt{\Omega}},\ \frac{-\lambda\sqrt{\Omega} - B}{2\lambda\sqrt{\Omega}};$$

which are the values given at the beginning of the paper, and which belong to the automorphic transformation

$$(a,\ b,\ c,\ d \between \alpha x + \beta y,\ \gamma x + \delta y)^3 = (a,\ b,\ c,\ d \between x,\ y)^3.$$

The *à priori* reason for the periodicity-equation $\phi^3x = x$, is best seen by expressing the cubic function as a product of factors

$$M\,(x - \mathrm{a}y)\,(x - \mathrm{b}y)\,(x - \mathrm{c}y).$$

The substitution must, it is clear, cyclically interchange these factors, and therefore, when performed three times in succession on any one of these factors, or consequently upon an arbitrary linear factor $x - \mathrm{f}y$, must leave the factor unaltered, and it must thus be a periodic substitution $\phi^3x = x$. But it was interesting to see how the condition for this, $\alpha^2 + \delta^2 + \alpha\delta + \beta\gamma = 0$, comes out from the consideration of the equation

$$(a,\ b,\ c,\ d \between \alpha x + \beta y,\ \gamma x + \delta y)^3 = (a,\ b,\ c,\ d \between x,\ y)^3.$$

782.

ON MONGE'S "MÉMOIRE SUR LA THÉORIE DES DÉBLAIS ET DES REMBLAIS."

[From the *Proceedings of the London Mathematical Society*, vol. XIV. (1883), pp. 139—142. Read March 8, 1883.]

THE Memoir referred to, published in the *Mémoires de l'Académie*, 1781, pp. 666—704, is a very remarkable one, as well for the problem of earthwork there considered as because the author was led by it to his capital discovery of the curves of curvature of a surface. The problem is, from a given area, called technically the Déblai, to transport the earth to a given equal area, called the Remblai, with the least amount of carriage. Taking the earth to be of uniform infinitesimal thickness over the whole of each area (and therefore of the same thickness for both areas), the problem is a plane one; viz. stating it in a purely geometrical form, the problem is: Given two equal areas, to transfer the elements of the first area to the second area in such wise that the sum of the products of each element into the traversed distance may be a minimum; the route of each element is, of course, a straight line. And we have the corresponding solid problem: Given two equal volumes, to transfer the elements of the first volume to the second volume in such wise that the sum of the products of each element into the traversed distance may be a minimum; the route of each element is, of course, a straight line. The Memoir is divided into two parts: the first relating to the plane problem (and to some variations of it): the second part contains a theorem as to congruences, the general theory of the curvature of surfaces, and finally a solution of the solid problem; in regard to this, I find a difficulty which will be referred to further on.

I have said that Monge gives a theorem as to congruences. This is not stated quite in the best form,—viz. instead of speaking of a singly infinite system of lines, or even of the lines drawn according to a given law from the several points of a *surface*, he speaks of the lines drawn according to a given law from the several points of a *plane* (but, of course, any congruence whatever of lines can be so represented); and he establishes the theorem that each line of the system is intersected by each of two consecutive lines,—viz. taking (x', y') as the coordinates of the point of intersection of any line with the plane of xy, he obtains, as the condition of intersection with the consecutive line a quadric equation in (dx', dy'). He then considers the normals of a surface, (which, as lines drawn according to a given law from any point of a *surface*, require a slightly different analytical investigation), establishes for them the like theorem, and shows moreover that the two directions of passage on the surface to a consecutive point are at right angles to each other; or, what is the same thing, that in the two sets of developable surfaces formed by the intersecting normals, each surface of the one set intersects each surface of the other set in a straight line, and at right angles. He speaks expressly of the lines of greatest and least curvature, and generally establishes the whole theory of the curvature of surfaces in a very complete and satisfactory manner; the particular case of surfaces of the second order is not considered. It may be remarked that, although not explicitly stating it, he must have seen that a congruence of lines is not, *in general*, a system of normals of a surface (that is, the lines of a congruence cannot be, in general, cut at right angles by any surface); he, in fact, assumes (quite correctly, but a proof should have been given) that a congruence of lines for which the two sets of developable surfaces intersect at right angles is a system of normals of a surface.

Reverting to the before-mentioned problem (plane or solid), I remark that this is a problem of minimum *sui generis*. Considering the first area or volume as divided in any manner into infinitesimal elements, we have to divide the second area or volume into corresponding equal elements, in such wise that the sum of the products of each element of the first area or volume into its distance from the corresponding element of the second area or volume may be a minimum; but, for doing this, we have no means of forming the analytical expression of any function which is to be, by the formulæ of the differential calculus or the calculus of variations, made a minimum.

For the plane problem, Monge obtains the solution by means of the very simple consideration that the routes of two elements must not cross each other; in fact, imagine an element A transferred to a, and an equal element B transferred to b: the lines Aa, Bb must not cross each other, for if they did, drawing the two lines

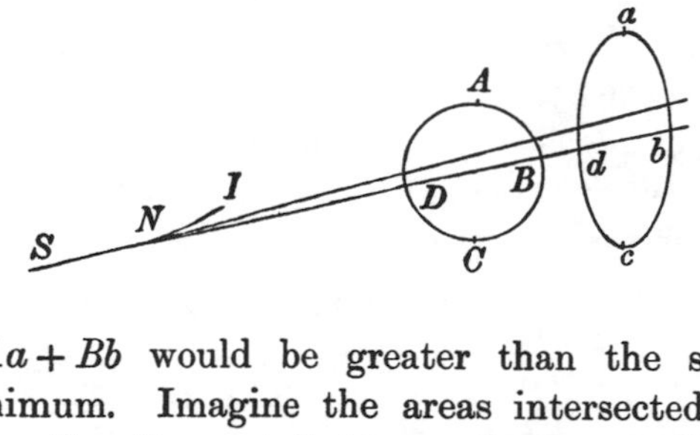

Ab and Ba, the sum $Aa + Bb$ would be greater than the sum $Ab + Ba$, contrary to the condition of the minimum. Imagine the areas intersected by two consecutive lines as shown in the figure: the filament between these two lines may be regarded as

a right line; and, assuming that some one element of the filament BD is transferred to a point of bd (that is, so as to coincide with an element of the filament bd), it follows that every other element of BD must be transferred so as to coincide with some other element of bd; and this obviously implies that the filaments BD and bd must be equal. Observe that, this being so, it is immaterial which element of BD is transferred to which element of bd; in whatever way this is done, the sum of the products will be the same*. The two lines may be regarded as the normals of a curve; and the problem thus is, to find a curve such that, drawing the normals thereof to intersect the two areas, then that the filaments BD and bd, cut off by consecutive normals on the two areas respectively, shall be equal. This leads to a differential equation of the second order for the normal curve; one of the constants of integration remains arbitrary, for the normal curve is any one of a system of parallel curves. It is to be observed that the filaments are the increments of the areas BCD and bcd; these increments are equal; a position of the line must be the common tangent Cc of the two areas (this, in fact, constitutes the condition for the determination of one of the arbitrary constants), and for this position the areas are each $=0$. Hence, in general, the areas must be equal; or the problem is, to find a curve such that any normal thereof cuts off equal areas BCD and bcd.

If, instead of the normal curve, we consider the curve which is the envelope of the several lines, or, what is the same thing, the locus of the point N, then we could, in like manner, obtain for this curve a differential equation of the first order: the constant of integration would be determined by the condition that Cc is a tangent. The curve in question is, of course, the evolute of the normal curve.

The several lines which intersect the two areas give rise to a finite arc IS of this evolute, and, as remarked by Monge, it is only when this arc IS lies (as in the figure) *outside* the two areas, that we have a true minimum.

Passing now to the solid problem, we may imagine a congruence of lines intersecting the two volumes; each line of the congruence is intersected by two consecutive lines, and the lines of the congruence thus form two sets of developable surfaces, each surface of the one set intersecting each surface of the other set. And, considering two consecutive surfaces of the one set, and two consecutive surfaces of the other set, these include between them a filament; and, treating the filament as a right line, it seems to follow (although it is more difficult to present the reasoning in a rigorous form) that, if any one element of the filament BD be transferred to any one element of the filament bd, then that every other element of the filament BD must be

* The most simple case is, take in the same straight line two equal segments AB, ab; it is immaterial how the elements of AB are transferred to ab, the sum of the products of each element into the traversed distance will be in every case the same. Analytically, if $dx=dx'$, then

$$\int (x'-x)\,dx = \int x'dx' - \int x\,dx,$$

the equation $dx'=dx$ meaning $x'=x+$a discontinuous constant. In the actual case of the filament, the formula is, if $rdr=r'dr'$, then

$$\int (r'-r)\,r\,dr = \int r'^2\,dr' - \int r^2\,dr.$$

transferred to some other element of the filament *bd*; and, this being so, the two filaments must be equal. But Monge goes on to argue that the condition of the minimum further requires that the developable surfaces shall cut at right angles, *and I cannot say that I see this.* He says (pp. 700, 701), "We know already that the routes must be the intersections of two sets of developable surfaces such that each surface of the first set intersects those of the second set in right lines; it remains to be found under what angles these surfaces must cut each other to satisfy the minimum. But it is evident that these angles must be right angles, for with these angles the elementary spaces comprised between four developable surfaces will be greater, and for equal distances the transported mass will be greater; therefore, in the case of a minimum, the routes must be the intersections of two sets of developable surfaces such that each surface of the one set cuts those of the second set in straight lines and at right angles." And, this being so, he infers, and it in fact follows, that the routes are the normals of a surface.

Admitting the conclusion, the problem becomes as follows:—Given two volumes, it is required to find a surface such that, drawing the normals thereof to intersect the two volumes, and considering the filament bounded by the developable surfaces which belong to two consecutive curves of curvature of the one set and those belonging to two consecutive curves of curvature of the other set, the portions cut off on the two volumes respectively may be equal. And we are thus led to a partial differential equation of the second order for determining the equation $z = f(x, y)$ of the required surface. As in the plane problem, it is immaterial how the elements of the one filament are transferred to the other filament.

783.

ON MR WILKINSON'S RECTANGULAR TRANSFORMATION.

[From the *Proceedings of the London Mathematical Society*, vol. XIV. (1883), pp. 222—229. Read May 10, 1883.]

CONSIDERING the three cones,

$$(p+\lambda) X^2 + (q+\lambda) Y^2 + (r+\lambda) Z^2 = 0,$$
$$(p+\mu) X^2 + (q+\mu) Y^2 + (r+\mu) Z^2 = 0,$$
$$(p+\nu) X^2 + (q+\nu) Y^2 + (r+\nu) Z^2 = 0,$$

where

$$p+q+r+\lambda+\mu+\nu=0,$$

it is easy to see that these contain a singly infinite system of rectangular axes, viz. we have in each cone one axis of a rectangular system, and for one of the cones the axis may be any line at pleasure of the cone. In fact, taking for the three axes (x, y, z), (x', y', z'), (x'', y'', z'') respectively, that is, for the first axis $X : Y : Z = x : y : z$, and so for each of the other two axes, then (x, y, z) being an arbitrary line on the first cone, we can find (x', y', z') and (x'', y'', z'') such that

$$(p+\lambda) x^2 + (q+\lambda) y^2 + (r+\lambda) z^2 = 0,$$
$$(p+\mu) x'^2 + (q+\mu) y'^2 + (r+\mu) z'^2 = 0,$$
$$(p+\nu) x''^2 + (q+\nu) y''^2 + (r+\nu) z''^2 = 0,$$
$$x' x'' + y' y'' + z' z'' = 0,$$
$$x'' x + y'' y + z'' z = 0,$$
$$x x' + y y' + z z' = 0.$$

For, eliminating (x'', y'', z'') from the third, fourth, and fifth equations, we have, first,

$$x'' : y'' : z'' = yz' - y'z : zx' - z'x : xy' - x'y,$$

and consequently

$$(p+\nu)(yz'-y'z)^2+(q+\nu)(zx'-z'x)^2+(r+\nu)(xy'-x'y)^2=0.$$

It is to be shown that this equation is implied in the remaining first, second, and third equations; for, this being so, (x, y, z), (x', y', z') satisfy only these equations; or (x, y, z) are any values whatever satisfying the first equation. The other two equations then determine (x', y', z'), and, these being known, (x'', y'', z'') are then determined as above.

In fact, attending to the sixth equation, the equation just obtained may be written in the form

$$(p+\nu)[(y^2+z^2)(y'^2+z'^2)-x^2x'^2]+(q+\nu)[(z^2+x^2)(z'^2+x'^2)-y^2y'^2]$$
$$+(r+\nu)[(x^2+y^2)(x'^2+y'^2)-z^2z'^2]=0,$$

or, what is the same thing, in the form

$$-(x^2+y^2+z^2)[(p+\mu)x'^2+(q+\mu)y'^2+(r+\mu)z'^2]$$
$$-(x'^2+y'^2+z'^2)[(p+\lambda)x^2+(q+\lambda)y^2+(r+\lambda)z^2]=0;$$

for, comparing in the two forms, first the coefficients of $x^2x'^2$, these are

$$(q+\nu)+(r+\nu)-(p+\nu) \quad \text{and} \quad -2p+\lambda+\mu,$$

which are equal in virtue of $p+q+r+\lambda+\mu+\nu=0$; and comparing next the coefficients of $y^2z'^2$, these are

$$p+\nu \quad \text{and} \quad -(r+\mu)-(q+\lambda),$$

which are equal in virtue of the same relation: and, similarly, the coefficients of the other terms $y^2y'^2$, &c., are equal in the two equations respectively.

Take now three arguments a_0, b_0, c_0, connected by the relation $a_0+b_0+c_0=0$, and write a, a, A for the sn, cn, and dn of a_0; and similarly b, b, B and c, c, C for those of b_0 and c_0 respectively: then we may write

$$p+\lambda,\ q+\lambda,\ r+\lambda= \left(1,\ \frac{\mathrm{a}}{\mathrm{bc}},\ \frac{A}{BC}\right),$$

$$p+\mu,\ q+\mu,\ r+\mu=\theta\left(1,\ \frac{\mathrm{b}}{\mathrm{ca}},\ \frac{B}{CA}\right),$$

$$p+\nu,\ q+\nu,\ r+\nu=\phi\left(1,\ \frac{\mathrm{c}}{\mathrm{ab}},\ \frac{C}{AB}\right);$$

for, starting from the first set of values, we have the second set if only

$$\mu-\lambda=\theta-1=\theta\frac{\mathrm{b}}{\mathrm{ca}}-\frac{\mathrm{a}}{\mathrm{bc}}=\theta\frac{B}{CA}-\frac{A}{BC}.$$

We thence obtain

$$\theta\left(1-\frac{\mathrm{b}}{\mathrm{ca}}\right)=1-\frac{\mathrm{a}}{\mathrm{bc}},\quad \theta\left(1-\frac{B}{CA}\right)=1-\frac{A}{BC};$$

and, in order to the identity of the two values of θ, we must have

$$\left(1-\frac{\mathrm{b}}{\mathrm{ca}}\right)\left(1-\frac{A}{BC}\right)-\left(1-\frac{\mathrm{a}}{\mathrm{bc}}\right)\left(1-\frac{B}{CA}\right)=0,$$

that is,

$$(\mathrm{abc}-\mathrm{b}^2)(ABC-A^2)-(\mathrm{abc}-\mathrm{a}^2)(ABC-B^2)=0,$$

or, reducing,

$$(\mathrm{a}^2-\mathrm{b}^2)ABC-(A^2-B^2)\,\mathrm{abc}+A^2\mathrm{b}^2-B^2\mathrm{a}^2=0.$$

But

$$\mathrm{a}^2-\mathrm{b}^2=-(a^2-b^2),\quad A^2-B^2=-k^2(a^2-b^2),\quad A^2\mathrm{b}^2-B^2\mathrm{a}^2=k'^2(a^2-b^2);$$

hence the whole equation divides by a^2-b^2, and, omitting this factor, it becomes

$$-ABC+k^2\mathrm{abc}+k'^2=0,$$

which is a known relation between the elliptic functions of the arguments a_0, b_0, c_0 connected by the equation $a_0+b_0+c_0=0$. Similarly, for ϕ, we have

$$\nu-\lambda=\phi-1=\phi\frac{\mathrm{c}}{\mathrm{ab}}-\frac{\mathrm{a}}{\mathrm{bc}}=\phi\frac{C}{AB}-\frac{A}{BC},$$

and, comparing the two values of ϕ, we have the same identical relation.

It thus appears that the three cones

$$X^2+\frac{\mathrm{a}}{\mathrm{bc}}Y^2+\frac{A}{BC}Z^2=0,$$

$$X^2+\frac{\mathrm{b}}{\mathrm{ca}}Y^2+\frac{B}{CA}Z^2=0,$$

$$X^2+\frac{\mathrm{c}}{\mathrm{ab}}Y^2+\frac{C}{AB}Z^2=0,$$

(the coefficients whereof depend on the elliptic functions sn, cn, and dn, of the arguments a_0, b_0, c_0 connected by the equation $a_0+b_0+c_0=0$) contain a singly infinite system of rectangular axes.

Considering an argument f_0, and denoting its sn, cn, dn by f, f, F respectively, we have, for an arbitrary line on the first cone, the values

$$x,\ y,\ z=M\sqrt{k'^2A\mathrm{a}},\quad M\sqrt{k^2A\mathrm{bc}}\,.\,\mathrm{f},\quad M\sqrt{-\mathrm{a}BC}\,.\,F.$$

In fact, substituting in the equation of the cone, we obtain the identity

$$k'^2+k^2\mathrm{f}^2-F^2=0;$$

and if we determine M by the condition that $x^2+y^2+z^2$ shall be $=1$, then we have

$$1=M^2\{k'^2A\mathrm{a}+k^2A\mathrm{bcf}^2-\mathrm{a}BCF^2\},$$

where the coefficient of M^2 is

$$=k'^2A\mathrm{a}+k^2A\mathrm{bc}(1-f^2)-\mathrm{a}BC(1-k^2f^2),$$

which is easily shown to be

$$= k^2k'^2bc\,(a^2-f^2),$$

so that the values of x, y, z are

$$= \{\sqrt{k'^2 A\mathrm{a}},\quad \sqrt{k^2 A\mathrm{bc}}\,.\,\mathrm{f},\quad \sqrt{-\mathrm{a}BC}\,.\,F\} \div \sqrt{k^2k'^2bc\,(a^2-f^2)},$$

and, similarly taking the arguments g_0, h_0, and denoting their elliptic functions by g, g, G, h, h, H, we have for a system of arbitrary lines in the three cones respectively, the values

$x,\ y,\ z\ =$	$\sqrt{k'^2 A\mathrm{a}}$	$\sqrt{k^2 A\mathrm{bc}}\,.\,f$	$\sqrt{-\mathrm{a}BC}\,.\,F$	$\div\sqrt{k^2k'^2bc\,(a^2-f^2)}$
$x',\ y',\ z'\ =$	$\sqrt{k'^2 B\mathrm{b}}$	$\sqrt{k^2 B\mathrm{ca}}\,.\,g$	$\sqrt{-\mathrm{b}CA}\,.\,G$	$\div\sqrt{k^2k'^2ca\,(b^2-g^2)}$
$x'',\ y'',\ z''=$	$\sqrt{k'^2 C\mathrm{c}}$	$\sqrt{k^2 C\mathrm{ab}}\,.\,h$	$\sqrt{-\mathrm{c}AB}\,.\,H$	$\div\sqrt{k^2k'^2ab\,(c^2-h^2)}$,

these values being such that $x^2+y^2+z^2$, $x'^2+y'^2+z'^2$, $x''^2+y''^2+z''^2$ are each $=1$. The radicals in the first line would be more correctly written, and may be understood as meaning $k'\sqrt{A}\sqrt{\mathrm{a}}$, $k\sqrt{A}\sqrt{\mathrm{b}}\sqrt{\mathrm{c}}$, $i\sqrt{\mathrm{a}}\sqrt{B}\sqrt{C}$, and similarly as regards the second and third lines respectively.

Taking now the arbitrary lines at right angles to each other, the condition for the second and third lines is

$$1 + k^2\mathrm{gh} - k'^2 AGH = 0,$$

which is satisfied if $a_0 = g_0 - h_0$; similarly the condition for the third and first lines is satisfied if $b_0 = h_0 - f_0$; and we then have $a_0 + b_0 = g_0 - f_0$; that is, $-c_0 = g_0 - f_0$ or $c_0 = f_0 - g_0$, which is the condition for the first and second lines; hence the arguments a_0, b_0, c_0, f_0, g_0, h_0 being such that

$$\begin{aligned} .\quad h_0 - g_0 + a_0 &= 0,\\ -h_0\quad .\quad + f_0 + b_0 &= 0,\\ g_0 - f_0\quad .\quad + c_0 &= 0,\\ -a_0 - b_0 - c_0\quad . \quad &= 0, \end{aligned}$$

or, what is the same thing, a_0, b_0, c_0, f_0, g_0, h_0 being the differences of any four arguments α, β, γ, δ, the foregoing values of (x, y, z), (x', y', z'), (x'', y'', z'') will satisfy the equations

$$\begin{aligned} x^2 + y^2 + z^2 &= 1,\\ x'^2 + y'^2 + z'^2 &= 1,\\ x''^2 + y''^2 + z''^2 &= 1,\\ x'x'' + y'y'' + z'z'' &= 0,\\ x''x + y''y + z''z &= 0,\\ xx' + yy' + zz' &= 0, \end{aligned}$$

for the transformation of a set of rectangular axes. These are, in fact, Mr Wilkinson's expressions, the a_0, b_0, c_0, f_0, g_0, h_0 being his $t-p$, $p-q$, $q-t$, t, p, q respectively.

Returning to the three cones, it is to be remarked that, taking in the first of them a line 1 at pleasure, then we have in the second of them *two* lines 2, 2′ each at right angles to the line 1, and such that the line 3 at right angles to the plane 12, and the line 3′ at right angles to the plane 12′, lie each of them in the third cone; or, what is the same thing, we have in the two cones respectively the rectangular lines 1 and 2, and also the rectangular lines 1 and 2′, such that the planes 12 and 12′ each of them envelope one and the same cone, the reciprocal of the third cone; where by the reciprocal cone of a given cone is meant the cone generated by the lines through the vertex at right angles to the tangent planes of the given cone. Introducing the notion of the absolute cone $X^2+Y^2+Z^2=0$, a line and plane through the vertex at right angles to each other are, in fact, reciprocal polars in regard to this absolute cone; and two lines at right angles to each other are reciprocals (or harmonics) in regard to this absolute cone; that is, the reciprocal plane of either of them passes through the other. The two cones are cones intersecting each other in four lines lying on the absolute cone; and in virtue of this relation they have the property in question, viz. taking in the first cone a line 1 at pleasure, then the reciprocal plane hereof in regard to the absolute cone meets the second cone in a pair of lines 2 and 2′ such that the planes 12 and 12′ each of them envelope one and the same cone; the reciprocal of this cone is then the third cone of the system, and as such it passes through the four lines on the absolute cone.

In verification, observe that the coefficients $p+\lambda$, $q+\lambda$, &c. of the equations of the three cones satisfy the equations

$$\left\|\begin{array}{lllll} 1, & p+\lambda, & p+\mu, & p+\nu, & (p+\lambda)(p+\mu)(p+\nu) \\ 1, & q+\lambda, & q+\mu, & q+\nu, & (q+\lambda)(q+\mu)(q+\nu) \\ 1, & r+\lambda, & r+\mu, & r+\nu, & (r+\lambda)(r+\mu)(r+\nu) \end{array}\right\| = 0.$$

This is obviously the case for each equation such as

$$|\,1,\ p+\lambda,\ p+\mu\,| = 0;$$

and any equation containing the fifth column is at once reducible to

$$|\,1,\ p,\ p^3+p^2(\lambda+\mu+\nu)\,| = 0,$$

that is,

$$|\,1,\ p,\ p^3\,| + (\lambda+\mu+\nu)\,|\,1,\ p,\ p^2\,| = 0;$$

or, dividing by $|\,1,\ p,\ p^2\,|$, this is $p+q+r+\lambda+\mu+\nu=0$, the equation connecting the coefficients.

Hence, representing the three cones by

$$\begin{aligned} p\,X^2+q\,Y^2+r\,Z^2&=0,\\ p'X^2+q'Y^2+r'Z^2&=0,\\ p''X^2+q''Y^2+r''Z^2&=0, \end{aligned}$$

and the absolute by

$$X^2+\ \ Y^2+\ \ Z^2=0,$$

the coefficients p, q, &c., are connected by the equations

$$\left\|\begin{matrix} 1, & p, & p', & p'', & pp'p'' \\ 1, & q, & q', & q'', & qq'q'' \\ 1, & r, & r', & r'', & rr'r'' \end{matrix}\right\| = 0;$$

among these are of course included the equation $|\, 1,\ p,\ p' \,| = 0$, which expresses that the first and second cones intersect on the absolute; (p, q, r), (p', q', r') are any quantities satisfying this relation, and, regarding them as given, we have then two independent equations determining the ratios $p'' : q'' : r''$. The theorem is that the planes 12 and 12′ envelope one and the same quadric cone

$$\frac{X^2}{p''} + \frac{Y^2}{q''} + \frac{Z^2}{r''} = 0.$$

The equations $|\, 1,\ p,\ p'' \,| = 0$ and $|\, 1,\ p,\ pp'p'' \,| = 0$ give

$$(q-r)\,p'' \quad + (r-p)\,q'' \quad + (p-q)\,r'' \quad = 0,$$
$$(q-r)\,pp'p'' + (r-p)\,qq'q'' + (p-q)\,rr'r'' = 0,$$

and thence

$$(q-r)\,p'' : (r-p)\,q'' : (p-q)\,r'' = qq' - rr' : rr' - pp' : pp' - qq';$$

or, observing that we have

$$q - r : r - p : p - q = qr' - q'r : rp' - r'p : pq' - p'q,$$

the equations may also be written

$$(qr' - q'r)\,p'' : (rp' - r'p)\,q'' : (pq' - p'q)\,r'' = qq' - rr' : rr' - pp' : pp' - qq'.$$

Starting with an arbitrary line (x, y, z) in the first cone, then the reciprocal plane thereof (in regard to the absolute cone) is the plane $Xx + Yy + Zz = 0$, which meets the second cone in two lines, say (2) and (2′), each of which is a line reciprocal to the line (1); and we have thus two planes (12) and (12′), each of which envelopes, as is to be shown, the same cone $q''r''X^2 + r''p''Y^2 + p''q''Z^2 = 0$.

Suppose, in general, that we have an arbitrary line (x, y, z) and an arbitrary plane $\alpha X + \beta Y + \gamma Z = 0$, and that it is required to find the equation of the two planes through the line (x, y, z), and the intersections of the plane $\alpha X + \beta Y + \gamma Z = 0$ with the cone $p'X^2 + q'Y^2 + r'Z^2 = 0$: the equation of the pair of planes is

$$\begin{aligned} &(\alpha X + \beta Y + \gamma Z)^2 (p'x^2 + q'y^2 + r'z^2) \\ &+ (\alpha x + \beta y + \gamma z)^2 (p'X^2 + q'Y^2 + r'Z^2) \\ &- 2\,(\alpha X + \beta Y + \gamma Z)(\alpha x + \beta y + \gamma z)(p'Xx + q'Yy + r'Zz) = 0. \end{aligned}$$

In the present case, the plane $\alpha X + \beta Y + \gamma Z = 0$ is the plane $xX + yY + zZ = 0$, which is the reciprocal of the line (x, y, z) in regard to the absolute cone, and the equation of the pair of planes is

$$\begin{aligned} &(xX + yY + zZ)^2 (p'x^2 + q'y^2 + r'z^2) \\ &+ (x^2 + y^2 + z^2)^2 (p'X^2 + q'Y^2 + r'Z^2) \\ &- 2\,(xX + yY + zZ)(x^2 + y^2 + z^2)(p'Xx + q'Yy + r'Zz) = 0, \end{aligned}$$

where the quantities (x, y, z), as belonging to a line on the first cone, satisfy the condition $px^2 + qy^2 + rz^2 = 0$. The equation may be written

$$(a, b, c, f, g, h \between yZ - zY, zX - xZ, xY - yX)^2 = 0,$$

where

$$a, b, c, f, g, h = q'z^2 + r'y^2, r'x^2 + p'z^2, p'y^2 + q'x^2, -p'yz, -q'zx, -r'xy,$$

and, as before, $px^2 + qy^2 + rz^2 = 0$; viz. this is the equation of the pair of planes (12) and (12′).

The equation of the pair of tangent planes through the line (x, y, z) to the cone $q''r''X^2 + r''p''Y^2 + p''q''Z^2 = 0$ is

$$(q''r''x^2 + r''p''y^2 + p''q''z^2)(q''r''X^2 + r''p''Y^2 + p''q''Z^2) - (q''r''xX + r''p''yY + p''q''zZ)^2 = 0;$$

viz. omitting a factor $p''q''r''$, this equation is

$$(p'', q'', r'', 0, 0, 0 \between yZ - zY, zX - xZ, xY - yX)^2 = 0.$$

And it is to be shown that this is equivalent to the former equation; viz. writing $yZ - zY, zX - xZ, xY - yX = \lambda, \mu, \nu$, then that the two equations

$$(q'z^2 + r'y^2, r'x^2 + p'z^2, p'y^2 + q'x^2, -p'yz, -q'zx, -r'xy \between \lambda, \mu, \nu)^2 = 0,$$
$$p''\lambda^2 + q''\mu^2 + r''\nu^2 = 0,$$

are equivalent to each other.

We have identically $\lambda x + \mu y + \nu z = 0$, and thence also

$$(\lambda x + \mu y + \nu z)[(p' - q' - r')\lambda x + (-p' + q' - r')\mu y + (-p' - q' + r')\nu z] = 0,$$

where, on the left-hand side, the terms in $\mu\nu$, $\nu\lambda$, and $\lambda\mu$ are

$$= -2p'yz\mu\nu - 2q'zx\nu\lambda - 2r'xy\lambda\mu.$$

Hence the first equation may be written

$$[q'z^2 + r'y^2 + (p' - q' - r')x^2]\lambda^2 + [r'x^2 + p'z^2 + (-p' + q' - r')y^2]\mu^2$$
$$+ [p'y^2 + q'x^2 + (-p' - q' + r')z^2]\nu^2 = 0,$$

and it is to be shown that this is equivalent to

$$p''\lambda^2 + q''\mu^2 + r''\nu^2 = 0;$$

viz. that we have $p'' : q'' : r'' =$

$$q'z^2 + r'y^2 - (\ \ p' - q' - r')x^2$$
$$: r'x^2 + p'z^2 - (-p' + q' - r')y^2$$
$$: p'y^2 + q'x^2 - (-p' - q' + r')z^2,$$

where $px^2 + qy^2 + rz^2 = 0$. Writing the equation in the form

$$p'' : q'' : r'' = A : B : C,$$

we have

$$\begin{aligned} A &= q'z^2 + r'y^2 - p'x^2 + q'x^2 + r'x^2 \\ &= -p'x^2 + q'(x^2+z^2) + r'(x^2+y^2) \\ &= -p'x^2 + (q'+r')(x^2+y^2+z^2) - q'y^2 - r'z^2. \end{aligned}$$

By what precedes, we have an identity of the form

$$x^2 + y^2 + z^2 = \alpha(p'x^2 + q'y^2 + r'z^2) + \beta(px^2 + qy^2 + rz^2),$$

where, determining α from the equations $1 = q'\alpha + q\beta$, $1 = r'\alpha + r\beta$, we find

$$\alpha = (q - r) \div (qr' - q'r);$$

but $px^2 + qy^2 + rz^2 = 0$, and the relation thus is

$$x^2 + y^2 + z^2 = \alpha(p'x^2 + q'y^2 + r'z^2);$$

hence

$$A = \{(q' + r')\alpha - 1\}(p'x^2 + q'y^2 + r'z^2),$$

or, substituting for α its value, this is

$$\begin{aligned} A &= \left\{\frac{(q'+r')(q-r) - qr' + q'r}{qr' - q'r}\right\}(p'x^2 + q'y^2 + r'z^2), \\ &= \frac{qq' - rr'}{qr' - q'r}(p'x^2 + q'y^2 + r'z^2); \end{aligned}$$

and, forming the like values of B and C, the relations to be verified become

$$p'' : q'' : r'' = \frac{qq' - rr'}{qr' - q'r} : \frac{rr' - pp'}{rp' - r'p} : \frac{pp' - qq'}{pq' - p'q},$$

which are, in fact, the values of the ratios $p'' : q'' : r''$ obtained above; and the theorem is thus seen to be true. It may be remarked that, if the first and second cones, instead of intersecting in four lines on the absolute cone, had been arbitrary cones; then, taking in the first cone a line (1) and in the second cone a line (2), the reciprocal of (1) in regard to the absolute, the envelope of the plane (12) would have been (instead of a quadric cone) a cone of the class 8.

784.

PRESIDENTIAL ADDRESS TO THE BRITISH ASSOCIATION, SEPTEMBER 1883.

[From the *Report of the British Association for the Advancement of Science*, (1883), pp. 3—37.]

SINCE our last meeting we have been deprived of three of our most distinguished members. The loss by the death of Professor Henry John Stephen Smith is a very grievous one to those who knew and admired and loved him, to his University, and to mathematical science, which he cultivated with such ardour and success. I need hardly recall that the branch of mathematics to which he had specially devoted himself was that most interesting and difficult one, the Theory of Numbers. The immense range of this subject, connected with and ramifying into so many others, is nowhere so well seen as in the series of reports on the progress thereof, brought up unfortunately only to the year 1865, contributed by him to the Reports of the Association; but it will still better appear when to these are united (as will be done in the collected works in course of publication by the Clarendon Press) his other mathematical writings, many of them containing his own further developments of theories referred to in the reports. There have been recently or are being published many such collected editions—Abel, Cauchy, Clifford, Gauss, Green, Jacobi, Lagrange, Maxwell, Riemann, Steiner. Among these the works of Henry Smith will occupy a worthy position.

More recently, General Sir Edward Sabine, K.C.B., for twenty-one years general secretary of the Association, and a trustee, President of the meeting at Belfast in the year 1852, and for many years treasurer and afterwards President of the Royal Society, has been taken from us, at an age exceeding the ordinary age of man. Born October 1788, he entered the Royal Artillery in 1803, and commanded batteries at the siege of Fort Erie in 1814; made magnetic and other observations in Ross and Parry's North Polar exploration in 1818–19, and in a series of other voyages. He contributed to the Association reports on Magnetic Forces in 1836–7–8, and about forty papers to the *Philosophical Transactions*; originated the system of Magnetic Observatories, and otherwise signally promoted the science of Terrestrial Magnetism.

There is yet a very great loss: another late President and trustee of the Association, one who has done for it so much, and has so often attended the meetings, whose presence among us at this meeting we might have hoped for—the President of the Royal Society, William Spottiswoode. It is unnecessary to say anything of his various merits: the place of his burial, the crowd of sorrowing friends who were present in the Abbey, bear witness to the esteem in which he was held.

I take the opportunity of mentioning the completion of a work promoted by the Association: the determination by Mr James Glaisher of the least factors of the missing three out of the first nine million numbers: the volume containing the sixth million is now published.

I wish to speak to you to-night upon Mathematics. I am quite aware of the difficulty arising from the abstract nature of my subject; and if, as I fear, many or some of you, recalling the Presidential Addresses at former meetings—for instance, the *résumé* and survey which we had at York of the progress, during the half century of the lifetime of the Association, of a whole circle of sciences—Biology, Palæontology, Geology, Astronomy, Chemistry—so much more familiar to you, and in which there was so much to tell of the fairy-tales of science; or at Southampton, the discourse of my friend who has in such kind terms introduced me to you, on the wondrous practical applications of science to electric lighting, telegraphy, the St Gothard Tunnel and the Suez Canal, gun-cotton, and a host of other purposes, and with the grand concluding speculation on the conservation of solar energy: if, I say, recalling these or any earlier Addresses, you should wish that you were now about to have, from a different President, a discourse on a different subject, I can very well sympathise with you in the feeling.

But be this as it may, I think it is more respectful to you that I should speak to you upon and do my best to interest you in the subject which has occupied me, and in which I am myself most interested. And in another point of view, I think it is right that the Address of a President should be on his own subject, and that different subjects should be thus brought in turn before the meetings. So much the worse, it may be, for a particular meeting; but the meeting is the individual, which on evolution principles must be sacrificed for the development of the race.

Mathematics connect themselves on the one side with common life and the physical sciences; on the other side with philosophy, in regard to our notions of space and time, and in the questions which have arisen as to the universality and necessity of the truths of mathematics, and the foundation of our knowledge of them. I would remark here that the connexion (if it exists) of arithmetic and algebra with the notion of time is far less obvious than that of geometry with the notion of space.

As to the former side, I am not making before you a defence of mathematics, but if I were I should desire to do it—in such manner as in the *Republic* Socrates was required to defend justice, quite irrespectively of the worldly advantages which may accompany a life of virtue and justice, and to show that, independently of all these, justice was a thing desirable in itself and for its own sake—*not* by speaking to you of the utility of mathematics in any of the questions of common life or of physical science. Still less would I speak of this utility before, I trust, a friendly audience, interested or willing to appreciate an interest in mathematics in itself and for its own sake. I would, on the contrary, rather consider the obligations of mathematics to these different subjects as the sources of mathematical theories now as remote from them, and in as different a region of thought—for instance, geometry from the measurement of land, or the Theory of Numbers from arithmetic—as a river at its mouth is from its mountain source.

On the other side, the general opinion has been and is that it is indeed by experience that we arrive at the truths of mathematics, but that experience is not their proper foundation: the mind itself contributes something. This is involved in the Platonic theory of reminiscence; looking at two things, trees or stones or anything else, which seem to us more or less equal, we arrive at the idea of equality: but we must have had this idea of equality before the time when first seeing the two things we were led to regard them as coming up more or less perfectly to this idea of equality; and the like as regards our idea of the beautiful, and in other cases.

The same view is expressed in the answer of Leibnitz, the *nisi intellectus ipse*, to the scholastic dictum, *nihil in intellectu quod non prius in sensu*: there is nothing in the intellect which was not first in sensation, except (said Leibnitz) the intellect itself. And so again in the *Critick of Pure Reason*, Kant's view is that while there is no doubt but that all our cognition begins with experience, we are nevertheless in possession of cognitions *a priori*, independent, not of this or that experience, but absolutely so of all experience, and in particular that the axioms of mathematics furnish an example of such cognitions *a priori*. Kant holds further that space is no empirical conception which has been derived from external experiences, but that in order that sensations may be referred to something external, the representation of space must already lie at the foundation; and that the external experience is itself first only possible by this representation of space. And in like manner time is no empirical conception which can be deduced from an experience, but it is a necessary representation lying at the foundation of all intuitions.

And so in regard to mathematics, Sir W. R. Hamilton, in an Introductory Lecture on Astronomy (1836), observes: "These purely mathematical sciences of algebra and geometry are sciences of the pure reason, deriving no weight and no assistance from experiment, and isolated or at least isolable from all outward and accidental phenomena. The idea of order with its subordinate ideas of number and figure, we must not indeed call innate ideas, if that phrase be defined to imply that all men must possess them with equal clearness and fulness: they are, however, ideas which seem to be so far born with us that the possession of them in any conceivable degree is only the development of our original powers, the unfolding of our proper humanity."

The general question of the ideas of space and time, the axioms and definitions of geometry, the axioms relating to number, and the nature of mathematical reasoning, are fully and ably discussed in Whewell's *Philosophy of the Inductive Sciences* (1840), which may be regarded as containing an exposition of the whole theory.

But it is maintained by John Stuart Mill that the truths of mathematics, in particular those of geometry, rest on experience; and as regards geometry, the same view is on very different grounds maintained by the mathematician Riemann.

It is not so easy as at first sight it appears to make out how far the views taken by Mill in his *System of Logic Ratiocinative and Inductive* (9th ed. 1879) are absolutely contradictory to those which have been spoken of; they profess to be so; there are most definite assertions (supported by argument), for instance, p. 263:—"It remains to enquire what is the ground of our belief in axioms, what is the evidence on which they rest. I answer, they are experimental truths, generalisations from experience. The proposition 'Two straight lines cannot enclose a space,' or, in other words, two straight lines which have once met cannot meet again, is an induction from the evidence of our senses." But I cannot help considering a previous argument (p. 259) as very materially modifying this absolute contradiction. After enquiring "Why are mathematics by almost all philosophers . . . considered to be independent of the evidence of experience and observation, and characterised as systems of necessary truth?" Mill proceeds (I quote the whole passage) as follows:—"The answer I conceive to be that this character of necessity ascribed to the truths of mathematics, and even (with some reservations to be hereafter made) the peculiar certainty ascribed to them, is a delusion, in order to sustain which it is necessary to suppose that those truths relate to and express the properties of purely imaginary objects. It is acknowledged that the conclusions of geometry are derived partly at least from the so-called definitions, and that these definitions are assumed to be correct representations, as far as they go, of the objects with which geometry is conversant. Now, we have pointed out that, from a definition as such, no proposition unless it be one concerning the meaning of a word can ever follow, and that what apparently follows from a definition, follows in reality from an implied assumption that there exists a real thing conformable thereto. This assumption in the case of the definitions of geometry is not strictly true: there exist no real things exactly conformable to the definitions. There exist no real points without magnitude, no lines without breadth, nor perfectly straight, no circles with all their radii exactly equal, nor squares with all their angles perfectly right. It will be said that the assumption does not extend to the actual but only to the possible existence of such things. I answer that according to every test we have of possibility they are not even possible. Their existence, so far as we can form any judgment, would seem to be inconsistent with the physical constitution of our planet at least, if not of the universal [*sic*]. To get rid of this difficulty and at the same time to save the credit of the supposed system of necessary truth, it is customary to say that the points, lines, circles and squares which are the subjects of geometry exist in our conceptions merely and are parts of our minds; which minds by working on their own materials construct an *a priori* science, the evidence of which is purely mental and has nothing to do with outward experience. By howsoever high authority this doctrine has been sanctioned, it appears to me psychologically incorrect. The points, lines and squares which anyone has in his mind are (as I apprehend) simply copies of the points, lines and squares which he has known in his experience. Our idea of a point I apprehend to be simply our idea of the *minimum visibile*, the small portion of surface which we can see. We can reason about a line as if it had no breadth, because we have a power which we can exercise over the operations of our minds: the power, when a perception is present to our senses or a conception to our intellects, of *attending* to a part only of that perception or conception instead of the whole. But we cannot *conceive* a line without breadth: we can form no mental picture of such a line; all the lines which we have in our mind are lines possessing breadth. If anyone doubt this, we may refer him to his own experience. I much question if anyone who fancies that he can conceive of a mathematical line thinks so from the evidence of his own consciousness. I suspect it is rather because he supposes that, unless such a perception be possible, mathematics could not exist as a science: a supposition which there will be no difficulty in showing to be groundless."

I think it may be at once conceded that the truths of geometry are truths precisely because they relate to and express the properties of what Mill calls "purely imaginary objects"; that these objects do not exist in Mill's sense, that they do not exist in nature, may also be granted; that they are "not even possible," if this means not possible in an existing nature, may also be granted. That we cannot "conceive" them depends on the meaning which we attach to the word conceive. I would myself say that the purely imaginary objects are the only realities, the *ὄντως ὄντα*, in regard to which the corresponding physical objects are as the shadows in the cave; and it is only by means of them that we are able to deny the existence of a corresponding physical object; if there is no conception of straightness, then it is meaningless to deny the existence of a perfectly straight line.

But at any rate the objects of geometrical truth are the so-called imaginary objects of Mill, and the truths of geometry are only true, and *a fortiori* are only necessarily true, in regard to these so-called imaginary objects; and these objects, points, lines, circles, &c., in the mathematical sense of the terms, have a likeness to and are represented more or less imperfectly, and from a geometer's point of view no matter how imperfectly, by corresponding physical points, lines, circles, &c. I shall have to return to geometry, and will then speak of Riemann, but I will first refer to another passage of the Logic.

Speaking of the truths of arithmetic, Mill says (p. 297) that even here there is one hypothetical element: "In all propositions concerning numbers a condition is implied without which none of them would be true, and that condition is an assumption which may be false. The condition is that $1 = 1$: that all the numbers are numbers of the same or of equal units." Here at least the assumption may be absolutely true; one shilling = one shilling in purchasing power, although they may not be absolutely of the same weight and fineness: but it is hardly necessary; one coin + one coin = two coins, even if the one be a shilling and the other a half-crown. In fact, whatever difficulty be raisable as to geometry, it seems to me that no similar difficulty applies to arithmetic; mathematician or not, we have each of us, in its most abstract form, the idea of a number; we can each of us appreciate the truth of a proposition in regard to numbers; and we cannot but see that a truth in regard to numbers is something different in kind from an experimental truth generalised from experience. Compare, for instance, the proposition that the sun, having already risen so many times, will rise to-morrow, and the next day, and the day after that, and so on; and the proposition that even and odd numbers succeed each other alternately *ad infinitum*: the latter at least seems to have the characters of universality and necessity. Or again, suppose a proposition observed to hold good for a long series of numbers, one thousand numbers, two thousand numbers, as the case may be: this is not only no proof, but it is absolutely no evidence, that the proposition is a true proposition, holding good for all numbers whatever; there are in the Theory of Numbers very remarkable instances of propositions observed to hold good for very long series of numbers and which are nevertheless untrue.

I pass in review certain mathematical theories.

In arithmetic and algebra, or say in analysis, the numbers or magnitudes which we represent by symbols are in the first instance ordinary (that is, positive) numbers or magnitudes. We have also in analysis and in analytical geometry *negative* magnitudes; there has been in regard to these plenty of philosophical discussion, and I might refer to Kant's paper, *Ueber die negativen Grössen in die Weltweisheit* (1763), but the notion of a negative magnitude has become quite a familiar one, and has extended itself into common phraseology. I may remark that it is used in a very refined manner in bookkeeping by double entry.

But it is far otherwise with the notion which is really the fundamental one (and I cannot too strongly emphasise the assertion) underlying and pervading the whole of modern analysis and geometry, that of imaginary magnitude in analysis and of imaginary space (or space as a *locus in quo* of imaginary points and figures) in geometry: I use in each case the word imaginary as including real. This has not been, so far as I am aware, a subject of philosophical discussion or enquiry. As regards the older metaphysical writers this would be quite accounted for by saying that they knew nothing, and were not bound to know anything, about it; but at present, and, considering the prominent position which the notion occupies—say even that the conclusion were that the notion belongs to mere technical mathematics, or has reference to nonentities in regard to which no science is possible, still it seems to me that (as a subject of philosophical discussion) the notion ought not to be thus ignored; it should at least be shown that there is a right to ignore it.

Although in logical order I should perhaps now speak of the notion just referred to, it will be convenient to speak first of some other quasi-geometrical notions; those of more-than-three-dimensional space, and of non-Euclidian two- and three-dimensional space, and also of the generalised notion of distance. It is in connexion with these that Riemann considered that our notion of space is founded on experience, or rather that it is only by experience that we know that our space is Euclidian space.

It is well known that Euclid's twelfth axiom, even in Playfair's form of it, has been considered as needing demonstration; and that Lobatschewsky constructed a perfectly consistent theory, wherein this axiom was assumed not to hold good, or say a system of non-Euclidian plane geometry. There is a like system of non-Euclidian solid geometry. My own view is that Euclid's twelfth axiom in Playfair's form of it does not need demonstration, but is part of our notion of space, of the physical space of our experience—the space, that is, which we become acquainted with by experience, but which is the representation lying at the foundation of all external experience. Riemann's view before referred to may I think be said to be that, having *in intellectu* a more general notion of space (in fact a notion of non-Euclidian space), we learn by experience that space (the physical space of our experience) is, if not exactly, at least to the highest degree of approximation, Euclidian space.

But suppose the physical space of our experience to be thus only approximately Euclidian space, what is the consequence which follows? *Not* that the propositions of geometry are only approximately true, but that they remain absolutely true in regard to that Euclidian space which has been so long regarded as being the physical space of our experience.

It is interesting to consider two different ways in which, without any modification at all of our notion of space, we can arrive at a system of non-Euclidian (plane or two-dimensional) geometry; and the doing so will, I think, throw some light on the whole question.

First, imagine the earth a perfectly smooth sphere; understand by a plane the surface of the earth, and by a line the apparently straight line (in fact, an arc of great circle) drawn on the surface; what experience would in the first instance teach would be Euclidian geometry; there would be intersecting lines which produced a few miles or so would seem to go on diverging: and apparently parallel lines which would exhibit no tendency to approach each other; and the inhabitants might very well conceive that they had by experience established the axiom that two straight lines cannot enclose a space, and the axiom as to parallel lines. A more extended experience and more accurate measurements would teach them that the axioms were each of them false; and that any two lines if produced far enough each way, would meet in two points: they would in fact arrive at a spherical geometry, accurately representing the properties of the two-dimensional space of their experience. But their original Euclidian geometry would not the less be a true system: only it would apply to an ideal space, not the space of their experience.

Secondly consider an ordinary, indefinitely extended plane; and let us modify only the notion of distance. We measure distance, say, by a yard measure or a foot rule, anything which is short enough to make the fractions of it of no consequence (in mathematical language, by an infinitesimal element of length); imagine, then, the length of this rule constantly changing (as it might do by an alteration of temperature), but under the condition that its actual length shall depend only on its situation on the plane and on its direction: viz. if for a given situation and direction it has a certain length, then whenever it comes back to the same situation and direction it must have the same length. The distance along a given straight or curved line between any two points could then be measured in the ordinary manner with this rule, and would have a perfectly determinate value: it could be measured over and over again, and would always be the same; but of course it would be the distance, not in the ordinary acceptation of the term, but in quite a different acceptation. Or in a somewhat different way: if the rate of progress from a given point in a given direction be conceived as depending only on the configuration of the ground, and the distance along a given path between any two points thereof be measured by the time required for traversing it, then in this way also the distance would have a perfectly determinate value; but it would be a distance, not in the ordinary acceptation of the term, but in quite a different acceptation. And corresponding to the new notion of distance we should have a new non-Euclidian system of plane geometry; all theorems involving the notion of distance would be altered.

We may proceed further. Suppose that as the rule moves away from a fixed central point of the plane it becomes shorter and shorter; if this shortening takes place with sufficient rapidity, it may very well be that a distance which in the ordinary sense of the word is finite will in the new sense be infinite; no number of repetitions of the length of the ever-shortening rule will be sufficient to cover it. There will be surrounding the central point a certain finite area such that (in the new acceptation of the term distance) each point of the boundary thereof will be at an infinite distance from the central point; the points outside this area you cannot by any means arrive at with your rule; they will form a *terra incognita,* or rather an unknowable land: in mathematical language, an imaginary or impossible space: and the plane space of the theory will be that within the finite area—that is, it will be finite instead of infinite.

We thus with a proper law of shortening arrive at a system of non-Euclidian geometry which is essentially that of Lobatschewsky. But in so obtaining it we put out of sight its relation to spherical geometry: the three geometries (spherical, Euclidian, and Lobatschewsky's) should be regarded as members of a system: viz. they are the geometries of a plane (two-dimensional) space of constant positive curvature, zero curvature, and constant negative curvature respectively; or again, they are the plane geometries corresponding to three different notions of distance; in this point of view they are Klein's elliptic, parabolic, and hyperbolic geometries respectively.

Next as regards solid geometry: we can by a modification of the notion of distance (such as has just been explained in regard to Lobatschewsky's system) pass from our present system to a non-Euclidian system; for the other mode of passing to a non-Euclidian system, it would be necessary to regard our space as a flat three-dimensional space existing in a space of four dimensions (i.e., as the analogue of a plane existing in ordinary space); and to substitute for such flat three-dimensional space a curved three-dimensional space, say of constant positive or negative curvature. In regarding the physical space of our experience as possibly non-Euclidian, Riemann's idea seems to be that of modifying the notion of distance, not that of treating it as a locus in four-dimensional space.

I have just come to speak of four-dimensional space. What meaning do we attach to it? Or can we attach to it any meaning? It may be at once admitted that we cannot conceive of a fourth dimension of space; that space as we conceive of it, and the physical space of our experience, are alike three-dimensional; but we can, I think, conceive of space as being two- or even one-dimensional; we can imagine rational beings living in a one-dimensional space (a line) or in a two-dimensional space (a surface), and conceiving of space accordingly, and to whom, therefore, a two-dimensional space, or (as the case may be) a three-dimensional space would be as inconceivable as a four-dimensional space is to us. And very curious speculative questions arise. Suppose the one-dimensional space a right line, and that it afterwards becomes a curved line: would there be any indication of the change? Or, if originally a curved line, would there be anything to suggest to them that it was not a right line? Probably not, for a one-dimensional geometry hardly exists. But let the space be two-dimensional, and imagine it originally a plane, and afterwards bent or converted into a curved surface (converted, that is, into some form of developable surface): or imagine it originally a developable or curved surface. In the former case there should be an indication of the change, for the geometry originally applicable to the space of their experience (our own Euclidian geometry) would cease to be applicable; but the change could not be apprehended by them as a bending or deformation of the plane, for this would imply the notion of a three-dimensional space in which this bending or deformation could take place. In the latter case their geometry would be that appropriate to the developable or curved surface which is their space: viz. this would be their Euclidian geometry: would they ever have arrived at our own more simple system? But take the case where the two-dimensional space is a plane, and imagine the beings of such a space familiar with our own Euclidian plane geometry; if, a third dimension being still inconceivable by them, they were by their geometry or otherwise led to the notion of it, there would be nothing to prevent them from forming a science such as our own science of three-dimensional geometry.

Evidently all the foregoing questions present themselves in regard to ourselves, and to three-dimensional space as we conceive of it, and as the physical space of our experience. And I need hardly say that the first step is the difficulty, and that granting a fourth dimension we may assume as many more dimensions as we please. But whatever answer be given to them, we have, as a branch of mathematics, potentially, if not actually, an analytical geometry of n-dimensional space. I shall have to speak again upon this.

Coming now to the fundamental notion already referred to, that of imaginary magnitude in analysis and imaginary space in geometry: I connect this with two great discoveries in mathematics made in the first half of the seventeenth century, Harriot's representation of an equation in the form $f(x) = 0$, and the consequent notion of the roots of an equation as derived from the linear factors of $f(x)$, (Harriot, 1560—1621: his *Algebra*, published after his death, has the date 1631), and Descartes' method of coordinates, as given in the *Géométrie*, forming a short supplement to his *Traité de la Méthode, etc.*, (Leyden, 1637).

Taking the coefficients of an equation to be real magnitudes, it at once follows from Harriot's form of an equation that an equation of the order n ought to have n roots. But it is by no means true that there are always n real roots. In particular, an equation of the second order, or quadric equation, may have no real root; but if we assume the existence of a root i of the quadric equation $x^2 + 1 = 0$, then the other root is $= -i$; and it is easily seen that every quadric equation (with real coefficients as before) has two roots, $a \pm bi$, where a and b are real magnitudes. We are thus led to the conception of an imaginary magnitude, $a + bi$, where a and b are real magnitudes, each susceptible of any positive or negative value, zero included. The general theorem is that, taking the coefficients of the equation to be imaginary magnitudes, then an equation of the order n has always n roots, each of them an imaginary magnitude, and it thus appears that the foregoing form $a + bi$ of imaginary magnitude is the only one that presents itself. Such imaginary magnitudes may be added or multiplied together or dealt with in any manner; the result is always a like imaginary magnitude. They are thus the magnitudes which are considered in analysis, and analysis is the science of such magnitudes. Observe the leading character that the imaginary magnitude $a + bi$ is a magnitude composed of the two real magnitudes a and b (in the case $b = 0$ it is the real magnitude a, and in the case $a = 0$ it is the pure imaginary magnitude bi). The idea is that of considering, in place of real magnitudes, these imaginary or complex magnitudes $a + bi$.

In the Cartesian geometry a curve is determined by means of the equation existing between the coordinates (x, y) of any point thereof. In the case of a right line, this equation is linear; in the case of a circle, or more generally of a conic, the equation is of the second order; and generally, when the equation is of the order n, the curve which it represents is said to be a curve of the order n. In the case of two given curves, there are thus two equations satisfied by the coordinates (x, y) of the several points of intersection, and these give rise to an equation of a certain order for the coordinate x or y of a point of intersection. In the case of a straight line and a circle, this is a quadric equation; it has two roots, real or imaginary. There are thus two values, say of x, and to each of these corresponds a single value of y. There are therefore two points of intersection—viz. a straight line and a circle intersect *always* in two points, real or imaginary. It is in this way that we are led analytically to the notion of imaginary points in geometry. The conclusion as to the two points of intersection cannot be contradicted by experience: take a sheet of paper and draw on it the straight line and circle, and try. But you might say, or at least be strongly tempted to say, that it is meaningless. The question of course arises, What is the meaning of an imaginary point? and further, In what manner can the notion be arrived at geometrically?

There is a well-known construction in perspective for drawing lines through the intersection of two lines, which are so nearly parallel as not to meet within the limits of the sheet of paper. You have two given lines which do not meet, and you draw a third line, which, when the lines are all of them produced, is found to pass through the intersection of the given lines. If instead of lines we have two circular arcs not meeting each other, then we can, by means of these arcs, construct a line; and if on completing the circles it is found that the circles intersect each other in two real points, then it will be found that the line passes through these two points: if the circles appear not to intersect, then the line will appear not to intersect either of the circles. But the geometrical construction being in each case the same, we say that in the second case also the line passes through the two intersections of the circles.

Of course it may be said in reply that the conclusion is a very natural one, provided we assume the existence of imaginary points; and that, this assumption not being made, then, if the circles do not intersect, it is meaningless to assert that the line passes through their points of intersection. The difficulty is not got over by the analytical method before referred to, for this introduces difficulties of its own: is there in a plane a point the coordinates of which have given imaginary values? As a matter of fact, we do consider in plane geometry imaginary points introduced into the theory analytically or geometrically as above.

The like considerations apply to solid geometry, and we thus arrive at the notion of imaginary space as a *locus in quo* of imaginary points and figures.

I have used the word imaginary rather than complex, and I repeat that the word has been used as including real. But, this once understood, the word becomes in many cases superfluous, and the use of it would even be misleading. Thus, "a problem has so many solutions": this means, so many imaginary (including real) solutions. But if it were said that the problem had "so many imaginary solutions," the word "imaginary" would here be understood to be used in opposition to real. I give this explanation the better to point out how wide the application of the notion of the imaginary is—viz. (unless expressly or by implication excluded), it is a notion implied and presupposed in all the conclusions of modern analysis and geometry. It is, as I have said, the fundamental notion underlying and pervading the whole of these branches of mathematical science.

I shall speak later on of the great extension which is thereby given to geometry, but I wish now to consider the effect as regards the theory of a function. In the original point of view, and for the original purposes, a function, algebraic or transcendental, such as $\sqrt{x}$, $\sin x$, or $\log x$, was considered as known, when the value was known for every real value (positive or negative) of the argument; or if for any such values the value of the function became imaginary, then it was enough to know that for such values of the argument there was no real value of the function. But now this is not enough, and to know the function means to know its value—of course, in general, an imaginary value $X+iY$,—for every imaginary value $x+iy$ whatever of the argument.

And this leads naturally to the question of the geometrical representation of an imaginary variable. We represent the imaginary variable $x+iy$ by means of a point in a plane, the coordinates of which are (x, y). This idea, due to Gauss, dates from about the year 1831. We thus picture to ourselves the succession of values of the imaginary variable $x+iy$ by means of the motion of the representative point: for instance, the succession of values corresponding to the motion of the point along a closed curve to its original position. The value $X+iY$ of the function can of course be represented by means of a point (taken for greater convenience in a different plane), the coordinates of which are X, Y.

We may consider in general two points, moving each in its own plane, so that the position of one of them determines the position of the other, and consequently the motion of the one determines the motion of the other: for instance, the two points may be the tracing-point and the pencil of a pentagraph. You may with the first point draw any figure you please, there will be a corresponding figure drawn by the second point: for a good pentagraph, a copy on a different scale (it may be); for a badly-adjusted pentagraph, a distorted copy: but the one figure will always be a sort of copy of the first, so that to each point of the one figure there will correspond a point of the other figure.

In the case above referred to, where one point represents the value $x+iy$ of the imaginary variable and the other the value $X+iY$ of some function $\phi(x+iy)$ of that variable, there is a remarkable relation between the two figures: this is the relation of orthomorphic projection, the same which presents itself between a portion of the earth's surface, and the representation thereof by a map on the stereographic projection or on Mercator's projection—viz. any indefinitely small area of the one figure is represented in the other figure by an indefinitely small area of the same shape. There will possibly be for different parts of the figure great variations of scale, but the shape will be unaltered; if for the one area the boundary is a circle, then for the other area the boundary will be a circle; if for one it is an equilateral triangle, then for the other it will be an equilateral triangle.

I have for simplicity assumed that to each point of either figure there corresponds one, and only one, point of the other figure; but the general case is that to each point of either figure there corresponds a determinate number of points in the other figure; and we have thence arising new and very complicated relations which I must just refer to. Suppose that to each point of the first figure there correspond in the second figure two points: say one of them is a red point, the other a blue point; so that, speaking roughly, the second figure consists of two copies of the first figure, a red copy and a blue copy, the one superimposed on the other. But the difficulty is that the two copies cannot be kept distinct from each other. If we consider in the first figure a closed curve of any kind—say, for shortness, an oval—this will be in the second figure represented in some cases by a red oval and a blue oval, but in other cases by an oval half red and half blue; or, what comes to the same thing, if in the first figure we consider a point which moves continuously in any manner, at last returning to its original position, and attempt to follow the corresponding points in the second figure, then it may very well happen that, for the corresponding point of either colour, there will be abrupt changes of position, or say jumps, from one position to another; so that, to obtain in the second figure a continuous path, we must at intervals allow the point to change from red to blue, or from blue to red. There are in the first figure certain critical points called branch-points (*Verzweigungspunkte*), and a system of lines connecting these, by means of which the colours in the second figure are determined; but it is not possible for me to go further into the theory at present. The notion of colour has of course been introduced only for facility of expression; it may be proper to add that in speaking of the two figures I have been following Briot and Bouquet rather than Riemann, whose representation of the function of an imaginary variable is a different one.

I have been speaking of an imaginary variable $(x+iy)$, and of a function $\phi(x+iy)=X+iY$ of that variable, but the theory may equally well be stated in regard to a plane curve: in fact, the $x+iy$ and the $X+iY$ are two imaginary variables connected by an equation; say their values are u and v, connected by an equation $F(u, v)=0$; then, regarding u, v as the coordinates of a point *in plano*, this will be a point on the curve represented by the equation. The curve, in the widest sense of the expression, is the whole series of points, real or imaginary, the coordinates of which satisfy the equation, and these are exhibited by the foregoing corresponding figures in two planes; but in the ordinary sense the curve is the series of real points, with coordinates u, v, which satisfy the equation.

In geometry it is the curve, whether defined by means of its equation, or in any other manner, which is the subject for contemplation and study. But we also use the curve as a representation of its equation—that is, of the relation existing between two magnitudes x, y, which are taken as the coordinates of a point on the curve. Such employment of a curve for all sorts of purposes—the fluctuations of the barometer, the Cambridge boat races, or the Funds—is familiar to most of you. It is in like manner convenient in analysis, for exhibiting the relations between any three magnitudes x, y, z, to regard them as the coordinates of a point in space; and, on the like ground, we should at least wish to regard any four or more magnitudes as the coordinates of a point in space of a corresponding number of dimensions. Starting with the hypothesis of such a space, and of points therein each determined by means of its coordinates, it is found possible to establish a system of n-dimensional geometry analogous in every respect to our two- and three-dimensional geometries, and to a very considerable extent serving to exhibit the relations of the variables. To quote from my memoir "On Abstract Geometry" (1869), [413]: "The science presents itself in two ways: as a legitimate extension of the ordinary two- and three-dimensional geometries, and as a need in these geometries and in analysis generally. In fact, whenever we are concerned with quantities connected in any manner, and which are considered as variable or determinable, then the nature of the connexion between the quantities is frequently rendered more intelligible by regarding them (if two or three in number) as the coordinates of a point in a plane or in space. For more than three quantities there is, from the greater complexity of the case, the greater need of such a representation; but this can only be obtained by means of the notion of a space of the proper dimensionality; and to use such representation we require a corresponding geometry. An important instance in plane geometry has already presented itself in the question of the number of curves which satisfy given conditions; the conditions imply relations between the coefficients in the equation of the curve; and for the better understanding of these relations it was expedient to consider the coefficients as the coordinates of a point in a space of the proper dimensionality."

It is to be borne in mind that the space, whatever its dimensionality may be, must always be regarded as an imaginary or complex space such as the two- or three-dimensional space of ordinary geometry; the advantages of the representation would otherwise altogether fail to be obtained.

I have spoken throughout of Cartesian coordinates; instead of these, it is in plane geometry not unusual to employ trilinear coordinates, and these may be regarded as absolutely undetermined in their magnitude—viz. we may take x, y, z to be, not equal, but only proportional to the distances of a point from three given lines; the ratios of the coordinates (x, y, z) determine the point; and so in one-dimensional geometry, we may have a point determined by the ratio of its two coordinates x, y, these coordinates being proportional to the distances of the point from two fixed points; and generally in n-dimensional geometry a point will be determined by the ratios of the $(n+1)$ coordinates $(x, y, z, \ldots)$. The corresponding analytical change is in the expression of the original magnitudes as fractions with a common denominator; we thus, in place of rational and integral non-homogeneous functions of the original variables, introduce rational and integral homogeneous functions (quantics) of the next succeeding number of variables—viz. we have binary quantics corresponding to one-dimensional geometry, ternary to two-dimensional geometry, and so on.

It is a digression, but I wish to speak of the representation of points or figures in space upon a plane. In perspective, we represent a point in space by means of the intersection with the plane of the picture (suppose a pane of glass) of the line drawn from the point to the eye, and doing this for each point of the object we obtain a representation or picture of the object. But such representation is an imperfect one, as not determining the object: we cannot by means of the picture alone find out the form of the object; in fact, for a given point of the picture the corresponding point of the object is not a determinate point, but it is a point anywhere in the line joining the eye with the point of the picture. To determine the object we need two pictures, such as we have in a plan and elevation, or, what is the same thing, in a representation on the system of Monge's descriptive geometry. But it is theoretically more simple to consider two projections on the same plane, with different positions of the eye: the point in space is here represented on the plane by means of two points which are such that the line joining them passes through a fixed point of the plane (this point is in fact the intersection with the plane of the picture of the line joining the two positions of the eye); the figure in space is thus represented on the plane by two figures, which are such that the lines joining corresponding points of the two figures pass always through the fixed point. And such two figures completely replace the figure in space; we can by means of them perform on the plane any constructions which could be performed on the figure in space, and employ them in the demonstration of properties relating to such figure. A curious extension has recently been made: two figures in space such that the lines joining corresponding points pass through a fixed point have been regarded by the Italian geometer Veronese as representations of a figure in four-dimensional space, and have been used for the demonstration of properties of such figure.

I referred to the connexion of Mathematics with the notions of space and time, but I have hardly spoken of time. It is, I believe, usually considered that the notion of number is derived from that of time; thus Whewell in the work referred to, p. xx, says number is a modification of the conception of repetition, which belongs to that of *time*. I cannot recognise that this is so: it seems to me that we have (independently, I should say, of space or time, and in any case not more depending on time than on space) the notion of plurality; we think of, say, the letters a, b, c, &c., and thence in the case of a finite set—for instance *a*, *b*, *c*, *d*, *e*—we arrive at the notion of number; coordinating them one by one with any other set of things, or, suppose, with the words first, second, &c., we find that the last of them goes with the word fifth, and we say that the number of things is = five: the notion of cardinal number would thus appear to be derived from that of ordinal number.

Questions of combination and arrangement present themselves, and it might be possible from the mere notion of plurality to develope a branch of mathematical science; this, however, would apparently be of a very limited extent, and it is difficult *not* to introduce into it the notion of number; in fact, in the case of a finite set of things, to avoid asking the question, How many? If we do this, we have a large enough subject, including the partition of numbers, which Sylvester has called Tactic.

From the notion thus arrived at of an integer number, we pass to that of a fractional number, and we see how by means of these the ratio of any two concrete magnitudes of the same kind can be expressed, not with absolute accuracy, but with any degree of accuracy we please: for instance, a length is so many feet, tenths of a foot, hundredths, thousandths, &c.; subdivide as you please, *non constat* that the length can be expressed accurately, we have in fact incommensurables; as to the part which these play in the Theory of Numbers, I shall have to speak presently: for the moment I am only concerned with them in so far as they show that we cannot from the notion of number pass to that which is required in analysis, the notion of an abstract (real and positive) magnitude susceptible of continuous variation. The difficulty is got over by a Postulate. We consider an abstract (real and positive) magnitude, and regard it as susceptible of continuous variation, without in anywise concerning ourselves about the actual expression of the magnitude by a numerical fraction or otherwise.

There is an interesting paper by Sir W. R. Hamilton, "Theory of Conjugate Functions, or Algebraical Couples: with a preliminary and elementary Essay on Algebra as the Science of Pure Time," 1833—35 (*Trans. R. I. Acad.* t. XVII.), in which, as appears by the title, he purposes to show that algebra is the science of pure time. He states there, in the General Introductory Remarks, his conclusions: first, that the notion of time is connected with existing algebra; second, that this notion or intuition of time may be unfolded into an independent pure science; and, third, that the science of pure time thus unfolded is coextensive and identical with algebra, so far as algebra itself is a science; and to sustain his first conclusion he remarks that "the history of algebraic science shows that the most remarkable discoveries in it have been made either expressly through the notion of *time*, or through the closely connected (and in some sort coincident) notion of continuous progression. It is the genius of algebra to consider what it reasons upon as *flowing*, as it was the genius of geometry to consider what it reasoned on as *fixed*. . . . And generally the revolution which Newton made in the higher parts of both pure and applied algebra was founded mainly on the notion of *fluxion*, which involves the notion of *time*." Hamilton uses the term algebra in a very wide sense, but whatever else he includes under it, he includes all that in contradistinction to the Differential Calculus would be called algebra. Using the word in this restricted sense, I cannot myself recognise the connexion of algebra with the notion of time: granting that the notion of continuous progression presents itself, and is of importance, I do not see that it is in anywise the fundamental notion of the science. And still less can I appreciate the manner in which the author connects with the notion of time his algebraical couple, or imaginary magnitude $a+bi$ ($a+b\sqrt{-1}$, as written in the memoir).

I would go further: the notion of continuous variation is a very fundamental one, made a foundation in the Calculus of Fluxions (if not always so in the Differential Calculus) and presenting itself or implied throughout in mathematics: and it may be said that a change of any kind takes place only in time; it seems to me, however, that the changes which we consider in mathematics are for the most part considered quite irrespectively of time.

It appears to me that we do not have in Mathematics the notion of time until we bring it there: and that even in kinematics (the science of motion) we have very little to do with it; the motion is a hypothetical one; if the system be regarded as actually moving, the rate of motion is altogether undetermined and immaterial. The relative rates of motion of the different points of the system are nothing else than the ratios of purely geometrical quantities, the indefinitely short distances simultaneously described, or which might be simultaneously described, by these points respectively. But whether the notion of time does or does not sooner enter into mathematics, we at any rate have the notion in Mechanics, and along with it several other new notions.

Regarding Mechanics as divided into Statics and Dynamics, we have in dynamics the notion of time, and in connexion with it that of velocity: we have in statics and dynamics the notion of force; and also a notion which in its most general form I would call that of corpus: viz. this may be, the material point or particle, the flexible inextensible string or surface, or the rigid body, of ordinary mechanics; the incompressible perfect fluid of hydrostatics and hydrodynamics; the ether of any undulatory theory; or any other imaginable corpus; for instance, one really deserving of consideration in any general treatise of mechanics is a developable or skew surface with absolutely rigid generating lines, but which can be bent about these generating lines, so that the element of surface between two consecutive lines rotates as a whole about one of them. We have besides, in dynamics necessarily, the notion of mass or inertia.

We seem to be thus passing out of pure mathematics into physical science; but it is difficult to draw the line of separation, or to say of large portions of the *Principia*, and the *Mécanique céleste*, or of the whole of the *Mécanique analytique*, that they are not pure mathematics. It may be contended that we first come to physics when we attempt to make out the character of the corpus as it exists in nature. I do not at present speak of any physical theories which cannot be brought under the foregoing conception of mechanics.

I must return to the Theory of Numbers; the fundamental idea is here integer number: in the first instance positive integer number, but which may be extended to include negative integer number and zero. We have the notion of a product, and that of a prime number, which is not a product of other numbers; and thence also that of a number as the product of a determinate system of prime factors. We have here the elements of a theory in many respects analogous to algebra: an equation is to be solved—that is, we have to find the integer values (if any) which satisfy the equation; and so in other cases: the congruence notation, although of the very highest importance, does not affect the character of the theory.

But as already noticed we have incommensurables, and the consideration of these gives rise to a new universe of theory. We may take into consideration any surd number such as $\sqrt{2}$, and so consider numbers of the form $a+b\sqrt{2}$, (a and b any positive or negative integer numbers not excluding zero); calling these integer numbers, every problem which before presented itself in regard to integer numbers in the original and ordinary sense of the word presents itself equally in regard to integer numbers in this new sense of the word; of course all definitions must be altered accordingly: an ordinary integer, which is in the ordinary sense of the word a prime number, may very well be the product of two integers of the form $a+b\sqrt{2}$, and consequently not a prime number in the new sense of the word. Among the incommensurables which can be thus introduced into the Theory of Numbers (and which was in fact *first* so introduced) we have the imaginary i of ordinary analysis: viz. we may consider numbers $a+bi$ (a and b ordinary positive or negative integers, not excluding zero), and, calling these integer numbers, establish in regard to them a theory analogous to that which exists for ordinary real integers. The point which I wish to bring out is that the imaginary i does not in the Theory of Numbers occupy a unique position, such as it does in analysis and geometry; it is in the Theory of Numbers one out of an indefinite multitude of incommensurables.

I said that I would speak to you, not of the utility of mathematics in any of the questions of common life or of physical science, but rather of the obligations of mathematics to these different subjects. The consideration which thus presents itself is in a great measure that of the history of the development of the different branches of mathematical science in connexion with the older physical sciences, Astronomy and Mechanics: the mathematical theory is in the first instance suggested by some question of common life or of physical science, is pursued and studied quite independently thereof, and perhaps after a long interval comes in contact with it, or with quite a different question. Geometry and algebra must, I think, be considered as each of them originating in connexion with objects or questions of common life—geometry, notwithstanding its name, hardly in the measurement of land, but rather from the contemplation of such forms as the straight line, the circle, the ball, the top (or sugar-loaf): the Greek geometers appropriated for the geometrical forms corresponding to the last two of these, the words *σφαῖρα* and *κῶνος*, our sphere and cone, and they extended the word cone to mean the complete figure obtained by producing the straight lines of the surface both ways indefinitely. And so algebra would seem to have arisen from the sort of easy puzzles in regard to numbers which may be made, either in the picturesque forms of the Bija-Ganita with its maiden with the beautiful locks, and its swarms of bees amid the fragrant blossoms, and the one queen-bee left humming around the lotus flower; or in the more prosaic form in which a student has presented to him in a modern text-book a problem leading to a simple equation.

The Greek geometry may be regarded as beginning with Plato (B.C. 430—347): the notions of geometrical analysis, loci, and the conic sections are attributed to him, and there are in his Dialogues many very interesting allusions to mathematical questions: in particular the passage in the *Theætetus*, where he affirms the incommensurability of the sides of certain squares. But the earliest extant writings are those of Euclid (B.C. 285): there is hardly anything in mathematics more beautiful than his wondrous fifth book; and he has also in the seventh, eighth, ninth and tenth books fully and ably developed the first principles of the Theory of Numbers, including the theory of incommensurables. We have next Apollonius (about B.C. 247), and Archimedes (B.C. 287—212), both geometers of the highest merit, and the latter of them the founder of the science of statics (including therein hydrostatics): his dictum about the lever, his "Εὕρηκα," and the story of the defence of Syracuse, are well known. Following these we have a worthy series of names, including the astronomers Hipparchus (B.C. 150) and Ptolemy (A.D. 125), and ending, say, with Pappus (A.D. 400), but continued by their Arabian commentators, and the Italian and other European geometers of the sixteenth century and later, who pursued the Greek geometry.

The Greek arithmetic was, from the want of a proper notation, singularly cumbrous and difficult; and it was for astronomical purposes superseded by the sexagesimal arithmetic, attributed to Ptolemy, but probably known before his time. The use of the present so-called Arabic figures became general among Arabian writers on arithmetic and astronomy about the middle of the tenth century, but was not introduced into Europe until about two centuries later. Algebra among the Greeks is represented almost exclusively by the treatise of Diophantus (A.D. 150), in fact a work on the Theory of Numbers containing questions relating to square and cube numbers, and other properties of numbers, with their solutions; this has no historical connexion with the later algebra, introduced into Italy from the East by Leonardi Bonacci of Pisa (A.D. 1202—1208) and successfully cultivated in the fifteenth and sixteenth centuries by Lucas Paciolus, or de Burgo, Tartaglia, Cardan, and Ferrari. Later on, we have Vieta (1540—1603), Harriot, already referred to, Wallis, and others.

Astronomy is of course intimately connected with geometry; the most simple facts of observation of the heavenly bodies can only be *stated* in geometrical language: for instance, that the stars describe circles about the pole-star, or that the different positions of the sun among the fixed stars in the course of the year form a circle. For astronomical calculations it was found necessary to determine the arc of a circle by means of its chord: the notion is as old as Hipparchus, a work of whom is referred to as consisting of twelve books on the chords of circular arcs; we have (A.D. 125) Ptolemy's *Almagest*, the first book of which contains a table of arcs and chords with the method of construction; and among other theorems on the subject he gives there the theorem afterwards inserted in Euclid (Book VI. Prop. D) relating to the rectangle contained by the diagonals of a quadrilateral inscribed in a circle. The Arabians made the improvement of using in place of the chord of an arc the sine, or half chord, of double the arc; and so brought the theory into the form in which it is used in modern trigonometry: the before-mentioned theorem of Ptolemy, or rather a particular case of it, translated into the notation of sines, gives the expression for the sine of the sum of two arcs in terms of the sines and cosines of the component arcs; and it is thus the fundamental theorem on the subject. We have in the fifteenth and sixteenth centuries a series of mathematicians who with wonderful enthusiasm and perseverance calculated tables of the trigonometrical or circular functions, Purbach, Müller or Regiomontanus, Copernicus, Reinhold, Maurolycus, Vieta, and many others; the tabulations of the functions tangent and secant are due to Reinhold and Maurolycus respectively.

Logarithms were invented, not exclusively with reference to the calculation of trigonometrical tables, but in order to facilitate numerical calculations generally; the invention is due to John Napier of Merchiston, who died in 1618 at 67 years of age; the notion was based upon refined mathematical reasoning on the comparison of the spaces described by two points, the one moving with a uniform velocity, the other with a velocity varying according to a given law. It is to be observed that Napier's logarithms were nearly but not exactly those which are now called (sometimes Napierian, but more usually) hyperbolic logarithms—those to the base e; and that the change to the base 10 (the great step by which the invention was perfected for the object in view) was indicated by Napier but actually made by Henry Briggs, afterwards Savilian Professor at Oxford (d. 1630). But it is the hyperbolic logarithm which is mathematically important. The direct function e^x or $\exp. x$, which has for its inverse the hyperbolic logarithm, presented itself, but not in a prominent way. Tables were calculated of the logarithms of numbers, and of those of the trigonometrical functions.

The circular functions and the logarithm were thus invented each for a practical purpose, separately and without any proper connexion with each other. The functions are connected through the theory of imaginaries and form together a group of the utmost importance throughout mathematics: but this is mathematical theory; the obligation of mathematics is for the discovery of the functions.

Forms of spirals presented themselves in Greek architecture, and the curves were considered mathematically by Archimedes; the Greek geometers invented some other curves, more or less interesting, but recondite enough in their origin. A curve which might have presented itself to anybody, that described by a point in the circumference of a rolling carriage-wheel, was first noticed by Mersenne in 1615, and is the curve afterwards considered by Roberval, Pascal, and others under the name of the Roulette, otherwise the Cycloid. Pascal (1623—1662) wrote at the age of seventeen his *Essais pour les Coniques* in seven short pages, full of new views on these curves, and in which he gives, in a paragraph of eight lines, his theorem of the inscribed hexagon.

Kepler (1571—1630) by his empirical determination of the laws of planetary motion, brought into connexion with astronomy one of the forms of conic, the ellipse, and established a foundation for the theory of gravitation. Contemporary with him for most of his life, we have Galileo (1564—1642), the founder of the science of dynamics; and closely following upon Galileo we have Isaac Newton (1643—1727): the *Philosophiæ naturalis Principia Mathematica* known as the *Principia* was first published in 1687.

The physical, statical, or dynamical questions which presented themselves before the publication of the *Principia* were of no particular mathematical difficulty, but it is quite otherwise with the crowd of interesting questions arising out of the theory of gravitation, and which, in becoming the subject of mathematical investigation, have contributed very much to the advance of mathematics. We have the problem of two bodies, or what is the same thing, that of the motion of a particle about a fixed centre of force, for any law of force; we have also the (mathematically very interesting) problem of the motion of a body attracted to two or more fixed centres of force; then, next preceding that of the actual solar system—the problem of three bodies; this has ever been and is far beyond the power of mathematics, and it is in the lunar and planetary theories replaced by what is mathematically a different problem, that of the motion of a body under the action of a principal central force and a disturbing force: or (in one mode of treatment) by the problem of disturbed elliptic motion. I would remark that we have here an instance in which an astronomical fact, the observed slow variation of the orbit of a planet, has directly suggested a mathematical method, applied to other dynamical problems, and which is the basis of very extensive modern investigations in regard to systems of differential equations. Again, immediately arising out of the theory of gravitation, we have the problem of finding the attraction of a solid body of any given form upon a particle, solved by Newton in the case of a homogeneous sphere, but which is far more difficult in the next succeeding cases of the spheroid of revolution (very ably treated by Maclaurin) and of the ellipsoid of three unequal axes: there is perhaps no problem of mathematics which has been treated by as great a variety of methods, or has given rise to so much interesting investigation as this last problem of the attraction of an ellipsoid upon an interior or exterior point. It was a dynamical problem, that of vibrating strings, by which Lagrange was led to the theory of the representation of a function as the sum of a series of multiple sines and cosines; and connected with this we have the expansions in terms of Legendre's functions P_n, suggested to him by the question just referred to of the attraction of an ellipsoid; the subsequent investigations of Laplace on the attractions of bodies differing slightly from the sphere led to the functions of two variables called Laplace's functions. I have been speaking of ellipsoids, but the general theory is that of attractions, which has become a very wide branch of modern mathematics; associated with it we have in particular the names of Gauss, Lejeune-Dirichlet, and Green; and I must not omit to mention that the theory is now one relating to n-dimensional space. Another great problem of celestial mechanics, that of the motion of the earth about its centre of gravity, in the most simple case, that of a body not acted upon by any forces, is a very interesting one in the mathematical point of view.

I may mention a few other instances where a practical or physical question has connected itself with the development of mathematical theory. I have spoken of two map projections—the stereographic, dating from Ptolemy; and Mercator's projection, invented by Edward Wright about the year 1600: each of these, as a particular case of the orthomorphic projection, belongs to the theory of the geometrical representation of an imaginary variable. I have spoken also of perspective, and of the representation of solid figures employed in Monge's descriptive geometry. Monge, it is well known, is the author of the geometrical theory of the curvature of surfaces and of curves of curvature: he was led to this theory by a problem of earthwork; from a given area, covered with earth of uniform thickness, to carry the earth and distribute it over an equal given area, with the least amount of cartage. For the solution of the corresponding problem in solid geometry he had to consider the intersecting normals of a surface, and so arrived at the curves of curvature. (See his "Mémoire sur les Déblais et les Remblais," *Mém. de l'Acad.*, 1781.) The normals of a surface are, again, a particular case of a doubly infinite system of lines, and are so connected with the modern theories of congruences and complexes.

The undulatory theory of light led to Fresnel's wave-surface, a surface of the fourth order, by far the most interesting one which had then presented itself. A geometrical property of this surface, that of having tangent planes each touching it along a plane curve (in fact, a circle), gave to Sir W. R. Hamilton the theory of conical refraction. The wave-surface is now regarded in geometry as a particular case of Kummer's quartic surface, with sixteen conical points and sixteen singular tangent planes.

My imperfect acquaintance as well with the mathematics as the physics prevents me from speaking of the benefits which the theory of Partial Differential Equations has received from the hydrodynamical theory of vortex motion, and from the great physical theories of heat, electricity, magnetism, and energy.

It is difficult to give an idea of the vast extent of modern mathematics. This word "extent" is not the right one: I mean extent crowded with beautiful detail—not an extent of mere uniformity such as an objectless plain, but of a tract of beautiful country seen at first in the distance, but which will bear to be rambled through and studied in every detail of hillside and valley, stream, rock, wood, and flower. But, as for anything else, so for a mathematical theory—beauty can be perceived, but not explained. As for mere extent, I can perhaps best illustrate this by speaking of the dates at which some of the great extensions have been made in several branches of mathematical science.

As regards geometry, I have already spoken of the invention of the Cartesian coordinates (1637). This gave to geometers the whole series of geometric curves of higher order than the conic sections: curves of the third order, or cubic curves; curves of the fourth order, or quartic curves; and so on indefinitely. The first fruits of it were Newton's *Enumeratio linearum tertii ordinis*, and the extremely interesting investigations of Maclaurin as to corresponding points on a cubic curve. This was at once enough to show that the new theory of cubic curves was a theory quite as beautiful and far more extensive than that of conics. And I must here refer to Euler's remark in the paper "Sur une contradiction apparente dans la théorie des courbes planes" (Berlin Memoirs, 1748), in regard to the nine points of intersection of two cubic curves (viz. that when eight of the points are given the ninth point is thereby completely determined): this is not only a fundamental theorem in cubic curves (including in itself Pascal's theorem of the hexagon inscribed in a conic), but it introduces into plane geometry a new notion—that of the point-system, or system of the points of intersection of two curves.

A theory derived from the conic, that of polar reciprocals, led to the general notion of geometrical duality—viz. that in plane geometry the point and the line are correlative figures; and founded on this we have Plücker's great work, the *Theorie der algebraischen Curven* (Bonn, 1839), in which he establishes the relation which exists between the order and class of a curve and the number of its different point- and line-singularities (Plücker's six equations). It thus appears that the true division of curves is not a division according to order only, but according to order and class, and that the curves of a given order and class are again to be divided into families according to their singularities: this is not a mere subdivision, but is really a widening of the field of investigation; each such family of curves is in itself a subject as wide as the totality of the curves of a given order might previously have appeared.

We *unite* families by considering together the curves of a given *Geschlecht*, or deficiency; and in reference to what I shall have to say on the Abelian functions, I must speak of this notion introduced into geometry by Riemann in the memoir "Theorie der Abel'schen Functionen," *Crelle*, t. LIV. (1857). For a curve of a given order, reckoning cusps as double points, the deficiency is equal to the greatest number $\frac{1}{2}(n-1)(n-2)$ of the double points which a curve of that order can have, less the number of double points which the curve actually has. Thus a conic, a cubic with one double point, a quartic with three double points, &c., are all curves of the deficiency 0; the general cubic is a curve, and the most simple curve, of the deficiency 1; the general quartic is a curve of deficiency 3; and so on. The deficiency is usually represented by the letter p. Riemann considers the general question of the rational transformation of a plane curve: viz. here the coordinates, assumed to be homogeneous or trilinear, are replaced by any rational and integral functions, homogeneous of the same degree in the new coordinates; the transformed curve is in general a curve of a different order, with its own system of double points; but the deficiency p remains unaltered; and it is on this ground that he unites together and regards as a single class the whole system of curves of a given deficiency p. It must not be supposed that all such curves admit of rational transformation the one into the other: there is the further theorem that any curve of the class depends, in the case of a cubic, upon one parameter, but for $p > 1$ upon $3p-3$ parameters, each such parameter being unaltered by the rational transformation; it is thus only the curves having the same one parameter, or $3p-3$ parameters, which can be rationally transformed the one into the other.

Solid geometry is a far wider subject: there are more theories, and each of them is of greater extent. The ratio is not that of the numbers of the dimensions of the spaces considered, or, what is the same thing, of the elementary figures—point and line in the one case; point, line and plane in the other case—belonging to these spaces respectively, but it is a very much higher one. For it is very inadequate to say that in plane geometry we have the curve, and in solid geometry the curve and surface: a more complete statement is required for the comparison. In plane geometry we have the curve, which may be regarded as a singly infinite system of points, and also as a singly infinite system of lines. In solid geometry we have, first, that which under one aspect is the curve, and under another aspect the developable, and which may be regarded as a singly infinite system of points, of lines, or of planes; secondly, the surface, which may be regarded as a doubly infinite system of points or of planes, and also as a special triply infinite system of lines (viz. the tangent-lines of the surface are a special complex): as distinct particular cases of the former figure, we have the plane curve and the cone; and as a particular case of the latter figure, the ruled surface or singly infinite system of lines; we have *besides* the congruence, or doubly infinite system of lines, and the complex, or triply infinite system of lines. But, even if in solid geometry we attend only to the curve and the surface, there are crowds of theories which have scarcely any analogues in plane geometry. The relation of a curve to the various surfaces which can be drawn through it, or of a surface to the various curves that can be drawn upon it, is different in kind from that which in plane geometry most nearly corresponds to it, the relation of a system of points to the curves through them, or of a curve to the points upon it. In particular, there is nothing in plane geometry corresponding to the theory of the curves of curvature of a surface. To the single theorem of plane geometry, a right line is the shortest distance between two points, there correspond in solid geometry two extensive and difficult theories—that of the geodesic lines upon a given surface, and that of the surface of minimum area for any given boundary. Again, in solid geometry we have the interesting and difficult question of the representation of a curve by means of equations; it is not every curve, but only a curve which is the complete intersection of two surfaces, which can be properly represented by two equations $(x, y, z, w)^m = 0$, $(x, y, z, w)^n = 0$, in quadriplanar coordinates; and in regard to this question, which may also be regarded as that of the classification of curves in space, we have quite recently three elaborate memoirs by Nöther, Halphen, and Valentiner respectively.

In n-dimensional geometry, only isolated questions have been considered. The field is simply too wide; the comparison with each other of the two cases of plane geometry and solid geometry is enough to show how the complexity and difficulty of the theory would increase with each successive dimension.

In Transcendental Analysis, or the Theory of Functions, we have all that has been done in the present century with regard to the general theory of the function of an imaginary variable by Gauss, Cauchy, Puiseux, Briot, Bouquet, Liouville, Riemann, Fuchs, Weierstrass, and others. The fundamental idea of the geometrical representation of an imaginary variable $x+iy$, by means of the point having x, y for its coordinates, belongs, as I mentioned, to Gauss; of this I have already spoken at some length. The notion has been applied to differential equations; in the modern point of view, the problem in regard to a given differential equation is, not so much to reduce the differential equation to quadratures, as to determine from it directly the course of the integrals for all positions of the point representing the independent variable: in particular, the differential equation of the second order leading to the hypergeometric series $F(\alpha, \beta, \gamma, x)$ has been treated in this manner, with the most interesting results; the function so determined for all values of the parameters (α, β, γ) is thus becoming a known function. I would here also refer to the new notion in this part of analysis introduced by Weierstrass—that of the one-valued integer function, as defined by an infinite series of ascending powers, convergent for all finite values, real or imaginary, of the variable x or $1/x-c$, and so having the one essential singular point $x=\infty$ or $x=c$, as the case may be: the memoir is published in the Berlin *Abhandlungen*, 1876.

But it is not only general theory: I have to speak of the various special functions to which the theory has been applied, or say the various known functions.

For a long time the only known transcendental functions were the circular functions sine, cosine, &c.; the logarithm—i.e. for analytical purposes the hyperbolic logarithm to the base e; and, as implied therein, the exponential function e^x. More completely stated, the group comprises the direct circular functions sin, cos, &c.; the inverse circular functions $\sin^{-1}$ or arc sin, &c.; the exponential function, exp.; and the inverse exponential, or logarithmic, function, log.

Passing over the very important Eulerian integral of the second kind or gamma-function, the theory of which has quite recently given rise to some very interesting developments—and omitting to mention at all various functions of minor importance,—we come (1811—1829) to the very wide groups, the elliptic functions and the single theta-functions. I give the interval of date so as to include Legendre's two systematic works, the *Exercices de Calcul Intégral* (1811—1816) and the *Théorie des Fonctions Elliptiques* (1825—1828); also Jacobi's *Fundamenta nova theoriæ Functionum Ellipticarum* (1829), calling to mind that many of Jacobi's results were obtained simultaneously by Abel. I remark that Legendre started from the consideration of the integrals depending on a radical $\sqrt{X}$, the square root of a rational and integral quartic function of a variable x; for this he substituted a radical $\Delta\phi$, $=\sqrt{1-k^2\sin^2\phi}$, and he arrived at his three kinds of elliptic integrals $F\phi$, $E\phi$, $\Pi\phi$, depending on the argument or amplitude ϕ, the modulus k, and also the last of them on a parameter n; the function F is properly an inverse function, and in place of it Abel and Jacobi each of them introduced the direct functions corresponding to the circular functions sine and cosine, Abel's functions called by him ϕ, f, F, and Jacobi's functions sinam, cosam, Δam, or as they are also written sn, cn, dn. Jacobi, moreover, in the development of his theory of transformation obtained a multitude of formulæ containing q, a transcendental function of the modulus defined by the equation $q=e^{-\pi K'/K}$, and he was also led by it to consider the two new functions H, Θ, which (taken each separately with two different arguments) are in fact the four functions called elsewhere by him $\Theta_1, \Theta_2, \Theta_3, \Theta_4$; these are the so-called theta-functions, or, when the distinction is necessary, the single theta-functions. Finally, Jacobi using the transformation $\sin\phi = \text{sinam}\, u$, expressed Legendre's integrals of the second and third kinds as integrals depending on the new variable u, denoting them by means of the letters Z, Π, and connecting them with his own functions H and Θ: and the elliptic functions sn, cn, dn are expressed with these, or say with $\Theta_1, \Theta_2, \Theta_3, \Theta_4$, as fractions having a common denominator.

It may be convenient to mention that Hermite in 1858, introducing into the theory in place of q the new variable ω connected with it by the equation $q=e^{i\pi\omega}$ (so that ω is in fact $=iK'/K$), was led to consider the three functions $\phi\omega$, $\psi\omega$, $\chi\omega$, which denote respectively the values of $\sqrt[4]{k}$, $\sqrt[4]{k'}$ and $\sqrt[12]{kk'}$ regarded as functions of ω.

A theta-function, putting the argument $=0$, and then regarding it as a function of ω, is what Professor Smith in a valuable memoir, left incomplete by his death, calls an omega-function, and the three functions $\phi\omega$, $\psi\omega$, $\chi\omega$ are his modular functions.

The proper elliptic functions sn, cn, dn form a system very analogous to the circular functions sine and cosine (say they are a sine and two separate cosines), having a like addition-theorem, viz. the form of this theorem is that the sn, cn and dn of $x+y$ are each of them expressible rationally in terms of the sn, cn and dn of x and of the sn, cn and dn of y; and, in fact, reducing itself to the system of the circular functions in the particular case $k=0$. But there is the important difference of form that the expressions for the sn, cn and dn of $x+y$ are fractional functions having a common denominator: this is a reason for regarding these functions as the ratios of four functions A, B, C, D, the absolute magnitudes of which are and remain indeterminate (the functions sn, cn, dn are in fact quotients $[\Theta_1, \Theta_2, \Theta_3] \div \Theta_4$ of the four theta-functions, but this is a further result in nowise deducible from the addition-equations, and which is intended to be for the moment disregarded; the remark has reference to what is said hereafter as to the Abelian functions). But there is in regard to the functions sn, cn, dn (what has no analogue for the circular functions), the whole theory of transformation of any order n prime or composite, and, as parts thereof, the whole theory of the modular and multiplier equations; and this theory of transformation spreads itself out in various directions, in geometry, in the Theory of Equations, and in the Theory of Numbers. Leaving the theta-functions out of consideration, the theory of the proper elliptic functions sn, cn, dn is at once seen to be a very wide one.

I assign to the Abelian functions the date 1826—1832. Abel gave what is called his theorem in various forms, but in its most general form in the *Mémoire sur une propriété générale d'une classe très-étendue de Fonctions Transcendantes* (1826), presented to the French Academy of Sciences, and crowned by them after the author's death, in the following year. This is in form a theorem of the integral calculus, relating to integrals depending on an irrational function y determined as a function of x by any algebraical equation $F(x, y)=0$ whatever: the theorem being that a sum of any number of such integrals is expressible by means of the sum of a determinate number p of like integrals, this number p depending on the form of the equation $F(x, y)=0$ which determines the irrational y (to fix the ideas, remark that considering this equation as representing a curve, then p is really the deficiency of the curve; but as already mentioned, the notion of deficiency dates only from 1857): thus in applying the theorem to the case where y is the square root of a function of the fourth order, we have in effect Legendre's theorem for elliptic integrals $F\phi + F\psi$ expressed by means of a single integral $F\mu$, and not a theorem applying in form to the elliptic functions sn, cn, dn. To be intelligible I must recall that the integrals belonging to the case where y is the square root of a rational and integral function of an order exceeding four are (in distinction from the general case) termed hyper-elliptic integrals: viz. if the order be 5 or 6, then these are of the class $p=2$; if the order be 7 or 8, then they are of the class $p=3$, and so on; the *general* Abelian integral of the class $p=2$ is a hyperelliptic integral: but if $p=3$, or any greater value, then the hyperelliptic integrals are only a particular case of the Abelian integrals of the same class. The further step was made by Jacobi in the short but very important memoir "Considerationes generales de transcendentibus Abelianis," *Crelle*, t. IX. (1832): viz. he there shows for the hyperelliptic integrals of any class (but the conclusion may be stated generally) that the direct functions to which Abel's theorem has reference are not functions of a single variable, such as the elliptic sn, cn, or dn, but functions of p variables. Thus, in the case $p=2$, which Jacobi specially considers, it is shown that Abel's theorem has reference to two functions $\lambda(u, v)$, $\lambda_1(u, v)$ each of two variables, and gives in effect an addition-theorem for the expression of the functions $\lambda(u+u', v+v')$, $\lambda_1(u+u', v+v')$ algebraically in terms of the functions $\lambda(u, v)$, $\lambda_1(u, v)$, $\lambda(u', v')$, $\lambda_1(u', v')$.

It is important to remark that Abel's theorem does not directly give, nor does Jacobi assert that it gives, the addition-theorem in a perfect form. Take the case $p=1$: the result from the theorem is that we have a function $\lambda(u)$, which is such that $\lambda(u+v)$ can be expressed algebraically in terms of $\lambda(u)$ and $\lambda(v)$. This is of course perfectly correct, $\operatorname{sn}(u+v)$ is expressible algebraically in terms of $\operatorname{sn} u$, $\operatorname{sn} v$, but the expression involves the radicals $\sqrt{1-\operatorname{sn}^2 u}$, $\sqrt{1-k^2\operatorname{sn}^2 u}$, $\sqrt{1-\operatorname{sn}^2 v}$, $\sqrt{1-k^2\operatorname{sn}^2 v}$; but it does not give the three functions sn, cn, dn, or in anywise amount to the statement that the sn, cn and dn u of $u+v$ are expressible rationally in terms of the sn, cn and dn of u and of v. In the case $p=1$, the right number of functions, each of one variable, is 3, but the three functions sn, cn and dn are properly considered as the ratios of 4 functions; and so, in general, the right number of functions, each of p variables, is 4^p-1, and these may be considered as the ratios of 4^p functions. But notwithstanding this last remark, it may be considered that the notion of the Abelian functions of p variables is established, and the addition-theorem for these functions in effect given by the memoirs (Abel 1826, Jacobi 1832) last referred to.

We have next for the case $p=2$, which is hyperelliptic, the two extremely valuable memoirs, Göpel, "Theoria transcendentium Abelianarum primi ordinis adumbratio læva," *Crelle*, t. XXXV. (1847), and Rosenhain, "Mémoire sur les fonctions de deux variables et à quatre périodes qui sont les inverses des intégrales ultra-elliptiques de la première classe" (1846), Paris, *Mém. Savans Étrang.* t. XI. (1851), each of them establishing on the analogy of the single theta-functions the corresponding functions of two variables, or double theta-functions, and in connexion with them the theory of the Abelian functions of two variables. It may be remarked that in order of simplicity the theta-functions certainly precede the Abelian functions.

Passing over some memoirs by Weierstrass which refer to the general hyperelliptic integrals, p any value whatever, we come to Riemann, who died 1866, at the age of forty: collected edition of his works, Leipzig, 1876. His great memoir on the Abelian and theta-functions is the memoir already incidentally referred to, "Theorie der Abel'schen Functionen," *Crelle*, t. LIV. (1857); but intimately connected therewith we have his Inaugural Dissertation (Göttingen, 1851), *Grundlagen für eine allgemeine Theorie der Functionen einer veränderlichen complexen Grösse*: his treatment of the problem of the Abelian functions, and establishment for the purpose of this theory of the multiple theta-functions, are alike founded on his general principles of the theory of the functions of a variable complex magnitude $x + iy$, and it is this which would have to be gone into for any explanation of his method of dealing with the problem.

Riemann, starting with the integrals of the most general form, and considering the inverse functions corresponding to these integrals—that is, the Abelian functions of p variables—defines a theta-function of p variables, or p-tuple theta-function, as the sum of a p-tuply infinite series of exponentials, the general term of course depending on the p variables; and he shows that the Abelian functions are algebraically connected with theta-functions of the proper arguments. The theory is presented in the broadest form; in particular as regards the theta-functions, the 4^p functions are not even referred to, and there is no development as to the form of the algebraic relations between the two sets of functions.

In the Theory of Equations, the beginning of the century may be regarded as an epoch. Immediately preceding it, we have Lagrange's *Traité des Équations Numériques* (1st ed. 1798), the notes to which exhibit the then position of the theory. Immediately following it, the great work by Gauss, the *Disquisitiones Arithmeticæ* (1801), in which he establishes the theory for the case of a prime exponent n, of the binomial equation $x^n - 1 = 0$: throwing out the factor $x - 1$, the equation becomes an equation of the order $n - 1$, and this is decomposed into equations the orders of which are the prime factors of $n - 1$. In particular, Gauss was thereby led to the remarkable geometrical result that it was possible to construct geometrically—that is, with only the ruler and compass—the regular polygons of 17 sides and 257 sides respectively. We have then (1826—1829) Abel, who, besides his demonstration of the impossibility of the solution of a quintic equation by radicals, and his very important researches on the general question of the algebraic solution of equations, established the theory of the class of equations since called Abelian equations. He applied his methods to the problem of the division of the elliptic functions, to (what is a distinct question) the division of the complete functions, and to the very interesting special case of the lemniscate. But the theory of algebraic solutions in its most complete form was established by Galois (born 1811, killed in a duel 1832), who for this purpose introduced the notion of a group of substitutions; and to him also are due some most valuable results in relation to another set of equations presenting themselves in the theory of elliptic functions—viz. the modular equations. In 1835 we have Jerrard's transformation of the general quintic equation. In 1870 an elaborate work, Jordan's *Traité des Substitutions et des équations algébriques*: a mere inspection of the table of contents of this would serve to illustrate my proposition as to the great extension of this branch of mathematics.

The Theory of Numbers was, at the beginning of the century, represented by Legendre's *Théorie des Nombres* (1st ed. 1798), shortly followed by Gauss' *Disquisitiones Arithmeticæ* (1801). This work by Gauss is, throughout, a theory of ordinary real numbers. It establishes the notion of a congruence; gives a proof of the theorem of reciprocity in regard to quadratic residues; and contains a very complete theory of binary quadratic forms $(a, b, c)(x, y)^2$, of negative and positive determinant, including the theory, there first given, of the composition of such forms. It gives also the commencement of a like theory of ternary quadratic forms. It contains also the theory already referred to, but which has since influenced in so remarkable a manner the whole theory of numbers—the theory of the solution of the binomial equation $x^n - 1 = 0$: it is, in fact, the roots or periods of roots derived from these equations which form the incommensurables, or unities, of the complex theories which have been chiefly worked at; thus, the i of ordinary analysis presents itself as a root of the equation $x^4 - 1 = 0$. It was Gauss himself who, for the development of a real theory—that of biquadratic residues—found it necessary to use complex numbers of the before-mentioned form, $a + bi$ (a and b positive or negative real integers, including zero), and the theory of these numbers was studied and cultivated by Lejeune-Dirichlet. We have thus a new theory of these complex numbers, side by side with the former theory of real numbers: everything in the real theory reproducing itself, prime numbers, congruences, theories of residues, reciprocity, quadratic forms, &c., but with greater variety and complexity, and increased difficulty of demonstration. But instead of the equation $x^4 - 1 = 0$, we may take the equation $x^3 - 1 = 0$: we have here the complex numbers $a + b\rho$ composed with an imaginary cube root of unity, the theory specially considered by Eisenstein: again a new theory, corresponding to but different from that of the numbers $a + bi$. The general case of any prime value of the exponent n, and with periods of roots, which here present themselves instead of single roots, was first considered by Kummer: viz. if $n - 1 = ef$, and $\eta_1, \eta_2, \ldots, \eta_e$ are the e periods, each of them a sum of f roots, of the equation $x^n - 1 = 0$, then the complex numbers considered are the numbers of the form $a_1\eta_1 + a_2\eta_2 + \ldots + a_e\eta_e$ ($a_1, a_2, \ldots, a_e$ positive or negative ordinary integers, including zero): f may be $= 1$, and the theory for the periods thus includes that for the single roots.

We have thus a new and very general theory, including within itself that of the complex numbers $a + bi$ and $a + b\rho$. But a new phenomenon presents itself; for these special forms the properties in regard to prime numbers corresponded precisely with those for real numbers; a non-prime number was in one way only a product of prime factors; the power of a prime number has only factors which are lower powers of the same prime number: for instance, if p be a prime number, then, excluding the obvious decomposition $p \cdot p^2$, we cannot have $p^3 =$ a product of two factors A, B. In the general case this is not so, but the exception first presents itself for the number 23; in the theory of the numbers composed with the 23rd roots of unity, we have prime numbers p, such that $p^3 = AB$. To restore the theorem, it is necessary to establish the notion of ideal numbers; a prime number p is by definition not the product of two actual numbers, but in the example just referred to the number p is the product of two ideal numbers having for their cubes the two actual numbers A, B, respectively, and we thus have $p^3 = AB$. It is, I think, in this way that we most easily get some notion of the meaning of an ideal number, but the mode of treatment (in Kummer's great memoir, "Ueber die Zerlegung der aus Wurzeln der Einheit gebildeten complexen Zahlen in ihre Primfactoren," *Crelle*, t. XXXV. 1847) is a much more refined one; an ideal number, without ever being isolated, is made to manifest itself in the properties of the prime number of which it is a factor, and without reference to the theorem afterwards arrived at, that there is always some power of the ideal number which is an actual number. In the still later developments of the Theory of Numbers by Dedekind, the units, or incommensurables, are the roots of any irreducible equation having for its coefficients ordinary integer numbers, and with the coefficient unity for the highest power of x. The question arises, What is the analogue of a whole number? thus, for the very simple case of the equation $x^2+3=0$, we have as a whole number the apparently fractional form $\frac{1}{2}(1+i\sqrt{3})$ which is the imaginary cube root of unity, the ρ of Eisenstein's theory. We have, moreover, the (as far as appears) wholly distinct complex theory of the numbers composed with the congruence-imaginaries of Galois: viz. these are imaginary numbers assumed to satisfy a congruence which is not satisfied by any real number; for instance, the congruence $x^2-2=0$ (mod 5) has no real root, but we assume an imaginary root i, the other root is then $=-i$, and we then consider the system of complex numbers $a+bi$ (mod 5), viz. we have thus the 5^2 numbers obtained by giving to each of the numbers a, b, the values 0, 1, 2, 3, 4, successively. And so in general, the consideration of an irreducible congruence $F(x)=0$ (mod p) of the order n, to any prime modulus p, gives rise to an imaginary congruence root i, and to complex numbers of the form $a+bi+ci^2+\ldots+ki^{n-1}$, where a, b, k, ... &c., are ordinary integers each $=0, 1, 2, \ldots, p-1$.

As regards the theory of forms, we have in the ordinary theory, in addition to the binary and ternary quadratic forms, which have been very thoroughly studied, the quaternary and higher quadratic forms (to these last belong, as very particular cases, the theories of the representation of a number as a sum of four, five or more squares), and also binary cubic and quartic forms, and ternary cubic forms, in regard to all of which something has been done; the binary quadratic forms have been studied in the theory of the complex numbers $a+bi$.

A seemingly isolated question in the Theory of Numbers, the demonstration of Fermat's theorem of the impossibility for any exponent λ greater than 3, of the equation $x^\lambda+y^\lambda=z^\lambda$, has given rise to investigations of very great interest and difficulty.

Outside of ordinary mathematics, we have some theories which must be referred to: algebraical, geometrical, logical. It is, as in many other cases, difficult to draw the line; we do in ordinary mathematics use symbols not denoting quantities, which we nevertheless combine in the way of addition and multiplication, $a+b$, and ab, and which may be such as not to obey the commutative law $ab=ba$: in particular, this is or may be so in regard to symbols of operation; and it could hardly be said that any development whatever of the theory of such symbols of operation did not belong to ordinary algebra. But I do separate from ordinary mathematics the system of multiple algebra or linear associative algebra, developed in the valuable memoir by the late Benjamin Peirce, *Linear Associative Algebra* (1870, reprinted 1881 in the *American Journal of Mathematics*, vol. IV., with notes and addenda by his son, C. S. Peirce); we here consider symbols A, B, &c. which are linear functions of a determinate number of letters or units i, j, k, l, &c., with coefficients which are ordinary analytical magnitudes, real or imaginary, viz. the coefficients are in general of the form $x+iy$, where

i is the before-mentioned imaginary or $\sqrt{-1}$ of ordinary analysis. The letters i, j, &c., are such that every binary combination i^2, ij, ji, &c., (the ij being in general not $= ji$), is equal to a linear function of the letters, but under the restriction of satisfying the associative law: viz. for each combination of three letters $ij \, . \, k$ is $= i \, . \, jk$, so that there is a determinate and unique product of three or more letters; or, what is the same thing, the laws of combination of the units i, j, k, are defined by a multiplication table giving the values of i^2, ij, ji, &c.; the original units may be replaced by linear functions of these units, so as to give rise, for the units finally adopted, to a multiplication table of the most simple form; and it is very remarkable, how frequently in these simplified forms we have nilpotent or idempotent symbols ($i^2 = 0$, or $i^2 = i$, as the case may be), and symbols i, j, such that $ij = ji = 0$; and consequently how simple are the forms of the multiplication tables which define the several systems respectively.

I have spoken of this multiple algebra before referring to various geometrical theories of earlier date, because I consider it as the general analytical basis, and the true basis, of these theories. I do not realise to myself directly the notions of the addition or multiplication of two lines, areas, rotations, forces, or other geometrical, kinematical, or mechanical entities; and I would formulate a general theory as follows: consider any such entity as determined by the proper number of parameters a, b, c (for instance, in the case of a finite line given in magnitude and position, these might be the length, the coordinates of one end, and the direction-cosines of the line considered as drawn from this end); and represent it by or connect it with the linear function $ai + bj + ck +$ &c., formed with these parameters as coefficients, and with a given set of units, i, j, k, &c. Conversely, any such linear function represents an entity of the kind in question. Two given entities are represented by two linear functions; the sum of these is a like linear function representing an entity of the same kind, which may be regarded as the sum of the two entities; and the product of them (taken in a determined order, when the order is material) is an entity of the same kind, which may be regarded as the product (in the same order) of the two entities. We thus establish by definition the notion of the sum of the two entities, and that of the product (in a determinate order, when the order is material) of the two entities. The value of the theory in regard to any kind of entity would of course depend on the choice of a system of units, i, j, k, ..., with such laws of combination as would give a geometrical or kinematical or mechanical significance to the notions of the sum and product as thus defined.

Among the geometrical theories referred to, we have a theory (that of Argand, Warren, and Peacock) of imaginaries in plane geometry; Sir W. R. Hamilton's very valuable and important theory of Quaternions; the theories developed in Grassmann's *Ausdehnungslehre*, 1841 and 1862; Clifford's theory of Biquaternions; and recent extensions of Grassmann's theory to non-Euclidian space, by Mr Homersham Cox. These different theories have of course been developed, not in anywise from the point of view in which I have been considering them, but from the points of view of their several authors respectively.

The literal symbols x, y, &c., used in Boole's *Laws of Thought* (1854) to represent things as subjects of our conceptions, are symbols obeying the laws of algebraic combination (the distributive, commutative, and associative laws) but which are such that for any one of them, say x, we have $x - x^2 = 0$, this equation not implying (as in ordinary algebra it would do) either $x = 0$ or else $x = 1$. In the latter part of the work relating to the Theory of Probabilities, there is a difficulty in making out the precise meaning of the symbols; and the remarkable theory there developed has, it seems to me, passed out of notice, without having been properly discussed. A paper by the same author, "Of Propositions numerically definite" (*Camb. Phil. Trans.* 1869), is also on the borderland of logic and mathematics. It would be out of place to consider other systems of mathematical logic, but I will just mention that Mr C. S. Peirce in his "Algebra of Logic," *American Math. Journal*, vol. III., establishes a notation for relative terms, and that these present themselves in connexion with the systems of units of the linear associative algebra.

Connected with logic, but primarily mathematical and of the highest importance, we have Schubert's *Abzählende Geometrie* (1878). The general question is, How many curves or other figures are there which satisfy given conditions? for example, How many conics are there which touch each of five given conics? The class of questions in regard to the conic was first considered by Chasles, and we have his beautiful theory of the characteristics μ, ν, of the conics which satisfy four given conditions; questions relating to cubics and quartics were afterwards considered by Maillard and Zeuthen; and in the work just referred to the theory has become a very wide one. The noticeable point is that the symbols used by Schubert are in the first instance, not numbers, but mere logical symbols: for example, a letter g denotes the condition that a line shall cut a given line; g^2 that it shall cut each of two given lines; and so in other cases; and these logical symbols are combined together by algebraical laws: they first acquire a numerical signification when the number of conditions becomes equal to the number of parameters upon which the figure in question depends.

In all that I have last said in regard to theories outside of ordinary mathematics, I have been still speaking on the text of the vast extent of modern mathematics. In conclusion I would say that mathematics have steadily advanced from the time of the Greek geometers. Nothing is lost or wasted; the achievements of Euclid, Archimedes, and Apollonius are as admirable now as they were in their own days. Descartes' method of coordinates is a possession for ever. But mathematics have never been cultivated more zealously and diligently, or with greater success, than in this century—in the last half of it, or at the present time: the advances made have been enormous, the actual field is boundless, the future full of hope. In regard to pure mathematics we may most confidently say:—

Yet I doubt not through the ages one increasing purpose runs,
And the thoughts of men are widened with the process of the suns.

785.

CURVE.

[From the *Encyclopædia Britannica, Ninth Edition*, vol. VI. (1877), pp. 716—728.]

THIS subject is treated here from an historical point of view, for the purpose of showing how the different leading ideas in the theory were successively arrived at and developed.

A curve is a line, or continuous singly infinite system of points. We consider in the first instance, and chiefly, a plane curve described according to a law. Such a curve may be regarded geometrically as actually described, or kinematically as in course of description by the motion of a point; in the former point of view, it is the locus of all the points which satisfy a given condition; in the latter, it is the locus of a point moving subject to a given condition. Thus the most simple and earliest known curve, the circle, is the locus of all the points at a given distance from a fixed centre, or else the locus of a point moving so as to be always at a given distance from a fixed centre. (The straight line and the point are not for the moment regarded as curves.)

Next to the circle we have the conic sections, the invention of them attributed to Plato (who lived 430 to 347 B.C.); the original definition of them as the sections of a cone was by the Greek geometers who studied them soon replaced by a proper definition *in plano* like that for the circle, viz. a conic section (or as we now say a "conic") is the locus of a point such that its distance from a given point, the focus, is in a given ratio to its (perpendicular) distance from a given line, the directrix; or it is the locus of a point which moves so as always to satisfy the foregoing condition. Similarly any other property might be used as a definition; an ellipse is the locus of a point such that the sum of its distances from two fixed points (the foci) is constant, &c., &c.

The Greek geometers invented other curves; in particular, the "conchoid," which is the locus of a point such that its distance from a given line, measured along the line drawn through it to a fixed point, is constant; and the "cissoid" which is the locus of a point such that its distance from a fixed point is always equal to the intercept (on the line through the fixed point) between a circle passing through the fixed point and the tangent to the circle at the point opposite to the fixed point. Obviously the number of such geometrical or kinematical definitions is infinite. In a machine of any kind, each point describes a curve; a simple but important instance is the "three-bar curve," or locus of a point in or rigidly connected with a bar pivotted on to two other bars which rotate about fixed centres respectively. Every curve thus arbitrarily defined has its own properties: and there was not any principle of classification.

The principle of classification first presented itself in the *Géométrie* of Descartes (1637). The idea was to represent any curve whatever by means of a relation between the coordinates (x, y) of a point of the curve, or say to represent the curve by means of its equation.

Descartes takes two lines xx', yy', called axes of coordinates, intersecting at a point O called the origin (the axes are usually at right angles to each other, and for the

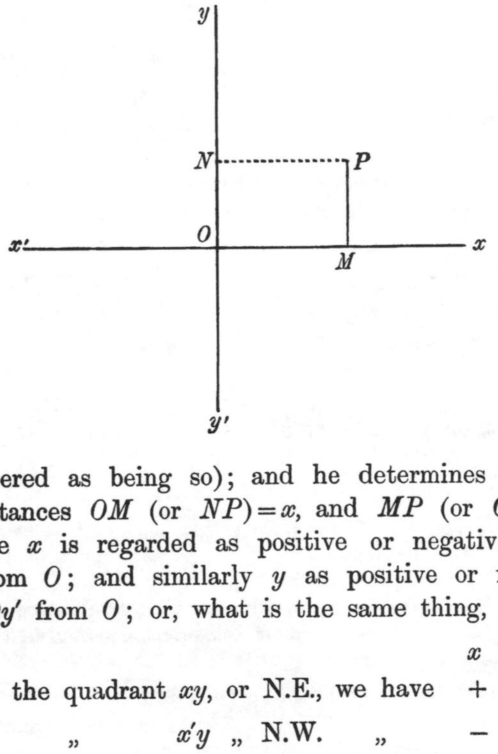

present they are considered as being so); and he determines the position of a point P by means of its distances OM (or NP) $= x$, and MP (or ON) $= y$, from these two axes respectively; where x is regarded as positive or negative according as it is in the sense Ox or Ox' from O; and similarly y as positive or negative according as it is in the sense Oy or Oy' from O; or, what is the same thing,

				x	y
In the quadrant	xy, or	N.E.,	we have	+	+
"	$x'y$ "	N.W.	"	−	+
"	xy' "	S.E.	"	+	−
"	$x'y'$ "	S.W.	"	−	−

Any relation whatever between (x, y) determines a curve, and conversely every curve whatever is determined by a relation between (x, y).

Observe that the distinctive feature is in the *exclusive* use of such determination of a curve by means of its equation. The Greek geometers were perfectly familiar with the property of an ellipse which in the Cartesian notation is $\frac{x^2}{a^2}+\frac{y^2}{b^2}=1$, the equation of the curve; but it was as one of a number of properties, and in no wise selected out of the others for the characteristic property of the curve *.

We obtain from the equation the notion of an algebraical or geometrical as opposed to a transcendental curve, viz. an algebraical or geometrical curve is a curve having an equation $F(x, y)=0$, where $F(x, y)$ is a rational and integral algebraical function of the coordinates (x, y); and in what follows we attend throughout (unless the contrary is stated) only to such curves. The equation is sometimes given, and may conveniently be used, in an irrational form, but we always imagine it reduced to the foregoing rational and integral form, and regard this as the equation of the curve. And we have hence the notion of a curve of a *given order*, viz. the order of the curve is equal to that of the term or terms of highest order in the coordinates (x, y) conjointly in the equation of the curve; for instance, $xy-1=0$ is a curve of the second order.

It is to be noticed here that the axes of coordinates may be any two lines at right angles to each other whatever; and that the equation of a curve will be different according to the selection of the axes of coordinates; but the order is independent of the axes, and has a determinate value for any given curve.

We hence divide curves according to their order, viz. a curve is of the first order, second order, third order, &c., according as it is represented by an equation of the first order, $ax+by+c=0$, or say $(*\between x, y, 1)=0$; or by an equation of the second order, $ax^2+2hxy+by^2+2fy+2gx+c=0$, say $(*\between x, y, 1)^2=0$; or by an equation of the third order, &c.; or, what is the same thing, according as the equation is linear, quadric, cubic, &c.

A curve of the first order is a right line; and conversely every right line is a curve of the first order.

* There is no exercise more profitable for a student than that of tracing a curve from its equation, or say rather that of so tracing a considerable number of curves. And he should make the equations for himself. The equation should be in the first instance a purely numerical one, where y is given or can be found as an explicit function of x; here, by giving different numerical values to x, the corresponding values of y may be found; and a *sufficient* number of points being thus determined, the curve is traced by drawing a continuous line through these points. The next step should be to consider an equation involving literal coefficients; thus, after such curves as $y=x^3$, $y=x(x-1)(x-2)$, $y=(x-1)\sqrt{x-2}$, &c., he should proceed to trace such curves as $y=(x-a)(x-b)(x-c)$, $y=(x-a)\sqrt{x-b}$, &c., and endeavour to ascertain for what different relations of equality or inequality between the coefficients the curve will assume essentially or notably distinct forms. The purely numerical equations will present instances of nodes, cusps, inflexions, double tangents, asymptotes, &c.,—specialities which he should be familiar with before he has to consider their general theory. And he may then consider an equation such that neither coordinate can be expressed as an explicit function of the other of them (practically, an equation such as $x^3+y^3-3xy=0$, which requires the solution of a cubic equation, belongs to this class); the problem of tracing the curve here frequently requires special methods, and it may easily be such as to require and serve as an exercise for the powers of an advanced algebraist.

A curve of the second order is a conic, or as it is also called a quadric; and conversely every conic, or quadric, is a curve of the second order.

A curve of the third order is called a cubic; one of the fourth order a quartic; and so on.

A curve of the order m has for its equation $(*\between x, y, 1)^m = 0$; and when the coefficients of the function are arbitrary, the curve is said to be the general curve of the order m. The number of coefficients is $\tfrac{1}{2}(m+1)(m+2)$; but there is no loss of generality if the equation be divided by one coefficient so as to reduce the coefficient of the corresponding term to unity, hence the number of coefficients may be reckoned as $\tfrac{1}{2}(m+1)(m+2)-1$, that is, $\tfrac{1}{2}m(m+3)$; and a curve of the order m may be made to satisfy this number of conditions; for example, to pass through $\tfrac{1}{2}m(m+3)$ points.

It is to be remarked that an equation may *break up;* thus a quadric equation may be $(ax+by+c)(a'x+b'y+c')=0$, breaking up into the two equations $ax+by+c=0$, $a'x+b'y+c'=0$, viz. the original equation is satisfied if either of these is satisfied. Each of these last equations represents a curve of the first order, or right line; and the original equation represents this pair of lines, viz. the pair of lines is considered as a quadric curve. But it is an *improper* quadric curve; and in speaking of curves of the second or any other given order, we frequently imply that the curve is a proper curve represented by an equation which does not break up.

The intersections of two curves are obtained by combining their equations; viz. the elimination from the two equations of y (or x) gives for x (or y) an equation of a certain order, say the resultant equation; and then to each value of x (or y) satisfying this equation there corresponds in general a single value of y (or x), and consequently a single point of intersection; the number of intersections is thus equal to the order of the resultant equation in x (or y).

Supposing that the two curves are of the orders m, n, respectively, then the order of the resultant equation is in general and at most $=mn$; in particular, if the curve of the order n is an arbitrary line $(n=1)$, then the order of the resultant equation is $=m$; and the curve of the order m meets therefore the line in m points. But the resultant equation may have all or any of its roots imaginary, and it is thus not always that there are m real intersections.

The notion of imaginary intersections, thus presenting itself, through algebra, in geometry, must be accepted in geometry—and it in fact plays an all-important part in modern geometry. As in algebra we say that an equation of the mth order has m roots, viz. we state this generally without in the first instance, or it may be without ever, distinguishing whether these are real or imaginary; so in geometry we say that a curve of the mth order is met by an arbitrary line in m points, or rather we thus, through algebra, obtain the proper geometrical definition of a curve of the mth order, as a curve which is met by an arbitrary line in m points (that is, of course, in m, and not more than m, points).

The theorem of the m intersections has been stated in regard to an *arbitrary* line; in fact, for particular lines the resultant equation may be or appear to be of an order less than m; for instance, taking $m = 2$, if the hyperbola $xy - 1 = 0$ be cut by the line $y = \beta$, the resultant equation in x is $\beta x - 1 = 0$, and there is apparently only the intersection $\left(x = \frac{1}{\beta},\ y = \beta\right)$; but the theorem is, in fact, true for every line whatever: a curve of the order m meets every line whatever in precisely m points. We have, in the case just referred to, to take account of a point at infinity on the line $y = \beta$; the two intersections are the point $\left(x = \frac{1}{\beta},\ y = \beta\right)$, and the point at infinity on the line $y = \beta$.

It is moreover to be noticed that the points at infinity may be all or any of them imaginary, and that the points of intersection, whether finite or at infinity, real or imaginary, may coincide two or more of them together, and have to be counted accordingly; to support the theorem in its universality, it is necessary to take account of these various circumstances.

The foregoing notion of a point at infinity is a very important one in modern geometry; and we have also to consider the paradoxical statement that in plane geometry, or say as regards the plane, infinity is a right line. This admits of an easy illustration in solid geometry. If with a given centre of projection, by drawing from it lines to every point of a given line, we project the given line on a given plane, the projection is a line, i.e., this projection is the intersection of the given plane with the plane through the centre and the given line. Say the projection is *always* a line, then if the figure is such that the two planes are parallel, the projection is the intersection of the given plane by a parallel plane, or it is the system of points at infinity on the given plane, that is, these points at infinity are regarded as situate on a given line, the line infinity of the given plane*.

Reverting to the purely plane theory, infinity is a line, related like any other right line to the curve, and thus intersecting it in m points, real or imaginary, distinct or coincident.

Descartes in the *Géométrie* defined and considered the remarkable curves called after him ovals of Descartes, or simply Cartesians, which will be again referred to. The next important work, founded on the *Géométrie*, was Sir Isaac Newton's *Enumeratio linearum tertii ordinis* (1706), establishing a classification of cubic curves founded chiefly on the nature of their infinite branches, which was in some details completed by Stirling, Murdoch, and Cramer; the work contains also the remarkable theorem (to be again referred to), that there are five kinds of cubic curves giving by their projections every cubic curve whatever.

Various properties of curves in general, and of cubic curves, are established in Maclaurin's memoir, "De linearum geometricarum proprietatibus generalibus Tractatus" (posthumous, say 1746, published in the 6th edition of his *Algebra*). We have in it a particular kind of *correspondence* of two points on a cubic curve, viz. two points correspond to each other when the tangents at the two points again meet the cubic in the same point.

* More generally, in solid geometry infinity is a plane,—its intersection with any given plane being the right line which is the infinity of this given plane.

The *Géométrie Descriptive* by Monge was written in the year 1794 or 1795 (7th edition, Paris, 1847), and in it we find stated, *in plano* with regard to the circle, and in three dimensions with regard to a surface of the second order, the fundamental theorem of reciprocal polars, viz. "Given a surface of the second order and a circumscribed conic surface which touches it.... then if the conic surface moves so that its summit is always in the same plane, the plane of the curve of contact passes always through the same point." The theorem is here referred to partly on account of its bearing on the theory of imaginaries in geometry. It is, in Brianchon's memoir "Sur les surfaces du second degré" (*Jour. Polyt.*, t. VI., 1806), shown how for any given position of the summit the plane of contact is determined, or reciprocally; say the plane XY is determined when the point P is given, or reciprocally; and it is noticed that when P is situate in the interior of the surface the plane XY does not cut the surface; that is, we have a real plane XY intersecting the surface in the imaginary curve of contact of the imaginary circumscribed cone having for its summit a given real point P inside the surface.

Stating the theorem in regard to a conic, we have a real point P (called the pole) and a real line XY (called the polar), the line joining the two (real or imaginary) points of contact of the (real or imaginary) tangents drawn from the point to the conic; and the theorem is that when the point describes a line the line passes through a point, this line and point being polar and pole to each other. The term "pole" was first used by Servois, and "polar" by Gergonne (*Gerg.*, t. I. and III., 1810—13); and from the theorem we have the method of reciprocal polars for the transformation of geometrical theorems, used already by Brianchon (in the memoir above referred to) for the demonstration of the theorem called by his name, and in a similar manner by various writers in the earlier volumes of Gergonne. We are here concerned with the method less in itself than as leading to the general notion of duality. And, bearing in a somewhat similar manner also on the theory of imaginaries in geometry (but the notion presents itself in a more explicit form), there is the memoir by Gaultier, on the graphical construction of circles and spheres (*Jour. Polyt.*, t. IX., 1813). The well-known theorem as to radical axes may be stated as follows. Consider two circles partially drawn so that it does not appear whether the circles, if completed, would or would not intersect in real points, say two arcs of circles; then we can, by means of a third circle drawn so as to intersect in two real points each of the two arcs, determine a right line, which, if the complete circles intersect in two real points, passes through the points, and which is on this account regarded as a line passing through two (real or imaginary) points of intersection of the two circles. The construction in fact is, join the two points in which the third circle meets the first arc, and join also the two points in which the third circle meets the second arc, and from the point of intersection of the two joining lines, let fall a perpendicular on the line joining the centre of the two circles; this perpendicular (considered as an indefinite line) is what Gaultier terms the "radical axis of the two circles"; it is a line determined by a real construction and itself always real; and by what precedes it is the line joining two (real or imaginary, as the case may be) intersections of the given circles.

The intersections which lie on the radical axis are two out of the four intersections of the two circles. The question as to the remaining two intersections did not present itself to Gaultier, but it is answered in Poncelet's *Traité des propriétés projectives* (1822), where we find (p. 49) the statement, "deux circles placés arbitrairement sur un plan...ont idéalement deux points imaginaires communs à l'infini"; that is, a circle *qua* curve of the second order is met by the line infinity in two points; but, more than this, they are the same two points for any circle whatever. The points in question have since been called (it is believed first by Dr Salmon) the circular points at infinity, or they may be called the circular points; these are also frequently spoken of as the points I, J; and we have thus the circle characterized as a conic which passes through the two circular points at infinity; the number of conditions thus imposed upon the conic is $=2$, and there remain three arbitrary constants, which is the right number for the circle. Poncelet throughout his work makes continual use of the foregoing theories of imaginaries and infinity, and also of the before-mentioned theory of reciprocal polars.

Poncelet's two memoirs "Sur les centres des moyennes harmoniques," and "Sur la théorie générale des polaires réciproques," although presented to the Paris Academy in 1824 were only published (*Crelle*, t. III. and IV., 1828, 1829), subsequent to the memoir by Gergonne, "Considérations philosophiques sur les élémens de la science de l'étendue" (*Gerg.*, t. XVI., 1825—26). In this memoir by Gergonne, the theory of duality is very clearly and explicitly stated; for instance, we find "dans la géométrie plane, à chaque théorème il en répond nécessairement un autre qui s'en déduit en échangeant simplement entre eux les deux mots *points* et *droites*; tandis que dans la géométrie de l'espace ce sont les mots *points* et *plans* qu'il faut échanger entre eux pour passer d'un théorème à son corrélatif"; and the plan is introduced of printing correlative theorems, opposite to each other, in two columns. There was a reclamation as to priority by Poncelet in the *Bulletin Universel* reprinted with remarks by Gergonne (*Gerg.*, t. XIX., 1827), and followed by a short paper by Gergonne, "Rectifications de quelques théorèmes, &c.," which is important as first introducing the word *class*. We find in it explicitly the two correlative definitions:—"a plane curve is said to be of the mth degree (order) when it has with a line m real or ideal intersections," and "a plane curve is said to be of the mth class when from any point of its plane there can be drawn to it m real or ideal tangents."

It may be remarked that in Poncelet's memoir on reciprocal polars, above referred to, we have the theorem that the number of tangents from a point to a curve of the order m, or say the class of the curve, is in general and at most $=m(m-1)$, and that he mentions that this number is subject to reduction when the curve has double points or cusps.

The theorem of duality as regards plane figures may be thus stated:—two figures may correspond to each other in such manner that to each point and line in either figure there corresponds in the other figure a line and point respectively. It is to be understood that the theorem extends to all points or lines, drawn or not drawn; thus if in the first figure there are any number of points on a line drawn or not drawn, the corresponding lines in the second figure, produced if necessary, must meet in a point. And we thus see how the theorem extends to curves, their points and tangents: if there is in the first figure a curve of the order m, any line meets it in m points; and hence from the corresponding point in the second figure there must be to the corresponding curve m tangents; that is, the corresponding curve must be of the class m.

Trilinear coordinates (to be again referred to) were first used by Bobillier in the memoir, "Essai sur un nouveau mode de recherche des propriétés de l'étendue" (*Gerg.*, t. XVIII., 1827—28). It is convenient to use these rather than Cartesian coordinates. We represent a curve of the order m by an equation $(*\between x, y, z)^m = 0$, the function on the left-hand being a homogeneous rational and integral function of the order m of the three coordinates (x, y, z); clearly the number of constants is the same as for the equation $(*\between x, y, 1)^m = 0$ in Cartesian coordinates.

The theory of duality is considered and developed, but chiefly in regard to its metrical applications, by Chasles in the "Mémoire de géométrie sur deux principes généraux de la science, la dualité et l'homographie," which forms a sequel to the "Aperçu historique sur l'origine et le développement des méthodes en géométrie" (*Mem. de Brux.*, t. XI., 1837).

We now come to Plücker; his "six equations" were given in a short memoir in *Crelle* (1842) preceding his great work, the *Theorie der algebraischen Curven* (1844).

Plücker first gave a scientific dual definition of a curve, viz. "A curve is a locus generated by a point, and enveloped by a line,—the point moving continuously along the line, while the line rotates continuously about the point"; the point is a point (ineunt) of the curve, the line is a tangent of the curve.

And, assuming the above theory of geometrical imaginaries, a curve such that m of its points are situate in an arbitrary line is said to be of the order m; a curve such that n of its tangents pass through an arbitrary point is said to be of the class n; as already appearing, this notion of the order and the class of a curve is, however, due to Gergonne. Thus the line is a curve of the order 1 and the class 0; and corresponding dually thereto, we have the point as a curve of the order 0 and the class 1.

Plücker moreover imagined a system of line-coordinates (tangential coordinates). The Cartesian coordinates (x, y) and trilinear coordinates (x, y, z) are point-coordinates for determining the position of a point; the new coordinates, say (ξ, η, ζ), are line-coordinates for determining the position of a line. It is possible, and (not so much for any application thereof as in order to more fully establish the analogy between the two kinds of coordinates) important, to give independent quantitative definitions of the two kinds of coordinates; but we may also derive the notion of line-coordinates from that of point-coordinates; viz. taking $\xi x + \eta y + \zeta z = 0$ to be the equation of a line, we say that (ξ, η, ζ) are the line-coordinates of this line. A linear relation $a\xi + b\eta + c\zeta = 0$ between these coordinates determines a point, viz. the point whose point-coordinates are (a, b, c); in fact, the equation in question $a\xi + b\eta + c\zeta = 0$ expresses that the equation $\xi x + \eta y + \zeta z = 0$, where (x, y, z) are current point-coordinates, is satisfied on writing therein $x, y, z = a, b, c$; or that the line in question passes through the point (a, b, c). Thus (ξ, η, ζ) are the line-coordinates of any line whatever; but when these, instead of being absolutely arbitrary, are subject to the restriction $a\xi + b\eta + c\zeta = 0$, this obliges the line to pass through a point (a, b, c); and the last-mentioned equation $a\xi + b\eta + c\zeta = 0$ is considered as the line-equation of this point.

A line has only a point-equation, and a point has only a line-equation; but any other curve has a point-equation and also a line-equation; the point-equation $(*\between x, y, z)^m = 0$ is the relation which is satisfied by the point-coordinates (x, y, z) of each point of the curve; and similarly the line-equation $(*\between \xi, \eta, \zeta)^n = 0$ is the relation which is satisfied by the line-coordinates (ξ, η, ζ) of each line (tangent) of the curve.

There is in analytical geometry little occasion for any explicit use of line-coordinates; but the theory is very important; it serves to show that, in demonstrating by point-coordinates any purely descriptive theorem whatever, we demonstrate the correlative theorem; that is, we do not demonstrate the one theorem, and then (as by the method of reciprocal polars) deduce from it the other, but we do at one and the same time demonstrate the two theorems; our (x, y, z) instead of meaning point-coordinates may mean line-coordinates, and the demonstration is then in every step of it a demonstration of the correlative theorem.

The above dual generation explains the nature of the singularities of a plane curve. The ordinary singularities, arranged according to a cross division, are

	Proper.	*Improper.*
Point-singularities—	1. The stationary point, cusp, or spinode;	2. The double point, or node;
Line-singularities—	3. The stationary tangent, or inflexion;	4. The double tangent:—

arising as follows:—

1. The cusp: the point as it travels along the line may come to rest, and then reverse the direction of its motion.

3. The stationary tangent: the line may in the course of its rotation come to rest, and then reverse the direction of its rotation.

2. The node: the point may in the course of its motion come to coincide with a former position of the point, the two positions of the line not in general coinciding.

4. The double tangent: the line may in the course of its motion come to coincide with a former position of the line, the two positions of the point not in general coinciding.

It may be remarked that we cannot with a real point and line obtain the node with two imaginary tangents (conjugate or isolated point, or acnode), nor again the real double tangent with two imaginary points of contact; but this is of little consequence, since in the general theory the distinction between real and imaginary is not attended to.

The singularities (1) and (3) have been termed proper singularities, and (2) and (4) improper; in each of the first-mentioned cases there is a real singularity, or peculiarity in the motion; in the other two cases there is not; in (2) there is not when the point is first at the node, or when it is secondly at the node, any peculiarity in the motion; the singularity consists in the point coming twice into the same position; and so in (4) the singularity is in the line coming twice into the same position. Moreover (1) and (2) are, the former a proper singularity, and the latter an improper singularity, *as regards the motion of the point;* and similarly (3) and (4) are, the former a proper singularity, and the latter an improper singularity, *as regards the motion of the line.*

But as regards the representation of a curve by an equation, the case is very different.

First, if the equation be in point-coordinates, (3) and (4) are in a sense not singularities at all. The curve $(*\between x, y, z)^m=0$, or general curve of the order m, has double tangents and inflexions; (2) presents itself as a singularity, for the equations $d_x(*\between x, y, z)^m=0$, $d_y(*\between x, y, z)^m=0$, $d_z(*\between x, y, z)^m=0$, implying $(*\between x, y, z)^m=0$, are not in general satisfied by any values (a, b, c) whatever of (x, y, z), but if such values exist, then the point (a, b, c) is a node or double point; and (1) presents itself as a further singularity or sub-case of (2), a cusp being a double point for which the two tangents become coincident.

In line-coordinates all is reversed:—(1) and (2) are not singularities; (3) presents itself as a sub-case of (4).

The theory of compound singularities will be referred to further on.

In regard to the ordinary singularities, we have

m, the order,

n " class,

δ " number of double points,

ι " " cusps,

τ " " double tangents,

κ " " inflexions;

and this being so, Plücker's "six equations" are

(1) $n = m(m-1)-2\delta-3\kappa$,

(2) $\iota = 3m(m-2)-6\delta-8\kappa$,

(3) $\tau = \frac{1}{2}m(m-2)(m^2-9)-(m^2-m-6)(2\delta+3\kappa)+2\delta(\delta-1)+6\delta\kappa+\frac{9}{2}\kappa(\kappa-1)$,

(4) $m = n(n-1)-2\tau-3\iota$,

(5) $\kappa = 3n(n-2)-6\tau-8\iota$,

(6) $\delta = \frac{1}{2}n(n-2)(n^2-9)-(n^2-n-6)(2\tau+3\iota)+2\tau(\tau-1)+6\tau\iota+\frac{9}{2}\iota(\iota-1)$.

It is easy to derive the further forms—

$$\begin{array}{lll}(7) & \iota - \kappa & = 3(n-m), \\ (8) & 2(\tau - \delta) & = \ (n-m)(n+m-9), \\ (9) & \tfrac{1}{2}m(m+3) - \delta - 2\kappa & = \tfrac{1}{2}n(n+3) - \tau - 2\iota, \\ (10) & \tfrac{1}{2}(m-1)(m-2) - \delta - \kappa & = \tfrac{1}{2}(n-1)(n-2) - \tau - \iota, \\ (11,\ 12) & m^2 - 2\delta - 3\kappa & = n^2 - 2\tau - 3\iota, \ = m + n,\end{array}$$

the whole system being equivalent to three equations only: and it may be added that, using α to denote the equal quantities $3m + \iota$ and $3n + \kappa$, everything may be expressed in terms of m, n, α. We have

$$\begin{aligned}\kappa &= \alpha - 3n, \\ \iota &= \alpha - 3m, \\ 2\delta &= m^2 - m + 8n - 3\alpha, \\ 2\tau &= n^2 - n + 8m - 3\alpha.\end{aligned}$$

It is implied in Plücker's theorem that, m, n, δ, κ, τ, ι signifying as above in regard to any curve, then in regard to the reciprocal curve n, m, τ, ι, δ, κ will have the same significations, viz. for the reciprocal curve these letters denote respectively the order, class, number of nodes, cusps, double tangents, and inflexions.

The expression $\tfrac{1}{2}m(m+3) - \delta - 2\kappa$ is that of the number of the disposable constants in a curve of the order m with δ nodes and κ cusps (in fact that there shall be a node is 1 condition, a cusp 2 conditions): and the equation (9) thus expresses that the curve and its reciprocal contain each of them the same number of disposable constants.

For a curve of the order m, the expression $\tfrac{1}{2}m(m-1) - \delta - \kappa$ is termed the "deficiency" (as to this more hereafter); the equation (10) expresses therefore that the curve and its reciprocal have each of them the same deficiency.

The relations $m^2 - 2\delta - 3\kappa = n^2 - 2\tau - 3\iota$, $= m + n$, present themselves in the theory of envelopes, as will appear further on.

With regard to the demonstration of Plücker's equations it is to be remarked that we are not able to write down the equation in point-coordinates of a curve of the order m, having the given numbers δ and κ of nodes and cusps. We can only use the general equation $(*\between x, y, z)^m = 0$, say for shortness $u = 0$, of a curve of the mth order, which equation, so long as the coefficients remain arbitrary, represents a curve without nodes or cusps. Seeking then, for this curve, the values n, ι, τ of the class, number of inflexions, and number of double tangents,—first, as regards the class, this is equal to the number of tangents which can be drawn to the curve from an arbitrary point, or what is the same thing, it is equal to the number of the points of contact of these tangents. The points of contact are found as the intersections of the curve $u = 0$ by a curve depending on the position of the arbitrary point, and called the "first polar" of this point; the order of the first polar is $= m - 1$, and the number of intersections is thus $=m(m-1)$. But it can be shown, analytically or geometrically, that if the given curve has a node, the first polar passes through this node, which therefore counts as two intersections: and that if the curve has a cusp, the first polar passes through the cusp, touching the curve there, and hence the cusp counts as three intersections. But, as is evident, the node or cusp is not a point of contact of a proper tangent from the arbitrary point; we have, therefore, for a node a diminution 2, and for a cusp a diminution 3, in the number of the intersections; and thus, for a curve with δ nodes and κ cusps, there is a diminution $2\delta+3\kappa$, and the value of n is $n=m(m-1)-2\delta-3\kappa$.

Secondly, as to the inflexions, the process is a similar one; it can be shown that the inflexions are the intersections of the curve by a derivative curve called (after Hesse, who first considered it) the Hessian, defined geometrically as the locus of a point such that its conic polar in regard to the curve breaks up into a pair of lines, and which has an equation $H=0$, where H is the determinant formed with the second differential coefficients of u in regard to the variables (x, y, z); $H=0$ is thus a curve of the order $3(m-2)$, and the number of inflexions is $=3m(m-2)$. But if the given curve has a node, then not only the Hessian passes through the node, but it has there a node the two branches at which touch respectively the two branches of the curve, and the node thus counts as six intersections; so if the curve has a cusp, then the Hessian not only passes through the cusp, but it has there a cusp through which it again passes, that is, there is a cuspidal branch touching the cuspidal branch of the curve, and besides a simple branch passing through the cusp, and hence the cusp counts as eight intersections. The node or cusp is not an inflexion, and we have thus for a node a diminution 6, and for a cusp a diminution 8, in the number of the intersections; hence for a curve with δ nodes and κ cusps, the diminution is $=6\delta+8\kappa$, and the number of inflexions is $\iota=3m(m-2)-6\delta-8\kappa$.

Thirdly, for the double tangents; the points of contact of these are obtained as the intersections of the curve by a curve $\Pi=0$, which has not as yet been geometrically defined, but which is found analytically to be of the order $(m-2)(m^2-9)$; the number of intersections is thus $=m(m-2)(m^2-9)$; but if the given curve has a node then there is a diminution $=4(m^2-m-6)$, and if it has a cusp then there is a diminution $=6(m^2-m-6)$, where, however, it is to be noticed that the factor (m^2-m-6) is in the case of a curve having only a node or only a cusp the number of the tangents which can be drawn from the node or cusp to the curve, and is used as denoting the number of these tangents, and ceases to be the correct expression if the number of nodes and cusps is greater than unity. Hence, in the case of a curve which has δ nodes and κ cusps, the apparent diminution $2(m^2-m-6)(2\delta+3\kappa)$ is too great, and it has in fact to be diminished by $2\{2\delta(\delta-1)+6\delta\kappa+\frac{9}{2}\kappa(\kappa-1)\}$, or the half thereof is 4 for each pair of nodes, 6 for each combination of a node and cusp, and 9 for each pair of cusps. We have thus finally an expression for 2τ, $=m(m-2)(m^2-9)-$&c.; or dividing the whole by 2, we have the expression for τ given by the third of Plücker's equations.

It is obvious that we cannot by consideration of the equation $u=0$ in point-coordinates obtain the remaining three of Plücker's equations; they might be obtained in a precisely analogous manner by means of the equation $v=0$ in line-coordinates, but they follow at once from the principle of duality, viz. they are obtained by the mere interchange of m, δ, κ with n, τ, ι respectively.

To complete Plücker's theory it is necessary to take account of compound singularities; it might be possible, but it is at any rate difficult, to effect this by considering the curve as in course of description by the point moving along the rotating line; and it seems easier to consider the compound singularity as arising from the variation of an actually described curve with ordinary singularities. The most simple case is when three double points come into coincidence, thereby giving rise to a triple point; and a somewhat more complicated one is when we have a cusp of the second kind, or node-cusp arising from the coincidence of a node, a cusp, an inflexion, and a double tangent, as shown in the annexed figure, which represents the singularities as on the

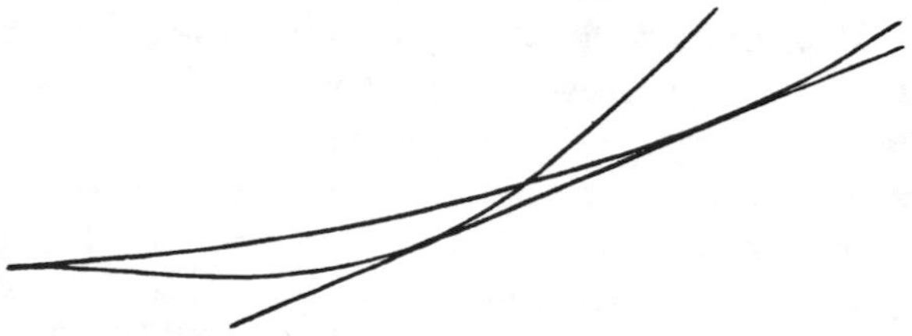

point of coalescing. The general conclusion (see Cayley, *Quart. Math. Jour.* t. VII., 1866, [374], "On the higher singularities of a plane curve") is that every singularity whatever may be considered as compounded of ordinary singularities, say we have a singularity $=\delta'$ nodes, κ' cusps, τ' double tangents, and ι' inflexions. So that, in fact, Plücker's equations properly understood apply to a curve with any singularities whatever.

By means of Plücker's equations we may form a table—

m	n	δ	κ	τ	ι
0	1	—	—	0	0
1	0	0	0	—	—
2	2	0	0	0	0
3	6	0	0	0	9
,,	4	1	0	0	3
,,	3	0	1	0	1
4	12	0	0	28	24
,,	10	1	0	16	18
,,	9	0	1	10	16
,,	8	2	0	8	12
,,	7	1	1	4	10
,,	6	0	2	1	8
,,	6	3	0	4	6
,,	5	2	1	2	4
,,	4	1	2	1	2
,,	3	0	3	1	0

The table is arranged according to the value of m; and we have $m=0$, $n=1$, the point; $m=1$, $n=0$, the line; $m=2$, $n=2$, the conic; of $m=3$, the cubic, there are three cases, the class being 6, 4, or 3, according as the curve is without singularities, or as it has 1 node, or 1 cusp; and so of $m=4$, the quartic, there are nine cases, where observe that in *two* of them the class is $=6$,—the reduction of class arising from two cusps or else from three nodes. The nine cases may be also grouped together into four, according as the number of nodes and cusps $(\delta+\kappa)$ is $=0$, 1, 2, or 3.

The cases may be divided into sub-cases, by the consideration of compound singularities; thus when $m=4$, $n=6$, $=3$, the three nodes may be all distinct, which is the general case, or two of them may unite together into the singularity called a tacnode, or all three may unite together into a triple point, or else into an oscnode.

We may further consider the inflexions and double tangents, as well in general as in regard to cubic and quartic curves.

The expression for the number of inflexions $3m(m-2)$ for a curve of the order m was obtained analytically by Plücker, but the theory was first given in a complete form by Hesse in the two papers "Ueber die Elimination, u.s.w.," and "Ueber die Wendepuncte der Curven dritter Ordnung" (*Crelle*, t. XXVIII., 1844); in the latter of these the points of inflexion are obtained as the intersections of the curve $u=0$ with the Hessian, or curve $\Delta=0$, where Δ is the determinant formed with the second derived functions of u. We have in the Hessian the first instance of a covariant of a ternary form. The whole theory of the inflexions of a cubic curve is discussed in a very interesting manner by means of the canonical form of the equation $x^3+y^3+z^3+6lxyz=0$; and in particular a proof is given of Plücker's theorem that the nine points of inflexion of a cubic curve lie by threes in twelve lines.

It may be noticed that the nine inflexions of a cubic curve are three real, six imaginary; the three real inflexions lie in a line, as was known to Newton and Maclaurin. For an acnodal cubic the six imaginary inflexions disappear, and there remain three real inflexions lying in a line. For a crunodal cubic, the six inflexions which disappear are two of them real, the other four imaginary, and there remain two imaginary inflexions and one real inflexion. For a cuspidal cubic the six imaginary inflexions and two of the real inflexions disappear, and there remains one real inflexion.

A quartic curve has 24 inflexions; it was conjectured by Salmon, and has been verified recently by Zeuthen, that at most 8 of these are real.

The expression $\frac{1}{2}m(m-2)(m^2-9)$ for the number of double tangents of a curve of the order m was obtained by Plücker only as a consequence of his first, second, fourth, and fifth equations. An investigation by means of the curve $\Pi=0$, which by its intersections with the given curve determines the points of contact of the double tangents, is indicated by Cayley, "Recherches sur l'élimination et la théorie des courbes", (*Crelle*, t. XXXIV., 1847), [53]: and in part carried out by Hesse in the memoir "Ueber Curven dritter Ordnung" (*Crelle*, t. XXXVI., 1848). A better process was indicated by Salmon in the "Note on the double tangents to plane curves," *Phil. Mag.* 1858; considering the $m-2$ points in which any tangent to the curve again meets the curve, he showed how to form the equation of a curve of the order $(m-2)$, giving by its intersection with the tangent the points in question; making the tangent touch this curve of the order $(m-2)$, it will be a double tangent of the original curve. See Cayley, "On the Double Tangents of a Plane Curve", (*Phil. Trans.* t. CXLVIII., 1859), [260], and Dersch (*Math. Ann.* t. VII., 1874). The solution is still in so far incomplete that we have no properties of the curve $\Pi=0$, to distinguish one such curve from the several other curves which pass through the points of contact of the double tangents.

A quartic curve has 28 double tangents, their points of contact determined as the intersections of the curve by a curve $\Pi=0$ of the order 14, the equation of which in a very elegant form was first obtained by Hesse (1849). Investigations in regard to them are given by Plücker in the *Theorie der algebraischen Curven*, and in two memoirs by Hesse and Steiner (*Crelle*, t. XLV., 1855), in respect to the triads of double tangents which have their points of contact on a conic, and other like relations. It was assumed by Plücker that the number of real double tangents might be 28, 16, 8, 4, or 0, but Zeuthen has recently found that the last case does not exist.

The Hessian Δ has just been spoken of as a covariant of the form u; the notion of invariants and covariants belongs rather to the form u than to the curve $u=0$ represented by means of this form; and the theory may be very briefly referred to. A curve $u=0$ may have some invariantive property, viz. a property independent of the particular axes of coordinates used in the representation of the curve by its equation; for instance, the curve may have a node, and in order to this, a relation, say $A=0$, must exist between the coefficients of the equation; supposing the axes of coordinates altered, so that the equation becomes $u'=0$, and writing $A'=0$ for the relation between the new coefficients, then the relations $A=0$, $A'=0$, as two different expressions of the same geometrical property, must each of them imply the other; this can only be the case when A, A' are functions differing only by a constant factor, or say, when A is an invariant of u. If, however, the geometrical property requires two or more relations between the coefficients, say $A=0$, $B=0$, &c., then we must have between the new coefficients the like relations, $A'=0$, $B'=0$, &c., and the two systems of equations must each of them imply the other; when this is so, the system of equations, $A=0$, $B=0$, &c., is said to be invariantive, but it does not follow that A, B, &c., are of necessity invariants of u. Similarly, if we have a curve $U=0$ derived from the curve $u=0$ in a manner independent of the particular axes of coordinates, then from the transformed equation $u'=0$ deriving in like manner the curve $U'=0$, the two equations $U=0$, $U'=0$ must each of them imply the other; and when this is so, U will be a covariant of u. The case is less frequent, but it may arise, that there are covariant systems $U=0$, $V=0$, &c., and $U'=0$, $V'=0$, &c., each implying the other, but where the functions U, V, &c., are not of necessity covariants of u.

The theory of the invariants and covariants of a ternary cubic function u has been studied in detail, and brought into connexion with the cubic curve $u=0$; but the theory of the invariants and covariants for the next succeeding case, the ternary quartic function, is still very incomplete.

In further illustration of the Plückerian dual generation of a curve, we may consider the question of the *envelope* of a variable curve. The notion is very probably older, but it is at any rate to be found in Lagrange's *Théorie des fonctions analytiques* (1798); it is there remarked that the equation obtained by the elimination of the parameter a from an equation $f(x, y, a) = 0$ and the derived equation in respect to a is a curve, the envelope of the series of curves represented by the equation $f(x, y, a) = 0$ in question. To develope the theory, consider the curve corresponding to any particular value of the parameter; this has with the consecutive curve (or curve belonging to the consecutive value of the parameter) a certain number of intersections, and of common tangents, which may be considered as the tangents at the intersections; and the so-called envelope is the curve which is at the same time generated by the points of intersection and enveloped by the common tangents; we have thus a dual generation. But the question needs to be further examined. Suppose that in general the variable curve is of the order m with δ nodes and κ cusps, and therefore of the class n with τ double tangents and ι inflexions, m, n, δ, κ, τ, ι being connected by the Plückerian equations,—the number of nodes or cusps may be greater for particular values of the parameter, but this is a speciality which may be here disregarded. Considering the variable curve corresponding to a given value of the parameter, or say simply the variable curve, the consecutive curve has then also δ and κ nodes and cusps, consecutive to those of the variable curve; and it is easy to see that among the intersections of the two curves we have the nodes each counting twice, and the cusps each counting three times; the number of the remaining intersections is $= m^2 - 2\delta - 3\kappa$. Similarly among the common tangents of the two curves we have the double tangents each counting twice, and the stationary tangents each counting three times, and the number of the remaining common tangents is $= n^2 - 2\tau - 3\iota$ $(= m^2 - 2\delta - 3\kappa$, inasmuch as each of these numbers is as was seen $= m + n)$. At any one of the $m^2 - 2\delta - 3\kappa$ points the variable curve and the consecutive curve have tangents distinct from yet infinitesimally near to each other, and each of these two tangents is also infinitesimally near to one of the $n^2 - 2\tau - 3\iota$ common tangents of the two curves; whence, attending only to the variable curve, and considering the consecutive curve as coming into actual coincidence with it, the $n^2 - 2\tau - 3\iota$ common tangents are the tangents to the variable curve at the $m^2 - 2\delta - 3\kappa$ points respectively, and the envelope is at the same time generated by the $m^2 - 2\delta - 3\kappa$ points, and enveloped by the $n^2 - 2\tau - 3\iota$ tangents; we have thus a dual generation of the envelope, which only differs from Plücker's dual generation, in that in place of a single point and tangent we have the group of $m^2 - 2\delta - 3\kappa$ points and $n^2 - 2\tau - 3\iota$ tangents.

The parameter which determines the variable curve may be given as a point upon a given curve, or say as a parametric point; that is, to the different positions of the parametric point on the given curve correspond the different variable curves, and the nature of the envelope will thus depend on that of the given curve; we have thus the envelope as a derivative curve of the given curve. Many well-known derivative curves present themselves in this manner; thus the variable curve may be the normal (or line at right angles to the tangent) at any point of the given curve; the intersection of the consecutive normals is the centre of curvature; and we have the evolute as at once the locus of the centre of curvature and the envelope of the normal. It may be added that the given curve is one of a series of curves, each cutting the several normals at right angles. Any one of these is a "parallel" of the given curve; and it can be obtained as the envelope of a circle of constant radius having its centre on the given curve. We have in like manner, as derivatives of a given curve, the caustic, catacaustic, or diacaustic, as the case may be, and the secondary caustic, or curve cutting at right angles the reflected or refracted rays.

We have in much that precedes disregarded, or at least been indifferent to, reality; it is only thus that the conception of a curve of the mth order, as one which is met by every right line in m points, is arrived at; and the curve itself, and the line which cuts it, although both are tacitly assumed to be real, may perfectly well be imaginary. For real figures we have the general theorem that imaginary intersections, &c., present themselves in conjugate pairs: hence, in particular, that a curve of an even order is met by a line in an even number (which may be $=0$) of points; a curve of an odd order in an odd number of points, hence in one point at least; it will be seen further on that the theorem may be generalized in a remarkable manner. Again, when there is in question only one pair of points or lines, these, if coincident, must be real; thus, a line meets a cubic curve in three points, one of them real, the other two real or imaginary; but if two of the intersections coincide they must be real, and we have a line cutting a cubic in one real point and touching it in another real point. It may be remarked that this is a limit separating the two cases where the intersections are all real, and where they are one real, two imaginary.

Considering always real curves, we obtain the notion of a branch; any portion capable of description by the continuous motion of a point is a branch; and a curve consists of one or more branches. Thus the curve of the first order or right line consists of one branch; but in curves of the second order, or conics, the ellipse and the parabola consist each of one branch, the hyperbola of two branches. A branch is either re-entrant, or it extends both ways to infinity, and in this case, we may regard it as consisting of two legs (*crura*, Newton), each extending one way to infinity, but without any definite separation. The branch, whether re-entrant or infinite, may have a cusp or cusps, or it may cut itself or another branch, thus having or giving rise to crunodes; an acnode is a branch by itself,—it may be considered as an indefinitely small re-entrant branch. A branch may have inflexions and double tangents, or there may be double tangents which touch two distinct branches; there are also double tangents with imaginary points of contact, which are thus lines having no visible connexion with the curve. A re-entrant branch not cutting itself may be everywhere convex, and it is then properly said to be an oval; but the term oval may be used more generally for any re-entrant branch not cutting itself; and we may thus speak of a once indented, twice indented oval, &c., or even of a cuspidate oval. Other descriptive names for ovals and re-entrant branches cutting themselves may be used when required; thus, in the last-mentioned case a simple form is that of a figure of eight; such a form may break up into two ovals, or into a doubly indented oval or hour-glass. A form which presents itself is when two ovals, one inside the other, unite, so as to give rise to a crunode—in default of a better name this may be called, after the curve of that name, a limaçon. Names may also be used for the different forms of infinite branches, but we have first to consider the distinction of hyperbolic and parabolic. The leg of an infinite branch may have at the extremity a tangent; this is an asymptote of the curve, and the leg is then hyperbolic; or the leg may tend to a fixed direction, but so that the tangent goes further and further off to infinity, and the leg is then parabolic; a branch may thus be hyperbolic or parabolic as to its two legs; or it may be hyperbolic as to one leg, and parabolic as to the other. The epithets hyperbolic and parabolic are of course derived from the conics hyperbola and parabola respectively. The nature of the two kinds of branches is best understood by considering them as projections, in the same way as we in effect consider the hyperbola and the parabola as projections of the ellipse. If a line Ω cut an arc aa', so that the two segments ab, ba' lie on opposite sides of the line, then projecting the figure so that the line Ω goes off to infinity, the tangent at b is projected into the asymptote, and the arc ab is projected into a hyperbolic leg touching the asymptote at one extremity; the arc ba' will at the same time be projected into a hyperbolic leg touching the same asymptote at the other extremity (and on the opposite side), but so that the two hyperbolic legs may or may not belong to one and the same branch. And we thus see that the two hyperbolic legs belong to a simple intersection of the curve by the line infinity. Next, if the line Ω touch at b the arc aa' so that the two portions ab', ba lie on the same side of the line Ω, then projecting the figure as before, the tangent at b, that is, the line Ω itself, is projected to infinity; the arc ab is projected into a parabolic leg, and at the same time the arc ba' is projected into a parabolic leg, having at infinity the same direction as the other leg, but so that the two legs may or may not belong to the same branch. And we thus see that the two parabolic legs represent a contact of the line infinity with the curve,—the point of contact being of course the point at infinity determined by the common direction of the two legs. It will readily be understood how the like considerations apply to other cases,—for instance, if the line Ω is a tangent at an inflexion, passes through a crunode, or touches one of the branches of a crunode, &c.; thus, if the line Ω passes through a crunode we have pairs of hyperbolic legs belonging to two parallel asymptotes. The foregoing considerations also show (what is very important) how different branches are connected together at infinity, and lead to the notion of a complete branch, or circuit.

The two legs of a hyperbolic branch may belong to different asymptotes, and in this case we have the forms which Newton calls inscribed, circumscribed, ambigene, &c.; or they may belong to the same asymptote, and in this case we have the serpentine form, where the branch cuts the asymptote, so as to touch it at its two extremities on opposite sides, or the conchoidal form, where it touches the asymptote on the same side. The two legs of a parabolic branch may converge to ultimate parallelism, as in the conic parabola, or diverge to ultimate parallelism, as in the semi-cubical parabola $y^2 = x^3$, and the branch is said to be convergent, or divergent, accordingly; or they may tend to parallelism in opposite senses, as in the cubical parabola $y = x^3$. As mentioned with regard to a branch generally, an infinite branch of any kind may have cusps, or, by cutting itself or another branch, may have or give rise to a crunode, &c.

We may now consider the various forms of cubic curves, as appearing by Newton's *Enumeratio*, and by the figures belonging thereto. The species are reckoned as 72, which are numbered accordingly 1 to 72; but to these should be added 10^a, 13^a, 22^a, and 22^b. It is not intended here to consider the division into species, nor even completely that into genera, but only to explain the principle of classification. It may be remarked generally that there are at most three infinite branches, and that there may besides be a re-entrant branch or oval.

The genera may be arranged as follows:—

1, 2, 3, 4	redundant hyperbolas,
5, 6	defective hyperbolas,
7, 8	parabolic hyperbolas,
9	hyperbolisms of hyperbola,
10	„ „ ellipse,
11	„ „ parabola,
12	trident curve,
13	divergent parabolas,
14	cubic parabola;

and, thus arranged, they correspond to the different relations of the line infinity to the curve. First, if the three intersections by the line infinity are all distinct, we have the hyperbolas; if the points are real, the redundant hyperbolas, with three hyperbolic branches; but if only one of them is real, the defective hyperbolas, with one hyperbolic branch. Secondly, if two of the intersections coincide, say if the line infinity meets the curve in a onefold point and a twofold point, both of them real, then there is always one asymptote: the line infinity may at the twofold point touch the curve, and we have the parabolic hyperbolas; or the twofold point may be a singular point,—viz. a crunode giving the hyperbolisms of the hyperbola; an acnode, giving the hyperbolisms of the ellipse; or a cusp, giving the hyperbolisms of the parabola. As regards the so-called hyperbolisms, observe that (besides the single asymptote) we have in the case of those of the hyperbola two parallel asymptotes; in the case of those of the ellipse the two parallel asymptotes become imaginary, that is, they disappear, and in the case of those of the parabola they become coincident, that is, there is here an ordinary asymptote, and a special asymptote answering to a cusp at infinity. Thirdly, the three intersections by the line infinity may be coincident and real; or say we have a threefold point: this may be an inflexion, a crunode, or a cusp, that is, the line infinity may be a tangent at an inflexion, and we have the divergent parabolas; a tangent at a crunode to one branch, and we have the trident curve; or lastly, a tangent at a cusp, and we have the cubical parabola.

It is to be remarked that the classification mixes together non-singular and singular curves, in fact, the five kinds presently referred to: thus the hyperbolas and the divergent parabolas include curves of every kind, the separation being made in the species; the hyperbolisms of the hyperbola and ellipse, and the trident curve, are nodal; the hyperbolisms of the parabola, and the cubical parabola, are cuspidal. The divergent parabolas are of five species which respectively belong to and determine the five kinds of cubic curves; Newton gives (in two short paragraphs without any development) the remarkable theorem that the five divergent parabolas by their shadows generate and exhibit all the cubic curves.

The five divergent parabolas are curves each of them symmetrical with regard to an axis. There are two non-singular kinds, the one with, the other without, an oval, but each of them has an infinite (as Newton describes it) *campaniform* branch; this cuts the axis at right angles, being at first convex, but ultimately concave, towards the axis, the two legs continually tending to become at right angles to the axis. The oval may unite itself with the infinite branch, or it may dwindle into a point, and we have the crunodal and the acnodal forms respectively; or if simultaneously the oval dwindles into a point and unites itself to the infinite branch, we have the cuspidal form. Drawing a line to cut any one of these curves and projecting the line to infinity, it would not be difficult to show how the line should be drawn in order to obtain a curve of any given species. We have herein a better principle of classification; considering cubic curves, in the first instance, according to singularities, the curves are non-singular, nodal (viz. crunodal or acnodal), or cuspidal; and we see further that there are two kinds of non-singular curves, the complex and the simplex. There is thus a complete division into the five kinds, the complex, simplex, crunodal, acnodal, and cuspidal. Each singular kind presents itself as a limit separating two kinds of inferior singularity; the cuspidal separates the crunodal and the acnodal, and these last separate from each other the complex and the simplex.

The whole question is discussed very fully and ably by Möbius in the memoir "Ueber die Grundformen der Linien dritter Ordnung" (*Abh. der K. Sachs. Ges. zu Leipzig*, t. I., 1852; *Ges. Werke*, t. I.). The author considers not only plane curves, but also cones, or, what is almost the same thing, the spherical curves which are their sections by a concentric sphere. Stated in regard to the cone, we have there the fundamental theorem that there are two different kinds of sheets: viz. the single sheet, not separated into two parts by the vertex (an instance is afforded by the plane considered as a cone of the first order generated by the motion of a line about a point), and the double or twin-pair sheet, separated into two parts by the vertex (as in the cone of the second order). And it then appears that there are two kinds of non-singular cubic cones, viz. the simplex, consisting of a single sheet, and the complex, consisting of a single sheet and a twin-pair sheet; and we thence obtain (as for cubic curves) the crunodal, the acnodal, and the cuspidal kinds of cubic cones. It may be mentioned that the single sheet is a sort of wavy form, having upon it three lines of inflexion, and which is met by any plane through the vertex in one or in three lines; the twin-pair sheet has no lines of inflexion, and resembles in its form a cone on an oval base.

In general a cone consists of one or more single or twin-pair sheets, and if we consider the section of the cone by a plane, the curve consists of one or more complete branches, or say circuits, each of them the section of one sheet of the cone;

thus, a cone of the second order is one twin-pair sheet, and any section of it is one circuit composed, it may be, of two branches. But although we thus arrive by projection at the notion of a circuit, it is not necessary to go out of the plane, and we may (with Zeuthen, using the shorter term *circuit* for his *complete branch*) define a circuit as any portion (of a curve) capable of description by the continuous motion of a point, it being understood that a passage through infinity is permitted. And we then say that a curve consists of one or more circuits; thus the right line, or curve of the first order, consists of one circuit; a curve of the second order consists of one circuit; a cubic curve consists of one circuit or else of two circuits.

A circuit is met by any right line always in an even number, or always in an odd number, of points, and it is said to be an even circuit or an odd circuit accordingly; the right line is an odd circuit, the conic an even circuit. And we have then the theorem, two odd circuits intersect in an odd number of points; an odd and an even circuit, or two even circuits, in an even number of points. An even circuit not cutting itself divides the plane into two parts, the one called the internal part, incapable of containing any odd circuit, the other called the external part, capable of containing an odd circuit.

We may now state in a more convenient form the fundamental distinction of the kinds of cubic curve. A non-singular cubic is simplex, consisting of one odd circuit, or it is complex, consisting of one odd circuit and one even circuit. It may be added that there are on the odd circuit three inflexions, but on the even circuit no inflexion; it hence also appears that from any point of the odd circuit there can be drawn to the odd circuit two tangents, and to the even circuit (if any) two tangents, but that from a point of the even circuit there cannot be drawn (either to the odd or the even circuit) any real tangent; consequently, in a simplex curve the number of tangents from any point is two; but in a complex curve the number is four, or none,—four if the point is on the odd circuit, none if it is on the even circuit. It at once appears from inspection of the figure of a non-singular cubic curve, which is the odd and which the even circuit. The singular kinds arise as before; in the crunodal and the cuspidal kinds the whole curve is an odd circuit, but in the acnodal kind the acnode must be regarded as an even circuit.

The analogous question of the classification of quartics (in particular non-singular quartics and nodal quartics) is considered in Zeuthen's memoir "Sur les différentes formes des courbes planes du quatrième ordre" (*Math. Ann.* t. VII., 1874). A non-singular quartic has only even circuits; it has at most four circuits external to each other, or two circuits one internal to the other, and in this last case the internal circuit has no double tangents or inflexions. A very remarkable theorem is established as to the double tangents of such a quartic:—distinguishing as a double tangent of the first kind a real double tangent which either twice touches the same circuit, or else touches the curve in two imaginary points, the number of the double tangents of the first kind of a non-singular quartic is $=4$; it follows that the quartic has at most 8 real inflexions. The forms of the non-singular quartics are very numerous, but it is not necessary to go further into the question.

We may consider in relation to a curve, not only the line infinity, but also the circular points at infinity; assuming the curve to be real, these present themselves always conjointly; thus a circle is a conic passing through the two circular points, and is thereby distinguished from other conics. Similarly a cubic through the two circular points is termed a circular cubic; a quartic through the two points is termed a circular quartic, and if it passes twice through each of them, that is, has each of them for a node, it is termed a bicircular quartic. Such a quartic is of course binodal ($m = 4$, $\delta = 2$, $\kappa = 0$); it has not in general, but it may have, a third node, or a cusp. Or again, we may have a quartic curve having a cusp at each of the circular points: such a curve is a "Cartesian," it being a complete definition of the Cartesian to say that it is a bicuspidal quartic curve ($m = 4$, $\delta = 0$, $\kappa = 2$), having a cusp at each of the circular points. The circular cubic and the bicircular quartic, together with the Cartesian (being in one point of view a particular case thereof), are interesting curves which have been much studied, generally, and in reference to their *focal* properties.

The points called *foci* presented themselves in the theory of the conic, and were well known to the Greek geometers, but the general notion of a focus was first established by Plücker, in the memoir "Ueber solche Puncte die bei Curven einer höheren Ordnung den Brennpuncten der Kegelschnitte entsprechen," (*Crelle*, t. x., 1833). We may from each of the circular points draw tangents to a given curve; the intersection of two such tangents (belonging of course to the two circular points respectively) is a focus. There will be from each circular point λ tangents (λ, a number depending on the class of the curve and its relation to the line infinity and the circular points, $= 2$ for the general conic, 1 for the parabola, 2 for a circular cubic or a bicircular quartic, &c.); the λ tangents from the one circular point and those from the other circular point intersect in λ real foci (viz. each of these is the only real point on each of the tangents through it), and in $\lambda^2 - \lambda$ imaginary foci; each pair of real foci determines a pair of imaginary foci (the so-called antipoints of the two real foci), and the $\frac{1}{2}\lambda(\lambda - 1)$ pairs of real foci thus determine the $\lambda^2 - \lambda$ imaginary foci. There are in some cases points termed çentres, or singular or multiple foci (the nomenclature is unsettled), which are the intersections of improper tangents from the two circular points respectively; thus, in the circular cubic, the tangents to the curve at the two circular points respectively (or two imaginary asymptotes of the curve) meet in a centre.

The notions of *distance* and of lines *at right angles* are connected with the circular points; and almost every construction of a curve by means of lines of a determinate length, or at right angles to each other, and (as such) mechanical constructions by means of linkwork, give rise to curves passing the same definite number of times through the two circular points respectively, or say to circular curves, and in which the fixed centres of the construction present themselves as ordinary, or as singular, foci. Thus the general curve of three-bar motion (or locus of the vertex of a triangle, the other two vertices whereof move on fixed circles) is a tricircular sextic, having besides three nodes ($m = 6$, $\delta = 3 + 3 + 3$, $= 9$), and having the centres of the fixed circles each for a singular focus; there is a third singular focus, and we have thus the remarkable theorem (due to Mr S. Roberts) of the triple generation of the curve by means of the three several pairs of singular foci.

Again, the normal, *qua* line at right angles to the tangent, is connected with the circular points, and these accordingly present themselves in the before-mentioned theories of evolutes and parallel curves.

We have several recent theories which depend on the notion of *correspondence:* two points whether in the same plane or in different planes, or on the same curve or in different curves, may determine each other in such wise that to any given position of the first point there correspond α' positions of the second point, and to any given position of the second point α positions of the first point; the two points have then an (α, α') correspondence; and if α, α' are each $=1$, then the two points have a (1, 1) or rational correspondence. Connecting with each theory the author's name, the theories in question are—Riemann, the rational transformation of a plane curve; Cremona, the rational transformation of a plane; and Chasles, correspondence of points on the same curve, and united points. The theory first referred to, with the resulting notion of Geschlecht, or *deficiency*, is more than the other two an essential part of the theory of curves, but they will all be considered.

Riemann's results are contained in the memoirs on "Theorie der Abel'schen Functionen," (*Crelle*, t. LIV., 1857); and we have next Clebsch, "Ueber die Singularitäten algebraischer Curven," (*Crelle*, t. LXV., 1865), and Cayley, "On the Transformation of Plane Curves," (*Proc. Lond. Math. Soc.* t. I., 1865, [384]). The fundamental notion of the rational transformation is as follows:—

Taking u, X, Y, Z to be rational and integral functions (X, Y, Z all of the same order) of the coordinates (x, y, z), and u', X', Y', Z' rational and integral functions (X', Y', Z' all of the same order) of the coordinates (x', y', z'), we transform a given curve $u=0$, by the equations $x' : y' : z' = X : Y : Z$, thereby obtaining a transformed curve $u'=0$, and a converse set of equations $x : y : z = X' : Y' : Z'$; viz. assuming that this is so, the point (x, y, z) on the curve $u=0$ and the point (x', y', z') on the curve $u'=0$ will be points having a (1, 1) correspondence. To show how this is, observe that to a given point (x, y, z) on the curve $u=0$ there corresponds a single point (x', y', z') determined by the equations $x' : y' : z' = X : Y : Z$; from these equations and the equation $u=0$ eliminating x, y, z we obtain the equation $u'=0$ of the transformed curve. To a given point (x', y', z') not on the curve $u'=0$ there corresponds, not a single point, but the system of points (x, y, z) given by the equations $x' : y' : z' = X : Y : Z$, viz. regarding x', y', z' as constants (and to fix the ideas, assuming that the curves $X=0$, $Y=0$, $Z=0$ have no common intersections), these are the points of intersection of the curves $X : Y : Z = x' : y' : z'$, but no one of these points is situate on the curve $u=0$. If, however, the point (x', y', z') is situate on the curve $u'=0$, then one point of the system of points in question is situate on the curve $u=0$, that is, to a given point of the curve $u'=0$ there corresponds a single point of the curve $u=0$; and hence also this point must be given by a system of equations such as $x : y : z = X' : Y' : Z'$.

It is an old and easily proved theorem that, for a curve of the order m, the number $\delta+\kappa$ of nodes and cusps is at most $=\frac{1}{2}(m-1)(m-2)$; for a given curve the deficiency of the actual number of nodes and cusps below this maximum number, viz.

$\frac{1}{2}(m-1)(m-2)-\delta-\kappa$, is the "Geschlecht," or "deficiency," of the curve, say this is $=D$. When $D=0$, the curve is said to be unicursal, when $=1$, bicursal, and so on.

The general theorem is that two curves corresponding rationally to each other have the same deficiency. In particular, a curve and its reciprocal have this rational or (1, 1) correspondence, and it has been already seen that a curve and its reciprocal have the same deficiency.

A curve of a given order can in general be rationally transformed into a curve of a lower order; thus a curve of any order for which $D=0$, that is, a unicursal curve, can be transformed into a line; a curve of any order having the deficiency 1 or 2 can be rationally transformed into a curve of the order $D+2$, deficiency D; and a curve of any order deficiency $=$ or >3 can be rationally transformed into a curve of the order $D+3$, deficiency D.

Taking x', y', z' as coordinates of a point of the transformed curve, and in its equation writing $x' : y' : z' = 1 : \theta : \phi$ we have ϕ a certain irrational function of θ, and the theorem is that the coordinates x, y, z of any point of the given curve can be expressed as proportional to rational and integral functions of θ, ϕ, that is, of θ and a certain irrational function of θ.

In particular, if $D=0$, that is, if the given curve be unicursal, the transformed curve is a line, ϕ is a mere linear function of θ, and the theorem is that the coordinates x, y, z of a point of the unicursal curve can be expressed as proportional to rational and integral functions of θ; it is easy to see that for a given curve of the order m, these functions of θ must be of the same order m.

If $D=1$, then the transformed curve is a cubic; it can be shown that in a cubic, the axes of coordinates being properly chosen, ϕ can be expressed as the square root of a quartic function of θ; and the theorem is that the coordinates x, y, z of a point of the bicursal curve can be expressed as proportional to rational and integral functions of θ, and of the square root of a quartic function of θ.

And so if $D=2$, then the transformed curve is a nodal quartic; ϕ can be expressed as the square root of a sextic function of θ, and the theorem is, that the coordinates x, y, z of a point of the tricursal curve can be expressed as proportional to rational and integral functions of θ, and of the square root of a sextic function of θ. But when $D=3$, we have no longer the like law, viz. ϕ is not expressible as the square root of an octic function of θ.

Observe that the radical, square root of a quartic function, is connected with the theory of elliptic functions, and the radical, square root of a sextic function, with that of the first kind of Abelian functions, but that the next kind of Abelian functions does not depend on the radical, square root of an octic function.

It is a form of the theorem for the case $D=1$, that the coordinates x, y, z of a point of the bicursal curve, or in particular the coordinates of a point of the cubic, can be expressed as proportional to rational and integral functions of the elliptic functions $\operatorname{sn} u$, $\operatorname{cn} u$, $\operatorname{dn} u$; in fact, taking the radical to be $\sqrt{1-\theta^2 . 1-k^2\theta^2}$, and writing

$\theta = \text{sn}\, u$, the radical becomes $= \text{cn}\, u\,.\,\text{dn}\, u$; and we have expressions of the form in question.

It will be observed that the equations $x' : y' : z' = X : Y : Z$ before-mentioned do not of themselves lead to the other system of equations $x : y : z = X' : Y' : Z'$, and thus that the theory does not in anywise establish a (1, 1) correspondence between the points (x, y, z) and (x', y', z') of two planes or of the same plane; this is the correspondence of Cremona's theory.

In this theory, given in the memoirs "Sulle trasformazioni geometriche delle figure piane," *Mem. di Bologna*, t. II. (1863), and t. V. (1865), we have a system of equations $x' : y' : z' = X : Y : Z$ which *does* lead to a system $x : y : z = X' : Y' : Z'$, where, as before, X, Y, Z denote rational and integral functions, all of the same order, of the coordinates x, y, z, and X', Y', Z' rational and integral functions, all of the same order, of the coordinates x', y', z', and there is thus a (1, 1) correspondence given by these equations between the two points (x, y, z) and (x', y', z'). To explain this, observe that starting from the equations $x' : y' : z' = X : Y : Z$, to a given point (x, y, z) there corresponds one point (x', y', z'), but that if n be the order of the functions X, Y, Z, then to a given point x', y', z' there would, if the curves $X = 0$, $Y = 0$, $Z = 0$ had no common intersections, correspond n^2 points (x, y, z). If, however, the functions are such that the curves $X = 0$, $Y = 0$, $Z = 0$ have k common intersections, then among the n^2 points are included these k points, which are fixed points independent of the point (x', y', z'); so that, disregarding these fixed points, the number of points (x, y, z) corresponding to the given point (x', y', z') is $= n^2 - k$; and in particular if $k = n^2 - 1$, then we have one corresponding point; and hence the original system of equations $x' : y' : z' = X : Y : Z$ must lead to the equivalent system $x : y : z = X' : Y' : Z'$; and in this system by the like reasoning the functions must be such that the curves $X' = 0$, $Y' = 0$, $Z' = 0$ have $n'^2 - 1$ common intersections. The most simple example is in the two systems of equations $x' : y' : z' = yz : zx : xy$ and $x : y : z = y'z' : z'x' : x'y'$; where $yz = 0$, $zx = 0$, $xy = 0$ are conics (pairs of lines) having three common intersections, and where obviously either system of equations leads to the other system. In the case where X, Y, Z are of an order exceeding 2, the required number $n^2 - 1$ of common intersections can only occur by reason of common multiple points on the three curves; and assuming that the curves $X = 0$, $Y = 0$, $Z = 0$ have $\alpha_1 + \alpha_2 + \alpha_3 + \ldots + \alpha_{n-1}$ common intersections, where the α_1 points are ordinary points, the α_2 points are double points, the α_3 points are triple points, &c., on each curve, we have the condition

$$\alpha_1 + 4\alpha_2 + 9\alpha_3 + \ldots + (n-1)^2\, \alpha_{n-1} = n^2 - 1\,;$$

but to this must be joined the condition

$$\alpha_1 + 3\alpha_2 + 6\alpha_3 + \ldots + \tfrac{1}{2}(n-1)(n-2)\, \alpha_{n-1} = \tfrac{1}{2}n(n+3) - 2,$$

(without which the transformation would be illusory); and the conclusion is that $\alpha_1, \alpha_2, \ldots, \alpha_{n-1}$ may be any numbers satisfying these two equations. It may be added that the two equations together give

$$\alpha_2 + 3\alpha_3 + \ldots + \tfrac{1}{2}(n-1)(n-2)\, \alpha_{n-1} = \tfrac{1}{2}(n-1)(n-2),$$

which expresses that the curves $X=0$, $Y=0$, $Z=0$ are unicursal. The transformation may be applied to any curve $u=0$, which is thus rationally transformed into a curve $u'=0$, by a rational transformation such as is considered in Riemann's theory; hence the two curves have the same deficiency.

Coming next to Chasles, the principle of correspondence is established and used by him in a series of memoirs relating to the conics which satisfy given conditions, and to other geometrical questions, contained in the *Comptes Rendus*, t. LVIII. *et seq.* (1864 to the present time). The theorem of united points in regard to points in a right line was given in a paper, June—July 1864, and it was extended to unicursal curves in a paper of the same series (March 1866), "Sur les courbes planes ou à double courbure dont les points peuvent se déterminer individuellement—application du principe de correspondance dans la théorie de ces courbes."

The theorem is as follows: if in a unicursal curve two points have an (α, β) correspondence, then the number of united points (or points each corresponding to itself) is $=\alpha+\beta$. In fact, in a unicursal curve the coordinates of a point are given as proportional to rational and integral functions of a parameter, so that any point of the curve is determined uniquely by means of this parameter; that is, to each point of the curve corresponds one value of the parameter, and to each value of the parameter one point on the curve; and the (α, β) correspondence between the two points is given by an equation of the form $(*\between\theta, 1)^{\alpha}(\phi, 1)^{\beta}=0$ between their parameters θ and ϕ; at a united point $\phi=\theta$, and the value of θ is given by an equation of the order $\alpha+\beta$. The extension to curves of any given deficiency D was made in the memoir of Cayley, "On the correspondence of two points on a curve,"—*Proc. Lond. Math. Soc.* t. I. (1866), [385],—viz. taking P, P' as the corresponding points in an (α, α') correspondence on a curve of deficiency D, and supposing that when P is given the corresponding points P' are found as the intersections of the curve by a curve Θ containing the coordinates of P as parameters, and having with the given curve k intersections at the point P, then the number of united points is $a=\alpha+\alpha'+2kD$; and more generally, if the curve Θ intersect the given curve in a set of points P' each p times, a set of points Q' each q times, &c., in such manner that the points (P, P'), the points (P, Q'), &c., are pairs of points corresponding to each other according to distinct laws; then if (P, P') are points having an (α, α') correspondence with a number $=a$ of united points, (P, Q') points having a (β, β') correspondence with a number $=b$ of united points, and so on, the theorem is that we have

$$p(a-\alpha-\alpha')+q(b-\beta-\beta')+\ldots=2kD.$$

The principle of correspondence, or say rather the theorem of united points, is a most powerful instrument of investigation, which may be used in place of analysis for the determination of the number of solutions of almost every geometrical problem. We can by means of it investigate the class of a curve, number of inflexions, &c.,—in fact, Plücker's equations; but it is necessary to take account of special solutions; thus, in one of the most simple instances, in finding the class of a curve, the cusps present themselves as special solutions.

Imagine a curve of order m, deficiency D, and let the corresponding points P, P' be such that the line joining them passes through a given point O; this is an $(m-1, m-1)$ correspondence, and the value of k is $=1$, hence the number of united points is $=2m-2+2D$; the united points are the points of contact of the tangents from O and (as special solutions) the cusps, and we have thus the relation $n+\kappa=2m-2+2D$; or, writing $D=\frac{1}{2}(m-1)(m-2)-\delta-\kappa$, this is $n=m(m-1)-2\delta-3\kappa$, which is right.

The principle in its original form as applying to a right line was used throughout by Chasles in the investigations on the number of the conics which satisfy given conditions, and on the number of solutions of very many other geometrical problems.

There is one application of the theory of the (α, α') correspondence between two planes which it is proper to notice.

Imagine a curve, real or imaginary, represented by an equation (involving, it may be, imaginary coefficients) between the Cartesian coordinates u, u'; then, writing $u=x+iy$, $u'=x'+iy'$, the equation determines real values of (x, y), and of (x', y'), corresponding to any given real values of (x', y') and (x, y) respectively; that is, it establishes a real correspondence (not of course a rational one) between the points (x, y) and (x', y'); for example in the imaginary circle $u^2+u'^2=(a+bi)^2$, the correspondence is given by the two equations $x^2-y^2+x'^2-y'^2=a^2-b^2$, $xy+x'y'=ab$. We have thus a means of geometrical representation for the portions, as well imaginary as real, of any real or imaginary curve. Considerations such as these have been used for determining the series of values of the independent variable, and the irrational functions thereof in the theory of Abelian integrals, but the theory seems to be worthy of further investigation.

The researches of Chasles (*Comptes Rendus*, t. LVIII., 1864, *et seq.*) refer to the conics which satisfy given conditions. There is an earlier paper by De Jonquières, "Théorèmes généraux concernant les courbes géométriques planes d'un ordre quelconque," *Liouv.* t. VI. (1861), which establishes the notion of a system of curves (of any order) of the index N, viz. considering the curves of the order n which satisfy $\frac{1}{2}n(n+3)-1$ conditions, then the index N is the number of these curves which pass through a given arbitrary point. But Chasles in the first of his papers (February 1864), considering the conics which satisfy four conditions, establishes the notion of the two characteristics (μ, ν) of such a system of conics, viz. μ is the number of the conics which pass through a given arbitrary point, and ν is the number of the conics which touch a given arbitrary line. And he gives the theorem, a system of conics satisfying four conditions, and having the characteristics (μ, ν) contains $2\nu-\mu$ line-pairs (that is, conics, each of them a pair of lines), and $2\mu-\nu$ point-pairs (that is, conics, each of them a pair of points,—coniques infiniment aplaties), which is a fundamental one in the theory. The characteristics of the system can be determined when it is known how many there are of these two kinds of degenerate conics in the system, and how often each is to be counted. It was thus that Zeuthen (in the paper *Nyt Bydrag*, "Contribution to the Theory of Systems of Conics which satisfy four Conditions," Copenhagen, 1865, translated with an addition in the *Nouvelles Annales*) solved the question of finding the characteristics of the systems of conics which satisfy four conditions of contact with a given curve or curves; and this led to the solution of the further problem of finding the number of the conics which satisfy five conditions of contact with a given curve or curves (Cayley, *Comptes Rendus*, t. LXIII., 1866, [377]), and "On the Curves which satisfy given Conditions" (*Phil. Trans.* t. CLVIII., 1868, [406]).

It may be remarked that although, as a process of investigation, it is very convenient to seek for the characteristics of a system of conics satisfying 4 conditions, yet what is really determined is in every case the number of the conics which satisfy 5 conditions; the characteristics of the system (4*p*) of the conics which pass through 4*p* points are (5*p*), (4*p*, 1*l*), the number of the conics which pass through 5 points, and which pass through 4 points and touch 1 line: and so in other cases. Similarly as regards cubics, or curves of any other order: a cubic depends on 9 constants, and the elementary problems are to find the number of the cubics (9*p*), (8*p*, 1*l*), &c., which pass through 9 points, pass through 8 points and touch 1 line, &c.; but it is in the investigation convenient to seek for the characteristics of the systems of cubics (8*p*), &c., which satisfy 8 instead of 9 conditions.

The elementary problems in regard to cubics are solved very completely by Maillard in his *Thèse, Recherche des caractéristiques des systèmes élémentaires des courbes planes du troisième ordre* (Paris, 1871). Thus, considering the several cases of a cubic

		No. of consts.
1.	With a given cusp	5,
2.	„ cusp on given line	6,
3.	„ cusp	7,
4.	„ a given node	6,
5.	„ node on given line	7,
6.	„ node	8,
7.	non-singular	9,

he determines in every case the characteristics (μ, ν) of the corresponding systems of cubics (4*p*), (3*p*, 1*l*), &c. The same problems, or most of them, and also the elementary problems in regard to quartics are solved by Zeuthen, who in the elaborate memoir "Almindelige Egenskaber, &c.," *Danish Academy*, t. X. (1873), considers the problem in reference to curves of any order, and applies his results to cubic and quartic curves.

The methods of Maillard and Zeuthen are substantially identical; in each case the question considered is that of finding the characteristics (μ, ν) of a system of curves by consideration of the special or degenerate forms of the curves included in the system. The quantities which have to be considered are very numerous. Zeuthen in the case of curves of any given order establishes between the characteristics μ, ν', and 18 other quantities, in all 20 quantities, a set of 24 equations (equivalent to 23 independent equations), involving (besides the 20 quantities) other quantities relating to the various forms of the degenerate curves, which supplementary terms he determines, partially for curves of any order, but completely only for quartic curves. It is in the discussion and complete enumeration of the special or degenerate forms of the curves, and of the supplementary terms to which they give rise, that the great difficulty of the question seems to consist; it would appear that the 24 equations are a complete system, and that (subject to a proper determination of the supplementary terms) they contain the solution of the general problem.

The remarks which follow have reference to the analytical theory of the degenerate curves which present themselves in the foregoing problem of the curves which satisfy given conditions.

A curve represented by an equation in point-coordinates may break up: thus if $P_1, P_2, \ldots$ be rational and integral functions of the coordinates (x, y, z) of the orders $m_1, m_2, \ldots$ respectively, we have the curve $P_1^{\alpha_1}P_2^{\alpha_2}\ldots = 0$, of the order $m, = \alpha_1 m_1 + \alpha_2 m_2 + \ldots$, composed of the curve $P_1 = 0$ taken α_1 times, the curve $P_2 = 0$ taken α_2 times, &c.

Instead of the equation $P_1^{\alpha_1}P_2^{\alpha_2}\ldots = 0$, we may start with an equation $u = 0$, where u is a function of the order m containing a parameter θ, and for a particular value say $\theta = 0$, of the parameter reducing itself to $P_1^{\alpha_1}P_2^{\alpha_2}\ldots$. Supposing θ indefinitely small, we have what may be called the penultimate curve, and when $\theta = 0$ the ultimate curve. Regarding the ultimate curve as derived from a given penultimate curve, we connect with the ultimate curve, and consider as belonging to it, certain points called "summits" on the component curves $P_1 = 0$, $P_2 = 0$, respectively; a summit Σ is a point such that, drawing from an arbitrary point O the tangents to the penultimate curve, we have $O\Sigma$ as the limit of one of these tangents. The ultimate curve together with its summits may be regarded as a degenerate form of the curve $u = 0$. Observe that the positions of the summits depend on the penultimate curve $u = 0$, viz. on the values of the coefficients in the terms multiplied by $\theta, \theta^2, \ldots$; they are thus in some measure arbitrary points as regards the ultimate curve $P_1^{\alpha_1}P_2^{\alpha_2}\ldots = 0$.

It may be added that we have summits only on the component curves $P_1 = 0$, of a multiplicity $\alpha_1 > 1$; the number of summits on such a curve is in general $= (\alpha_1^2 - \alpha_1) m_1^2$. Thus assuming that the penultimate curve is without nodes or cusps, the number of the tangents to it is $= m^2 - m, = (\alpha_1 m_1 + \alpha_2 m_2 + \ldots)^2 - (\alpha_1 m_1 + \alpha_2 m_2 + \ldots)$, taking $P_1 = 0$ to have δ_1 nodes and κ_1 cusps, and therefore its class n_1 to be $= m_1^2 - m_1 - 2\delta_1 - 3\kappa_1$, &c., the expression for the number of tangents to the penultimate curve is

$$= (\alpha_1^2 - \alpha_1) m_1^2 + (\alpha_2^2 - \alpha_2) m_2^2 + \ldots + 2\alpha_1\alpha_2 m_1 m_2 + \ldots + \alpha_1 (n_1 + 2\delta_1 + 3\kappa_1) + \alpha_2 (n_2 + 2\delta_2 + 3\kappa_2) + \ldots$$

where a term $2\alpha_1\alpha_2 m_1 m_2$ indicates tangents which are in the limit the lines drawn to the intersections of the curves $P_1 = 0$, $P_2 = 0$ each line $2\alpha_1\alpha_2$ times; a term $\alpha_1(n_1 + 2\delta_1 + 3\kappa_1)$ tangents which are in the limit the proper tangents to $P_1 = 0$ each α_1 times, the lines to its nodes each $2\alpha_1$ times, and the lines to its cusps each $3\alpha_1$ times; the remaining terms $(\alpha_1^2 - \alpha_1) m_1^2 + (\alpha_2^2 - \alpha_2) m_2^2 + \ldots$ indicate tangents which are in the limit the lines drawn to the several summits, that is, we have $(\alpha_1^2 - \alpha_1) m_1^2$ summits on the curve $P_1 = 0$, &c.

There is of course a precisely similar theory as regards line-coordinates; taking Π_1, Π_2, &c., to be rational and integral functions of the coordinates (ξ, η, ζ), we connect with the ultimate curve $\Pi_1^{\alpha_1}\Pi_2^{\alpha_2}\ldots = 0$, and consider as belonging to it certain lines, which for the moment may be called "axes," tangents to the component curves

$\Pi_1 = 0$, $\Pi_2 = 0$ respectively. Considering an equation in point-coordinates, we may have among the component curves right lines; and, if in order to put these in evidence, we take the equation to be $L_1^{\gamma_1} \ldots P_1^{\alpha_1} \ldots = 0$, where $L_1 = 0$ is a right line, $P_1 = 0$ a curve of the second or any higher order, then the curve will contain as part of itself summits not exhibited in this equation, but the corresponding line-equation will be $\Lambda_1^{\delta_1} \ldots \Pi_1^{\alpha_1} \ldots = 0$, where $\Lambda_1 = 0, \ldots$ are the equations of the summits in question, $\Pi_1 = 0$, &c., are the line-equations corresponding to the several point-equations $P_1 = 0$, &c.; and this curve will contain as part of itself axes not exhibited by this equation, but which are the lines $L_1 = 0, \ldots$ of the equation in point-coordinates.

In conclusion a little may be said as to curves of double curvature, otherwise twisted curves, or curves in space. The analytical theory by Cartesian coordinates was first considered by Clairaut, *Recherches sur les courbes à double courbure* (Paris, 1731). Such a curve may be considered as described by a point, moving in a line which at the same time rotates about the point in a plane which at the same time rotates about the line; the point is a point, the line a tangent, and the plane an osculating plane, of the curve; moreover the line is a generating line, and the plane a tangent plane, of a developable surface or torse, having the curve for its edge of regression. Analogous to the order and class of a plane curve we have the order, rank, and class, of the system (assumed to be a geometrical one), viz. if an arbitrary plane contains m points, an arbitrary line meets r lines, and an arbitrary point lies in n planes, of the system, then m, r, n are the order, rank, and class respectively. The system has singularities, and there exist between m, r, n and the numbers of the several singularities equations analogous to Plücker's equations for a plane curve.

It is a leading point in the theory that a curve in space cannot in general be represented by means of two equations $U = 0$, $V = 0$; the two equations represent surfaces, intersecting in a curve; but there are curves which are not the complete intersection of any two surfaces; thus we have the cubic in space, or skew cubic, which is the residual intersection of two quadric surfaces which have a line in common; the equations $U = 0$, $V = 0$ of the two quadric surfaces represent the cubic curve, not by itself, but together with the line.

786.

EQUATION.

[From the *Encyclopædia Britannica, Ninth Edition,* vol. VIII. (1878), pp. 497—509.]

THE present article includes Determinant and Theory of Equations; and it may be proper to explain the relation to each other of the two subjects. Theory of Equations is used in its ordinary conventional sense to denote the theory of a single equation of any order in one unknown quantity; that is, it does not include the theory of a system or systems of equations of any order between any number of unknown quantities. Such systems occur very frequently in analytical geometry and other parts of mathematics, but they are hardly as yet the subject-matter of a distinct theory; and even Elimination, the transition-process for passing from a system of any number of equations involving the same number of unknown quantities to a single equation in one unknown quantity, hardly belongs to the Theory of Equations in the above restricted sense. But there is one case of a system of equations which precedes the Theory of Equations, and indeed presents itself at the outset of algebra, that of a system of simple (or linear) equations. Such a system gives rise to the function called a Determinant, and it is by means of these functions that the solution of the equations is effected. We have thus the subject Determinant as nearly equivalent to (but somewhat more extensive than) that of a system of linear equations; and we have the other subject, Theory of Equations, used in the restricted sense above referred to, and as not including Elimination.

Determinant.

1. A sketch of the history of determinants is given under [the Article] Algebra; it thereby appears that the algebraical function called a determinant presents itself in the solution of a system of simple equations, and we have herein a natural source of the theory. Thus, considering the equations

$$\begin{aligned} a\ x+b\ y+c\ z&=d\ ,\\ a'x+b'y+c'z&=d',\\ a''x+b''y+c''z&=d'', \end{aligned}$$

and proceeding to solve them by the so-called method of cross multiplication, we multiply the equations by factors selected in such a manner that, upon adding the results, the whole coefficient of y becomes $=0$ and the whole coefficient of z becomes $=0$; the factors in question are $b'c''-b''c'$, $b''c-bc''$, $bc'-b'c$ (values which, as at once seen, have the desired property); we thus obtain an equation which contains on the left-hand side only a multiple of x, and on the right-hand side a constant term; the coefficient of x has the value

$$a(b'c''-b''c')+a'(b''c-bc'')+a''(bc'-b'c),$$

and this function, represented in the form

$$\begin{vmatrix} a, & b, & c \\ a', & b', & c' \\ a'', & b'', & c'' \end{vmatrix},$$

is said to be a determinant; or, the number of elements being 3^2, it is called a determinant of the third order. It is to be noticed that the resulting equation is

$$\begin{vmatrix} a, & b, & c \\ a', & b', & c' \\ a'', & b'', & c'' \end{vmatrix} x = \begin{vmatrix} d, & b, & c \\ d', & b', & c' \\ d'', & b'', & c'' \end{vmatrix},$$

where the expression on the right-hand side is the like function with d, d', d'' in place of a, a', a'' respectively, and is of course also a determinant. Moreover, the functions $b'c''-b''c'$, $b''c-bc''$, $bc'-b'c$ used in the process are themselves the determinants of the second order

$$\begin{vmatrix} b', & c' \\ b'', & c'' \end{vmatrix}, \quad \begin{vmatrix} b'', & c'' \\ b, & c \end{vmatrix}, \quad \begin{vmatrix} b, & c \\ b', & c' \end{vmatrix}.$$

We have herein the suggestion of the rule for the derivation of the determinants of the orders 1, 2, 3, 4, &c., each from the preceding one, viz. we have

$$|a| = a,$$

$$\begin{vmatrix} a, & b \\ a', & b' \end{vmatrix} = a\,|b'| - a'\,|b|,$$

$$\begin{vmatrix} a, & b, & c \\ a', & b', & c' \\ a'', & b'', & c'' \end{vmatrix} = a\begin{vmatrix} b', & c' \\ b'', & c'' \end{vmatrix} + a'\begin{vmatrix} b'', & c'' \\ b, & c \end{vmatrix} + a''\begin{vmatrix} b, & c \\ b', & c' \end{vmatrix},$$

$$\begin{vmatrix} a, & b, & c, & d \\ a', & b', & c', & d' \\ a'', & b'', & c'', & d'' \\ a''', & b''', & c''', & d''' \end{vmatrix} = a\begin{vmatrix} b', & c', & d' \\ b'', & c'', & d'' \\ b''', & c''', & d''' \end{vmatrix} - a'\begin{vmatrix} b'', & c'', & d'' \\ b''', & c''', & d''' \\ b, & c, & d \end{vmatrix} + a''\begin{vmatrix} b''', & c''', & d''' \\ b, & c, & d \\ b', & c', & d' \end{vmatrix} - a'''\begin{vmatrix} b, & c, & d \\ b', & c', & d' \\ b'', & c'', & d'' \end{vmatrix},$$

and so on, the terms being all + for a determinant of an odd order, but alternately + and − for a determinant of an even order.

2. It is easy, by induction, to arrive at the general results:—

A determinant of the order n is the sum of the $1.2.3...n$ products which can be formed with n elements out of n^2 elements arranged in the form of a square, no two of the n elements being in the same line or in the same column, and each such product having the coefficient $\pm$ unity.

The products in question may be obtained by permuting in every possible manner the columns (or the lines) of the determinant, and then taking for the factors the n elements in the dexter diagonal. And we thence derive the rule for the signs, viz. considering the primitive arrangement of the columns as positive, then an arrangement obtained therefrom by a single interchange (inversion, or derangement) of two columns is regarded as negative; and so in general an arrangement is positive or negative according as it is derived from the primitive arrangement by an even or an odd number of interchanges. This implies the theorem that a given arrangement can be derived from the primitive arrangement only by an odd number, or else only by an even number of interchanges,—a theorem the verification of which may be easily obtained from the theorem (in fact, a particular case of the general one), an arrangement can be derived from itself only by an even number of interchanges. And this being so, each product has the sign belonging to the corresponding arrangement of the columns; in particular, a determinant contains with the sign + the product of the elements in its dexter diagonal. It is to be observed that the rule gives as many positive as negative arrangements, the number of each being $=\frac{1}{2}.1.2...n$.

The rule of signs may be expressed in a different form. Giving to the columns in the primitive arrangement the numbers $1, 2, 3, ..., n$, to obtain the sign belonging to any other arrangement we take, as often as a lower number succeeds a higher one, the sign $-$, and, compounding together all these minus signs, obtain the proper sign, $+$ or $-$ as the case may be.

Thus, for three columns, it appears by either rule that 123, 231, 312 are positive; 132, 321, 213 are negative; and the developed expression of the foregoing determinant of the third order is

$$=ab'c'' - ab''c' + a'b''c - a'bc'' + a''bc' - a''b'c.$$

3. It further appears that a determinant is a linear function* of the elements of each column thereof, and also a linear function of the elements of each line thereof; moreover, that the determinant retains the same value, only its sign being altered, when any two columns are interchanged, or when any two lines are interchanged; more generally, when the columns are permuted in any manner, or when the lines are permuted in any manner, the determinant retains its original value, with the sign $+$ or $-$ according as the new arrangement (considered as derived from the primitive arrangement) is positive or negative according to the foregoing rule of signs.

* The expression, a linear function, is here used in its narrowest sense, a linear function without constant term; what is meant is, that the determinant is in regard to the elements $a, a', a'', ...$ of any column or line thereof, a function of the form $Aa + A'a' + A''a'' + ...$, without any term independent of $a, a', a'', ...$.

It at once follows that, if two columns are identical, or if two lines are identical, the value of the determinant is $=0$. It may be added that, if the lines are converted into columns, and the columns into lines, in such a way as to leave the dexter diagonal unaltered, the value of the determinant is unaltered; the determinant is in this case said to be *transposed*.

4. By what precedes it appears that there exists a function of the n^2 elements, linear as regards the terms of each column (or say, for shortness, linear as to each column), and such that only the sign is altered when any two columns are interchanged; these properties completely determine the function, except as to a common factor which may multiply all the terms. If, to get rid of this arbitrary common factor, we assume that the product of the elements in the dexter diagonal has the coefficient $+1$, we have a complete definition of the determinant; and it is interesting to show how from these properties, assumed for the definition of the determinant, it at once appears that the determinant is a function serving for the solution of a system of linear equations. Observe that the properties show at once that if any column is $=0$ (that is, if the elements in the column are each $=0$), then the determinant is $=0$; and further that, if any two columns are identical, then the determinant is $=0$.

5. Reverting to the system of linear equations written down at the beginning of this article, consider the determinant

$$\begin{vmatrix} a\,x+b\,y+c\,z-d\,, & b\,, & c \\ a'x+b'y+c'z-d', & b', & c' \\ a''x+b''y+c''z-d'', & b'', & c'' \end{vmatrix};$$

it appears that this is

$$=x\begin{vmatrix} a\,, & b\,, & c \\ a', & b', & c' \\ a'', & b'', & c'' \end{vmatrix}+y\begin{vmatrix} b\,, & b\,, & c \\ b', & b', & c' \\ b'', & b'', & c'' \end{vmatrix}+z\begin{vmatrix} c\,, & b\,, & c \\ c', & b', & c' \\ c'', & b'', & c'' \end{vmatrix}-\begin{vmatrix} d\,, & b\,, & c \\ d', & b', & c' \\ d'', & b'', & c'' \end{vmatrix},$$

viz. the second and the third terms each vanishing, it is

$$=x\begin{vmatrix} a\,, & b\,, & c \\ a', & b', & c' \\ a'', & b'', & c'' \end{vmatrix}-\begin{vmatrix} d\,, & b\,, & c \\ d', & b', & c' \\ d'', & b'', & c'' \end{vmatrix}.$$

But if the linear equations hold good, then the first column of the original determinant is $=0$, and therefore the determinant itself is $=0$; that is, the linear equations give

$$x\begin{vmatrix} a\,, & b\,, & c \\ a', & b', & c' \\ a'', & b'', & c'' \end{vmatrix}-\begin{vmatrix} d\,, & b\,, & c \\ d', & b', & c' \\ d'', & b'', & c'' \end{vmatrix}=0;$$

which is the result obtained above.

We might in a similar way find the values of y and z, but there is a more symmetrical process. Join to the original equations the new equation

$$\alpha x + \beta y + \gamma z = \delta;$$

a like process shows that, the equations being satisfied, we have

$$\begin{vmatrix} \alpha, & \beta, & \gamma, & \delta \\ a, & b, & c, & d \\ a', & b', & c', & d' \\ a'', & b'', & c'', & d'' \end{vmatrix} = 0;$$

or, as this may be written,

$$\begin{vmatrix} \alpha, & \beta, & \gamma & \\ a, & b, & c, & d \\ a', & b', & c', & d' \\ a'', & b'', & c'', & d'' \end{vmatrix} - \delta \begin{vmatrix} a, & b, & c \\ a', & b', & c' \\ a'', & b'', & c'' \end{vmatrix} = 0;$$

which, considering δ as standing herein for its value $\alpha x + \beta y + \gamma z$, is a consequence of the original equations only. We have thus an expression for $\alpha x + \beta y + \gamma z$, an arbitrary linear function of the unknown quantities x, y, z; and by comparing the coefficients of α, β, γ on the two sides respectively, we have the values of x, y, z; in fact, these quantities, each multiplied by

$$\begin{vmatrix} a, & b, & c \\ a', & b', & c' \\ a'', & b'', & c'' \end{vmatrix},$$

are in the first instance obtained in the forms

$$\begin{vmatrix} 1 & & & \\ a, & b, & c, & d \\ a', & b', & c', & d' \\ a'', & b'', & c'', & d'' \end{vmatrix}, \quad \begin{vmatrix} & 1 & & \\ a, & b, & c, & d \\ a', & b', & c', & d' \\ a'', & b'', & c'', & d'' \end{vmatrix}, \quad \begin{vmatrix} & & 1 & \\ a, & b, & c, & d \\ a', & b', & c', & d' \\ a'', & b'', & c'', & d'' \end{vmatrix};$$

but these are

$$= \begin{vmatrix} b, & c, & d \\ b', & c', & d' \\ b'', & c'', & d'' \end{vmatrix}, \quad - \begin{vmatrix} c, & d, & a \\ c', & d', & a' \\ c'', & d'', & a'' \end{vmatrix}, \quad \begin{vmatrix} d, & a, & b \\ d', & a', & b' \\ d'', & a'', & b'' \end{vmatrix},$$

or, what is the same thing,

$$= \begin{vmatrix} b, & c, & d \\ b', & c', & d' \\ b'', & c'', & d'' \end{vmatrix}, \quad \begin{vmatrix} c, & a, & d \\ c', & a', & d' \\ c'', & a'', & d'' \end{vmatrix}, \quad \begin{vmatrix} a, & b, & d \\ a', & b', & d' \\ a'', & b'', & d'' \end{vmatrix}$$

respectively.

6. *Multiplication of two determinants of the same order.*—The theorem is obtained very easily from the last preceding definition of a determinant. It is most simply expressed thus—

$$\begin{array}{c|ccc|} & (\alpha,\ \alpha',\ \alpha''), & (\beta,\ \beta',\ \beta''), & (\gamma,\ \gamma',\ \gamma'') \\ \hline (a\ ,\ b\ ,\ c\ ) & ,, & ,, & ,, \\ (a',\ b',\ c') & ,, & ,, & ,, \\ (a'',\ b'',\ c'') & ,, & ,, & ,, \end{array} = \begin{vmatrix} a\ , & b\ , & c \\ a', & b', & c' \\ a'', & b'', & c'' \end{vmatrix} \cdot \begin{vmatrix} \alpha\ , & \beta\ , & \gamma \\ \alpha', & \beta', & \gamma' \\ \alpha'', & \beta'', & \gamma'' \end{vmatrix},$$

where the expression on the left side stands for a determinant, the terms of the first line being $(a,\ b,\ c)(\alpha,\ \alpha',\ \alpha'')$, that is, $a\alpha + b\alpha' + c\alpha''$, $(a,\ b,\ c)(\beta,\ \beta',\ \beta'')$, that is, $a\beta + b\beta' + c\beta''$, $(a,\ b,\ c)(\gamma,\ \gamma',\ \gamma'')$, that is, $a\gamma + b\gamma' + c\gamma''$; and similarly the terms in the second and third lines are the like functions with $(a',\ b',\ c')$ and $(a'',\ b'',\ c'')$ respectively.

There is an apparently arbitrary transposition of lines and columns; the result would hold good if on the left-hand side we had written $(\alpha,\ \beta,\ \gamma)$, $(\alpha',\ \beta',\ \gamma')$, $(\alpha'',\ \beta'',\ \gamma'')$, or what is the same thing, if on the right-hand side we had transposed the second determinant; and either of these changes would, it might be thought, increase the elegance of the form, but, for a reason which need not be explained*, the form actually adopted is the preferable one.

To indicate the method of proof, observe that the determinant on the left-hand side, *qua* linear function of its columns, may be broken up into a sum of $(3^3 =)$ 27 determinants, each of which is either of some such form as

$$\pm\,\alpha\beta\gamma' \begin{vmatrix} a\ , & a\ , & b \\ a', & a', & b' \\ a'', & a'', & b'' \end{vmatrix},$$

where the term $\alpha\beta\gamma'$ is not a term of the $\alpha\beta\gamma$-determinant, and its coefficient (as a determinant with two identical columns) vanishes; or else it is of a form such as

$$\pm\,\alpha\beta'\gamma'' \begin{vmatrix} a\ , & b\ , & c \\ a', & b', & c' \\ a'', & b'', & c'' \end{vmatrix},$$

that is, every term which does not vanish contains as a factor the abc-determinant last written down; the sum of all other factors $\pm\,\alpha\beta'\gamma''$ is the $\alpha\beta\gamma$-determinant of the formula; and the final result then is, that the determinant on the left-hand side is equal to the product on the right-hand side of the formula.

7. *Decomposition of a determinant into complementary determinants.*—Consider, for simplicity, a determinant of the fifth order, $5 = 2 + 3$, and let the top two lines be

$$\begin{matrix} a\ , & b\ , & c\ , & d\ , & e\ , \\ a'\ , & b'\ , & c'\ , & d'\ , & e'; \end{matrix}$$

* The reason is the connexion with the corresponding theorem for the multiplication of two matrices.

then, if we consider how these elements enter into the determinant, it is at once seen that they enter only through the determinants of the second order $\begin{vmatrix} a, & b \\ a', & b' \end{vmatrix}$, &c., which can be formed by selecting any two columns at pleasure. Moreover, representing the remaining three lines by

$$\begin{matrix} a'', & b'', & c'', & d'', & e'', \\ a''', & b''', & c''', & d''', & e''', \\ a'''', & b'''', & c'''', & d'''', & e'''', \end{matrix}$$

it is further seen that the factor which multiplies the determinant formed with any two columns of the first set is the determinant of the third order formed with the complementary three columns of the second set; and it thus appears that the determinant of the fifth order is a sum of all the products of the form

$$\pm \begin{vmatrix} a, & b \\ a', & b' \end{vmatrix} \begin{vmatrix} c'', & d'', & e'' \\ c''', & d''', & e''' \\ c'''', & d'''', & e'''' \end{vmatrix},$$

the sign $\pm$ being in each case such that the sign of the term $\pm ab' . c''d'''e''''$ obtained from the diagonal elements of the component determinants may be the actual sign of this term in the determinant of the fifth order; for the product written down the sign is obviously $+$.

Observe that for a determinant of the nth order, taking the decomposition to be $1+(n-1)$, we fall back upon the equations given at the commencement, in order to show the genesis of a determinant.

8. Any determinant $\begin{vmatrix} a, & b \\ a', & b' \end{vmatrix}$ formed out of the elements of the original determinant, by selecting the lines and columns at pleasure, is termed a *minor* of the original determinant; and when the number of lines and columns, or order of the determinant, is $n-1$, then such determinant is called a *first minor;* the number of the first minors is $=n^2$, the first minors, in fact, corresponding to the several elements of the determinant—that is, the coefficient therein of any term whatever is the corresponding first minor. The first minors, each divided by the determinant itself, form a system of elements *inverse* to the elements of the determinant.

A determinant is *symmetrical* when every two elements symmetrically situated in regard to the dexter diagonal are equal to each other; if they are equal and opposite (that is, if the sum of the two elements be $=0$), this relation not extending to the diagonal elements themselves, which remain arbitrary, then the determinant is *skew;* but if the relation does extend to the diagonal terms (that is, if these are each $=0$), then the determinant is *skew symmetrical;* thus the determinants

$$\begin{vmatrix} a, & h, & g \\ h, & b, & f \\ g, & f, & c \end{vmatrix}, \quad \begin{vmatrix} a, & \nu, & -\mu \\ -\nu, & b, & \lambda \\ \mu, & -\lambda, & c \end{vmatrix}, \quad \begin{vmatrix} 0, & \nu, & -\mu \\ -\nu, & 0, & \lambda \\ \mu, & -\lambda, & 0 \end{vmatrix},$$

are respectively symmetrical, skew, and skew symmetrical.

The theory admits of very extensive algebraic developments, and applications in algebraical geometry and other parts of mathematics; but the fundamental properties of the functions may fairly be considered as included in what precedes.

Theory of Equations.

9. In the subject "Theory of Equations," the term *equation* is used to denote an equation of the form $x^n - p_1x^{n-1} + \ldots \pm p_n = 0$, where $p_1, p_2, \ldots, p_n$ are regarded as known, and x as a quantity to be determined; for shortness, the equation is written $f(x) = 0$.

The equation may be *numerical;* that is, the coefficients $p_1, p_2, \ldots, p_n$ are then numbers,—understanding by number a quantity of the form $\alpha + \beta i$ (α and β having any positive or negative real values whatever, or say each of these is regarded as susceptible of continuous variation from an indefinitely large negative to an indefinitely large positive value), and i denoting $\sqrt{-1}$.

Or the equation may be *algebraical;* that is, the coefficients are not then restricted to denote, or are not explicitly considered as denoting, numbers.

I. We consider first numerical equations. (Real theory, 10 to 14; Imaginary theory, 15 to 18.)

10. Postponing all consideration of imaginaries, we take in the first instance the coefficients to be real, and attend only to the real roots (if any); that is, $p_1, p_2, \ldots, p_n$ are real positive or negative quantities, and a root a, if it exists, is a positive or negative quantity such that $a^n - p_1a^{n-1} + \ldots \pm p_n = 0$, or say, $f(a) = 0$. The fundamental theorems are given in the article Algebra, sections X., XIII., XIV.; but there are various points in the theory which require further development.

It is very useful to consider the curve $y = f(x)$,—or, what would come to the same, the curve $Ay = f(x)$,—but it is better to retain the first-mentioned form of equation, drawing, if need be, the ordinate y on a reduced scale. For instance, if the given equation be $x^3 - 6x^2 + 11x - 6{\cdot}06 = 0$,* then the curve $y = x^3 - 6x^2 + 11x - 6{\cdot}06$ is as shown in the figure at page 501, without any reduction of scale for the ordinate.

It is clear that, in general, y is a continuous one-valued function of x, finite for every finite value of x, but becoming infinite when x is infinite; i.e. assuming throughout that the coefficient of x^n is $+1$, then when $x = \infty$, $y = +\infty$; but when $x = -\infty$, then $y = +\infty$ or $-\infty$, according as n is even or odd; the curve cuts any line whatever, and in particular it cuts the axis of x, in at most n points; and the value of x, at any point of intersection with the axis, is a root of the equation $f(x) = 0$.

If β, α are any two values of x ($\alpha > \beta$, that is, α nearer $+\infty$), then if $f(\beta)$, $f(\alpha)$ have opposite signs, the curve cuts the axis an odd number of times, and therefore at least once, between the points $x = \beta$, $x = \alpha$; but if $f(\beta)$, $f(\alpha)$ have the same sign, then between these points the curve cuts the axis an even number of times, or it may be not at all. That is, $f(\beta)$, $f(\alpha)$ having opposite signs, there are between the limits β, α an odd number of real roots, and therefore at least one real

* The coefficients were selected so that the roots might be nearly 1, 2, 3.

root; but $f(\beta)$, $f(\alpha)$ having the same sign, there are between these limits an even number of real roots, or it may be there is no real root. In particular, by giving to β, α the values $-\infty$, $+\infty$ (or, what is the same thing, any two values sufficiently near to these values respectively) it appears that an equation of an odd order has always an odd number of real roots, and therefore at least one real root; but that an equation of an even order has an even number of real roots, or it may be no real root.

If α be such that for $x =$ or $> \alpha$ (that is, x nearer to $+\infty$) $f(x)$ is always $+$, and β be such that for $x =$ or $< \beta$ (that is, x nearer to $-\infty$) $f(x)$ is always $-$, then the real roots (if any) lie between these limits $x = \beta$, $x = \alpha$; and it is easy to find by trial such two limits including between them all the real roots (if any).

11. Suppose that the positive value δ is an inferior limit to the difference between two real roots of the equation; or rather (since the foregoing expression would imply the existence of real roots) suppose that there are not two real roots such that their difference taken positively is $=$ or $< \delta$; then, γ being any value whatever, there is clearly at most one real root between the limits γ and $\gamma + \delta$; and by what precedes there is such real root or there is not such real root, according as $f(\gamma)$, $f(\gamma + \delta)$ have opposite signs or have the same sign. And by dividing in this manner the interval β to α into intervals each of which is $=$ or $< \delta$, we should not only ascertain the number of the real roots (if any), but we should also *separate* the real roots, that is, find for each of them limits γ, $\gamma + \delta$ between which there lies this one, and only this one, real root.

In particular cases it is frequently possible to ascertain the number of the real roots, and to effect their separation by trial or otherwise, without much difficulty; but the foregoing was the general process as employed by Lagrange even in the second edition (1808) of the *Traité de la résolution des Équations Numériques**; the determination of the limit δ had to be effected by means of the "equation of differences" or equation of the order $\frac{1}{2}n(n-1)$, the roots of which are the squares of the differences of the roots of the given equation, and the process is a cumbrous and unsatisfactory one.

12. The great step was effected by Sturm's theorem (1835)—viz. here starting from the function $f(x)$, and its first derived function $f'(x)$, we have (by a process which is a slight modification of that for obtaining the greatest common measure of these two functions) to form a series of functions

$$f(x),\ f'(x),\ f_2(x), \ldots,\ f_n(x)$$

of the degrees n, $n-1$, $n-2, \ldots, 0$ respectively,—the last term $f_n(x)$ being thus an absolute constant. These lead to the immediate determination of the number of real roots (if any) between any two given limits β, α; viz. supposing $\alpha > \beta$ (that is, α nearer to $+\infty$), then substituting successively these two values in the series of functions, and attending only to the signs of the resulting values, the number of the changes of sign lost in passing from β to α is the required number of real roots

* The third edition (1826) is a reproduction of that of 1808; the first edition has the date 1798, but a large part of the contents is taken from memoirs of 1767—68 and 1770—71.

between the two limits. In particular, taking β, $\alpha = -\infty, +\infty$ respectively, the signs of the several functions depend merely on the signs of the terms which contain the highest powers of x, and are seen by inspection, and the theorem thus gives at once the whole number of real roots.

And although theoretically, in order to complete by a finite number of operations the separation of the real roots, we still need to know the value of the before-mentioned limit δ; yet in any given case the separation may be effected by a limited number of repetitions of the process. The practical difficulty is when two or more roots are very near to each other. Suppose, for instance, that the theorem shows that there are two roots between 0 and 10; by giving to x the values 1, 2, 3,... successively, it might appear that the two roots were between 5 and 6; then again that they were between 5·3 and 5·4, then between 5·34 and 5·35, and so on until we arrive at a separation; say it appears that between 5·346 and 5·347 there is one root, and between 5·348 and 5·349 the other root. But in the case in question δ would have a very small value, such as ·002, and even supposing this value known, the direct application of the first-mentioned process would be still more laborious.

13. Supposing the separation once effected, the determination of the single real root which lies between the two given limits may be effected to any required degree of approximation either by the processes of Horner and Lagrange (which are in principle a carrying out of the method of Sturm's theorem), or by the process of Newton, as perfected by Fourier (which requires to be separately considered).

First as to Horner and Lagrange. We know that between the limits β, α there lies one, and only one, real root of the equation; $f(\beta)$ and $f(\alpha)$ have therefore opposite signs. Suppose any intermediate value is θ; in order to determine by Sturm's theorem whether the root lies between β, θ, or between θ, α, it would be quite unnecessary to calculate the signs of $f(\theta)$, $f'(\theta)$, $f_2(\theta)$, ...; only the sign of $f(\theta)$ is required: for, if this has the same sign as $f(\beta)$, then the root is between β, θ; if the same sign as $f(\alpha)$, then the root is between θ, α. We want to make θ increase from the inferior limit β, at which $f(\theta)$ has the sign of $f(\beta)$, so long as $f(\theta)$ retains this sign, and then to a value for which it assumes the opposite sign; we have thus two nearer limits of the required root, and the process may be repeated indefinitely.

Horner's method (1819) gives the root as a decimal, figure by figure; thus, if the equation be known to have one real root between 0 and 10, it is in effect shown say that 5 is too small (that is, the root is between 5 and 6); next that 5·4 is too small (that is, the root is between 5·4 and 5·5); and so on to any number of decimals. Each figure is obtained, *not* by the successive trial of all the figures which precede it, but (as in the ordinary process of the extraction of a square root, which is in fact Horner's process applied to this particular case) it is given presumptively as the first figure of a quotient; such value may be too large, and then the next inferior integer must be tried instead of it, or it may require to be further diminished. And it is to be remarked that the process not only gives the approximate value α of the root, but (as in the extraction of a square root) it includes the calculation of the function $f(\alpha)$ which should be, and approximately is, $=0$. The arrangement of the calculations is very elegant, and forms an integral part of the actual method. It is to be observed that after a certain number of decimal places have been obtained, a good many more can be found by a mere division. It is in the progress tacitly assumed that the roots have been first separated.

Lagrange's method (1767) gives the root as a continued fraction $a+\frac{1}{b+}\frac{1}{c+}\cdots$, where a is a positive or negative integer (which may be $=0$), but $b, c, \ldots$ are positive integers. Suppose the roots have been separated; then (by trial if need be of consecutive integer values) the limits may be made to be consecutive integer numbers: say they are $a, a+1$; the value of x is therefore $=a+\frac{1}{y}$, where y is positive and greater than 1; from the given equation for x, writing therein $x=a+\frac{1}{y}$, we form an equation of the same order for y, and this equation will have one, and only one, positive root greater than 1; hence finding for it the limits $b, b+1$ (where b is $=$ or >1), we have $y=b+\frac{1}{z}$, where z is positive and greater than 1; and so on—that is, we thus obtain the successive denominators $b, c, d, \ldots$ of the continued fraction. The method is theoretically very elegant, but the disadvantage is that it gives the result in the form of a continued fraction, which for the most part must ultimately be converted into a decimal. There is one advantage in the method, that a commensurable root (that is, a root equal to a rational fraction) is found accurately, since, when such root exists, the continued fraction terminates.

14. Newton's method (1711), as perfected by Fourier (1831), may be roughly stated as follows. If $x=\gamma$ be an approximate value of any root, and $\gamma+h$ the correct value, then $f(\gamma+h)=0$, that is,

$$f(\gamma)+\frac{h}{1}f'(\gamma)+\frac{h^2}{1\,.\,2}f''(\gamma)+\ldots=0;$$

and then, if h be so small that the terms after the second may be neglected, $f(\gamma)+hf'(\gamma)=0$, that is, $h=-\frac{f(\gamma)}{f'(\gamma)}$, or the new approximate value is $x=\gamma-\frac{f(\gamma)}{f'(\gamma)}$; and so on, as often as we please. It will be observed that so far nothing has been assumed as to the separation of the roots, or even as to the existence of a real root; γ has been taken as the approximate value of a root, but no precise meaning has been attached to this expression. The question arises, what are the conditions to be satisfied by γ in order that the process may by successive repetitions actually lead to a certain real root of the equation; or say that, γ being an approximate value of a certain real root, the new value $\gamma-\frac{f(\gamma)}{f'(\gamma)}$ may be a more approximate value.

Referring to the figure, it is easy to see that, if OC represent the assumed value γ, then, drawing the ordinate CP to meet the curve in P, and the tangent PC' to meet the axis in C', we shall have OC' as the new approximate value of the root. But observe that there is here a real root OX, and that the curve beyond X

is convex to the axis; under these conditions the point C' is nearer to X than was C; and, starting with C' instead of C, and proceeding in like manner to draw a new ordinate and tangent, and so on as often as we please, we approximate continually, and that with great rapidity, to the true value OX. But if C had been taken on the other side of X, where the curve is concave to the axis, the new point C' might or might not be nearer to X than was the point C; and in this

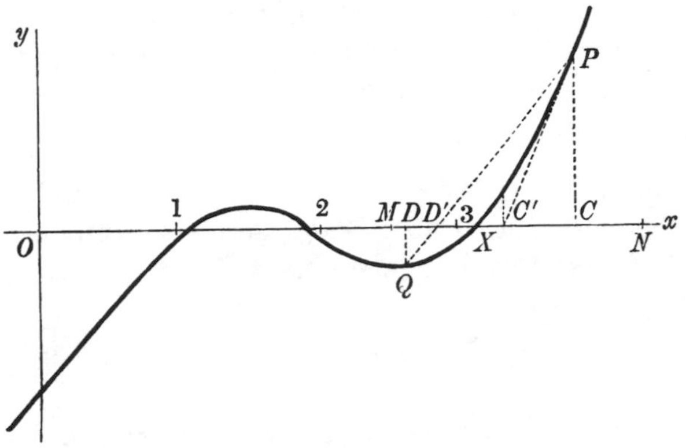

case the method, if it succeeds at all, does so by accident only, i.e., it may happen that C' or some subsequent point comes to be a point C, such that OC is a *proper* approximate value of the root, and then the subsequent approximations proceed in the same manner as if this value had been assumed in the first instance, all the preceding work being wasted. It thus appears that for the proper application of the method we require *more* than the mere separation of the roots. In order to be able to approximate to a certain root a, $= OX$, we require to know that, between OX and some value ON, the curve is always convex to the axis: analytically, between the two values, $f(x)$ and $f''(x)$ must have always the same sign. When this is so, the point C may be taken anywhere on the proper side of X, and within the portion XN of the axis; and the process is then the one already explained. The approximation is in general a very rapid one. If we know for the required root OX the two limits OM, ON such that from M to X the curve is always *concave* to the axis, while from X to N it is always convex to the axis,—then, taking D anywhere in the portion MX and (as before) C in the portion XN, drawing the ordinates DQ, CP, and joining the points P, Q by a line which meets the axis in D', also constructing the point C' by means of the tangent at P as before, we have for the required root the new limits OD', OC'; and proceeding in like manner with the points D', C', and so on as often as we please, we obtain at each step two limits approximating more and more nearly to the required root OX. The process as to the point D', translated into analysis, is the ordinate process of interpolation. Suppose $OD = \beta$, $OC = \alpha$, we have approximately

$$f(\beta + h) = f(\beta) + \frac{h\{f(\alpha) - f(\beta)\}}{\alpha - \beta},$$

whence, if the root is $\beta + h$, then

$$h = -\frac{(\alpha - \beta) f(\beta)}{f(\alpha) - f(\beta)}.$$

Returning for a moment to Horner's method, it may be remarked that the correction h, to an approximate value α, is therein found as a quotient, the same or such as the quotient $f(\alpha) \div f'(\alpha)$ which presents itself in Newton's method. The difference is that with Horner the integer part of this quotient, is taken as the presumptive value of h, and the figure is verified at each step. With Newton the quotient itself, developed to the proper number of decimal places, is taken as the value of h; if too many decimals are taken, there would be a waste of work; but the error would correct itself at the next step. Of course the calculation should be conducted without any such waste of work.

Next as to the theory of imaginaries.

15. It will be recollected that the expression *number* and the correlative epithet *numerical* were at the outset used in a wide sense, as extending to imaginaries. This extension arises out of the theory of equations by a process analogous to that by which number, in its original most restricted sense of positive integer number, was extended to have the meaning of a real positive or negative magnitude susceptible of continuous variation.

If for a moment number is understood in its most restricted sense as meaning positive integer number, the solution of a simple equation leads to an extension; $ax - b = 0$, gives $x = \frac{b}{a}$, a positive fraction, and we can in this manner represent, not accurately, but as nearly as we please, any positive magnitude whatever; so an equation $ax + b = 0$ gives $x = -\frac{b}{a}$, which (approximately as before) represents any negative magnitude. We thus arrive at the extended signification of number as a continuously varying positive or negative magnitude. Such numbers may be added or subtracted, multiplied or divided one by another, and the result is always a number. Now from a quadric equation we derive, in like manner, the notion of a complex or imaginary number such as is spoken of above. The equation $x^2 + 1 = 0$ is not (in the foregoing sense, number = real number) satisfied by any numerical value whatever of x; but we assume that there is a number which we call i, satisfying the equation $i^2 + 1 = 0$; and then taking a and b any real numbers, we form an expression such as $a + bi$, and use the expression number in this extended sense: any two such numbers may be added or subtracted, multiplied or divided one by the other, and the result is always a number. And if we consider first a quadric equation $x^2 + px + q = 0$ where p and q are real numbers, and next the like equation, where p and q are any numbers whatever, it can be shown that there exists for x a numerical value which satisfies the equation; or, in other words, it can be shown that the equation has a numerical root. The like theorem, in fact, holds good for an equation of any order whatever. But suppose for a moment that this was not the case: say that there was a cubic equation $x^3 + px^2 + qx + r = 0$, with numerical coefficients, not satisfied by any numerical value of x, we should have to establish a new imaginary j satisfying some such equation, and should then have to consider numbers of the form $a + bj$, or perhaps $a + bj + cj^2$ (a, b, c numbers $\alpha + \beta i$ of the kind heretofore considered),—first we should be thrown back on the quadric equation $x^2 + px + q = 0$, p and q being now numbers of the last-mentioned extended form—*non constat* that every such equation has a numerical root—and if not, we might be led to *other* imaginaries k, l, &c., and so on *ad infinitum* in inextricable confusion.

But in fact a numerical equation of any order whatever has always a numerical root, and thus numbers (in the foregoing sense, number = quantity of the form $\alpha+\beta i$) form (*what real numbers do not*) a universe complete in itself, such that starting in it we are never led out of it. There may very well be, and perhaps are, numbers in a more general sense of the term (quaternions are not a case in point, as the ordinary laws of combination are not adhered to): but in order to have to do with such numbers (if any), we must start with them.

16. The capital theorem as regards numerical equations thus is, every numerical equation has a numerical root; or for shortness (the meaning being as before), every equation has a root. Of course the theorem is the reverse of self-evident, and it requires proof; but provisionally assuming it as true, we derive from it the general theory of numerical equations. As the term root was introduced in the course of an explanation, it will be convenient to give here the formal definition.

A number a such that substituted for x it makes the function $x^n - p_1x^{n-1} + \ldots \pm p_n$ to be $=0$, or say such that it satisfies the equation $f(x)=0$, is said to be a root of the equation; that is, a being a root, we have

$$a^n - p_1a^{n-1} + \ldots \pm p_n = 0, \text{ or say } f(a)=0;$$

and it is then easily shown that $x-a$ is a factor of the function $f(x)$, viz. that we have $f(x)=(x-a)f_1(x)$, where $f_1(x)$ is a function $x^{n-1} - q_1x^{n-2} + \ldots \pm q_{n-1}$ of the order $n-1$, with numerical coefficients $q_1, q_2, \ldots, q_{n-1}$.

In general, a is not a root of the equation $f_1(x)=0$, but it may be so—i.e., $f_1(x)$ may contain the factor $x-a$; when this is so, $f(x)$ will contain the factor $(x-a)^2$; writing then $f(x)=(x-a)^2 f_2(x)$, and assuming that a is not a root of the equation $f_2(x)=0$, $x=a$ is then said to be a double root of the equation $f(x)=0$; and similarly $f(x)$ may contain the factor $(x-a)^3$ and no higher power, and $x=a$ is then a triple root; and so on.

Supposing, in general, that $f(x)=(x-a)^\alpha F(x)$, α being a positive integer which may be $=1$, $(x-a)^\alpha$ the highest power of $x-a$ which divides $f(x)$, and $F(x)$ being of course of the order $n-\alpha$, then the equation $F(x)=0$ will have a root b which will be different from a; $x-b$ will be a factor, in general a simple one, but it may be a multiple one, of $F(x)$, and $f(x)$ will in this case be $=(x-a)^\alpha (x-b)^\beta \Phi(x)$, β a positive integer which may be $=1$, $(x-b)^\beta$ the highest power of $x-b$ in $F(x)$ or $f(x)$, and $\Phi(x)$ being of course of the order $n-\alpha-\beta$. The original equation $f(x)=0$ is in this case said to have α roots each $=a$, β roots each $=b$; and so on for any other factors $(x-c)^\gamma$, &c.

We have thus the *theorem*—A numerical equation of the order n has in every case n roots, viz. there exist n numbers $a, b, \ldots$, in general all distinct, but which may arrange themselves in any sets of equal values, such that $f(x)=(x-a)(x-b)(x-c)\ldots$ identically.

If the equation has equal roots, these can in general be determined: and the case is at any rate a special one which may be in the first instance excluded from consideration. It is therefore, in general, assumed that the equation $f(x) = 0$ has all its roots unequal.

If the coefficients p_1, p_2, ... are all or any one or more of them imaginary, then the equation $f(x) = 0$, separating the real and imaginary parts thereof, may be written $F(x) + i\Phi(x) = 0$, where $F(x)$, $\Phi(x)$ are each of them a function with real coefficients; and it thus appears that the equation $f(x) = 0$, with imaginary coefficients, has not in general any real root; supposing it to have a real root a, this must be at once a root of each of the equations $F(x) = 0$ and $\Phi(x) = 0$.

But an equation with real coefficients may have as well imaginary as real roots, and we have further the theorem that for any such equation the imaginary roots enter in pairs, viz. $\alpha + \beta i$ being a root, then $\alpha - \beta i$ will be also a root. It follows that, if the order be odd, there is always an odd number of real roots, and therefore at least one real root.

17. In the case of an equation with real coefficients, the question of the existence of real roots, and of their separation, has been already considered. In the general case of an equation with imaginary (it may be real) coefficients, the like question arises as to the situation of the (real or imaginary) roots; thus if, for facility of conception, we regard the constituents α, β of a root $\alpha + \beta i$ as the coordinates of a point *in plano*, and accordingly represent the root by such point, then drawing in the plane any closed curve or "contour," the question is how many roots lie within such contour.

This is solved theoretically by means of a theorem of Cauchy's (1837), viz. writing in the original equation $x + iy$ in place of x, the function $f(x + iy)$ becomes $= P + iQ$, where P and Q are each of them a rational and integral function (with real coefficients) of (x, y). Imagining the point (x, y) to travel along the contour, and considering the number of changes of sign from $-$ to $+$ and from $+$ to $-$ of the fraction corresponding to passages of the fraction through zero, that is, to values for which P becomes $= 0$, disregarding those for which Q becomes $= 0$, the difference of these numbers gives the number of roots within the contour.

It is important to remark that the demonstration does not presuppose the existence of any root; the contour may be the infinity of the plane (such infinity regarded as a contour, or closed curve), and in this case it can be shown (and that very easily) that the difference of the numbers of changes of sign is $= n$; that is, there are within the infinite contour, or (what is the same thing) there are in all, n roots; thus Cauchy's theorem contains really the proof of the fundamental theorem that a numerical equation of the nth order (not only has a numerical root, but) has precisely n roots. It would appear that this proof of the fundamental theorem in its most complete form is in principle identical with Gauss's last proof (1849) of the theorem, in the form—A numerical equation of the nth order has always a root*.

---

* The earlier demonstrations by Euler, Lagrange, &c., relate to the case of a numerical equation with real coefficients; and they consist in showing that such equation has always a real quadratic divisor, furnishing two roots, which are either real or else conjugate imaginaries $\alpha + \beta i$: see Lagrange's *Équations Numériques*.

But in the case of a finite contour, the actual determination of the difference which gives the number of real roots can be effected only in the case of a rectangular contour, by applying to each of its sides separately a method such as that of Sturm's theorem; and thus the actual determination ultimately depends on a method such as that of Sturm's theorem.

Very little has been done in regard to the calculation of the imaginary roots of an equation by approximation; and the question is not here considered.

18. A class of numerical equations which needs to be considered is that of the binomial equations $x^n - a = 0$ ($a = \alpha + \beta i$, a complex number). The foregoing conclusions apply, viz. there are always n roots, which, it may be shown, are all unequal. And these can be found numerically by the extraction of the square root, and of an nth root, of *real* numbers, and by the aid of a table of natural sines and cosines*. For writing

$$\alpha + \beta i = \sqrt{\alpha^2 + \beta^2}\left\{\frac{\alpha}{\sqrt{\alpha^2+\beta^2}} + \frac{\beta}{\sqrt{\alpha^2+\beta^2}}\,i\right\},$$

there is always a real angle λ (positive and less than 2π), such that its cosine and sine are $=\frac{\alpha}{\sqrt{\alpha^2+\beta^2}}$ and $\frac{\beta}{\sqrt{\alpha^2+\beta^2}}$ respectively; that is, writing for shortness $\sqrt{\alpha^2+\beta^2} = \rho$, we have $\alpha + \beta i = \rho(\cos\lambda + i\sin\lambda)$, or the equation is $x^n = \rho(\cos\lambda + i\sin\lambda)$; hence observing that $\left(\cos\frac{\lambda}{n} + i\sin\frac{\lambda}{n}\right)^n = \cos\lambda + i\sin\lambda$, a value of x is $=\sqrt[n]{\rho}\left(\cos\frac{\lambda}{n} + i\sin\frac{\lambda}{n}\right)$. The formula really gives all the roots, for instead of λ we may write $\lambda + 2s\pi$, s a positive or negative integer, and then we have

$$x = \sqrt[n]{\rho}\left(\cos\frac{\lambda+2s\pi}{n} + i\sin\frac{\lambda+2s\pi}{n}\right),$$

which has the n values obtained by giving to s the values $0, 1, 2, \ldots, n-1$ in succession; the roots are, it is clear, represented by points lying at equal intervals on a circle. But it is more convenient to proceed somewhat differently; taking one of the roots to be θ, so that $\theta^n = a$, then assuming $x = \theta y$, the equation becomes $y^n - 1 = 0$, which equation, like the original equation, has precisely n roots (one of them being of course $=1$). And the original equation $x^n - a = 0$ is thus reduced to the more simple equation $x^n - 1 = 0$; and although the theory of this equation is included in the preceding one, yet it is proper to state it separately.

The equation $x^n - 1 = 0$ has its several roots expressed in the form $1, \omega, \omega^2, \ldots, \omega^{n-1}$, where ω may be taken $=\cos\frac{2\pi}{n} + i\sin\frac{2\pi}{n}$; in fact, ω having this value, any integer power ω^k is $=\cos\frac{2\pi k}{n} + i\sin\frac{2\pi k}{n}$, and we thence have $(\omega^k)^n = \cos 2\pi k + i\sin 2\pi k$, $=1$, that is, ω^k is a root of the equation. The theory will be resumed further on.

* The square root of $\alpha + \beta i$ can be determined by the extraction of square roots of positive real numbers, without the trigonometrical tables.

By what precedes, we are led to the notion (a numerical) of the radical $a^{\frac{1}{n}}$ regarded as an n-valued function; any one of these being denoted by $\sqrt[n]{a}$, then the series of values is $\sqrt[n]{a}$, $\omega\sqrt[n]{a}, \ldots, \omega^{n-1}\sqrt[n]{a}$; or we may, if we please, use $\sqrt[n]{a}$ instead of $a^{\frac{1}{n}}$ as a symbol to denote the n-valued function.

As the coefficients of an algebraical equation may be numerical, all which follows in regard to algebraical equations is (with, it may be, some few modifications) applicable to numerical equations; and hence, concluding for the present this subject, it will be convenient to pass on to algebraical equations.

II. We consider, secondly, algebraical equations (19 to 34).

19. The equation is

$$x^n - p_1x^{n-1} + \ldots \pm p_n = 0,$$

and we here *assume* the existence of roots, viz. we assume that there are n quantities $a, b, c, \ldots$ (in general all of them different, but which in particular cases may become equal in sets in any manner), such that

$$x^n - p_1x^{n-1} + \ldots \pm p_n = 0;$$

or looking at the question in a different point of view, and starting with the roots $a, b, c, \ldots$ as given, we express the product of the n factors $x-a$, $x-b, \ldots$ in the foregoing form, and thus arrive at an equation of the order n having the n roots $a, b, c, \ldots$. In either case we have

$$p_1 = \Sigma a, \quad p_2 = \Sigma ab, \ldots, \quad p_n = abc\ldots;$$

i.e., regarding the coefficients $p_1, p_2, \ldots, p_n$ as given, then we assume the existence of roots $a, b, c, \ldots$ such that $p_1 = \Sigma a$, &c.; or, regarding the roots as given, then we write p_1, p_2, &c., to denote the functions Σa, Σab, &c.

As already explained, the epithet algebraical is not used in opposition to numerical; an algebraical equation is merely an equation wherein the coefficients are not restricted to denote, or are not explicitly considered as denoting, numbers. That the abstraction is legitimate, appears by the simplest example; in saying that the equation $x^2 - px + q = 0$ has a root $x = \frac{1}{2}(p + \sqrt{p^2 - 4q})$, we mean that writing this value for x the equation becomes an identity, $\{\frac{1}{2}(p + \sqrt{p^2 - 4q})\}^2 - p\{\frac{1}{2}(p + \sqrt{p^2 - 4q})\} + q = 0$; and the verification of this identity in nowise depends upon p and q meaning numbers. But if it be asked what there is beyond numerical equations included in the term algebraical equation, or, again, what is the full extent of the meaning attributed to the term—the latter question at any rate it would be very difficult to answer; as to the former one, it may be said that the coefficients may, for instance, be symbols of operation. As regards such equations, there is certainly no proof that every equation has a root, or that an equation of the nth order has n roots; nor is it in any wise clear what the precise signification of the statement is. But it is found that the assumption of the existence of the n roots can be made without contradictory results; conclusions derived from it, if they involve the roots, rest on the same ground as the original assumption; but the conclusion may be independent of the roots altogether, and in this case it is undoubtedly valid; the reasoning, although actually conducted by aid of the assumption (and, it may be, most easily and elegantly in this manner), is really independent of the assumption. In illustration, we observe that it is allowable to express a function of p and q as follows,—that is, by means of a rational symmetrical function of a and b; this can, as a fact, be expressed as a rational function of $a+b$ and ab; and if we prescribe that $a+b$ and ab shall then be changed into p and q respectively, we have the required function of p, q. That is, we have $F(\alpha, \beta)$ as a representation of $f(p, q)$, obtained as if we had $p=a+b$, $q=ab$, but without in any wise assuming the existence of the a, b of these equations.

20. Starting from the equation

$$x^n - p_1 x^{n-1} + \ldots = x - a \,.\, x - b \,.\, \&\text{c.},$$

or the equivalent equations $p_1 = \Sigma a$, &c., we find

$$\begin{aligned} a^n - p_1 a^{n-1} + \ldots &= 0, \\ b^n - p_1 b^{n-1} + \ldots &= 0; \\ \vdots \qquad \vdots \qquad &\vdots \end{aligned}$$

(it is as satisfying these equations that $a, b, \ldots$ are said to be the roots of $x^n - p_1 x^{n-1} + \ldots = 0$); and conversely from the last-mentioned equations, assuming that $a, b, \ldots$ are all different, we deduce

$$p_1 = \Sigma a, \quad p_2 = \Sigma ab, \ \&\text{c.},$$

and

$$x^n - p_1 x^{n-1} + \ldots = x - a \,.\, x - b \,.\, \&\text{c.}$$

Observe that if, for instance, $a = b$, then the equations $a^n - p_1 a^{n-1} + \ldots = 0$, $b^n - p_1 b^{n-1} + \ldots = 0$ would reduce themselves to a single relation, which would not of itself express that a was a double root,—that is, that $(x-a)^2$ was a factor of $x^n - p_1 x^{n-1} + \&\text{c.}$; but by considering b as the limit of $a+h$, h indefinitely small, we obtain a second equation

$$na^{n-1} - (n-1)\, p_1 a^{n-2} + \ldots = 0,$$

which, with the first, expresses that a is a double root; and then the whole system of equations leads as before to the equations $p_1 = \Sigma a$, &c. But the existence of a double root implies a certain relation between the coefficients; the general case is when the roots are all unequal.

We have then the *theorem* that every rational symmetrical function of the roots is a rational function of the coefficients. This is an easy consequence from the less general theorem, every rational and integral symmetrical function of the roots is a rational and integral function of the coefficients.

In particular, the sums of the powers Σa^2, Σa^3, &c., are rational and integral functions of the coefficients.

The process originally employed for the expression of other functions $\Sigma a^\alpha b^\beta$, &c., in terms of the coefficients is to make them depend upon the sums of powers: for instance, $\Sigma a^\alpha b^\beta = \Sigma a^\alpha \Sigma a^\beta - \Sigma a^{\alpha+\beta}$; but this is very objectionable; the true theory consists in showing that we have systems of equations

$$p_1 = \Sigma a,$$

$$\begin{cases} p_2 = \Sigma ab, \\ p_1^2 = \Sigma a^2 + 2\Sigma ab, \end{cases}$$

$$\begin{cases} p_3 = \Sigma abc, \\ p_1 p_2 = \Sigma a^2 b + 3\Sigma abc, \\ p_1^3 = \Sigma a^3 + 3\Sigma a^2 b + 6\Sigma abc, \end{cases}$$

where in each system there are precisely as many equations as there are root-functions on the right-hand side—e.g. 3 equations and 3 functions Σabc, $\Sigma a^2 b$, Σa^3. Hence in each system the root-functions can be determined linearly in terms of the powers and products of the coefficients:

$$\begin{cases} \Sigma ab = p_2, \\ \Sigma a^2 = p_1^2 - 2p_2, \end{cases}$$

$$\begin{cases} \Sigma abc = p_3, \\ \Sigma a^2 b = p_1 p_2 - 3p_3, \\ \Sigma a^3 = p_1^3 - 3p_1 p_2 + 3p_3, \end{cases}$$

and so on. The older process, if applied consistently, would derive the originally assumed value Σab, $= p_2$, from the two equations $\Sigma a = p_1$, $\Sigma a^2 = p_1^2 - 2p_2$; i.e. we have $2\Sigma ab = \Sigma a \,.\, \Sigma a - \Sigma a^2$, $= p_1^2 - (p_1^2 - 2p_2)$, $= 2p_2$.

21. It is convenient to mention here the theorem that, x being determined as above by an equation of the order n, any rational and integral function whatever of x, or more generally any rational function which does not become infinite in virtue of the equation itself, can be expressed as a rational and integral function of x, of the order $n-1$, the coefficients being rational functions of the coefficients of the equation. Thus the equation gives x^n a function of the form in question; multiplying each side by x, and on the right-hand side writing for x^n its foregoing value, we have x^{n+1}, a function of the form in question; and the like for any higher power of x, and therefore also for any rational and integral function of x. The proof in the case of a rational non-integral function is somewhat more complicated. The final result is of the form $\frac{\phi(x)}{\psi(x)} = I(x)$, or say $\phi(x) - \psi(x) I(x) = 0$, where ϕ, ψ, I are rational and integral functions; in other words, this equation, being true if only $f(x) = 0$, can only be so by reason that the left-hand side contains $f(x)$ as a factor, or we must have identically $\phi(x) - \psi(x) I(x) = M(x) f(x)$. And it is, moreover, clear that the equation $\frac{\phi(x)}{\psi(x)} = I(x)$, being satisfied if only $f(x) = 0$, must be satisfied by each root of the equation.

From the theorem that a rational symmetrical function of the roots is expressible in terms of the coefficients, it at once follows that it is possible to determine an equation (of an assignable order) having for its roots the several values of any given (unsymmetrical) function of the roots of the given equation. For example, in the case of a quartic equation, having the roots (a, b, c, d), it is possible to find an equation having the roots ab, ac, ad, bc, bd, cd, being therefore a sextic equation: viz. in the product

$$(y-ab)(y-ac)(y-ad)(y-bc)(y-bd)(y-cd),$$

the coefficients of the several powers of y will be symmetrical functions of a, b, c, d and therefore rational and integral functions of the coefficients of the quartic equation; hence, supposing the product so expressed, and equating it to zero, we have the required sextic equation. In the same manner can be found the sextic equation having the roots $(a-b)^2$, $(a-c)^2$, $(a-d)^2$, $(b-c)^2$, $(b-d)^2$, $(c-d)^2$, which is the equation of differences previously referred to; and similarly we obtain the equation of differences for a given equation of any order. Again, the equation sought for may be that having for its n roots the given rational functions $\phi(a)$, $\phi(b)$, ... of the several roots of the given equation. Any such rational function can (as was shown) be expressed as a rational and integral function of the order $n-1$; and, retaining x in place of any one of the roots, the problem is to find y from the equations $x^n - p_1x^{n-1} + \ldots = 0$, and $y = M_0x^{n-1} + M_1x^{n-2} + \ldots$, or, what is the same thing, from these two equations to eliminate x. This is, in fact, Tschirnhausen's transformation (1683).

22. In connexion with what precedes, the question arises as to the number of values (obtained by permutations of the roots) of given unsymmetrical functions of the roots, or say of a given set of letters: for instance, with roots or letters (a, b, c, d) as before, how many values are there of the function $ab + cd$, or better, how many functions are there of this form? The answer is 3, viz. $ab + cd$, $ac + bd$, $ad + bc$; or again we may ask whether, in the case of a given number of letters, there exist functions with a given number of values, 3-valued, 4-valued functions, &c.

It is at once seen that for any given number of letters there exist 2-valued functions; the product of the differences of the letters is such a function; however the letters are interchanged, it alters only its sign; or say the two values are Δ, $-\Delta$. And if P, Q are symmetrical functions of the letters, then the general form of such a function is $P + Q\Delta$; this has only the two values $P + Q\Delta$, $P - Q\Delta$.

In the case of 4 letters there exist (as appears above) 3-valued functions: but in the case of 5 letters there does not exist any 3-valued or 4-valued function; and the only 5-valued functions are those which are symmetrical in regard to four of the letters, and can thus be expressed in terms of one letter and of symmetrical functions of all the letters. These last theorems present themselves in the demonstration of the non-existence of a solution of a quintic equation by radicals.

The theory is an extensive and important one, depending on the notions of *substitutions* and of *groups* *.

* A substitution is the operation by which we pass from the primitive arrangement of n letters to any other arrangement of the same letters: for instance, the substitution $\binom{bcda}{abcd}$ means that a is to be changed

23. Returning to equations, we have the very important theorem that, given the value of any unsymmetrical function of the roots, e.g. in the case of a quartic equation, the function $ab+cd$, it is in general possible to determine rationally the value of any similar function, such as $(a+b)^3+(c+d)^3$.

The *a priori* ground of this theorem may be illustrated by means of a numerical equation. Suppose that the roots of a quartic equation are 1, 2, 3, 4, then if it is given that $ab+cd=14$, this in effect determines a, b to be 1, 2 and c, d to be 3, 4 (viz. $a=1$, $b=2$ or $a=2$, $b=1$, and $c=3$, $d=4$ or $c=4$, $d=3$) or else a, b to be 3, 4 and c, d to be 1, 2; and it therefore in effect determines $(a+b)^3+(c+d)^3$ to be $=370$, and not any other value; that is, $(a+b)^3+(c+d)^3$, as having a single value, must be determinable rationally. And we can in the same way account for cases of failure as regards particular equations; thus, the roots being 1, 2, 3, 4 as before, $a^2b=2$ determines a to be $=1$ and b to be $=2$; but if the roots had been 1, 2, 4, 16 then $a^2b=16$ does not uniquely determine a, b but only makes them to be 1, 16 or 2, 4 respectively.

As to the *a posteriori* proof, assume, for instance,

$$t_1=ab+cd,\quad y_1=(a+b)^3+(c+d)^3,$$
$$t_2=ac+bd,\quad y_2=(a+c)^3+(b+d)^3,$$
$$t_3=ad+bc,\quad y_3=(a+d)^3+(b+c)^3:$$

then

$$y_1+y_2+y_3,\quad t_1y_1+t_2y_2+t_3y_3,\quad t_1^2y_1+t_2^2y_2+t_3^2y_3,$$

will be respectively symmetrical functions of the roots of the quartic, and therefore rational and integral functions of the coefficients; that is, they will be known.

Suppose for a moment that t_1, t_2, t_3 are *all* known; then the equations being linear in y_1, y_2, y_3 these can be expressed rationally in terms of the coefficients and of t_1, t_2, t_3; that is, y_1, y_2, y_3 will be known. But observe further that y_1 is obtained as a function of t_1, t_2, t_3 symmetrical as regards t_2, t_3; it can therefore be expressed

into b, b into c, c into d, d into a. Substitutions may, of course, be represented by single letters α, β, . . ; $\binom{abcd}{abcd}$, $=1$, is the substitution which leaves the letters unaltered. Two or more substitutions may be compounded together and give rise to a substitution; i.e., performing upon the primitive arrangement first the substitution β and then upon the result the substitution α, we have the substitution $\alpha\beta$. Substitutions are not commutative; thus, $\alpha\beta$ is not in general $=\beta\alpha$; but they are associative, $\alpha\beta.\gamma=\alpha.\beta\gamma$, so that $\alpha\beta\gamma$ has a determinate meaning. A substitution may be compounded any number of times with itself, and we thus have the powers α^2, α^3, . . , &c. Since the number of substitutions is limited, some power α^ν must be $=1$: or, as this may be expressed, every substitution is a root of unity. A group of substitutions is a set such that each two of them compounded together in either order gives a substitution belonging to the set; every group includes the substitution unity, so that we may in general speak of a group 1, α, β, ... (the number of terms is the order of the group). The whole system of the $1.2.3...n$ substitutions which can be performed upon the n letters is obviously a group: the order of every other group which can be formed out of these substitutions is a submultiple of this number; but it is not conversely true that a group exists the order of which is any given submultiple of this number. In the case of a determinant the substitutions which give rise to the positive terms form a group the order of which is $=\frac{1}{2}.1.2.3...n$. For any function of the n letters, the whole series of substitutions which leave the value of the functions unaltered form a group; and thence also the number of values of the function is $=1.2.3...n$ divided by the order of the group.

as a rational function of t_1 and of t_2+t_3, t_2t_3, and thence as a rational function of t_1 and of $t_1+t_2+t_3$, $t_1t_2+t_1t_3+t_2t_3$, $t_1t_2t_3$; but these last are symmetrical functions of the roots, and as such they are expressible rationally in terms of the coefficients; that is, y_1 will be expressed as a rational function of t_1 and of the coefficients; or t_1 (alone, not t_2 or t_3) being known, y_1 will be rationally determined.

24. We now consider the question of the algebraical solution of equations, or, more accurately, that of the *solution of equations by radicals*.

In the case of a quadric equation $x^2-px+q=0$, we can by the assistance of the sign $\sqrt{(\ )}$ or $(\ )^{\frac{1}{2}}$ find an expression for x as a two-valued function of the coefficients p, q such that, substituting this value in the equation, the equation is thereby identically satisfied; it has been found that this expression is

$$x=\tfrac{1}{2}\{p\pm\sqrt{p^2-4q}\},$$

and the equation is on this account said to be algebraically solvable, or more accurately solvable by radicals. Or we may by writing $x=-\frac{1}{2}p+z$, reduce the equation to $z^2=\frac{1}{4}(p^2-4q)$ viz. to an equation of the form $z^2=a$; and in virtue of its being thus reducible we say that the original equation is solvable by radicals. And the question for an equation of any higher order, say of the order n, is, can we by means of radicals, that is, by aid of the sign $\sqrt[m]{(\ )}$ or $(\ )^{\frac{1}{m}}$, using as many as we please of such signs and with any values of m, find an n-valued function (or any function) of the coefficients which substituted for x in the equation shall satisfy it identically.

It will be observed that the coefficients $p, q, \ldots$ are not explicitly considered as numbers, but even if they do denote numbers, the question whether a numerical equation admits of solution by radicals is wholly unconnected with the before-mentioned theorem of the existence of the n roots of such an equation. It does not even follow that in the case of a numerical equation solvable by radicals the algebraical solution gives the numerical solution, but this requires explanation. Consider first a numerical quadric equation with imaginary coefficients. In the formula $x=\frac{1}{2}(p\pm\sqrt{p^2-4q})$, substituting for p, q their given numerical values, we obtain for x an expression of the form $x=\alpha+\beta i\pm\sqrt{\gamma+\delta i}$, where α, β, γ, δ are real numbers. This expression substituted for x in the quadric equation would satisfy it identically, and it is thus an algebraical solution; but there is no obvious *a priori* reason why $\sqrt{\gamma+\delta i}$ should have a value $=c+di$, where c and d are real numbers calculable by the extraction of a root or roots of real numbers; however the case is (what there was no *a priori* right to expect) that $\sqrt{\gamma+\delta i}$ has such a value calculable by means of the radical expressions $\sqrt{\{\sqrt{\gamma^2+\delta^2}\pm\gamma\}}$: and hence the algebraical solution of a numerical quadric equation does in every case give the numerical solution. The case of a numerical cubic equation will be considered presently.

25. A cubic equation can be solved by radicals. Taking for greater simplicity the cubic in the reduced form $x^3+qx-r=0$, and assuming $x=a+b$, this will be a solution if only $3ab=q$ and $a^3+b^3=r$, equations which give $(a^3-b^3)^2=r^2-\frac{4}{27}q^3$, a

quadric equation solvable by radicals, and giving $a^3 - b^3 = \sqrt{r^2 - \frac{4}{27}q^3}$, a two-valued function of the coefficients: combining this with $a^3 + b^3 = r$, we have $a^3 = \frac{1}{2}(r + \sqrt{r^2 - \frac{4}{27}q^3})$, a two-valued function: we then have a by means of a cube root, viz.

$$a = \sqrt[3]{\{\tfrac{1}{2}(r + \sqrt{r^2 - \tfrac{4}{27}q^3})\}},$$

a six-valued function of the coefficients; but then, writing $q = \dfrac{b}{3a}$, we have, as may be shown, $a + b$ a three-valued function of the coefficients; and $x = a + b$ is the required solution by radicals. It would have been wrong to complete the solution by writing

$$b = \sqrt[3]{\{\tfrac{1}{2}(r - \sqrt{r^2 - \tfrac{4}{27}q^3})\}},$$

for then $a + b$ would have been given as a 9-valued function having only 3 of its values roots, and the other 6 values being irrelevant. Observe that in this last process we make no use of the equation $3ab = q$, in its original form, but use only the derived equation $27a^3b^3 = q^3$, implied in, but not implying, the original form.

An interesting variation of the solution is to write $x = ab(a + b)$, giving $a^3b^3(a^3 + b^3) = r$ and $3a^3b^3 = q$, or say $a^3 + b^3 = \dfrac{3r}{q}$, $a^3b^3 = \frac{1}{3}q$; and consequently

$$a^3 = \frac{\frac{3}{2}}{q}(r + \sqrt{r^2 - \tfrac{4}{27}q^3}),\ b^3 = \frac{\frac{3}{2}}{q}(r - \sqrt{r^2 - \tfrac{4}{27}q^3}),$$

i.e., here a^3, b^3 are each of them a two-valued function, but as the only effect of altering the sign of the quadric radical is to interchange a^3, b^3, they may be regarded as each of them one-valued; a and b are each of them 3-valued (for observe that here only a^3b^3, not ab, is given); and $ab(a + b)$ thus is in appearance a 9-valued function, but it can easily be shown that it is (as it ought to be) only 3-valued.

In the case of a numerical cubic, even when the coefficients are real, substituting their values in the expression

$$x = \sqrt[3]{\{\tfrac{1}{2}(r + \sqrt{r^2 - \tfrac{4}{27}q^3})\}} + \tfrac{1}{3}q \div \sqrt[3]{\{\tfrac{1}{2}(r + \sqrt{r^2 - \tfrac{4}{27}q^3})\}},$$

this may depend on an expression of the form $\sqrt[3]{\gamma + \delta i}$, where γ and δ are real numbers (it will do so if $r^2 - \frac{4}{27}q^3$ is a negative number), and then we *cannot* by the extraction of any root or roots of real positive numbers reduce $\sqrt[3]{\gamma + \delta i}$ to the form $c + di$, c and d real numbers; hence here the algebraical solution does *not* give the numerical solution, and we have here the so-called "irreducible case" of a cubic equation. By what precedes, there is nothing in this that might not have been expected; the algebraical solution makes the solution depend on the extraction of the cube root of a negative number, and there was no reason for expecting this to be a real number. It is well known that the case in question is that wherein the three roots of the numerical cubic equation are all real; if the roots are two imaginary, one real, then contrariwise the quantity under the cube root is real; and the algebraical solution gives the numerical one.

The irreducible case is solvable by a trigonometrical formula, but this is not a solution by radicals: it consists, in effect, in reducing the given numerical cubic (not to a cubic of the form $z^3 = a$, solvable by the extraction of a cube root, but) to a cubic of the form $4x^3 - 3x = a$, corresponding to the equation $4\cos^3\theta - 3\cos\theta = \cos 3\theta$ which serves to determine $\cos\theta$ when $\cos 3\theta$ is known. The theory is applicable to an algebraical cubic equation; say that such an equation, if it can be reduced to the form $4x^3 - 3x = a$, is solvable by "trisection"—then the general cubic equation is solvable by trisection.

26. A quartic equation is solvable by radicals: and it is to be remarked that the existence of such a solution depends on the existence of 3-valued functions such as $ab + cd$ of the four roots (a, b, c, d): by what precedes, $ab + cd$ is the root of a cubic equation, which equation is solvable by radicals: hence $ab + cd$ can be found by radicals; and since $abcd$ is a given function, ab and cd can then be found by radicals. But by what precedes, if ab be known then any similar function, say $a + b$, is obtainable rationally; and then from the values of $a + b$ and ab we may by radicals obtain the value of a or b, that is, an expression for the root of the given quartic equation: the expression ultimately obtained is 4-valued, corresponding to the different values of the several radicals which enter therein, and we have thus the expression by radicals of each of the four roots of the quartic equation. But when the quartic is numerical the same thing happens as in the cubic, and the algebraical solution does not in every case give the numerical one.

It will be understood, from the foregoing explanation as to the quartic, how in the next following case, that of the quintic, the question of the solvability by radicals depends on the existence or non-existence of k-valued functions of the five roots (a, b, c, d, e); the fundamental theorem is the one already stated, a rational function of five letters, if it has less than 5, cannot have more than 2 values, that is, there are no 3-valued or 4-valued functions of 5 letters: and by reasoning depending in part upon this theorem, Abel (1824) showed that a general quintic equation is not solvable by radicals; and *a fortiori* the general equation of any order higher than 5 is not solvable by radicals.

27. The general theory of the solvability of an equation by radicals depends fundamentally on Vandermonde's remark (1770) that, supposing an equation is solvable by radicals, and that we have therefore an algebraical expression of x in terms of the coefficients, then substituting for the coefficients their values in terms of the roots, the resulting expression must reduce itself to any one at pleasure of the roots $a, b, c, ..$; thus in the case of the quadric equation, in the expression $x = \frac{1}{2}(p + \sqrt{p^2 - 4q})$, substituting for p and q their values, and observing that $(a + b)^2 - 4ab = (a - b)^2$, this becomes $x = \frac{1}{2}\{a + b + \sqrt{(a - b)^2}\}$, the value being a or b according as the radical is taken to be $+(a - b)$ or $-(a - b)$.

So in the cubic equation $x^3 - px^2 + qx - r = 0$, if the roots are a, b, c, and if ω is used to denote an imaginary cube root of unity, $\omega^2 + \omega + 1 = 0$, then writing for shortness $p = a + b + c$, $L = a + \omega b + \omega^2 c$, $M = a + \omega^2 b + \omega c$, it is at once seen that LM,

L^3+M^3, and therefore also $(L^3-M^3)^2$ are symmetrical functions of the roots, and consequently rational functions of the coefficients: hence

$$\tfrac{1}{2}\{L^3+M^3+\sqrt{(L^3-M^3)^2}\}$$

is a rational function of the coefficients, which when these are replaced by their values as functions of the roots becomes, according to the sign given to the quadric radical, $=L^3$ or M^3: taking it $=L^3$, the cube root of the expression has the three values $L, \omega L, \omega^2 L$; and LM divided by the same cube root has therefore the values $M, \omega^2 M, \omega M$; whence finally the expression

$$\tfrac{1}{3}[p \quad +\sqrt[3]{\{\tfrac{1}{2}(L^3+M^3+\sqrt{(L^3-M^3)^2})\}}$$
$$+LM\div\sqrt[3]{\{\tfrac{1}{2}(L^3+M^3+\sqrt{(L^3-M^3)^2})\}}]$$

has the three values

$$\tfrac{1}{3}(p+L+M),\ \tfrac{1}{3}(p+\omega L+\omega^2 M),\ \tfrac{1}{3}(p+\omega^2 L+\omega M);$$

that is, these are $=a$, b, c respectively. If the value M^3 had been taken instead of L^3, then the expression would have had the same three values a, b, c. Comparing the solution given for the cubic $x^3+qx-r=0$, it will readily be seen that the two solutions are identical, and that the function $r^2-\tfrac{4}{27}q^3$ under the radical sign must (by aid of the relation $p=0$ which subsists in this case) reduce itself to $(L^3-M^3)^2$; it is only by each radical being equal to a rational function of the roots that the final expression *can* become equal to the roots a, b, c respectively.

28. The formulæ for the cubic were obtained by Lagrange (1770—71) from a different point of view. Upon examining and comparing the principal known methods for the solution of algebraical equations, he found that they all ultimately depended upon finding a "resolvent" equation of which the root is $a+\omega b+\omega^2 c+\omega^3 d+\dots$, ω being an imaginary root of unity, of the same order as the equation; e.g., for the cubic the root is $a+\omega b+\omega^2 c$, ω an imaginary cube root of unity. Evidently the method gives for L^3 a quadric equation, which is the "resolvent" equation in this particular case.

For a quartic the formulæ present themselves in a somewhat different form, by reason that 4 is not a prime number. Attempting to apply it to a quintic, we seek for the equation of which the root is $(a+\omega b+\omega^2 c+\omega^3 d+\omega^4 e)$, ω an imaginary fifth root of unity, or rather the fifth power thereof $(a+\omega b+\omega^2 c+\omega^3 d+\omega^4 e)^5$; this is a 24-valued function, but if we consider the four values corresponding to the roots of unity ω, ω^2, ω^3, ω^4, viz. the values

$$(a+\omega\, b+\omega^2 c+\omega^3 d+\omega^4 e)^5,$$
$$(a+\omega^2 b+\omega^4 c+\omega\, d+\omega^3 e)^5,$$
$$(a+\omega^3 b+\omega\, c+\omega^4 d+\omega^2 e)^5,$$
$$(a+\omega^4 b+\omega^3 c+\omega^2 d+\omega\, e)^5,$$

any symmetrical function of these, for instance their sum, is a six-valued function of the roots, and may therefore be determined by means of a sextic equation, the coefficients whereof are rational functions of the coefficients of the original quintic equation; the conclusion being that the solution of an equation of the fifth order is made to depend upon that of an equation of the sixth order. This is, of course, useless for the solution of the quintic equation, which, as already mentioned, does not admit of solution by radicals; but the equation of the sixth order, Lagrange's resolvent sextic, is very important, and is intimately connected with all the later investigations in the theory.

29. It is to be remarked, in regard to the question of solvability by radicals, that not only the coefficients are taken to be arbitrary, but it is assumed that they are represented each by a single letter, or say rather that they are not so expressed in terms of other arbitrary quantities as to make a solution possible. If the coefficients are not all arbitrary, for instance, if some of them are zero, a sextic equation might be of the form $x^6+bx^4+cx^2+d=0$, and so be solvable as a cubic; or if the coefficients of the sextic are given functions of the six arbitrary quantities a, b, c, d, e, f, such that the sextic is really of the form

$$(x^2+ax+b)(x^4+cx^3+dx^2+ex+f)=0,$$

then it breaks up into the equations $x^2+ax+b=0$, $x^4+cx^3+dx^2+ex+f=0$, and is consequently solvable by radicals; so also if the form is

$$(x-a)(x-b)(x-c)(x-d)(x-e)(x-f)=0,$$

then the equation is solvable by radicals,—in this extreme case rationally. Such cases of solvability are self-evident; but they are enough to show that the general theorem of the non-solvability by radicals of an equation of the fifth or any higher order does not in any wise exclude for such orders the existence of particular equations solvable by radicals, and there are, in fact, extensive classes of equations which are thus solvable; the binomial equations $x^n-1=0$ present an instance.

30. It has already been shown how the several roots of the equation $x^n-1=0$ can be expressed in the form $\cos\frac{2s\pi}{n}+i\sin\frac{2s\pi}{n}$, but the question is now that of the algebraical solution (or solution by radicals) of this equation. There is always a root $=1$; if ω be any other root, then obviously $\omega, \omega^2, \ldots, \omega^{n-1}$ are all of them roots; x^n-1 contains the factor $x-1$, and it thus appears that $\omega, \omega^2, \ldots, \omega^{n-1}$ are the $n-1$ roots of the equation

$$x^{n-1}+x^{n-2}+\ldots+x+1=0;$$

we have, of course,

$$\omega^{n-1}+\omega^{n-2}+\ldots+\omega+1=0.$$

It is proper to distinguish the cases n prime and n composite; and in the latter case there is a distinction according as the prime factors of n are simple or multiple. By way of illustration, suppose successively $n=15$ and $n=9$; in the former case, if α be an imaginary root of $x^3-1=0$ (or root of $x^2+x+1=0$), and β an imaginary root of $x^5-1=0$ (or root of $x^4+x^3+x^2+x+1=0$), then ω may be taken $=\alpha\beta$; the successive powers thereof, $\alpha\beta$, $\alpha^2\beta^2$, β^3, $\alpha\beta^4$, α^2, β, $\alpha\beta^2$, $\alpha^2\beta^3$, β^4, α,

$\alpha^2\beta$, β^2, $\alpha\beta^3$, $\alpha^2\beta^4$, are the roots of $x^{14}+x^{13}+\ldots+x+1=0$; the solution thus depends on the solution of the equations $x^3-1=0$ and $x^5-1=0$. In the latter case, if α be an imaginary root of $x^3-1=0$ (or root of $x^2+x+1=0$), then the equation $x^9-1=0$ gives $x^3=1$, α, or α^2; $x^3=1$ gives $x=1$, α, or α^2; and the solution thus depends on the solution of the equations $x^3-1=0$, $x^3-\alpha=0$, $x^3-\alpha^2=0$. The first equation has the roots 1, α, α^2; if β be a root of either of the others, say if $\beta^3=\alpha$, then assuming $\omega=\beta$, the successive powers are β, β^2, α, $\alpha\beta$, $\alpha\beta^2$, α^2, $\alpha^2\beta$, $\alpha^2\beta^2$, which are the roots of the equation $x^8+x^7+\ldots+x+1=0$.

It thus appears that the only case which need be considered is that of n a prime number, and writing (as is more usual) r in place of ω, we have $r, r^2, r^3, \ldots, r^{n-1}$ as the $(n-1)$ roots of the reduced equation

$$x^{n-1}+x^{n-2}+\ldots+x+1=0;$$

then not only $r^n-1=0$, but also $r^{n-1}+r^{n-2}+\ldots+r+1=0$.

31. The process of solution due to Gauss (1801) depends essentially on the arrangement of the roots in a certain order, viz. not as above, with the indices of r in arithmetical progression, but with their indices in geometrical progression; the prime number n has a certain number of prime roots g, which are such that g^{n-1} is the lowest power of g, which is $\equiv 1$ to the modulus n; or, what is the same thing, that the series of powers $1, g, g^2, \ldots, g^{n-2}$, each divided by n, leave (in a different order) the remainders $1, 2, 3, \ldots, n-1$; hence giving to r in succession the indices $1, g, g^2, \ldots, g^{n-2}$, we have, in a different order, the whole series of roots $r, r^2, r^3, \ldots, r^{n-1}$.

In the most simple case, $n=5$, the equation to be solved is $x^4+x^3+x^2+x+1=0$; here 2 is a prime root of 5, and the order of the roots is r, r^2, r^4, r^3. The Gaussian process consists in forming an equation for determining the periods $P_1, P_2, =r+r^4$ and r^2+r^3 respectively,—these being such that the symmetrical functions P_1+P_2, P_1P_2 are rationally determinable: in fact,

$$P_1+P_2=-1,\quad P_1P_2=(r+r^4)(r^2+r^3),\ =r^3+r^4+r^6+r^7,\ =r^3+r^4+r+r^2,\ =-1.$$

P_1, P_2 are thus the roots of $u^2+u-1=0$; and taking them to be known, they are themselves broken up into subperiods, in the present case single terms, r and r^4 for P_1, r^2 and r^3 for P_2; the symmetrical functions of these are then rationally determined in terms of P_1 and P_2; thus $r+r^4=P_1$, $r.r^4=1$, or r, r^4 are the roots of $u^2-P_1u+1=0$. The mode of division is more clearly seen for a larger value of n; thus, for $n=7$ a prime root is $=3$, and the arrangement of the roots is $r, r^3, r^2, r^6, r^4, r^5$. We may form either 3 periods each of 2 terms,

$$P_1, P_2, P_3, =r+r^6,\ r^3+r^4,\ r^2+r^5,$$

respectively; or else 2 periods each of 3 terms, P_1, $P_2=r+r^2+r^4$, $r^3+r^6+r^5$ respectively; in each case the symmetrical functions of the periods are rationally determinable; thus in the case of the two periods $P_1+P_2=-1$, $P_1P_2=3+r+r^2+r^3+r^4+r^5+r^6$, $=2$;

and, the periods being known, the symmetrical functions of the several terms of each period are rationally determined in terms of the periods, thus

$$r + r^2 + r^4 = P_1, \quad r \,.\, r^2 + r \,.\, r^4 + r^2 \,.\, r^4 = P_2, \quad r \,.\, r^2 \,.\, r^4 = 1.$$

The theory was further developed by Lagrange (1808), who, applying his general process to the equation in question, $x^{n-1} + x^{n-2} + \ldots + x + 1 = 0$, the roots $a, b, c, \ldots$ being the several powers of r, the indices in geometrical progression as above, showed that the function $(a + \omega b + \omega^2 c + \ldots)^{n-1}$ was in this case a given function of ω with integer coefficients. Reverting to the before-mentioned particular equation $x^4 + x^3 + x^2 + x + 1 = 0$, it is very interesting to compare the process of solution with that for the solution of the general quartic the roots whereof are a, b, c, d.

Take ω, a root of the equation $\omega^4 - 1 = 0$ (whence ω is $= 1, -1, i,$ or $-i$, at pleasure), and consider the expression

$$(a + \omega b + \omega^2 c + \omega^3 d)^4.$$

The developed value of this is

$$\begin{aligned} = \quad & a^4 + b^4 + c^4 + d^4 + 6(a^2c^2 + b^2d^2) + 12(a^2bd + b^2ca + c^2db + d^2ac) \\ + \omega \quad & \{4(a^3b + b^3c + c^3d + d^3a) + 12(a^2cd + b^2da + c^2ab + d^2bc)\} \\ + \omega^2 & \{6(a^2b^2 + b^2c^2 + c^2d^2 + d^2a^2) + 4(a^3c + b^3d + c^3a + d^3b) + 24abcd\} \\ + \omega^3 & \{4(a^3d + b^3a + c^3b + d^3c) + 12(a^2bc + b^2cd + c^2da + d^2ab)\}; \end{aligned}$$

that is, this is a 6-valued function of a, b, c, d, the root of a sextic (which is, in fact, solvable by radicals; but this is not here material).

If, however, a, b, c, d denote the roots r, r^2, r^4, r^3 of the special equation, then the expression becomes

$$\begin{aligned} & r^4 + r^3 + r + r^2 + 6(1 + 1) + 12(r^2 + r^4 + r^3 + r) \\ + \omega \; & \{4(1 + 1 + 1 + 1) + 12(r^4 + r^3 + r + r^2)\} \\ + \omega^2 & \{6(r + r^2 + r^4 + r^3) + 4(r^2 + r^4 + r^3 + r)\} \\ + \omega^3 & \{4(r + r^2 + r^4 + r^3) + 12(r^3 + r + r^2 + r^4)\}; \end{aligned}$$

viz. this is

$$= -1 + 4\omega + 14\omega^2 - 16\omega^3,$$

a completely determined value. That is, we have

$$(r + \omega r^2 + \omega^2 r^4 + \omega^3 r^3)^4 = -1 + 4\omega + 14\omega^2 - 16\omega^3,$$

which result contains the solution of the equation. If $\omega = 1$, we have $(r + r^2 + r^4 + r^3)^4 = 1$, which is right; if $\omega = -1$, then $(r + r^4 - r^2 - r^3)^4 = 25$; if $\omega = i$, then we have $\{r - r^4 + i(r^2 - r^3)\}^4 = -15 + 20i$; and if $\omega = -i$, then $\{r - r^4 - i(r^2 - r^3)\}^4 = -15 - 20i$; the solution may be completed without difficulty.

The result is perfectly general, thus:—n being a prime number, r a root of the equation $x^{n-1} + x^{n-2} + \ldots + x + 1 = 0$, ω a root of $\omega^{n-1} - 1 = 0$, and g a prime root of $g^{n-1} \equiv 1$ (mod. n), then

$$(r + \omega r^g + \ldots + \omega^{n-2} r^{g^{n-2}})^{n-1}$$

is a given function $M_0 + M_1\omega + \ldots + M_{n-2}\omega^{n-2}$ with integer coefficients, and by the extraction of $(n-1)$th roots of this and similar expressions we ultimately obtain r in terms of ω, which is taken to be known; the equation $x^n - 1 = 0$, n a prime number, is thus solvable by radicals. In particular, if $n-1$ be a power of 2, the solution (by either process) requires the extraction of square roots only; and it was thus that Gauss discovered that it was possible to construct geometrically the regular polygons of 17 sides and 257 sides respectively. Some interesting developments in regard to the theory were obtained by Jacobi (1837); see the memoir "Ueber die Kreistheilung, u.s.w.," *Crelle*, t. XXX. (1846).

The equation $x^{n-1} + \ldots + x + 1 = 0$ has been considered for its own sake, but it also serves as a specimen of a class of equations solvable by radicals, considered by Abel (1828), and since called Abelian equations, viz., for the Abelian equation of the order n, if x be any root, the roots are $x, \theta x, \theta^2 x, \ldots, \theta^{n-1}x$ (θx being a rational function of x, and $\theta^n x = x$); the theory is, in fact, very analogous to that of the above particular case. A more general theorem obtained by Abel is as follows:—If the roots of an equation of any order are connected together in such wise that *all* the roots can be expressed rationally in terms of any one of them, say x; if, moreover, θx, $\theta_1 x$ being any two of the roots, we have $\theta\theta_1 x = \theta_1\theta x$, the equation will be solvable algebraically. It is proper to refer also to Abel's definition of an *irreducible* equation:—an equation $\phi x = 0$, the coefficients of which are rational functions of a certain number of known quantities $a, b, c, \ldots$, is called irreducible when it is impossible to express its roots by an equation of an inferior degree, the coefficients of which are also rational functions of $a, b, c, \ldots$ (or, what is the same thing, when ϕx does not break up into factors which are rational functions of $a, b, c, \ldots$). Abel applied his theory to the equations which present themselves in the division of the elliptic functions, but not to the modular equations.

32. But the theory of the algebraical solution of equations in its most complete form was established by Galois (born October 1811, killed in a duel May 1832; see his collected works, *Liouville*, t. XI., 1846). The definition of an irreducible equation resembles Abel's,—an equation is reducible when it admits of a rational divisor, irreducible in the contrary case; only the word *rational* is used in this extended sense that, in connexion with the coefficients of the given equation, or with the irrational quantities (if any) whereof these are composed, he considers any number of other irrational quantities called "adjoint radicals," and he terms rational any rational function of the coefficients (or the irrationals whereof they are composed) and of these adjoint radicals; the epithet irreducible is thus taken either absolutely or in a relative sense, according to the system of adjoint radicals which are taken into account. For instance, the equation $x^4 + x^3 + x^2 + x + 1 = 0$; the left-hand side has here no rational divisor, and the equation is irreducible; but this function is $= (x^2 + \frac{1}{2}x + 1)^2 - \frac{5}{4}x^2$, and it has thus the irrational divisors $x^2 + \frac{1}{2}(1 + \sqrt{5})\,x + 1$, $x^2 + \frac{1}{2}(1 - \sqrt{5})\,x + 1$; and these, if we *adjoin* the radical $\sqrt{5}$, are rational, and the equation is no longer irreducible. In the case of a given equation, assumed to be irreducible, the problem to solve the equation is, in fact, that of finding radicals by the adjunction of which the equation

becomes reducible; for instance, the general quadric equation $x^2 + px + q = 0$ is irreducible, but it becomes reducible, breaking up into rational linear factors, when we adjoin the radical $\sqrt{\tfrac{1}{4}p^2 - q}$.

The fundamental theorem is the Proposition I. of the "Mémoire sur les conditions de résolubilité des équations par radicaux"; viz. given an equation of which $a, b, c, \ldots$ are the m roots, there is always a group of permutations of the letters $a, b, c, \ldots$ possessed of the following properties:—

1. Every function of the roots invariable by the substitutions of the group is rationally known.

2. Reciprocally, every rationally determinable function of the roots is invariable by the substitutions of the group.

Here by an invariable function is meant not only a function of which the form is invariable by the substitutions of the group, but further, one of which the value is invariable by these substitutions: for instance, if the equation be $\phi x = 0$, then ϕx is a function of the roots invariable by any substitution whatever. And in saying that a function is rationally known, it is meant that its value is expressible rationally in terms of the coefficients and of the adjoint quantities.

For instance, in the case of a general equation, the group is simply the system of the $1.2.3 \ldots n$ permutations of all the roots, since, in this case, the only rationally determinable functions are the symmetric functions of the roots.

In the case of the equation $x^{n-1} + \ldots + x + 1 = 0$, n a prime number,

$$a, b, c, \ldots, k = r, r^g, r^{g^2}, \ldots, r^{g^{n-2}},$$

where g is a prime root of n, then the group is the cyclical group $abc \ldots k$, $bc \ldots ka, \ldots, kab \ldots j$, that is, in this particular case the number of the permutations of the group is equal to the order of the equation.

This notion of the group of the original equation, or of the group of the equation as varied by the adjunction of a series of radicals, seems to be the fundamental one in Galois's theory. But the problem of solution by radicals, instead of being the sole object of the theory, appears as the first link of a long chain of questions relating to the transformation and classification of irrationals.

Returning to the question of solution by radicals, it will be readily understood that by the adjunction of a radical the group may be diminished; for instance, in the case of the general cubic, where the group is that of the six permutations, by the adjunction of the square root which enters into the solution, the group is reduced to abc, bca, cab; that is, it becomes possible to express rationally, in terms of the coefficients and of the adjoint square root, any function such as $a^2b + b^2c + c^2a$ which is not altered by the cyclical substitution a into b, b into c, c into a. And hence, to determine whether an equation of a given form is solvable by radicals, the course of investigation is to inquire whether, by the successive adjunction of radicals, it is possible to reduce the original group of the equation so as to make it ultimately consist of a single permutation.

The condition in order that an equation of a given prime order n may be solvable by radicals was in this way obtained—in the first instance in the form, scarcely intelligible without further explanation, that every function of the roots $x_1, x_2, \ldots, x_n$, invariable by the substitutions x_{ak+b} for x_k, must be rationally known; and then in the equivalent form that the resolvent equation of the order $1.2\ldots\overline{n-2}$ must have a rational root. In particular, the condition in order that a quintic equation may be solvable is that Lagrange's resolvent of the order 6 may have a rational factor, a result obtained from a direct investigation in a valuable memoir by E. Luther, *Crelle*, t. XXXIV. (1847).

Among other results demonstrated or announced by Galois may be mentioned those relating to the modular equations in the theory of elliptic functions; for the transformations of the orders 5, 7, 11, the modular equations of the orders 6, 8, 12 are depressible to the orders 5, 7, 11 respectively; but for the transformation, n a prime number greater than 11, the depression is impossible.

The general theory of Galois in regard to the solution of equations was completed, and some of the demonstrations supplied, by Betti (1852). See also Serret's *Cours d'Algèbre supérieure*, 2nd ed. 1854; 4th ed. 1877—78.

33. Returning to quintic equations, Jerrard (1835) established the theorem that the general quintic equation is, by the extraction of only square and cubic roots, reducible to the form $x^5 + ax + b = 0$, or what is the same thing, to $x^5 + x + b = 0$. The actual reduction by means of Tschirnhausen's theorem was effected by Hermite in connexion with his elliptic-function solution of the quintic equation (1858) in a very elegant manner. It was shown by Cockle and Harley (1858—59) in connexion with the Jerrardian form, and by Cayley (1861), that Lagrange's resolvent equation of the sixth order can be replaced by a more simple sextic equation occupying a like place in the theory.

The theory of the modular equations, more particularly for the case $n = 5$, has been studied by Hermite, Kronecker, and Brioschi. In the case $n = 5$, the modular equation of the order 6 depends, as already mentioned, on an equation of the order 5; and conversely the general quintic equation may be made to depend upon this modular equation of the order 6; that is, assuming the solution of this modular equation, we can solve (not by radicals) the general quintic equation; this is Hermite's solution of the general quintic equation by elliptic functions (1858); it is analogous to the before-mentioned trigonometrical solution of the cubic equation. The theory is reproduced and developed in Brioschi's memoir, "Ueber die Auflösung der Gleichungen vom fünften Grade," *Math. Annalen*, t. XIII. (1877—78).

34. The great modern work, reproducing the theories of Galois, and exhibiting the theory of algebraic equations as a whole, is Jordan's *Traité des Substitutions et des Équations Algébriques*, Paris, 1870. The work is divided into four books—book I.,

preliminary, relating to the theory of congruences; book II. is in two chapters, the first relating to substitutions in general, the second to substitutions defined analytically, and chiefly to linear substitutions; book III. has four chapters, the first discussing the principles of the general theory, the other three containing applications to algebra, geometry, and the theory of transcendents; lastly, book IV., divided into seven chapters, contains a determination of the general types of equations solvable by radicals, and a complete system of classification of these types. A glance through the index will show the vast extent which the theory has assumed, and the form of general conclusions arrived at; thus, in book III., the algebraical applications comprise Abelian equations, equations of Galois; the geometrical ones comprise Hesse's equation, Clebsch's equations, lines on a quartic surface having a nodal line, singular points of Kummer's surface, lines on a cubic surface, problems of contact; the applications to the theory of transcendents comprise circular functions, elliptic functions (including division and the modular equation), hyperelliptic functions, solution of equations by transcendents. And on this last subject, solution of equations by transcendents, we may quote the result,—"the solution of the general equation of an order superior to five cannot be made to depend upon that of the equations for the division of the circular or elliptic functions"; and again (but with a reference to a possible case of exception), "the general equation cannot be solved by aid of the equations which give the division of the hyperelliptic functions into an odd number of parts."

787.

FUNCTION.

[From the *Encyclopædia Britannica, Ninth Edition,* vol. IX. (1879), pp. 818—824.]

FUNCTIONALITY, in Analysis, is dependence on a variable or variables; in the case of a single variable u, it is the same thing to say that v depends upon u, or to say that v is a function of u, only in the latter form of expression the mode of dependence is embodied in the term "function." We have given or known functions such as u^2 or $\sin u$, and the general notation of the form ϕu, where the letter ϕ is used as a functional symbol to denote a function of u, known or unknown as the case may be: in each case u is the independent variable or argument of the function, but it is to be observed that, if v be a function of u, then v like u is a variable, the values of v regarded as known serve to determine those of u; that is, we may conversely regard u as a function of v. In the case of two or more independent variables, say when w depends on or is a function of u, v, &c., or $w = \phi(u, v, \ldots)$, then $u, v, \ldots$ are the independent variables or arguments of the function; frequently when one of these variables, say u, is principally or alone attended to, it is regarded as the independent variable or argument of the function, and the other variables v, &c., are regarded as parameters, the values of which serve to complete the definition of the function. We may have a set of quantities $w, t, \ldots$ each of them a function of the same variables $u, v, \ldots$; and this relation may be expressed by means of a single functional symbol ϕ, $(w, t, \ldots) = \phi(u, v, \ldots)$; but, as to this, more hereafter.

The notion of a function is applicable in geometry and mechanics as well as in analysis; for instance, a point Q, the position of which depends upon that of a variable point P, may be regarded as a function of the point P; but here, substituting for the points themselves the coordinates (of any kind whatever) which determine their positions, we may say that the coordinates of Q are each of them a function of the coordinates of P, and we thus return to the analytical notion of a function. And in what follows a function is regarded exclusively in this point of view, viz. the variables are regarded as numbers; and we attend to the case of a function of one variable $v=fu$. But it has been remarked (see Equation) that it is not allowable to confine the attention to *real* numbers; a number u must in general be taken to be a complex number $u=x+iy$, x and y being real numbers, each susceptible of continuous variation between the limits $-\infty$, $+\infty$, and i denoting $\sqrt{-1}$. In regard to any particular function, fu, although it may for some purposes be sufficient to know the value of the function for any real value whatever of u, yet to attend only to the real values of u is an essentially incomplete view of the question; to properly know the function, it is necessary to consider u under the aforesaid imaginary or complex form $u=x+iy$.

To a given value $x+iy$ of u there corresponds in general for v a value or values of the like form $v=x'+iy'$, and we obtain a geometrical notion of the meaning of the functional relation $v=fu$ by regarding x, y as rectangular coordinates of a point P in a plane Π, and x', y' as rectangular coordinates of a point P' in a plane (for greater convenience a different plane) Π'; P, P' are thus the geometrical representations, or representative points, of the variables $u=x+iy$ and $u'=x'+iy'$ respectively; and, according to a locution above referred to, the point P' might be regarded as a function of the point P; a given value of $u=x+iy$ is thus represented by a point P in the plane Π, and corresponding hereto we have a point or points P' in the plane Π', representing (if more than one, each of them) a value of the variable $v=x'+iy'$. And, if we attend only to the values of u as corresponding to a series of positions of the representative point P, we have the notion of the "path" of a complex variable $u=x+iy$.

Known Functions.

1. The most simple kind of function is the rational and integral function. We have the series of powers u^2, u^3, ... each calculable not only for a real but also for a complex value of u, $(x+iy)^2=x^2-iy^2+2ixy$, $(x+iy)^3=x^3-3xy^2+i(3x^2y-y^3)$, &c., and thence, if a, b, ... be real or complex numbers, the general form $a+bu+cu^2+\ldots+ku^m$, of a rational and integral function of the order m. And taking two such functions, say of the orders m and n respectively, the quotient of one of these by the other represents the general form of a rational function of u.

The function which next presents itself is the algebraical function, and in particular the algebraical function expressible by radicals. To take the most simple case, suppose (m being a positive integer) that $v^m=u$; v is here the irrational function $=u^{\frac{1}{m}}$. Obviously, if u is real and positive, there is always a real and positive value of v, calculable to any extent of approximation from the equation $v^m=u$, which serves as the definition of $u^{\frac{1}{m}}$; but it is known (see Equation) that, as well in this case as in the general case where u is a complex number, there are in fact m values of the function $u^{\frac{1}{m}}$; and that for their determination we require the theory of the so-called circular functions sine and cosine; and these depend on the exponential function $\exp u$, or, as it is commonly written, e^u, which has for its inverse the logarithmic function $\log u$; these are all of them transcendental functions.

2. In a rational and integral function $a + bu + cu^2 + \ldots + ku^m$, the number of terms is finite, and the coefficients $a, b, \ldots. k$ may have any values whatever, but if we imagine a like series $a + bu + cu^2 + \ldots$ extending to infinity, *non constat* that such an expression has any calculable value,—that is, any meaning at all; the coefficients $a, b, c, \ldots$ must be such as, either for every value whatever of u (that is, for every finite value) or for values included within certain limits, to make the series *convergent.* It is easy to see that the values of $a, b, c, \ldots$ may be such as to make the series always convergent; for instance, this is the case for the exponential function,

$$\exp u = 1 + \frac{u}{1} + \frac{u^2}{1.2} + \frac{u^3}{1.2.3} + \&c.;$$

taking for the moment u to be real and positive, then it is evident that however large u may be, the successive terms will become ultimately smaller and smaller, and the series will have a determinate calculable value. A function thus expressed by means of a convergent infinite series is not in general algebraical, and when it is not so, it is said to be transcendental (but observe that it is in nowise true that we have thus the most general form of a transcendental function); in particular, the exponential function above written down is not an algebraical function.

By forming the expression of $\exp v$, and multiplying together the two series, we derive the fundamental property

$$\exp u \exp v = \exp(u + v);$$

whence also

$$\exp x \exp iy = \exp(x + iy),$$

so that $\exp(x + iy)$ is given as the product of the two series $\exp x$ and $\exp iy$. As regards this last, if in place of u we actually write the value iy, we find

$$\exp iy = \left(1 - \frac{y^2}{1.2} + \frac{y^4}{1.2.3.4} - \ldots\right) + i\left(y - \frac{y^3}{1.2.3} + \ldots\right),$$

where obviously each series is convergent and actually calculable for any real value whatever of y. Calling the two series cosine y and sine y respectively, or in the ordinary abbreviated notation $\cos y$ and $\sin y$, the equation is

$$\exp iy = \cos y + i \sin y;$$

and if we herein for y write z, and multiply the two expressions together, observing that the product will be $= \exp i(y + z)$, we obtain the fundamental equations

$$\cos(y + z) = \cos y \cos z - \sin y \sin z,$$

$$\sin(y + z) = \sin y \cos z + \sin z \cos y,$$

for the functions sine and cosine.

Taking y as an angle, and defining as usual the sine and cosine as the ratios of the perpendicular and base respectively to the radius, the sine and cosine will be functions of y; and we obtain geometrically the foregoing fundamental equations for the sine and cosine; but in order to the truth of the foregoing equation $\exp iy = \cos y + i \sin y$, it is further necessary that the angle should be measured in circular measure, that is, by the ratio of the arc to the radius; so that π denoting as usual the number 3·14159..., the measure of a right angle is $= \frac{1}{2}\pi$. And this being so, the functions sine and cosine, obtained as above by consideration of the exponential function, have their ordinary geometrical significations.

3. The foregoing investigation was given in detail in order to the completion of the theory of the irrational function $u^{\frac{1}{m}}$. We henceforth take the theory of the circular functions as known, and speak of $\tan x$, &c., as the occasion may arise.

We have

$$x + iy = r(\cos\theta + i\sin\theta),$$

where, writing $\sqrt{x^2+y^2}$ to denote the positive value of the square root, we have

$$r = \sqrt{x^2+y^2},\quad \cos\theta = \frac{x}{\sqrt{x^2+y^2}},\quad \sin\theta = \frac{y}{\sqrt{x^2+y^2}},$$

and therefore also

$$\tan\theta = \frac{y}{x}.$$

Treating x, y as the rectangular coordinates of a point P, r is the distance (regarded as positive) of this point from the origin, and θ is the inclination of r to the positive part of the axis of x; to fix the ideas θ may be regarded as lying within the limits 0, π, or 0, $-\pi$, according as y is positive or negative; θ is thus completely determinate, except in the case, x negative, $y = 0$, for which θ is $= \pi$ or $-\pi$ indifferently.

And if $u = x + iy$, we hence have

$$u^{\frac{1}{m}} = (x+iy)^{\frac{1}{m}} = r^{\frac{1}{m}}\left(\cos\frac{\theta + 2s\pi}{m} + i\sin\frac{\theta+2s\pi}{m}\right),$$

where $r^{\frac{1}{m}}$ is real and positive and s has any positive or negative integer value whatever: but we thus obtain for $u^{\frac{1}{m}}$ only the m values corresponding to the values $0, 1, 2, \ldots, m-1$ of s. More generally we may, instead of the index $\frac{1}{m}$, take the index to be any rational fraction $\frac{n}{m}$. Supposing this to be in its least terms, and m to be positive, the number of distinct values is always $= m$. If instead of $\frac{n}{m}$ we take the index to be the general real or complex quantity m, we have u^m, no longer an algebraical function of u, and having in general an infinity of values.

4. The foregoing equation $\exp(x+y)=\exp x\,.\exp y$ is, in fact, the equation of indices, $a^{x+y}=a^x\,.\,a^y$; $\exp x$ is thus the same thing as e^x, where e denotes a properly determined number, and putting e^x equal to the series, and then writing $x=1$, we have $e=1+\frac{1}{1}+\frac{1}{1\,.\,2}+\frac{1}{1\,.\,2\,.\,3}+\&c.$, that is, $e=2{\cdot}7128\ldots$ But as well theoretically as for convenience of printing, there is considerable advantage in the use of the notation $\exp u$.

From the equation, $\exp iy=\cos y+i\sin y$, we deduce $\exp(-iy)=\cos y-i\sin y$, and thence

$$\cos y=\tfrac{1}{2}\{\exp(iy)+\exp(-iy)\},$$

$$\sin y=\frac{1}{2i}\{\exp(iy)-\exp(-iy)\};$$

if we write herein ix instead of y we have

$$\cos ix=\tfrac{1}{2}\{\exp x+\exp(-x)\},$$

$$\sin ix=\frac{i}{2}\{\exp x-\exp(-x)\},$$

viz. these values are

$$\cos ix=1+\frac{x^2}{1\,.\,2}+\frac{x^4}{1\,.\,2\,.\,3\,.\,4}+\ldots$$

$$\frac{1}{i}\sin ix=x+\frac{x^3}{1\,.\,2\,.\,3}+\ldots$$

each of them real when x is real. The functions in question $1+\frac{x^2}{1\,.\,2}+\frac{x^4}{1\,.\,2\,.\,3\,.\,4}+\ldots$ and $x+\frac{x^3}{1\,.\,2\,.\,3}+\ldots$, regarded as functions of x, are termed the hyperbolic cosine and sine, and are represented by the notations $\cosh x$ and $\sinh x$ respectively; and similarly we have the hyperbolic tangent $\tanh x$, &c.: although it is easy to remember that $\cos ix$, $\frac{1}{i}\sin ix$, are, in fact, real functions of x, and to understand accordingly the formulæ wherein they occur, yet the use of these notations of the hyperbolic functions is often convenient.

5. Writing $u=\exp v$, then v is conversely a function of u which is called the logarithm (hyperbolic logarithm, to distinguish it from the tabular or Briggian logarithm), and we write $v=\log u$, or what is the same thing, we have $u=\exp(\log u)$: and it is clear that if u be real and positive there is always a real and positive value of $\log u$, in particular the real logarithm of e is $=1$; it is however to be observed that the logarithm is not a one-valued function, but has an infinity of values corresponding to the different integer values of a constant s; in fact, if $\log u$ be any one of its values, then $\log u+2s\pi i$ is also a value, for we have $\exp(\log u+2s\pi i)=\exp\log u\exp 2s\pi i$, or since $\exp 2s\pi i$ is $=1$, this is $=u$; that is, $\log u+2s\pi i$ is a value of the logarithm of u.

We have

$$uv=\exp(\log uv)=\exp\log u\,.\,\exp\log v,$$

and hence the equation which is commonly written

$$\log uv = \log u + \log v,$$

but which requires the addition on one side of a term $2s\pi i$. And reverting to the equation $x + iy = r(\cos\theta + i\sin\theta)$, or as it is convenient to write it, $x + iy = r \exp i\theta$, we hence have

$$\log(x + iy) = \log r + i(\theta + 2s\pi),$$

where $\log r$ may be taken to denote the real logarithm of the real positive quantity r, and θ the completely determinate angle defined as already mentioned.

Reverting to the function u^m, we have $u = \exp\log u$, and thence $u^m = \exp(m\log u)$, which, on account of the infinity of values of $\log u$, has in general (as before remarked) an infinity of values; if $u = e$, then e^m, $= \exp(m\log e)$, has in general in like manner an infinity of values, but in regarding e^m as identical with the one-valued function $\exp m$, we take $\log e$ to be $=$ its real value, 1.

The inverse functions $\cos^{-1} x$, $\sin^{-1} x$, $\tan^{-1} x$, are in fact logarithmic functions; thus in the equation $\exp ix = \cos x + i\sin x$, writing first $\cos x = u$, the equation becomes $\exp i\cos^{-1} u = u + i\sqrt{1-u^2}$, or we have $\cos^{-1} u = \frac{1}{i}\log(u + i\sqrt{1-u^2})$, and from the same equation, writing secondly $\sin x = u$, we have $\sin^{-1} u = \frac{1}{i}\log(\sqrt{1-u^2} + iu)$. But the formula for $\tan^{-1} u$ is a more elegant one, as not involving the radical $\sqrt{1-u^2}$; we have

$$i\tan x = \frac{\exp ix - \exp(-ix)}{\exp ix + \exp(-ix)}, \; = \frac{\exp 2ix - 1}{\exp 2ix + 1},$$

and thence

$$\exp 2ix = \frac{1 + i\tan x}{1 - i\tan x},$$

that is,

$$x = \frac{1}{2i}\log\frac{1 + i\tan x}{1 - i\tan x},$$

or, if $\tan x = u$, then

$$\tan^{-1} u = \frac{1}{2i}\log\frac{1 + iu}{1 - iu}.$$

The logarithm (or inverse exponential function) and the inverse circular functions present themselves as the integrals of algebraic functions

$$\int\frac{dx}{x} = \log x,$$

whence also

$$\int\frac{dx}{1 + x^2} = \frac{1}{2i}\log\frac{1 + ix}{1 - ix} = \tan^{-1} x,$$

and

$$\int\frac{dx}{\sqrt{1 - x^2}} = \sin^{-1} x.$$

6. Each of the functions $\exp u$, $\sin u$, $\cos u$, $\tan u$, &c., as a one-valued function of u, is in this respect analogous to a rational function of u; and there are further analogies of $\exp u$, $\sin u$, $\cos u$, to a rational and integral function; and of $\tan u$, $\sec u$, &c., to a rational non-integral function.

A rational and integral function has a certain number of roots, or zeros, each of a given multiplicity, and is completely determined (except as to a constant factor) when the several roots and the multiplicity of each of them is given; i.e., if $a, b, c, \ldots$ be the roots, $p, q, r, \ldots$ their multiplicities, then the form is $A\left(1-\frac{u}{a}\right)^p\left(1-\frac{u}{b}\right)^q\ldots$; a rational (non-integral) function has a certain number of infinities, or poles, each of them of a given multiplicity, viz. the infinities are the roots or zeros of the rational and integral function which is its denominator.

The function $\exp u$ has no finite roots, but an infinity of roots each $=-\infty$; this appears from the equation $\exp u=\left(1+\frac{u}{n}\right)^n$, where n is indefinitely large and positive. The function $\sin u$ has the roots $s\pi$ where s is any positive or negative integer, zero included; or, what is the same thing, its roots are 0 and $\pm s\pi$, s now denoting any positive integer from 1 to ∞; each of these is a simple root, and we in fact have $\sin u=u\Pi\left(1-\frac{u^2}{s^2\pi^2}\right)$. Similarly the roots of $\cos u$ are $(s+\frac{1}{2})\pi$, s denoting any positive or negative integer, zero included, or, what is the same thing, they are $\pm(s+\frac{1}{2})\pi$, s now denoting any positive integer from 0 to ∞; each root is simple, and we have $\cos u=\Pi\left(1+\frac{u^2}{(s+\frac{1}{2})^2\pi^2}\right)$. Obviously $\tan u$, as the quotient $\sin u\div\cos u$, has both roots and infinities, its roots being the roots of $\sin u$, its infinities the roots of $\cos u$; $\sec u$ as the reciprocal of $\cos u$ has infinities only, these being the roots of $\cos u$, &c.

In the foregoing expression $\sin u=u\Pi\left(1-\frac{u^2}{s^2\pi^2}\right)$, the product must be understood to mean the limit of $\Pi_1^n\left(1-\frac{u^2}{s^2\pi^2}\right)$ for an indefinitely large positive integer value of n, viz. the product is first to be formed for the values $s=1, 2, 3, \ldots$ up to a determinate number n, and then n is to be taken indefinitely large. If, separating the positive and the negative values of s, we consider the product $u\Pi_1^n\left(1+\frac{u}{s\pi}\right)\Pi_1^m\left(1-\frac{u}{s\pi}\right)$, (where in the first product s has all the positive integer values from 1 to n, and in the second product s has all the positive integer values from 1 to m), then by making m and n each of them indefinitely large, the function does *not* approximate to $\sin u$, unless $m:n$ be a ratio of equality*. And similarly as regards $\cos u$, the product $\Pi_0^n\left(1+\frac{u}{(s+\frac{1}{2})\pi}\right)\Pi\left(\frac{u}{(s+\frac{1}{2})\pi}\right)$, m and n indefinitely large, does not approximate to $\cos u$, unless $m:n$ be a ratio of equality.

* The value of the function in question $u\Pi_1^n\left(1+\frac{u}{s\pi}\right)\Pi_1^m\left(1-\frac{u}{s\pi}\right)$, when m, n are each indefinitely large, but $\frac{m}{n}$ not $=1$, is $=\left(\frac{n}{m}\right)^{\frac{u}{\pi}}\sin u$.

7. The functions $\sin u$, $\cos u$, are *periodic*, having the period 2π, $\frac{\sin}{\cos}(u+2\pi)=\frac{\sin}{\cos}(u)$; and the half-period π, $\frac{\sin}{\cos}(u+\pi)=-\frac{\sin}{\cos}u$; the periodicity may be verified by means of the foregoing fractional forms, but some attention is required; thus writing, as we may do, $\sin u=\frac{\Pi\,(u+s\pi)}{\Pi s\pi}$, where s extends from $-n$ to n, n ultimately infinite, if for u we write $u+\pi$, each factor of the numerator is changed into the following one and the numerator is unaltered, save only that there is an introduced factor $u+(n+1)\pi$ at the superior limit, and an omitted factor $u-n\pi$ at the inferior limit; the ratio of these, $(u+\overline{n+1}\pi)\div(u-n\pi)$, for n infinite is $=-1$, and we thus have, as we should have, $\sin(u+\pi)=-\sin u$.

The most general periodic function having no infinities, and each root a simple root, and having a given period a, has the form $A\sin\frac{2\pi u}{a}+B\cos\frac{2\pi u}{a}$, or, what is the same thing, $L\frac{\sin}{\cos}\left(\frac{2\pi u}{a}+\lambda\right)$.

8. We come now to the Elliptic Functions. These arose from the consideration of the integral $\int\frac{Rdx}{\sqrt{X}}$, where R is a rational function of x, and X is the general rational and integral quartic function

$$\alpha x^4+\beta x^3+\gamma x^2+\delta x+\epsilon\,;$$

a form arrived at was

$$\int\frac{dx}{\sqrt{1-x^2\,.\,1-k^2x^2}},\quad=\int\frac{d\phi}{\sqrt{1-k^2\sin^2\phi}},$$

on putting therein $x=\sin\phi$, and this last integral was represented by $F\phi$, and called the elliptic integral of the first kind. In the particular case $k=0$, the integral is $\int\frac{dx}{\sqrt{1-x^2}}=\sin^{-1}x$, and it thus appears that $F\phi$ is of the nature of an inverse function; for passing to the direct functions we write $F\phi=u$, and consider ϕ as hereby determined as a function of u, $\phi=$ amplitude of u, or for shortness $\text{am}\,u$. And the functions $\sin\phi$, $\cos\phi$, and $\sqrt{1-k^2\sin^2\phi}$ were then considered as functions of the amplitude, and written $\sin\text{am}\,u$, $\cos\text{am}\,u$, $\Delta\,\text{am}\,u$; these were afterwards written $\text{sn}\,u$, $\text{cn}\,u$, $\text{dn}\,u$, which may be regarded either as mere abbreviations of the former functional symbols, or (in a different point of view) as functions, no longer of $\text{am}\,u$, but of u itself as the argument of the functions; sn is thus a function in some respects analogous to a sine, and cn and dn functions analogous to a cosine; they have the corresponding property that the three functions of $u+v$ are expressible in terms of the functions of u and of v. The following formulæ may be mentioned:

$$\text{cn}^2u=1-\text{sn}^2u,\quad \text{dn}^2u=1-k^2\,\text{sn}^2u,$$

$$\text{sn}'u=\text{cn}\,u\,\text{dn}\,u,\quad \text{cn}'u=-\text{sn}\,u\,\text{dn}\,u,\quad \text{dn}'u=-k^2\,\text{sn}\,u\,\text{cn}\,u,$$

where the accent denotes differentiation in regard to u; and the addition-formulæ:

$$\begin{aligned} \text{sn}\,(u+v) &= \text{sn}\,u\,\text{cn}\,v\,\text{dn}\,v + \text{sn}\,v\,\text{cn}\,u\,\text{dn}\,u, && (\div), \\ \text{cn}\,(u+v) &= \text{cn}\,u\,\text{cn}\,v - \text{sn}\,u\,\text{dn}\,u\,\text{sn}\,v\,\text{dn}\,v, && (\div), \\ \text{dn}\,(u+v) &= \text{dn}\,u\,\text{dn}\,v - k^2\,\text{sn}\,u\,\text{cn}\,u\,\text{sn}\,v\,\text{cn}\,v, && (\div), \end{aligned}$$

each of the expressions on the right-hand side being the numerator of a fraction of which

$$\text{Denom.} = 1 - k^2\,\text{sn}^2\,u\,\text{sn}^2\,v.$$

It may be remarked that any one of the fractional expressions, differentiated in regard to u and to v respectively, gives the same result; such expression is therefore a function of $u+v$, and the addition-formulæ can be thus directly verified.

9. The existence of a denominator in the addition-formulæ suggests that sn, cn, dn are not, like the sine and cosine, functions having zeros only, without infinities; they are in fact functions, having each its own zeros, but having a common set of infinities; moreover, the zeros and the infinities are simple zeros and infinities respectively. And this further suggests, what in fact is the case, that the three functions are quotients having each its own numerator but a common denominator, say they are the quotients of four θ-functions, each of them having zeros only (and these simple zeros) but no infinities.

The functions sn, cn, dn, but not the θ-functions, are moreover *doubly periodic;* that is, there exist values 2ω, 2υ, $=4K$ and $4(K+iK')$ in the ordinary notation, such that the sn, cn, or dn of $u+2\omega$, and the sn, cn, and dn of $u+2\upsilon$ are equal to the sn, cn, and dn respectively of u; or say that $\phi(u+2\omega)=\phi(u+2\upsilon)=\phi u$, where ϕ is any one of the three functions.

As regards this double periodicity, it is to be observed that the equations $\phi(u+2\omega)=\phi u$, $\phi(u+2\upsilon)=\phi u$, imply $\phi(u+2m\omega+2n\upsilon)=\phi u$, and hence it easily follows that if ω, υ were commensurable, say if they were multiples of some quantity α, we should have $\phi(u+2\alpha)=\phi u$, an equation which would replace the original two equations $\phi(u+2\omega)=\phi u$, $\phi(u+2\upsilon)=\phi u$, or there would in this case be only the single period α; ω and υ must therefore be incommensurable. And this being so, they cannot have a real ratio, for if they had, the integer values m, n could be taken such as to make $2m\omega+2n\upsilon=k$ times a given real or imaginary value, k *as small as we please;* the ratio $\omega:\upsilon$ must be therefore imaginary, as is in fact the case when the values are $4K$ and $4(K+iK')$.

10. The function $\text{sn}\,u$ has the zero 0 and the zeros $m\omega+n\upsilon$, m and n any positive or negative integers whatever; and this suggests that the numerator of $\text{sn}\,u$ is equal to a doubly infinite product, (Cayley, "On the Inverse Elliptic Functions," *Camb. Math. Jour.* t. IV., 1845, [24]; and "Mémoire sur les fonctions doublement périodiques," *Liouville,* t. X., 1845, [25]). The numerator is equal to

$$u\Pi\Pi\left(1+\frac{u}{m\omega+n\upsilon}\right),$$

m and n having any positive or negative integer values whatever, including zero, except that m, n must not be simultaneously $=0$, these values being taken account of in the factor u outside the product. But until further defined, such a product has no definite value, and consequently no meaning whatever. Imagine m, n to be coordinates, and suppose that we have, surrounding the origin, a closed curve having the origin for its centre, i.e. the curve is such that, if α, β be the coordinates of a point thereof, then $-\alpha$, $-\beta$ are also the coordinates of a point thereof; suppose further that the form of the curve is given, but that its magnitude depends upon a parameter h, and that the curve is such that, when h is indefinitely large, each point of the curve is at an indefinitely large distance from the origin; for instance, the curve might be a circle or ellipse, or a parallelogram, the origin being in each case the centre. Then if in the double product, taking the value of h as given, we first give to m, n all the positive or negative integer values (the simultaneous values 0, 0 excluded) which correspond to points within the curve, and then make h indefinitely large, the product will thus have a definite value; *but this value will still be dependent on the form of the curve.* Moreover, varying in any manner the form of the curve, the ratio of the two values of the double product will be $=\exp \beta u^2$, where β is a determinate value depending only on the forms of the two curves; or, what is the same thing, if we first give to the curve a certain form, say we take it to be a circle, and then give it any other form, the product in the latter case is equal to its former value multiplied by $\exp \beta u^2$, where β depends only upon the form of the curve in the latter case.

Considering the form of the bounding curve as given, and writing the double product in the form

$$\Pi\Pi\left(\frac{u+m\omega+n\upsilon}{m\omega+n\upsilon}\right),$$

the simultaneous values $m=0$, $n=0$ being now admitted in the numerator, although still excluded from the denominator, then if we write for instance $u+2\omega$ instead of u, each factor in the numerator is changed into a contiguous factor, and the numerator remains unaltered, except that we introduce certain factors which lie outside the bounding curve, and omit certain factors which lie inside the bounding curve; we, in fact, affect the result by a singly infinite series of factors belonging to points adjacent to the bounding curve; and it appears on investigation that we thus introduce a constant factor $\exp \gamma(u+\omega)$. The final result thus is that the product

$$u\Pi\Pi\left(1+\frac{u}{m\omega+n\upsilon}\right)$$

does not remain unaltered when u is changed into $u+2\omega$, but that it becomes therefore affected with a constant factor, $\exp \gamma(u+\omega)$. And similarly the function does not remain unaltered when u is changed into $u+2\upsilon$, but it becomes affected with a factor, $\exp \delta(u+\upsilon)$. The bounding curve may however be taken such that the function is unaltered when u is changed into $u+2\omega$: this will be the case if the curve is a rectangle such that the length in the direction of the axis of m is infinitely great in comparison of that in the direction of the axis of n; or it may be taken such that the function is unaltered when u is changed into $u+2\upsilon$: this will be so if the curve be a rectangle such that the length in the direction of the axis of n is indefinitely great in comparison with that in the direction of the axis of m; but the two conditions cannot be satisfied simultaneously.

11. We have three other like functions, viz. writing for shortness $\bar{m}$, $\bar{n}$ to denote $m+\frac{1}{2}$, $n+\frac{1}{2}$ respectively, and (m, n) to denote $m\omega+nv$, then the four functions are

$$u\Pi\Pi\left(1+\frac{u}{(m,\ n)}\right),\quad \Pi\Pi\left(1+\frac{u}{(\bar{m},\ n)}\right),\quad \Pi\Pi\left(1+\frac{u}{(m,\ \bar{n})}\right),\quad \Pi\Pi\left(1+\frac{u}{(\bar{m},\ \bar{n})}\right),$$

the bounding curve being in each case the same; and, dividing the first three of these each by the last, we have (except as to constant factors) the three functions sn u, cn u, dn u; writing in each of the four functions $u+2\omega$ or $u+2v$ in place of u, the functions acquire each of them the *same* exponential factor $\exp\gamma(u+\omega)$, or $\exp\delta(u+v)$, and the quotient of any two of them, and therefore each of the functions sn u, cn u, dn u, remains unaltered.

It is easily seen that, disregarding constant factors, the four θ-functions are in fact one and the same function with different arguments, or they may be written θu, $\theta(u+\frac{1}{2}\omega)$, $\theta(u+\frac{1}{2}v)$, $\theta(u+\frac{1}{2}\omega+\frac{1}{2}v)$; by what precedes, the functions may be so determined that they shall remain unaltered when u is changed into $u+2\omega$, that is, be singly periodic, but that the change u into $u+2v$ shall affect them each with the same exponential factor $\exp\delta(u+v)$.

12. Taking the last-mentioned property as a definition of the function θ, it appears that θu may be expressed as a sum of exponentials

$$\theta u = A\Sigma \exp\frac{\pi i}{\omega}(vm^2+um),$$

where the summation extends to all positive and negative integer values of m, including zero. In fact, if we first write herein $u+2\omega$ instead of u, then in each term the index of the exponential is altered by $\frac{\pi i}{\omega}2\omega m$, $=2m\pi i$, and the term itself thus remains unaltered; that is, $\theta(u+2\omega)=\theta u$. But writing $u+2v$ in place of u, each term is changed into the succeeding term, multiplied by the factor $\exp\frac{\pi i}{\omega}(u+v)$; in fact, making the change in question u into $u+2v$, and writing also $m-1$ in place of m, vm^2+um becomes $v(m-1)^2+(u+2v)(m-1)$, $=vm^2+um-u-v$, and we thus have $\theta(u+2v)=\exp\left\{-\frac{\pi i}{\omega}(u+v)\right\}.\theta u$. In order to the convergency of the series it is necessary that $\exp\frac{\pi i v m^2}{\omega}$ should vanish for indefinitely large values of m, and this will be so if $\frac{iv}{\omega}$ be a complex quantity of the form $\alpha+\beta i$, α *negative*; for instance, this will be the case if ω be real and positive and v be $=i$ multiplied by a real and positive quantity. The original definition of θ as a double product seems to put more clearly in evidence the real nature of this function, but the new definition has the advantage that it admits of extension to the θ-functions of two or more variables.

The elliptic functions $\text{sn}\,u$, $\text{cn}\,u$, $\text{dn}\,u$, have thus been expressed each of them as the quotient of two θ-functions, but the question arises to express conversely a θ-function by means of the elliptic functions; the form is found to be

$$\theta u = C \exp\left(Au^2 + B\int_0\int_0 \text{sn}^2 u\, du^2\right),$$

viz. θu is expressible as an exponential, the index of which depends on the double integral

$$\int_0\int_0 \text{sn}^2 u\, du^2.$$

The object has been to explain the general nature of the elliptic functions $\text{sn}\,u$, $\text{cn}\,u$, $\text{dn}\,u$, and of the θ-functions with which they are thus intimately connected; it would be out of place to go into the theories of the multiplication, division, and transformation of the elliptic functions, or into the theory of the elliptic integrals, and of the application of the θ-functions to the representation of the elliptic integrals of the second and third kinds.

13. The reasoning which shows that for a doubly periodic function the ratio of the two periods 2ω, 2υ is imaginary shows that we cannot have a function *of a single variable*, which shall be *triply* periodic, or of any higher order of periodicity. For if the periods of a triply periodic function $\phi(u)$ were 2ω, 2υ, 2χ, then m, n, p being any positive or negative integer values, we should have $\phi(u + 2m\omega + 2n\upsilon + 2p\chi) = \phi u$; ω, υ, χ must be incommensurable, for if not, the three periods would really reduce themselves to two periods, or to a single period; and being incommensurable, it would be possible to determine the integers m, n, p in such manner that the real part and also the coefficient of i of the expression $m\omega + n\upsilon + p\chi$ shall be each of them as small as we please, say $\phi(u + \epsilon) = \phi u$, and thence $\phi(u + k\epsilon) = \phi u$ (k an integer), and $k\epsilon$ as near as we please to any given real or imaginary value whatever. We have thus the nugatory result ϕu = a constant, or at least the function if not a constant is a function of an infinitely and perpetually discontinuous kind, a conception of which can hardly be formed. But a function of two variables may be triply or quadruply periodic—viz. we may have a function $\phi(u, v)$ having for u, v the simultaneous periods 2ω, $2\omega'$; 2υ, $2\upsilon'$; 2χ, $2\chi'$; 2ψ, $2\psi'$; or, what is the same thing, it may be such that, m, n, p, q being any integers whatever, we have

$$\phi(u + 2m\omega + 2n\upsilon + 2p\chi + 2q\psi,\ v + 2m\omega' + 2n\upsilon' + 2p\chi' + 2q\psi') = \phi(u, v);$$

and similarly a function of $2n$ variables may be $2n$-tuply periodic.

It is, in fact, in this manner that we pass from the elliptic functions and the single θ-functions to the hyperelliptic or Abelian functions and the multiple θ-functions; the case next succeeding the elliptic functions is when we have X, Y the same rational and integral sextic functions of x, y respectively, and then writing

$$\frac{dx}{\sqrt{X}} + \frac{dy}{\sqrt{Y}} = du, \quad \frac{x\,dx}{\sqrt{X}} + \frac{y\,dy}{\sqrt{Y}} = dv,$$

we regard certain symmetrical functions of x, y, in fact, the ratios of $(2^4 =)$ 16 such symmetrical functions, as functions of (u, v); say we thus have 15 hyperelliptic functions $f(u, v)$, analogous to the 3 elliptic functions $\operatorname{sn} u$, $\operatorname{cn} u$, $\operatorname{dn} u$, and being quadruply periodic. And these are the quotients of 16 double θ-functions $\theta(u, v)$, the general form being

$$\theta(u, v) = A\Sigma\Sigma \exp\{\tfrac{1}{2}(a, h, b)(m, n)^2 + mu + nv\},$$

where the summations extend to all positive and negative integer values of (m, n); and we thus see the form of the θ-function for any number of variables whatever. The epithet "hyperelliptic" is used in the case where the differentials are of the form just mentioned $\frac{dx}{\sqrt{X}}$, where X is a rational and integral function of x; the epithet "Abelian" extends to the more general case where the differential involves the irrational function of x, determined by any rational and integral equation $\phi(x, y) = 0$ whatever.

As regards the literature of the subject, it may be noticed that the various memoirs by Riemann, 1851—1866, are republished in the collected edition of his works, Leipsic, 1876; and shortly after his death we have the *Theorie der Abel'schen Functionen*, by Clebsch and Gordan, Leipsic, 1866. Preceding this, we have by MM. Briot and Bouquet, the *Théorie des Fonctions doublement périodiques et en particulier des Fonctions Elliptiques*, Paris, 1859, the results of which are reproduced and developed in their larger work, *Théorie des Fonctions Elliptiques*, 2nd ed., Paris, 1875.

14. It is proper to mention the gamma (Γ) or Π function, $\Gamma(n+1) = \Pi n$, $= 1.2.3\ldots n$, when n is a positive integer. In the case just referred to, n a positive integer, this presents itself almost everywhere in analysis,—for instance, the binomial coefficients, and the coefficients of the exponential series are expressible by means of such functions of a number n. The definition for any real positive value of n is taken to be

$$\Gamma n = \int_0^\infty x^{n-1} e^{-x}\, dx;$$

it is then shown that, n being real and positive, $\Gamma(n+1) = n\Gamma n$, and by assuming that this equation holds good for positive or negative real values of n, the definition is extended to real negative values; the equation gives $\Gamma 1 = 0\Gamma 0$, that is, $\Gamma 0 = \infty$, and similarly $\Gamma(-n) = \infty$, where $-n$ is any negative integer. The definition by the definite integral has been or may be extended to imaginary values of n, but the theory is not an established one. A definition extending to all values of n is that of Gauss

$$\Pi n = \text{limit}\, \frac{1\ .\ 2\ .\ 3\ \ldots\ k}{n+1\,.\,n+2\,.\,n+3\ldots n+k}\, k^n,$$

the ultimate value of k being $= \infty$; but the function is chiefly considered for real values of the variable.

A formula for the calculation, when x has a large real and positive value, is

$$\Pi x = \sqrt{2\pi}\, x^{x+\frac{1}{2}} \exp\left(-x + \frac{1}{12x} + \ldots\right),$$

or as this may also be written, neglecting the negative powers of x,

$$\Pi x = \sqrt{2\pi} \exp \{(x+\tfrac{1}{2})\log x - x\}.$$

Another formula is $\Gamma x \Gamma(1-x) = \dfrac{\pi}{\sin \pi x}$: or, as this may also be written,

$$\Pi(x-1)\,\Pi(-x) = \frac{\pi}{\sin \pi x}.$$

It is to be observed that the function Π serves to express the product of a set of factors in arithmetical progression; we have

$$(x+a)(x+2a)\ldots(x+ma) = a^m\left(\frac{x}{a}+1\right)\left(\frac{x}{a}+2\right)\ldots\left(\frac{x}{a}+m\right) = a^m \Pi\left(\frac{x}{a}+m\right) \div \Pi\frac{x}{a}.$$

We can consequently express by means of it the product of any number of the factors which present themselves in the factorial expression of $\sin u$. Starting from the form

$$u\Pi_1^{\,m}\left(1+\frac{u}{s\pi}\right)\Pi_1^{\,n}\left(1-\frac{u}{s\pi}\right),$$

where Π is here as before the sign of a product of factors corresponding to the different integer values of s, this is thus converted into

$$u\Pi\left(\frac{u}{\pi}+m\right)\Pi\left(-\frac{u}{\pi}+n\right) \div \Pi\left(\frac{u}{\pi}\right)\Pi\left(-\frac{u}{\pi}\right)\Pi m\,\Pi n,$$

or as this may also be written,

$$\pi\Pi\left(\frac{u}{\pi}+m\right)\Pi\left(-\frac{u}{\pi}+n\right) \div \Pi\left(\frac{u}{\pi}-1\right)\Pi\left(-\frac{u}{\pi}\right)\Pi m\,\Pi n,$$

which, in virtue of

$$\Pi\left(\frac{u}{\pi}-1\right)\Pi\left(-\frac{u}{\pi}\right) = \frac{\pi}{\sin u},$$

becomes

$$= \sin u\Pi\left(\frac{u}{\pi}+m\right)\Pi\left(-\frac{u}{\pi}+n\right) \div \Pi m\,\Pi n.$$

Here m and n are large and positive; calculating the second factor by means of the formula for Πx, in this case we have the before-mentioned formula

$$u\Pi_1^{\,m}\left(1+\frac{u}{s\pi}\right)\Pi_1^{\,n}\left(1-\frac{u}{s\pi}\right) = \left(\frac{n}{m}\right)^{\frac{u}{\pi}} \sin u.$$

The gamma or Π function is the so-called second Eulerian integral; the first Eulerian integral

$$\int_0^1 x^{p-1}(1-x)^{q-1}\,dx, \ = \Gamma p \Gamma q \div \Gamma(p+q),$$

is at once expressible in terms of Γ, and is therefore not a new function to be considered.

15. We have the function defined by its expression as a hypergeometric series

$$F(\alpha,\ \beta,\ \gamma,\ u) = 1 + \frac{\alpha\,.\,\beta}{1\,.\,\gamma}u + \frac{\alpha\,.\,\alpha+1\,.\,\beta\,.\,\beta+1}{1\,.\,2\,.\,\gamma\,.\,\gamma+1}u^2 + \&c.,$$

i.e., this expression of the function serves as a definition, if the series be finite or if, being infinite, it is convergent. The function may also be defined as a definite integral; in other words, if, in the integral

$$\int_0^1 x^{\alpha'-1}(1-x)^{\beta'-1}(1-ux)^{-\gamma'}\,dx,$$

we expand the factor $(1-ux)^{-\gamma'}$ in powers of ux, and then integrate each term separately by the formula for the second Eulerian integral, the result is

$$= \frac{\Gamma\alpha'\,.\,\Gamma\beta'}{\Gamma(\alpha'+\beta')} + \frac{\Gamma(\alpha'+1)\,.\,\Gamma\beta'}{\Gamma(\alpha'+\beta'+1)}\frac{\gamma'}{1}u + \&c.,$$

which is

$$= \frac{\Gamma\alpha'\,.\,\Gamma\beta'}{\Gamma(\alpha'+\beta')}\left\{1 + \frac{\alpha'\,.\,\gamma'}{\alpha'+\beta'\,.\,1}u + \frac{\alpha'\,.\ \ \alpha'+1\,.\ \gamma'\,.\,\gamma'+1}{\alpha'+\beta'\,.\,\alpha'+\beta'+1\,.\,1\,.\,2}u^2 + \ldots\right\}$$

or writing $\alpha',\ \beta',\ \gamma' = \alpha,\ \gamma-\alpha,\ \beta$ respectively, this is

$$= \frac{\Gamma\alpha\Gamma(\gamma-\alpha)}{\Gamma\gamma}F(\alpha,\ \beta,\ \gamma,\ u),$$

so that the new definition is

$$F(\alpha,\ \beta,\ \gamma,\ u) = \frac{\Gamma\alpha\Gamma(\gamma-\alpha)}{\Gamma\gamma}\int_0^1 x^{\alpha-1}(1-x)^{\beta-1}(1-ux)^{-\beta}\,dx\,;$$

but this is in like manner only a definition under the proper limitations as to the values of $\alpha,\ \beta,\ \gamma,\ u$. It is not here considered how the definition is to be extended so as to give a meaning to the function $F(\alpha,\ \beta,\ \gamma,\ u)$ for all values, say of the parameters $\alpha,\ \beta,\ \gamma$, and of the variable u. There are included a large number of special forms which are either algebraic or circular or exponential, for instance $F(\alpha,\ \beta,\ \beta,\ u) = (1-u)^{-\alpha}$, &c.; or which are special transcendents which have been separately studied, for instance, Bessel's functions, the Legendrian functions X_n presently referred to, series occurring in the development of the reciprocal of the distance between two planets, &c.

16. There is a class of functions depending upon a variable or variables $x,\ y, \ldots$ and a parameter n, say the function for the parameter n is X_n such that the product of two functions having the same variables, multiplied it may be by a given function of the variables, and integrated between given limits, gives a result $=0$ or not $=0$, according as the parameters are unequal or equal; $\int UX_mX_n\,dx = 0$, but $\int UX_n{}^2dx$ not $=0$; the admissible values of the parameters being either any integer values, or it may be the roots of a determinate algebraical or transcendental equation; and the functions X_n may be either algebraical or transcendental. For instance, such a function is $\cos nx$; m and n being integers, we have $\int_0^\pi \cos mx\,.\,\cos nx\,dx = 0$, but $\int_0^\pi \cos^2 nx\,dx = \tfrac{1}{2}\pi$.

Assuming the existence of the expansion of a function fx, in a series of multiple cosines, we thus obtain at once the well-known Fourier series, wherein the coefficient of $\cos mx$ is $=\frac{1}{2}\pi\int_0^\pi \cos mx . fx dx$. The question whether the process is applicable is elaborately discussed in Riemann's memoir (1854), *Ueber die Darstellbarkeit einer Function durch eine trigonometrische Reihe*, No. XII. in the collected works. And again we have the Legendrian functions, which present themselves as the coefficients of the successive powers of α in the development of $(1-2\alpha x+\alpha^2)^{-\frac{1}{2}}$, $X_0=1$, $X_1=x$, $X_2=\frac{3}{2}(x^2-\frac{1}{3})$, &c.: here m, n being any positive integers, $\int_{-1}^{1} X_m X_n \, dx = 0$, but $\int_{-1}^{1} X_n^2 \, dx = \frac{2}{2n+1}$. And we have also Laplace's functions, &c.

Functions in General.

17. In what precedes, a review has been given, not by any means an exhaustive one, but embracing the most important kinds of known functions; but there are questions to be considered in regard to functions in general.

A function of $x+iy$ has been built up by means of analytical operations performed upon $x+iy$, $(x+iy)^2=x^2-y^2+i.2xy$, &c., and the question next referred to has not arisen. But observe that, knowing $x+iy$, we know x and y, and therefore any two given functions $\phi(x, y)$, $\psi(x, y)$ of x and y: we therefore also know $\phi(x, y)+i\psi(x, y)$, and the question is, whether such a function of x, y (being known when $x+iy$ is known) is to be regarded as a function of $x+iy$; and if not, what is the condition to be satisfied in order that $\phi(x, y)+i\psi(x, y)$ may be a function of $x+iy$. Cauchy at one time considered that the general form was to be regarded as a function of $x+iy$, and he introduced the expression "fonction monogène," monogenous function, to denote the more restricted form which is the proper function of $x+iy$.

Consider for a moment the above general form, say $x'+iy'=\phi(x, y)+i\psi(x, y)$, where ϕ, ψ are any real functions of the real variables (x, y); or what is the same thing, assume $x'=\phi(x, y)$, $y'=\psi(x, y)$; if these functions have each or either of them more than one value, we attend only to one value of each of them. We may then as before take x, y to be the coordinates of a point P in a plane Π, and x', y' to be the coordinates of a point P' in a plane Π'. If, for any given values of x, y, the increments of $\phi(x, y)$, $\psi(x, y)$ corresponding to the indefinitely small real increments h, k of x, y be $Ah+Bk$, $Ch+Dk$, A, B, C, D being functions of x, y, then if the new coordinates of P are $x+h$, $y+k$, the new coordinates of P' will be $x'+Ah+Bk$, $y'+Ch+Dk$; or P, P' will respectively describe the indefinitely small straight paths at the inclinations $\tan^{-1}\frac{k}{h}$, $\tan^{-1}\frac{Ch+Dk}{Ah+Bk}$ to the axes of x, x' respectively; calling these angles θ, θ', we have therefore $\tan\theta'=\frac{C+D\tan\theta}{A+B\tan\theta}$. Now in order that $x'+iy'$ may be $=\phi(x+iy)$, a function of $x+iy$, the condition to be satisfied is that the increment of $x'+iy'$ shall be proportional to the increment $h+ik$ of $x+iy$, or say that it shall be $=(\lambda+i\mu)(h+ik)$, λ, μ being functions of x, y, but independent of

h, k; we must therefore have $Ah+Bk$, $Ch+Dk=\lambda h-\mu k$, $\mu h+\lambda k$ respectively, that is $A, B, C, D=\lambda, -\mu, \mu, \lambda$ respectively, and the equation for $\tan\theta'$ thus becomes $\tan\theta'=\frac{\mu+\lambda\tan\theta}{\lambda-\mu\tan\theta}$; hence writing $\frac{\mu}{\lambda}=\tan\alpha$, where α is a function of x, y, but independent of h, k, we have $\tan\theta'=\frac{\tan\alpha+\tan\theta}{1-\tan\alpha\tan\theta}$, that is, $\theta'=\alpha+\theta$; or for the given points (x, y), (x', y'), the path of P being at any inclination whatever θ to the axis of x, the path of P' is at the inclination θ + constant angle α to the axis of x'; also $(\lambda h-\mu k)^2+(\mu h+\lambda k)^2=(\lambda^2+\mu^2)(h^2+k^2)$, i.e., the lengths of the paths are in a constant ratio.

The condition may be written $\delta(x'+iy')=(\lambda+i\mu)(\delta x+i\delta y)$; or what is the same thing

$$\left(\frac{dx'}{dx}+i\frac{dy'}{dx}\right)\delta x+\left(\frac{dx'}{dy}+i\frac{dy'}{dy}\right)\delta y=(\lambda+i\mu)(\delta x+i\delta y),$$

that is,

$$\frac{dx'}{dx}+i\frac{dy'}{dx}=(\lambda+i\mu),\quad \frac{dx'}{dy}+i\frac{dy'}{dy}=i(\lambda+i\mu);$$

consequently

$$\frac{dx'}{dy}+i\frac{dy'}{dy}=i\left(\frac{dx'}{dx}+i\frac{dy'}{dx}\right);$$

that is,

$$\frac{dx'}{dy}=-\frac{dy'}{dx},\quad \frac{dy'}{dy}=\frac{dx'}{dx},$$

as the analytical conditions in order that $x'+iy'$ may be a function of $x+iy$. They obviously imply

$$\frac{d^2x'}{dx^2}+\frac{d^2x'}{dy^2}=0,\quad \frac{d^2y'}{dx^2}+\frac{d^2y'}{dy^2}=0;$$

and if x' be a function of x, y, satisfying the first of these conditions, then

$$-\frac{dx'}{dy}dx+\frac{dx'}{dx}dy$$

is a complete differential, and

$$y'=\int\left(-\frac{dx'}{dy}dx+\frac{dx'}{dx}dy\right).$$

18. We have, in what just precedes, the ordinary behaviour of a function $\phi(x+iy)$ in the neighbourhood of the value $x+iy$ of the argument or point $x+iy$; or say the behaviour in regard to a point $x+iy$ such that the function is in the neighbourhood of this point a *continuous* function of $x+iy$ (or that the point is not a point of *discontinuity*): the correlative definition of continuity will be that the function $\phi(x+iy)$, assumed to have at the given point $x+iy$ a single finite value, is continuous in the neighbourhood of this point, when the point $x+iy$ describing continuously a straight infinitesimal element $h+ik$, the point $\phi(x+iy)$ describes continuously a straight infinitesimal element $(\lambda+i\mu)(h+ik)$; or what is really the same thing, when the function $(x+iy)$ has at the point $x+iy$ a differential coefficient.

19. It would doubtless be possible to give for the continuity of a function $\phi(x+iy)$ a less stringent definition not implying the existence of a differential coefficient; but we have this theory only in regard to the functions ϕx of a real variable in memoirs by Riemann, Hankel, du Bois Reymond, Schwarz, Gilbert, Klein, and Darboux. The last-mentioned geometer, in his "Mémoire sur les fonctions discontinues," *Jour. de l'École Normale*, t. IV. (1875), pp. 57—112, gives (after Bonnet) the following definition of a continuous function (observe that we are now dealing with real quantities only):—the function $f(x)$ is continuous for the value $x=x_0$ when, h and ϵ being positive quantities as small as we please and θ any positive quantity at pleasure between 0 and 1, we have, for all the values of θ, $f(x_0 \pm \theta h)-f(x)$ less in absolute magnitude than ϵ; and moreover $f(x)$ is continuous through the interval x_0, x_1 ($x_1 > x_0$, that is, nearer $+\infty$) when $f(x)$ is continuous for every value of x between x_0 and x_1, and, h tending to zero through positive values, $f(x_0+h)$ and $f(x_0-h)$ tend to the limits $f(x_0)$, $f(x_1)$ respectively. It is possible, consistently with this definition, to form continuous functions not having in any proper sense a differential coefficient, and having other anomalous properties; thus if $a_1, a_2, a_3, \ldots$ be an infinite series of real positive or negative quantities, such that the series Σa_n is absolutely convergent (i.e. the sum $\Sigma \pm a_n$, each term being made positive, is convergent), then the function $\Sigma a_n (\sin n\pi x)^{\frac{2}{3}}$ is a continuous function actually calculable for any assumed value of x; but it is shown in the memoir that, taking $x =$ any commensurable value $\frac{p}{q}$ whatever, and then writing $x=\frac{p}{q}+h$, h indefinitely small, the increment of the function is of the form $(k+\epsilon)h^{\frac{2}{3}}$, k finite, ϵ an indefinitely small quantity vanishing with h; there is thus no term varying with h, nor consequently any differential coefficient. See also Riemann's Memoir, *Ueber die Darstellbarkeit*, &c. (No. XII. in the collected works), already referred to.

20. It was necessary to allude to the foregoing theory of (as they may be termed) infinitely discontinuous functions; but the ordinary and most important functions of analysis are those which are continuous, except for a finite number (or it may be an infinite number) of points of discontinuity. It is to be observed that a point at which the function becomes infinite is *ipso facto* a point of discontinuity; a value of the variable for which the function becomes infinite is, as already mentioned, said to be an "infinity" (or a "pole") of the function; thus, in the case of a rational function expressed as a fraction in its least terms, if the denominator contains a factor $(x-a)^m$, a a real or imaginary value, m a positive integer, then a is said to be an infinity of the mth order (and in the particular case $m=1$, it is said to be a simple infinity). The circular functions $\tan x$, $\sec x$ are instances of a function having an infinite number of simple infinities.

A rational function is a one-valued function, and in regard to a rational function the infinities are the only points of discontinuity; but a one-valued function may have points of discontinuity of a character quite distinct from an infinity: for instance, in the exponential function $\exp\left(\frac{1}{u-a}\right)$ where a is real or imaginary, the value $u\,(=x+iy)=a$, is a point of discontinuity but not an infinity; taking $u=a+\rho e^{\alpha i}$, where ρ is an indefinitely small real positive quantity, the value of the function is $\exp\left(\frac{1}{\rho}e^{-\alpha i}\right)$, $=\exp\frac{1}{\rho}(\cos\alpha - i\sin\alpha)$, which is indefinitely large or indefinitely small according as $\cos\alpha$ is positive or negative, and in the separating case $\cos\alpha = 0$, and therefore $\sin\alpha = \pm 1$, it is $=\cos\frac{1}{\rho} \pm i\sin\frac{1}{\rho}$ which is indeterminate. If, instead of $\exp\frac{1}{u-a}$, we consider a linear function

$$\left\{A + B\exp\frac{1}{u-a}\right\} \div \left\{C + D\exp\frac{1}{u-a}\right\},$$

then writing as before $u = a + \rho e^{\alpha i}$, the value is $= A \div C$, or $= B \div D$, according as $\cos\alpha$ is negative or positive. As regards the theory of one-valued functions in general, the memoir by Weierstrass, "Zur Theorie der eindeutigen analytischen Functionen," *Berl. Abh.* 1876, pp. 11—60, may be referred to.

21. A one-valued function *ex vi termini* cannot have a point of discontinuity of the kind next referred to; if the representative point P, moving in any manner whatever, returns to its original position, the corresponding point P' cannot but return to its original position. But consider a many-valued function, say an n-valued function $x' + iy'$, of $x + iy$; to each position of P there correspond n positions, in general all of them different, of P'. But the point P may be such that (to take the most simple case) two of the corresponding points P' coincide with each other, say such a position of P is at V, then (using for greater distinctness a different letter W' instead of V') corresponding thereto we have two coincident points (W'), and $n-2$ other points W'; V is then a branch-point (Verzweigungspunkt). Taking for P any point which is not a branch-point, then in the neighbourhood of this value each of the n functions $x' + iy'$ is a continuous function of $x + iy$, and by what precedes, if P describing an infinitely small closed curve (or oval) return to its original position, then each of the corresponding points P' describing a corresponding indefinitely small oval will return to its original position. But imagine the oval described by P to be gradually enlarged, so that it comes to pass through a branch-point V; the ovals described by two of the corresponding points P' will gradually approach each other, and will come to unite at the point (W'), each oval then sharpening itself out so that the two form together a figure of eight. And if we imagine the oval described by P to be still further enlarged so as to include within it the point V, then the figure of eight, losing the crossing, will be at first an hour-glass form, or twice-indented oval, and ultimately in form an ordinary oval, but having the character of a twofold oval; i.e. to the oval described by P (and which surrounds the branch-point V) there will correspond this twofold oval, and $n-2$ onefold ovals, in such wise that to a given position of P on its oval there correspond two points, say P_1', P_2', on the twofold oval, and $n-2$ points $P_3', \ldots, P_n'$, each on its own onefold oval. And then as P describing its oval returns to its original position, the point P_1' describing a portion only of the twofold oval, will pass to the original position of P_2', while the point P_2' describing the remaining portion of the twofold oval will pass to the original position of P_1'; the other points $P_3', \ldots, P_n'$, describing each of them its own onefold oval, will

return each of them to its original position. And it is easy to understand how, when the oval described by P surrounds two or more of the branch-points V, the corresponding curves for P' may be a system of manifold ovals, such that the sum of the manifoldness is always $=n$, and to conceive in a general way the behaviour of the corresponding points P and P'.

Writing for a moment $x+iy=u$, $x'+iy'=v$, the branch-points are the points of contact of parallel tangents to the curve $\phi(u, v)=0$, a line through a cusp (but not a line through a node), being reckoned as a tangent; that is, if this be a curve of the order n and class m, with δ nodes and κ cusps, the number of branch-points is $=m+\kappa$, that is, it is $=n^2-n-2\delta-2\kappa$, or if $p, =\frac{1}{2}(n-1)(n-2)-\delta-\kappa$, be the *deficiency*, then the number is $=2n-2+2p$.

To illustrate the theory of the n-valued algebraical function $x'+iy'$ of the complex variable $x+iy$, Riemann introduces the notion of a surface composed of n coincident planes or sheets, such that the transition from one sheet to another is made at the branch-points, and that the n sheets form together a multiply-connected surface, which can be by cross-cuts (Querschnitte) converted into a simply-connected surface; the n-valued function $x'+iy'$ becomes thus a one-valued function of $x+iy$, considered as belonging to a point on some determinate sheet of the surface: and upon such considerations he founds the whole theory of the functions which arise from the *integration* of the differential expressions depending on the n-valued algebraical function (that is, any irrational algebraical function whatever) of the independent variable, establishing as part of the investigation the theory of the multiple θ-functions. But it would be difficult to give a further account of these investigations.

The Calculus of Functions.

22. The so-called Calculus of Functions, as considered chiefly by Herschel and Babbage and De Morgan, is not so much a theory of functions as a theory of the solution of functional equations; or, as perhaps should rather be said, the solution of functional equations by means of known functions, or symbols,—the epithet known being here used in reference to the actual state of analysis. Thus for a functional equation $\phi x+\phi y=\phi(xy)$, taking the logarithm as a known function, the solution is $\phi x=c\log x$; or if the logarithm is not taken to be a known function, then a solution may be obtained by means of the sign of integration $\phi x=c\int\frac{dx}{x}$; but the establishment of the properties of the function logarithm (assumed to be previously unknown) would not be considered as coming within the theory. A class of equations specially considered is where $\alpha x, \beta x, \ldots$ being given functions of x, the unknown function ϕ is to be determined by means of a given relation between $x, \phi x, \phi\alpha x, \phi\beta x, \ldots$; in particular the given relation may be between $x, \phi x, \phi\alpha x$; this can be at once reduced to equations of finite differences; for writing $x=u_n$, $\alpha x=u_{n+1}$, we have $u_{n+1}=\alpha u_n$, giving u_n, and therefore also x, each of them as a function of n; and then writing $\phi x=v_n$, $\phi\alpha x$ will be the same function of $n+1$, $=v_{n+1}$, and the given relation is again an equation of finite differences in v_{n+1}, v_n, and n; we have thus $v_n, =\phi x$,

as a function of n, that is, of x. As regards the equation $u_{n+1} = \alpha u_n$, considered in itself apart from what precedes, observe that this is satisfied by writing $u_n = \alpha^n(x)$, or the question of solving this equation of finite differences is, in fact, identical with that of finding the nth function $\alpha^n(x)$, where $\alpha(x)$ is a given function of x. It of course depends on the form of $\alpha(x)$ whether this question admits of solution in any proper sense; thus, for a function such as $\log x$, the nth logarithm is expressible in its original function $\log^n x$, ($= \log\log \ldots x$), and not in any other form. But there are forms, for instance $\alpha x = \dfrac{a + bx}{c + dx}$, where the nth function $\alpha^n x$ is a function of the like form $\alpha^n x = \dfrac{A + Bx}{C + Dx}$, in which the actual value can be expressed as a function of n; if α be such a form, then $\phi\alpha\phi^{-1}$, whatever ϕ may be, is a like form, for we obviously have $(\phi\alpha\phi^{-1})^n = \phi\alpha^n\phi^{-1}$. The determination of the nth function is, in fact, a leading question in the calculus of functions.

It is to be observed that considering the case of two variables, if for instance $\alpha(x, y)$ denote a given function of x, y, the notation $\alpha^2(x, y)$ is altogether meaningless; in order to generalize the question, we require an extended notation wherein a single functional symbol is used to denote two functions of the two variables. Thus $\phi(x, y) = \alpha(x, y)$, $\beta(x, y)$, α and β given functions; writing for shortness $x_1 = \alpha(x, y)$, $y_1 = \beta(x, y)$, then $\phi^2(x, y)$ will denote $\phi(x_1, y_1)$, that is, two functions $\alpha(x_1, y_1)$, $\beta(x_1, y_1)$, say these are x_2, y_2; $\phi^3(x, y)$ will denote $\phi(x_2, y_2)$, and so on, so that $\phi^n(x, y)$ will have a determinate meaning. And the like is obviously the case in regard to any number of variables, the single functional symbol denoting in each case a set of functions equal in number to the variables.

788.

GALOIS.

[From the *Encyclopædia Britannica, Ninth Edition*, vol. x. (1879), p. 48.]

GALOIS, EVARISTE (1811—1832), an eminently original and profound French mathematician, born 26th October 1811, killed in a duel May 1832. A necrological notice by his friend M. Auguste Chevalier appeared in the *Revue Encyclopédique*, September 1832, p. 744; and his collected works are published, *Liouville*, t. XI. (1846), pp. 381—444, about fifty of these pages being occupied by researches on the resolubility of algebraic equations by radicals. But these researches, crowning as it were the previous labours of Lagrange, Gauss, and Abel, have in a signal manner advanced the theory, and it is not too much to say that they are the foundation of all that has since been done, or is doing, in the subject. The fundamental notion consists in the establishment of a group of permutations of the roots of an equation, such that every function of the roots invariable by the substitutions of the group is rationally known, and reciprocally that every rationally determinable function of the roots is invariable by the substitutions of the group; some further explanation of the theorem, and in connexion with it an explanation of the notion of an adjoint radical, is given under Equation, No. 32, [786]. As part of the theory (but the investigation has a very high independent value as regards the Theory of Numbers, to which it properly belongs), Galois introduces the notion of the imaginary roots of an irreducible congruence of a degree superior to unity; i.e., such a congruence, $F(x) \equiv 0$ (mod. a prime number p), has no integer root; but what is done is to introduce a quantity i subjected to the condition of verifying the congruence in question, $F(i) \equiv 1$ (mod. p), which quantity i is an imaginary of an entirely new kind, occupying in the theory of numbers a position analogous to that of $\sqrt{-1}$ in algebra.

789.

GAUSS.

[From the *Encyclopædia Britannica, Ninth Edition*, vol. x. (1879), p. 116.]

GAUSS, CARL FRIEDRICH (1777—1855), an eminent German mathematician, was born of humble parents at Brunswick, April 23, 1777, and was indebted for a liberal education to the notice which his talents procured him from the reigning duke. His name became widely known by the publication, in his twenty-fifth year (1801), of the *Disquisitiones Arithmeticæ*. In 1807 he was appointed director of the Göttingen observatory, an office which he retained to his death: it is said that he never slept away from under the roof of his observatory, except on one occasion, when he accepted an invitation from Humboldt to attend a meeting of natural philosophers at Berlin. In 1809 he published at Hamburg his *Theoria Motus Corporum Cœlestium*, a work which gave a powerful impulse to the true methods of astronomical observation; and his astronomical workings, observations, calculations of orbits of planets and comets, &c., are very numerous and valuable. He continued his labours in the theory of numbers and other analytical subjects, and communicated a long series of memoirs to the Royal Society of Sciences at Göttingen. His first memoir on the theory of magnetism, *Intensitas vis magneticæ terrestris ad mensuram absolutam revocata*, was published in 1833, and he shortly afterwards proceeded, in conjunction with Professor Wilhelm Weber, to invent new apparatus for observing the earth's magnetism and its changes; the instruments devised by them were the declination instrument and the bifilar magnetometer. With Weber's assistance he erected in 1833 at Göttingen a magnetic observatory free from iron (as Humboldt and Arago had previously done on a smaller scale), where he made magnetic observations, and from this same observatory he sent telegraphic signals to the neighbouring town, thus showing the practicability of an electromagnetic telegraph. He further instituted an association (Magnetische Verein), composed at first almost entirely of Germans, whose continuous observations on fixed term-days extended from Holland to Sicily. The volumes of their publication, *Resultate aus den Beobachtungen des magnetischen Vereins*, extend from 1836 to 1839; and in those for 1838 and 1839 are contained the two important memoirs by Gauss, *Allgemeine Theorie des Erdmagnetismus*, and the *Allgemeine Lehrsätze*—on the theory of forces attracting according to the inverse square of the distance. The instruments and methods thus due to him are substantially those employed in the magnetic observatories throughout the world. He co-operated in the Danish and Hanoverian measurements of an arc and trigonometrical operations (1821—48), and wrote (1843, 1846) the two memoirs *Ueber Gegenstände der höhern Geodäsie.* Connected with observations in general we have (1812—26) the memoir *Theoria combinationis observationum erroribus minimis obnoxia,* with a second part and a supplement. Another memoir of applied mathematics is the *Dioptrische Untersuchungen,* 1840. Gauss was well versed in general literature and the chief languages of modern Europe, and was a member of nearly all the leading scientific societies in Europe. He died at Göttingen early in the spring of 1855. The centenary of his birth was celebrated (1877) at his native place, Brunswick.

Gauss's collected works have been recently published by the Royal Society of Göttingen, in 7 vols. 4to, Gött., 1863—71, edited by E. J. Schering,—(1) the *Disquisitiones Arithmeticæ,* (2) *Theory of Numbers,* (3) *Analysis,* (4) *Geometry and Method of Least Squares,* (5) *Mathematical Physics,* (6) *Astronomy,* and (7) the *Theoria Motus Corporum Cœlestium.* They include, besides his various works and memoirs, notices by him of many of these, and of works of other authors in the *Göttingische gelehrte Anzeigen,* and a considerable amount of previously unpublished matter, *Nachlass.* Of the memoirs in pure mathematics, comprised for the most part in vols. II., III., and IV. (but to these must be added those on *Attractions* in vol. V.), it may be safely said there is not one which has not signally contributed to the progress of the branch of mathematics to which it belongs, or which would not require to be carefully analysed in a history of the subject. Running through these volumes in order, we have in the second the memoir, *Summatio quarundam serierum singularium,* the memoirs on the theory of biquadratic residues, in which the notion of complex numbers of the form $a + bi$ was first introduced into the theory of numbers; and included in the *Nachlass* are some valuable tables. That for the conversion of a fraction into decimals (giving the complete period for all the prime numbers up to 997) is a specimen of the extraordinary love which Gauss had for long arithmetical calculations; and the amount of work gone through in the construction of the table of the number of the classes of binary quadratic forms must also have been tremendous. In vol. III. we have memoirs relating to the proof of the theorem that every numerical equation has a real or imaginary root, the memoirs on the *Hypergeometric Series,* that on *Interpolation,* and the memoir *Determinatio Attractionis*—in which a planetary mass is considered as distributed over its orbit according to the time in which each portion of the orbit is described, and the question (having an implied reference to the theory of secular perturbations) is to find the attraction of such a ring. In the solution the value of an elliptic function is found by means of the *arithmetico-geometrical mean.* The *Nachlass* contains further researches on this subject, and also researches (unfortunately very fragmentary) on the lemniscate-function, &c., showing that Gauss was, even before 1800, in possession of many of the discoveries which have made the names of Abel and Jacobi illustrious. In vol. IV. we have the memoir *Allgemeine Auflösung...,* on the graphical representation of one surface upon another, and the *Disquisitiones generales circa superficies curvas.* And in vol. V. we have a memoir *On the Attraction of Homogeneous Ellipsoids,* and the already mentioned memoir *Allgemeine Lehrsätze...,* on the theory of forces attracting according to the inverse square of the distance.

790.

GEOMETRY (ANALYTICAL).

[From the *Encyclopædia Britannica, Ninth Edition*, vol. x. (1879), pp. 408—420.]

THIS will be here treated as a method. The science is Geometry; and it would be possible, analytically, or by the method of coordinates, to develope the truths of geometry in a systematic course. But it is proposed not in any way to attempt this, but simply to explain the method, giving such examples, interesting (it may be) in themselves, as are suitable for showing how the method is employed in the demonstration and solution of theorems or problems.

Geometry is one-, two-, or three-dimensional, or, what is the same thing, it is lineal, plane, or solid, according as the space dealt with is the line, the plane, or ordinary (three-dimensional) space. No more general view of the subject need here be taken:—but in a certain sense one-dimensional geometry does not exist, inasmuch as the geometrical constructions for points in a line can only be performed by travelling out of the line into other parts of a plane which contains it, and conformably to the usual practice Analytical Geometry will be treated under the two divisions, Plane and Solid.

It is proposed to consider Cartesian coordinates almost exclusively; for the proper development of the science homogeneous coordinates (three and four in plane and solid geometry respectively) are required; and it is moreover necessary to have the correlative line- and plane-coordinates; and in solid geometry to have the *six* coordinates of the line. The most comprehensive English works are those of Dr Salmon, *Conic Sections* (5th edition, 1869), *Higher Plane Curves* (2nd edition, 1873), and *Geometry of Three Dimensions* (3rd edition, 1874); we have also, on plane geometry, Clebsch's *Vorlesungen über Geometrie*, posthumous, edited by Dr F. Lindemann, Leipsic, 1875, not yet complete.

I. PLANE ANALYTICAL GEOMETRY (§§ 1—25).

1. It is assumed that the points, lines, and figures considered exist in one and the same plane, which plane, therefore, need not be in any way referred to. The position of a point is determined by means of its (Cartesian) coordinates; i.e. as

explained under the article Curve, [785], we take the two lines $x'Ox$ and $y'Oy$, called the axes of x and y respectively, intersecting in a point O called the origin, and determine the position of any other point P by means of its coordinates $x = OM$ (or NP), and

Fig. 1.

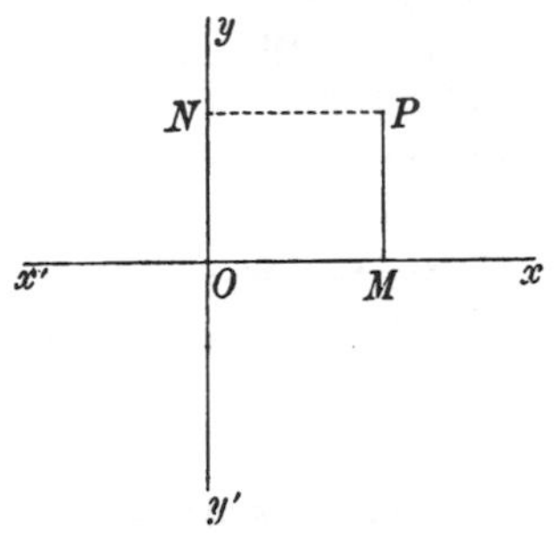

$y = MP$ (or ON). The two axes are usually (as in fig. 1) at right angles to each other, and the lines PM, PN are then at right angles to the axes of x and y respectively. Assuming a scale at pleasure, the coordinates x, y of a point have numerical values.

It is necessary to attend to the signs: x has opposite signs according as the point is on one side or the other of the axis of y, and similarly y has opposite signs according as the point is on the one side or the other of the axis of x. Using the letters N., E., S., W. as in a map, and considering the plane as divided into four quadrants by the axes, the signs are usually taken to be—

x	y	for quadt.
+	+	N. E.
+	−	S. E.
−	+	N. W.
−	−	S. W.

A point is said to have the coordinates (a, b), and is referred to as the point (a, b), when its coordinates are $x = a$, $y = b$; the coordinates x, y of a variable point, or of a point which is for the time being regarded as variable, are said to be current coordinates.

2. It is sometimes convenient to use oblique coordinates; the only difference is that the axes are not at right angles to each other; the lines PM, PN are drawn parallel to the axes of y and x respectively, and the figure $OMPN$ is thus a parallelogram. But in all that follows, the Cartesian coordinates are taken to be rectangular; polar coordinates and other systems will be briefly referred to in the sequel.

3. If the coordinates (x, y) of a point are not given, but only a relation between them $f(x, y) = 0$, then we have a curve. For, if we consider x as a real quantity varying continuously from $-\infty$ to $+\infty$, then, for any given value of x, y has a value

or values. If these are all imaginary, there is not any real point; but if one or more of them be real, we have a real point or points, which (as the assumed value of x varies continuously) varies or vary continuously therewith; and the locus of all these real points is a curve. The equation completely defines the curve; to trace the curve directly from the equation, nothing else being known, we obtain as above a series of points sufficiently near to each other, and draw the curve through them. For instance, let this be done in a simple case. Suppose $y = 2x - 1$; it is quite easy to obtain and lay down a series of points as near to each other as we please, and the application of a ruler would show that these were in a line; that the curve is a line depends upon something more than the equation itself, viz. the theorem that every equation of the form $y = ax + b$ represents a line; supposing this known, it will be at once understood how the process of tracing the curve may be abbreviated; we have $x = 0$, $y = -1$, and $x = \frac{1}{2}$, $y = 0$; the curve is thus the line passing through these two points. But in the foregoing example the notion of a line is taken to be a known one, and such notion of a line does in fact precede the consideration of any equation of a curve whatever, since the notion of the coordinates themselves rests upon that of a line. In other cases it may very well be that the equation is the definition of the curve; the points laid down, although (as finite in number) they do not actually determine the curve, determine it to any degree of accuracy; and the equation thus enables us to construct the curve.

A curve may be determined in another way; viz. the coordinates x, y may be given each of them as a function of the same variable parameter θ; $x, y = f(\theta), \phi(\theta)$ respectively. Here, giving to θ any number of values in succession, these equations determine the values of x, y, that is, the positions of a series of points on the curve. The ordinary form $y = \phi(x)$, where y is given explicitly as a function of x, is a particular case of each of the other two forms: we have $f(x, y), = y - \phi(x), = 0$; and $x = \theta$, $y = \phi(\theta)$.

4. As remarked under Curve, [785], it is a useful exercise to trace a considerable number of curves, first taking equations which are purely numerical, and then equations which contain literal constants (representing numbers); the equations most easily dealt with are those wherein one coordinate is given as an explicit function of the other, say $y = \phi(x)$ as above. A few examples are here given, with such explanations as seem proper.

(i) $y = 2x - 1$, as before; it is at once seen that this is a line; and taking it to be so, any two points, for instance, $(0, -1)$ and $(\frac{1}{2}, 0)$, determine the line.

(ii) $y = x^2$. The equation shows that x may be positive or negative, but that y is always positive, and has the same values for equal positive and negative values of x: the curve passes through the origin, and through the points $(\pm 1, 1)$. It is already known that the curve lies wholly above the axis of x. To find its form in the neighbourhood of the origin, give x a small value, $x = \pm 0{\cdot}1$ or $\pm 0{\cdot}01$, then y is very much smaller, $= 0{\cdot}01$ and $0{\cdot}0001$ in the two cases respectively; this shows that the curve touches the axis of x at the origin. Moreover, x may be as large as we please, but when it is large, y is much larger; for instance, $x=10$, $y=100$. The curve is a parabola (fig. 2).

Fig. 2.

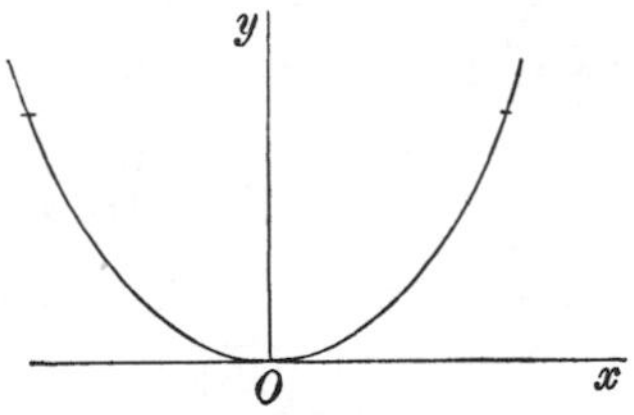

(iii) $y=x^3$. Here x being positive y is positive, but x being negative y is also negative: the curve passes through the origin, and also through the points (1, 1) and (−1, −1). Moreover, when x is small, $=0{\cdot}1$ for example, then not only is y, $=0{\cdot}001$, very much smaller than x, but it is also very much smaller than y was for the last-mentioned curve $y=x^2$, that is, in the neighbourhood of the origin the present curve approaches more closely the axis of x. The axis of x is a tangent at the origin, but it is a tangent of a peculiar kind (a stationary or inflexional tangent), cutting the curve at the origin, which is an inflexion. The curve is the cubical parabola (fig. 3).

Fig. 3.

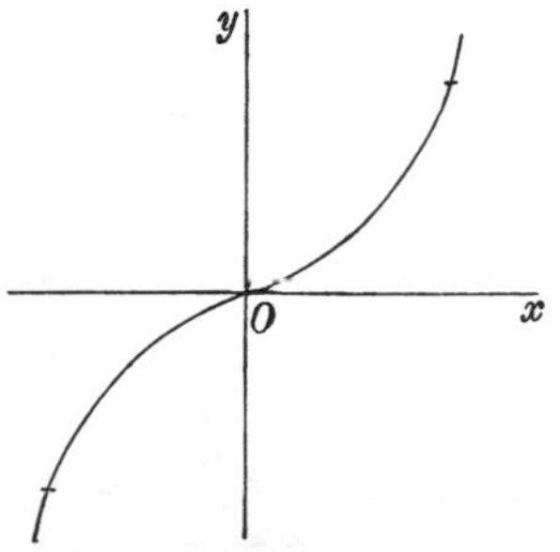

(iv) $y^2=x-1\,.\,x-3\,.\,x-4$. Here $y=0$ for $x=1$, $=3$, $=4$. Whenever $x-1.x-3.x-4$ is positive, y has two equal and opposite values; but when $x-1\,.\,x-3\,.\,x-4$ is negative, then y is imaginary. In particular, for x less than 1, or between 3 and 4, y is imaginary, but for x between 1 and 3, or greater than 4, y has two values. It

Fig. 4.

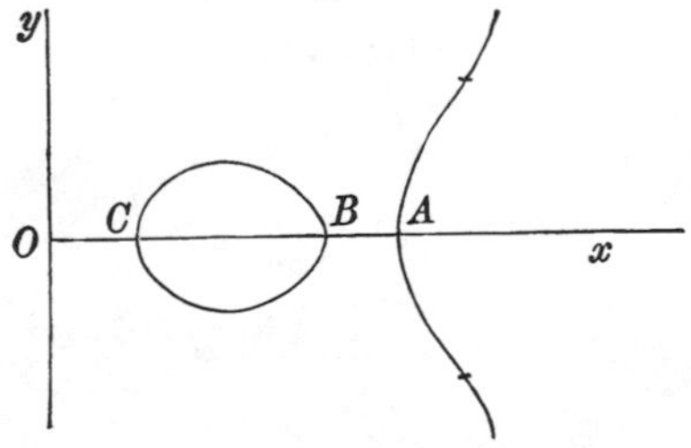

is clear that for x somewhere between 1 and 3, y will attain a maximum: the values of x and y may be found approximately by trial. The curve will consist of an oval and infinite branch, and it is easy to see that, as shown in fig. 4, the curve where it cuts the axis of x cuts it at right angles. It may be further remarked that, as

x increases from 4, the value of y will increase more and more rapidly; for instance, $x=5$, $y^2=8$, $x=10$, $y^2=378$, &c., and it is easy to see that this implies that the curve has on the infinite branch two inflexions as shown.

(v) $y^2=x-c\,.\,x-b\,.\,x-a$, where $a>b>c$ (that is, a nearer to $+\infty$, c to $-\infty$). The curve has the same general form as in the last figure, the oval extending between the limits $x=c$, $x=b$, the infinite branch commencing at the point $x=a$.

(vi) $y^2=(x-c)^2(x-a)$. Suppose that in the last-mentioned curve, $y^2=x-c.x-b.x-a$, b gradually diminishes, and becomes ultimately $=c$. The infinite branch (see fig. 5) changes its form, but not in a very marked manner, and it retains the two inflexions.

Fig. 5.

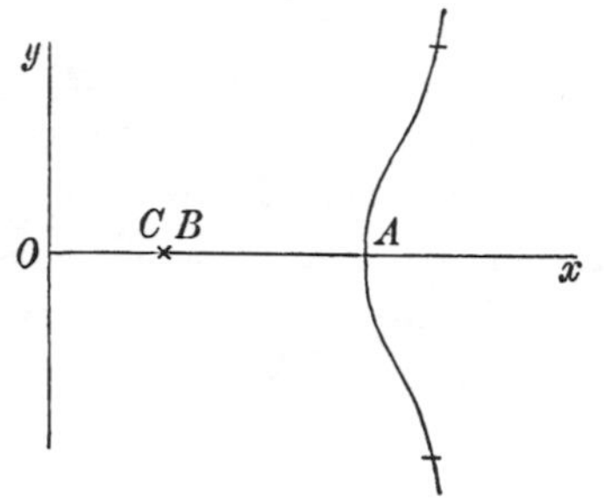

The oval lies always between the values $x=c$, $x=b$, and therefore its length continually diminishes; it is easy to see that its breadth will also continually diminish; ultimately it shrinks up into a mere point. The curve has thus a conjugate or isolated point, or acnode. For a direct verification observe that $x=c$, $y=0$, so that $(c, 0)$ is a point of the curve, but if x is either less than c, or between c and a, y^2 is negative, and y is imaginary.

(vii) $y^2=(x-c)(x-a)^2$. If in the same curve b gradually increases and becomes ultimately $=a$, the oval and the infinite branch change each of them its form, the oval extending always between the values $x=c$, $x=b$, and thus continually approaching the infinite branch, which begins at $x=a$. The consideration of a few numerical examples, *with careful drawing*, would show that the oval and the infinite branch as they approach sharpen out each towards the other, the two inflexions on the infinite branch coming always nearer to the point $(a, 0)$,—so that finally, when b becomes

Fig. 6.

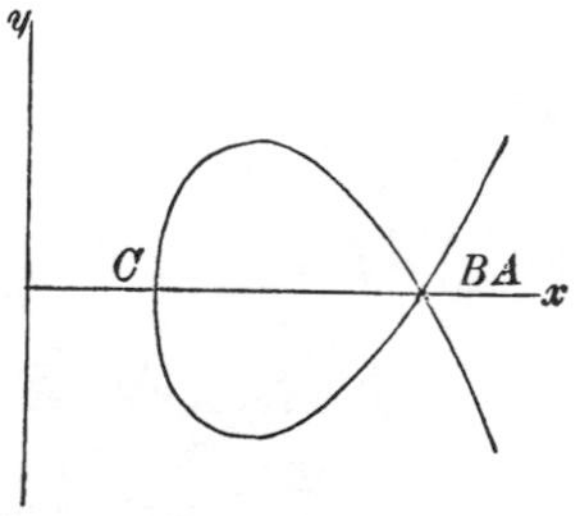

$=a$, the curve has the form shown in fig. 6, there being now a double point or node (crunode) at A, and the inflexions on the infinite branch having disappeared.

In the last four examples, the curve is one of the cubical curves called the divergent parabolas: (iv) is a mere numerical example of (v), and (vi), (vii), (viii) are in Newton's language the parabola *cum ovali*, *punctata*, and *nodata* respectively. When a, b, c are all equal, or the form is $y^2=(x-c)^3$, we have a cuspidal form, Newton's parabola *cuspidata*, otherwise the semicubical parabola.

(viii) As an example of a curve given by an implicit equation, suppose the equation is

$$x^3+y^3-3xy=0;$$

this is a nodal cubic curve, the node at the origin, and the axes touching the two branches respectively (fig. 7). An easy mode of tracing it is to express x, y each of them

Fig. 7.

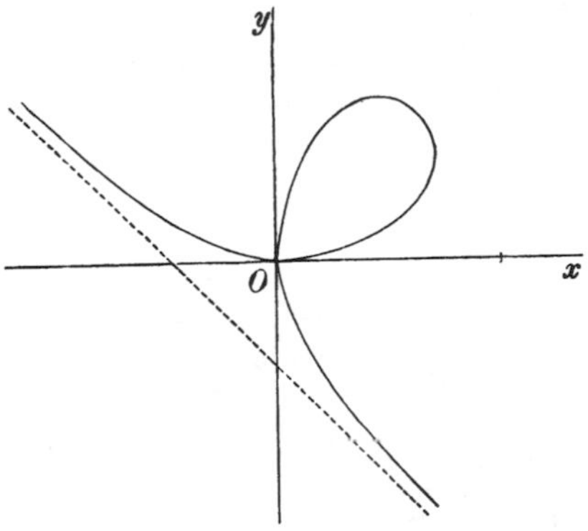

in terms of a variable θ, $x=\dfrac{3\theta}{1+\theta^3}$, $y=\dfrac{3\theta^2}{1+\theta^3}$; but it is instructive to trace the curve directly from its equation.

5. It may be remarked that the purely algebraical process, which is, in fact, that employed in finding a differential coefficient $\dfrac{dy}{dx}$, if applied directly to the equation of the curve, determines the point consecutive to any given point of the curve, that is, the direction of the curve at such given point, or, what is the same thing, the direction of the tangent at that point. In fact, if α, β are the coordinates of any point on a curve $f(x, y)=0$, then writing in the equation of the curve $x=\alpha+h$, $y=\beta+k$, and in the resulting equation $f(\alpha+h, \beta+k)=0$, developed in powers of h and k, omitting the term $f(\alpha, \beta)$, which vanishes, and the terms containing the second and higher powers of h, k, we have a linear equation $Ah+Bk=0$, which determines the ratio of the increments h, k. Of course, in the analytical development of the theory, we translate this into the notation of the differential calculus; but the question presents itself, and is thus seen to be solvable, as soon as it is attempted to trace a curve from its equation.

Geometry is Descriptive, or Metrical.

6. A geometrical proposition is either *descriptive* or *metrical*: in the former case it is altogether independent of the idea of magnitude (length, inclination, &c.); in the latter case it has reference to this idea. It is to be noticed that, although the method of coordinates seems to be by its inception essentially metrical, and we can hardly, except by metrical considerations, connect an equation with the curve which it represents (for instance, even assuming it to be known that an equation $Ax+By+C=0$ represents a line, yet if it be asked what line, the only form of answer is, that it is the line cutting the axes at *distances* from the origin $-C\div A$, $-C\div B$ respectively), yet in dealing by this method with descriptive propositions, we are, in fact, eminently free from all metrical considerations.

7. It is worth while to illustrate this by the instance of the well-known theorem of the radical centre of three circles. The theorem is that, given any three circles A, B, C (fig. 8), the common chords $\alpha\alpha'$, $\beta\beta'$, $\gamma\gamma'$ of the three pairs of circles meet in a point.

Fig. 8.

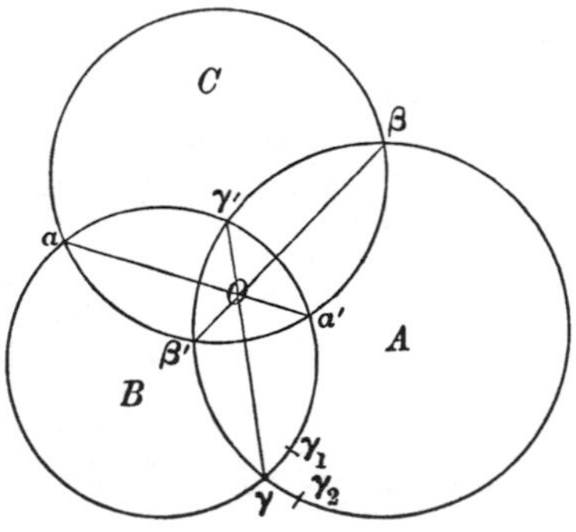

The geometrical proof is metrical throughout:—

Take O the point of intersection of $\alpha\alpha'$, $\beta\beta'$, and joining this with γ', suppose that $\gamma'O$ does not pass through γ, but that it meets the circles A, B in two distinct points γ_1, γ_2 respectively. We have then the known metrical property of intersecting chords of a circle; viz. in the circle C, where $\alpha\alpha'$, $\beta\beta'$ are chords meeting at a point O,

$$O\alpha\,.\,O\alpha'=O\beta\,.\,O\beta',$$

where, as well as in what immediately follows $O\alpha$, &c., denote, of course, *lengths* or *distances*.

Similarly in the circle A,

$$O\beta\,.\,O\beta'=O\gamma_1\,.\,O\gamma',$$

and in the circle B,

$$O\alpha\,.\,O\alpha'=O\gamma_2\,.\,O\gamma'.$$

Consequently $O\gamma_1\,.\,O\gamma'=O\gamma_2\,.\,O\gamma'$, that is, $O\gamma_1=O\gamma_2$, or the points γ_1 and γ_2 coincide; that is, they each coincide with γ.

We contrast this with the analytical method.

Here it only requires to be known that an equation $Ax + By + C = 0$ represents a line, and an equation $x^2 + y^2 + Ax + By + C = 0$ represents a circle. A, B, C have, in the two cases respectively, metrical significations; but these we are not concerned with. Using S to denote the function $x^2 + y^2 + Ax + By + C$, the equation of a circle is $S = 0$, where S stands for its value; more briefly, we say the equation is $S, = x^2 + y^2 + Ax + By + C, = 0$. Let the equation of any other circle be $S', = x^2 + y^2 + A'x + B'y + C' = 0$; the equation $S - S' = 0$ is a linear equation: $S - S'$ is, in fact, $= (A - A')x + (B - B')y + C - C'$: and it thus represents a line; this equation is satisfied by the coordinates of each of the points of intersection of the two circles (for at each of these points $S = 0$ and $S' = 0$, therefore also $S - S' = 0$); hence the equation $S - S' = 0$ is that of the line joining the two points of intersection of the two circles, or say it is the equation of the common chord of the two circles. Considering then a third circle $S'', = x^2 + y^2 + A''x + B''y + C'' = 0$, the equations of the common chords are $S - S' = 0$, $S - S'' = 0$, $S' - S'' = 0$ (each of these a linear equation); at the intersection of the first and second of these lines $S = S'$ and $S = S''$, therefore also $S' = S''$, or the equation of the third line is satisfied by the coordinates of the point in question; that is, the three chords intersect in a point O, the coordinates of which are determined by the equations $S = S' = S''$.

It further appears that, if the two circles $S = 0$, $S' = 0$ do not intersect in any real points, they must be regarded as intersecting in two imaginary points, such that the line joining them is the real line represented by the equation $S - S' = 0$; or that two circles, whether their intersections be real or imaginary, have always a real common chord (or radical axis), and that for *any* three circles the common chords intersect in a point (of course real) which is the radical centre. And by this very theorem, given two circles with imaginary intersections, we can, by drawing circles which meet each of them in real points, construct the radical axis of the first-mentioned two circles.

8. The principle employed in showing that the equation of the common chord of two circles is $S - S' = 0$ is one of very extensive application, and some more illustrations of it may be given.

Suppose $S = 0$, $S' = 0$ are lines, that is, let S, S' now denote linear functions $Ax + By + C$, $A'x + B'y + C'$, then $S - kS' = 0$ (k an arbitrary constant) is the equation of *any* line passing through the point of intersection of the two given lines. Such a line may be made to pass through any given point, say the point (x_0, y_0); i.e., if S_0, S_0' are what S, S' respectively become on writing for (x, y) the values (x_0, y_0), then the value of k is $k = S_0 \div S_0'$. The equation in fact is $SS_0' - S_0S' = 0$; and starting from this equation we at once verify it *à posteriori*; the equation is a linear equation satisfied by the values of (x, y) which make $S = 0$, $S' = 0$; and satisfied also by the values (x_0, y_0); and it is thus the equation of the line in question.

If, as before, $S = 0$, $S' = 0$ represent circles, then (k being arbitrary) $S - kS' = 0$ is the equation of any circle passing through the two points of intersection of the two circles; and to make this pass through a given point (x_0, y_0) we have again $k = S_0 \div S_0'$. In the particular case $k = 1$, the circle becomes the common chord; more accurately, it becomes the common chord together with the line infinity, but this is a question which is not here gone into.

If S denote the general quadric function,

$$S = ax^2 + 2hxy + by^2 + 2fy + 2gx + c, \quad = (a, b, c, f, g, h)(x, y, 1)^2,$$

then the equation $S=0$ represents a conic; assuming this, then, if $S'=0$ represents another conic, the equation $S - kS' = 0$ represents *any* conic through the four points of intersection of the two conics.

Returning to the equation $Ax + By + C = 0$ of a line, if this pass through two given points (x_1, y_1), (x_2, y_2), then we must have $Ax_1 + By_1 + C = 0$, $Ax_2 + By_2 + C = 0$, equations which determine the ratios $A : B : C$, and it thus appears that the equation of the line through the two given points is

$$x(y_1 - y_2) - y(x_1 - x_2) + x_1y_2 - x_2y_1 = 0;$$

or, what is the same thing,

$$\begin{vmatrix} x, & y, & 1 \\ x_1, & y_1, & 1 \\ x_2, & y_2, & 1 \end{vmatrix} = 0.$$

9. The object still being to illustrate the mode of working with coordinates, we consider the theorem of the polar of a point in regard to a circle. Given a circle and a point O (fig. 9), we draw through O any two lines meeting the circle in the points

Fig. 9.

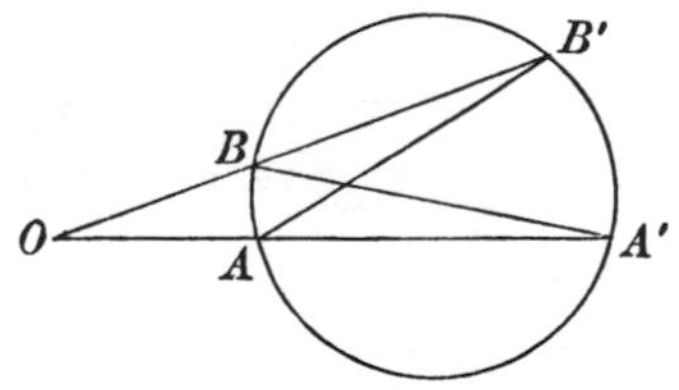

A, A' and B, B' respectively, and then taking Q as the intersection of the lines AB' and $A'B$, the theorem is that the locus of the point Q is a right line depending only upon O and the circle, but independent of the particular lines OAA' and OBB'.

Taking O as the origin, and for the axes any two lines through O at right angles to each other, the equation of the circle will be

$$x^2 + y^2 + 2Ax + 2By + C = 0;$$

and if the equation of the line OAA' is taken to be $y = mx$, then the points A, A' are found as the intersections of the straight line with the circle; or to determine x we have

$$x^2(1 + m^2) + 2x(A + Bm) + C = 0.$$

If (x_1, y_1) are the coordinates of A, and (x_2, y_2) of A', then the roots of this equation are x_1, x_2, whence easily

$$\frac{1}{x_1}+\frac{1}{x_2}=-2\frac{A+Bm}{C}.$$

And similarly, if the equation of the line OBB' is taken to be $y=m'x$, and the coordinates of B, B' to be (x_3, y_3) and (x_4, y_4) respectively, then

$$\frac{1}{x_3}+\frac{1}{x_4}=-2\frac{A+Bm'}{C}.$$

We have then

$$x(y_1-y_4)-y(x_1-x_4)+x_1y_4-x_4y_1=0,$$

$$x(y_2-y_3)-y(x_2-x_3)+x_2y_3-x_3y_2=0,$$

as the equations of the lines AB' and $A'B$ respectively; for the first of these equations, being satisfied if we write therein (x_1, y_1) or (x_4, y_4) for (x, y), is the equation of the line AB': and similarly the second equation is that of the line $A'B$. Reducing by means of the relations $y_1-mx_1=0$, $y_2-mx_2=0$, $y_3-m'x_3=0$, $y_4-m'x_4=0$, the two equations become

$$x(mx_1-m'x_4)-y(x_1-x_4)+(m'-m)x_1x_4=0,$$

$$x(mx_2-m'x_3)-y(x_2-x_3)+(m'-m)x_2x_3=0;$$

and if we divide the first of these equations by m_1m_4, and the second by m_2m_3, and then add, we obtain

$$x\left\{m\left(\frac{1}{x_3}+\frac{1}{x_4}\right)-m'\left(\frac{1}{x_1}+\frac{1}{x_2}\right)\right\}-y\left\{\frac{1}{x_3}+\frac{1}{x_4}-\left(\frac{1}{x_1}+\frac{1}{x_2}\right)\right\}+2m'-2m=0,$$

or, what is the same thing,

$$\left(\frac{1}{x_1}+\frac{1}{x_2}\right)(y-m'x)-\left(\frac{1}{x_3}+\frac{1}{x_4}\right)(y-mx)+2m'-2m=0,$$

which by what precedes is the equation of a line through the point Q. Substituting herein for $\frac{1}{x_1}+\frac{1}{x_2}$, $\frac{1}{x_3}+\frac{1}{x_4}$ their foregoing values, the equation becomes

$$-(A+Bm)(y-m'x)+(A+Bm')(y-mx)+m'-m=0;$$

that is,

$$(m-m')(Ax+By+C)=0;$$

or finally it is $Ax+By+C=0$, showing that the point Q lies in a line the position of which is independent of the particular lines OAA', OBB' used in the construction. It is proper to notice that there is no correspondence to each other of the points A, A' and B, B'; the grouping might as well have been A, A' and B', B; and it thence appears that the line $Ax+By+C=0$ just obtained is in fact the line joining the point Q with the point R which is the intersection of AB and $A'B'$.

10. The equation $Ax+By+C=0$ of a line contains in appearance 3, but really only 2 constants (for one of the constants can be divided out), and a line depends

accordingly upon 2 parameters, or can be made to satisfy 2 conditions. Similarly, the equation $(a, b, c, f, g, h\between x, y, 1)^2 = 0$ of a conic contains really 5 constants, and the equation $(*\between x, y, 1)^3 = 0$ of a cubic contains really 9 constants. It thus appears that a cubic can be made to pass through 9 given points, and that the cubic so passing through 9 given points is completely determined. There is, however, a remarkable exception. Considering two given cubic curves $S = 0$, $S' = 0$, these intersect in 9 points, and through these 9 points we have the whole series of cubics $S - kS' = 0$, where k is an arbitrary constant: k may be determined so that the cubic shall pass through a given tenth point, viz. $k = S_0 \div S_0'$, if the coordinates are (x_0, y_0), and S_0, S_0' denote the corresponding values of S, S'. The resulting curve $SS_0' - S'S_0 = 0$ may be regarded as the cubic determined by the conditions of passing through 8 of the 9 points and through the given point (x_0, y_0); and from the equation it thence appears that the curve passes through the remaining one of the 9 points. In other words, we thus have the theorem, any cubic curve which passes through 8 of the 9 intersections of two given cubic curves passes through the 9th intersection.

The applications of this theorem are very numerous; for instance, we derive from it Pascal's theorem of the inscribed hexagon. Consider a hexagon inscribed in a conic. The three alternate sides constitute a cubic, and the other three alternate sides another cubic. The cubics intersect in 9 points, being the 6 vertices of the hexagon, and the 3 Pascalian points, or intersections of the pairs of opposite sides of the hexagon. Drawing a line through two of the Pascalian points, the conic and this line constitute a cubic passing through 8 of the 9 points of intersection, and it therefore passes through the remaining point of intersection—that is, the third Pascalian point; and since obviously this does not lie on the conic, it must lie on the line—that is, we have the theorem that the three Pascalian points (or points of intersection of the pairs of opposite sides) lie on a line.

Metrical Theory.

11. The foundation of the metrical theory consists in the simple theorem that if a finite line PQ (fig. 10) be projected upon any other line OO' by lines perpendicular

Fig. 10.

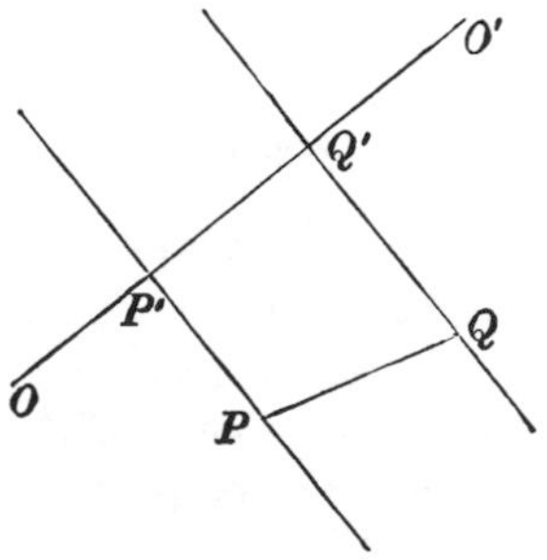

to OO', then the length of the projection $P'Q'$ is equal to the length of PQ multiplied by the cosine of its inclination to $P'Q'$; or, what is the same thing, that the perpendicular distance $P'Q'$ of any two parallel lines is equal to the inclined distance PQ

multiplied by the cosine of the inclination. It at once follows that the algebraical sum of the projections of the sides of a closed polygon upon any line is $=0$; or, reversing the signs of certain sides, and considering the polygon as consisting of two broken lines, each extending from the same initial to the same terminal point, the sum of the projections of the lines of the first set upon any line is equal to the sum of the projections of the lines of the second set. Observe that, if any line be perpendicular to the line on which the projection is made, then its projection is $=0$.

Thus, if we have a right-angled triangle PQR (fig. 11), where QR, RP, QP are

Fig. 11.

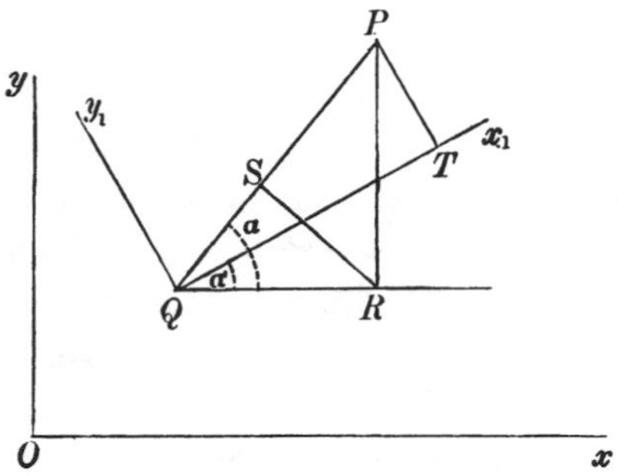

$=\xi, \eta, \rho$ respectively, and whereof the base-angle is $=\alpha$, then projecting successively on the three sides, we have

$$\xi=\rho\cos\alpha,\quad \eta=\rho\sin\alpha,\quad \rho=\xi\cos\alpha+\eta\sin\alpha;$$

and we thence obtain

$$\rho^2=\xi^2+\eta^2;\quad \cos^2\alpha+\sin^2\alpha=1.$$

And again, by projecting on a line Qx_1, inclined at the angle α' to QR, we have

$$\rho\cos(\alpha-\alpha')=\xi\cos\alpha'+\eta\sin\alpha';$$

and by substituting for ξ, η their foregoing values,

$$\cos(\alpha-\alpha')=\cos\alpha\cos\alpha'+\sin\alpha\sin\alpha'.$$

It is to be remarked that, assuming only the theory of similar triangles, we have herein a *proof* of Euclid, Book I., Prop. 47; in fact, the same as is given Book VI., Prop. 31; and also a *proof* of the trigonometrical formula for $\cos(\alpha-\alpha')$. The formulæ for $\cos(\alpha+\alpha')$ and $\sin(\alpha\pm\alpha')$ could be obtained in the same manner.

Draw PT at right angles to Qx_1, and suppose QT, $TP=\xi_1, \eta_1$ respectively, so that we have now the quadrilateral $QRPTQ$, or, what is the same thing, the two broken lines QRP and QTP, each extending from Q to P. Projecting on the four sides successively, we have

$$\begin{aligned}\xi &= \xi_1\cos\alpha'-\eta_1\sin\alpha',\\ \eta &= \xi_1\sin\alpha'+\eta_1\cos\alpha',\\ \xi_1 &= \xi\cos\alpha'+\eta\sin\alpha',\\ \eta_1 &= -\xi\sin\alpha'+\eta\cos\alpha',\end{aligned}$$

where the third equation is that previously written

$$\rho\cos(\alpha-\alpha')=\xi\cos\alpha+\eta\sin\alpha.$$

Equations of Right Line and Circle.—Transformation of Coordinates.

12. The required formulæ are really contained in the foregoing results. For, in fig. 11, supposing that the axis of x is parallel to QR, and taking a, b for the coordinates of Q, and (x, y) for those of P, then we have $\xi,\ \eta = x-a,\ y-b$ respectively; and therefore

$$x - a = \rho\cos\alpha,\quad y - b = \rho\sin\alpha,$$
$$\rho^2 \quad = (x-a)^2 + (y-b)^2.$$

Writing the first two of these in the form

$$\frac{x-a}{\cos\alpha} = \frac{y-b}{\sin\alpha}\ (=\rho),$$

we may regard Q as a fixed point, but P as a point moving in the direction Q to P, so that α remains constant, and then, omitting the equation $(=\rho)$, we have a relation between the coordinates x, y of the point P thus moving in a right line,—that is, we have the equation of the line through the given point (a, b) at a given inclination α to the axis of x. And, moreover, if, using this equation $(=\rho)$, we write $x = a + \rho\cos\alpha$, $y = b + \rho\sin\alpha$, then we have expressions for the coordinates x, y of a point of this line, in terms of the variable parameter ρ.

Again, take the point T to be fixed, but consider the point P as moving in the line TP at right angles to QT. If instead of ξ_1 we take p for the distance QT, then the equation $\xi_1 = \xi\cos\alpha' + \eta\sin\alpha'$ will be

$$(x-a)\cos\alpha' + (y-b)\sin\alpha' = p;$$

that is, this will be the equation of a line such that its perpendicular distance from the point (a, b) is $= p$, and that the inclination of this distance to the axis of x is $= \alpha'$.

From either form it appears that the equation of a line is, in fact, a linear equation of the form $Ax + By + C = 0$. It is important to notice that, starting from this equation, we can determine conversely the α but not the (a, b) of the form of equation which contains these quantities; and in like manner the α' but not the (a, b) or p of the other form of equation. The reason is obvious. In each case (a, b) denote the coordinates of a point, fixed indeed, but which is in the first form any point of the line, and in the second form any point whatever. Thus, in the second form the point from which the perpendicular is let fall may be the origin. Here $(a, b) = (0, 0)$, and the equation is $x\cos\alpha' + y\sin\alpha' - p = 0$. Comparing this with $Ax + By + C = 0$, we have the values of $\cos\alpha'$, $\sin\alpha'$, and p.

13. The equation

$$\rho^2 = (x-a)^2 + (y-b)^2$$

is an expression for the squared distance of the two points (a, b) and (x, y). Taking as before the point Q, coordinates (a, b), as a fixed point, and writing c in the place of ρ, the equation

$$(x-a)^2 + (y-b)^2 = c^2$$

expresses that the point (x, y) is always at a given distance c from the given point (a, b); viz. this is the equation of a circle, having (a, b) for the coordinates of its centre, and c for its radius.

The equation is of the form

$$x^2 + y^2 + 2Ax + 2By + C = 0,$$

and here, the number of constants being the same, we can identify the two equations; we find $a = -A$, $b = -B$, $c^2 = A^2 + B^2 - C$, or the last equation is that of a circle having $-A$, $-B$ for the coordinates of its centre, and $\sqrt{A^2 + B^2 - C}$ for its radius.

14. Drawing (fig. 11) Qy_1 at right angles at Qx_1, and taking Qx_1, Qy_1 as a new set of rectangular axes, if instead of ξ_1, η_1 we write x_1, y_1, we have x_1, y_1 as the new coordinates of the point P; and writing also α in place of α', α now denoting the inclination of the axes Qx_1 and Ox, we have the formulæ for transformation between two sets of rectangular axes. These are

$$x - a = x_1 \cos\alpha - y_1 \sin\alpha,$$
$$y - b = x_1 \sin\alpha + y_1 \cos\alpha,$$

and

$$x_1 = (x - a)\cos\alpha + (y - b)\sin\alpha,$$
$$y_1 = -(x - a)\sin\alpha + (y - b)\cos\alpha,$$

each set being obviously at once deducible from the other one. In these formulæ (a, b) are the xy-coordinates of the new origin Q_1, and α is the inclination of Qx_1 to Ox. It is to be noticed that Qx_1, Qy_1 are so placed that, by moving O to Q, and then turning the axes Ox_1, Oy_1 round Q (through an angle α measured in the sense Ox to Oy), the original axes Ox, Oy will come to coincide with Qx_1, Qy_1 respectively. This could not have been done if Qy_1 had been drawn (at right angles always to Qx_1) in the reverse direction: we should then have had in the formulæ $-y_1$ instead of y_1. The new formulæ which would be thus obtained are of an essentially distinct form: the analytical test is that in the formulæ as written down we can, by giving to α a proper value (in fact, $\alpha = 0$), make the $(x - a)$ and $(y - b)$ equal to x_1 and y_1 respectively; in the other system we could only make them equal to x_1, $-y_1$, or $-x_1$, y_1 respectively. But for the very reason that the second system can be so easily derived from the first, it is proper to attend exclusively to the first system,—that is, always to take the new axes so that the two sets admit of being brought into coincidence.

In the foregoing system of two pairs of equations, the first pair give the original coordinates x, y in terms of the new coordinates x_1, y_1; the second pair the new coordinates x_1, y_1 in terms of the original coordinates x, y. The formulæ involve (a, b), the original coordinates of the new origin; it would be easy, instead of these, to introduce (a_1, b_1), the new coordinates of the origin. Writing $(a, b) = (0, 0)$, we have, of course, the formulæ for transformation between two sets of rectangular axes *having the same origin*, and it is as well to write the formulæ in this more simple form; the subsequent transformation to a new origin, but with axes parallel to the original axes, can then be effected without any difficulty.

15. All questions in regard to the line may be solved by means of one or other of the foregoing forms—

$$Ax + By + C = 0,$$

$$y = Ax + B,$$

$$\frac{x-a}{\cos\alpha} = \frac{y-b}{\sin\alpha},$$

$$(x-a)\cos\alpha' + (y-b)\sin\alpha' - p = 0;$$

or it may be by a comparison of these different forms: thus, using the first form, it has been already shown that the equation of the line through two given points (x_1, y_1), (x_2, y_2) is

$$x(y_1 - y_2) - y(x_1 - x_2) + x_1 y_2 - x_2 y_1 = 0,$$

or, as this may be written,

$$y - y_1 = \frac{y_2 - y_1}{x_2 - x_1}(x - x_1).$$

A particular case is the equation

$$\frac{x}{a} + \frac{y}{b} = 1,$$

representing the line through the points $(a, 0)$ and $(0, b)$, or, what is the same thing, the line meeting the axes of x and y at the distances from the origin a and b respectively. It may be noticed that, in the form $Ax + By + C = 0$, $-\frac{A}{B}$ denotes the tangent of the inclination to the axis of x, or we may say that $B \div \sqrt{A^2 + B^2}$ and $-A \div \sqrt{A^2 + B^2}$ denote respectively the cosine and the sine of the inclination to the axis of x. A better form is this: $A \div \sqrt{A^2 + B^2}$ and $B \div \sqrt{A^2 + B^2}$ denote respectively the cosine and the sine of inclination to the axis of x of the perpendicular upon the line. So, of course, in regard to the form $y = Ax + B$, A is here the tangent of the inclination to the axis of x; $1 \div \sqrt{A^2 + 1}$ and $A \div \sqrt{A^2 + 1}$ are the cosine and sine of this inclination, &c. It thus appears that the condition, in order that the lines $Ax + By + C = 0$ and $A'x + B'y - C' = 0$ may meet at right angles, is $AA' + BB' = 0$; so when the equations are $y = Ax + B$, $y = A'x + B'$, the condition is $AA' + 1 = 0$, or say the value of A' is $= -1 \div A$.

The perpendicular distance of the point (a, b) from the line $Ax + By + C = 0$ is $(Aa + Bb + C) \div \sqrt{A^2 + B^2}$. In all the formulæ involving $\sqrt{A^2 + B^2}$ or $\sqrt{A^2 + 1}$, the radical should be written with the sign $\pm$, which is essentially indeterminate: the like indeterminateness of sign presents itself in the expression for the distance of two points $\rho = \pm\sqrt{(x-a)^2 + (y-b)^2}$; if, as before, the points are Q, P, and the indefinite line through these is $z'QPz$, then it is the same thing whether we measure off from Q along this line, considered as drawn from z' towards z, a positive distance k, or along the line considered as drawn reversely from z towards z', the equal negative distance $-k$, and the expression for the distance ρ is thus properly of the form $\pm k$. It is interesting to compare expressions which do not involve a radical: thus, in seeking for the expression for the perpendicular distance of the point (a, b) from a given line, let the equation of the given line be taken in the form, $x\cos\alpha+y\sin\alpha-p=0$, p being the perpendicular distance from the origin, α its inclination to the axis of x: the equation of the line may also be written $(x-a)\cos\alpha+(y-b)\sin\alpha-p_1=0$, and we have thence $p_1=p-a\cos\alpha-b\sin\alpha$, the required expression for the distance p_1: it is here assumed that p_1 is drawn from (a, b) in the same sense as p is drawn from the origin, and the indeterminateness of sign is thus removed.

16. As an instance of the mode of using the formulæ, take the problem of finding the locus of a point such that its distance from a given point is in a given ratio to its distance from a given line.

We take (a, b) as the coordinates of the given point, and it is convenient to take (x, y) as the coordinates of the variable point, the locus of which is required: it thus becomes necessary to use other letters, say (X, Y), for current coordinates in the equation of the given line. Suppose this is a line such that its perpendicular distance from the origin is $=p$, and that the inclination of p to the axis of x is $=\alpha$; the equation is $X\cos\alpha+Y\sin\alpha-p=0$. In the result obtained in § 15, writing (x, y) in place of (a, b), it appears that the perpendicular distance of this line from the point (x, y) is

$$=p-x\cos\alpha-y\sin\alpha;$$

hence the equation of the locus is

$$\sqrt{(x-a)^2+(y-b)^2}=e(p-x\cos\alpha-y\sin\alpha),$$

or say

$$(x-a)^2+(y-b)^2-e^2(x\cos\alpha+y\sin\alpha-p)^2=0,$$

an equation of the second order.

The Conics (Parabola, Ellipse, Hyperbola).

17. The conics or, as they were called, conic sections were originally defined as the sections of a right circular cone; but Apollonius substituted a definition, which is, in fact, that of the last example: the curve is the locus of a point such that its

Fig. 12.

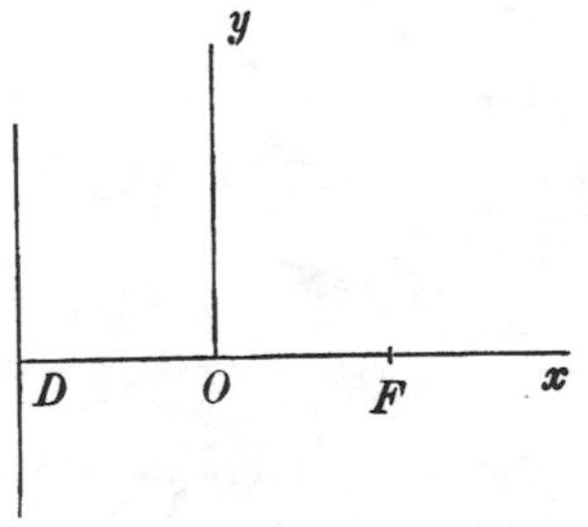

distance from a given point (called the focus) is in a given ratio to its distance from a given line (called the directrix); and taking the ratio as $e:1$, then e is called the eccentricity.

Take FD for the perpendicular from the focus F upon the directrix, and the given ratio being that of $e:1$ ($e >, =,$ or <1, but positive), and let the distance FD

be divided at O in the given ratio, say we have $OD=m$, $OF=em$, where m is positive;—then the origin may be taken at O, the axis Ox being in the direction OF (that is, from O to F), and the axis Oy at right angles to it. The distance of the point (x, y) from F is $=\sqrt{(x-em)^2+y^2}$, its distance from the directrix is $=x+m$; the equation therefore is

$$(x-em)^2+y^2=e^2(x+m)^2;$$

or, what is the same thing, it is

$$(1-e^2)x^2-2me(1+e)x+y^2=0.$$

If $e^2=1$, or, since e is taken to be positive, if $e=1$, this is

$$y^2-4mx=0,$$

which is the parabola.

If e^2 not $=1$, then the equation may be written

$$(1-e^2)\left(x-\frac{me}{1-e}\right)^2+y^2=\frac{m^2e^2(1+e)}{1-e}.$$

Supposing e positive and <1, then, writing $m=\dfrac{a(1-e)}{e}$, the equation becomes

$$(1-e^2)(x-a)^2+y^2=a^2(1-e^2),$$

that is,

$$\frac{(x-a)^2}{a^2}+\frac{y^2}{a^2(1-e^2)}=1;$$

or, changing the origin and writing $b^2=a^2(1-e^2)$, this is

$$\frac{x^2}{a^2}+\frac{y^2}{b^2}=1,$$

which is the ellipse.

And similarly if e be positive and >1, then writing $m=\dfrac{a(e-1)}{e}$, the equation becomes

$$(1-e^2)(x+a)^2+y^2=a^2(1-e^2),$$

that is,

$$\frac{(x+a)^2}{a^2}+\frac{y^2}{a^2(1-e^2)}=1,$$

or changing the origin and writing $b^2=a^2(e^2-1)$, this is

$$\frac{x^2}{a^2}-\frac{y^2}{b^2}=1,$$

which is the hyperbola.

18. The general equation $ax^2+2hxy+by^2+2fy+2gx+c=0$, or as it is written $(a, b, c, f, g, h)(x, y, 1)^2=0$, may be such that the quadric function breaks up into factors, $=(\alpha x+\beta y+\gamma)(\alpha' x+\beta' y+\gamma')$; and in this case the equation represents a pair of lines, or (it may be) two coincident lines. When it does not so break up, the function can be put in the form $\lambda\{(x-a')^2+(y-b')^2-e^2(x\cos\alpha+y\sin\alpha-p)^2\}$, or, equating the two expressions, there will be six equations for the determination of $\lambda, a', b', e, p, \alpha$; and by what precedes, if a', b', e, p, α are real, the curve is either

a parabola, ellipse, or hyperbola. The original coefficients (a, b, c, f, g, h) may be such as not to give any system of real values for a', b', e, p, α; but when this is so the equation $(a, b, c, f, g, h)(x, y, 1)^2 = 0$ does not represent a real curve*; the imaginary curve which it represents is, however, regarded as a conic. Disregarding the special cases of the pair of lines and the twice repeated line, it thus appears that the only real curves represented by the general equation $(a, b, c, f, g, h)(x, y, 1)^2 = 0$ are the parabola, the ellipse, and the hyperbola. The circle is considered as a particular case of the ellipse.

The same result is obtained by transforming the equation $(a, b, c, f, g, h)(x, y, 1)^2 = 0$ to new axes. If in the first place the origin be unaltered, then the directions of the new (rectangular) axes Ox_1, Oy_1 can be found so that h_1 (the coefficient of the term x_1y_1) shall be $=0$; when this is done, then either one of the coefficients of x_1^2, y_1^2 is $=0$, and the curve is then a parabola, or neither of these coefficients is $=0$, and the curve is then an ellipse or hyperbola, according as the two coefficients are of the same sign or of opposite signs.

19. The curves can be at once traced from their equations:—

$$y^2 = 4mx, \text{ for the parabola (fig. 13),}$$

$$\frac{x^2}{a^2} + \frac{y^2}{b^2} = 1, \text{ for the ellipse (fig. 14),}$$

$$\frac{x^2}{a^2} - \frac{y^2}{b^2} = 1, \text{ for the hyperbola (fig. 15);}$$

Fig. 13.

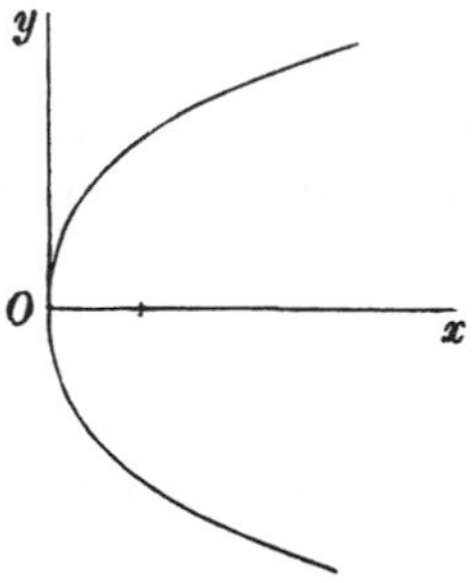

Fig. 14.

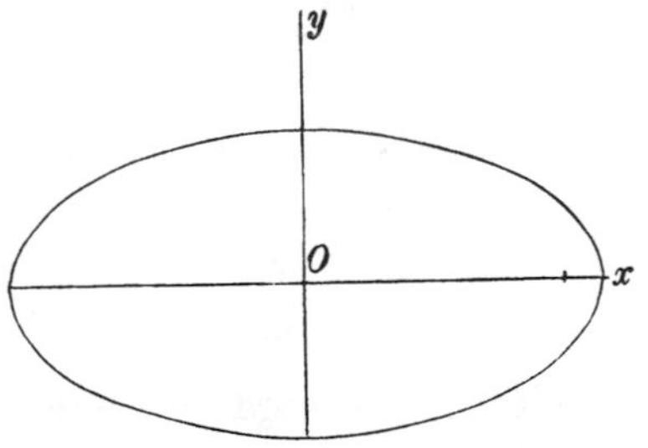

Fig. 15.

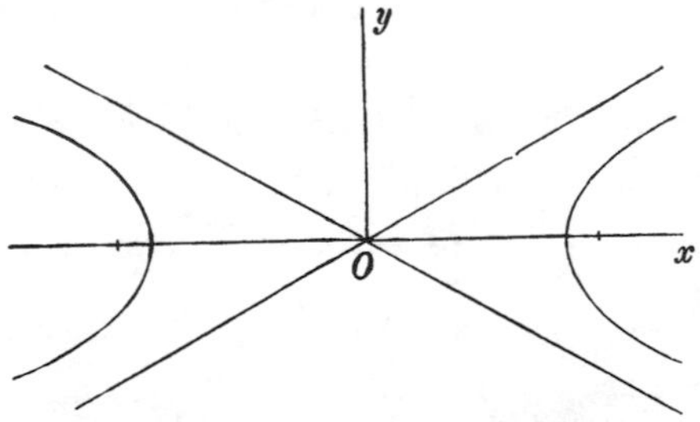

* It is proper to remark that, when $(a, b, c, f, g, h)(x, y, 1)^2=0$ *does* represent a real curve, there are, in fact, four systems of values of a', b', e, p, α, two real, the other two imaginary; we have thus two real equations and two imaginary equations, each of them of the form $(x-a')^2+(y-b')^2=e^2(x\cos\alpha+y\cos\beta-p)^2$, representing each of them one and the same real curve. This is consistent with the assertion of the text that the real curve is in every case represented by a real equation of this form.

and it will be noticed how the form of the last equation puts in evidence the two asymptotes $\frac{x}{a} = \pm \frac{y}{b}$ of the hyperbola. Referred to the asymptotes (as a set of oblique axes) the equation of the hyperbola takes the form $xy = c$; and in particular, if in this equation the axes are at right angles, then the equation represents the rectangular hyperbola referred to its asymptotes as axes.

Tangent, Normal, Circle and Radius of Curvature, &c.

20. There is great convenience in using the language and notation of the infinitesimal analysis; thus we consider on a curve a point with coordinates (x, y), and a consecutive point the coordinates of which are $(x + dx, y + dy)$, or again a second consecutive point with coordinates $(x + dx + \frac{1}{2}d^2x, y + dy + \frac{1}{2}d^2y)$, &c.; and in the final results the ratios of the infinitesimals must be replaced by differential coefficients in the proper manner; thus, if x, y are considered as given functions of a parameter θ, then dx, dy have in fact the values $\frac{dx}{d\theta}d\theta$, $\frac{dy}{d\theta}d\theta$, and (only the ratio being really material) they may in the result be replaced by $\frac{dx}{d\theta}, \frac{dy}{d\theta}$. This includes the case where the equation of the curve is given in the form $y = \phi(x)$; θ is here $= x$, and the increments dx, dy are in the result to be replaced by $1, \frac{dy}{dx}$. So also with the infinitesimals of the higher orders d^2x, &c.

21. The tangent at the point (x, y) is the line through this point and the consecutive point $(x + dx, y + dy)$; hence, taking ξ, η as current coordinates, the equation is

$$\frac{\xi - x}{dx} = \frac{\eta - y}{dy},$$

an equation which is satisfied on writing therein $\xi, \eta = (x, y)$ or $= (x + dx, y + dy)$. The equation may be written

$$\eta - y = \frac{dy}{dx}(\xi - x),$$

$\frac{dy}{dx}$ being now the differential coefficient of y in regard to x; and this form is applicable whether y is given directly as a function of x, or in whatever way y is in effect given as a function of x: if as before x, y are given each of them as a function of θ, then the value of $\frac{dy}{dx}$ is $= \frac{dy}{d\theta} \div \frac{dx}{d\theta}$, which is the result obtained from the original form on writing therein $\frac{dx}{d\theta}, \frac{dy}{d\theta}$, for dx, dy respectively.

So again, when the curve is given by an equation $u = 0$ between the coordinates (x, y), then $\frac{dy}{dx}$ is obtained from the equation $\frac{du}{dx} + \frac{du}{dy}\frac{dy}{dx} = 0$. But here it is more elegant, using the original form, to eliminate dx, dy by the formula $\frac{du}{dx}dx + \frac{du}{dy}dy$; we thus obtain the equation of the tangent in the form

$$\frac{du}{dx}(\xi - x) + \frac{du}{dy}(\eta - y) = 0.$$

For example, in the case of the ellipse $\frac{x^2}{a^2} + \frac{y^2}{b^2} = 1$, the equation is

$$\frac{x}{a^2}(\xi - x) + \frac{y}{b^2}(\eta - y) = 0;$$

or reducing by means of the equation of the curve the equation of the tangent is

$$\frac{\xi x}{a^2} + \frac{\eta y}{b^2} = 1.$$

The normal is a line through the point at right angles to the tangent; the equation therefore is

$$(\xi - x)\,dx + (\eta - y)\,dy = 0,$$

where dx, dy are to be replaced by their proportional values as before.

22. The circle of curvature is the circle through the point and two consecutive points of the curve. Taking the equation to be

$$(\xi - \alpha)^2 + (\eta - \beta)^2 = \gamma^2,$$

the values of α, β are given by

$$x - \alpha = \frac{dy\,(dx^2 + dy^2)}{dx\,d^2y - dy\,d^2x}, \quad y - \beta = \frac{-\,dx\,(dx^2 + dy^2)}{dx\,d^2y - dy\,d^2x},$$

and we then have

$$\gamma^2, = (x - \alpha)^2 + (y - \beta)^2, = \frac{(dx^2 + dy^2)^3}{(dx\,d^2y - dy\,d^2x)^2}.$$

In the case where y is given directly as a function of x, then, writing for shortness $p = \frac{dy}{dx}$, $q = \frac{d^2y}{dx^2}$, this is $\gamma^2 = \frac{(1 + p^2)^3}{q^2}$, or, as the equation is usually written, $\gamma = \frac{(1 + p^2)^{\frac{3}{2}}}{-q}$, the radius of curvature, considered to be positive or negative according as the curve is concave or convex to the axis of x.

It may be added that the centre of curvature is the intersection of the normal by the consecutive normal.

The locus of the centre of curvature is the evolute. If from the expressions of α, β regarded as functions of x we eliminate x, we have thus an equation between (α, β), which is the equation of the evolute.

Polar Coordinates.

23. The position of a point may be determined by means of its distance from a fixed point and the inclination of this distance to a fixed line through the fixed point. Say we have r the distance from the origin, and θ the inclination of r to the axis of x; r and θ are then the polar coordinates of the point, r the radius vector, and θ the inclination. These are immediately connected with the Cartesian coordinates x, y by the formulæ $x = r\cos\theta$, $y = r\sin\theta$; and the transition from either set of coordinates to the other can thus be made without difficulty. But the use of polar coordinates is very convenient, as well in reference to certain classes of questions relating to curves of any kind—for instance, in the dynamics of central forces—as in relation to curves having in regard to the origin the symmetry of the regular polygon (curves such as that represented by the equation $r = \cos m\theta$), and also in regard to the class of curves called spirals, where the radius vector r is given as an algebraical or exponential function of the inclination θ.

Trilinear Coordinates.

24. Consider a fixed triangle ABC, and (regarding the sides as indefinite lines) suppose for a moment that p, q, r denote the distances of a point P from the sides BC, CA, AB respectively,—these distances being measured either perpendicularly to the several sides, or each of them in a given direction. To fix the ideas each distance may be considered as positive for a point inside the triangle, and the sign is thus fixed for any point whatever. There is then an identical relation between p, q, r: if a, b, c are the lengths of the sides, and the distances are measured perpendicularly thereto, the relation is $ap + bq + cr =$ twice the area of triangle. But taking x, y, z proportional to p, q, r, or if we please proportional to given multiples of p, q, r, then only the ratios of x, y, z are determined; their absolute values remain arbitrary. But the ratios of p, q, r, and consequently also the ratios of x, y, z determine, and that uniquely, the point; and it being understood that only the ratios are attended to, we say that (x, y, z) are the coordinates of the point. The equation of a line has thus the form $ax + by + cz = 0$, and generally that of a curve of the nth order is a homogeneous equation of this order between the coordinates, $(*\between x, y, z)^n = 0$. The advantage over Cartesian coordinates is in the greater symmetry of the analytical forms, and in the more convenient treatment of the line infinity and of points at infinity. The method includes that of Cartesian coordinates, the homogeneous equation in x, y, z is, in fact, an equation in $\frac{x}{z}$, $\frac{y}{z}$, which two quantities may be regarded as denoting Cartesian coordinates; or, what is the same thing, we may in the equation write $z = 1$. It may be added that, if the trilinear coordinates (x, y, z) are regarded as the Cartesian coordinates of a point of space, then the equation is that of a cone having the origin for its vertex; and conversely that such equation of a cone may be regarded as the equation in trilinear coordinates of a plane curve.

General Point-Coordinates.—Line-Coordinates.

25. All the coordinates considered thus far are point-coordinates. More generally, any two quantities (or the ratios of three quantities) serving to determine the position

of a point in the plane may be regarded as the coordinates of the point; or, if instead of a single point they determine a system of two or more points, then as the coordinates of the system of points. But, as noticed under Curve, [785], there are also line-coordinates serving to determine the position of a line; the ordinary case is when the line is determined by means of the ratios of three quantities ξ, η, ζ (correlative to the trilinear coordinates x, y, z). A linear equation $a\xi + b\eta + c\zeta = 0$ represents then the system of lines such that the coordinates of each of them satisfy this relation, in fact, all the lines which pass through a given point; and it is thus regarded as the line-equation of this point; and generally a homogeneous equation $(*\between \xi, \eta, \zeta)^n = 0$ represents the curve which is the envelope of all the lines the coordinates of which satisfy this equation, and it is thus regarded as the line-equation of this curve.

II. Solid Analytical Geometry (§§ 26—40).

26. We are here concerned with points in space,—the position of a point being determined by its three coordinates x, y, z. We consider three coordinate planes, at right angles to each other, dividing the whole of space into eight portions called octants, the coordinates of a point being the perpendicular distances of the point from the three planes respectively, each distance being considered as positive or negative according as it lies on the one or the other side of the plane. Thus the coordinates in the eight octants have respectively the signs

x,	y,	z
+	+	+
+	−	+
−	+	+
−	−	+
+	+	−
+	−	−
−	+	−
−	−	−

Fig. 16.

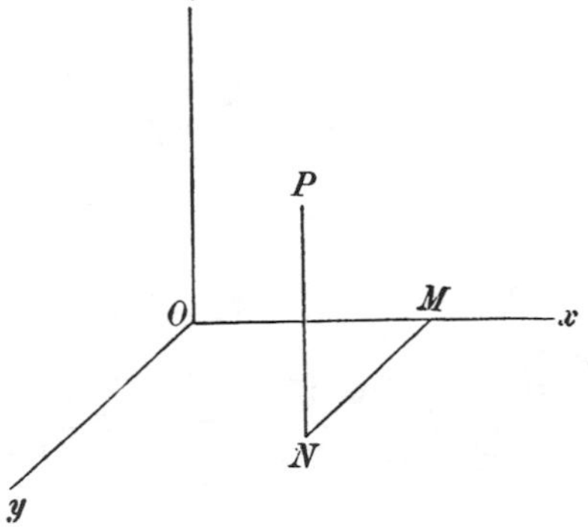

The positive parts of the axes are usually drawn as in fig. 16, which represents a point P, the coordinates of which have the positive values OM, MN, NP.

27. It may be remarked, as regards the delineation of such solid figures, that if we have in space three lines at right angles to each other, say Oa, Ob, Oc, of equal lengths, then it is possible to project these by parallel lines upon a plane in such wise that the projections Oa', Ob', Oc' shall be at given inclinations to each other, and that these lengths shall be to each other in given ratios: in particular, the two lines Oa', Oc' may be at right angles to each other, and their lengths equal, the direction of Ob', and its proportion to the two equal lengths Oa', Oc', being arbitrary. It thus appears that we may as in the figure draw Ox, Oz at right angles to each other, and Oy in an arbitrary direction; and moreover represent the coordinates x, z on equal scales, and the remaining coordinate y on an arbitrary scale (which may be that of the other two coordinates x, z, but is in practice usually smaller). The advantage, of course, is that a figure in *one* of the coordinate planes xz is represented in its proper form without distortion; but it may be in some cases preferable to employ the isometrical projection, wherein the three axes are represented by lines inclined to each other at angles of 120°, and the scales for the coordinates are equal (fig. 17).

Fig. 17.

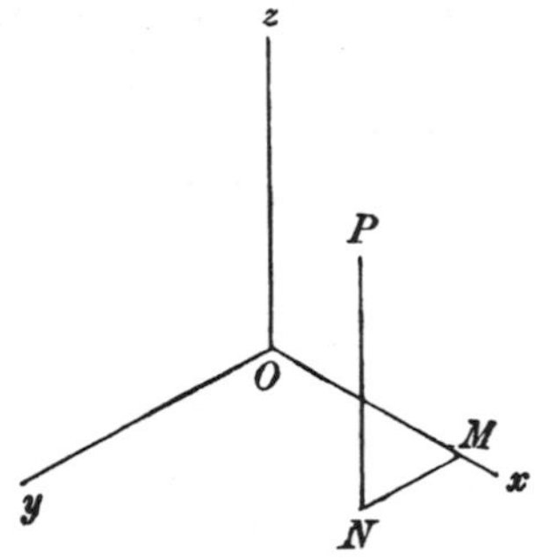

For the delineation of a surface of a tolerably simple form, it is frequently sufficient to draw (according to the foregoing projection) the sections by the coordinate planes; and in particular, when the surface is symmetrical in regard to the

Fig. 18.

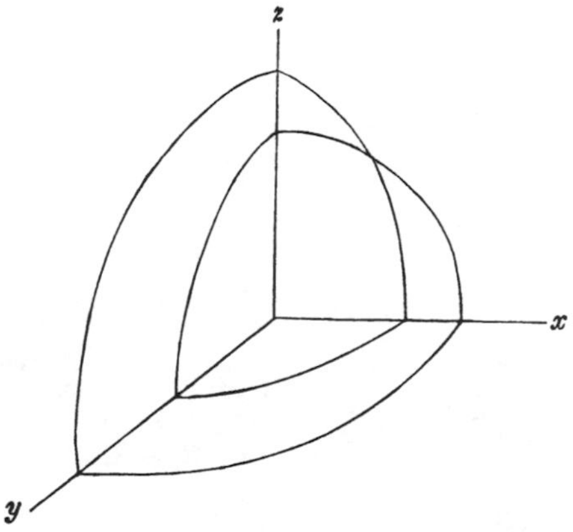

coordinate planes, it is sufficient to draw the quarter-sections belonging to a single octant of the surface; thus fig. 18 is a convenient representation of an octant of the wave surface. Or a surface may be delineated by means of a series of parallel sections, or (taking these to be the sections by a series of horizontal planes) say by a series of contour lines. Of course, other sections may be drawn or indicated, if necessary. For the delineation of a curve, a convenient method is to represent, as above, a series of the points P thereof, each point P being accompanied by the ordinate PN, which serves to refer the point to the plane of xy; this is in effect a representation of each point P of the curve, by means of two points P, N such that the line PN has a fixed direction. Both as regards curves and surfaces, the employment of stereographic representations is very interesting.

28. In plane geometry, reckoning the line as a curve of the first order, we have only the point and the curve. In solid geometry, reckoning a line as a curve of the first order, and the plane as a surface of the first order, we have the point, the curve, and the surface; but the increase of complexity is far greater than would hence at first sight appear. In plane geometry a curve is considered in connexion with lines (its tangents); but in solid geometry the curve is considered in connexion with lines and planes (its tangents and osculating planes), and the surface also in connexion with lines and planes (its tangent lines and tangent planes); there are surfaces arising out of the line—cones, skew surfaces, developables, doubly and triply infinite systems of lines, and whole classes of theories which have nothing analogous to them in plane geometry: it is thus a very small part indeed of the subject which can be even referred to in the present article.

In the case of a surface, we have between the coordinates (x, y, z) a single, or say a onefold relation, which can be represented by a single relation $f(x, y, z) = 0$; or we may consider the coordinates expressed each of them as a given function of two variable parameters p, q; the form $z = f(x, y)$ is a particular case of each of these modes of representation; in other words, we have in the first mode $f(x, y, z) = z - f(x, y)$, and in the second mode $x = p$, $y = q$ for the expression of two of the coordinates in terms of the parameters.

In the case of a curve, we have between the coordinates (x, y, z) a twofold relation: two equations $f(x, y, z) = 0$, $\phi(x, y, z) = 0$ give such a relation; i.e., the curve is here considered as the intersection of two surfaces (but the curve is not always the complete intersection of two surfaces, and there are hence difficulties); or, again, the coordinates may be given each of them as a function of a single variable parameter. The form $y = \phi x$, $z = \psi x$, where two of the coordinates are given in terms of the third, is a particular case of each of these modes of representation.

29. The remarks under plane geometry as to descriptive and metrical propositions, and as to the non-metrical character of the method of coordinates when used for the proof of a descriptive proposition, apply also to solid geometry; and they might be illustrated in like manner by the instance of the theorem of the radical centre of four spheres. The proof is obtained from the consideration that S and S' being each of them a function of the form $x^2 + y^2 + z^2 + ax + by + cz + d$, the difference $S - S'$ is a mere linear function of the coordinates, and consequently that $S - S' = 0$ is the equation of the plane containing the circle of intersection of the two spheres $S = 0$ and $S' = 0$.

Metrical Theory.

30. The foundation in solid geometry of the metrical theory is, in fact, the before-mentioned theorem that, if a finite right line PQ be projected upon any other line OO' by lines perpendicular to OO', then the length of the projection $P'Q'$ is equal to the length of PQ multiplied by the cosine of its inclination to $P'Q'$—or (in the form in which it is now convenient to state the theorem) the perpendicular distance $P'Q'$ of two parallel planes is equal to the inclined distance PQ into the cosine of the inclination. Hence also the algebraical sum of the projections of the sides of a closed polygon upon any line is $=0$; or, reversing the signs of certain sides and considering the polygon as made up of two broken lines each extending from the same initial to the same terminal point, the sum of the projections of the one set of lines upon any line is equal to the sum of the projections of the other set of lines upon the same line. When any of the lines are at right angles to the given line (or, what is the same thing, in a plane at right angles to the given line), the projections of these lines severally vanish.

31. Consider the skew quadrilateral $QMNP$, the sides QM, MN, NP being respectively parallel to the three rectangular axes Ox, Oy, Oz; let the lengths of these sides be ξ, η, ζ, and that of the side QP be $=\rho$; and let the cosines of the inclinations (or say the cosine-inclinations) of ρ to the three axes be α, β, γ; then projecting successively on the three sides and on QP, we have

$$\xi,\ \eta,\ \zeta = \rho\alpha,\ \rho\beta,\ \rho\gamma,$$

and

$$\rho = \alpha\xi + \beta\eta + \gamma\zeta,$$

whence $\rho^2 = \xi^2 + \eta^2 + \zeta^2$, which is the relation between a distance ρ and its projections ξ, η, ζ upon three rectangular axes. And from the same equations we obtain $\alpha^2 + \beta^2 + \gamma^2 = 1$, which is a relation connecting the cosine-inclinations of a line to three rectangular axes.

Suppose we have through Q any other line QT, and let the cosine-inclinations of this to the axes be α', β', γ', and δ be its cosine-inclination to QP; also let p be the length of the projection of QP upon QT; then projecting on QT, we have

$$p = \alpha'\xi + \beta'\eta + \gamma'\zeta, \ = \rho\delta.$$

And in the last equation substituting for ξ, η, ζ their values $\rho\alpha$, $\rho\beta$, $\rho\gamma$, we find

$$\delta = \alpha\alpha' + \beta\beta' + \gamma\gamma',$$

which is an expression for the mutual cosine-inclination of two lines, the cosine-inclinations of which to the axes are α, β, γ and α', β', γ' respectively. We have of course $\alpha^2 + \beta^2 + \gamma^2 = 1$, and $\alpha'^2 + \beta'^2 + \gamma'^2 = 1$, and hence also

$$\begin{aligned} 1 - \delta^2 &= (\alpha^2 + \beta^2 + \gamma^2)(\alpha'^2 + \beta'^2 + \gamma'^2) - (\alpha\alpha' + \beta\beta' + \gamma\gamma')^2 \\ &= (\beta\gamma' - \beta'\gamma)^2 + (\gamma\alpha' - \gamma'\alpha)^2 + (\alpha\beta' - \alpha'\beta)^2; \end{aligned}$$

so that the sine of the inclination can only be expressed as a square root. These formulæ are the foundation of spherical trigonometry.

The Line, Plane, and Sphere.

32. The foregoing formulæ give at once the equations of these loci.

For first, taking Q to be a fixed point, coordinates (a, b, c) and the cosine-inclinations (α, β, γ) to be constant, then P will be a point in the line through Q in the direction thus determined; or, taking (x, y, z) for its coordinates, these will be the current coordinates of a point in the line. The values of ξ, η, ζ then are $x-a$, $y-b$, $z-c$, and we thus have

$$\frac{x-a}{\alpha}=\frac{y-b}{\beta}=\frac{z-c}{\gamma}(=\rho),$$

which (omitting the last equation, $=\rho$) are the equations of the line through the point (a, b, c), the cosine-inclinations to the axes being α, β, γ, and these quantities being connected by the relation $\alpha^2+\beta^2+\gamma^2=1$. This equation may be omitted, and then α, β, γ, instead of being equal, will only be proportional to the cosine-inclinations.

Using the last equation, and writing

$$x,\ y,\ z=a+\alpha\rho,\ \ b+\beta\rho,\ \ c+\gamma\rho,$$

these are expressions for the current coordinates in terms of a parameter ρ, which is in fact the distance from the fixed point (a, b, c).

It is easy to see that, if the coordinates (x, y, z) are connected by any two linear equations, these equations can always be brought into the foregoing form, and hence that the two linear equations represent a line.

Secondly, taking for greater simplicity the point Q to be coincident with the origin, and $\alpha', \beta', \gamma', p$ to be constant, then p is the perpendicular distance of a plane from the origin, and α', β', γ' are the cosine-inclinations of this distance to the axes $(\alpha'^2+\beta'^2+\gamma'^2=1)$. Now P is any point in this plane; taking its coordinates to be (x, y, z), then (ξ, η, ζ) are $=(x, y, z)$, and the foregoing equation $p=\alpha'\xi+\beta'\eta+\gamma'\zeta$ becomes

$$\alpha'x+\beta'y+\gamma'z=p,$$

which is the equation of the plane in question.

If, more generally, Q is not coincident with the origin, then, taking its coordinates to be (a, b, c), and writing p_1 instead of p, the equation is

$$\alpha'(x-a)+\beta'(y-b)+\gamma'(z-c)=p_1;$$

and we thence have $p_1=p-(a\alpha'+b\beta'+c\gamma')$, which is an expression for the perpendicular distance of the point (a, b, c) from the plane in question.

It is obvious that any linear equation $Ax+By+Cz+D=0$ between the coordinates can always be brought into the foregoing form, and hence that such equation represents a plane.

Thirdly, supposing Q to be a fixed point, coordinates (a, b, c) and the distance $QP, =\rho$, to be constant, say this is $=d$, then, as before, the values of ξ, η, ζ are $x-a, y-b, z-c$, and the equation $\xi^2+\eta^2+\zeta^2=\rho^2$ becomes

$$(x-a)^2+(y-b)^2+(z-c)^2=d^2,$$

which is the equation of the sphere, coordinates of the centre $=(a, b, c)$ and radius $=d$.

A quadric equation wherein the terms of the second order are $x^2+y^2+z^2$, viz. an equation

$$x^2+y^2+z^2+Ax+By+Cz+D=0,$$

can always, it is clear, be brought into the foregoing form; and it thus appears that this is the equation of a sphere, coordinates of the centre $-\tfrac{1}{2}A, -\tfrac{1}{2}B, -\tfrac{1}{2}C$, and squared radius $=\tfrac{1}{4}(A^2+B^2+C^2)-D$.

Cylinders, Cones, Ruled Surfaces.

33. A singly infinite system of lines, or a system of lines depending upon one variable parameter, forms a surface; and the equation of the surface is obtained by eliminating the parameter between the two equations of the line.

If the lines all pass through a given point, then the surface is a cone; and, in particular, if the lines are all parallel to a given line, then the surface is a cylinder.

Beginning with this last case, suppose the lines are parallel to the line $x=mz$, $y=nz$, the equations of a line of the system are $x=mz+a$, $y=nz+b$,—where a, b are supposed to be functions of the variable parameter, or, what is the same thing, there is between them a relation $f(a, b)=0$: we have $a=x-mz$, $b=y-nz$, and the result of the elimination of the parameter therefore is $f(x-mz, y-nz)=0$, which is thus the general equation of the cylinder the generating lines whereof are parallel to the line $x=mz$, $y=nz$. The equation of the section by the plane $z=0$ is $f(x, y)=0$, and conversely if the cylinder be determined by means of its curve of intersection with the plane $z=0$, then, taking the equation of this curve to be $f(x, y)=0$, the equation of the cylinder is $f(x-mz, y-nz)=0$. Thus, if the curve of intersection be the circle $(x-\alpha)^2+(y-\beta)^2=\gamma^2$, we have $(x-mz-\alpha)^2+(y-nz-\beta)^2=\gamma^2$ as the equation of an oblique cylinder on this base, and thus also $(x-\alpha)^2+(y-\beta)^2=\gamma^2$ as the equation of the right cylinder.

If the lines all pass through a given point (a, b, c), then the equations of a line are $x-a=\alpha(z-c)$, $y-b=\beta(z-c)$, where α, β are functions of the variable parameter, or, what is the same thing, there exists between them an equation $f(\alpha, \beta)=0$; the elimination of the parameter gives, therefore, $f\left(\dfrac{x-a}{z-c}, \dfrac{y-b}{z-c}\right)=0$; and this equation, or, what is the same thing, any homogeneous equation $f(x-a, y-b, z-c)=0$, or, taking f to be a rational and integral function of the order n, say $(*)(x-a, y-b, z-c)^n=0$, is the general equation of the cone having the point (a, b, c) for its vertex. Taking the vertex to be at the origin, the equation is $(*)(x, y, z)^n=0$; and, in particular, $(*)(x, y, z)^2=0$ is the equation of a cone of the second order, or quadricone, having the origin for its vertex.

34. In the general case of a singly infinite system of lines, the locus is a ruled surface (or *regulus*). If the system be such that a line does not intersect the consecutive line, then the surface is a skew surface, or scroll; but if it be such that each line intersects the consecutive line, then it is a developable, or torse.

Suppose, for instance, that the equations of a line (depending on the variable parameter θ) are

$$\frac{x}{a}+\frac{z}{c}=\theta\left(1+\frac{y}{b}\right),\quad \frac{x}{a}-\frac{z}{c}=\frac{1}{\theta}\left(1-\frac{y}{b}\right),$$

then, eliminating θ, we have $\frac{x^2}{a^2}-\frac{z^2}{c^2}=1-\frac{y^2}{b^2}$, or say $\frac{x^2}{a^2}+\frac{y^2}{b^2}-\frac{z^2}{c^2}=1$, the equation of a quadric surface, afterwards called the hyperboloid of one sheet; this surface is consequently a scroll. It is to be remarked that we have upon the surface a second singly infinite series of lines; the equations of a line of this second system (depending on the variable parameter ϕ) are

$$\frac{x}{a}+\frac{z}{c}=\phi\left(1-\frac{y}{b}\right),\quad \frac{x}{a}-\frac{z}{c}=\frac{1}{\phi}\left(1+\frac{y}{b}\right).$$

It is easily shown that any line of the one system intersects every line of the other system.

Considering any curve (of double curvature) whatever, the tangent lines of the curve form a singly infinite system of lines, each line intersecting the consecutive line of the system,—that is, they form a developable, or torse; the curve and torse are thus inseparably connected together, forming a single geometrical figure. A plane through three consecutive points of the curve (or osculating plane of the curve) contains two consecutive tangents, that is, two consecutive lines of the torse, and is thus a tangent plane of the torse along a generating line.

Transformation of Coordinates.

35. There is no difficulty in changing the origin, and it is for brevity assumed that the origin remains unaltered. We have, then, two sets of rectangular axes, Ox, Oy, Oz, and Ox_1, Oy_1, Oz_1, the mutual cosine-inclinations being shown by the diagram—

	x	y	z
x_1	α	β	γ
y_1	α'	β'	γ'
z_1	α''	β''	γ''

that is, α, β, γ are the cosine-inclinations of Ox_1 to Ox, Oy, Oz; α', β', γ' those of Oy_1, &c.

And this diagram gives also the linear expressions of the coordinates (x_1, y_1, z_1) or (x, y, z) of either set in terms of those of the other set; we thus have

$$\begin{aligned} x_1 &= \alpha\, x + \beta\, y + \gamma\, z, & x &= \alpha x_1 + \alpha' y_1 + \alpha'' z_1, \\ y_1 &= \alpha' x + \beta' y + \gamma' z, & y &= \beta x_1 + \beta' y_1 + \beta'' z_1, \\ z_1 &= \alpha'' x + \beta'' y + \gamma'' z, & z &= \gamma x_1 + \gamma' y_1 + \gamma'' z_1, \end{aligned}$$

which are obtained by projection, as above explained. Each of these equations is, in fact, nothing else than the before-mentioned equation $p = \alpha'\xi + \beta'\eta + \gamma'\zeta$, adapted to the problem in hand.

But we have to consider the relations between the nine coefficients. By what precedes, or by the consideration that we must have identically $x^2 + y^2 + z^2 = x_1^2 + y_1^2 + z_1^2$, it appears that these satisfy the relations—

$$\begin{aligned} \alpha^2 + \beta^2 + \gamma^2 &= 1, & \alpha^2 + \alpha'^2 + \alpha''^2 &= 1, \\ \alpha'^2 + \beta'^2 + \gamma'^2 &= 1, & \beta^2 + \beta'^2 + \beta''^2 &= 1, \\ \alpha''^2 + \beta''^2 + \gamma''^2 &= 1, & \gamma^2 + \gamma'^2 + \gamma''^2 &= 1, \\ \alpha'\alpha'' + \beta'\beta'' + \gamma'\gamma'' &= 0, & \beta\gamma + \beta'\gamma' + \beta''\gamma'' &= 0, \\ \alpha''\alpha + \beta''\beta + \gamma''\gamma &= 0, & \gamma\alpha + \gamma'\alpha' + \gamma''\alpha'' &= 0, \\ \alpha\alpha' + \beta\beta' + \gamma\gamma' &= 0, & \alpha\beta + \alpha'\beta' + \alpha''\beta'' &= 0, \end{aligned}$$

either set of six equations being implied in the other set.

It follows that the square of the determinant

$$\begin{vmatrix} \alpha, & \beta, & \gamma \\ \alpha', & \beta', & \gamma' \\ \alpha'', & \beta'', & \gamma'' \end{vmatrix}$$

is $=1$; and hence that the determinant itself is $=\pm 1$. The distinction of the two cases is an important one: if the determinant is $=+1$, then the axes Ox_1, Oy_1, Oz_1 are such that they can by a rotation about O be brought to coincide with Ox, Oy, Oz respectively; if it is $=-1$, then they cannot. But in the latter case, by measuring x_1, y_1, z_1 in the opposite directions we change the signs of all the coefficients and so make the determinant to be $=+1$; hence this case need alone be considered, and it is accordingly assumed that the determinant is $=+1$. This being so, it is found that we have a further set of nine equations, $\alpha = \beta'\gamma'' - \beta''\gamma'$, &c.; that is, the coefficients arranged as in the diagram have the values

$\beta'\gamma'' - \beta''\gamma'$	$\gamma'\alpha'' - \gamma''\alpha'$	$\alpha'\beta'' - \alpha''\beta'$
$\beta''\gamma - \beta\gamma''$	$\gamma''\alpha - \gamma\alpha''$	$\alpha''\beta - \alpha\beta''$
$\beta\gamma' - \beta'\gamma$	$\gamma\alpha' - \gamma'\alpha$	$\alpha\beta' - \alpha'\beta$

36. It is important to express the nine coefficients in terms of three independent quantities. A solution which, although unsymmetrical, is very convenient in Astronomy and Dynamics is to use for the purpose the three angles θ, ϕ, τ of fig. 19; say θ = longitude of the node; ϕ = inclination; and τ = longitude of x_1 from node.

Fig. 19.

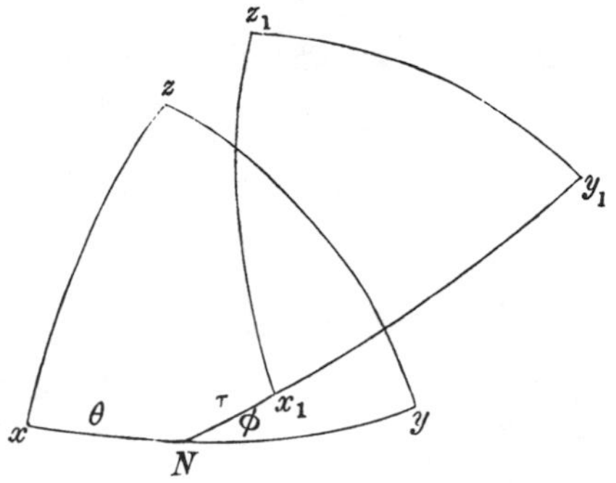

The diagram of transformation then is

	x	y	z
x_1	$\cos\tau\cos\theta - \sin\tau\sin\theta\cos\phi$	$\cos\tau\sin\theta + \sin\tau\cos\theta\cos\phi$	$\sin\tau\sin\phi$
y_1	$-\sin\tau\cos\theta - \cos\tau\sin\theta\cos\phi$	$-\sin\tau\sin\theta + \cos\tau\cos\theta\cos\phi$	$\cos\tau\sin\phi$
z_1	$\sin\theta\sin\phi$	$-\cos\theta\sin\phi$	$\cos\phi$

But a more elegant solution (due to Rodrigues) is that contained in the diagram

	x	y	z
x_1	$1+\lambda^2-\mu^2-\nu^2$	$2(\lambda\mu-\nu)$	$2(\lambda\nu+\mu)$
y_1	$2(\lambda\mu+\nu)$	$1-\lambda^2+\mu^2-\nu^2$	$2(\mu\nu-\lambda)$
z_1	$2(\nu\lambda-\mu)$	$2(\mu\nu+\lambda)$	$1-\lambda^2-\mu^2+\nu^2$

$$\div(1+\lambda^2+\mu^2+\nu^2).$$

The nine coefficients of transformation are the nine functions of the diagram, each divided by $1+\lambda^2+\mu^2+\nu^2$; the expressions contain as they should do the three arbitrary quantities λ, μ, ν; and the identity $x_1^2+y_1^2+z_1^2=x^2+y^2+z^2$ can be at once verified. It may be added that the transformation can be expressed in the quaternion form

$$ix_1+jy_1+kz_1=(1+\Lambda)(ix+jy+kz)(1+\Lambda)^{-1},$$

where Λ denotes the vector $i\lambda+j\mu+k\nu$.

Quadric Surfaces (Paraboloids, Ellipsoid, Hyperboloids).

37. It appears, by a discussion of the general equation of the second order $(a,\ldots\between x, y, z, 1)^2=0$, that the proper quadric surfaces * represented by such an equation are the following five surfaces (a and b positive):—

$$(1)\quad z=\frac{x^2}{2a}+\frac{y^2}{2b},\ \text{elliptic paraboloid.}$$

$$(2)\quad z=\frac{x^2}{2a}-\frac{y^2}{2b},\ \text{hyperbolic paraboloid.}$$

$$(3)\quad \frac{x^2}{a^2}+\frac{y^2}{b^2}+\frac{z^2}{c^2}=\ \ 1,\ \text{ellipsoid.}$$

$$(4)\quad \frac{x^2}{a^2}+\frac{y^2}{b^2}-\frac{z^2}{c^2}=\ \ 1,\ \text{hyperboloid of one sheet.}$$

$$(5)\quad \frac{x^2}{a^2}+\frac{y^2}{b^2}-\frac{z^2}{c^2}=-1,\ \text{hyperboloid of two sheets.}$$

It is at once seen that these are distinct surfaces; and the equations also show very readily the general form and mode of generation of the several surfaces.

Fig. 20.

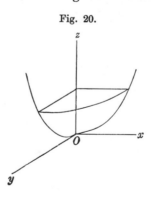

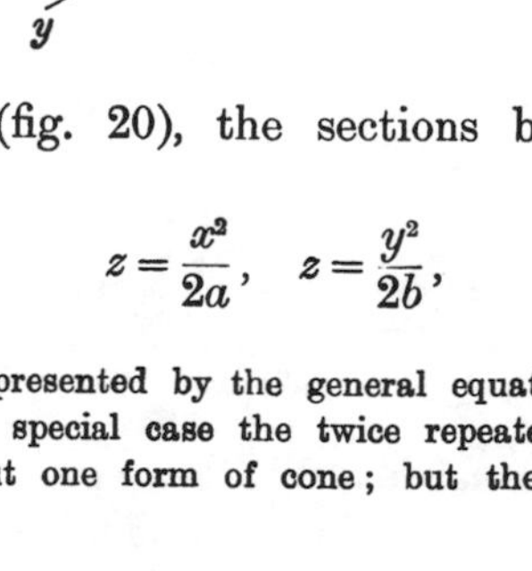

In the elliptic paraboloid (fig. 20), the sections by the planes of zx and zy are the parabolas

$$z=\frac{x^2}{2a},\quad z=\frac{y^2}{2b},$$

* The improper quadric surfaces represented by the general equation of the second order are (1) the pair of planes or plane-pair, including as a special case the twice repeated plane, and (2) the cone, including as a special case the cylinder. There is but one form of cone; but the cylinder may be parabolic, elliptic, or hyperbolic.

having the common axis Oz; and the section by any plane $z=\gamma$ parallel to that of xy is the ellipse

$$\gamma = \frac{x^2}{2a} + \frac{y^2}{2b};$$

so that the surface is generated by a variable ellipse moving parallel to itself along the parabolas as directrices.

In the hyperbolic paraboloid (fig. 21), the sections by the planes of zx, zy are the parabolas

$$z = \frac{x^2}{2a}, \quad z = -\frac{y^2}{2b},$$

Fig. 21.

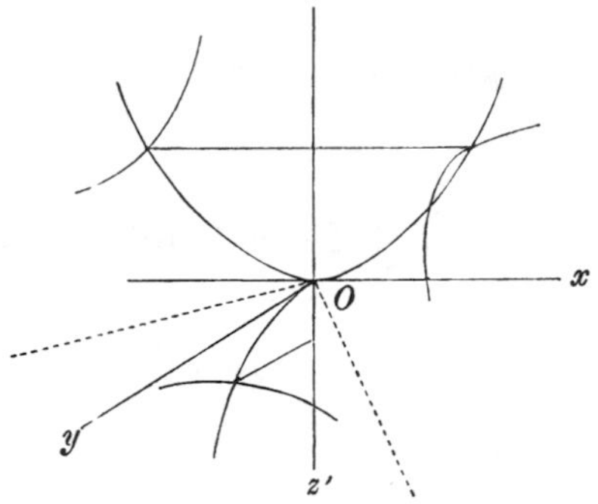

having the opposite axes Oz, Oz'; and the section by a plane $z=\gamma$ parallel to that of xy is the hyperbola

$$\gamma = \frac{x^2}{2a} - \frac{y^2}{2b},$$

which has its transverse axis parallel to Ox or Oy according as γ is positive or negative. The surface is thus generated by a variable hyperbola moving parallel to

Fig. 22.

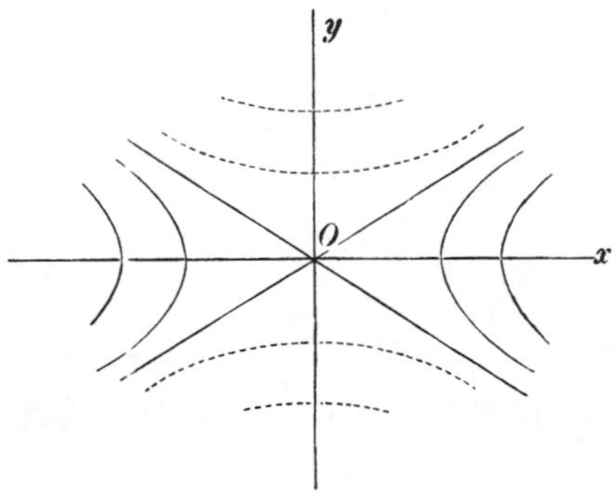

itself along the parabolas as directrices. The form is best seen from fig. 22, which represents the sections by planes parallel to the plane of xy, or say the contour lines;

the continuous lines are the sections above the plane of xy, and the dotted lines the sections below this plane. The form is, in fact, that of a saddle.

In the ellipsoid (fig. 23), the sections by the planes of zx, zy, and xy are each of them an ellipse, and the section by any parallel plane is also an ellipse. The

Fig. 23.

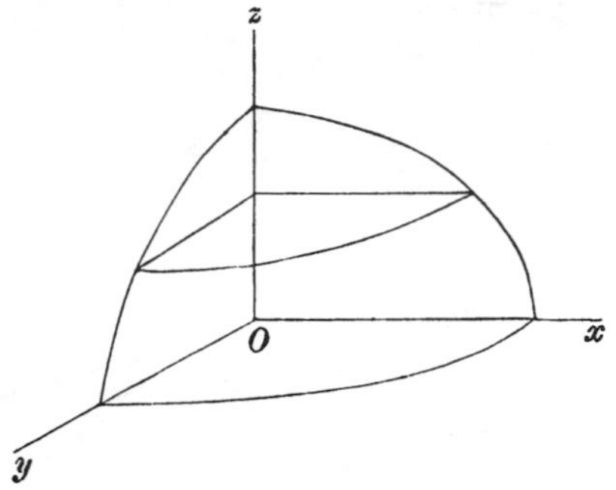

surface may be considered as generated by an ellipse moving parallel to itself along two ellipses as directrices.

In the hyperboloid of one sheet (fig. 24), the sections by the planes of zx, zy are the hyperbolas

$$\frac{x^2}{a^2} - \frac{z^2}{c^2} = 1, \quad \frac{y^2}{b^2} - \frac{z^2}{c^2} = 1,$$

having a common conjugate axis zOz'; the section by the plane of xy, and that by

Fig. 24.

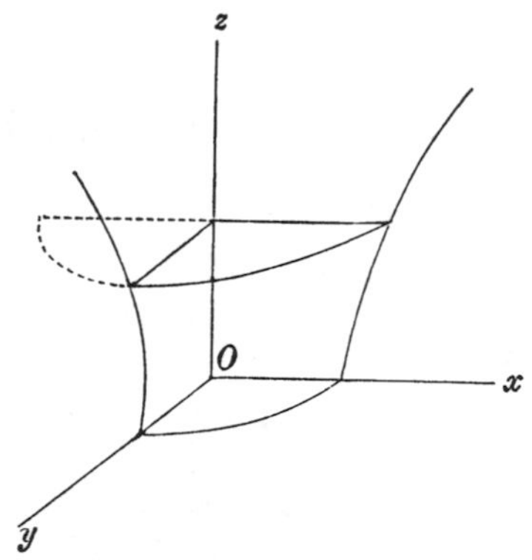

any parallel plane, is an ellipse; and the surface may be considered as generated by a variable ellipse moving parallel to itself along the two hyperbolas as directrices.

In the hyperboloid of two sheets (fig. 25), the sections by the planes of zx and zy are the hyperbolas

$$\frac{z^2}{c^2} - \frac{x^2}{a^2} = 1, \quad \frac{z^2}{c^2} - \frac{y^2}{b^2} = 1,$$

having the common transverse axis zOz'; the section by any plane $z = \pm\gamma$ parallel to that of xy, γ being in absolute magnitude $> c$, is the ellipse

$$\frac{x^2}{a^2} + \frac{y^2}{b^2} = \frac{\gamma^2}{c^2} - 1;$$

and the surface, consisting of two distinct portions or sheets, may be considered as

Fig. 25.

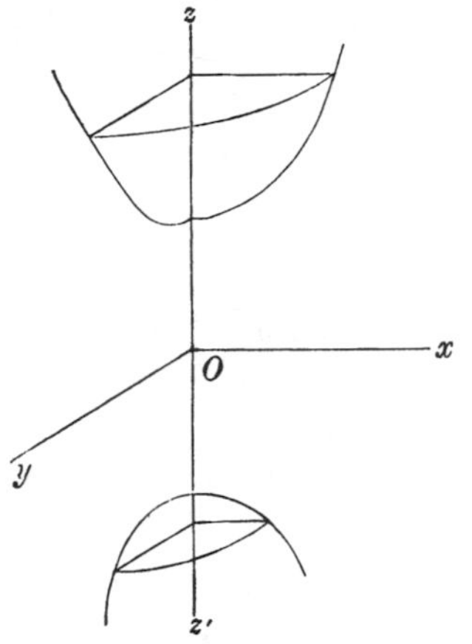

generated by a variable ellipse moving parallel to itself along the hyperbolas as directrices.

The hyperbolic paraboloid is such (and it is easy from the figure to understand how this may be the case) that there exist upon it two singly infinite series of right lines. The same is the case with the hyperboloid of one sheet (ruled or skew hyperboloid, as with reference to this property it is termed). If we imagine two equal and parallel circular disks, their points connected by strings of equal length, so that these are the generating lines of a right circular cylinder, then by turning one of the disks about its centre through the same angle in one or the other direction, the strings will in each case generate one and the same hyperboloid, and will in regard to it be the two systems of lines on the surface, or say the two systems of generating lines; and the general configuration is the same when instead of circles we have ellipses. It has been already shown analytically that the equation $\frac{x^2}{a^2} + \frac{y^2}{b^2} - \frac{z^2}{c^2} = 1$ is satisfied by each of two pairs of linear relations between the coordinates.

Curves; Tangent, Osculating Plane, Curvature, &c.

38. It will be convenient to consider the coordinates (x, y, z) of the point on the curve as given in terms of a parameter θ, so that dx, dy, dz, d^2x, &c., will be proportional to $\frac{dx}{d\theta}, \frac{dy}{d\theta}, \frac{dz}{d\theta}, \frac{d^2x}{d\theta^2}$, &c. But only a part of the analytical formulæ will be given; in them ξ, η, ζ are used as current coordinates.

The tangent is the line through the point (x, y, z) and the consecutive point $(x+dx, y+dy, z+dz)$; its equations therefore are

$$\frac{\xi - x}{dx} = \frac{\eta - y}{dy} = \frac{\zeta - z}{dz}.$$

The osculating plane is the plane through the point and two consecutive points, and contains therefore the tangent; its equation is

$$\begin{vmatrix} \xi - x, & \eta - y, & \zeta - z \\ dx\ , & dy\ , & dz \\ d^2x\ , & d^2y\ , & d^2z \end{vmatrix} = 0,$$

or, what is the same thing,

$$(\xi - x)(dyd^2z - dzd^2y) + (\eta - y)(dzd^2x - dxd^2z) + (\zeta - z)(dxd^2y - dyd^2x) = 0.$$

The normal plane is the plane through the point at right angles to the tangent. It meets the osculating plane in a line called the principal normal; and drawing through the point a line at right angles to the osculating plane, this is called the binormal. We have thus at the point a set of three rectangular axes—the tangent, the principal normal, and the binormal.

We have through the point and three consecutive points a sphere of spherical curvature,—the centre and radius thereof being the centre, and radius, of spherical curvature. The sphere is met by the osculating plane in the circle of absolute curvature,—the centre and radius thereof being the centre, and radius, of absolute curvature. The centre of absolute curvature is also the intersection of the principal normal by the normal plane at the consecutive point.

Surfaces; Tangent Lines and Plane, Curvature, &c.

39. It will be convenient to consider the surface as given by an equation $f(x, y, z) = 0$ between the coordinates; taking (x, y, z) for the coordinates of a given point, and $(x+dx, y+dy, z+dz)$ for those of a consecutive point, the increments dx, dy, dz satisfy the condition

$$\frac{df}{dx}dx + \frac{df}{dy}dy + \frac{df}{dz}dz = 0,$$

but the ratio of two of the increments, suppose $dx : dy$, may be regarded as arbitrary. Only a part of the analytical formulæ will be given; in them ξ, η, ζ are used as current coordinates.

We have through the point a singly infinite series of right lines, each meeting the surface in a consecutive point, or say having each of them two-point intersection with the surface. These lines lie all of them in a plane which is the tangent plane; its equation is

$$\frac{df}{dx}(\xi - x) + \frac{df}{dy}(\eta - y) + \frac{df}{dz}(\zeta - z) = 0,$$

as is at once verified by observing that this equation is satisfied (irrespectively of the value of $dx : dy$) on writing therein $\xi, \eta, \zeta = x + dx, y + dy, z + dz$.

The line through the point at right angles to the tangent plane is called the normal; its equations are

$$\frac{\xi - x}{\frac{df}{dx}} = \frac{\eta - y}{\frac{df}{dy}} = \frac{\zeta - z}{\frac{df}{dz}}.$$

In the series of tangent lines there are in general two (real or imaginary) lines, each of which meets the surface in a second consecutive point, or say it has three-point intersection with the surface; these are called the chief-tangents (Haupt-tangenten). The tangent-plane cuts the surface in a curve, having at the point of contact a node (double point), the tangents to the two branches being the chief-tangents.

In the case of a quadric surface the curve of intersection, *qua* curve of the second order, can only have a node by breaking up into a pair of lines; that is, every tangent-plane meets the surface in a pair of lines, or we have on the surface two singly infinite systems of lines; these are real for the hyperbolic paraboloid and the hyperboloid of one sheet, imaginary in other cases.

At each point of a surface the chief-tangents determine two directions; and passing along one of them to a consecutive point, and thence (without abrupt change of direction) along the new chief-tangent to a consecutive point, and so on, we have on the surface a chief-tangent curve; and there are, it is clear, two singly infinite series of such curves. In the case of a quadric surface, the curves are the right lines on the surface.

40. If at the point we draw in the tangent-plane two lines bisecting the angles between the chief-tangents, these lines (which are at right angles to each other) are called the principal tangents*. We have thus at each point of the surface a set of rectangular axes, the normal and the two principal tangents.

Proceeding from the point along a principal tangent to a consecutive point on the surface, and thence (without abrupt change of direction) along the new principal tangent to a consecutive point, and so on, we have on the surface a curve of curvature; there are, it is clear, two singly infinite series of such curves, cutting each other at right angles at each point of the surface.

Passing from the given point in an arbitrary direction to a consecutive point on the surface, the normal at the given point is not intersected by the normal at the consecutive point; but passing to the consecutive point along a curve of curvature (or, what is the same thing, along a principal tangent) the normal at the given point is intersected by the normal at the consecutive point; we have thus on the normal two centres of curvature, and the distances of these from the point on the surface are the two principal radii of curvature of the surface at that point; these are also the radii of curvature of the sections of the surface by planes through the normal and the two principal tangents respectively; or say they are the radii of curvature of the

* The point on the surface may be such that the directions of the principal tangents become arbitrary; the point is then an umbilicus. It is in the text assumed that the point on the surface is not an umbilicus.

normal sections through the two principal tangents respectively. Take at the point the axis of z in the direction of the normal, and those of x and y in the directions of the principal tangents respectively, then, if the radii of curvature be a, b (the signs being such that the coordinates of the two centres of curvature are $z=a$ and $z=b$ respectively), the surface has in the neighbourhood of the point the form of the paraboloid

$$z=\frac{x^2}{2a}+\frac{y^2}{2b},$$

and the chief-tangents are determined by the equation $0=\frac{x^2}{2a}+\frac{y^2}{2b}$. The two centres of curvature may be on the same side of the point or on opposite sides; in the former case a and b have the same sign, the paraboloid is elliptic, and the chief-tangents are imaginary; in the latter case a and b have opposite signs, the paraboloid is hyperbolic, and the chief-tangents are real.

The normal sections of the surface and the paraboloid by the same plane have the same radius of curvature; and it thence readily follows that the radius of curvature of a normal section of the surface by a plane inclined at an angle θ to that of zx is given by the equation

$$\frac{1}{\rho}=\frac{\cos^2\theta}{a}+\frac{\sin^2\theta}{b}.$$

The section in question is that by a plane through the normal and a line in the tangent plane inclined at an angle θ to the principal tangent along the axis of x. To complete the theory, consider the section by a plane having the same trace upon the tangent plane, but inclined to the normal at an angle ϕ; then it is shown without difficulty (Meunier's theorem) that the radius of curvature of this inclined section of the surface is $=\rho\cos\phi$.

791.

LANDEN.

[From the *Encyclopædia Britannica, Ninth Edition,* vol. XIV. (1882), p. 271.]

LANDEN, JOHN, a distinguished mathematician of the 18th century, was born at Peakirk near Peterborough in Northamptonshire in 1719, and died 15th January 1790 at Milton in the same county. Most of his time was spent in the pursuits of active life, but he early showed a strong talent for mathematical study, which he eagerly cultivated in his leisure hours. In 1762 he was appointed agent to the Earl Fitzwilliam, and held that office to within two years of his death. He lived a very retired life, and saw little or nothing of society; when he did mingle in it, his dogmatism and pugnacity caused him to be generally shunned. He was first known as a mathematician by his essays in the *Ladies' Diary* for 1744. In 1766 he was elected a Fellow of the Royal Society. He was well acquainted and *au courant* with the works of the mathematicians of his own time, and has been called the English D'Alembert. In his *Discourse* on the "Residual Analysis," in which he proposes to substitute for the method of fluxions a purely algebraical method, he says, "It is by means of the following theorem, viz.

$$\frac{x^{\frac{m}{n}} - v^{\frac{m}{n}}}{x - v} = x^{\frac{m}{n}-1} \times 1 + \frac{v}{x} + \left(\frac{v}{x}\right)^2 + \dots (m \text{ terms})$$

$$\div 1 + \left(\frac{v}{x}\right)^{\frac{m}{n}} + \left(\frac{v}{x}\right)^{\frac{2m}{n}} + \dots (n \text{ terms})$$

(where m and n are integers), that we are able to perform all the principal operations in our said analysis; and I am not a little surprised that a theorem so obvious, and of such vast use, should so long escape the notice of algebraists." The idea is of course a perfectly legitimate one, and may be compared with that of Lagrange's *Calcul des Fonctions.* His memoir (1775) on the rotatory motion of a body contains (as the author was aware) conclusions at variance with those arrived at by D'Alembert and

Euler in their researches on the same subject. He reproduces and further develops and defends his own views in his *Mathematical Memoirs,* and in his paper in the *Philosophical Transactions* for 1785. But Landen's capital discovery is that of the theorem known by his name (obtained in its complete form in the memoir of 1775, and reproduced in the first volume of the *Mathematical Memoirs*) for the expression of the arc of an hyperbola in terms of two elliptic arcs. To find this, he integrates a differential equation derived from the equation

$$t = gx\sqrt{\frac{m^2 - x^2}{m^2 - gx^2}},$$

interpreting geometrically in an ingenious and elegant manner three integrals which present themselves. If in the foregoing equation we write $m = 1$, $g = k^2$, and instead of t consider the new variable $y = t \div (1 - k')$, then

$$y = (1 + k')\, x\sqrt{\frac{1 - x^2}{1 - k^2x^2}},$$

which is the form known as Landen's transformation in the theory of elliptic functions; but his investigation does not lead him to obtain the equivalent of the resulting differential equation

$$\frac{dy}{\sqrt{1 - y^2 \,.\, 1 - \lambda^2 y^2}} = \frac{(1 + k')\, dx}{\sqrt{1 - x^2 \,.\, 1 - k^2x^2}}, \text{ where } \lambda = \frac{1 - k'}{1 + k'},$$

due it would appear to Legendre, and which (over and above Landen's own beautiful result) gives importance to the theorem as leading directly to the quadric transformation of an elliptic integral in regard to the modulus.

The list of his writings is as follows:—*Ladies' Diary,* various communications, 1744—1760; papers in the *Phil. Trans.,* 1754, 1760, 1768, 1771, 1775, 1777, 1785; *Mathematical Lucubrations,* 1755; *A Discourse concerning the Residual Analysis,* 1758; *The Residual Analysis,* book I., 1764; *Animadversions on Dr Stewart's Method of computing the Sun's Distance from the Earth,* 1771; *Mathematical Memoirs,* 1780, 1789.

792.

LOCUS.

[From the *Encyclopædia Britannica, Ninth Edition,* vol. XIV. (1882), pp. 764, 765.]

LOCUS, in Greek τόπος, a geometrical term, the invention of the notion of which is attributed to Plato. It occurs in such statements as these:—the locus of the points which are at the same distance from a fixed point, or of a point which moves so as to be always at the same distance from a fixed point, is a circle; conversely a circle is the locus of the points at the same distance from a fixed point, or of a point moving so as to be always at the same distance from a fixed point; and so, in general, a curve of any given kind is the locus of the points which satisfy, or of a point moving so as always to satisfy, a given condition. The theory of loci is thus identical with that of curves; and it is in fact in this very point of view that a curve is considered in the article Curve, [785]; see that article, and also Geometry (Analytical), [790]. It is only necessary to add that the notion of a locus is useful as regards determinate problems or theorems: thus, to find the centre of the circle circumscribed about a given triangle ABC, we see that the circumscribed circle must pass through the two vertices A, B, and the locus of the centres of the circles which pass through these two points is the straight line at right angles to the side AB at its mid-point; similarly the circumscribed circle must pass through A, C, and the locus of the centres of the circles through these two points is the line at right angles to the side AC at its mid-point; thus we get the ordinary construction, and also the theorem that the lines at right angles to the sides, at their mid-points respectively, meet in a point. The notion of a locus applies, of course, not only to plane but also to solid geometry. Here the locus of the points satisfying a single (or onefold) condition is a surface; the locus of the points satisfying two conditions (or a twofold condition) is a curve in space, which is in general a twisted curve or curve of double curvature.

793.

MONGE.

[From the *Encyclopædia Britannica, Ninth Edition,* vol. XVI. (1883), pp. 738, 739.]

MONGE, GASPARD (1746—1818), a French mathematician, the inventor of descriptive geometry, was born at Beaune on the 10th May 1746. He was educated first at the college of the Oratorians at Beaune, and then in their college at Lyons,—where, at sixteen, the year after he had been learning physics, he was made a teacher of it. Returning to Beaune for a vacation, he made, on a large scale, a plan of the town, inventing the methods of observation and constructing the necessary instruments; the plan was presented to the town, and preserved in their library. An officer of engineers seeing it wrote to recommend Monge to the commandant of the military school at Mézières, and he was received as draftsman and pupil in the practical school attached to that institution; the school itself was of too aristocratic a character to allow of his admission to it. His manual skill was duly appreciated: "I was a thousand times tempted," he said long afterwards, "to tear up my drawings in disgust at the esteem in which they were held, as if I had been good for nothing better." An opportunity, however, presented itself: being required to work out from data supplied to him the "défilement" of a proposed fortress (an operation then only performed by a long arithmetical process), Monge, substituting for this a geometrical method, obtained the result so quickly that the commandant at first refused to receive it—the time necessary for the work had not been taken; but upon examination the value of the discovery was recognized, and the method was adopted. And Monge, continuing his researches, arrived at that general method of the application of geometry to the arts of construction which is now called descriptive geometry. But such was the system in France before the Revolution that the officers instructed in the method were strictly forbidden to communicate it even to those engaged in other branches of the public service; and it was not until many years afterwards that an account of it was published. The method consists, as is well known, in the use of the two halves of a sheet of paper to represent say the planes of xy and xz at right angles to each other, and the consequent representation of points, lines, and figures in space by means of their plan and elevation, placed in a determinate relative position.

In 1768 Monge became professor of mathematics, and in 1771 professor of physics, at Mézières; in 1778 he married Madame Horbon, a young widow whom he had previously defended in a very spirited manner from an unfounded charge; in 1780 he was appointed to a chair of hydraulics at the Lyceum in Paris (held by him together with his appointments at Mézières), and was received as a member of the Academy; his intimate friendship with Berthollet began at this time. In 1783, quitting Mézières, he was, on the death of Bezout, appointed examiner of naval candidates. Although pressed by the minister to prepare for them a complete course of mathematics, he declined to do so, on the ground that it would deprive Madame Bezout of her only income, arising from the sale of the works of her late husband; he wrote, however (1786), his *Traité élémentaire de la Statique.*

Monge contributed (1770—1790) to the *Memoirs* of the Academy of Turin, the *Mémoires des Savants Étrangers* of the Academy of Paris, the *Mémoires* of the same Academy, and the *Annales de Chimie,* various mathematical and physical papers. Among these may be noticed the memoir "Sur la théorie des déblais et des remblais" (*Mém. de l'Acad. de Paris,* 1781), which, while giving a remarkably elegant investigation in regard to the problem of earthwork referred to in the title, establishes in connexion with it his capital discovery of the curves of curvature of a surface. Euler, in his paper on curvature in the Berlin *Memoirs* for 1760, had considered, not the normals of the surface, but the normals of the plane sections through a particular normal, so that the question of the intersection of successive normals of the surface had never presented itself to him. Monge's memoir just referred to gives the ordinary differential equation of the curves of curvature, and establishes the general theory in a very satisfactory manner; but the application to the interesting particular case of the ellipsoid was first made by him in a later paper in 1795. A memoir in the volume for 1783 relates to the production of water by the combustion of hydrogen; but Monge's results in this matter had been anticipated by Watts and Cavendish.

In 1792, on the creation by the Legislative Assembly of an executive council, Monge accepted the office of minister of the marine, but retained it only until April 1793. When the Committee of Public Safety made an appeal to the savants to assist in producing the *matériel* required for the defence of the republic, he applied himself wholly to these operations, and distinguished himself by his indefatigable activity therein; he wrote at this time his *Description de l'art de fabriquer les canons,* and his *Avis aux ouvriers en fer sur la fabrication de l'acier.* He took a very active part in the measures for the establishment of the Normal School (which existed only during the first four months of the year 1795), and of the School for Public Works, afterwards the Polytechnic School, and was at each of them professor of descriptive geometry; his methods in that science were first published in the form in which the shorthand writers took down his lessons given at the Normal School in 1795, and again in 1798—99. In 1796 Monge was sent into Italy with Berthollet and some artists to receive the pictures and statues levied from several Italian towns, and made there the acquaintance of General Bonaparte. Two years afterwards he was sent to Rome on a political mission, which terminated in the establishment, under Massena, of the shortlived Roman republic; and he thence joined the expedition to Egypt, taking part with his friend Berthollet as well in various operations of the war as in the scientific labours of the Egyptian Institute of Sciences and Arts; they accompanied Bonaparte to Syria, and returned with him in 1798 to France. Monge was appointed president of the Egyptian commission, and he resumed his connexion with the Polytechnic School. His later mathematical papers are published (1794—1816) in the *Journal* and the *Correspondance* of the Polytechnic School. On the formation of the Senate he was appointed a member of that body, with an ample provision and the title of count of Pelusium; but on the fall of Napoleon he was deprived of all his honours, and even excluded from the list of members of the reconstituted Institute. He died at Paris on the 28th July 1818.

For further information see B. Brisson, *Notice historique sur Gaspard Monge*; Dupin, *Essai historique sur les services et les travaux scientifiques de Gaspard Monge,* Paris, 1819, which contains (pp. 162—166) a list of Monge's memoirs and works; and the biography by Arago (*Œuvres,* t. II., 1854).

Monge's various mathematical papers are to a considerable extent reproduced in the *Application de l'Analyse à la Géométrie,* 4th edition (last revised by the author), Paris, 1819—the pure text of this is reproduced in the 5th edition (revue, corrigée et annotée par M. Liouville), Paris, 1850, which contains also Gauss's Memoir, "Disquisitiones generales circa superficies curvas," and some valuable notes by the editor. The other principal separate works are *Traité élémentaire de la Statique, 8ᵉ édition, conformée à la précédente, par M. Hachette, et suivie d'une Note etc., par M. Cauchy,* Paris, 1846; and the *Géométrie Descriptive* (originating, as mentioned above, in the lessons given at the Normal School). The 4th edition, published shortly after the author's death, seems to have been substantially the same as the 7th (*Géométrie Descriptive, par G. Monge, suivie d'une théorie des Ombres et de la Perspective, extraite des papiers de l'auteur, par M. Brisson,* Paris, 1847).

794.

NUMBERS, PARTITION OF.

[From the *Encyclopædia Britannica, Ninth Edition*, vol. XVII. (1884), p. 614.]

THIS subject, created by Euler, though relating essentially to positive integer numbers, is scarcely regarded as a part of the Theory of Numbers. We consider in it a number as made up by the *addition* of other numbers: thus the partitions of the successive numbers 1, 2, 3, 4, 5, 6, &c., are as follows:—

1;

2, 11;

3, 21, 111;

4, 31, 22, 211, 1111;

5, 41, 32, 311, 221, 2111, 11111;

6, 51, 42, 411, 33, 321, 3111, 222, 2211, 21111, 111111.

These are formed each from the preceding ones; thus, to form the partitions of 6 we take first 6; secondly, 5 prefixed to each of the partitions of 1 (that is, 51); thirdly, 4 prefixed to each of the partitions of 2 (that is, 42, 411); fourthly, 3 prefixed to each of the partitions of 3 (that is, 33, 321, 3111); fifthly, 2 prefixed, not to each of the partitions of 4, but only to those partitions which begin with a number not exceeding 2 (that is, 222, 2211, 21111); and lastly, 1 prefixed to all the partitions of 5 which begin with a number not exceeding 1 (that is, 111111); and so in other cases.

The method gives *all* the partitions of a number, but we may consider different classes of partitions: the partitions into a given number of parts, or into not more than a given number of parts; or the partitions into given parts, either with repetitions or without repetitions, &c. It is possible, for any particular class of partitions, to obtain methods more or less easy for the formation of the partitions either of a given number or of the successive numbers 1, 2, 3, &c. And of course in any case, having obtained the partitions, we can count them and so obtain the number of partitions.

Another method is by Arbogast's rule of the last and the last but one; in fact, taking the value of a to be unity, and, understanding this letter in each term, the rule gives b; c, b^2; d, bc, b^3; e, bd, c^2, b^2c, b^4, &c., which, if b, c, d, e, &c., denote 1, 2, 3, 4, &c., respectively, are the partitions of 1, 2, 3, 4, &c., respectively.

An important notion is that of conjugate partitions. Thus a partition of 6 is 42; writing this in the form $\begin{cases}1111\\11\end{cases}$, and summing the columns instead of the lines, we obtain the conjugate partition 2211; evidently, starting from 2211, the conjugate partition is 42. If we form all the partitions of 6 into not more than three parts, these are

6, 51, 42, 33, 411, 321, 222,

and the conjugates are

111111, 21111, 2211, 222, 3111, 321, 33,

where no part is greater than 3; and so, in general, we have the theorem, the number of partitions of n into not more than k parts is equal to the number of partitions of n with no part greater than k.

We have for the number of partitions an analytical theory depending on generating functions; thus for the partitions of a number n with the parts 1, 2, 3, 4, 5, &c., without repetitions, writing down the product

$$1+x\,.\,1+x^2\,.\,1+x^3\,.\,1+x^4\,\ldots,\ =1+x+x^2+2x^3+\ldots+Nx^n+\ldots,$$

it is clear that, if x^α, x^β, $x^\gamma, \ldots$ are terms of the series x, x^2, $x^3, \ldots$ for which $\alpha+\beta+\gamma+\ldots=n$, then we have in the development of the product a term x^n, and hence that, in the term Nx^n of the product, the coefficient N is equal to the number of partitions of n with the parts 1, 2, 3, ..., without repetitions; or say that the product is the generating function (G. F.) for the number of such partitions. And so in other cases we obtain a generating function.

Thus for the function

$$\frac{1}{1-x\,.\,1-x^2\,.\,1-x^3\ldots},\ =1+x+2x^2+\ldots+Nx^n+\ldots,$$

observing that any factor $1/1-x^l$ is $=1+x^l+x^{2l}+\ldots$, we see that, in the term Nx^n, the coefficient is equal to the number of partitions of n, with the parts 1, 2, 3, ..., with repetitions.

Introducing another letter z, and considering the function

$$1+xz\,.\,1+x^2z\,.\,1+x^3z\ldots,\ =1+z(x+x^2+\ldots)+\ldots+Nx^nz^k+\ldots,$$

we see that, in the term Nx^nz^k of the development, the coefficient N is equal to the number of partitions of n into k parts, with the parts 1, 2, 3, 4, ..., without repetitions.

And similarly, considering the function

$$\frac{1}{1-xz\,.\,1-x^2z\,.\,1-x^3z\ldots},\ =1+z(x+x^2+\ldots)+\ldots+Nx^nz^k+\ldots,$$

we see that, in the term Nx^nz^k of the development, the coefficient N is equal to the number of partitions of n into k parts, with the parts 1, 2, 3, 4, ..., with repetitions.

We have such analytical formulæ as

$$\frac{1}{1-xz\,.\,1-x^2z\,.\,1-x^3z\,\ldots}=1+\frac{zx}{1-x}+\frac{z^2x^2}{1-x\,.\,1-x^2}+\ldots,$$

which lead to theorems in the Partition of Numbers. A remarkable theorem is

$$1-x\,.\,1-x^2\,.\,1-x^3\,.\,1-x^4\ldots=1-x-x^2+x^5+x^7-x^{12}-x^{15}+\ldots,$$

where the only terms are those with an exponent $\frac{1}{2}(3n^2\pm n)$, and for each such pair of terms the coefficient is $(-)^n 1$. The formula shows that, except for numbers of the form $\frac{1}{2}(3n^2\pm n)$, the number of partitions without repetitions into an odd number of parts is equal to the number of partitions without repetitions into an even number of parts, whereas for the excepted numbers these numbers differ by unity. Thus for the number 11, which is not an excepted number, the two sets of partitions are

11, 821, 731, 641, 632, 542,

10.1, 92, 83, 74, 65, 5321,

in each set 6.

We have

$$1-x\,.\,1+x\,.\,1+x^2\,.\,1+x^4\,.\,1+x^8\ldots=1;$$

or, as this may be written,

$$1+x\,.\,1+x^2\,.\,1+x^4\,.\,1+x^8\ldots=\frac{1}{1-x},\ =1+x+x^2+x^3+\ldots,$$

showing that a number n can always be made up, and in one way only, with the parts 1, 2, 4, 8,.... The product on the left-hand side may be taken to k terms only: thus if $k=4$, we have

$$1+x\,.\,1+x^2\,.\,1+x^4\,.\,1+x^8,\ =\frac{1-x^{16}}{1-x},\ =1+x+x^2+\ldots+x^{15},$$

that is, any number from 1 to 15 can be made up, and in one way only, with the parts 1, 2, 4, 8; and similarly any number from 1 to 2^k-1 can be made up, and in one way only, with the parts 1, 2, 4, ..., 2^{k-1}. A like formula is

$$\frac{1-x^3}{x\,.\,1-x}\cdot\frac{1-x^9}{x^3\,.\,1-x^3}\cdot\frac{1-x^{27}}{x^9\,.\,1-x^9}\cdot\frac{1-x^{81}}{x^{27}\,.\,1-x^{27}}=\frac{1-x^{81}}{x^{40}\,.\,1-x};$$

that is,

$$x^{-1}+1+x\,.\,x^{-3}+1+x^3\,.\,x^{-9}+1+x^9\,.\,x^{-27}+1+x^{27}=x^{-40}+x^{-39}+\ldots+1+x+\ldots+x^{39}+x^{40},$$

showing that any number from -40 to $+40$ can be made up, and that in one way only, with the parts 1, 3, 9, 27 taken positively or negatively; and so in general any number from $-\frac{1}{2}(3^k-1)$ to $+\frac{1}{2}(3^k-1)$ can be made up, and that in one way only, with the parts 1, 3, 9, ..., 3^{k-1} taken positively or negatively.

795.

NUMBERS, THEORY OF.

[From the *Encyclopædia Britannica, Ninth Edition*, vol. XVII. (1884), pp. 614—624.]

THE Theory of Numbers, or higher arithmetic, otherwise arithmology, is a subject which, originating with Euclid, has in modern times, in the hands of Legendre, Gauss, Lejeune-Dirichlet, Kummer, Kronecker, and others, been developed into a most extensive and interesting branch of mathematics. We distinguish between the ordinary (or say the simplex) theory and the various complex theories.

In the ordinary theory we have, in the first instance, positive integer numbers, the unit or unity 1, and the other numbers 2, 3, 4, 5, &c. We introduce the zero 0, which is a number *sui generis*, and the negative numbers -1, -2, -3, -4, &c., and we have thus the more general notion of integer numbers, 0, ± 1, ± 2, ± 3, &c.; $+1$ and -1 are units or unities. The sum of any two or more numbers is a number; conversely, any number is a sum of two or more parts; but even when the parts are positive a number cannot be, in a determinate manner, represented as a sum of parts. The product of two or more numbers is a number; but (disregarding the unities $+1$, -1, which may be introduced as factors at pleasure) it is not conversely true that every number is a product of numbers. A number such as 2, 3, 5, 7, 11, &c., which is not a product of numbers, is said to be a prime number; and a number which is not prime is said to be composite. A number other than zero is thus either prime or composite; and we have the theorem that every composite number is, in a determinate way, a product of prime factors.

We have complex theories in which all the foregoing notions (integer, unity, zero, prime, composite) occur; that which first presented itself was the theory with the unit i $(i^2=-1)$; we have here complex numbers, $a+bi$, where a and b are in the before-mentioned (ordinary) sense positive or negative integers, not excluding zero; we have the zero 0, $=0+0i$, and the four units $1, -1, i, -i$. A number other than zero is here either prime or else composite; for instance, 3, 7, 11, are prime numbers, and 5, $=(2+i)(2-i)$, 9, $=3.3$, 13, $=(3+2i)(3-2i)$, are composite numbers (generally any positive real prime of the form $4n+3$ is prime, but any positive real prime of the form $4n+1$ is a sum of two squares, and is thus composite). And disregarding unit factors we have, as in the ordinary theory, the theorem that every composite number is, in a determinate way, a product of prime factors.

There is, in like manner, a complex theory involving the cube roots of unity—if α be an imaginary cube root of unity $(\alpha^2+\alpha+1=0)$, then the integers of this theory are $a+b\alpha$ (a and b real positive or negative integers, including zero); a complex theory with the fifth roots of unity—if α be an imaginary fifth root of unity $(\alpha^4+\alpha^3+\alpha^2+\alpha+1=0)$, then the integers of the theory are $a+b\alpha+c\alpha^2+d\alpha^3$ ($a, b, c, d,$ real positive or negative integers, including zero); and so on for the roots of the orders 7, 11, 13, 17, 19. In all these theories, or at any rate for the orders 3, 5, 7 (see No. 37, *post*), we have the foregoing theorem: disregarding unit factors, a number other than zero is either prime or composite, and every composite number is, in a determinate way, a product of prime factors. But coming to the 23rd roots of unity the theorem ceases to be true. Observe that it is a particular case of the theorem that, if N be a prime number, any integer power of N has for factors only the lower powers of N,—for instance, $N^3=N.N^2$; there is no other decomposition $N^3=AB$. This is obviously true in the ordinary theory, and it is true in the complex theories preceding those for the 3rd, 5th, and 7th roots of unity, and probably in those for the other roots preceding the 23rd roots; but it is not true in the theory for the 23rd roots of unity. We have, for instance, 47, a number not decomposable into factors, but 47^3, $=AB$, is a product of two numbers each of the form $a+b\alpha+\ldots+k\alpha^{21}$ (α a 23rd root). The theorem recovers its validity by the introduction into the theory of Kummer's notion of an ideal number.

The complex theories above referred to would be more accurately described as theories for the complex numbers involving the periods of the roots of unity: the units are the roots either of the equation $x^{p-1}+x^{p-2}+\ldots+x+1=0$ (p a prime number) or of any equation $x^{\frac{p-1}{e}}+\ldots\pm1=0$ belonging to a factor of the function of the order $p-1$: in particular, this may be the quadric equation for the periods each of $\tfrac{1}{2}(p-1)$ roots; they are the theories which were first and have been most completely considered, and which led to the notion of an ideal number. But a yet higher generalization which has been made is to consider the complex theory, the units whereof are the roots of any given irreducible equation which has integer numbers for its coefficients.

There is another complex theory the relation of which to the foregoing is not very obvious, viz. Galois's theory of the numbers composed with the imaginary roots of an irreducible congruence, $F(x)\equiv0$ (modulus a prime number p); the nature of this will be indicated in the sequel.

In any theory, ordinary or complex, we have a first part, which has been termed (but the name seems hardly wide enough) the theory of congruences; a second part, the theory of homogeneous forms: this includes in particular the theory of the binary quadratic forms $(a, b, c)(x, y)^2$; and a third part, comprising those miscellaneous investigations which do not come properly under either of the foregoing heads.

Ordinary Theory, First Part.

1. We are concerned with the integer numbers 0, ±1, ±2, ±3, &c., or in the first place with the positive integer numbers 1, 2, 3, 4, 5, 6, &c. Some of these, 1, 2, 3, 5, 7, &c., are prime, others, 4, $=2^2$, 6, $=2.3$, &c., are composite; and we have the fundamental theorem that a composite number is expressible, and that in one way only, as a product of prime factors, $N=a^\alpha b^\beta c^\gamma \ldots$ (a, b, c, ... primes other than 1; α, β, γ, ... positive integers).

Gauss makes the proof to depend on the following steps: (i) the product of two numbers each smaller than a given prime number is not divisible by this number; (ii) if neither of two numbers is divisible by a given prime number the product is not so divisible; (iii) the like as regards three or more numbers; (iv) a composite number cannot be resolved into factors in more than one way.

2. Proofs will in general be only indicated or be altogether omitted, but, as a specimen of the reasoning in regard to whole numbers, the proofs of these fundamental propositions are given at length. (i) Let p be the prime number, a a number less than p, and if possible let there be a number b less than p, and such that ab is divisible by p; it is further assumed that b is the only number, or, if there is more than one, then that b is the least number having the property in question; b is greater than 1, for a being less than p is not divisible by p. Now p as a prime number is not divisible by b, but must lie between two consecutive multiples mb and $(m+1)b$ of b. Hence, ab being divisible by p, mab is also divisible by p; moreover, ap is divisible by p, and hence the difference of these numbers, $=a(p-mb)$, must also be divisible by p, or, writing $p-mb=b'$, we have ab' divisible by p, where b' is less than b; so that b is not the least number for which ab is divisible by p. (ii) If a and b are neither of them divisible by p, then a divided by p leaves a remainder α which is less than p, say we have $a=mp+\alpha$; and similarly b divided by p leaves a remainder β which is less than p, say we have $b=np+\beta$; then

$$ab=(mp+\alpha)(np+\beta), \quad =(mnp+n\alpha+m\beta)p+\alpha\beta,$$

and $\alpha\beta$ is not divisible by p, therefore ab is not divisible by p. (iii) The like proof applies to the product of three or more factors a, b, c, ... (iv) Suppose that the number N, $=a^\alpha b^\beta c^\gamma \ldots$ (a, b, c, ... prime numbers other than 1), is decomposable in some other way into prime factors; we can have no prime factor p, other than a, b, c, ..., for no such number can divide $a^\alpha b^\beta c^\gamma \ldots$; and we must have each of the numbers a, b, c, ..., for if any one of them, suppose a, were wanting, the number N would not be divisible by a. Hence the new decomposition if it exists must be a decomposition $N=a^{\alpha'} b^{\beta'} c^{\gamma'} \ldots$; and here, if any two corresponding indices, say α, α', are different from each other, then one of them, suppose α', is the greater, and we have $N \div p^\alpha = b^\beta c^\gamma \ldots = a^{\alpha'-\alpha} b^{\beta'} c^{\gamma'} \ldots$ That is, we have the number $N \div p^\alpha$ expressed in two different ways as a product, the number a being a factor in the one case, but not a factor in the other case. Thus the two exponents cannot be unequal, that is, we must have $\alpha=\alpha'$, and similarly we have $\beta=\beta'$, $\gamma=\gamma'$, ...; that is, there is *only* the original decomposition $N=a^\alpha b^\beta c^\gamma \ldots$

3. The only numbers divisible by a number $N = a^\alpha b^\beta c^\gamma \ldots$ are the numbers $a^{\alpha'} b^{\beta'} c^{\gamma'} \ldots$, where each exponent α' is equal to or greater than the corresponding exponent α. And conversely the only numbers which divide N are those of the form $a^{\alpha'} b^{\beta'} c^{\gamma'} \ldots$, where each index α' is at most equal to the corresponding index α; and in particular each or any of the indices α' may be $= 0$. Again, the least common multiple of two numbers $N = a^\alpha b^\beta c^\gamma \ldots$ and $N' = a^{\alpha'} b^{\beta'} c^{\gamma'} \ldots$ is $a^{\alpha''} b^{\beta''} c^{\gamma''} \ldots$, where each index α'' is equal to the largest of the corresponding indices α, α';—observe that any one or more of the indices $\alpha, \beta, \gamma, \ldots, \alpha', \beta', \gamma', \ldots$, may be $= 0$, so that the theorem extends to the case where either of the numbers N, N', has prime factors which are not factors of the other number. And so the greatest common measure of two numbers $N = a^\alpha b^\beta c^\gamma \ldots$ and $N' = a^{\alpha'} b^{\beta'} c^{\gamma'} \ldots$ is $a^{\alpha''} b^{\beta''} c^{\gamma''} \ldots$, where each index α'' is equal to the least of the corresponding indices α and α'.

4. The divisors of $N = a^\alpha b^\beta c^\gamma \ldots$ are the several terms of the product

$$(1 + a + \ldots + a^\alpha)(1 + b + \ldots + b^\beta)(1 + c + \ldots + c^\gamma),$$

where unity and the number N itself are reckoned each of them as a divisor. Hence the number of divisors is $= (\alpha + 1)(\beta + 1)(\gamma + 1) \ldots$, and the sum of the divisors is

$$= \frac{(a^{\alpha+1} - 1)(b^{\beta+1} - 1)(c^{\gamma+1} - 1) \ldots}{(a - 1)(b - 1)(c - 1) \ldots}.$$

5. In $N = a^\alpha b^\beta c^\gamma \ldots$ the number of integers less than N and prime to it is

$$\phi(N), = N\left(1 - \frac{1}{a}\right)\left(1 - \frac{1}{b}\right)\left(1 - \frac{1}{c}\right) \ldots$$

To find the numbers in question write down the series of numbers $1, 2, 3, \ldots, N$; strike out all the numbers divisible by a, then those divisible by b, then those divisible by c, and so on; there will remain only the numbers prime to N. For actually finding the numbers we may of course in striking out those divisible by b disregard the numbers already struck out as divisible by a, and in striking out with respect to c disregard the numbers already struck out as divisible by a or b, and so on; but in order to count the remaining numbers it is more convenient to ignore the previous strikings out. Suppose, for a moment, there are only two prime factors a and b, then the number of terms struck out as divisible by a is $= N \cdot \frac{1}{a}$, and the number of terms struck out as divisible by b is $= N \cdot \frac{1}{b}$; but then each term divisible by ab will have been twice struck out; the number of these is $= N \cdot \frac{1}{ab}$, and thus the number of the remaining terms is $N\left(1 - \frac{1}{a} - \frac{1}{b} + \frac{1}{ab}\right)$, which is $= N\left(1 - \frac{1}{a}\right)\left(1 - \frac{1}{b}\right)$. By treating in like manner the case of three or more prime factors $a, b, c, \ldots$ we arrive at the general theorem. The formula gives $\phi(1) = 1$ viz. when $N = 1$, there is no factor $1 - \frac{1}{a}$; and it is necessary to consider $\phi(1)$ as being $= 1$. The explanation is that $\phi(N)$

properly denotes the number of integers not greater than N and prime to it; so that, when $N=1$, we have 1 an integer not greater than N and prime to it; but in every other case the two definitions agree.

6. If N, N', are numbers prime to each other, then $\phi(NN')=\phi(N)\phi(N')$, and so also for any number of numbers having no common divisor; in particular,

$$\phi(a^\alpha b^\beta c^\gamma \ldots)=\phi(a^\alpha)\phi(b^\beta)\phi(c^\gamma)\ldots; \quad \phi(a^\alpha)=a^\alpha\left(1-\frac{1}{a}\right),$$

and the theorem is at once verified. We have $N=\Sigma\phi(N')$, where the summation extends to all the divisors N' of N, unity and the number N itself being included; thus $15=\phi(15)+\phi(5)+\phi(3)+\phi(1), =8+4+2+1$.

7. The prime factor of the binomial function x^N-1 is

$$=\frac{(x^N-1)(x^{N/ab}-1)\ldots}{(x^{N/a}-1)(x^{N/b}-1)\ldots},$$

a rational and integral function of the degree $\phi(N)$; say this is called $[x^N-1]$, and we have $x^N-1=\Pi[x^{N'}-1]$, where the product extends to all the divisors N' of N, unity and the number N included. For instance

$$[x^{15}-1]=\frac{(x^{15}-1)(x-1)}{(x^5-1)(x^3-1)}, \quad =x^8-x^7+x^5-x^4+x^3-x+1;$$

and we have

$$x^{15}-1=[x^{15}-1][x^5-1][x^3-1][x-1],$$
$$=(x^8-x^7+\ldots-x+1)(x^4+x^3+x^2+x+1)(x^2+x+1)(x-1).$$

8. Congruence to a given modulus. A number x is congruent to 0, to the modulus N, $x\equiv 0 \,(\text{mod. } N)$, when x is divisible by N; two numbers x, y are congruent to the modulus N, $x\equiv y\,(\text{mod. } N)$, when their difference $x-y$ divides by N, or, what is the same thing, if $x-y\equiv 0\,(\text{mod. } N)$. Observe that, if $xy\equiv 0\,(\text{mod. } N)$, and x be prime to N, then $y\equiv 0\,(\text{mod. } N)$.

9. Residues to a given modulus. For a given modulus N we can always find, and that in an infinity of ways, a set of N numbers, say N residues, such that every number whatever is, to the modulus N, congruent to one and only one of these residues. For instance, the residues may be 0, 1, 2, 3, ..., $N-1$ (the residue of a given number is here simply the positive remainder of the number when divided by N); or, N being odd, the system may be

$$0, \pm 1, \pm 2, \ldots, \pm \tfrac{1}{2}(N-1),$$

and N even,

$$0, \pm 1, \pm 2, \ldots, \pm \tfrac{1}{2}(N-2), +\tfrac{1}{2}N.$$

10. Prime residues to a given modulus. Considering only the numbers which are prime to a given modulus N, we have here a set of $\phi(N)$ numbers, say $\phi(N)$ prime residues, such that every number prime to N is, to the modulus N, congruent to

one and only one of these prime residues. For instance, the prime residues may be the numbers less than N and prime to it. In particular, if N is a prime number p, then the residues may be the $p-1$ numbers, $1, 2, 3, \ldots, p-1$.

In all that follows, the letter p, in the absence of any statement to the contrary, will be used to denote an *odd* prime other than unity. A theorem for p may hold good for the even prime 2, but it is in general easy to see whether this is so or not.

11. Fermat's theorem, $x^{p-1}-1\equiv 0$ (mod. p). The generalized theorem is $x^{\phi(N)}-1\equiv 0$ (mod. N). The proof of the generalized theorem is as easy as that of the original theorem. Consider the series of the $\phi(N)$ numbers $a, b, c, \ldots$, each less than N and prime to it; let x be any number prime to N, then each of the numbers $xa, xb, xc, \ldots$, is prime to N, and no two of them are congruent to the modulus N, that is, we cannot have $x(a-b)\equiv 0$ (mod. N); in fact, x is prime to N, and the difference $a-b$ of two positive numbers each less than N will be less than N. Hence the numbers $xa, xb, xc, \ldots$, are in a different order congruent to the numbers $a, b, c, \ldots$; and multiplying together the numbers of each set we have $x^{\phi(N)}\, abc \ldots \equiv abc \ldots$ (mod. N), or, since $a, b, c, \ldots$, are each prime to N, and therefore also the product $abc\ldots$ is prime to N, we have $x^{\phi(N)}\equiv 1$, or say $x^{\phi(N)}-1\equiv 0$ (mod. N).

In particular, if N be a prime number $=p$, then $\phi(N)$ is $=p-1$, and the theorem is $x^{p-1}-1\equiv 0$ (mod. p), x being now any number not divisible by p.

12. The general congruence $f(x)\equiv 0$ (mod. p). $f(x)$ is written to denote a rational and integral function with integer coefficients which may without loss of generality be taken to be each of them less than p; it is assumed that the coefficient A of the highest power of x is not $=0$. If there is for x an integer value a such that $f(a)\equiv 0$ (mod. p, throughout), then a is said to be a root of the congruence $f(x)\equiv 0$; we may, it is clear, for a substitute any value whatever $a'=a+kp$, or say any value a' which is $\equiv a$, but such value a' is considered not as a different root but as the same root of the congruence. We have thus $f(a)\equiv 0$; and therefore $f(x)\equiv f(x)-f(a)$, $=(x-a)f_1(x)$, where $f_1(x)$ is a function of like form with $f(x)$, that is, with integer coefficients, but of the next inferior order $n-1$. Suppose there is another root b of the congruence, that is, an integer value b such that $f(b)\equiv 0$; we have then $(b-a)f_1(b)\equiv 0$, and $b-a$ is not $\equiv 0$ (for then b would be the same root as a). Hence $f_1(b)\equiv 0$, and $f(x)=(x-a)\{f_1(x)-f_1(b)\}$, $=(x-a)(x-b)f_2(x)$, where $f_2(x)$ is an integral function such as $f(x)$, but of the order $n-2$; and so on, that is, if there exist n different (non-congruent) roots of the congruence $f(x)\equiv 0$, then $f(x)=A(x-a)(x-b)\ldots(x-k)$, and the congruence may be written $A(x-a)(x-b)\ldots(x-k)\equiv 0$. And this cannot be satisfied by any other value l; for if so we should have $A(l-a)(l-b)\ldots(l-k)\equiv 0$, that is, some one of the congruences $(l-a)\equiv 0$, &c., would have to be satisfied, and l would be the same as one of the roots $a, b, c, \ldots, k$. That is, a congruence of the order n cannot have more than n roots, and if it have precisely n roots $a, b, c, \ldots, k$, then the form is $f(x)\equiv A(x-a)(x-b)\ldots(x-k)$, $\equiv 0$.

Observe that a congruence may have equal roots, viz. if the form be

$$f(x)\equiv A(x-a)^{\alpha}(x-b)^{\beta}\ldots, \equiv 0,$$

then the roots a, b, ... are to be counted α times, β times, ... respectively; but clearly the whole number of roots $\alpha+\beta+\ldots$ is at most $=n$.

It is hardly necessary to remark that this theory of a congruence of the order n is precisely analogous to that of an equation of the order n, when only real roots are attended to. The theory of the imaginary roots of a congruence will be considered further on (see No. 41).

13. The linear congruence $ax \equiv c$ (mod. b). This is equivalent to the indeterminate equation $ax+by=c$; if a and b are not prime to each other, but have a greatest common measure q, this must also divide c; supposing the division performed, the equation becomes $a'x+b'y=c'$, where a' and b' are prime to each other, or, what is the same thing, we have the congruence $a'x \equiv c'$ (mod. b'). This can always be solved, for, if we consider the b' numbers $0, 1, 2, \ldots, b'-1$, one and only one of these will be $\equiv c'$ (mod. b'). Multiplying these by any number a' prime to b', and taking the remainders in regard to b', we reproduce in a different order the same series of numbers $0, 1, 2, \ldots, b'-1$; that is, in the series a', $2a'$, ..., $(b'-1)a'$ there will be one and only one term $\equiv c'$ (mod. b'), or, calling the term in question α, we have $x=\alpha$ as the solution of the congruence $a'x \equiv c'$ (mod. b'); $a'\alpha - c'$ is then a multiple of b', say it is $=-b'\beta$, and the corresponding value of y is $y=\beta$. We may for α write $\alpha+mb'$, m being any positive or negative integer, not excluding zero (but, as already remarked, this is not considered as a distinct solution of the congruence); the corresponding value of y is clearly $=\beta-ma'$.

The value of x can be found by a process similar to that for finding the greatest common measure of the two numbers a' and c'; this is what is really done in the apparently tentative process which at once presents itself for small numbers, thus $6x \equiv 9$ (mod. 35), we have $36x \equiv 54$, or, rejecting multiples of 35, $x \equiv 19$, or, if we please, $x=35m+19$.

In particular, we can always find a number ξ such that $a'\xi \equiv 1$ (mod. b'); and we have then $x=c'\xi$ as the solution of the congruence $a'x \equiv c'$. The value of ξ may be written $\xi \equiv \frac{1}{a'}$ (mod. b'), where $\frac{1}{a'}$ stands for that integer value ξ which satisfies the original congruence $a'\xi \equiv 1$ (mod. b'); and the value of x may then be written $x \equiv \frac{c'}{a'}$ (mod. b'). Another solution of the linear congruence is given in No. 21.

14. Wilson's theorem, $1.2.3\ldots\overline{p-1}+1 \equiv 0$ (mod. p). It has been seen that, for any prime number p, the congruence $x^{p-1}-1 \equiv 0$ (mod. p) of the order $p-1$ has the $p-1$ roots $1, 2, \ldots, p-1$; we have therefore

$$x^{p-1}-1 \equiv (x-1)(x-2)\ldots(x-\overline{p-1}),$$

or, comparing the terms independent of x, it appears that $1.2.3\ldots p-1 \equiv -1$, that is, $1.2.3\ldots\overline{p-1}+1 \equiv 0$ (mod. p),—the required theorem. For instance, where $p=5$, then $1.2.3.4+1 \equiv 0$ (mod. 5), and where $p=7$, then $1.2.3.4.5.6+1 \equiv 0$ (mod. 7).

15. A proof on wholly different principles may be given. Suppose, to fix the ideas, $p=7$; consider on a circle 7 points, the summits of a regular heptagon, and join

these in any manner so as to form a heptagon; the whole number of heptagons is $\frac{1}{2}.1.2.3.4.5.6$. Now of these we have $\frac{1}{2}(7-1)$, $=3$, which are regular heptagons (convex or stellated); the number of remaining heptagons must be divisible by 7, for with any one such heptagon we connect the 6 heptagons which can be obtained from it by making it rotate through successive angles of $\frac{1}{7}360°$. That is, $\frac{1}{2}.1.2.3.4.5.6-\frac{1}{2}(7-1)\equiv 0$ (mod. 7), whence $1.2.3.4.5.6-7+1\equiv 0$, or finally $1.2.3.4.5.6+1\equiv 0$ (mod. 7). It is clear that the proof applies without alteration to the case of any prime number p.

If p is not a prime number, then $1.2.3\ldots\overline{p-1}\equiv 0$ (mod. p); hence the theorem shows directly whether a number p is or is not a prime number; but it is not of any practical utility for this purpose.

16. Prime roots of a prime number—application to the binomial equation $x^p-1=0$. Take, for instance, $p=7$. By what precedes, we have

$$x^6-1=[x^6-1][x^3-1][x^2-1][x-1], \ =(x^2-x+1)(x^2+x+1)(x+1)(x-1);$$

and we have

$$x^6-1\equiv(x-1)(x-2)(x-3)(x-4)(x-5)(x-6) \text{ (mod. 7)};$$

whence also

$$(x^2-x+1)(x^2+x+1)(x+1)(x-1)\equiv(x-1)(x-2)(x-3)(x-4)(x-5)(x-6).$$

These two decompositions must agree together, and we in fact have

$$x^2-x+1\equiv(x-3)(x-5),\quad x^2+x+1\equiv(x-2)(x-3),\quad x+1\equiv x-6,\quad x-1\equiv x-1.$$

In particular, we thus have 3, 5, as the roots of the congruence $x^2-x+1\equiv 0$, that is, $[x^6-1]\equiv 0$, and these roots 3, 5, are not the roots of any other of the congruences $[x^3-1]\equiv 0$, $[x^2-1]\equiv 0$, $[x-1]\equiv 0$; that is, writing $a=3$ or 5 in the series of numbers a, a^2, a^3, a^4, a^5, a^6, we have a^6 as the first term which is $\equiv 1$ (mod. 7); the series in fact are

$$3,\ 9\ ,\ 27\ ,\ 81\ ,\ 243\ ,\ 729\quad \equiv 3,\ 2,\ 6,\ 4,\ 5,\ 1,$$

$$5,\ 25,\ 125,\ 625,\ 3125,\ 15625\equiv 5,\ 4,\ 6,\ 2,\ 3,\ 1.$$

And so, in general, the congruence $x^{p-1}-1\equiv 0$ (mod. p) has the $p-1$ real roots $1, 2, 3, \ldots, p-1$; hence the congruence $[x^{p-1}-1]\equiv 0$, which is of the order $\phi(p-1)$, has this number $\phi(p-1)$ of real roots; and, calling any one of these g, then in the series of powers $g, g^2, g^3, \ldots, g^{p-1}$, the first term which is $\equiv 1$ (mod. p) is g^{p-1}, that is, we have $g, g^2, g^3, \ldots, g^{p-1}=1, 2, 3, \ldots, p-1$ in a different order. Any such number g is said to be a prime root of p, and the number of prime roots is $\phi(p-1)$, the number of integers less than and prime to $p-1$.

The notion of a prime root was applied by Gauss to the solution of the binomial equation $x^p-1=0$, or, what is the same thing, to the question of the division of the circle (Kreistheilung), see Equation, Nos. 30 and 31, [786]; and, as remarked in the introduction to the present article, the roots or periods of roots of this equation present themselves as the units of a complex theory in the Theory of Numbers.

17. Any number x less than p is $\equiv g^m$, and, if m is not prime to $p-1$, but has with it a greatest common measure e, suppose $m=ke$, $p-1=ef$, then

$$x \equiv g^{ke},\quad x^f \equiv g^{kef} \equiv g^{k(p-1)} \equiv 1,$$

that is, $x^f \equiv 1$; and it is easily seen that in the series of powers $x, x^2, \ldots, x^f$, we have x^f as the first term which is $\equiv 1 \pmod{p}$. A number $\equiv g^m$, where m is not prime to $p-1$, is thus not a prime root; and it further appears that, g being any particular prime root, the $\phi(p-1)$ prime roots are $\equiv$ the numbers g^m, where m is any number less than $p-1$ and prime to it. Thus in the foregoing example $p=7$, where the prime roots were 3 and 5, the integers less than 6 and prime to it are 1, 5; and we, in fact, have $5 \equiv 3^5$ and $3 \equiv 5^5 \pmod{7}$.

18. Integers belonging to a given exponent; index of a number. If, as before, $p-1=ef$, that is, if f be a submultiple of $p-1$, then any integer x such that x^f is the lowest power of x which is $\equiv 1 \pmod{p}$ is said to belong to the exponent f. The number of residues, or terms of the series $1, 2, 3, \ldots, p-1$, which belong to the exponent f is $\phi(f)$, the number of integers less than f and prime to it; these are the roots of the congruence $[x^f-1] \equiv 0$ of the order $\phi(f)$. It is hardly necessary to remark that the prime roots belong to the exponent $p-1$.

A number $x \equiv g^m$ is said to have the index m; observe the distinction between the two terms exponent and index; and, further, that the index is dependent on the selected prime root g.

19. Special forms of composite modulus. If instead of a prime modulus p we have a modulus p^m which is the power of an odd prime, or a modulus $2p$ or $2p^m$ which is twice an odd prime or a power of an odd prime, then there is a theory analogous to that of prime roots, viz. the numbers less than the modulus and prime to it are congruent to successive powers of a prime root g; thus,

if $p^m = 3^2$, we have

$$2,\ 4,\ 8,\ 16,\ 32,\ 64 \equiv 2,\ 4,\ 8,\ 7,\ 5,\ 1 \pmod{9},$$

and if $2p^m = 2 \,.\, 3^2$, we have

$$5,\ 25,\ 125,\ 625,\ 3125,\ 15625 \equiv 5,\ 7,\ 11,\ 13,\ 17,\ 1 \pmod{18}.$$

As regards the even prime 2 and its powers—for the modulus 2 or 4 the theory of prime roots does not come into existence, and for the higher powers it is not applicable; thus with modulus $=8$ the numbers less than 8 and prime to it are 1, 3, 5, 7; and we have $3^2 \equiv 5^2 \equiv 7^2 \equiv 1 \pmod{8}$.

20. Composite modulus $N = a^\alpha b^\beta c^\gamma \ldots$—no prime roots—irregularity. In the general case of a composite modulus it has been seen that, if x is any number less than N and prime to it, then $x^{\phi(N)} - 1 \equiv 0 \pmod{N}$. But, except in the above-mentioned cases p^m, $2p^m$, 2 or 4, there is not any number a such that $a^{\phi(N)}$ is the first power of a which is $\equiv 1$; there is always some submultiple $i = \frac{1}{\theta}\phi(N)$ such that a^i is the first

power which is $\equiv 1$. For instance, say $N = 24$, $\phi(N) = 8$, then the numbers less than 24 and prime to it are 1, 5, 7, 11, 13, 17, 19, 23; and we have

$$1^1 \equiv 1,\quad 5^2 \equiv 7^2 \equiv 13^2 \equiv 17^2 \equiv 19^2 \equiv 23^2 \equiv 1 \text{ (mod. 24)},$$

that is, 1 has the exponent 1, but all the other numbers have the exponent 2. So again where $N = 48$, the 16 numbers less than 48 and prime to it have, 1 the exponent 1, and 7, 13, 17, 23, 25, 31, 35, 41, 47 each the exponent 2, and the remaining numbers 5, 11, 19, 29, 37, 43 each the exponent 4. We cannot in this case by means of any single root or of any two roots express all the numbers, but we can by means of three roots, for instance, 5, 7, 13, express all the numbers less than 48 and prime to it; the numbers are in fact $\equiv 5^\alpha 7^\beta 13^\gamma$, where $\alpha = 0$, 1, 2, or 3, and β and γ each $= 0$ or 1.

Comparing with the theorem for a prime number p, where the several numbers $1, 2, 3, \ldots, p-1$, are expressed by means of a single prime root, $\equiv g^\alpha$, where $\alpha = 0, 1, 2, \ldots, p-1$, we have the analogue of a case presenting itself in the theory of quadratic forms,—the "irregularity" of a determinant (*post*, No. 31); the difference is that here (the law being known, $N =$ a composite number) the case is not regarded as an irregular one, while the irregular determinants do not present themselves according to any apparent law.

21. Maximum indicator—application to solution of a linear congruence. In the case $N = 48$ it was seen that the exponents were 1, 2, 4, the largest exponent 4 being divisible by each of the others, and this property is a general one, viz. if $N = a^\alpha b^\beta c^\gamma \ldots$ in the series of exponents (or, as Cauchy calls them, indicators) of the numbers less than N and prime to it, the largest exponent I is a multiple of each of the other exponents, and this largest exponent Cauchy calls the maximum indicator; the maximum indicator I is thus a submultiple of $\phi(N)$, and it is the smallest number such that for every number x less than N and prime to it we have $x^I - 1 \equiv 0$ (mod. N). The values of I have been tabulated from $N = 2$ to 1000.

Reverting to the linear congruence $ax \equiv c$ (mod. b), where a and b are prime to each other, then, if I is the maximum indicator for the modulus b, we have $a^I \equiv 1$, and hence it at once appears that the solution of the congruence is $x \equiv ca^{I-1}$.

22. Residues of powers for an odd prime modulus. For the modulus p, if g be a prime root, then every number not divisible by p is $\equiv$ one of the series of numbers $g, g^2, \ldots, g^{p-1}$; and, if k be any positive number prime to $p-1$, then raising each of these to the power k we reproduce in a different order the same series of numbers $g, g^2, \ldots, g^{p-1}$, which numbers are in a different order $\equiv 1, 2, \ldots, p-1$, that is, the residue of a kth power may be any number whatever of the series $1, 2, \ldots, p-1$.

But, if k is not prime to $p-1$, say their greatest common measure is e, and that we have $p - 1 = ef$, $k = me$, then for any number not divisible by p the kth power is $\equiv$ one of the series of f numbers $g^e, g^{2e}, \ldots, g^{fe}$; there are thus only f, $= \frac{1}{e}(p-1)$, out of the $p-1$ numbers $1, 2, 3, \ldots, p-1$, which are residues of a kth power.

23. Quadratic residues for an odd prime modulus. In particular, if $k=2$, then $e=2$, $f=\frac{1}{2}(p-1)$, and the square of every number not divisible by p is $\equiv$ one of the $\frac{1}{2}(p-1)$ numbers $g^2, g^4, \ldots, g^{p-1}$; that is, there are only $\frac{1}{2}(p-1)$ numbers out of the series $1, 2, 3, \ldots, p-1$ which are residues of a square number, or say quadratic residues, and the remaining $\frac{1}{2}(p-1)$ numbers are said to be quadratic non-residues of the modulus p,—we may say simply, residues and non-residues. But this result can be obtained more easily without the aid of the theory of prime roots. Every number not divisible by p is, to the modulus p, $\equiv$ one of the series of numbers $\pm 1, \pm 2, \pm 3, \ldots, \pm \frac{1}{2}(p-1)$; hence every square number is $\equiv$ one of the series of numbers $1^2, 2^2, 3^2, \ldots, \frac{1}{4}(p-1)^2$; and thus the $p-1$ numbers $1, 2, 3, \ldots, p-1$, are one-half of them residues and the other half non-residues of p. Thus, in the case $p=11$, every number not divisible by 11 is, to this modulus, $\equiv$ one of the series $\pm 1, \pm 2, \pm 3, \pm 4, \pm 5$; whence the square of any such number is $\equiv$ one of the series 1, 4, 9, 16, 25, or say the series 1, 4, 9, 5, 3; that is, we have

residues	1,	.	3,	4,	5,	.	.	.	9,	.
non-residues	.	2,	.	.	.	6,	7,	8,	.	10

Calling as usual the residues a and the non-residues b, we have in this case

$$\tfrac{1}{11}(\Sigma b - \Sigma a) = \tfrac{1}{11}(33-22), \ = 1,$$

a positive integer; this is a property true for any prime number of the form $4n+3$, but for a prime number of the form $4n+1$ we have $\Sigma b - \Sigma a = 0$; the demonstration belongs to a higher part of the theory.

It is easily shown that the product of two residues or of two non-residues is a residue; but the product of a residue and a non-residue is a non-residue.

24. The law of reciprocity—Legendre's symbol. The question presents itself, given that P is a residue or a non-residue of Q, can we thence infer whether Q is a residue or a non-residue of P? In particular, if P, Q, are the odd primes p, q, for instance, given that $13 = R(17)$, can we thence infer that $17 = R(13)$, or that $17 = NR(13)$? The answer is contained in the following theorem: If p, q, are odd primes each or one of them of the form $4n+1$, then p, q, are each of them a residue or each of them a non-residue of the other; but, if p, q, are each of them of the form $4n+3$, then, according as p is a residue or a non-residue of q, we have q a non-residue or a residue of p.

The theorem is conveniently expressed by means of Legendre's symbol, viz. p being a positive odd prime, and Q any positive or negative number not divisible by p, then $\left(\frac{Q}{p}\right)$ denotes $+1$ or -1, according as Q is or is not a residue of p; if, as before, q is (as p) a positive odd prime, then the foregoing theorem is

$$\left(\frac{p}{q}\right)\left(\frac{q}{p}\right) = (-1)^{\frac{1}{4}(p-1)(q-1)}.$$

The denominator symbol may be negative, say it is $-p$, we then have as a definition $\left(\frac{Q}{-p}\right) = \left(\frac{Q}{p}\right)$—observe that $\left(\frac{-Q}{p}\right)$ is not $= \left(\frac{Q}{-p}\right)$—and we have further the theorems

$$\left(\frac{-1}{p}\right) = (-1)^{\frac{1}{2}(p-1)}, \quad \left(\frac{2}{p}\right) = (-1)^{\frac{1}{8}(p^2-1)},$$

viz. -1 is a residue or a non-residue of p according as $p \equiv 1$ or $\equiv 3$ (mod. 4), and 2 is a residue or a non-residue of p according as $p \equiv 1$ or 7, or $\equiv 3$ or 5 (mod. 8). If, as definitions, $\left(\frac{p}{-1}\right) = +1$ and $\left(\frac{p}{2}\right) = +1$, these may be written

$$\left(\frac{-1}{p}\right)\left(\frac{p}{-1}\right) = (-1)^{\frac{1}{2}(p-1)}, \text{ and } \left(\frac{2}{p}\right)\left(\frac{p}{2}\right) = (-1)^{\frac{1}{8}(p^2-1)}.$$

We have also, what is in fact a theorem given at the end of No. 23,

$$\left(\frac{QQ'}{p}\right) = \left(\frac{Q}{p}\right)\left(\frac{Q'}{p}\right).$$

The further definition is sometimes convenient—

$$\left(\frac{Q}{p}\right) = 0, \text{ when } p \text{ divides } Q.$$

The law of reciprocity, as contained in the theorem

$$\left(\frac{p}{q}\right)\left(\frac{q}{p}\right) = (-1)^{\frac{1}{4}(p-1)(q-1)},$$

is a fundamental theorem in the whole theory; it was enunciated by Legendre, but first proved by Gauss, who gave no less than six demonstrations of it.

25. Jacobi's generalized symbol. Jacobi defined this as follows: The symbol $\left(\frac{Q}{\pm pp'p''\ldots}\right)$, where $p, p', p'', \ldots$ are positive odd primes equal or unequal, and Q is any positive or negative odd number prime to $pp'p''\ldots$, denotes $+1$ or -1 according to the definition

$$\left(\frac{Q}{\pm pp'p''\ldots}\right) = \left(\frac{Q}{p}\right)\left(\frac{Q}{p'}\right)\left(\frac{Q}{p''}\right)\ldots,$$

the symbols on the right-hand side being Legendre's symbols. But the definition may be regarded as extending to the case where Q is not prime to $pp'p''\ldots$: then we have Q divisible by some factor p, and by the definition of Legendre's symbol in this case we have $\left(\frac{Q}{p}\right) = 0$; hence in the case in question of Q not being prime to $pp'p''\ldots$, the value of Jacobi's symbol is $= 0$.

We may further extend the definition of the symbol to the case where the numerator and the denominator of the symbol are both or one of them even, and present the definition in the most general form, as follows: suppose that $p, p', p'', \ldots$

being positive or negative even or odd primes, equal or unequal, and similarly $q, q', q'', \ldots$ being positive or negative even or odd primes, equal or unequal, we have $P = pp'p''\ldots$ and $Q = qq'q''\ldots$, then the symbol $\left(\frac{Q}{P}\right)$ will denote $+1$, -1, or 0, according to the definition

$$\left(\frac{Q}{P}\right) = \left(\frac{q}{p}\right)\left(\frac{q}{p'}\right)\left(\frac{q}{p''}\right)\cdots\left(\frac{q'}{p}\right)\left(\frac{q'}{p'}\right)\left(\frac{q'}{p''}\right)\cdots,$$

the symbols on the right-hand being Legendre's symbols. If P and Q are not prime to each other, then for some pair of factors p and q we have $p = \pm q$, and the corresponding Legendrian symbol $\left(\frac{q}{p}\right)$ is $= 0$, whence in this case $\left(\frac{Q}{P}\right) = 0$.

It is important to remark that $\left(\frac{Q}{P}\right) = +1$ is not a sufficient condition in order that Q may be a residue of P; if $P = 2^{\alpha}pp'p''\ldots$, $p, p', p'', \ldots$ being positive odd primes, then, in order that Q may be a residue of P, it must be a residue of each of the prime factors $p, p', p'', \ldots$, that is, we must have

$$\left(\frac{Q}{p}\right) = +1,\quad \left(\frac{Q}{p'}\right) = +1,\quad \left(\frac{Q}{p''}\right) = +1, \ldots,$$

as many equations as there are unequal factors $p, p', p'', \ldots$ of the modulus P.

Ordinary Theory, Second Part,—Theory of Forms.

26. Binary quadratic (or quadric) forms—transformation and equivalence. We consider a form

$$ax^2 + 2bxy + cy^2, \quad = (a, b, c)(x, y)^2,$$

or when, as usual, only the coefficients are attended to, $= (a, b, c)$. The coefficients (a, b, c) and the variables (x, y) are taken to be positive or negative integers, not excluding zero. The discriminant $ac - b^2$ taken negatively, that is, $b^2 - ac$, is said to be the determinant of the form: and we thus distinguish between forms of a positive and of a negative determinant.

Considering new variables, $\alpha x + \beta y$, $\gamma x + \delta y$, where $\alpha, \beta, \gamma, \delta$, are positive or negative integers, not excluding zero, we have identically

$$(a, b, c)(\alpha x + \beta y, \gamma x + \delta y)^2 = (a', b', c')(x, y)^2,$$

where

$$\begin{aligned} a' &= (a, b, c)(\alpha, \gamma)^2, && = a\alpha^2 + 2b\alpha\gamma + c\gamma^2, \\ b' &= (a, b, c)(\alpha, \gamma)(\beta, \delta), && = a\alpha\beta + b(\alpha\delta + \beta\gamma) + c\gamma\delta, \\ c' &= (a, b, c)(\beta, \delta)^2, && = a\beta^2 + 2b\beta\delta + c\delta^2; \end{aligned}$$

and thence

$$b'^2 - a'c' = (\alpha\delta - \beta\gamma)^2(b^2 - ac).$$

The form (a', b', c') is in this case said to be contained in the form (a, b, c); and a condition for this is obviously that the determinant D' of the contained form shall be equal to the determinant D of the containing form multiplied by a square number; in particular, the determinants must be of the same sign. If the determinants are equal, then $(\alpha\delta-\beta\gamma)^2=1$, that is, $\alpha\delta-\beta\gamma=\pm 1$. Assuming in this case that the transformation exists, and writing $\alpha\delta-\beta\gamma=\epsilon$, and writing also

$$x'=\alpha x+\beta y,$$
$$y'=\gamma x+\delta y,$$

then conversely

$$x=\frac{1}{\epsilon}(\ \delta x'-\beta y'),\ =\alpha'x'+\beta'y',$$
$$y=\frac{1}{\epsilon}(-\gamma x'+\alpha y'),\ =\gamma'x'+\delta'y',$$

suppose, where α', β', γ', δ' are integers; and we have, moreover,

$$\alpha'\delta'-\beta'\gamma'=\frac{1}{\epsilon^2}(\alpha\delta-\beta\gamma),\ =\frac{1}{\epsilon},\ =\epsilon,$$

that is, $\alpha'\delta'-\beta'\gamma'=+1$ or -1, according as $\alpha\delta-\beta\gamma$ is $=+1$ or -1. The two forms (a, b, c), (a', b', c') are in this case said to be equivalent, and to be, in regard to the particular transformation, equivalent properly or improperly according as $\alpha\delta-\beta\gamma(=\alpha'\delta'-\beta'\gamma')$ is $=+1$ or $=-1$. We have, therefore, as a condition for the equivalence of two forms, that their determinants shall be equal; but this is not a sufficient condition. It is to be remarked also that two forms of the same determinant may be equivalent properly and also improperly; there may exist a transformation for which $\alpha\delta-\beta\gamma$ is $=+1$, and also a transformation for which $\alpha\delta-\beta\gamma$ is $=-1$. But this is only the case when each of the forms is improperly equivalent to itself; for instance, a form x^2-Dy^2, which remains unaltered by the change x, y, into x, $-y$ (that is, α, β, γ, $\delta=1$, 0, 0, -1, and therefore $\alpha\delta-\beta\gamma=-1$), is a form improperly equivalent to itself. A form improperly equivalent to itself is said to be an ambiguous form. In what follows, equivalent means always properly equivalent.

27. Forms for a given determinant—classes, &c. In the case where D, $=b^2-ac$, is a square, the form $(a, b, c)(x, y)^2$ is a product of two rational factors; this case may be excluded from consideration, and we thus assume that the determinant D is either negative, or, being positive, that it is not a square. The forms (a, b, c) of a given positive or negative determinant are each of them equivalent to some one out of a finite number of non-equivalent forms which may be considered as representing so many distinct classes. For instance, every form of the determinant -1 is equivalent to $(1, 0, 1)$, that is, given any form (a, b, c) for which $b^2-ac=-1$, it is possible to find integer values α, β, γ, δ, such that $\alpha\delta-\beta\gamma=+1$, and $(a, b, c)(\alpha x+\beta y, \gamma x+\delta y)^2=(1, 0, 1)(x, y)^2$, that is, $=x^2+y^2$. Or, to take a less simple example, every form of the determinant -35 is equivalent to one of the following forms: $(1, 0, 35)$, $(5, 0, 7)$, $(3, \pm 1, 12)$, $(4, \pm 1, 18)$,—$(2, 1, 8)$, $(6, 1, 6)$; for the first six forms, the numbers a, $2b$, c have no common factor, and these are said to be *properly primitive* forms, or to belong to the properly primitive order; for the last two forms, the numbers a, b, c have no common factor, but, a and c being each even, the numbers a, $2b$, c have a common factor 2,

and these are said to be *improperly primitive* forms, or to belong to the improperly primitive order. The properly primitive forms are thus the six forms (1, 0, 35), (5, 0, 7), (3, ± 1, 12), (4, ± 1, 18); or we may say that there are represented hereby six properly primitive classes. Derived forms, or forms which belong to a derived order, present themselves in the case of a determinant D having a square factor or factors, and it is not necessary to consider them here.

It is not proposed to give here the rules for the determination of the system of non-equivalent forms; it will be enough to state that this depends on the determination in the first instance of a system of *reduced* forms, that is, forms for which the coefficients a, b, c, taken positively satisfy certain numerical inequalities admitting only of a finite number of solutions. In the case of a negative determinant, the reduced forms are no two of them equivalent, and we thus have the required system of non-equivalent forms; in the case of a positive determinant, the reduced forms group themselves together in *periods* in such wise that the forms belonging to a period are equivalent to each other, and the required system of non-equivalent forms is obtained by selecting one form out of each such period. The principal difference in the theory of the two cases of a positive and a negative determinant consists in these periods; the system of non-equivalent forms once arrived at, the two theories are nearly identical.

28. Characters of a form or class—division into genera. Attending only to the properly primitive forms: for instance, those mentioned above for the determinant -35: the form (1, 0, 35) represents only numbers f which are residues of 5, and also residues of 7; we have, in fact, $f = x^2 + 35y^2$, $\equiv x^2 \pmod{5}$, and also $\equiv x^2 \pmod{7}$. Using the Legendrian symbols $\left(\frac{f}{5}\right)$ and $\left(\frac{f}{7}\right)$, we say that the form (1, 0, 35) has the characters $\left(\frac{f}{5}\right)$, $\left(\frac{f}{7}\right) = ++$. Each of the other forms has in like manner a determinate character $+$ or $-$ in regard to $\left(\frac{f}{5}\right)$ and also in regard to $\left(\frac{f}{7}\right)$; and it is found that for each of them the characters are $+\,+$ or else $-\,-$ (that is, they are never $+\,-$ or $-\,+$). We, in fact, have

	$\left(\frac{f}{5}\right)$	$\left(\frac{f}{7}\right)$
(1, 0, 35)	+	+
(4, ± 1, 9)		
(5, 0, 7)	−	−
(3, ± 1, 12)		

and we thus arrange the six forms into genera, viz. we have three forms belonging to the genus $\left(\frac{f}{5}\right)$, $\left(\frac{f}{7}\right) = +\,+$, and three to the genus $\left(\frac{f}{5}\right)$, $\left(\frac{f}{7}\right) = -\,-$, these characters $+\,+$ and $-\,-$ of genera being one-half of all the combinations $+\,+$, $-\,-$, $+\,-$, $-\,+$.

The like theory applies to any other negative or positive determinant; the several characters have reference in some cases not only to the odd prime factors of D but also to the numbers 4 and 8, that is, there is occasion to consider also the Legendrian symbols $\left(\frac{-1}{f}\right)$, $=(-1)^{\frac{1}{2}(f-1)}$, and $\left(\frac{2}{f}\right)$, $=(-1)^{\frac{1}{8}(f^2-1)}$, and there are various cases to be considered according to the form of D in regard to its simple and squared factors respectively; but in every case there are certain combinations of characters (in number one-half of all the combinations) which correspond to genera, and the properly primitive forms belong to different genera accordingly, the number of forms being the same in each genus.

The form $(1, 0, -D)$ has the characters all $+$, and this is said to be the principal form, and the genus containing it the principal genus. For a given determinant, the characters of two genera may be compounded together according to the ordinary rule of signs, giving the characters of a new genus; in particular, if the characters of a genus are compounded with themselves, then we have the characters of the principal genus.

29. Composition of quadratic forms. Considering X, Y, as given lineo-linear functions of (x, y), (x', y'), defined by the equations

$$X = p_0xx' + p_1xy' + p_2yx' + p_3yy',$$
$$Y = q_0xx' + q_1xy' + q_2yx' + q_3yy',$$

the coefficients p_0, p_1, p_2, p_3, q_0, q_1, q_2, q_3, may be so connected with the coefficients (A, B, C), (a, b, c), (a', b', c'), of three quadratic forms as to give rise to the identity

$$(A, B, C)(X, Y)^2 = (a, b, c)(x, y)^2 \,.\, (a', b', c')(x', y')^2;$$

and, this being so, the form (A, B, C) is said to be compounded of the two forms (a, b, c) and (a', b', c'), the order of composition being indifferent.

The necessary and sufficient condition, in order that it may be possible to compound together two given forms (a, b, c), (a', b', c'), is that their determinants shall be to each other in the proportion of two square numbers; in particular, the two forms may have the same determinant D; and when this is so the compound form (A, B, C) will also have the same determinant D. The rules for this composition of two forms of the same determinant have been (as part of the general theory) investigated and established. The forms compounded of equivalent forms are equivalent to each other; we thus in effect compound *classes*, viz. considering any two classes, the composition of their representative forms gives a form which is the representative of a new class, and the composition of any two forms belonging to the two classes respectively gives a form belonging to the new class. But, this once understood, it is more simple to speak of the composition of forms, that is, of the forms belonging to the finite system of representative forms for a given determinant; and it will be enough to consider the properly primitive forms.

30. The principal form $(1, 0, D)$, compounded with any other form (a, b, c), gives rise to this same form (a, b, c); the principal form is on this account denoted by 1, viz. denoting the other form by ϕ, and expressing composition in like manner with

multiplication, we have $1 . \phi = \phi$. The form ϕ may be compounded with itself, giving a form denoted by ϕ^2; compounding this again with ϕ, we have a form denoted by ϕ^3; and so on. Since the whole number of forms is finite, we must in this manner arrive at the principal form, say we have $\phi^n = 1$, n being the least exponent for which this equation is satisfied. In particular, if the form ϕ belong to the principal genus, then the forms $\phi^2, \phi^3, \ldots, \phi^{n-1}$ will all belong to the principal genus, or the principal genus will include the forms $1, \phi, \phi^2, \ldots, \phi^{n-1}$, the powers of a form ϕ having the exponent n.

31. Regular and irregular determinants. The principal genus may consist of such a series of forms, and the determinant is then said to be *regular*; in particular, for a negative determinant D, $= -1$ to -1000, the determinant is always regular except in the thirteen cases $-D = 243$, 307, 339, 459, 576, 580, 675, 755, 820, 884, 891, 900, 974 (and, Perott, in *Crelle*, vol. XCV., 1883, except also for $-D = 468$, 931); the determinant is here said to be irregular. Thus for each of the values $-D = 576$, 580, 820, 900, the principal genus consists of four forms, not $1, \phi, \phi^2, \phi^3$, where $\phi^4 = 1$, but $1, \phi, \phi_1, \phi\phi_1$, where $\phi^2 = 1$, $\phi_1^2 = 1$, and therefore also $(\phi\phi_1)^2 = 1$.

Compounding together any two forms, we have a form with the characters compounded of the characters of the two forms; and in particular, combining a form with itself, we have a form with the characters of the principal form. Or, what is the same thing, any two genera compounded together give rise to a determinate genus, viz. the genus having the characters compounded of the characters of the two genera; and any genus compounded with itself gives rise to the principal genus.

Considering any regular determinant, suppose that there is more than one genus, and that the number of forms in each genus is $= n$; then, except in the case $n = 2$, it can be shown that there are always forms having the exponent $2n$. For instance, in the case $D = -35$, we have two genera each of three forms; there will be a form g having the exponent 6, or $g^6 = 1$; and the forms are $1, g, g^2, g^3, g^4, g^5$, where $1, g^2, g^4$, belong to the principal genus, and g, g^3, g^5, to the other genus. The characters refer to $\left(\frac{f}{5}\right)$, $\left(\frac{f}{7}\right)$, and the forms are

+ +,	(1, 0, 35)	1		− −,	(3, − 1, 12)	g
	(4, 1, 9)	g^2			(5, 0, 7)	g^3
	(4, − 1, 9)	g^4			(3, 1, 12)	g^5.

An instance of the case $n = 1$ is $D = -21$, there are here four genera each of a single form $1, c, c_1, cc_1$, where $c^2 = 1$, $c_1^2 = 1$; an instance of the case $n = 2$ is $D = -88$, there are here two genera each of two forms $1, c$, and c_1, cc_1, where $c^2 = 1$, $c_1^2 = 1$, thus there is here no form having the exponent $2n$. (See Cayley, *Tables, &c.*, in *Crelle*, t. LX., 1862, pp. 357—372, [335].) We may have 2^{k+1} genera, each of n forms, viz. such a system may be represented by $(1, \phi^2, \ldots, \phi^{2n-2}; \phi, \phi^3, \ldots, \phi^{2n-1})(1, c)(1, c_1)\ldots(1, c_{k-1})$, where $\phi^{2n} = 1$, $c^2 = 1$, $c_1^2 = 1, \ldots, c^2_{k-1} = 1$; there is no peculiarity in the form ϕ: we may instead of it take any form such as $c\phi$, $cc_1\phi$, &c., for each of these is like ϕ, a form belonging to the exponent $2n$, and such that the even powers give the principal genus.

32. Ternary and higher quadratic forms—cubic forms, &c. The theory of the ternary quadratic forms

$$(a,\ b,\ c,\ a',\ b',\ c')\,(x,\ y,\ z)^2, \ = ax^2 + by^2 + cz^2 + 2a'yz + 2b'zx + 2c'xy,$$

or when only the coefficients are attended to, $\begin{pmatrix} a, & b, & c \\ a', & b', & c' \end{pmatrix}$, has been studied in a very complete manner; and those of the quaternary and higher quadratic forms have also been studied; in particular, the forms $x^2 + y^2 + z^2$, $x^2 + y^2 + z^2 + w^2$ composed of three or four squares; and the like forms with five, six, seven, and eight squares. The binary cubic forms $(a, b, c, d)(x, y)^3$, $= ax^3 + 3bx^2y + 3cxy^2 + dy^3$, or when only the coefficients are attended to, (a, b, c, d), have also been considered, though the higher binary forms have been scarcely considered at all. The special ternary cubic forms $ax^3+by^3+cz^3+6lxyz$ have been considered. Special forms of the degree n with n variables, the products of linear factors, present themselves in the theory of the division of the circle (the Kreistheilung) and of the complex numbers connected therewith; but it can hardly be said that these have been studied as a part of the general theory of forms.

Complex Theories.

33. The complex theory which first presented itself is that of the numbers $a + bi$ composed with the imaginary i, $=\sqrt{-1}$; here if a and b are ordinary, or say simplex positive or negative integers, including zero, we regard $a + bi$ as an integer number, or say simply as a number in this complex theory. We have here a zero 0 $(a = 0,$ $b = 0)$ and the units $1, i, -1, -i$, or as these may be written, $1, i, i^2, i^3$ $(i^4 = 1)$; the numbers $a + bi$, $a - bi$, are said to be conjugate numbers, and their product $(a + bi)(a - bi)$, $= a^2 + b^2$, is the norm of each of them. And so the norm of the real number a is $= a^2$, and that of the pure imaginary number bi is $= b^2$. Denoting the norm by the letter N, $N(a \pm bi) = a^2 + b^2$.

Any simplex prime number, $\equiv 1 \pmod{4}$, is the sum of two squares $a^2 + b^2$, for instance $13 = 9 + 4$, and it is thus a product $(a + bi)(a - bi)$, that is, it is not a prime number in the present theory, but each of these factors (or say any number $a + bi$, where $a^2 + b^2$ is a prime number in the simplex theory) is a prime; and any simplex prime number, $\equiv 3 \pmod{4}$, is also a prime in the present theory. The number 2, $= (1 + i)(1 - i)$, is not a prime, but the factors $1 + i$, $1 - i$ are each of them prime; these last differ only by a unit factor $i—1 = i(1 + i)$—so that 2, $= -i(1 + i)^2$, contains a square factor.

In the simplex theory we have numbers, for instance 5, -5, differing from each other only by a unit factor, but we can out of these select one, say the positive number, and attend by preference to this number of the pair. It is in this way—viz. by restricting $a, b, c, \ldots$ to denote terms of the series $2, 3, 5, 7, \ldots$ of positive primes other than unity—that we are enabled to make the definite statement, a positive number N is, and that in one way only, $= a^\alpha b^\beta c^\gamma \ldots$; if N be a positive or negative number, then the theorem of course is, N is, and that in one way only, $= (-1)^m a^\alpha b^\beta c^\gamma \ldots$, where $m = 0$, or 1, and $a, b, c, \ldots, \alpha, \beta, \gamma, \ldots$ are as before. To obtain a like definite statement in the present theory, we require to distinguish between the four numbers $a+bi$, $-a-bi$, $-b+ai$, $b-ai$, which differ from each other only by a unit factor -1, $\pm i$. Consider a number $a+bi$ where a and b are the one of them odd and the other even (a and b may be either of them $=0$, the other is then odd), every prime number $a+bi$ other than $\pm 1 \pm i$ is necessarily of this form: for if a and b were both even, the number would be divisible by 2, or say by $(1+i)^2$, and if a and b were both odd, it would be divisible by $1+i$; then of the four associated numbers $a+bi$, $-a-bi$, $-b+ai$, $b-ai$, there is one and only one, $a+bi$, such that b is even and $a+b-1$ is evenly even; or say one and only one which is $\equiv 1$ (mod. $2(1+i)$). We distinguish such one of the four numbers from the other three and call it a *primary* number; the units ± 1, $\pm i$, and the numbers $\pm 1 \pm i$, are none of them primary numbers. We have then the theorem, a number N is in one way only $= i^m (1+i)^n A^\alpha B^\beta \ldots$, where $m=0$, 1, 2, or 3, n is $=0$ or a positive integer, A, B,... are primary primes, α, β,... positive integers. Here i is a unit of the theory, $1+i$ is a special prime having reference to the number 2, but which might, by an extension of the definition, be called a primary prime, and so reckoned as one of the numbers A, B,...; the theorem stated broadly still is that the number N is, and that in one way only, a product of prime factors, but the foregoing complete statement shows the precise sense in which this theorem must be understood. A like explanation is required in other complex theories; we have to select out of each set of primes differing only by unit factors some one number as a primary prime, and the general theorem then is that every number N is, and that in one way only, $= P \,.\, A^\alpha B^\beta C^\gamma \ldots$, where P is a product of unities, and A, B, C,... are primary primes.

34. We have in the simplex theory (*ante*, No. 10) the theorem that, p being an odd prime, there exists a system of $p-1$ residues, that is, that any number not divisible by p is, to the modulus p, congruent to one, and only one, of the $p-1$ numbers 1, 2, 3,..., $p-1$. The analogous theorem in the complex theory is that, for any prime number p other than $\pm 1 \pm i$, there exists a system of $N(p)-1$ residues, that is, that every number not divisible by p is, to the modulus p, congruent to one of these $N(p)-1$ numbers.

But p may be a real prime such as 3, or a complex prime such as $3+2i$; and the system of residues presents itself naturally under very different forms in the two cases respectively. Thus in the case $p=3$, $N(3)=9$, the residues may be taken to be

$$\begin{array}{ccc} & 1\ , & 2\ , \\ i, & 1+i\ , & 2+i\ , \\ 2i, & 1+2i, & 2+2i, \end{array}$$

being in number $N(3)-1=8$. And for $p=3+2i$, $N(3+2i)=13$, they may be taken to be the system of residues of 13 in the simplex theory, viz. the real numbers 1, 2, 3,..., 12. We have in fact $5+i=(2+3i)(1-i)$, that is, $5+i\equiv 0\,(\text{mod. } 2+3i)$, and consequently $a+bi\equiv a-5b$, a real number which, when $a+bi$ is not divisible by $3+2i$, may have any one of the foregoing values 1, 2, 3,..., 12.

Taking then any number x not divisible by p, the $N(p)-1$ residues each multiplied by x are, to the modulus p, congruent to the series of residues in a different order; and we thus have,—say this is Fermat's theorem for the complex theory—$x^{N(p)-1}-1\equiv 0$ (mod. p), with all its consequences, in particular, the theory of prime roots.

In the case of a complex modulus such as $3+2i$, the theory is hardly to be distinguished from its analogue in the ordinary theorem; a prime root is $=2$, and the series of powers is 2, 4, 8, 3, 6, 12, 11, 9, 5, 10, 7, 1, for the modulus $3+2i$ as for the modulus 13. But for a real prime such as 3 the prime root is a complex number; taking it to be $=2+i$, we have $(2+i)^8-1\equiv 0$ (mod. 3), and the series of powers in fact is $2+i$, i, $2+2i$, 2, $1+2i$, $2i$, $1+i$, 1, viz. we thus have the system of residues (mod. 3).

We have in like manner a theory of quadratic residues; a Legendrian symbol $\left[\frac{p}{q}\right]$ (which, if p, q, are uneven primes not necessarily primary but subject to the condition that their imaginary parts are even, denotes $+1$ or -1 according as $p^{\frac{1}{2}(Nq-1)}$ is $\equiv 1$ or $\equiv -1$ (mod. q), so that $\left[\frac{p}{q}\right]=+1$ or -1 according as p is or is not a residue of q), a law of reciprocity expressed by the very simple form of equation $\left[\frac{p}{q}\right]=\left[\frac{q}{p}\right]$, and generally a system of properties such as that which exists in the simplex theory.

The theory of quadratic forms (a, b, c) has been studied in this complex theory; the results correspond to those of the simplex theory.

35. The complex theory with the imaginary cube root of unity has also been studied; the imaginary element is here γ, $=\frac{1}{2}(-1+\sqrt{-3})$, a root of the equation $\gamma^2+\gamma+1=0$; the form of the complex number is thus $a+b\gamma$, where a and b are any positive or negative integers, including zero. The conjugate number is $a+b\gamma^2$, $=a-b-b\gamma$, and the product $(a+b\gamma)(a+b\gamma^2)$, $=a^2-ab+b^2$, is the norm of each of the factors $a+b\gamma$, $a+b\gamma^2$. The whole theory corresponds very closely to, but is somewhat more simple than, that of the complex numbers $a+bi$.

36. The last-mentioned theory is a particular case of the complex theory for the imaginary λth roots of unity, λ being an odd prime. Here α is determined by the equation $\frac{\alpha^\lambda-1}{\alpha-1}=0$, that is, $\alpha^{\lambda-1}+\alpha^{\lambda-2}+\ldots+\alpha+1=0$, and the form of the complex number is $f(\alpha)$, $=a+b\alpha+c\alpha^2+\ldots+k\alpha^{\lambda-2}$, where $a, b, c, \ldots, k$, are any positive or negative integers, including zero. We have $\lambda-1$ conjugate forms, viz. $f(\alpha), f(\alpha^2), \ldots, f(\alpha^{\lambda-1})$, and the product of these is the norm of each of the factors $Nf(\alpha)$, $=Nf(\alpha^2)$, $=\ldots$, $=Nf(\alpha^{\lambda-1})$. Taking g any prime root of λ, $g^{\lambda-1}-1\equiv 0$ (mod. λ), the roots $\alpha, \alpha^2, \ldots, \alpha^{\lambda-1}$, may be arranged in the order $\alpha, \alpha^g, \alpha^{g^2}, \ldots, \alpha^{g^{\lambda-2}}$; and we have thence a grouping of the roots in periods, viz. if $\lambda-1$ be in any manner whatever expressed as a product of two factors, $\lambda-1=ef$, we may with the $\lambda-1$ roots form e periods $\eta_0, \eta_1, \ldots, \eta_{e-1}$, each of

f roots. For instance, when $\lambda = 13$, a prime root is $g = 2$, and $\lambda - 1 = ef = 3.4$; then the three periods each of four roots are

$$\eta_0 = \alpha + \alpha^8 + \alpha^{12} + \alpha^5,$$
$$\eta_1 = \alpha^2 + \alpha^3 + \alpha^{11} + \alpha^{10},$$
$$\eta_2 = \alpha^4 + \alpha^6 + \alpha^9 + \alpha^7.$$

So also, if $ef = 2.6$, then the 2 periods each of 6 roots are

$$\eta_0 = \alpha + \alpha^4 + \alpha^3 + \alpha^{12} + \alpha^9 + \alpha^{10},$$
$$\eta_1 = \alpha^2 + \alpha^8 + \alpha^6 + \alpha^{11} + \alpha^5 + \alpha^7;$$

and so in other cases. In particular, if $f = 1$ and consequently $e = \lambda - 1$, the e periods each of f roots are, in fact, the single roots $\alpha, \alpha^g, \ldots, \alpha^{g^{\lambda-2}}$. We may, in place of the original form of the complex number

$$f(\alpha) = a + b\alpha + c\alpha^2 + \ldots + k\alpha^{\lambda-2},$$

consider the new form $f(\eta) = a\eta + b\eta_1 + \ldots + l\eta_{e-1}$, which when $f = 1$ is equivalent to the original form, but in any other case denotes a special form of complex number; instead of $\lambda - 1$ we have only e conjugate numbers, and the product of these e numbers may be regarded as the norm of $f(\eta)$.

37. The theory for the roots α includes as part of itself the theory for the periods corresponding to every decomposition whatever $\lambda - 1 = ef$ of $\lambda - 1$ into two factors, but each of these may be treated apart from the others as a theory complete in itself. In particular, a simple case is that of the half-periods $e = 2$, $f = \frac{1}{2}(\lambda - 1)$; and, inasmuch as the characteristic phenomenon of ideal numbers presents itself in this theory of the half-periods (first for the value $\lambda = 23$), it will be sufficient, by way of illustration of the general theory, to consider only this more special and far easier theory; we may even assume $\lambda = 23$.*

For the case in question, $\lambda - 1 = ef = 2 . \frac{1}{2}(\lambda - 1)$, we have the two periods η_0, η_1, each of $\frac{1}{2}(\lambda - 1)$ roots; from the expressions for η_0, η_1, in terms of the roots we obtain at once $\eta_0 + \eta_1 = -1$, and with a little more difficulty $\eta_0\eta_1 = -\frac{1}{4}(\lambda - 1)$ or $\frac{1}{4}(\lambda + 1)$, according as λ is $\equiv 1$ or $3 \pmod{4}$, that is, in the two cases respectively η_0, η_1, are the roots of the equation $\eta^2 + \eta - \frac{1}{4}(\lambda - 1) = 0$, and $\eta^2 + \eta + \frac{1}{4}(\lambda + 1) = 0$. And this equation once obtained, there is no longer any occasion to consider the original equation of the order $\lambda - 1$, but the theory is that of the complex numbers $a\eta_0 + b\eta_1$, or

* In the theory of the roots α, ideal numbers do not present themselves for the values $\lambda = 3$, 5, or 7; they do for the value $\lambda = 23$. It is stated in Smith's "Report on the Theory of Numbers," *Brit. Assoc. Report* for 1860, p. 136, [Collected Works, vol. I. p. 114], that "for $\lambda = 11$, $\lambda = 13$, $\lambda = 17$, and $\lambda = 19$, it is not possible to say whether this is or is not the case for these values also." The writer is not aware whether this question has been settled; but in Reuschle's *Tafeln*, 1875, no ideal factors present themselves for these values of λ; and it is easy to see that, in the theory of the half-periods, the ideal factors first present themselves for the value $\lambda = 23$. It may be remarked that the solution of the question depends on the determination of a system of fundamental units for the values in question $\lambda = 11$, 13, 17, and 19; the theory of the units in the several complex theories is an important and difficult part of the theory, not presenting itself in the theory of the half-periods, which is alone attended to in the text.

if we please $a+b\eta$, composed with the roots of this quadric equation,—say the complex numbers $a+b\eta$, where a and b are any positive or negative integer numbers, including zero. In the case $\lambda=23$, the quadric equation is $\eta^2+\eta+6=0$. We have $N(a+b\eta)=(a+b\eta_0)(a+b\eta_1)=a^2-ab+\frac{1}{4}(\lambda+1)b^2$; and for $\lambda=23$, this is $N(a+b\eta)=a^2-ab+6b^2$. It may be remarked that there is a connexion with the theory of the quadratic forms of the determinant -23, viz. there are here the three improperly primitive forms (2, 1, 12), (4, 1, 6), (4, −1, 6), 23 being the smallest prime number for which there exists more than one improperly primitive form.

38. Considering then the case $\lambda=23$, we have η_0, η_1, the roots of the equation $\eta^2+\eta+6=0$; and a real number P is composite when it is $=(a+b\eta_0)(a+b\eta_1)$, $=a^2-ab+6b^2$, viz. if $4P=(2a-b)^2+23b^2$. Hence no number, and in particular no positive real prime P, can be composite unless it is a (quadratic) residue of 23; the residues of 23 are 1, 2, 3, 4, 6, 8, 9, 12, 13, 16, 18; and we have thus, for instance, 5, 7, 11, as numbers which are *not* composite, while 2, 3, 13, are numbers which are not by the condition precluded from being composite: they are not, according to the foregoing signification of the word, composite (for 8, 12, 52, are none of them of the form x^2+23y^2), but *some* such numbers, residues that is of 23, are composite, for instance 59, $=(5-2\eta_0)(5-2\eta_1)$. And we have an indication, so to speak, of the composite nature of *all* such numbers; take for instance 13, we have $(\eta-4)(\eta+5)=-2.13$, where 13 does not divide either $\eta-4$ or $\eta+5$, and we are led to conceive it as the product of two ideal factors, one of them dividing $\eta-4$, the other dividing $\eta+5$. It appears, moreover, that a power 13^3 is in fact composite, viz. we have

$$13^3=(31-12\eta_0)(31-12\eta_1),\quad (2197=961+372+864);$$

and writing $13=\sqrt[3]{31-12\eta_0}.\sqrt[3]{31-12\eta_1}$ we have 13 as the product of two ideal numbers each represented as a cube root; it is to be observed that, 13 being in the simplex theory a prime number, these are regarded as prime ideal numbers. We have in like manner

$$2=\sqrt[3]{1-\eta_0}.\sqrt[3]{1-\eta_1},\quad 3=\sqrt[3]{1-2\eta_0}.\sqrt[3]{1-2\eta_1},\ \&c.;$$

every positive real prime which is a residue of 23 is thus a product of two factors ideal or actual. And, reverting to the equation $(\eta-4)(\eta+5)=-2.13$, or as this may be written

$$(\eta_1-4)(\eta_1+5)=-\sqrt[3]{1-\eta_0}\sqrt[3]{1-\eta_1}\sqrt[3]{31-12\eta_0}\sqrt[3]{31-12\eta_1},$$

we have $(\eta_1-4)^3$ and $(1-\eta_0)(31-12\eta_0)$ each $=14+55\eta_1$, or say

$$\eta_1-4=\ \sqrt[3]{1-\eta_0}\sqrt[3]{31-12\eta_0},$$

and similarly

$$\eta_1+5=-\sqrt[3]{1-\eta_1}\sqrt[3]{31-12\eta_1},$$

so that we verify that η_1-4, η_1+5, do thus in fact each of them contain an ideal factor of 13.

39. We have $2=\sqrt[3]{1-\eta_0}\sqrt[3]{1-\eta_1}$, viz. the ideal multiplier $\sqrt[3]{1-\eta_0}$ renders actual one of the ideal factors $\sqrt[3]{1-\eta_1}$ of 2, and it is found that this same ideal multiplier

$\sqrt[3]{1-\eta_0}$ renders actual one of the two ideal factors of any other decomposable number 3, 13, &c.,

$$\sqrt[3]{1-2\eta_0}\sqrt[3]{1-\eta_0}=1+\eta_0,\quad \sqrt[3]{31-12\eta_0}\sqrt[3]{1-\eta_0}=-5-\eta_0,\ \text{\&c.}$$

Similarly the conjugate multiplier $\sqrt[3]{1-\eta_1}$ renders actual the other ideal factor of any number 2, 3, 13, &c. We have thus two classes, or, reckoning also actual numbers, three classes of prime numbers, viz. (i) ideal primes rendered actual by the multiplier $\sqrt[3]{1-\eta_0}$, (ii) ideal primes rendered actual by the multiplier $\sqrt[3]{1-\eta_1}$, (iii) actual primes. This is a general property in the several complex theories; there is always a finite number of classes of ideal numbers, distinguished according to the multipliers by which they are rendered actual; the actual numbers form a "principal" class.

40. General theory of congruences—irreducible functions. In the complex theory relating to the roots of the equation $\eta^2+\eta+6=0$, there has just been occasion to consider the equation $(\eta-4)(\eta+5)=-2.13$, or say the congruence $(\eta-4)(\eta+5)\equiv 0\ (\text{mod.}\ 13)$; in this form the relation $\eta^2+\eta+6=0$ is presupposed, but if, dropping this equation, η be regarded as arbitrary, then there is the congruence $\eta^2+\eta+6\equiv(\eta-4)(\eta+5)\ (\text{mod.}\ 13)$. For a different modulus, for instance 11, there is not any such congruence exhibiting a decomposition of $\eta^2+\eta+6$ into factors. The function $\eta^2+\eta+6$ is irreducible, that is, it is not a product of factors with integer coefficients; in respect of the modulus 13 it becomes reducible, that is, it breaks up into factors having integer coefficients, while for the modulus 11 it continues irreducible. And there is a like general theory in regard to any rational and integral function $F(x)$ with integer coefficients; such function, assumed to be irreducible, may for a given prime modulus p continue irreducible, that is, it may not admit of any decomposition into factors with integer coefficients; or it may become reducible, that is, admit of a decomposition $F(x)\equiv\phi(x)\psi(x)\chi(x)\ldots(\text{mod.}\ p)$. And, when this is so, it is thus a product, in one way only, of factors $\phi(x)$, $\psi(x)$, $\chi(x)$, ..., which are each of them irreducible in regard to the same modulus p; any such factor may be a linear function of x, and as such irreducible; or it may be an irreducible function of the second or any higher degree. It is hardly necessary to remark that, in this theory, functions which are congruent to the modulus p are regarded as identical, and that in the expression of $F(x)$ an irreducible function $\phi(x)$ may present itself either as a simple factor, or as a multiple factor, with any exponent. The decomposition is analogous to that of a number into its prime factors; and the whole theory of the rational and integral function $F(x)$ in regard to the modulus p is in many respects analogous to that of a prime number regarded as a modulus. The theory has also been studied where the modulus is a power p^ν.

41. The congruence-imaginaries of Galois. If $F(x)$ be an irreducible function to a given prime modulus p, this implies that there is no integer value of x satisfying the congruence $F(x)\equiv 0\ (\text{mod.}\ p)$; we assume such a value and call it i, that is, we assume $F(i)\equiv 0\ (\text{mod.}\ p)$; the step is exactly analogous to that by which, starting from the notion of a real root, we introduce into algebra the ordinary imaginary $i=\sqrt{-1}$. For instance, x^2-x+3 is an irreducible function to the modulus 7: there is no integer solution of the congruence $x^2-x+3\equiv 0\ (\text{mod.}\ 7)$. Assuming a solution i such that

$i^2 - i + 3 \equiv 0$ (mod. 7), we have, always to this modulus, $i^2 = i - 3$, and thence i^3, i^4, &c., each of them equal to a linear function of i. We consider the numbers of the form $a + bi$, where a and b are ordinary integers which may be regarded as having each of them the values 0, 1, 2, 3, 4, 5, or 6; there are thus 7^2, $=49$, such numbers, or, excluding zero, 48 numbers; and it is easy to verify that these are, in fact, the numbers $i, i^2, \ldots, i^{47}, i^{48}, = 1$, that is, we have i a prime root of the congruence $x^{48} - 1 \equiv 0$ (mod. 7). The irreducible function may be of the third or any higher degree; thus for the same modulus 7 there is the cubic function $x^3 - x + 2$, giving rise to a theory of the numbers of the form $a + bi + ci^2$, where i is a congruence-imaginary such that $i^3 - i + 2 \equiv 0$ (mod. 7); and instead of 7 the modulus may be any other odd prime p.

Ordinary Theory, Third Part.

42. In what precedes, no mention has been made of the so-called Pellian equation $x^2 - Dy^2 = 1$ (where D is a given positive number), and of the allied equations $x^2 - Dy^2 = -1$, or $= \pm 4$. The equations with the sign + have always a series of solutions, those with the sign − only for certain values of D; in every case where the solutions exist, a least solution is obtainable by a process depending on the expression of $\sqrt{D}$ as a continued fraction, and from this least solution the whole series of solutions can be obtained without difficulty. The equations are very interesting, as well for their own sake as in connexion with the theory of the binary quadratic forms of a positive non-square determinant.

43. The theory of the expression of a number as a sum of squares or polygonal numbers has been developed apart from the general theory of the binary, ternary, and other quadratic forms to which it might be considered as belonging. The theorem for two squares, that every prime number of the form $4n + 1$ is, and that in one way only, a sum of two squares, is a fundamental theorem in relation to the complex numbers $a + bi$. A sum of two squares multiplied by a sum of two squares is always a sum of two squares, and hence it appears that every number of the form $2^a(4n + 1)$ is (in general, in a variety of ways) a sum of two squares.

Every number of the form $4n + 2$ or $8n + 3$ is a sum of three squares; even in the case of a prime number $8n + 3$ there is in general more than one decomposition, thus $59 = 25 + 25 + 9$ and $= 49 + 9 + 1$. Since a sum of three squares multiplied by a sum of three squares is not a sum of three squares, it is not enough to prove the theorem in regard to the primes of the form $8n + 3$.

Every prime number is (in general, in more than one way) a sum of four squares; and therefore every number is (in general, in more than one way) a sum of four squares, for a sum of four squares multiplied by a sum of four squares is always a sum of four squares.

Every number is (in general, in several ways) a sum of $m + 2$ $(m + 2)$gonal numbers, that is, of numbers of the form $\frac{1}{2}m(x^2 - x) + x$; and of these $m - 2$ may be at pleasure equal to 0 or 1; in particular, every number is a sum of three triangular numbers (a theorem of Fermat's).

The theorems in regard to three triangular numbers and to four square numbers are exhibited by certain remarkable identities in the Theory of Elliptic Functions; and generally there is in this subject a great mass of formulæ connected with the theory of the representation of numbers by quadratic forms. The various theorems in regard to the number of representations of a number as the sum of a definite number of squares cannot be here referred to.

44. The equation $x^\lambda + y^\lambda = z^\lambda$, where λ is any positive integer greater than 2, is not resoluble in whole numbers (a theorem of Fermat's). The general proof depends on the theory of the complex numbers composed of the λth roots of unity, and presents very great difficulty; in particular, distinctions arise according as the number λ does or does not divide certain of Bernoulli's numbers.

45. Lejeune-Dirichlet employs, for the determination of the number of quadratic forms of a given positive or negative determinant, a remarkable method depending on the summation of a series Σf^{-s}, where the index s is greater than but indefinitely near to unity.

46. Very remarkable formulæ have been given by Legendre, Tchebycheff, and Riemann for the approximate determination of the number of prime numbers less than a given large number x. Factor tables have been formed for the first nine million numbers, and the number of primes counted for successive intervals of 50,000; and these are found to agree very closely with the numbers calculated from the approximate formulæ. Legendre's expression is of the form $\dfrac{x}{\log x - A}$, where A is a constant not very different from unity; Tchebycheff's depends on the logarithm-integral $\mathrm{li}(x)$; and Riemann's, which is the most accurate, but is of a much more complicated form, contains a series of terms depending on the same integral.

The classical works on the Theory of Numbers are Legendre, *Théorie des Nombres*, 1st ed. 1798, 3rd ed. 1830; Gauss, *Disquisitiones Arithmeticæ*, Brunswick, 1801 (reprinted in the collected works, vol. I., Göttingen, 1863; French translation, under the title *Recherches Arithmétiques*, by Poullet-Delisle, Paris, 1807); and Lejeune-Dirichlet, *Vorlesungen über Zahlentheorie*, 3rd ed., with extensive and valuable additions by Dedekind, Brunswick, 1879—81. We have by the late Prof. H. J. S. Smith the extremely valuable series of "Reports on the Theory of Numbers," Parts I. to VI., *British Association Reports*, 1859—62, 1864—65, which, with his own original researches, [are] printed in the [first volume of the] collected works [published in 1894] by the Clarendon Press. See also Cayley, "Report of the Mathematical Tables Committee," *Brit. Assoc. Report*, 1875, pp. 305—336, [611], for a list of tables relating to the Theory of Numbers, and Mr J. W. L. Glaisher's introduction to the *Factor Table for the Sixth Million*, London, 1883, in regard to the approximate formulæ for the number of prime numbers.

796.

SERIES.

[From the *Encyclopædia Britannica, Ninth Edition,* vol. XXI. (1886), pp. 677—682.]

A SERIES is a set of terms considered as arranged in order. Usually the terms are or represent numerical magnitudes, and we are concerned with the sum of the series. The number of terms may be limited or without limit; and we have thus the two theories, finite series and infinite series. The notions of convergency and divergency present themselves only in the latter theory.

Finite Series.

1. Taking the terms to be numerical magnitudes, or say numbers, if there be a definite number of terms, then the sum of the series is nothing else than the number obtained by the addition of the terms; e.g. $4+9+10=23$, $1+2+4+8=15$. In the first example there is no apparent law for the successive terms; in the second example there is an apparent law. But it is important to notice that in neither case is there a determinate law: we can in an infinity of ways form series beginning with the apparently irregular succession of terms 4, 9, 10, or with the apparently regular succession of terms 1, 2, 4, 8. For instance, in the latter case we may have a series with the general term 2^n, when for $n=0, 1, 2, 3, 4, 5, \ldots$ the series will be $1, 2, 4, 8, 16, 32, \ldots$; or a series with the general term $\frac{1}{6}(n^3+5n+6)$, where for the same values of n the series will be $1, 2, 4, 8, 15, 26, \ldots$ The series may contain negative terms, and in forming the sum each term is of course to be taken with the proper sign.

2. But we may have a given law, such as either of those just mentioned, and the question then arises, to find the sum of an indefinite number of terms, or say of n terms (n standing for any positive integer number at pleasure) of the series. The expression for the sum cannot in this case be obtained by actual addition; the formation by addition of the sum of two terms, of three terms, &c., will, it may be, suggest (but it cannot do more than suggest) the expression for the sum of n terms

of the series. For instance, for the series of odd numbers $1+3+5+7+\ldots$, we have $1=1$, $1+3=4$, $1+3+5=9$, &c. These results at once suggest the law, $1+3+5+\ldots+(2n-1)=n^2$, which is in fact the true expression for the sum of n terms of the series; and this general expression, once obtained, can afterwards be verified.

3. We have here the theory of finite series: the general problem is, u_n being a given function of the positive integer n, to determine as a function of n the sum $u_0+u_1+u_2+\ldots+u_n$, or, in order to have n instead of $n+1$ terms, say the sum $u_0+u_1+u_2+\ldots+u_{n-1}$.

Simple cases are the three which follow.

(i) The arithmetic series,

$$a+(a+b)+(a+2b)+\ldots+(a+\overline{n-1}\,)b;$$

writing here the terms in the reverse order, it at once appears that twice the sum is $=2a+\overline{n-1}b$ taken n times: that is, the sum $=na+\tfrac{1}{2}n(n-1)b$. In particular, we have an expression for the sum of the natural numbers

$$1+2+3+\ldots+n=\tfrac{1}{2}n(n+1),$$

and an expression for the sum of the odd numbers

$$1+3+5+\ldots+(2n-1)=n^2.$$

(ii) The geometric series,

$$a+ar+ar^2+\ldots+ar^{n-1};$$

here the difference between the sum and r times the sum is at once seen to be $=a-ar^n$, and the sum is thus $=a\dfrac{1-r^n}{1-r}$; in particular, the sum of the series

$$1+r+r^2+\ldots+r^{n-1}=\frac{1-r^n}{1-r}.$$

(iii) But the harmonic series,

$$\frac{1}{a}+\frac{1}{a+b}+\frac{1}{a+2b}+\ldots+\frac{1}{a+(n-1)b},$$

or say $\frac{1}{1}+\frac{1}{2}+\frac{1}{3}\ldots+\dfrac{1}{n}$, does not admit of summation; there is no algebraical function of n which is equal to the sum of the series.

4. If the general term be a given function u_n, and we can find v_n a function of n such that $v_{n+1}-v_n=u_n$, then we have $u_0=v_1-v_0$, $u_1=v_2-v_1$, $u_2=v_3-v_2,\ldots$, $u_n=v_{n+1}-v_n$; and hence $u_0+u_1+u_2+\ldots+u_n=v_{n+1}-v_0$,—an expression for the required sum. This is in fact an application of the Calculus of Finite Differences. In the notation of this calculus $v_{n+1}-v_n$ is written Δv_n; and the general inverse problem, or problem of integration, is from the equation of differences $\Delta v_n=u_n$ (where u_n is a given function of n) to find v_n. The general solution contains an arbitrary constant, $v_n=V_n+C$; but this disappears in the difference $v_{n+1}-v_0$. As an example consider the series

$$u_0+u_1+\ldots+u_n=0+1+3+\ldots+\tfrac{1}{2}n(n+1);$$

here, observing that

$$n(n+1)(n+2)-(n-1)n(n+1)=n(n+1)(\overline{n+2}-\overline{n-1}),\ =3n(n+1),$$

we have

$$v_{n+1}=\tfrac{1}{6}n(n+1)(n+2);$$

and hence

$$1+3+6+\ldots+\tfrac{1}{2}n(n+1)=\tfrac{1}{6}n(n+1)(n+2),$$

as may be at once verified for any particular value of n.

Similarly, when the general term is a factorial of the order r, we have

$$1+\frac{r+1}{1}+\ldots+\frac{n(n+1)\ldots(n+r-1)}{1.2\ \ldots\ r}=\frac{n(n+1)\ldots(n+r)}{1.2\ \ldots(r+1)}.$$

5. If the general term u_n be any rational and integral function of n, we have

$$u_n=u_0+\frac{n}{1}\Delta u_0+\frac{n(n-1)}{1.2}\Delta^2u_0+\ldots+\frac{n(n-1)\ldots(n-p+1)}{1.2\ldots p}\Delta^p u_0,$$

where the series is continued only up to the term depending on p, the degree of the function u_n, for all the subsequent terms vanish. The series is thus decomposed into a set of series which have each a factorial for the general term, and which can be summed by the last formula; thus we obtain

$$u_0+u_1\ldots+u_n=(n+1)u_0+\frac{(n+1)n}{1.2}\Delta u_0+\ldots+\frac{(n+1)n(n-1)\ldots(n-p+1)}{1.2.3\ldots(p+1)}\Delta^p u_0,$$

which is a function of the degree $p+1$.

Thus for the before-mentioned series $1+2+4+8+\ldots$, if it be assumed that the general term u_n is a cubic function of n, and writing down the given terms and forming the differences, 1, 2, 4, 8; 1, 2, 4; 1, 2; 1, we have

$$u_n=1+\frac{n}{1}+\frac{n(n-1)}{1.2}+\frac{n(n-1)(n-2)}{1.2.3}\ \{=\tfrac{1}{6}(n^3+5n+6),\text{ as above}\};$$

and the sum

$$u_0+u_1+\ldots+u_n=n+1+\frac{(n+1)n}{1.2}+\frac{(n+1)n(n-1)}{1.2.3}+\frac{(n+1)n(n-1)(n-2)}{1.2.3.4},$$

$$=\tfrac{1}{24}(n^4+2n^3+11n^2+34n+24).$$

As particular cases we have expressions for the sums of the powers of the natural numbers—

$$1^2+2^2+\ldots+n^2=\tfrac{1}{6}n(n+1)(2n+1);\quad 1^3+2^3+\ldots+n^3=\tfrac{1}{4}n^2(n+1)^2:$$

observe that the latter $=(1+2\ldots+n)^2$; and so on.

6. We may, from the expression for the sum of the geometric series, obtain by differentiation other results: thus

$$1+r+r^2+\ldots+r^{n-1}=\frac{1-r^n}{1-r}$$

gives

$$1+2r+3r^2+\dots+(n-1)\,r^{n-2}=\frac{d}{dr}\,\frac{1-r^n}{1-r},\quad =\frac{1-nr^{n-1}+(n-1)\,r^n}{(1-r)^2};$$

and we might in this way find the sum $u_0+u_1r+\dots+u_nr^n$, where u_n is any rational and integral function of n.

7. The expression for the sum $u_0+u_1+\dots+u_n$ of an indefinite number of terms will in many cases lead to the sum of the infinite series $u_0+u_1+\dots$; but the theory of infinite series requires to be considered separately. Often in dealing apparently with an infinite series $u_0+u_1+\dots$ we consider rather an indefinite than an infinite series, and are not in any wise really concerned with the sum of the series or the question of its convergency: thus the equation

$$\left(1+mx+\frac{m(m-1)}{1.2}x^2+\dots\right)\left(1+nx+\frac{n(n-1)}{1.2}x^2+\dots\right)$$
$$=1+(m+n)\,x+\frac{(m+n)(m+n-1)}{1.2}x^2+\dots$$

really means the series of identities

$$(m+n)=m+n,$$
$$\frac{(m+n)(m+n-1)}{1.2}=\frac{m(m-1)}{1.2}+2\frac{m}{1}\frac{n}{1}+\frac{n(n-1)}{1.2},\ \text{\&c.},$$

obtained by multiplying together the two series of the left-hand side. Again, in the method of generating functions we are concerned with an equation $\phi(t)=A_0+A_1t+\dots+A_nt^n+\dots$, where the function $\phi(t)$ is used only to express the law of formation of the successive coefficients.

It is an obvious remark that, although according to the original definition of a series the terms are considered as arranged in a determinate order, yet in a finite series (whether the number of terms be definite or indefinite) the sum is independent of the order of arrangement.

Infinite Series.

8. We consider an infinite series $u_0+u_1+u_2+\dots$ of terms proceeding according to a given law, that is, the general term u_n is given as a function of n. To fix the ideas the terms may be taken to be positive numerical magnitudes, or say numbers continually diminishing to zero; that is, $u_n>u_{n+1}$, and u_n is, moreover, such a function of n that, by taking n sufficiently large, u_n can be made as small as we please.

Forming the successive sums $S_0=u_0$, $S_1=u_0+u_1$, $S_2=u_0+u_1+u_2,\dots$, these sums S_0, S_1, $S_2,\dots$ will be a series of continually increasing terms, and if they increase up to a determinate finite limit S (that is, if there exists a determinate numerical magnitude S such that, by taking n sufficiently large, we can make $S-S_n$ as small as we please), S is said to be the sum of the infinite series. To show that we can actually have an infinite series with a given sum S, take u_0 any number less than S, then $S-u_0$ is positive, and taking u_1 any numerical magnitude less than $S-u_0$, then $S-u_0-u_1$ is positive. And going on continually in this manner we obtain a series $u_0+u_1+u_2+\ldots$, such that for any value of n however large $S-u_0-u_1\ldots-u_n$ is positive; and if as n increases this difference diminishes to zero, we have $u_0+u_1+u_2+\ldots$, an infinite series having S for its sum. Thus, if $S=2$, and we take $u_0<2$, say $u_0=1$; $u_1<2-1$, say $u_1=\frac{1}{2}$; $u_2<2-1-\frac{1}{2}$, say $u_2=\frac{1}{4}$; and so on, we have $1+\frac{1}{2}+\frac{1}{4}+\ldots=2$; or, more generally, if r be any positive number less than 1, then $1+r+r^2+\ldots=\dfrac{1}{1-r}$, that is, the infinite geometric series with the first term $=1$, and with a ratio $r<1$, has the finite sum $\dfrac{1}{1-r}$. This, in fact, follows from the expression $1+r+r^2\ldots+r^{n-1}=\dfrac{1-r^n}{1-r}$ for the sum of the finite series; taking $r<1$, then as n increases r^n decreases to zero, and the sum becomes more and more nearly $=\dfrac{1}{1-r}$.

9. An infinite series of positive numbers can, it is clear, have a sum only if the terms continually diminish to zero; but it is not conversely true that, if this condition be satisfied, there will be a sum. For instance, in the case of the harmonic series $1+\frac{1}{2}+\frac{1}{3}+\ldots$, it can be shown that by taking a sufficient number of terms the sum of the finite series may be made as large as we please. For, writing the series in the form $1+\frac{1}{2}+(\frac{1}{3}+\frac{1}{4})+(\frac{1}{5}+\frac{1}{6}+\frac{1}{7}+\frac{1}{8})+\ldots$, the number of terms in the brackets being doubled at each successive step, it is clear that the sum of the terms in any bracket is always $>\frac{1}{2}$; hence by sufficiently increasing the number of brackets the sum may be made as large as we please. In the foregoing series, by grouping the terms in a different manner $1+(\frac{1}{2}+\frac{1}{3})+(\frac{1}{4}+\frac{1}{5}+\frac{1}{6}+\frac{1}{7})+\ldots$, the sum of the terms in any bracket is always <1; we thus arrive at the result that ($n=3$ at least) the sum of 2^n terms of the series is $>1+\frac{1}{2}n$ and $<n$.

10. An infinite series may contain negative terms; suppose in the first instance that the terms are alternately positive and negative. Here the absolute magnitudes of the terms must decrease down to zero, but this is a sufficient condition in order that the series may have a sum. The case in question is that of a series $v_0-v_1+v_2-\ldots$, where $v_0, v_1, v_2,\ldots$ are all positive and decrease down to zero. Here, forming the successive sums $S_0=v_0$, $S_1=v_0-v_1$, $S_2=v_0-v_1+v_2,\ldots$, $S_0, S_1, S_2,\ldots$ are all positive, and we have $S_0>S_1$, $S_1<S_2$, $S_2>S_3,\ldots$, and $S_{n+1}-S_n$ tends continually to zero. Hence the sums $S_0, S_1, S_2,\ldots$ tend continually to a positive limit S in such wise that $S_0, S_2, S_4,\ldots$ are each of them greater and $S_1, S_3, S_5,\ldots$ are each of them less than S; and we thus have S as the sum of the series. The series $1-\frac{1}{2}+\frac{1}{3}-\frac{1}{4}+\ldots$ will serve as an example. The case just considered includes the apparently more general one where the series consists of alternate groups of positive and negative terms respectively; the terms of the same group may be united into a single term $\pm v_n$, and the original series will have a sum only if the resulting series $v_0-v_1+v_2\ldots$ has a sum, that is, if the positive partial sums $v_0, v_1, v_2,\ldots$ decrease down to zero.

The terms at the beginning of a series may be irregular as regards their signs; but, when this is so, all the terms in question (assumed to be finite in number) may be united into a single term, which is of course finite, and instead of the original series only the remaining terms of the series need be considered. Every infinite series whatever is thus substantially included under the two forms,—terms all positive and terms alternately positive and negative.

11. In brief, the sum (if any) of the infinite series $u_0 + u_1 + u_2 + \dots$ is the finite limit (if any) of the successive sums u_0, $u_0 + u_1$, $u_0 + u_1 + u_2, \dots$; if there is no such limit, then there is no sum. Observe that the assumed order $u_0, u_1, u_2, \dots$ of the terms is part of and essential to the definition; the terms in any other order may have a different sum, or may have no sum. A series having a sum is said to be "convergent"; a series which has no sum is "divergent."

If a series of positive terms be convergent, the terms cannot, it is elear, continually increase, nor can they tend to a fixed limit: the series $1 + 1 + 1 + \dots$ is divergent. For the convergency of the series it is necessary (but, as has been shown, not sufficient) that the terms shall decrease to zero. So, if a series with alternately positive and negative terms be convergent, the absolute magnitudes cannot, it is clear, continually increase. In reference to such a series Abel remarks, "Peut-on imaginer rien de plus horrible que de débiter $0 = 1^n - 2^n + 3^n - 4^n +$, &c., où n est un nombre entier positif?" Neither is it allowable that the absolute magnitudes shall tend to a fixed limit. The so-called "neutral" series $1 - 1 + 1 - 1 \dots$ is divergent: the successive sums do not tend to a determinate limit, but are alternately $+1$ and 0; it is necessary (and also sufficient) that the absolute magnitudes shall decrease to zero.

In the so-called semi-convergent series, we have an equation of the form

$$S = U_0 - U_1 + U_2 - \dots,$$

where the positive values U_0, U_1, $U_2, \dots$ decrease to a minimum value, suppose U_p, and afterwards increase; the series is divergent and has no sum, and thus S is not the sum of the series. S is only a number or function calculable approximately by means of the series regarded as a finite series terminating with the term $\pm U_p$. The successive sums U_0, $U_0 - U_1$, $U_0 - U_1 + U_2, \dots$ up to that containing $\pm U_p$, give alternately superior and inferior limits of the number or function S.

12. The condition of convergency may be presented under a different form: let the series $u_0 + u_1 + u_2 + \dots$ be convergent, then, taking m sufficiently large, the sum is the limit not only of $u_0 + u_1 + \dots + u_m$ but also of $u_0 + u_1 + \dots + u_{m+r}$, where r is any number as large as we please. The difference of these two expressions must therefore be indefinitely small; by taking m sufficiently large the sum $u_{m+1} + u_{m+2} + \dots + u_{m+r}$ (where r is any number however large) can be made as small as we please; or, as this may also be stated, the sum of the infinite series $u_{m+1} + u_{m+2} + \dots$ can be made as small as we please. If the terms are all positive (but not otherwise), we may take, instead of the entire series $u_{m+1} + u_{m+2} + \dots$, any set of terms (not of necessity consecutive terms) subsequent to u_m; that is, for a convergent series of positive terms the sum of any set of terms subsequent to u_m can, by taking m sufficiently large, be made as small as we please.

13. It follows that, in a convergent series of positive terms, the terms may be grouped together in any manner so as to form a finite number of partial series which will be each of them convergent, and such that the sum of their sums will be the sum of the given series. For instance, if the given series be $u_0 + u_1 + u_2 + \dots$, then the two series $u_0 + u_2 + u_4 + \dots$ and $u_1 + u_3 + \dots$ will each be convergent and the sum of their sums will be the sum of the original series.

14. Obviously the conclusion does not hold good in general for series of positive and negative terms: for instance, the series $1 - \frac{1}{2} + \frac{1}{3} - \frac{1}{4} + \dots$ is convergent, but the two series $1 + \frac{1}{3} + \frac{1}{5} + \dots$ and $-\frac{1}{2} - \frac{1}{4} - \dots$ are each divergent, and thus without a sum. In order that the conclusion may be applicable to a series of positive and negative terms the series must be "absolutely convergent," that is, it must be convergent when all the terms are made positive. This implies that the positive terms taken by themselves are a convergent series, and also that the negative terms taken by themselves are a convergent series. It is hardly necessary to remark that a convergent series of positive terms is absolutely convergent. The question of the convergency or divergency of a series of positive and negative terms is of less importance than the question whether it is or is not absolutely convergent. But in this latter question we regard the terms as all positive, and the question in effect relates to series containing positive terms only.

15. Consider, then, a series of positive terms $u_0 + u_1 + u_2 + \dots$; if they are increasing—that is, if in the limit u_{n+1}/u_n be greater than 1—the series is divergent, but if less than 1 the series is convergent. This may be called a first criterion; but there is the doubtful case where the limit $=1$. A second criterion was given by Cauchy and Raabe; but there is here again a doubtful case when the limit considered $= 1$. A succession of criteria was established by De Morgan, which it seems proper to give in the original form; but the equivalent criteria established by Bertrand are somewhat more convenient. In what follows lx is for shortness written to denote the logarithm of x, no matter to what base. De Morgan's form is as follows:—Writing $u_n = \frac{1}{\phi(n)}$, put $p_0 = \frac{x\phi' x}{\phi x}$; if for $x = \infty$ the limit a_0 of p_0 be greater than 1 the series is convergent, but if less than 1 it is divergent. If the limit $a_0 = 1$, seek for the limit of p_1, $= (p_0 - 1)lx$; if this limit a_1 be greater than 1 the series is convergent, but if less than 1 it is divergent. If the limit $a_1 = 1$, seek for the limit p_2, $= (p_1 - 1)\,llx$; if this limit a_2 be greater than 1 the series is convergent, but if less than 1 it is divergent. And so on indefinitely.

16. Bertrand's form is:—If, in the limit for $n = \infty$, $l\frac{1}{u_n}\Big/ln$ be negative or less than 1 the series is divergent, but if greater than 1 it is convergent. If it $= 1$, then if $l\frac{1}{nu_n}\Big/lln$ be negative or less than 1 the series is divergent, but if greater than 1 it is convergent. If it $= 1$, then if $l\frac{1}{nu_n ln}\Big/llln$ be negative or less than 1 the series is divergent, but if greater than 1 it is convergent. And so on indefinitely.

The last-mentioned criteria follow at once from the theorem that the several series having the general terms $\frac{1}{n^\alpha}$, $\frac{1}{n\,(ln)^\alpha}$, $\frac{1}{nln\,(lln)^\alpha}$, $\frac{1}{nlnlln\,(llln)^\alpha}$, ... respectively are each of them convergent if α be greater than 1, but divergent if α be negative or less than 1 or $=1$. In the simplest case, the series having the general term $\frac{1}{n^\alpha}$, the theorem may be proved nearly in the manner in which it is shown above (cf. § 9) that the harmonic series is divergent.

17. Two or more absolutely convergent series may be added together,

$$\{u_0 + u_1 + u_2 + \ldots\} + \{v_0 + v_1 + v_2 + \ldots\} = (u_0 + v_0) + (u_1 + v_1) + \ldots;$$

that is, the resulting series is absolutely convergent and has for its sum the sum of the two sums. And similarly two or more absolutely convergent series may be multiplied together

$$\{u_0 + u_1 + u_2 + \ldots\} \times \{v_0 + v_1 + v_2 + \ldots\} = u_0v_0 + (u_0v_1 + u_1v_0) + (u_0v_2 + u_1v_1 + u_2v_0) + \ldots;$$

that is, the resulting series is absolutely convergent and has for its sum the product of the two sums. But more properly the multiplication gives rise to a doubly infinite series—

$$\begin{matrix} u_0v_0, & u_0v_1, & u_0v_2 \ldots \\ u_1v_0, & u_1v_1, & u_1v_2 \\ \vdots & & \end{matrix}$$

—which is a kind of series which will be presently considered.

18. But it is, in the first instance, proper to consider a single series extending backwards and forwards to infinity, or say a back-and-forwards infinite series $\ldots + u_{-2} + u_{-1} + u_0 + u_1 + u_2 + \ldots$; such a series may be absolutely convergent, and the sum is then independent of the order of the terms, and in fact equal to the sum of the sums of the two series $u_0 + u_1 + u_2 + \ldots$ and $u_{-1} + u_{-2} + u_{-3} + \ldots$ respectively. But, if not absolutely convergent, the expression has no definite meaning until it is explained in what manner the terms are intended to be grouped together; for instance, the expression may be used to denote the foregoing sum of two series, or to denote the series $u_0 + (u_1 + u_{-1}) + (u_2 + u_{-2}) + \ldots$ and the sum may have different values, or there may be no sum, accordingly. Thus, if the series be $\ldots - \frac{1}{2} - \frac{1}{1} + 0 + \frac{1}{1} + \frac{1}{2} + \ldots$, in the former meaning the two series $0 + \frac{1}{1} + \frac{1}{2} + \ldots$ and $-\frac{1}{1} - \frac{1}{2} - \ldots$ are each divergent, and there is not any sum. But in the latter meaning the series is $0 + 0 + 0 + \ldots$, which has a sum $= 0$. So, if the series be taken to denote the limit of

$$(u_0 + u_1 + u_2 + \ldots + u_m) + (u_{-1} + u_{-2} + \ldots + u_{-m'}),$$

where m, m' are each of them ultimately infinite, there may be a sum depending on the ratio $m : m'$, which sum consequently acquires a determinate value only when this ratio is given.

19. In a singly infinite series we have a general term u_n, where n is an integer positive in the case of an ordinary series, and positive or negative in the case of a

back-and-forwards series. Similarly for a doubly infinite series, we have a general term $u_{m,n}$, where m, n are integers which may be each of them positive, and the form of the series is then

$$\begin{array}{lll} u_{0,0}, & u_{0,1}, & u_{0,2} \ldots \\ u_{1,0}, & u_{1,1}, & u_{1,2} \\ \vdots & & ; \end{array}$$

or m, n may be each of them positive or negative. The latter is the more general supposition, and includes the former, since $u_{m,n}$ may $=0$ for m or n each or either of them negative. To put a definite meaning on the notion of a sum, we may regard m, n as the rectangular coordinates of a point in a plane; that is, if m, n are each of them positive, we attend only to the positive quadrant of the plane, but otherwise to the whole plane; and we have thus a doubly infinite system or lattice-work of points. We may imagine a boundary depending on a parameter T which for $T=\infty$ is at every point thereof at an infinite distance from the origin; for instance, the boundary may be the circle $x^2+y^2=T$, or the four sides of a rectangle, $x=\pm\alpha T$, $y=\pm\beta T$. Suppose the form is given and the value of T, and let the sum $\Sigma u_{m,n}$ be understood to denote the sum of those terms $u_{m,n}$ which correspond to points within the boundary, then, if as T increases without limit the sum in question continually approaches a determinate limit (dependent, it may be, on the form of the boundary), *for such form of boundary* the series is said to be convergent, and the sum of the doubly infinite series is the aforesaid limit of the sum $\Sigma u_{m,n}$. The condition of convergency may be otherwise stated: it must be possible to take T so large that the sum $\Sigma u_{m,n}$ for all terms $u_{m,n}$ which correspond to points outside the boundary shall be as small as we please.

It is easy to see that, if the terms $u_{m,n}$ be all of them positive, and the series be convergent for any particular form of boundary, it will be convergent for any other form of boundary, and the sum will be the same in each case. Thus, let the boundary be in the first instance the circle $x^2+y^2=T$; by taking T sufficiently large the sum $\Sigma u_{m,n}$ for points outside the circle may be made as small as we please. Consider any other form of boundary—for instance, an ellipse of given eccentricity,—and let such an ellipse be drawn including within it the circle $x^2+y^2=T$. Then the sum $\Sigma u_{m,n}$ for terms $u_{m,n}$ corresponding to points outside the ellipse will be smaller than the sum for points outside the circle, and the difference of the two sums—that is, the sum for points outside the circle and inside the ellipse—will also be less than that for points outside the circle, and can thus be made as small as we please. Hence finally the sum $\Sigma u_{m,n}$, whether restricted to terms $u_{m,n}$ corresponding to points inside the circle or to terms corresponding to points inside the ellipse, will have the same value, or the sum of the series is independent of the form of the boundary. Such a series, viz. a doubly infinite convergent series of positive terms, is said to be absolutely convergent; and similarly a doubly infinite series of positive and negative terms which is convergent when the terms are all taken as positive is absolutely convergent.

20. We have in the preceding theory the foundation of the theorem (§ 17) as to the product of two absolutely convergent series. The product is in the first instance

expressed as a doubly infinite series; and, if we sum this for the boundary $x+y=T$, this is in effect a summation of the series $u_0v_0+(u_0v_1+u_1v_0)+\ldots$, which is the product of the two series. It may be further remarked that, starting with the doubly infinite series and summing for the rectangular boundary $x=\alpha T$, $y=\beta T$, we obtain the sum as the product of the sums of the two single series. For series not absolutely convergent, the theorem is not true. A striking instance is given by Cauchy: the series $1-\frac{1}{\sqrt{2}}+\frac{1}{\sqrt{3}}-\frac{1}{\sqrt{4}}+\ldots$ is convergent and has a calculable sum, but it can be shown without difficulty that its square, viz. the series $1-\frac{2}{\sqrt{2}}+\left(\frac{2}{\sqrt{3}}+\frac{1}{2}\right)-\ldots$, is divergent.

21. The case where the terms of a series are imaginary comes under that where they are real. Suppose the general term is p_n+q_ni, then the series will have a sum, or will be convergent, if and only if the series having for its general term p_n and the series having for its general term q_n be each convergent; then the sum = sum of first series + i multiplied by sum of second series. The notion of absolute convergence will of course apply to each of the series separately; further, if the series having for its general term the modulus $\sqrt{p^2_n+q^2_n}$ be convergent (that is, absolutely convergent, since the terms are all positive), each of the component series will be absolutely convergent; but the condition is not necessary for the convergence, or the absolute convergence, of the two component series respectively.

22. In the series thus far considered, the terms are actual numbers, or are at least regarded as constant; but we may have a series $u_0+u_1+u_2+\ldots$, where the successive terms are functions of a parameter z; in particular, we may have a series $a_0+a_1z+a_2z^2+\ldots$ arranged in powers of z. It is in view of a complete theory *necessary* to consider z as having the imaginary value $x+iy=r(\cos\phi+i\sin\phi)$. The two component series will then have the general terms $a_nr^n\cos n\phi$ and $a_nr^n\sin n\phi$ respectively; accordingly each of these series will be absolutely convergent for any value whatever of ϕ, provided the series with the general term a_nr^n be absolutely convergent. Moreover, the series, if thus absolutely convergent for any particular value R of r, will be absolutely convergent for any smaller value of r, that is, for any value of $x+iy$ having a modulus not exceeding R; or, representing as usual $x+iy$ by the point whose rectangular coordinates are x, y, the series will be absolutely convergent for any point whatever inside or on the circumference of the circle having the origin for centre and its radius $=R$. The origin is of course an arbitrary point: or, what is the same thing, instead of a series in powers of z, we may consider a series in powers of $z-c$ (where c is a given imaginary value $=\alpha+\beta i$). Starting from the series, we may within the aforesaid limit of absolute convergency consider the series as the definition of a function of the variable z; in particular, the series may be absolutely convergent for every finite value of the modulus, and we have then a function defined for every finite value whatever $x+iy$ of the variable. Conversely, starting from a given function of the variable, we may inquire under what conditions it admits of expansion in a series of powers of z (or $z-c$), and seek to determine the expansion of the function in a series of this form. But in all this, however, we are travelling out of the theory of series into the general theory of functions.

23. Considering the modulus r as a given quantity and the several powers of r as included in the coefficients, the component series are of the forms $a_0+a_1\cos\phi+a_2\cos 2\phi+\ldots$ and $a_1\sin\phi+a_2\sin 2\phi+\ldots$ respectively. The theory of these trigonometrical or multiple sine and cosine series, and of the development, under proper conditions, of an arbitrary function in series of these forms, constitutes an important and interesting branch of analysis.

24. In the case of a real variable z, we may have a series $a_0+a_1z+a_2z^2+\ldots$, where the series $a_0+a_1+a_2+\ldots$ is a divergent series of decreasing positive terms (or as a limiting case where this series is $1+1+1+\ldots$). For a value of z inferior but indefinitely near to ± 1, say $z=\pm(1-\epsilon)$, where ϵ is indefinitely small and positive, the series will be convergent and have a determinate sum $\phi(z)$, and we may write $\phi(\pm 1)$ to denote the limit of $\phi(\pm(1-\epsilon))$ as ϵ diminishes to zero; but unless the series be convergent for the value $z=\pm 1$, it cannot for this value have a sum, nor consequently a sum $=\phi(\pm 1)$. For instance, let the series be $z+\frac{z^2}{2}+\frac{z^3}{3}+\ldots$, which for values of z between the limits ± 1 (both limits excluded) $=-\log(1-z)$. For $z=+1$ the series is divergent and has no sum; but for $z=1-\epsilon$, as ϵ diminishes to zero, we have $-\log\epsilon$ and $(1-\epsilon)+\frac{1}{2}(1-\epsilon)^2+\ldots$, each positive and increasing without limit; for $z=-1$, the series $1-\frac{1}{2}+\frac{1}{3}-\frac{1}{4}+\ldots$ is convergent, and we have *at the limit* $\log 2=1-\frac{1}{2}+\frac{1}{3}-\frac{1}{4}+\ldots$. As a second example, consider the series $1+z+z^2+\ldots$, which for values of z between the limits ± 1 (both limits excluded) $=\frac{1}{1-z}$. For $z=+1$, the series is divergent and has no sum; but for $z=1-\epsilon$, as ϵ diminishes to zero, we have $\frac{1}{\epsilon}$ and $1+(1-\epsilon)+(1-\epsilon)^2+\ldots$, each positive and increasing without limit; for $z=-1$ the series is divergent and has no sum; the equation $\frac{1}{2-\epsilon}=1-(1-\epsilon)+(1-\epsilon)^2-\ldots$ is true for any positive value of ϵ however small, but *not* for the value $\epsilon=0$.

The following memoirs and works may be consulted:—Cauchy, *Cours d'Analyse de l'École Polytechnique*—part I., *Analyse Algébrique*, 8vo. Paris, 1821; Abel, "Untersuchungen über die Reihe $1+\frac{m}{1}x+\frac{m(m-1)}{1.2}x^2+\ldots$," in Crelle's *Journ. de Math.*, vol. I. (1826), pp. 211—239, and *Œuvres* (French trans.), vol. I.; De Morgan, *Treatise on the Differential and Integral Calculus*, 8vo. London, 1842; Id., "On Divergent Series and various Points of Analysis connected with them" (1844), in *Camb. Phil. Trans.*, vol. VIII. (1849), and other memoirs in *Camb. Phil. Trans.*; Bertrand, "Règles sur la Convergence des Séries," in *Liouv. Journ. de Math.*, vol. VII. (1842), pp. 35—54; Cayley, "On the Inverse Elliptic Functions," *Camb. Math. Journ.*, vol. IV. (1845), pp. 257—277, [24], and "Mémoire sur les Fonctions doublement périodiques," in *Liouv. Journ. de Math.*, vol. X. (1845), pp. 385—420, [25], (as to the boundary for a doubly infinite series); Riemann, "Ueber die Darstellbarkeit einer Function durch eine trigonometrische Reihe," in *Gött. Abh.*, vol. XIII. (1854), and *Werke*, Leipsic, 1876, pp. 213—253 (contains an account of preceding researches by Euler, D'Alembert, Fourier, Lejeune-Dirichlet, &c.); Catalan, *Traité Élémentaire des Séries*, 8vo. Paris, 1860; Boole, *Treatise on the Calculus of Finite Differences*, 2nd ed. by Moulton, 8vo. London, 1872.

797.

SURFACE, CONGRUENCE, COMPLEX.

[From the *Encyclopædia Britannica, Ninth Edition*, vol. XXII. (1887), pp. 668—672.]

IN the article Curve [785], the subject was treated from an historical point of view for the purpose of showing how the leading ideas of the theory were successively arrived at. These leading ideas apply to surfaces, but the ideas peculiar to surfaces are scarcely of the like fundamental nature, being rather developments of the former set in their application to a more advanced portion of geometry; there is consequently less occasion for the historical mode of treatment. Curves in space were briefly considered in the same article, and they will not be discussed here; but it is proper to refer to them in connexion with the other notions of solid geometry. In plane geometry the elementary figures are the point and the line; and we then have the curve, which may be regarded as a singly infinite system of points, and also as a singly infinite system of lines. In solid geometry the elementary figures are the point, the line, and the plane; we have, moreover, first, that which under one aspect is the curve and under another aspect the developable (or torse), and which may be regarded as a singly infinite system of points, of lines, or of planes; and secondly, the surface, which may be regarded as a doubly infinite system of points or of planes, and also as a special triply infinite system of lines. (The tangent lines of a surface are a special complex.) As distinct particular cases of the first figure, we have the plane curve and the cone: and as a particular case of the second figure, the ruled surface, regulus, or singly infinite system of lines; we have, besides, the congruence or doubly infinite system of lines, and the complex or triply infinite system of lines. And thus crowds of theories arise which have hardly any analogues in plane geometry; the relation of a curve to the various surfaces which can be drawn through it, and that of a surface to the various curves which can be drawn upon it, are different in kind from those which in plane geometry most nearly correspond to them,—the relation of a system of points to the different curves through them and that of a curve to the systems of points upon it. In particular, there is nothing in plane geometry to correspond to the theory of the curves of curvature of a surface. Again, to the single theorem of plane geometry, that a line is the shortest distance between two points, there correspond in solid geometry two extensive and difficult theories,—that of the geodesic lines on a surface and that of the minimal surface, or surface of minimum area, for a given boundary. And it would be easy to say more in illustration of the great extent and complexity of the subject.

Surfaces in General; Torses, &c.

1. A surface may be regarded as the locus of a doubly infinite system of points,—that is, the locus of the system of points determined by a single equation $U = (*\between x, y, z, 1)^n, = 0$, between the Cartesian coordinates (to fix the ideas, say rectangular coordinates) x, y, z; or, if we please, by a single homogeneous relation $U = (*\between x, y, z, w)^n, = 0$, between the quadriplanar coordinates x, y, z, w. The degree n of the equation is the order of the surface; and this definition of the order agrees with the geometrical one, that the order of the surface is equal to the number of the intersections of the surface by an arbitrary line. Starting from the foregoing point definition of the surface, we might develop the notions of the tangent line and the tangent plane; but it will be more convenient to consider the surface *ab initio* from the more general point of view in its relation to the point, the line, and the plane.

2. Mention has been made of the plane curve and the cone; it is proper to recall that the *order* of a plane curve is equal to the number of its intersections by an arbitrary line (in the plane of the curve), and that its *class* is equal to the number of tangents to the curve which pass through an arbitrary point (in the plane of the curve). The cone is a figure correlative to the plane curve: corresponding to the plane of the curve we have the vertex of the cone, to its tangents the generating lines of the cone, and to its points the tangent planes of the cone. But from a different point of view, we may consider the generating lines of the cone as corresponding to the points of the curve and its tangent planes as corresponding to the tangents of the curve. From this point of view, we define the order of the cone as equal to the number of its intersections (generating lines) by an arbitrary plane through the vertex, and its class as equal to the number of the tangent planes which pass through an arbitrary line through the vertex. And in the same way that a plane curve has singularities (singular points and singular tangents), so a cone has singularities (singular generating lines and singular tangent planes).

3. Consider now a surface in connexion with an arbitrary line. The line meets the surface in a certain number of points, and, as already mentioned, the *order* of the surface is equal to the number of these intersections. We have through the line a certain number of tangent planes of the surface, and the *class* of the surface is equal to the number of these tangent planes.

But, further, through the line imagine a plane; this meets the surface in a curve the order of which is equal (as is at once seen) to the order of the surface. Again, on the line imagine a point; this is the vertex of a cone circumscribing the surface, and the class of this cone is equal (as is at once seen) to the class of the surface.

The tangent lines of the surface, which lie in the plane, are nothing else than the tangents of the plane section, and thus form a singly infinite series of lines; similarly, the tangent lines of the surface, which pass through the point, are nothing else than the generating lines of the circumscribed cone, and thus form a singly infinite series of lines. But, if we consider those tangent lines of the surface which are at once in the plane and through the point, we see that they are finite in number; and we define the *rank* of a surface as equal to the number of tangent lines which lie in a given plane and pass through a given point in that plane. It at once follows that the class of the plane section and the order of the circumscribed cone are each equal to the rank of the surface, and are thus equal to each other. It may be noticed that for a general surface $(*\between x, y, z, w)^n, =0$, of order n without point singularities the rank is $a, =n(n-1)$, and the class is $n', =n(n-1)^2$; this implies (what is, in fact, the case) that the circumscribed cone has line singularities, for otherwise its class, that is, the class of the surface, would be $a(a-1)$, which is not $=n(n-1)^2$.

4. In the last preceding number, the notions of the tangent line and the tangent plane have been assumed as known, but they require to be further explained in reference to the original point definition of the surface. Speaking generally, we may say that the points of the surface consecutive to a given point on it lie in a plane which is the tangent plane at the given point, and conversely the given point is the point of contact of this tangent plane, and that any line through the point of contact and in the tangent plane is a tangent line touching the surface at the point of contact. Hence we see at once that the tangent line is any line meeting the surface in two consecutive points, or—what is the same thing—a line meeting the surface in the point of contact, counting as two intersections, and in $\overline{n-2}$ other points. But, from the foregoing notion of the tangent plane as a plane containing the point of contact and the consecutive points of the surface, the passage to the true definition of the tangent plane is not equally obvious. A plane in general meets the surface of the order n in a curve of that order without double points; but the plane may be such that the curve has a double point, and when this is so the plane is a tangent plane having the double point for its point of contact. The double point is either an acnode (isolated point), then the surface at the point in question is convex towards (that is, concave away from) the tangent plane; or else it is a crunode, and the surface at the point in question is then concavo-convex, that is, it has its two curvatures in opposite senses (see *infra*, No. 16). Observe that, in either case, any line whatever in the plane and through the point meets the surface in the points in which it meets the plane curve, namely, in the point of contact, which *qua* double point counts as two intersections, and in $\overline{n-2}$ other points; that is, we have the preceding definition of the tangent line.

5. The complete enumeration and discussion of the singularities of a surface is a question of extreme difficulty which has not yet been solved*. A plane curve has

* In a plane curve, the only singularities which need to be considered are those that present themselves in Plücker's equations: for every higher singularity whatever is equivalent to a certain number of nodes, cusps, inflexions, and double tangents. As regards a surface, no such reduction of the higher singularities has as yet been made.

point singularities and line singularities; corresponding to these, we have for the surface isolated point singularities and isolated plane singularities, but there are besides continuous singularities applying to curves on or torses circumscribed to the surface, and it is among these that we have the non-special singularities which play the most important part in the theory. Thus the plane curve represented by the general equation $(*\between x, y, z)^n = 0$, of any given order n, has the non-special line singularities of inflexions and double tangents; corresponding to this, the surface represented by the general equation $(*\between x, y, z, w)^n = 0$, of any given order n, has, not the isolated plane singularities, but the continuous singularities of the spinode curve or torse and the node-couple curve or torse. A plane may meet the surface in a curve having (1) a cusp (spinode) or (2) a pair of double points; in each case, there is a singly infinite system of such singular tangent planes, and the locus of the points of contact is the curve, the envelope of the tangent planes the torse. The reciprocal singularities to these are the nodal curve and the cuspidal curve: the surface may intersect or touch itself along a curve in such wise that, cutting the surface by an arbitrary plane, the curve of intersection has, at each intersection of the plane with the curve on the surface, (1) a double point (node) or (2) a cusp. Observe that these are singularities not occurring in the surface represented by the general equation $(*\between x, y, z, w)^n = 0$ of any order; observe further that, in the case of both or either of these singularities, the definition of the tangent plane must be modified. A tangent plane is a plane such that there is in the plane section a double point in addition to the nodes or cusps at the intersections with the singular lines on the surface.

6. As regards isolated singularities, it will be sufficient to mention the point singularity of the conical point (or cnicnode) and the corresponding plane singularity of the conic of contact (or cnictrope). In the former case, we have a point such that the consecutive points, instead of lying in a tangent plane, lie on a quadric cone, having the point for its vertex; in the latter case, we have a plane touching the surface along a conic, that is, the complete intersection of the surface by the plane is made up of the conic taken twice and of a residual curve of the order $n-4$.

7. We may, in the general theory of surfaces, consider either a surface and its reciprocal surface, the reciprocal surface being taken to be the surface enveloped by the polar planes (in regard to a given quadric surface) of the points of the original surface; or—what is better—we may consider a given surface in reference to the reciprocal relations of its order, rank, class, and singularities. In either case, we have a series of unaccented letters and a corresponding series of accented letters, and the relations between them are such that we may in any equation interchange the accented and the unaccented letters; in some cases, an unaccented letter may be equal to the corresponding accented letter. Thus, let n, n' be as before the order and the class of the surface, but, instead of immediately defining the rank, let a be used to denote the class of the plane section and a' the order of the circumscribed cone; also let S, S' be numbers referring to the singularities. The form of the relations is $a = a'$ (= rank of surface); $a' = n(n-1) - S$; $n' = n(n-1)^2 - S$; $a = n'(n'-1) - S'$; $n = n'(n'-1)^2 - S'$. In these last equations S, S' are merely written down to denote proper corresponding combinations of the several numbers referring to the singularities collectively denoted

by S, S' respectively. The theory, as already mentioned, is a complex and difficult one, and it is not the intention to further develop it here.

8. A developable or torse corresponds to a curve in space, in the same manner as a cone corresponds to a plane curve: although capable of representation by an equation $U = (*\between x, y, z, w)^n, = 0$, and so of coming under the foregoing point definition of a surface, it is an entirely distinct geometrical conception. We may indeed, *qua* surface, regard it as a surface characterized by the property that each of its tangent planes touches it, not at a single point, but along a line; this is equivalent to saying that it is the envelope, not of a doubly infinite series of planes, as is a proper surface, but of a singly infinite system of planes. But it is perhaps easier to regard it as the locus of a singly infinite system of lines, each line meeting the consecutive line, or, what is the same thing, the lines being tangent lines of a curve in space. The tangent plane is then the plane through two consecutive lines, or, what is the same thing, an osculating plane of the curve, whence also the tangent plane intersects the surface in the generating line counting twice, and in a residual curve of the order $n-2$. The curve is said to be the edge of regression of the developable, and it is a cuspidal curve thereof; that is to say, any plane section of the developable has at each point of intersection with the edge of regression a cusp. A sheet of paper bent in any manner without crumpling gives a developable; but we cannot with a single sheet of paper properly exhibit the form in the neighbourhood of the edge of regression: we need two sheets connected along a plane curve, which, when the paper is bent, becomes the edge of regression and appears as a cuspidal curve on the surface.

It may be mentioned that the condition which must be satisfied in order that the equation $U = 0$ shall represent a developable is $H(U) = 0$; that is, the Hessian or functional determinant formed with the second differential coefficients of U must vanish in virtue of the equation $U = 0$, or—what is the same thing—$H(U)$ must contain U as a factor. If in Cartesian coordinates the equation is taken in the form $z - f(x, y) = 0$, then the condition is $rt - s^2 = 0$ identically, where r, s, t denote as usual the second differential coefficients of z in regard to x, y respectively.

9. A ruled surface or regulus is the locus of a singly infinite system of lines, where the consecutive lines do not intersect; this is a true surface, for there is a doubly infinite series of tangent planes,—in fact, any plane through any one of the lines is a tangent plane of the surface, touching it at a point on the line, and in such wise that, as the tangent plane turns about the line, the point of contact moves along the line. The complete intersection of the surface by the tangent plane is made up of the line counting once and of a residual curve of the order $n-1$. A quadric surface is a regulus in a twofold manner, for there are on the surface two systems of lines each of which is a regulus. A cubic surface may be a regulus (see No. 11 *infra*).

Surfaces of the Orders 2, 3, *and* 4.

10. A surface of the second order or a quadric surface is a surface such that every line meets it in two points, or—what comes to the same thing—such that every plane section thereof is a conic or quadric curve. Such surfaces have been studied

from every point of view. The only singular forms are when there is (i) a conical point (cnicnode), when the surface is a cone of the second order or quadricone; (ii) a conic of contact (cnictrope), when the surface is this conic; from a different point of view it is a *surface aplatie* or flattened surface. Excluding these degenerate forms, the surface is of the order, rank, and class each $=2$, and it has no singularities. Distinguishing the forms according to *reality*, we have the ellipsoid, the hyperboloid of two sheets, the hyperboloid of one sheet, the elliptic paraboloid, and the hyperbolic paraboloid (see Geometry, Analytical, [790]). A particular case of the ellipsoid is the sphere; in abstract geometry, this is a quadric surface passing through a given quadric curve, the circle at infinity. The tangent plane of a quadric surface meets it in a quadric curve having a node, that is, in a pair of lines; hence there are on the surface two singly infinite sets of lines. Two lines of the same set do not meet, but each line of the one set meets each line of the other set; the surface is thus a regulus in a twofold manner. The lines are real for the hyperboloid of one sheet and for the hyperbolic paraboloid; for the other forms of surface they are imaginary.

11. We have next the surface of the third order or cubic surface, which has also been very completely studied. Such a surface may have isolated point singularities (cnicnodes or points of higher singularity), or it may have a nodal line; we have thus $21+2$, $=23$ cases. In the general case of a surface without any singularities, the order, rank, and class are $=3$, 6, 12 respectively. The surface has upon it 27 lines, lying by threes in 45 planes, which are triple tangent planes. Observe that the tangent plane is a plane meeting the surface in a curve having a node. For a surface of any given order n there will be a certain number of planes each meeting the surface in a curve with 3 nodes, that is, triple tangent planes; and, in the particular case where $n=3$, the cubic curve with 3 nodes is of course a set of 3 lines; it is found that the number of triple tangent planes is, as just mentioned, $=45$. This would give 135 lines, but through each line we have 5 such planes, and the number of lines is thus $=27$. The theory of the 27 lines is an extensive and interesting one; in particular, it may be noticed that we can, in thirty-six ways, select a system of 6×6 lines, or "double sixer," such that no 2 lines of the same set intersect each other, but that each line of the one set intersects each line of the other set.

A cubic surface having a nodal line is a ruled surface or regulus; in fact, any plane through the nodal line meets the surface in this line counting twice and in a residual line, and there is thus on the surface a singly infinite set of lines. There are two forms; but the distinction between them need not be referred to here.

12. As regards quartic surfaces, only particular forms have been much studied. A quartic surface can have at most 16 conical points (cnicnodes); an instance of such a surface is Fresnel's wave surface, which has 4 real cnicnodes in one of the principal planes, 4×2 imaginary ones in the other two principal planes, and 4 imaginary ones at infinity,—in all 16 cnicnodes; the same surface has also 4 real $+$ 12 imaginary planes each touching the surface along a circle (cnictropes),—in all 16 cnictropes. It was easy by a mere homographic transformation to pass to the more general surface called the tetrahedroid; but this was itself only a particular form of the general surface

with 16 cnicnodes and 16 cnictropes first studied by Kummer. Quartic surfaces with a smaller number of cnicnodes have also been considered.

Another very important form is the quartic surface having a nodal conic; the nodal conic may be the circle at infinity, and we have then the so-called anallagmatic surface, otherwise the cyclide (which includes the particular form called Dupin's cyclide). These correspond to the bicircular quartic curve of plane geometry. Other forms of quartic surface might be referred to.

Congruences and Complexes.

13. A congruence is a doubly infinite system of lines. A line depends on four parameters and can therefore be determined so as to satisfy four conditions; if only two conditions are imposed on the line, we have a doubly infinite system of lines or a congruence. For instance, the lines meeting each of two given lines form a congruence. It is hardly necessary to remark that, imposing on the line one more condition, we have a ruled surface or regulus; thus we can in an infinity of ways separate the congruence into a singly infinite system of reguli or of torses (see *infra*, No. 16).

Considering in connexion with the congruence two arbitrary lines, there will be in the congruence a determinate number of lines which meet each of these two lines; and the number of lines thus meeting the two lines is said to be the *order-class* of the congruence. If the two arbitrary lines are taken to intersect each other, the congruence lines which meet each of the two lines separate themselves into two sets,—those which lie in the plane of the two lines and those which pass through their intersection. There will be in the former set a determinate number of congruence lines which is the *order* of the congruence, and in the latter set a determinate number of congruence lines which is the *class* of the congruence. In other words, the order of the congruence is equal to the number of congruence lines lying in an arbitrary plane, and its class to the number of congruence lines passing through an arbitrary point.

The following systems of lines form each of them a congruence:—(A) lines meeting each of two given curves; (B) lines meeting a given curve twice; (C) lines meeting a given curve and touching a given surface; (D) lines touching each of two given surfaces; (E) lines touching a given surface twice, or, say, the bitangents of a given surface.

The last case is the most general one; and conversely, for a given congruence, there will be in general a surface having the congruence lines for bitangents. This surface is said to be the *focal surface* of the congruence; the general surface with 16 cnicnodes first presented itself in this manner as the focal surface of a congruence. But the focal surface may degenerate into the forms belonging to the other cases A, B, C, D.

14. A complex is a triply infinite system of lines,—for instance, the tangent lines of a surface. Considering an arbitrary point in connexion with the complex, the com-

plex lines which pass through the point form a cone; considering a plane in connexion with it, the complex lines which lie in the plane envelope a curve. It is easy to see that the class of the curve is equal to the order of the cone; in fact, each of these numbers is equal to the number of complex lines which lie in an arbitrary plane and pass through an arbitrary point of that plane; and we then say *order* of complex = order of curve; *rank* of complex = class of curve = order of cone; *class* of complex = class of cone. It is to be observed that, while for a congruence there is in general a surface having the congruence lines for bitangents, for a complex there is not in general any surface having the complex lines for tangents; the tangent lines of a surface are thus only a special form of complex. The theory of complexes first presented itself in the researches of Malus on systems of rays of light in connexion with double refraction.

15. The analytical theory as well of congruences as of complexes is most easily carried out by means of the six coordinates of a line; viz. there are coordinates (a, b, c, f, g, h) connected by the equation $af + bg + ch = 0$, and therefore such that the ratios $a : b : c : f : g : h$ constitute a system of four arbitrary parameters. We have thus a congruence of the order n represented by a single homogeneous equation of that order $(*\between a, b, c, f, g, h)^n = 0$ between the six coordinates; two such relations determine a congruence. But we have in regard to congruences the same difficulty as that which presents itself in regard to curves in space: it is not every congruence which can be represented completely and precisely by *two* such equations.

The linear equation $(*\between a, b, c, f, g, h) = 0$ represents a congruence of the first order or linear congruence; such congruences are interesting both in geometry and in connexion with the theory of forces acting on a rigid body.

Curves of Curvature; Asymptotic Lines.

16. The normals of a surface form a congruence. In any congruence, the lines consecutive to a given congruence line do not in general meet this line; but there is a determinate number of consecutive lines which do meet it; or, attending for the moment to only one of these, say the congruence line is met by a consecutive congruence line. In particular, each normal is met by a consecutive normal; this again is met by a consecutive normal, and so on. That is, we have a singly infinite system of normals each meeting the consecutive normal, and so forming a torse; starting from different normals successively, we obtain a singly infinite system of such torses. But each normal is in fact met by two consecutive normals, and, using in the construction first the one and then the other of these, we obtain two singly infinite systems of torses each intersecting the given surface at right angles. In other words, if in place of the normal we consider the point on the surface, we obtain on the surface two singly infinite systems of curves such that for any curve of either system the normals at consecutive points intersect each other; moreover, for each normal the torses of the two systems intersect each other at right angles; and therefore for each point of the surface the curves of the two systems intersect each other at right angles. The two systems of curves are said to be the curves of curvature of the surface.

The normal is met by the two consecutive normals in two points which are the centres of curvature for the point on the surface; these lie either on the same side of the point or on opposite sides, and the surface has at the point in question like curvatures or opposite curvatures in the two cases respectively (see *supra*, No. 4).

17. In immediate connexion with the curves of curvature, we have the so-called asymptotic curves (Haupt-tangenten-linien). The tangent plane at a point of the surface cuts the surface in a curve having at that point a node. Thus we have at the point of the surface two directions of passage to a consecutive point, or, say, two elements of arc; and, passing along one of these to the consecutive point, and thence to a consecutive point, and so on, we obtain on the surface a curve. Starting successively from different points of the surface we thus obtain a singly infinite system of curves; or, using first one and then the other of the two directions, we obtain two singly infinite systems of curves, which are the curves above referred to. The two curves at any point are equally inclined to the two curves of curvature at that point, or—what is the same thing—the supplementary angles formed by the two asymptotic lines are bisected by the two curves of curvature. In the case of a quadric surface, the asymptotic curves are the two systems of lines on the surface.

Geodesic Lines.

18. A geodesic line (or curve) is a shortest curve on a surface; more accurately, the element of arc between two consecutive points of a geodesic line is a shortest arc on the surface. We are thus led to the fundamental property that, at each point of the curve, the osculating plane of the curve passes through the normal of the surface; in other words, any two consecutive arcs PP', $P'P''$ are *in plano* with the normal at P'. Starting from a given point P on the surface, we have a singly infinite system of geodesics proceeding along the surface in the direction of the several tangent lines at the point P; and, if the direction PP' is given, the property gives a construction by successive elements of arc for the required geodesic line.

Considering the geodesic lines which proceed from a given point P of the surface, any particular geodesic line is or is not again intersected by the consecutive generating line; if it is thus intersected, the generating line is a shortest line on the surface up to, but not beyond, the point at which it is first intersected by the consecutive generating line; if it is not intersected, it continues a shortest line for the whole course.

In the analytical theory both of geodesic lines and of the curves of curvature, and in other parts of the theory of surfaces, it is very convenient to consider the rectangular coordinates x, y, z of a point of the surface as given functions of two independent parameters p, q; the form of these functions of course determines the surface, since by the elimination of p, q from the three equations we obtain the equation in the coordinates x, y, z. We have for the geodesic lines a differential equation of the second order between p and q; the general solution contains two arbitrary constants, and is thus capable of representing the geodesic line which can be drawn from a given point in a given direction on the surface. In the case of a quadric surface, the solution involves hyperelliptic integrals of the first kind, depending on the square root of a sextic function.

Curvilinear Coordinates.

19. The expressions of the coordinates x, y, z in terms of p, q may contain a parameter r, and, if this is regarded as a given constant, these expressions will as before refer to a point on a given surface. But, if p, q, r are regarded as three independent parameters, x, y, z will be the coordinates of a point in space, determined by means of the three parameters p, q, r; these parameters are said to be the curvilinear coordinates, or (in a generalized sense of the term) simply the coordinates of the point. We arrive otherwise at the notion by taking p, q, r each as a given function of x, y, z; say we have $p=f_1(x, y, z)$, $q=f_2(x, y, z)$, $r=f_3(x, y, z)$, which equations of course lead to expressions for p, q, r each as a function of x, y, z. The first equation determines a singly infinite set of surfaces: for any given value of p we have a surface; and similarly the second and third equations determine each a singly infinite set of surfaces. If, to fix the ideas, f_1, f_2, f_3 are taken to denote each a rational and integral function of x, y, z, then two surfaces of the same set will not intersect each other, and through a given point of space there will pass one surface of each set; that is, the point will be determined as a point of intersection of three surfaces belonging to the three sets respectively; moreover, the whole of space will be divided by the three sets of surfaces into a triply infinite system of elements, each of them being a parallelepiped.

Orthotomic Surfaces; Parallel Surfaces.

20. The three sets of surfaces may be such that the three surfaces through any point of space whatever intersect each other at right angles; and they are in this case said to be orthotomic. The term curvilinear coordinates was almost appropriated by Lamé, to whom this theory is chiefly due, to the case in question: assuming that the equations $p=f_1(x, y, z)$, $q=f_2(x, y, z)$, $r=f_3(x, y, z)$ refer to a system of orthotomic surfaces, we have in the restricted sense p, q, r as the curvilinear coordinates of the point.

An interesting special case is that of confocal quadric surfaces. The general equation of a surface confocal with the ellipsoid

$$\frac{x^2}{a^2}+\frac{y^2}{b^2}+\frac{z^2}{c^2}=1 \text{ is } \frac{x^2}{a^2+\theta}+\frac{y^2}{b^2+\theta}+\frac{z^2}{c^2+\theta}=1;$$

and, if in this equation we consider x, y, z as given, we have for θ a cubic equation with three real roots p, q, r, and thus we have through the point three real surfaces, one an ellipsoid, one a hyperboloid of one sheet, and one a hyperboloid of two sheets.

21. The theory is connected with that of curves of curvature by Dupin's theorem. Thus in any system of orthotomic surfaces, each surface of any one of the three sets is intersected by the surfaces of the other two sets in its curves of curvature.

22. No one of the three sets of surfaces is altogether arbitrary: in the equation $p=f_1(x, y, z)$, p is not an arbitrary function of x, y, z, but it must satisfy a certain partial differential equation of the third order. Assuming that p has this value, we have $q=f_2(x, y, z)$ and $r=f_3(x, y, z)$ determinate functions of x, y, z, such that the three sets of surfaces form an orthotomic system.

23. Starting from a given surface, it has been seen (No. 16) that the normals along the curves of curvature form two systems of torses intersecting each other, and also the given surface, at right angles. But there are, intersecting the two systems of torses at right angles, not only the given surface, but a singly infinite system of surfaces. If at each point of the given surface we measure off along the normal one and the same distance at pleasure, then the locus of the points thus obtained is a surface cutting all the normals of the given surface at right angles, or, in other words, having the same normals as the given surface; and it is therefore a parallel surface to the given surface. Hence the singly infinite system of parallel surfaces and the two singly infinite systems of torses form together a set of orthotomic surfaces.

The Minimal Surface.

24. This is the surface of minimum area—more accurately, a surface such that, for any indefinitely small closed curve which can be drawn on it round any point, the area of the surface is less than it is for any other surface whatever through the closed curve. It at once follows that the surface at every point is concavo-convex; for, if at any point this was not the case, we could, by cutting the surface by a plane, describe round the point an indefinitely small closed plane curve, and the plane area within the closed curve would then be less than the area of the element of surface within the same curve. The condition leads to a partial differential equation of the second order for the determination of the minimal surface: considering z as a function of x, y, and writing as usual p, q, r, s, t for the first and second differential coefficients of z in regard to x, y respectively, the equation (as first shown by Lagrange) is $(1+q^2)r-2pqs+(1+p^2)t=0$, or, as this may also be written,

$$\frac{d}{dy}\frac{q}{\sqrt{1+p^2+q^2}}+\frac{d}{dx}\frac{p}{\sqrt{1+p^2+q^2}}=0.$$

The general integral contains of course arbitrary functions, and, if we imagine these so determined that the surface may pass through a given closed curve, and if, moreover, there is but one minimal surface passing through that curve, we have the solution of the problem of finding the surface of minimum area within the same curve. The surface continued beyond the closed curve is a minimal surface, but it is not of necessity or in general a surface of minimum area for an arbitrary bounding curve not wholly included within the given closed curve. It is hardly necessary to remark that the plane is a minimal surface, and that, if the given closed curve is a plane curve, the plane is the proper solution; that is, the plane area within the given closed curve is less than the area for any other surface through the same curve. The given closed curve is not of necessity a single curve: it may be, for instance, a skew polygon of four or more sides.

The partial differential equation was dealt with in a very remarkable manner by Riemann. From the second form given above it appears that we have $\frac{q\,dx - p\,dy}{\sqrt{1+p^2+q^2}} =$ a complete differential, or, putting this $= d\zeta$, we introduce into the solution a variable ζ, which combines with z in the forms $z \pm i\zeta$ ($i = \sqrt{-1}$ as usual). The boundary conditions have to be satisfied by the determination of the conjugate variables η, η' as functions of $z + i\zeta$, $z - i\zeta$, or, say, of Z, Z' respectively. By writing S, S' to denote $x + iy$, $x - iy$ respectively, Riemann obtains finally two ordinary differential equations of the first order in S, S', η, η', Z, Z', and the results are completely worked out in some very interesting special cases.

The memoirs on various parts of the general subject are very numerous; references to many of them will be found in Salmon's *Treatise on the Analytic Geometry of Three Dimensions*, 4th ed., Dublin, 1882 (the most comprehensive work on solid geometry); for the minimal surface (which is not considered there) see Memoirs XVII. and XXVI. in Riemann's *Gesammelte mathematische Werke*, Leipsic, 1876; the former—"Ueber die Fläche vom kleinsten Inhalt bei gegebener Begrenzung," as published in *Gött. Abhandl.*, vol. XIII. (1866—67)—contains an introduction by Hattendorff giving the history of the question.

798.

WALLIS (JOHN).

[From the *Encyclopædia Britannica, Ninth Edition*, vol. XXIV. (1888), pp. 331, 332.]

WALLIS, JOHN (1616—1703), an eminent English mathematician, logician, and grammarian, was born on the 23rd November 1616 at Ashford, in Kent, of which parish his father was then incumbent. Having been previously instructed in Latin, Greek, and Hebrew, he was in 1632 sent to Emmanuel College, Cambridge, and afterwards was chosen fellow of Queens' College. Having been admitted to holy orders, he left the university in 1641 to act as chaplain to Sir William Darley, and in the following year accepted a similar appointment from the widow of Sir Horatio Vere. It was about this period that he displayed surprising talents in deciphering the intercepted letters and papers of the Royalists. His adherence to the Parliamentary party was in 1643 rewarded by the living of St Gabriel, Fenchurch Street, London. In 1644 he was appointed one of the scribes or secretaries of the Assembly of Divines at Westminster. During the same year he married Susanna Glyde, and thus vacated his fellowship; but the death of his mother had left him in possession of a handsome fortune. In 1645 he attended those scientific meetings which led to the establishment of the Royal Society. When the Independents obtained the superiority, Wallis adhered to the Solemn League and Covenant. The living of St Gabriel he exchanged for that of St Martin, Ironmonger Lane; and, as rector of that parish, he in 1648 subscribed the Remonstrance against putting Charles I. to death. Notwithstanding this act of opposition, he was in June 1649 appointed Savilian professor of geometry at Oxford. In 1654 he there took the degree of D.D., and four years later succeeded Dr Langbaine as keeper of the archives. After the Restoration, he was named one of the king's chaplains in ordinary. While complying with the terms of the Act of Uniformity, Wallis seems always to have retained moderate and rational notions of ecclesiastical polity. He died at Oxford on the 28th of October 1703, in the eighty-seventh year of his age.

The works of Wallis are numerous, and relate to a multiplicity of subjects. His *Institutio Logicæ*, published in 1687, was very popular, and in his *Grammatica Linguæ Anglicanæ* we find indications of an acute and philosophic intellect. The mathematical works are published some of them in a small 4to volume, Oxford, 1657, and a complete collection in three thick folio volumes, Oxford, 1695–93–99. The third volume includes, however, some theological treatises, and the first part of it is occupied with editions of treatises on harmonics and other works of Greek geometers, some of them first editions from the MSS., and in general with Latin versions and notes (Ptolemy, Porphyrius, Briennius, Archimedes, Eutocius, Aristarchus, and Pappus). The second and third volumes include also two collections of letters to and from Brouncker, Frenicle, Leibnitz, Newton, Oldenburg, Schooten, and others; and there is a tract on trigonometry by Caswell. Excluding all these, the mathematical works contained in the first and second volumes occupy about 1800 pages. The titles in the order adopted, but with date of publication, are as follows:—"Oratio Inauguralis," on his appointment (1649) as Savilian professor, 1657; "Mathesis Universalis, seu Opus Arithmeticum Philologice et Mathematice Traditum, Arithmeticam Numerosam et Speciosam Aliaque Continens," 1657; "Adversus Meibomium, de Proportionibus Dialogus," 1657; "De Sectionibus Conicis Nova Methodo Expositis," 1655; "Arithmetica Infinitorum, sive Nova Methodus Inquirendi in Curvilineorum Quadraturam Aliaque Difficiliora Matheseos Problemata," 1655; "Eclipsis Solaris Observatio Oxonii Habita 2nd Aug. 1654," 1655; "Tractatus Duo, prior de Cycloide, posterior de Cissoide et de Curvarum tum Linearum Εὐθύνσει tum Superficierum Πλατυσμῷ," 1659; "Mechanica, sive de Motu Tractatus Geometricus," three parts, 1669–70–71; "De Algebra Tractatus Historicus et Practicus, ejusdem originem et progressus varios ostendens," English, 1685; "De Combinationibus Alternationibus et Partibus Aliquotis Tractatus," English, 1685; "De Sectionibus Angularibus Tractatus," English, 1685; "De Angulo Contactus et Semicirculi Tractatus," 1656; "Ejusdem Tractatus Defensio," 1685; "De Postulato Quinto, et Quinta Definitione, Lib. VI. Euclidis, Disceptatio Geometrica," ?1663; "Cuno-Cuneus, seu Corpus partim Conum partim Cuneum Representans Geometrice Consideratum," English, 1685; "De Gravitate et Gravitatione Disquisitio Geometrica," 1662 (English, 1674); "De Æstu Maris Hypothesis Nova," 1666—69.

The *Arithmetica Infinitorum* relates chiefly to the quadrature of curves by the so-called method of indivisibles established by Cavalieri, 1629, and cultivated in the interval by him, Fermat, Descartes, and Roberval. The method is substantially that of the integral calculus; thus, e.g., for the curve $y = x^2$ to find the area from $x = 0$ to $x = 1$, the base is divided into n equal parts, and the area is obtained as

$$= \frac{1}{n^3}(1^2 + 2^2 + \ldots + n^2), \quad = \frac{1}{6n^3}n(n+1)(2n+1),$$

which, taking n indefinitely large, is $= \frac{1}{3}$. The case of the general parabola $y = x^m$ (m a positive integer or fraction), where the area is $\frac{1}{m+1}$, had been previously solved. Wallis made the important remark that the reciprocal of such a power of x could be regarded as a power with a negative exponent $\left(\frac{1}{x^m} = x^{-m}\right)$, and he was thus enabled to extend the theorem to certain hyperbolic curves, but the case m a negative value larger than 1 presented a difficulty which he did not succeed in overcoming. It should be noticed that Wallis, although not using the notation x^m in the case of a positive or negative fractional value, nor indeed in the case of a negative integer value of m, deals continually with such powers, and speaks of the positive or negative integer or fractional value of m as the index of the power. The area of a curve, $y =$ sum of a finite number of terms Ax^m, was at once obtained from that for the case of a single term; and Wallis, after thus establishing the several results which would now be written $\int_0^1 (x - x^2)^0 \, dx = 1$, $\int_0^1 (x - x^2)^1 \, dx = \frac{1}{6}$, $\int_0^1 (x - x^2)^2 \, dx = \frac{1}{30}$, $\int_0^1 (x - x^2)^3 \, dx = \frac{1}{140}$, &c., proposed to himself to interpolate from these the value of $\int_0^1 (x - x^2)^{\frac{1}{2}} \, dx$, which is the expression for the area $(= \frac{1}{8}\pi)$ of a semicircle, diameter $= 1$; making a slight transformation, the actual problem was to find the value of $\square \left(= \frac{4}{\pi}\right)$, the term halfway between 1 and 2, in the series of terms 1, 2, 6, 20, 70, ...; and he thus obtained the remarkable expression $\pi = \frac{2.4.4.6.6.8.8 \ldots}{3.3.5.5.7.7.9 \ldots}$, together with a succession of superior and inferior limits for the number π.

In the same work, Wallis obtained the expression which would now be written $ds = dx \sqrt{1 + \left(\frac{dy}{dx}\right)^2}$ for the length of the element of a curve, thus reducing the problem of rectification to that of quadrature. An application of this formula to an algebraical curve was first made a few years later by W. Neil; the investigation is reproduced in the "Tractatus de Cissoide, &c." (1659, as above), and Wallis adds the remark that the curve thus rectified is in fact the semicubical parabola.

The *Mathesis Universalis* is a more elementary work intended for learners. It contains copious dissertations on fundamental points of algebra, arithmetic, and geometry, and critical remarks.

The *De Algebra Tractatus* contains (chapters 66—69) the idea of the interpretation of imaginary quantities in geometry. This is given somewhat as follows: the distance represented by the square root of a negative quantity cannot be measured in the line backwards or forwards, but can be measured in the same plane above the line, or (as appears elsewhere) at right angles to the line either in the plane, or in the plane at right angles thereto. Considered as a history of algebra, this work is strongly objected to by Montucla, on the ground of its unfairness as against the early Italian algebraists and also Viéta and Descartes, and in favour of Harriot; but De Morgan, while admitting this, attributes to it considerable merit.

The two treatises on the cycloid and on the cissoid, &c., and the *Mechanica* contain many results which were then new and valuable. The latter work contains elaborate investigations in regard to the centre of gravity, and it is remarkable also for the employment of the principle of virtual velocities. The cuno-cuneus is a highly interesting surface; it is a ruled quartic surface, the equation of which may be written $c^2y^2 = (c-z)^2(a^2-x^2)$.

Among the letters in volume III., there is one to the editor of the Leipsic *Acts*, giving the decipherment of two letters in secret characters. The ciphers are different, but on the same principle: the characters in each are either single digits or combinations of two or three digits, standing some of them for letters, others for syllables or words,—the number of distinct characters which had to be deciphered being thus very considerable.

For the prolonged conflict between Hobbes and Wallis, see the article Hobbes, [*Encyclopædia Britannica, ninth edition,*] vol. XII. pp. 36—38.

END OF VOL. XI.

CAMBRIDGE: PRINTED BY J. AND C. F. CLAY, AT THE UNIVERSITY PRESS.

Printed in the United States
By Bookmasters